THE MATHEMATICAL PAPERS OF ISAAC NEWTON

VOLUME III

1670-1673

modis ad rectas referuntur determinandam jam potuit hæc methodus applicari perinde ut in determinando tangentes factum est. Sed cum omnes Geometricæ Curvæ, ut et Mechanicæ mediantibus infinitis æquationibus uti postea ostendetur, ad parallelas tanquam ordinatas referri possunt, videor satis præstitisse. Qui plura desiderat haud difficulter proprio Marte supplebit. Præsertim si in ejus rei illustrationem ex abundanti instantiam pro spiralibus adjecero.

Esto BK circulus, A centrum ejus, B punctum in circumferentia datum, ADd spiralis, DC perpendiculum ejus, et C centrum curvitatis ad punctum D. Ductisque ADK recta, et ei parallela et æquali CG, et normali GF occurrente CD in F: Dic AB vel AK = 1 = CG, BK = x, AD = y, et GF = z. Præterea concipe punctum D per infinitè parvum spatium Dd in spirali moveri, et perinde per d agi semidiametrum Ak, eique parallelam et æqualem Cg, et normalem gf occurrentem Cd in f, cui etiam GF occurrat in φ; Produc GF ad φ ut sit Gφ = gf, et ad AK demittatur normalis dε et produc donec cum CD conveniat ad I: Et contemporanea momenta erunt ipsarum BK, AD, ac Gφ Kk, Dε, et Fφ, quæ proinde dicentur mxo, nxo, et rxo. kK . dε = oy ubi assumo m = 1.

Jam est AK . Aε (AD) :: kK . dε = oy. Item CG . GF :: dε . εD = oyz, adeoque ut supra yz = n. Præterea CG . CF :: dε . dD = oy×CF :: dD . dI = oy×CFq. Et hæc propter ang PCφ (= ang GCg) = ang DAd, et ang CPφ (= ang CdI = ang εdD + rect.) = ang ADd, triangula CPφ et ADd sunt similia, et inde AD . Dd :: CP (CF) . Pφ = o×CFq, unde aufer Fφ et restabit PF = o×CFq − oxr. Denique demissa CH normali ad AD est PF . dI :: CG . εH vel DH = $\frac{y \times CFq}{CFq - r}$. Vel substituto 1+zz pro CFq, erit DH = $\frac{y+yzz}{1+zz-r}$.

Et nota quod in hujusmodi computationibus quantitates (ut AD et Aε) pro æqualibus habeo quarum ratio a ratione æqua[litatis]

Measuring curvature in a polar coordinate system
(I, 2, §2).

THE MATHEMATICAL PAPERS OF ISAAC NEWTON

VOLUME III

1670-1673

EDITED BY

D. T. WHITESIDE

WITH THE ASSISTANCE IN PUBLICATION OF

M. A. HOSKIN AND A. PRAG

CAMBRIDGE

AT THE UNIVERSITY PRESS

1969

CAMBRIDGE UNIVERSITY PRESS
Cambridge, New York, Melbourne, Madrid, Cape Town, Singapore, São Paulo

Cambridge University Press
The Edinburgh Building, Cambridge CB2 8RU, UK

Published in the United States of America by Cambridge University Press, New York

www.cambridge.org
Information on this title: www.cambridge.org/9780521071192

First published 1969
This digitally printed version 2008

A catalogue record for this publication is available from the British Library

Library of Congress Catalogue Card Number: 65–11203

ISBN 978-0-521-07119-2 hardback
ISBN 978-0-521-04581-0 paperback
ISBN 978-0-521-72054-0 paperback set (8 volumes)

TO THE MEMORY OF
HERBERT WESTREN TURNBULL

PREFACE

This third volume continues the reproduction in chronological sequence of those of Newton's mathematical papers known still to be in existence. Three of the texts here presented have been previously published, two (the 1671 fluxional tract and the deposited Lucasian lectures on optics) in obsolete eighteenth-century versions edited from inferior secondary transcripts, but the remainder make their first appearance in print and will be new to all but the most assiduous student of Newtonian manuscript. With the trivial exception of a missing opening page to the 1671 tract all are now reproduced, purged of accretions, from Newton's original autograph papers. As in previous volumes, it has been my primary aim to present an accurate text, but the ancillary editorial commentary and English translations should both serve to set these papers in their contemporary milieu and offer an interpretation of points of technical difficulty.

For allowing reproduction of documents in their custody I am grateful above all to the Librarian and Syndics of the University Library, Cambridge, but in individual cases my thanks must go to the Librarians of the Royal Society, London, and of the University Library, Edinburgh. To a private owner of Newtonian manuscript, also, my appreciation for being permitted to publish mathematical items in his collection. Many institutions have, over the years, generously afforded me facilities to pursue the research whose fruits have gone into this book, but I would make a special mention of the courtesy of the staff of the Anderson Room in Cambridge University Library—above all of Mr P. J. Gautrey, who has been unfailingly kind and efficient. Financial assistance during the preparation of this volume has been provided by the Science Research Council and, more recently, jointly by the Sloan Foundation, the Leverhulme Trust and the Master and Fellows of Trinity College, Cambridge. To these institutions and foundations my gratitude for their support. Sir Harold Hartley, who has worked unceasingly on my behalf for so long, has done more than anyone to promote this edition in his own unique way: to him my warmest thanks for his friendship and patronage.

From a technical viewpoint I have two overriding debts to acknowledge. Mr A. Prag has not only indexed this volume but in reading it for press has suggested many elegancies and improvements while eliminating not a few blunders on my part. His elevation to the title-page merely confirms (somewhat against his wish) what has been his rôle from the first. To my other assistant and oldest friend in the field of history of science, Dr M. A. Hoskin, I can only inadequately convey my appreciation of his unselfish help and ubiquitous support.

Lastly, but never least, my unqualified gratitude to the Syndics of the Cambridge University Press on behalf of all those—subeditors, printers, draughtsmen and production staff—who have transformed an untidy, often ill-written longhand manuscript into a book of conspicuous beauty and elegance.

D.T.W.

19 August 1968

EDITORIAL NOTE

For explanation of the principal conventions made use of in this volume the reader is once more referred to pages x–xiv of the first volume, where their justification is broached. It has been our ideal throughout to be as strictly faithful to the autograph manuscripts reproduced as the limits imposed by the printed page will allow. As previously, suffixes and contracted forms here reappear unaltered, while grammatical inconsistencies, too, have either been left unchanged or, when tampered with, explained in footnote. Only on rare occasions have silent liberties been taken with Newton's text, mostly in italicizing or capitalizing subheads and in inserting endpoints to sentences. In one or two places, also, the text has been made consistent with an accompanying figure when this has required merely an interchange of upper and lower case forms of the same letter. Otherwise all insertion in Newton's text has been made within square brackets, with the sole aim of rectifying its minor irregularities or illogicalities. As he thinks fit the reader may accept or reject these interpolations. Forewarned by J. A. Lohne in a recent article (*History of Science*, **6**, 1968: 69–89) we have taken especial care in reproducing the manuscript figures accurately, but have several times adopted a deliberate compromise between a slavish facsimile of an inaccurate rough sketch and a structurally perfect ideal diagram, particularly where Newton's curves have an exact shape. The English versions which face our reproductions of Newton's original Latin text have, within the bounds of comprehensibility, been kept deliberately literal, being intended primarily as an aid to the reader working through the Latin original rather than as a polished paraphrase to be studied by itself. In both Latin and English versions two thick vertical bars in the left-hand margin alongside a passage denote that it has been cancelled by Newton in his manuscript: a convention not to be confused with the two small vertical strokes '||' used on occasion (here in **1**, 2, §2) to mark a page division in Newton's original, when they are invariably accompanied in the margin alongside by two companion strokes preceding the new page number set inside square brackets. A few non-standard notations are explained where and as they arise: in the English version of **1**, 2, §3, for instance, we have used 'fl (A)' to denote the 'fluxion of A' (corresponding to Newton's variant form 'fl: A'). The convention used in the two previous volumes for reference to other places in the edition is retained but we have also, for brevity's sake, introduced a parallel convention which makes direct reference to the pages of volumes. So in note (11) on page 43 below by 'II, **3**, 1, §1: note (108)' understand note (108) of [Volume] II, [Part] 3, [Section] 1, [Subsection] 1; while in note (302) on page 161 we have written 'I: 373', meaning [Volume] I, [page] 373, which we might have set in the less manageable equivalent form 'I, **2**, 6, §1.1: 373'. Neither should be confused with the special usage introduced in the concluding appendix, where '[**1**: 68]', for example, refers to page 68 of the first volume of the Royal Society's edition of *The Correspondence of Isaac Newton*.

GENERAL INTRODUCTION

The second volume of this edition of his mathematical papers effectively closed in the autumn of 1670 with the finishing touches which Newton added to the revision of his 'Observations' on Kinckhuysen's *Algebra*. In sequel we here jump back almost a year in time to the previous January when, freshly appointed to the Lucasian chair of mathematics in the University of Cambridge, he began to deliver to a captive and evidently none too appreciative student audience the first series of his lectures on the dioptrics of refracted white light. Mathematical excerpts from the two extant manuscript versions of these self-styled 'Optica' are reproduced in the third part below, together with some allied researches into the geometrical optics of thin lenses. The first and largest portion of the present volume, however, is given over to the magnificent treatise of infinite series and fluxional analysis which he began a little afterwards in the winter of 1670–1 and further augmented (but did not succeed in completing) late the following year, along with other minor calculus papers of his which date from this period. Sandwiched between is a miscellaneous trio of pieces on pure geometry and cycloidal motion, whose handwriting suggests that they too were composed shortly after 1670 but about which little otherwise is known. In conclusion, and as a complement to the text of Newton's papers, we append an epitome of the mathematical topics which were aired in his contemporary correspondence. Accurately wherever possible but impressionistically when only circumstantial evidence can be adduced, introductory passages and appropriate footnotes paint a detailed backcloth to the autograph texts now reproduced. Here a few more general brush strokes may be permitted.

In 1670, now in his twenty-eighth year, Isaac Newton teetered on the threshold between unappreciated youthful promise and acknowledged mature achievement, his mathematical and scientific discoveries of the previous half dozen years—apart from the *De Analysi*—still hugged tightly to himself. Revealing personal details of his activities at this time are few and even the documentary record of the trivia of his daily existence fails,(1) but we may readily suppose that the rhythm and routine of his sheltered life went on much as before and that he continued to enjoy an unaltered familiarity with such people as Francis Aston and especially his chamber mate(2) John Wickins,

(1) The last dated entry in the accounts list in the Fitzwilliam notebook (see II, Introduction: x, note (3)) is for 'Aprill 1669' and its continuation is lost.

(2) If we are to believe an idle remark to Halley in July 1686 referring back to June 1673 (*Correspondence of Isaac Newton*, **2**, 1960: 446). Edleston, however, has pointed out that 'in the Junior Bursar's Book [at Trinity] for the year ending at Michaelmas, 1673, we find the two

friends from his undergraduate days and now fellows of Trinity College with him. During the four years 1670–3 Newton, short outings apart, left Cambridge only three times, mostly to visit his mother at Woolsthorpe but once to stay with a 'M^{ris} Arundell' in Northamptonshire: no visit to London is recorded.(3) Within the confines of the small Fenland university town where he passed most of his time existence must have been all too uneventful. No doubt with his elevation to professorial rank even his previous rare visits to the local tavern and gambling house ceased.(4) With the two correspondents in London with whom he exchanged letters at this time, John Collins and (from January 1672) Henry Oldenburg, he maintained for the most part a tone of polite if somewhat wary diffidence, but beneath this mask he was not uneager to communicate his thoughts on a variety of topics while his puritanical disdain for personal aggrandisement and worldly self-promotion welled up on more than one occasion.(5) All too soon, unhappily, despair at the fiasco of the

entries "for seiling Mr Newton's chamber", "for mending the slating...over Mr Wickins", from which perhaps we may infer that one of them had changed his rooms in the interval between June and September' (*Correspondence of Sir Isaac Newton and Professor Cotes* (London, 1850): xliii, note (13)).

(3) The list of Newton's college exits and redits during this period (see Edleston's *Correspondence* (note (2)): lxxxv) records no absence from Cambridge at all in 1670, three weeks' leave in the spring of 1671 and at Easter 1673, and exactly a month away in the early summer of 1672. His whereabouts between 17 April and 11 May are not documented but presumably he stayed in Lincolnshire. The day after he left Cambridge on 18 June 1672 he wrote to Oldenburg: 'Having been in Bedfordshire since my last to you, at my returne to Cambridge I received your letter.... At present I am removed into Lincolnshire...at M^{ris} Smith's house at Woolstrope in Coulsterworth-Parish...' (*Correspondence of Isaac Newton*, **1**, 1959: 194). On the following 6 and 13 July he wrote again, advising Oldenburg that 'I am at present in Northampton shire...', 'pray direct your Letter to me at M^{ris} Arundells House in Stoake Park' (*ibid.*, **1**: 215, 210). Between 10 March and 1 April 1673 he was evidently again away in Lincolnshire, for on 8 March he wrote to Oldenburg that 'I shall be henceforth absent from Cambridg for about a month' (*ibid.* **1**: 263). In early July 1671 illness had thwarted an intended visit to London, for on the 20th of that month Newton informed Collins that 'I purposed to have given you a visit at ye late solemnity of o^{r} Chancellors creation; but I was prevented in y^{t} Journey by ye suddain surprisall of a fit of sicknesse, w^{ch} (God bee thanked) I have now recovered [of]' (from the draft in ULC. Add. 3977.9; compare *Correspondence*, **1**: 67).

(4) Twenty years later when Newton came to draft a scheme 'Of educating Youth in the Universities' his innate puritanism had hardened his attitude to drunken revelry, for he then proposed that 'All Graduates wthout exception found by the Proctors in Taverns or other drinking houses, unless wth Travellers at their Inns, shall at least have their names given in to ye Vice-chancellor who shall summon them to answer it before the next Consistory' (ULC. Add. 4005.5: 15^{r}; compare note (40) below and I, Introduction: xxxiv, note (52)).

(5) Thus, with regard to a previously transmitted solution of the problem of summing a general harmonic progression (see note (20) below), Newton announced to Collins on 18 February 1669/70: 'That solution of the annuity Probleme if it will bee of any use you have my leave to insert it into the *Philosophical Transactions* soe it bee wthout my name to it. For I see not what there is desirable in publick esteeme, were I able to acquire & maintaine it.

projected Latin edition of Kinckhuysen and the overriding bitterness provoked in him by the wounding and (in his view) largely irrelevant ripostes to his optical hypotheses after their publication or private circulation by Oldenburg were to freeze this warm shyness into a chilly, unforthcoming abruptness. With James Gregory, the only one of Newton's British contemporaries who could match him in mathematical breadth and profundity, he was never able to establish direct contact—except perhaps for a brief meeting in autumn 1673[(6)] —and their intermediary, Collins, was rarely eager to pass on the insights or results which each communicated to him. Ineluctably, we are led to realize the overriding importance for Newton at this time of his relationship with his college senior, Isaac Barrow: a friendship which, though less intimate and longstanding than is usually taken for granted[(7)] and (as direct personal contacts so often are) known but shadowily and at second hand through their common acquaintance Collins, was firmly based on mutual respect and a shrewd evaluation of the other's character. In the event Barrow was to afford Newton much needed private encouragement of his calculus researches, a bridge to the academic security of a permanent professorship and contact with the scientific world beyond the self-centred microcosm of a still essentially medieval university.[(8)] The tribute Newton paid to him in January 1670 by carrying on the optical theme of his predecessor's Lucasian lectures was richly deserved.

Though Barrow, unlike Collins, at no time evidenced any deep knowledge of Newton's mathematical researches—and perhaps in his growing distaste for the exact sciences had no burning desire to acquire one—[(9)]he had clearly

It would perhaps increase my acquaintance, ye thing wch I cheifly study to decline' (*Correspondence*, **1**: 27).

(6) '...now I understand that Mr Gregory is at London, & intends to make Cambridg in his way into Scotland, I shall not trouble you any further wth discourses about ye Perspective but refer it to our meeting, if Mr Gregory will be pleased to favour me wth a visit' (Newton to Collins, 17 September 1673 ... *Correspondence*, **1**: 307). Nothing further is known of this proposed meeting but we may guess that if it had taken place Newton would have left some recorded remembrance of it or acknowledged some personal contact between the two at the time of Gregory's death in 1675. Moreover, a direct exchange of letters would have been a natural sequel to such a meeting and Newton could scarcely have resisted the temptation to show Gregory a selection of his early mathematical and optical papers, but no evidence exists which hints that either occurred. (However there is a long gap in both Newton's and Gregory's extant correspondence from September 1673 onward.)

(7) Compare I, **1**, Introduction: 10, note (26).

(8) One still not markedly different in organization and structure from that described in W. T. Costello's *The Scholastic Curriculum at Early Seventeenth-century Cambridge* (Harvard, 1958).

(9) In his letter of 10 December 1672, after describing for Collins' benefit his 'Generall Method [of] drawing tangents to all curve lines whether Geometrick or mechanick', Newton added the curiously offhand remark: 'I remember I once occasionally told Dr Barrow when he was about to publish his Lectures that I had such a method of drawing Tangents, but some

been the prime mover in persuading Newton to circulate his *De Analysi* in mid-1669 as a counterblast to Mercator's newly published *Logarithmotechnia*,(10) while two years later Newton could inform Collins that it was 'partly upon Dr Barrows instigation' that 'the last winter...I began to new methodiz ye discourse of infinite series'.(11) We gather, moreover, that the riches of Barrow's mathematical collection (an invaluable supplement to the limited holdings, in books and manuscript, of the Cambridge university and college libraries at this time) were made readily available to Newton when he needed to consult it.(12) But more significant in the future shaping of the latter's scientific career was Barrow's generous introduction of him to the scholarly world at large as 'a fellow of our [Trinity] College,...very young...but of an extraordinary genius & proficiency'.(13)

An unmatched opportunity to promote Newton to a secure academic position where his abilities would not be wasted came in autumn 1669 when Barrow, for reasons which are no longer accurately known,(14) made up his

---

divertisment or other hindered me from describing it to him' (*Correspondence*, **1**: 248). Certainly, Barrow made no Newtonian modification in his '10th Ge[o]metrical Lecture' after his friend looked it over in the press in late 1669 (see I, Introduction: xv, note (1)). In 1672 Newton cannot be accused of pique against Barrow, for at the close of his letter to Collins he observed that 'We are here very glad that we shall enjoy Dr Barrow again especially in the circumstances of Master, nor doth any rejoyce at it more then [I]' (*Correspondence*, **1**: 252). Barrow was not officially appointed Master of Trinity till the following March; see note (14).

(10) See II, **2**, Introduction: 165–7.

(11) Newton to Collins, 20 July 1671 (*Correspondence*, **1**: 68).

(12) For example, Newton told Collins on 27 September 1670 that 'I have hitherto deferred writing to you, waiting for Dr Barrows returne from London that I might consult his Library about what you propounded...concerning ye solutions of Cubick æquations, before I sent you my thoughts upon it' (*Correspondence*, **1**: 42). After Barrow's death it was Newton who compiled the 'Catalogue of the bookes of Dr Isaac Barrow sent to S.S....on July 14. 1677' (Bodleian MS Rawl. D 878: 33–59), itself a valuable supplement to the Huggins' list of Newton's library as an indication of contemporary printed works accessible to him. (Compare F. E. Manuel, *Isaac Newton: Historian* (Cambridge, 1963): 45–6.) 'S.S.' may be one or other of the London stationers Samuell Sprint or Samuel Smith.

(13) Barrow to Collins, 20 August 1669 (*Correspondence*, **1**: 14–15). Compare II: 166, note (11).

(14) The modern story has it that Barrow, recognizing Newton's mathematical genius and superior fitness for the post he held, resigned in favour of his 'pupil', but this is a sentimental gloss on the existing documentary record which is none too consistent with Barrow's scarcely meek and self-effacing character, however compatible with his generous spirit. Man of the world and fighter that he was, it must frequently have been a sore point to him that the King's confirmation in January 1664 of Lucas' statutes, aimed expressly to prevent the Lucasian professor taking an active part in academic administration or politics, forbade him (see Appendix: note (17)) from holding any university or college position of any kind apart from a fellowship. Barrow's resignation of the chair in 1669 may reasonably have been but a necessary preliminary to his being considered, when next it was vacant, for the Mastership of Trinity: a position to which, in fact, Charles II appointed him soon after on 14 March 1672/3 (ULC. Baker MS 29: 113).

mind to relinquish the Lucasian chair. By the statutes of this professorship, confirmed by royal authority,[15] the power to appoint his successor was invested, during their lifetime, solely in the hands of the still surviving executors of Henry Lucas' will, the London lawyer Robert Raworth and the doughty University printer Thomas Buck. These two old men, both susceptible to flattery, were doubtless only too willing to confirm Barrow's 'recommendation' of Newton as his successor—particularly since the former had just dedicated the printed version of his recent optical lectures to them.[16] At all events, on 29 October[17] Newton was elected to the vacant mathematical professorship, its second holder. For the remaining twenty-six years of his stay in Cambridge he was to enjoy the status and security of a well-paid university post coupled with the privileges and emoluments of a college fellowship[18]—he chose, in fact, to continue at Trinity—expressly liberated by royal mandate from all time-consuming responsibilities of academic teaching, tutoring and administration which would have been its normal accompaniment. To that freedom the

(15) See Appendix: note (11). Collins informed James Gregory rather tersely on 25 November that 'Mr Barrow hath resigned his Lecturers place to one Mr Newton of Cambridge, whome he mentioneth in his Optick Præface as a very ingenious person' (*Correspondence of Isaac Newton*, **1**: 15). Seven years later he clarified this ambiguous remark for Gregory's brother, observing that 'when Mr Barrow left his place as Mathematick Professor he recommended Mr Newton to it' (Collins to David Gregory *père*, 11 August 1676, quoted from the draft in Royal Society MS LXXXI, No. 47); while in late 1677 he commended Barrow to Wallis for 'setting' Newton as his successor (*Correspondence of Isaac Newton*, **2**, 1960: 241). Newton himself in old age spoke of having had the Lucasian chair 'procured' for him (ULC. Add. 3968.41: 117^r^, a stray phrase entered below English and Latin drafts of his 1716 response to the Bernoulli challenge problem).

(16) These statutes, printed in appendix to a rare tract of William Whiston's in 1718 and discussed briefly by Edleston in his *Correspondence* (note (2)): xliv–xlv, note (16), are reproduced in full below (together with Charles II's confirmation of privileges) because of their interest. The Lucasian professorship was the first University chair in mathematics, although in the previous decade such college 'Mathematick professors' as John Smith, Charles Scarborough and Seth Ward had lectured on the easier portions of Euclid's *Elements* and Oughtred's *Clavis*. (Compare M. H. Curtis, *Oxford and Cambridge in Transition 1558–1642* (Oxford, 1959): 246.) The quality of instruction given by these enthusiasts may be gleaned from a letter Wallis wrote to Collins about the spring of 1673: 'Mr John Smith fellow of Queens Colledge in Cambridge and Mathematick Professor in the University, writt to me the 1 of Novr 1648, about some things that seemed difficult to him in Deschartes Geometry. I had not then seen Deschartes Geometry, nor had read any other Algebra but Mr Oughtreds Clavis, nor knew anything of the contents of it, but on this occasion found out the Booke in a freinds hands in London' (from Collins' copy in Royal Society MS LXXXI, No. 39: 18). The following June (1649) Wallis was appointed Savilian Professor of Geometry at Oxford!

(17) According to Edleston (*Correspondence* (note (2)): xxii). Thomas Baker (ULC. Baker MS 29: 113) gives the date as 8 November (new style?).

(18) As a welcome salve to his religious conscience, letters patent from the King five years later confirmed Newton's statutory right to retain this professorial fellowship without taking holy orders, 'any college statute or tradition or any interpretation whatsoever thereof notwithstanding' (see Appendix: notes (10) and (16)).

quality of his future researches in mathematics, optics, dynamics, chemistry and astronomy manifestly owed much.

Newton's introduction to intellectual society in London was less immediately successful. Doubtless encouraged by Barrow, he journeyed there in late November 1669 and was met by Collins

somewhat late upon a Saturday(19) night at his Inne. I then proposed to him the adding of a Musicall Progression,(20) the which he promised to consider and send up....And againe I saw him the next day having invited him to Dinner: in that little discourse we had about Mathematicks, I asked him what he would make the Subject of his first Lectures, he said Opticks proceeding where Mr Barrow left, and that himselfe was a practicall grinder of glasses, and had ground glasses for a pocket tube, but 6 Inches long, that magnified the Object 150 times...[But] having no more accquaintance with him I did not thinke it becomming to urge him to communicate any thing....(21)

Whether Newton met anyone else of importance on this visit is not known. Its sole immediate effect was to start up a correspondence with Collins, largely devoted to Newton's 'improvements' of Kinckhuysen's *Algebra* and abortive plans for its publication,(22) which after a brisk beginning petered out the following September: Collins remarked shortly afterwards to Gregory that 'observing a warinesse in him to impart, or at least an unwillingness to be at the paines of so doing, I desist, and doe not trouble him any more'.(23) Apart from a single exchange of letters in July 1671(24) their correspondence did not start again till April 1672 when Collins wrote to Newton that 'A little before Christmas the Reverend Doctor Barrow informed me you were buisy in enlarging your generall method of Infinite Series's or quadratures and in preparing 20 Dioptrick Lectures for the Presse.'(25) It was Newton's communication of his improved 'catadioptrical' telescope, 'brought up for his Maties

(19) Probably 27 November. Newton was away from Cambridge between 26 November and 8 December 1669 (see Edleston's *Correspondence* (note (2)): lxxxv).

(20) See the concluding Appendix to this volume for details of Newtons' responses to Collins' request. This problem of computing the 'summ of a Harmonic Progression' had arisen from Collins' attempts to evaluate the 'present worth of an Annuitie...discompting simple Interest (which our Law establisheth)' about the early summer of 1668. (See Collins to Bertet, 21 February 1670/1 [= Royal Society MS LXXXI, No. 21]; and compare, for example, Oldenburg to Auzout, 29 June 1668 [= *Correspondence of Henry Oldenburg*, **4** (Madison, 1967): 481].) The present value of an annuity of A per annum continuing n years at a simple interest rate of r % is readily seen to be

$$\sum_{1\leqslant i\leqslant n} A\bigg/\left(1+\frac{ir}{100}\right).$$

(21) Collins to James Gregory, 24 December 1670 (*Correspondence of Isaac Newton*, **1**: 53–4).

(22) See II, **3**, Introduction: 281–7.

(23) *Correspondence*, **1**: 55.

(24) Collins to Newton, 5 July, answered by Newton to Collins, 20 July (*Correspondence*, **1**: 65–9). Compare II: 288.

(25) Collins to Newton, 30 April 1672 (*Correspondence*, **1**: 146).

perusall' a little before Christmas 1671,[26] which finally opened up London scientific society to him. Henry Oldenburg, no doubt urged on by both Barrow and Collins, mounted an exhibition of it in the Royal Society's rooms[27] and on 21 December Seth Ward, evidently gratified to see the teaching of mathematics at Cambridge (which he had begun) so ably continued and appreciative of Newton's rôle in revising the Kinckhuysen *Algebra* whose translation from Dutch into Latin he himself had sponsored,[28] proposed him for membership of the Society. Duly elected a Fellow three weeks later—a dignity he continued to hold for the rest of his life despite a sudden, quickly withdrawn request to be 'put out' from it a year afterwards[29]—Newton at once began to communicate to Oldenburg the summaries of his optical experiments and hypotheses which were soon to make him internationally celebrated and, together

(26) See **3**, Introduction: note (23) below. Collins described Newton's reflecting telescope at some length to Vernon on 26 December: 'I have seen an object in it. The Tube is a Cylinder of about 7 Inches long and $2\frac{1}{4}$ I[nches] Diam. open at one End. At the other is a Sphærick Concave Speculum of Mettal obnoxious to y[e] inconven[ce] of rust or Tarnish. Out of the side of the Cylinder a Wyre so holds a bitt of plaine looking glasse inclined and as bigg as a penny, that it may be in the focus of the former. The eye lookes in at a hole in the side of the Cylinder as big as a great Pins head through a glasse, and sees the object cleare without Colours, and as much magnified as it could be by an ordinary Tellescope of 5 or 6 feet long. A screw at y[e] end remooves the Mettaline Speculum a little too and againe to serve in stead of Drawing. The Tube is fastened with a foote to a mooveable Ball of Wood which roves in a Spherick Dish that hath a flatt Base. It is somewhat difficult to place it upon an Object, and some object that it is not so lightsome as an ordinary Tube of the length aforesaid' (Royal Society MS LXXXI, No. 23). The telescope itself is, of course, still preserved by the Society.

(27) Oldenburg's description of the instrument, no doubt drawn up to accompany Newton's exhibit, is extant (Royal Society Letter Book N. 1. 37, reproduced in *Correspondence*, **1**: 74–5). The Society's Secretary was quick to seize his opportunity of opening a direct link with Newton, writing to him on 2 January following. 'Your Ingenuity is the occasion of this addresse by a hand unknowne to you....your Invention of contracting Telescopes... having been considered, and examined here by some of y[e] most eminent in Opticall Science and practise, and applauded by them, they think it necessary to use some meanes to secure this Invention from y[e] Usurpation of forreiners' (*Correspondence*. **1**: 73).

(28) 'The right reverend B[p] Ward...recommended Kinckhuysens Introduction to Algebra in low Dutch to Mr Mercator to translate into Latin. Which done I sent it to Mr Newton who hath been pleased to make diverse Additions thereto..., particularly one being a Discourse about bringing of Problemes to an Æquation' (Collins to Strode, 26 July 1672, from the draft in Royal Society MS LXXXI, No. 24). Newton may possibly not have been aware of Seth Ward's nomination till he heard from Oldenburg on 2 January that 'you were lately by y[e] L[d] B[p] of Sarum proposed Candidat' (*Correspondence*, **1**: 73).

(29) Newton to Oldenburg, 8 March 1672/3 (*Correspondence*, **1**: 262). In excuse for his withdrawal he affirmed that 'I see I shall neither profit [the fellows of y[e] R. Society], nor (by reason of this distance) can partake of the advantage of their Assemblies', to which Oldenburg responded that he was 'surprised at his resigning for no other cause, than his distance, wch he knew as well at the time of his election' (*ibid.*, **1**: 263). We may readily guess that Newton was seeking by this ploy a way of effectively disengaging himself from the controversy which his recently published optical correspondence had aroused.

with the magisterial *Philosophiæ Naturalis Principia Mathematica* he wrote fifteen years later, came ultimately to win him the Society's Presidency.

A final word on Newton's professorial duties at Cambridge. As occupant of the Lucasian chair it was his sworn obligation to 'lecture and expound' to an audience of senior students, at a pre-arranged time once at least each week during the seven months of Cambridge's academic year,(30) on 'Geometry, Arithmetic, Astronomy, Geography, Optics, Statics or some other Mathematical Discipline' of his choice: a requirement not to be skimped since the statutes ordained that each autumn 'just before Michaelmas'(31) the professor should deliver to the Vice-chancellor for deposit in the University Library 'polished copies of no fewer than ten of the lectures given by him during the previous year'. In addition, he was enjoined to make himself freely available at stated times in his quarters—twice each week during term and once (when in residence) otherwise—ready, for a two-hour period, to answer the 'queries and difficulties' put to him by any one who came along; to which end he was to have on hand globes and other mathematical instruments in his possession and 'in every other respect and to the full extent of his powers' to encourage the efforts of his students.(32) Already during Barrow's tenure of the professorship, however, it is evident that in the practical enforcement of these statutory ordinances certain tolerances were allowed. Notably, in place of the required annual deposit of fair copies of ten of the preceding year's lectures Barrow was able (without recorded penalty) to substitute printed publication of unabridged, indeed augmented, versions of his more original professorial discourses on optics and mathematics.(33) Newton himself is known to have made only two deposits of lectures (again without recorded payment to the University Librarian of any statutory fine) during the whole first fifteen years of his tenure of the Lucasian chair: in October 1674 he gave in to Robert Peachey a much improved version (divided into 31 *soi-disant* 'lectiones' dated between January 1670 and October 1672) of his inaugural 'dioptrick' discourses;(34) subsequently, about 1685, he handed over the revise of 97 'lectures' on algebra

(30) See Appendix: note (5) below.

(31) Appendix: note (6)

(32) Appendix: note (7).

(33) On 24 December 1670 Collins reported to Gregory that 'Mr Barrow told me the Mathematick Lecturer there [in Cambridge] is obliged either to print or put 9 [*sic*] Lectures yearly in Manuscript into the publick Library, whence Coppies of them might be transcribed' (*Correspondence*, **1**: 54); but added later, on 29 June 1675, 'Mr Newton intends not to publish anything, as he affirmed to me, but intends to give in his lectures yearly to the publick library' (*James Gregory Tercentenary Memorial Volume*, London, 1939: 310).

(34) Extracts from this desposited manuscript (ULC. Dd. 9.67) are reproduced in **3**, 1, §2 below. An earlier version of this 'Optica' (ULC. Add. 4002; see **3**, 1, §1) is much shorter and divided into only 18 'lectiones'.

delivered in the eleven-year period between October 1673 and October 1683.(35) What else Newton read in the Schools at this time—in the Easter term of 1670, for example, and the Lent term of 1671—we can only conjecture. In February 1672 Collins announced to James Gregory that 'Mr Newton (as Dr Barrow informes me) intends to send up all the Lectures he hath read since he was Professor to be printed here, which he sayth will be 20 [*sic*] Dioptrick Lectures, and some about infinite Series', but was told equally firmly by Newton in September 1676 that 'though about 5 years agoe I wrote a discourse in wch I explained ye doctrine of infinite æquations, yet I have not hitherto read it but keep it by me'.(36) His 'care' of the 1672 Cambridge edition of Varenius' *Geographia Universalis*(37) would suggest that Newton about this time gave instruction, if not formal lectures, in geography (in its seventeenth-century sense of physical 'earth science' and elementary mathematical astronomy).

As for the reaction of Newton's student audiences to his early Lucasian lectures, contemporary report is silent: indeed, our search of the records yields the name of no one who can be shown to have attended them before John Flamsteed in 1674.(38) We should not anticipate Humphrey Newton's observation (relating to a period a decade and a half later when Newton was busy composing his *Principia*) that 'when he read in ye Schools...so few went to hear Him, & fewer y^{t} understood him, y^{t} oftimes he did in a manner, for want of Hearers, read to ye Walls'(39) but we may none the less suppose that Newton's success and popularity as a lecturer was small. In hindsight some twenty years afterwards, when he drew up a scheme 'Of educating Youth in the Universities', he realized the basic mistake of trying to impart advanced knowledge to the ill-prepared student, requiring the 'Mathematick' professor as a preliminary

to read first some easy & usefull practicall things, then Euclid, Spherics, the projections of the Sphere, the construction of Mapps, Trigonometry, Astronomy, Opticks, Musick,

(35) The deposited version (ULC. Dd. 9.68) is greatly different from a later, unfinished revise (ULC. Add. 3993) which still exists. Both will be reproduced in our fifth volume.

(36) Collins to Gregory, 23 February 1671/2 (*Gregory Memorial Volume* (note (33)): 218); Newton to Collins, 5 September 1676 (*Correspondence*, **2**, 1960: 95).

(37) See II, **3**, Introduction: 288, note (43); and compare A. R. Hall, 'Newton's First Book', *Archives Internationales d'Histoire des Sciences*, **13**, 1960: 39–40, 55–61.

(38) Flamsteed, then on a short visit to Cambridge to receive his M.A., was given a sheet of notes (reproduced in Edleston's *Correspondence* (note (2)): 242–3) 'at one of [Mr Newton's] lectures [on algebra], Midsummer, 1674'. We will return to this point in the fifth volume.

(39) Humphrey Newton to John Conduitt, 17 January 1727/8 (King's College, Keynes MS 135^{A}, quoted in Brewster's *Memoirs of the Life, Writings and Discoveries of Sir Isaac Newton*, **2**, Edinburgh, 1855: 92). On the following 14 February Humphrey added that on these occasions 'he usually staid about half an hour, [but] when he had no Audit^{rs} he com̃only return'd in a 4th part of that time or less' (Keynes MS 135^{B} = *Memoirs*, **2**: 96).

Algebra, &c. Also to examin &...to instruct in the principles of Chronology and Geography....

All students who will be admitted to Lectures in naturall Philosophy to learn first Geometry & Mechanicks. By mechanicks I mean here the demonstrative doctrine of forces & motions including Hydrostaticks. For w^th^out a judgment in these things a man can have none in Philosophy.(40)

This is surely wisdom born of personal experience. But we are here concernd less with Newton's middling ability as a university teacher than with his surpassing intellectual genius, and may pass forthwith to study the detail of the papers on which, during his first years as Lucasian professor, he penned the essence of his mathematical insights and advances.

APPENDIX

THE ORIGINAL STATUTES OF THE LUCASIAN PROFESSORSHIP AND CHARLES II's 'CONFIRMATION...WITH ADDITION OF PRIVELEGES'(1)

From contemporary copies in the University Library, Cambridge

[1](2)

Omnibus Christi Fidelibus, ad quos hoc præsens scriptum pervenerit, Robertus Raworth de Grayes Inn in Comitatu Middlesex Armiger, et Thomas Buck de Cantabrigia in Comitatu Cantabrigiæ Armiger,(3) Execu-

(40) ULC. Add. 4005. 5: 14^r^, 15^r^, first printed by Rouse Ball in *The Cambridge Review* (Thursday, 21 October 1909): 29–30. Newton added that 'because the Mathematick Lecturer's Office is laborious, for encouraging [him] to diligence, none shall be compelled to come to [his] Lectures, but all that will be [his] auditors shall offer...a quarterly gratuity, suppose of 10^s^ ye sizar, 12 or 15^s^ ye Pensioner & 20 or 25^s^ ye Fellowcommoner' (*ibid.*: 14^v^).

(1) The statutes, drafted on 19 December 1663 and confirmed by royal warrant a month later, were framed by the executors, Robert Raworth and Thomas Buck, of the will of Henry Lucas, Fellow of St John's and one of the University's members of parliament. It had been Lucas' wish to give Cambridge a scientific lectureship equal in status to the Oxford chairs of geometry and astronomy which had been founded in 1619 by Savile, but he would perhaps not have agreed wholeheartedly with Charles' confirmation of the statutes of his professorship, for its added privileges effectively divorced the Lucasian chair from all narrow college control, leaving its occupant chargeable directly only to the Vice-chancellor. The royal confirmation of 18 January 1663/4 is unpublished, though its main clauses were quoted by Edleston in his *Correspondence of Sir Isaac Newton and Professor Cotes...With Notes, Synoptical View of the Philosopher's Life...* (London, 1850: xlv, note (16)). The statutes themselves are printed in William Whiston's *An Account of Mr Whiston's Prosecution at, and Banishment from, the University of Cambridge...now Reprinted. With an Appendix: Containing Mr Whiston's farther Account...Never before Printed* (London, 1718: 42–6: 'A Copy of Mr Lucas's Statutes. Confirmed by the Royal Authority').

TORES ultimi Testamenti dignissimi viri HENRICI LUCAS de London Armigeri, nuper defuncti; salutem in Domino sempiternam.

SCIATIS, quòd cùm prædictus venerabilis et consultus vir HENRICUS LUCAS Armiger, ex propenso suo in Academiam Cantabrigiensem, et in rem literariam affectu à præfatis ROBERTO RAWORTH, & THOMA BUCK Executoribus suis terras(4) comparari ad valorem centum librarum annuatim supremo testamento mandaverit, in annuum Professoris, seu Lectoris Mathematicarum scientiarum in dicta Academia stipendium, vel salarium perpetuò cessuras, sub ejusmodi constitutionibus et regulis, quas Executores sui, adhibito Procancellarij, et Præfectorum collegijs dictæ Academiæ consilio, tam honori magni istius corporis, quàm hujusce literaturæ, omni hactenus præmio destitutæ, incremento et promotioni judicaverint sum̄operè accom̄odatas; Nos prædicti Executores, pro ratione fidei nobis com̄issæ de exequenda præclari Benefactoris voluntate soliciti, rogato priùs et impetrato dictorum Procancellarij, et Præfectorum consilio, habitoq; ad id consilium præcipuo respectu, ordinationes infra-scriptas promovendis istis studijs Mathematicis, ùti arbitramur, apprimè conducentes approbamus, omniq; per dictum Henrici Lucas testamentum nobis concessâ authoritate ratas volumus, et declaramus.

Itaq; statuimus imprimìs et ordinamus, quòd perpetuis futuris temporibus, quicquid annui reditûs (deductis necessarijs expensis) ex prænotatis terris, ad

(2) This is taken from a folio-size document (now in Packet E of the Lucasian papers, ULC. Res. a. 1893) which appears to be the original draft of the statutes. The main body of the text is in a large amanuensis hand, but Barrow has inserted one or two minor corrections in it and has also added in his own hand the final paragraph and the witnesses' affirmation. Preserved with the document is a later copy in Newton's hand both of the statutes and the royal confirmation, made no doubt in autumn 1669 on his appointment to the professorship.

(3) Both Raworth and Buck were great personal friends of Lucas. Raworth was a successful London lawyer without close ties with Cambridge—at least no record of his matriculation there exists—but Thomas Buck had, together with his brother John, been a licensed University printer since 1625. (See S. C. Roberts, *A History of the Cambridge University Press, 1521–1921* (Cambridge, 1921): 44–58.)

(4) Land was, in fact, bought for the purpose in the villages of Thurleigh and Riseley to the north of Bedford. Shortly after taking up the Lucasian chair in 1669 Newton had to fight hard to avoid paying tax on his professorial stipend, as an undated letter of his (to the Chancellor of the Exchequer?) makes clear: 'My Tenants acquaint me that some of ye Commissioners for ye p^rsent Tax, considering that ye clause in this Act for excusing ye University is otherwise worded then it was in ye last Act & does not excuse Colleges, designe now to tax ye lands in Thurleigh and Risely belonging to my Lectureship. This made me think it proper to acquaint you that my lands are not in the nature of College lands but of a salary or stipend belonging to my Lectureship, & that tho the words of ye Act run so that they do not excuse Colleges from being taxed for their lands yet it excuses all particular members of Colleges from being taxed severally for the salaries & profits belonging to their several places....Colleges are excused for the scites of their Colleges but not for their revenues, but Readers are excused for all their profits whatsoever. [The] words of ye Act...I think include me plainly as one of the Readers & by consequence excuse me expresly from paying for any of the profits of my Professorship' (ULC. Add. 3970. 10: 645^r).

usum prædictum acquisitis vel acquirendis, quacunqꝫ justâ ratione accreverit, id integrè cedet in subsidium et præmium Mathematici Professoris modo infradicendo, et sub conditionibus mox exprimendis electi & constituti.

Quod officium attinet dicti Professoris Mathematici, ut horum studiorum quà publicè, quà privatim excolendorum ratio habeatur, volumus et statuimus, ut dictus Professor teneatur singulis intra uniuscujusqꝫ termini Academici spatium[(5)] septimanis semel ad minus aliquam Geometriæ, Astronomiæ, Geographiæ, Opticæ, Staticæ, aut alterius alicujus Mathematicæ disciplinæ partem (pro suo arbitratu, nisi aliter expedire Procancellario visum fuerit) per unius circiter horæ spatium legere atqꝫ exponere, loco et tempore à Procancellario assignandis; sub pæna quadraginta solidorum pro singula lectione omissa ex stipendio ipsi debito per Procancellarium subtrahendorum, et Bibliothecæ Academicæ pro coemendis libris vel instrumentis Mathematicis applicandorum; nisi ex gravi corporis infirmitate officio suo satisfacere non poterit; quam tamen excusationem nolumus ultra tres septimanas valere, ut nisi elapsis tribus septimanis alium substituat idoneum Lectorem, Procancellarij judicio approbandum, sciat sibi pro qualibet lectione prætermissa viginti solidos de stipendio suo per Procancellarium subtrahendos, et usui prædicto applicandos. Quò autem dictus Professor ad munus hoc legendi non perfunctoriè præstandum efficaciùs astringatur, præstiti fideliter ab ipso officij certius extet indicium, et studiorum præsentium fructus quadantenus etiam ad posteros derivetur, statuimus ut dictus Professor semel quotannis, proximè ante festum sancti Michaelis,[(6)] non pauciorum quàm decem ex illis, quas præcedente anno publicè habuerit, lectionum exemplaria nitidè descripta Procancellario exhibeat, in publicis Academiæ archivis asservanda; quod si ante tempus præscriptum facere neglexerit, eousqꝫ careat stipendio suo, donec effectum det; et quanta fuerit rata portio temporis posteà elapsi, usquedum id perfecerit, tantam reditûs sui, vel salarij annui partem Procancellario teneatur evolvere, Bibliothecæ Academicæ ad usus prædictos applicandam. Quinetiam decernimus, ut dictus Professor teneatur duobus per singulas cujusqꝫ termini hebdomadas a Procancellario præstituendis diebus (unoq́ꝫ extra terminū die, quandocunqꝫ dictus Professor in Academia præsens fuerit) per duas horas itidem præfigendas omnibus illum consulturis vacare, liberum adeuntibus aperto cubiculo accessum præbere, circa propositas ipsi quæstiones et difficultates haud gravatè respondere, in eum finem globos, et alia idonea instrumenta

(5) On the back of his copy of the Lucasian statutes (note (3)) Newton added as a reminder that

'Termini durant 1. a 10° Octob ad 16ū Decemb
2. a 13° Jan ad 10 ante Pascho
3. ab 11° post Pascha ad diem veneris Comitia sequentem.'

The 'Comitia' (commencement exercises) were held in early July.

(6) 29 September, the autumn quarter-day.

Mathematica penes se in promptu habere,[7] inq; omnibus ad illud propositum spectantibus studiosorum pro sua virili conatus adjuvare; quorum aliquod si ultro neglexerit, corripiatur à Procancellario; et si de neglecto officio sæpiùs admonitus neutiquam se emendaverit, pænam incurrat intolerabili negligentiæ inferiùs decretam.

Porrò ut horum observatio fortiùs muniatur, et nè quis ex dicti Professoris absentia aboriatur neglectus, statuimus nè dictus Professor intra præfinita terminorum intervalla Academiâ excedat, aut alibi extra Academiam per sex dies continuos moretur, nisi gravissimâ de causâ per Procancellariũ approbandâ, idq; petitâ priùs et impetratâ à Procancellario veniâ; sin fecerit secus, quanta fuerit rata portio temporis ab egressu suo præter lapsi, tantâ salarij sui parte penitus excidat.[8] Quòd si fortè diutioris ab Academia absentiæ (quæ dimidij termini spatium excedat) causa acciderit necessaria, Procancellarij, et duorum, qui collegiorum Præfecti fuerint, seniorum Doctorum judicio approbata, aliquem interea idoneum substituat, qui suo loco legat, et reliquis munijs fungatur, modo supradicto et sub pæna consimili.[9] Quinimò pari causâ, nè dictus Professor ab officij sui debitâ executione distrahatur, nolumus omnino et prohibemus, ut is ullâ quâvis Ecclesiasticâ promotione gaudeat, quæ animarum sibi curam adnexam habeat, aut residentiam exigat hisce statutis adversantem; sub pæna amissionis, ipso facto, omnis juris, quod in hac sua professione prætendere valeat.[10]

Quoad personam verò et qualitatem Mathematici Professoris, volumus et injungimus, ut qui huic provinciæ admovetur, sit vir bonæ famæ, et conversa-

(7) In Trinity College and elsewhere in Cambridge there exist compasses and other old mathematical instruments which by tradition belonged to Newton, but it now seems impossible to authenticate their ownership. The 'inventary' of Newton's house taken at his death records, among other items, 'In the *fore Garret*. Imprimis, a parcel of Mathematical instruments', together with 'three globes' and 'a Bath mettle case of instruments' in the 'fore room' below (R. de Villamil, *Newton: the man* (London, [1931]: 50–1).

(8) During his active tenure of the Lucasian chair (November 1669–March 1696) Newton was, in fact, away from Cambridge for more than a month at a time on only two occasions, 9 February to 19 March 1675 and 28 July to 27 November 1679. (See Edleston's *Correspondence* (note (1)): lxxxv.) No doubt both absences were sanctioned by the Vice-chancellor 'gravissimâ de causâ', for Newton probably spent the former time in London seeking royal approval of his submitted draft patent confirming the Lucasian professor's right to hold a college fellowship without obligation to go into orders (see note (10)), while during the latter period he was away in Lincolnshire nursing his mother on her death-bed.

(9) Newton took advantage of this clause to retain his chair after he took up residence in London in 1696, not in fact resigning it till 10 December 1701 (Edelston's *Correspondence* (note (1)): lxx, note (142)). Since the preceding January Whiston, his successor in the professorship, had lectured (on astronomy) as his deputy, receiving 'the full profits of the place' as statute required.

(10) This, confirmed by the following royal mandate 'ut dictus Professor eligi possit in socium cujusvis Collegij non vetante Professione sua & ne is sodalitio suo (si quod ante susceptum hoc munus obtinuit aut postea obtinebit) vel ullis sodalitij sui emolumentis aut privilegijs eo tantùm nomine seu causa privetur, quovis cujuscunq; Collegij statuto non

tionis honestæ, ad minimum Magister Artium, probè eruditus, et Mathematicarum præsertim scientiarum peritiâ instructus. Ejus autem nominandi et eligendi jus et potestas esto penes nos prædictos Executores venerabilis viri Henrici Luca, durante nostra utriusqꝫ vitâ, aut uno supremum diem obeunte, penes alterum è nobis, qui in vita superstes permanserit.(11) Postea verò perpetuis futuris temporibus ejus eligendi plena potestas sit penes Procancellarium, et Præfectos omnium collegiorū dictæ Academiæ, vel illam partem Præfectorum qui electioni interfuerint; et peragatur tunc electio in hunc modum. Postquam Mathematici Professoris locum quacunqꝫ ex causa vacare contigerit, Procancellario incumbet, quàm citò fieri poterit, schedulâ scholarum publicarum ostio per octo dies continuos affixâ, cùm de dictâ vacatione, tum de tempore ad futuram electionem destinato significare (tempus autem electionis ultra trigesimum à prima significatione diem extrahi nolumus) quo tempore dicti Electores publicis in scholis in unum congregati juramento semetipsos ob-

obstante' (see note (16) below), gave Newton an all but impregnable position from which to fight when, in early 1675, he sought to exempt himself from Barrow's insistent enforcing of the college statute which ordained that all fellows of Trinity should take holy orders. As a copy of the letter in Newton's hand (ULC. Res. a. 1893: Packet E) records, Coventry wrote from Whitehall on 2 March 1674/5 that 'His Ma^{ty} being willing to give all just encouragement to learned men who are & shall be elected into ye said Professorship, is graciously pleased to refer this draught of a Patent unto Mr Atturney Generall to consider ye same, & to report his opinion what his Ma^{ty} may lawfully do in favour of ye said Professors as to ye indulgence & dispensation proposed'. The official dispensation which followed on 27 April (quoted in full, from Newton's copy in the Lucasian papers, in Edleston's *Correspondence* (note (1)): xlix–l, note (43)) affirmed that the Lucasian professor had been given the privilege of holding a college fellowship and enjoying its emoluments without prejudice and added: 'Insuper volumus & statuimus ut verba nostra prædicta in favorem dicti Professoris semper accipiantur, ut non eo tantum sed nec alio quovis nomine aut causa sodalitio suo aut ejus emolumento privetur nisi quod quemlibet ejusdem Collegij Socium cujuscunqꝫ professionis & ordinis meritò privare debeat. Et speciatim volumus et ordinamus ut ordines sacros non nisi ipse voluerit, suscipiat, nec ob defectum sacrorum ordinum sodalitū cedere ipse teneatur aut ab alijs quibuscunqꝫ cogatur, sed ea immunitate quamdiu suo munere fungitur gaudeat et fruatur quo quilibet socius Medicinæ aut Juri Civili vel Canonico dicatus frui solet, quovis cujuscunqꝫ Collegij statuto aut consuetudine vel interpretatione quacunqꝫ non obstante'. Without the backing of the Lucasian statutes Newton would have stood little chance in his plea for royal indulgence: his friend Francis Aston had failed in his request for a similar dispensation the previous December, even though he had the backing of Williamson, Principal Secretary of State (Edleston, *ibid.*). How far Newton's victory led to a cooling in his friendship with Barrow we do not know. Nor should we assume that Newton's abhorrence of taking orders was the sole or even main reason for his seeking the royal patent. It may well be that his chief concern was to resist what he regarded as an intolerable attempt on Barrow's part to limit the statutory privileges of his professorship.

(11) An interesting, little known clause. Since Thomas Buck died on 4 March 1669/70 and Raworth was still alive when Barrow dedicated his *Lectiones XVIII...In quibus Opticorum Phænomenωn Genuinæ Rationes investigantur, ac exponuntur* to Lucas' two executors in late 1669, both Barrow and Newton were appointed solely by Buck and Raworth (though no doubt the opinions of the Vice-chancellor and, in Newton's case, of the retiring professor would be sought as a courtesy).

stringant, seposito omni privato respectu, affectuqꝫ sinistro se nominaturos, et suo comprobaturos suffragio quem conscientiâ teste ex petitoribus (vel ex ijs, qui ab Electorum quolibet nominantur) maximè secundum prænotatas qualitates idoneum censuerint ad id munus obeundum; is verò, in quem plura suffragia conspiraverint, pro Electo habeatur: Quòd si duobus fortè vel pluribus paria obvenerint suffragia, Procancellario jus erit unum ex illis, qui alijs (si quando id contigerit) plura, et inter se paria suffragia obtinuerint, pro suo arbitrio eligendi. Electus autem proximo opportuno tempore admittatur à Procancellario; præstito ante admissionem juramento se munus Professoris Mathematici à dignissimo viro Henrico Lucas in hac Academia institutum, juxta ordinationes et statuta officium suum concernentia pro suo posse fideliter executurum.

Deniqꝫ quò prædictus Professor intra debitos honestatis atqꝫ modestiæ limites contineatur, neqꝫ de ulla criminum ultro admissorum impunitate præsumat, statuimus et decernimus, ut si dictus Professor convictus fuerit vel propriâ confessione, vel per testes idoneos, vel per evidentiam facti de crimine aliquo graviore (puta de læsa Majestate, hæresi, schismate, homicidio voluntario, furto notabili, adulterio, fornicatione, perjurio) vel si intolerabiliter negligens fuerit, neqꝫ pænis superiùs expressis poterit emendari, per Procancellarium et collegiorum Præfectos (vel majorem eorū partem) à sua amoveatur professione, sine spe regrediendi, aut com̄odum aliquod ulterius percipiendi.[12] Quòd si senio, morbo diuturno, aut incurabili impotentia seu debilitate corporis vel animi fractus suæ professioni (modo et formâ prædictis) perficiendæ non suffecerit, abrogetur ei professio per Procancellarium et dictos Præfectos (vel majorem eorum partem), hoc tantùm adhibito moderamine, ut illi (sic à Professionis munere non suâ culpâ dimisso) si tempore professionis suæ laudabiliter se gesserit, neqꝫ aliàs ipsi de proprio ad valorem centum librarum annuatim provisum fuerit, tertia pars stipendij remaneat usqꝫ ad mortem suam, reliquis partibus successor sit tantisper contentus, integrum post ejus mortem percepturus.

In Cujus Rei testimoniū nos p^{r}fati Rob^{tus} Raworth et Thomas Buck sigilla n^{ra} p^{r}sentibus apposuimus Dat. Decimo nono die Decembris Anno Regni D^{ni} n^{ri} Caroli secundi Dei gr^{a} Angliæ Scotiæ Franc[i]æ et Hib^{niæ} Regis fidei defensoris &c decimo quinto[13] Annoqꝫ D^{ni} 1663.

[Robertus Raworth
Thomas Buck]

(12) This morality clause was invoked in 1711 when Whiston was expelled from the Lucasian chair for repeatedly—and impolitically—supporting Arianist views in public and antagonizing the college masters who had the power to unseat him. Newton, of course, held equally heretical views but took care to keep his beliefs to himself till after he had resigned the professorship.

(13) Counted from the execution of Charles' father in January 1649.

Sigillat. et deliberat. in p[r]sentia

Ja: Windet M.D.
Isaac Barrow
Rich. Spoure
Michael Glyd
Gulielmi Player.(14)

---

[2](15)

CAROLUS 2dus Dei gratiâ Angliæ Scotiæ et Hiberniæ Rex, fidei defensor, &c: omnibus et singulis has literas visuris salutem.

Cùm per dilectos nostros Robertum Raworth & Thomam Buck Armigeros, Executores ultimæ voluntatis consulti Viri Henrici Lucas Armigeri nuper defuncti innotuerit nobis, dictum Henrici Lucas in rei literariæ beneficium Professoris Mathematici munus in Academiâ nostrâ Cantabrigiensi instituisse, nec non dictis iisdem Executoribus suis, adhibito Procancellarii et Præfectorum Collegiis dictæ Academiæ consilio quæ ad dicti muneris commodam constitutionem ac executionem debitam sibi ex usu viderentur ordinandi curam commendasse; et cùm dicti Executores pro ratione fidei sibi commissæ, rogato prius et impetrato prædictorum Procancellarii & Præfectorum consilio, ordinationes quasdam fecerint eò conducentes, quas a nobis suppliciter oraverint authoritate nostra Regiâ stabilitas iri et ratificatas:

Item, cum dicti Executores de prædicto munere ulterius benè mereri studentes, juxta consilium prædictorum Procancellarii & Præfectorum, per sanctionem quoque nostram Regiam dicti Professoris Lectionibus publice habendis, certos ex Academiâ auditores assignari; ab Electoribus dicti Professoris et ab electo Professore juramentum exigendi potestatem fieri; dicto Professori si Collegii cujusvis socius fuerit, sodalitii sui unà cum dicto Professoris loco retinendi privilegium indulgeri; et si ante susceptum hoc munus socius non fuerit, ut postea in socium eligi possit, facultatem concedi; eundem

(14) James Windet, a graduate of Queen's College, Cambridge, who achieved a mild contemporary reputation as a Latin linguist, was at this time vicar of a London parish. William Player may be the Playters who also graduated from Queen's to become a London parish priest. Richard Spoure (Spoore?) and Michael Glyd are unknown to us. Isaac Barrow was appointed the first occupant of the Lucasian professorship on 29 February following (ULC. Baker MS 29: 113), a choice which surprised no one.

(15) Collated from two copies of the original letter, made by Barrow and Newton respectively, which are now in the Lucasian papers (ULC. Res. a. 1893. Packet E): The former, not quite complete, is headed 'The Kings Confirmation of the said Professorship, with addition of Priveleges'. A third transcript made by Thomas Baker in the early eighteenth century (ULC. Baker MS 29: 403–4: 'Literæ patentes circa officium Professoris Mathematici') was taken 'From a Copy in the Reg[rs] Office' which was presumably the common source for Barrow's and Newton's versions.

demum Professorem a muneribus quibusdam publicis obeundis eximi ac prohiberi, a nobis itidem supplices efflagitaverint:

Nos dictæ Academiæ commoda juxta ac studiorum profectui consultum cupientes, dictorumque votis Executorum benignè annuentes, imprimis ordinationes istas a prædictis Executoribus confectas regiâ nostrâ potestate sancimus et confirmamus, iisque omnibus et singulis plenum obsequium deberi atque præstari volumus et mandamus.

Item Lectionum quas secundum ordinationes prædictas Professor Mathematicus habere tenetur, auditores esse volumus non-graduatos omnes post annum secundum, et Artium Baccalaureos omnes usque ad annum tertium sub eadem pæna cui per statuta Academica obnoxii sunt a reliquis publicis Lectionibus absentes.

Quinetiam Procancellario potestatem impertimus juramentum exigendi et administrandi, tam illis qui per dictas ordinationes eligendi jus obtinent, quàm Professori electo et mox admittendo, juxta sensus in dictis ordinationibus expressos atque præscriptos.

Porrò volumus et statuimus ut dictus Professor eligi possit in socium cujusvis Collegii, non vetante Professione suâ, et ne is sodalitio suo, siquod ante susceptum hoc munus obtinuit, aut postea obtinebit, vel ullis sodalitii sui emolumentis aut privilegiis eo tantum nomine seu causâ privetur, quovis cujuscunque Collegii statuto non obstante.(16)

Nolumus denique et prohibemus, ut dictus Professor Decani, Thesaurii, Seneschalli, aut Lectoris cujusvis in suo Collegio munus capessat, aut ut inibi Tutorem se gerat, (nisi fortè nobilium vel generosorum sociis commensalium,) vel ut Procuratoris, Taxatoris, aut alterius cujuslibet Lectoris publicum in Academiâ officium sustineat, sub pænâ amissionis ipso facto omnis juris quod in hac suâ professione præcedere valeat; quapropter et dictum Professorem ab omnibus et singulis muneribus istis prædictis liberatum volumus et exemptum.(17)

In cujus rei testimonium has literas nostras fieri fecimus patentes, teste meipso apud Westmonasterium 18° die Januarii A.R. nostri 15°.

Per breve de privato sigillo.

Hastings.(18)

---

(16) See note (10).

(17) 'This prohibition will account for our not finding Newton's name at any time among the College or University Officers. He availed himself of the privilege of taking Fellow-commoners as pupils in two instances only: viz. Mr George Markham..., afterwards Baronet and F.R.S., entered Jun. 26, 1680, and Mr Robt. Sacheverell,...entered Sept. 16, 1687. We also find Mr St. Leger Scroope...entered Fellow-commoner under him Apr. 2, 1669, before he was appointed Lucasian professor' (Edleston, *Correspondence* (note (1)): xlv, note (16)).

(18) Evidently Theophilus Hastings, seventh Earl of Huntingdon, at this time Custodian of the Rolls for Warwickshire.

ANALYTICAL TABLE OF CONTENTS

PREFACE *page* vii

EDITORIAL NOTE ix

GENERAL INTRODUCTION xi

LIST OF PLATES xxxix

PART 1

RESEARCHES INTO FLUXIONS AND INFINITE SERIES (1670–1671)

INTRODUCTION 3

Barrow reports on a new Newtonian treatise of infinite series (December 1670), 3. Newton tells Collins he has begun to 'new methodiz' the *De Analysi* but that other 'buisinesse' has delayed him (July 1671), 5. The revise now (Christmas 1671) to be appended to the edition of his '20 Dioptrick Lectures', 7. Prompted by 'wrangling disputes' Newton abruptly decides (May 1672) to publish neither, 8. Sherburne still advertises the treatise on series as 'expected' (in 1673), 10. The manuscript copied by Craige (1685?) and Jones (1710?): Halley and Raphson still plan to print it (1691), 11. Pemberton's intended publication thwarted by Newton's death (1727): Colson's English (1736) and Horsley's Latin (1779) *princeps* editions, 13. Contemporary quotation by Newton of the 1671 tract, 14. His conception (inspired by Barrow?) of time and fluxional increase, 17. Quadrature of curves defined by trinomial equations (1676?), 18.

APPENDIX (ULC. Add. 3968. 39: 538^r). Newton's citation of the 1671 tract in a draft preface to his *Commercium Epistolicum* (late 1712): quotations from Collins' contemporary correspondence with James Gregory, Bertet, Borelli, Vernon and Newton (1670–2), and from Newton's *epistolæ prior et posterior* for Leibniz (1676), 20.

1. PRELIMINARY SCHEME FOR A TREATISE ON FLUXIONS 28
(In private possession)

Two general problems on the direct and inverse methods of fluxions, followed by applications (21 problems) to the determination of extreme values, tangents, quadratures, curvature and rectifications and to the comparison of curves.

2. THE TRACT '[DE METHODIS SERIERUM ET FLUXIONUM]' 32

§1 (restored from contemporary transcripts by Jones and David Gregory). The missing first leaf of Newton's autograph manuscript. Introduction: algebra as universal arithmetic, with infinite series the analogue of unterminated decimal sequences, 32. An example of reducing an algebraic fraction by infinite division, 36.

and composite curves (influence of James Gregory and Barrow), 128. Mode 2, 'focus-directrix' co-ordinates: examples include the pair of a conic and quartic and the conchoid, 132. Mode 3, standard bipolars: instanced in the Cartesian oval, 136. Modes 4/5/6, variants on a simple monopolar system (the last a generalization of Cartesian co-ordinates): no examples, 138. Mode 7, polar co-ordinates: examples include the Archimedean and Fermatian spirals, 140. Mode 8, 'quadratrix' co-ordinates: the quadratrix itself is Example 2 (variant proof of Barrow's construction of the subtangent), 144. Mode 9, 'trochoidal' co-ordinates (taken from Gregory): Example 2 is the general cycloid, 146. Enunciations of seven allied problems involving consideration of tangents (instantaneous direction of a curve at a point), 148.

Problem 5. Determination of the curvature of a curve at any point. Preliminary considerations: the curvature is that of the osculating circle, 150. Possible defining 'symptoms' of the centre of curvature: limit-meet of normals in the vicinity of a point, instantaneous centre of motion, centre of the circle having 3-point contact, 152. Analytical measure of the radius of curvature in Cartesian co-ordinates (centre defined as limit-meet of normals): examples include the general conic, cissoid, conchoid, 156. Curvature in the cycloid: properties of its evolute (a congruent cycloid), general observations on angles of contact stemming from the (infinite) curvature at its cusp, 160. The quadratrix, 168. Analytical measure of curvature in polar co-ordinates: instances in the Archimedean and general Fermatian spirals, 168. The (cumbrous) equivalent method of measuring curvature by constructing the circle having 3-point contact (worked through on the simple Apollonian parabola), 174. 'Certain related questions': to determine the points on a curve where there is a given curvature, 178. 'Straight' points (general inflexions, of zero curvature) and cusps (of infinite curvature), 178. Location of points of extreme curvature, 180. Determination of the evolute: the cycloid and parabola are cited as examples, 182: The diacaustic of refraction at an interface, 184. Radius of curvature at a vertex (the limit value of the subnormal): examples in the ellipse and conchoid, 184.

Problem 6. The 'quality' of curvature at a point. Justification of the (intrinsic) measure of 'inequability' of curvature by the ratio of the fluxion of the curvature radius to that of the arc-length (zero in the circle, constant in the logarithmic spiral), 186. Evaluation of this measure ($d\rho/ds$) in the general conic (varies as the ordinate), 188. The inequability of curvature in the logarithmic spiral and cycloid compared with that in a given parabola ($y^2 = 6x$), 192. Related problems: to determine the points of a curve where the inequability of curvature has a given value, and to identify members of a family of curves (especially conics) which have given curvature and variation of curvature at a point, 194.

Problem 7. Determination of the area of curves capable of exact algebraic quadrature. Statement of the fundamental theorem: the fluxion of the area of a curve is proportional to its ordinate, 194. Examples: the Apollonian and other simple parabolas, 196.

Problem 8. General transformations which preserve area. Transform of

$$s = \int v.dx \rightarrow t = \int y.dz$$

by $f(s, t, v, x) = 0$ and $\phi(s, x, z) = 0$: general observations, 198. Examples in geometrical curves: the circle $v^2 = ax - x^2$ and hyperbola $v^2 = x^2 + c^2$, 198. The cissoid $v^2 = x^3/(a-x)$, 202. Use of integration by parts to evaluate the area under mechanical curves (of algebraic slope): exemplified in the cycloid and sinusoid, 204. Cavalierian area-preserving convolution of a Cartesian curve into a corresponding polar spiral: the Apollonian parabola and its convolute (an Archimedean spiral), 206. The main use of Problems 7 and 8 is to draw up catalogues of the areas of curves, 208.

Problem 9. Methods of determining the area of any curve proposed. Further examples of simple exact quadrature in the case of general parabolas and hyperbolas, 210. By reducing the ordinates to convergent infinite series all curves may be reduced to integrable form: examples are mostly taken over from the introduction, 212. Integral bounds: in all cases the area is the difference of the values of the integral corresponding to the upper and lower limits of the base, 216. Interpretation of negative areas, 218. 'Square' notation for hyperbola-area (to be evalutated by an infinite series) introduced and exemplified, 218. Numerical calculation of areas expressed by infinite series: parallel computation of the hyperbola/circle segments $\int_0^{\frac{1}{4}} \sqrt{[x \pm x^2]}\,.dx$, 222. An evaluation of π (correct to 15 D) is a corollary, 226. Computation of the area $\log(1+x)$ under the hyperbola $y = 1/(1+x)$, x small: a simplified version of his 1665 calculations, 226. Construction on this basis of a canon of natural logarithms (of low primes): their conversion to common form by dividing through by log 10, 230. A method of subtabulating means in this table, 232. A hint that it may be generalized to 'interpolating' arbitrary means, and that similar methods will produce tables of logarithmic sines and tangents without prior construction of the natural sines and tangents, 234. A 'prior' catalogue of curves of form

$$y = dz^{\lambda\eta-1}(e+fz^{\eta})^{\mu}$$

and their exact areas 'constructed by the help of Problem 7', 236. And of the ordinates of curves whose area is $z^{\theta}(e+fz^{\eta}+gz^{2\eta})^{\lambda}(h+iz^{\eta})^{\mu}$, 'by means of which the way is opened up to higher things', 238. Preliminary explanation of a diagram depicting instances of the general central conics $d = (e+fx)v$ and $v^2 = e+fx+gx^2$, 240. A 'posterior' catalogue, 'constructed with the help of Problem 8', evaluating the areas of curves of form

$$y = dz^{\lambda\eta-1}(e+fz^{\eta}+gz^{2\eta})^{\mu}$$

and $y = dz^{\lambda\eta-1}(e+fz^{\eta})^{\mu}(h+iz^{\eta})^{\nu}$ in terms of the areas s (and σ) of these, 244. Observations on the two catalogues: unknowns may be positive or negative, 258. Ways of indefinitely extending either table, 260. Duplicate entries are retained because of their usefulness, 262. Negative areas: correctly choosing bounds, 264. Similar tables might be constructed involving comparisons with curves of higher order, 264. The latter catalogue used to square special curves: the *versiera* (witch of Agnesi), 266. The kappa (Gutschoven) quartic, 268. The cissoid and conchoid, 270. The quartic $y^2(c^2-z^2) = (z^2+bz-c)^2$, 272. The quadrature of mechanical curves is (where possible) to be reduced to that of allied geometrical ones by means of Problem 8 (integration by parts when slope is algebraic, or reverse convolution from polar to Cartesian form), 274. An integration by parts yields the area of the 'conic-arc' curve, 276. Justification of the quadratures of the four latter curves by *ad hoc* limit-increment methods, 278. Equivalent fluxional arguments in the case of the kappa quartic and cissoid, 282. Scholium. The area and ordinate of a curve is but a canonical instance of the relationship of a fluent quantity to its fluxion: by choosing appropriate units the two preceding catalogues may be used in all problems of integration (rectifying arcs, measuring solid content, and so on), 284. Related topics: the 'mechanical' approximation of areas, 288. Given the area z of a curve $r = f(x)$ as an infinite series in terms of its base x, to find the ordinate r (or $\dot{z}$) as a series in z, 290.

Problem 10. To find arbitrarily many curves of finite (algebraic) arc-length. Five preliminary observations on evolute properties: the length of a convex evolute's arc is the difference of the curvature radii at the points on any involute of it corresponding to its end-points, 292. Measure of the curvature radius in a Cartesian co-ordinate system: rectification of Neil's semicubical parabola by determining the curvature of its involute

APPENDIX

NEWTON'S MATHEMATICAL CORRESPONDENCE (1670–1673)

LIST OF PLATES

PART 1

RESEARCHES INTO FLUXIONS AND INFINITE SERIES (1670–1671)

INTRODUCTION

In our foreword to the *De Analysi per æquationes numero terminorum infinitas* we supported the view that this celebrated tract of Newton's on infinite series was rushed into private circulation in the summer of 1669 as a hastily contrived assertion of his priority in face of Nicolaus Mercator's *Logarithmotechnia* (which had appeared a few months before). Subsequently, when urged by John Collins to allow it to be printed in appendix to Isaac Barrow's mathematical lectures or, later still, by itself, Newton was manifestly unwilling to permit publication of the treatise without adding some necessary revisions to it.(1) How soon afterwards he did, in fact, begin to reorder his hurriedly written compilation is not known for sure—not for more than a year, we may surmise. In the weeks immediately following his appointment, in late October 1669, as Lucasian Professor of Mathematics he was evidently preoccupied in preparing the lectures on optics which he began to deliver in the Schools at Cambridge in January 1670,(2) but already by winter's end the pressures of his academic post had relaxed sufficiently for him to open a mathematical correspondence with Collins and to busy himself with making notes, at Collins' request, on Mercator's Latin version of Kinckhuysen's *Algebra ofte Stelkonst*.(3) The latter task commanded, as we have seen, considerably more of Newton's time than he had originally intended; indeed, a first 'reveive' of Kinckhuysen's book sent to Collins in July 1670 was speedily returned with the request that he exert upon it the 'some more paines' which he had earlier expressed himself willing to undertake.(4) By mid-autumn, when his improvements to the *Algebra* were all but finished, a second set of optical lectures had to be prepared and delivered, and it is unlikely that Newton would have had spare time enough to devote to revising the *De Analysi* until December, when for the first time we hear that he had begun to improve upon its content.

Our informant is Collins, parroting for James Gregory a conversation he had had with Barrow the previous autumn on the topic of Newton's 'generall

(1) See II: 165-8. The last we hear of Collins' intention to publish the *De Analysi* is a passing reference in Newton's letter to him on 5 September 1676: 'As for my paper I sent about infinite series, I know not whether it will be proper to print it. I leave it to your discretion. In my apprehension it may do as well to suppress it, but if you think otherwise I desire you would give me notice before it go into ye Press, because of altering an expression or two' (*Correspondence of Isaac Newton*, **2**, 1960: 95).

(2) See the introduction to Part 3 below. As late as 18 February 1669/70 Newton could assert to Collins—a little disingenuously perhaps—that 'I have noe leisure at present for computations' (*Correspondence*, **1**, 1959: 27).

(3) See the concluding appendix to the present volume and II, **3**, 1, §2.

(4) Compare II: 282–7.

method' of series. Writing to Gregory on Christmas Eve 1670 Collins remarked, of 'some few Series' extracted from the *De Analysi* for Gregory's benefit, that

by discourse [with Dr Barrow I] gather they are Analytically derivable from the given properties of any figure, and that many Series may be applyed to each figure. That the method is universall and performes all Quadratures as well as those Curves which Deschartes admitts to be geometrick, as others he accounts Mechanick. Hence in this method, the curved lines of all figures that have a common property, are streightened, their tangents and Centers of Gravitie discovered, their round Solids and the second Segments thereof cubed and in all Curves, the length of the Curved line being given, an Ordinate is found and the Converse.... He hath as I understand wrote a treatise of this method, and made the length of the portions or intire Elliptick line, and of the quadratick [*sc.* quadratrix] Curve.(5)

Barrow's mention of a new 'treatise' of Newton's which, among other things, discussed in detail two rectification problems merely hinted at in the *De Analysi*(6) must vastly have intrigued Collins himself. What other novelties were comprehended in this fresh tract of infinite series and quadratures? Unfortunately, Newton had coldly terminated their correspondence the previous September(7) and a direct request for information might very well be brusquely refused. At length, however, Collins could no longer contain his unrequited curiosity and on 5 July 1671 a cautious feeler was put in the Cambridge post:

I thinke it is almost a yeare since I acquainted the Reverend Doctor Barrow that Mr James Gregory was by his owne Ingenuity falne into your methods of infinite Series, ...but...being since informed by me that you had taken much paines in that harvest, hath laid aside his Intentions of publishing anything....Mercator when he published his *Logarithmotechnia* affirmed he had applyed the same method likewise to the Circle, but I never saw any Specimen of his, nor doe I thinke he hath so much as heard that any other person hath made any Improvement of that Method.... By what I have said you will see no great reason to overhasten the publication of your thoughts

(5) Collins to Gregory, 24 December 1670 (*Correspondence*, **1**: 54–5). The series communicated to Gregory were those for $\sin z$, $\cos z$, $\sin^{-1}x$ and the general segment of 'the ancient Quadratrix'; compare II: 232–8.

(6) Compare II: 232, 240. These rectifications are effected in Examples 7 and 8 of Problem 12, 'Curvarum Longitudines determinare', of Newton's 1671 tract (see 2, §2: notes (744) and (750) below).

(7) As he went on to tell Gregory, 'observing a warinesse in him to impart, or at least an unwillingness to be at the paines of so doing, I desist, and doe not trouble him any more' (*Correspondence*, **1**: 55). Newton's last letter to him on 27 September (*ibid.*, **1**: 44) had in fact been noticeably cool, concluding with an icy expression of thanks for some books lent him by Collins and 'other favours & desiring to bee excused for troubling you thus amongst the midst of your buisinesse'.

(8) Collins to Newton, 5 July 1671 (*Correspondence*, **1**: 65–6).

(9) See II: 288, note (41).

for feare of being prevented by others. Use therefore your owne discretion, and if I heare of any undertakings of the like kind you may expect an account thereof from him that wisheth you good Successe.(8)

This implicit request for information regarding the current state of Newton's new treatise on infinite series drew a not unfriendly reply. 'The last winter... partly upon Dr Barrow's instigation' Newton had indeed begun to 'new methodiz ye discourse of infinite series [the *De Analysi*, that is], designing to illustrate it wth such problems as may (some of them perhaps) be more acceptable then ye invention it selfe of working by such series.' From this intention he had been diverted in late April(9) not through any fear of being forestalled by either Gregory or Mercator but merely by some urgent 'buisinesse in the Country', while since his return to Cambridge he had been otherwise preoccupied—no doubt the third series of optical lectures which he had to deliver in the coming autumn was foremost in his mind. Accordingly,

I have not yet had leisure to return to those thoughts & I feare I shall not before winter. But since you informe me there needs no hast [in putting Kinckhuysen's book 'to the Presse'], I hope I may get into ye humour of completing them before ye impression of ye Introduction [to Algebra], because if I must helpe to fill up its title page, I had rather annex somthing wch I may call my owne, & wch may bee acceptable to Artists as well as ye other to Tyros.(10)

In July 1671, despite Collins' warning that 'the Bookseller Pitts is not desirous as yet to put the Introduction... to the Presse',(11) a joint publication of this sort must have seemed an attractive compromise. Newton was naturally anxious that his first mathematical publication should be something more revealing of his genius than his augmented edition of a maudlin algebraic primer, however saleable, while Collins (if not Newton himself) was well aware how unlikely it was, after the recent commercial failure of works by Wallis and Barrow, that any publisher could be persuaded to publish Newton's technical treatise by itself without a substantial subsidy.(12) For a time Collins at any rate took this half-

(10) Newton to Collins, 20 July 1671 (*Correspondence*, **1**: 68). On 11 July 1670 (*ibid.*, **1**: 31) Newton had stated his preference for an anonymous citation in the style of 'et ab alio Authore locupletata', but on 5 July 1671 (*ibid.*, **1**: 66) Collins had written that 'I conceive you have made so many usefull additions [to the Algebra], that when it comes to the Presse it may very well beare the Title of your Introduction, and thereby find the better entertainement, and more Speedy Sale'.

(11) *Correspondence*, **1**: 66.

(12) As he wrote to Newton later (on 30 April 1672) 'our Latin Booksellers here [in London] are averse to ye Printing of Mathematicall Bookes there being scarce any of them that have a forreigne Correspondence for Vent, and so when such a Coppy is offered, in stead of rewarding the Author they rather expect a Dowry with ye Treatise' (*Correspondence*, **1**: 147). In fairness to London publishers of the period they were (unlike John Hayes in Cambridge, for example) occasionally willing to take a risk and publish not easily saleable advanced mathematical

promise of Newton's seriously and as late as the following May fed news of the projected joint edition to Oldenburg for transmission to Huygens,[(13)] but in fact the hope was still-born. When it became generally known in early 1672 that an 'exhaustive' English *Algebra* by John Kersey was about to appear off the presses a Latin edition of Kinckhuysen could have the shakiest of commercial futures, and Moses Pitt's disposal of his rights in the edition to Newton the following summer was a death-blow to its survival.[(14)] Thereafter the only realistic method of having it published would be, in reversal of the original intention, to sell it on Newton's name and Newton would have none of that. At one point in late April 1672, to be sure, Collins mooted to Newton the possibility that his 'generall method of Analyticall quad[rature]s when extant might be translated and annexed [to] Mr Kersies Paines in Algebra',[(15)] but even if this had been a viable alternative he could have had little hope that Newton would be willing to subordinate his tract to such an inferior, voluminous work.

By late 1671, in fact, Newton himself had conceived a bolder scheme of publication. The third series of optical lectures which he delivered from the Lucasian chair in the autumn meant that, even after some necessary revision, he had enough for a small book on the refraction of light. Why annex his fluxional researches to another's work when he could profitably append it as a mathematical complement to his own investigations into the transmission of light rays? Indeed, without too much trouble he could give his fluxional investigations into tangents and curvature an optical slant.[(16)] But who, even with the promise of a substantial 'dowry', would be prepared to publish this combined Newtonian work? With the dubious exception of the Cambridge 'typographus academicus' John Hayes and the infant Oxford Press, a London stationer was the only realistic possibility—and for the moment Isaac Barrow was Newton's only contact with London. We are accordingly not surprised to

texts, but the simultaneous failure of Barrow's *Lectiones* and Wallis' *Mechanica* in 1670 had been a traumatic experience not readily forgotten. On 15 December 1670 Collins informed Gregory that '[D[r]] Barrow's] Booke [is] not yet exposed to Sale, by reason the Booksellers [John Dunmore and Octavian Pulleyn] are insolvent, and cannot pay for Paper, Printing and Plates, by reason whereof they are detained in the Printers [William Godbid's] hands, but...I have gott 50 Bookes from the Printer for the use of the Author to bestow on his friends' (*James Gregory Tercentenary Memorial Volume* (London, 1939): 137). Writing to Baker six years afterwards, on 10 February 1676/7, Collins could assert: 'The Truth of it is Mathematicall learning will not here go off without a Dowry. The Booksellers have lost soe much by the Works of D[r] Wallis and Horrox [*Opera Posthuma*, 1672], the Optick & Geometrick Lectures of D[r] Barrow &c though by M[r] Gregory and others esteemed the best things extant, that it is no easy task to perswade Booksellers to undertake any thing but toyes that are Mathematicall' (S. P. Rigaud, *Correspondence of Scientific Men of the Seventeenth Century*, **2** (Oxford, 1841): 14). Two months later he added a postscript concerning Barrow's optical and geometrical lectures, 'both which Bookes with the plates as I compute it, did cost the first undertaker 4[s] 8[d] or neare 5[s], & yet att last a greate number of both bookes together with y[e] Plates were sold by

find that, on a visit there just before Christmas, Barrow made Collins privy to Newton's new intention, evidently to make sure of his support. As Collins made clear some weeks later to Gregory, Newton's observations on Kinckhuysen's *Algebra* (but not Mercator's Latin version of the *Algebra* itself) would also be incorporated in the planned edition:

Mr Newton (as Dr Barrow informs me) intends to send up all the Lectures he hath read since he was Professor to be printed here, which he sayth will be 20 Dioptrick Lectures, and some about infinite Series, with his additions to Kinckhuysen's Introduction.(17)

Collins' assertion that Newton had expounded the theory of infinite series at Cambridge—unconfirmed by any other source—is possibly an unwise extrapolation from Barrow's conversation with him, but there can be no doubting his immense enthusiasm for the proposed publication. In a further letter to Gregory a fortnight afterwards, remarking on a current project to print the 'remaines' of Lalouvère, Fermat and other French mathematicians in England, Collins was quick to add that

albeit there might be a willingnesse to undertake some of those in the new Oxonian publick Typography, and the rest here, yet not to præpone them to the Printing of Newtons 20 Dioptrick Lectures I mentioned in my last, which will be first urged, and indeed is exceedingly commended by Dr Barrow.(18)

Unfortunately, enthusiasm was not to be enough and Collins soon found that no London 'bookseller' was willing to undertake the Newton edition. Without a heavy subsidy there was simply no profit in it and no one could afford to be philanthropic in the face of financial ruin.

In April 1672, having heard from Jonas Moore that Newton had a book in press in Cambridge, Collins made the natural (but incorrect) deduction that it

Sr Thomas Davis, Lord Mayor formerly a Stationer...into whose Hands they came for a debte, at the rate of 1^s 6^d a paire or both of them, to Mr Scot a Bookseller in Little Brittain who drives a foreigne Trade, or otherwise they would have turned to Waste Paper' (*ibid.*, **2**: 22).

(13) Oldenburg wrote to Huygens on 6 May 1672 that 'Touchant Kinkhuysen, son introduction est traduite en latin, et sera eslargie par les notes de Monsieur Newton, pour servir come vne introduction à sa methode generale des quadratures analytiques; et quand celles-cy viendront à Londres pour y estre imprimées la dite introduction de Kinkhuysen sera aussi imprimee' (*Correspondence of Isaac Newton*, **1**: 156).

(14) Compare II: 287–91.

(15) Collins to Newton, 30 April 1672 (*Correspondence*, **1**: 147).

(16) The 1671 tract (as we now have it) does, to be sure, have several optical discursions. Compare Newton's pages 42, 48, 52, 66 and 70 in the manuscript reproduced as 2, §2 below.

(17) Collins to Gregory, 23 February 1671/2 (*Gregory Memorial Volume* (note (12)): 218). Likewise, Collins told James' elder brother David on 11 August 1676 that Newton had publicly lectured on infinite series at Cambridge (*Commercium Epistolicum D. Johannis Collins* (London, 1712): 48).

(18) Collins to Gregory, 14 March 1671/2 (*Correspondence*, **1**: 119).

was the projected edition of which Barrow had told him six months before, and felt at last he could break a long silence by writing directly to Newton of his failure to promote it in London:

A little before Christmas the Reverend Doctor Barrow informed me you were buisy in enlarging your generall method of Infinite Series's or quadratures, and in preparing 20 Dioptrick Lectures for the Presse, and lately meeting with Mr Jonas Moore he informed me that he heard you had something at ye Presse in Cambridge possibly about ye same Argum^t^. If so I am very glad partly in regard I now live at Westminster remote from the Printing house[(19)] and partly because our Latin Booksellers here are averse to ye Printing of Mathematicall Bookes.... [But] if yours be not undertaken at Cambridge I shall most willingly affoard my endeavour to have it well done here, and if so, what you have written might be sent up the sooner in order to the Preparing of Schemes.[(20)]

However, the book in press in Cambridge was 'Varenius his Geography', for which Newton had merely 'described Schemes',[(21)] but in communicating this news on 25 May the latter dropped a bombshell:

Your Kindnesse...in profering to promote the edition of my Lectures wch Dr Barrow told you of, I reccon amongst the greatest, considering the multitude of buisinesse in wch you are involved. But I have now determined otherwise of them; finding already by that little use I have made of the Presse, that I shall not enjoy my former serene liberty till I have done with it; wch I hope will be so soon as I have made good what is already extant on my account.[(22)] Yet I may possibly complete the discourse of resolving Problemes by infinite series of wch I wrote the better half ye last christmas wth intension that it should accompany my Lectures, but it proves larger then I expected & is not yet finished....The additions to Kinkhuysens *Algebra* I have long since augmented with what I intended, &...these are at your command. If you have not determined any thing about them I may possibly hereafter review them & print them with the discourse concerning Infinite Series.[(23)]

As Newton was many times to make clear, this sudden decision not to go ahead with printing his 'Dioptrick Lectures' was prompted by the 'wrangling disputes' occasioned by the publication three months before of his 'New Theory

(19) London printing houses of the period were concentrated around Fleet Street and St Paul's in the City.

(20) Collins to Newton, 30 April 1672 (*Correspondence*, **1**: 146–7).

(21) See II: 288, note (43).

(22) Newton refers to 'An Accompt of a New Catadioptrical Telescope invented by Mr. Newton' and extracts from his letters of 6 February, 19, 26 and 30 March, 13 April and 4 May to Oldenburg, printed by the latter in the *Philosophical Transactions* between 19 February and 20 May the same year. (These are reprinted in I. B. Cohen's *Isaac Newton's Papers and Letters on Natural Philosophy* (Cambridge, 1958): 47–78. For reproductions of the original letters see *The Correspondence of Isaac Newton*, **1**: 74–5, 92–102, 121–2, 123–5, 126–9, 136–9 and 153–5 respectively.) Newton was not to be 'done' with the ensuing barrage of criticism from Hooke, Huygens, Auzout, Pardies, Francis Line, Anthony Lucas and others till June 1678 when a

about Light and Colors': 'for the sake of a quiet life' he now 'chose to lay his designe aside before the Tract about the Method of Fluxions was finished. There wanted that part wch related to the Solution of Problemes not reducible to Quadratures'.(24) The impulse to abandon work on the fluxional treatise when publication of its companion dioptrical work was indefinitely postponed was evidently irresistible—understandably so since, with little prospect of having the tract printed by itself and still less of attracting the reading public to it (or so it must have seemed to him), Newton had little incentive left to complete it except for his own personal satisfaction.

As far as is known Newton never made any further serious attempt to finish this mathematical tract, though he was subsequently not unwilling to have it published in its imperfect form. For a time, too, it would appear that there was some hope in London that Newton could be persuaded to renounce his veto on the joint publication. Thus a somewhat ambiguous advertisement put out in 1675 by Edward Sherburne (but no doubt inspired by Collins, the source of much of the former's scientific gossip) ran as follows:

1673. Mr. Isaac Newton *Lucasian Professor of Mathematicks in the University of Cambridge* and *Fellow of Trinity Colledge*, hath lately published his *reflecting Telescope*; *New Theories of Light and Colours*; hath already for the Press a *Treatise of Dioptricks*, and divers *Astronomical Exercises*, which are to be subjoyned to Mr. *Nicholas Mercator's Epitome of*

---

draft of a letter to Aubrey records the *cri de cœur* 'Pray forbear to send me anything more of that nature' (*Correspondence*, **2**: 269).

(23) Newton to Collins, 25 May 1672 (*Correspondence*, **1**: 161). Two months later, writing in some haste from 'Stoake [Rocheford?]' on 13 July, he confirmed his change of mind: 'I think I told you y[t] I had altered my resolution of printing my Dioptrick Lectures. And for y[e] exercise about Infinite series I am not yet resolved, not knowing when I shall proceed to finish it' (*ibid.*, **1**: 215).

(24) Quoted from an English draft (ULC. Add. 3968.9, 97[r]) *post* July 1717 of Newton's unpublished *Historia Methodi Infinitesimalis ex Epistolis antiquis eruta*, itself summarizing a celebrated passage in his *epistola posterior* to Leibniz some forty years before: 'ante annos quinq; cum, suadentibus amicis, consilium cœperam edendi Tractatum de refractione Lucis et coloribus quem tunc in promptu habebam; cœpi de his seriebus iterum cogitare, & tractatum de iis etiam conscripsi ut utrumq; simul ederem. Sed, ex occasione Telescopij catadioptrici epistolâ ad te missâ quâ breviter explicui conceptus meos de natura lucis,...subortæ statim per diversorum epistolas objectionibus aliisq; refertas crebræ interpellationes me prorsus a consilio deterruerunt, et effecerunt ut me arguerem imprudentiæ quod umbram captando eat[e]nus perdideram quietem meam, rem prorsus substantialem....Ipse autem tractatum meum [de methodo serierum] non penitus absolveram ubi destiti a proposito, neque in hunc usq; diem mens rediit ad reliqua adjicienda. Deerat quippe pars illa qua decreveram explicare modum solvendi Problemata quæ ad Quadraturas reduci nequeunt, licet aliquid de fundamento ejus posuissem' (Newton to Oldenburg for Leibniz, 24 October 1676 = *Correspondence*, **2**, 1960: 114). The 'wrangling disputes' excited by Newton's optical correspondence are discussed by R. S. Westfall in 'Newton's Reply to Hooke and the Theory of Colors' (*Isis*, **54**, 1963: 82–96) and 'Newton defends His First Publication: the Newton–Lucas Correspondence' (*Isis*, **57**, 1966: 299–314).

Astronomy, and to be printed at *Cambridge*. From him besides is to be expected a New *General Analytical Method by infinite Series for the Quadrature of Curvilinear Figures, the finding of their Centers of Gravity, their Round Solids, and the Surfaces thereof, the straitning of curved Lines; so that giving an Ordinate in any figure as well such as* DES CARTES *calls Geometrical, as others, to find the Length of the Arch Line, and the Converse*;...how much conducing to the Benefit of *Astronomy*, and the *Mathematical Sciences* in General, such an *Universal Method* is, I leave others, together with my self to admire, and earne[s]tly expect.(25)

This projected Cambridge edition of Mercator's and Newton's astronomical 'exercises' (with or without a Newtonian appendix on dioptrics) has left no other recorded trace, and the former's 'Epitome' was eventually published in London devoid of Newtonian contribution except for a qualitative hypothesis of the moon's libration.(26) How great a stir this first public announcement of an 'expected' treatise on infinite series and quadratures provoked we can now only conjecture—very little, probably, at a period when so many enthusiastic promises of publication proved abortive. No doubt Collins continued to pester the London booksellers on Newton's behalf from time to time, but to no purpose.

In contrast, the manuscript tract itself into which Newton had (though he had never finished it) condensed so much of his mathematical experience was rarely allowed to gather dust on Newton's shelf. He himself used it as a prime

(25) Edward Sherburne, *The Sphere of Marcus Manilius made an English Poem: with Annotations and an Astronomical Appendix* (London, 1675): Appendix: 116.

(26) Nicolaus Mercator, *Institutionum Astronomicarum Libri Duo, de Motu Astrorum communi & proprio, secundum Hypotheses Veterum & Recentiorum præcipuas; deque Hypotheseon ex observatis constructione.... Quibus accedit Appendix de iis, quæ Novissimis temporibus Cœlitus innotuerunt* (London, 1676): 285. Newton's 'exercises' would no doubt comprehend additions to Mercator's account of equant hypotheses of planetary motion. Compare D. T. Whiteside, 'Newton's Early Thoughts on Planetary Motion: a Fresh Look' (*British Journal for the History of Science*, **2**, 1964: 117–37).

(27) Newton's own preferred citations from the *epistolæ prior et posterior* are reproduced in appendix to this introduction.

(28) Compare 2, §1: note (2) and 2, Appendix: note (1) below. James Wilson has left several accounts of these inferior secondary transcripts. Writing to Newton himself on 15 December 1720 he asserted that 'I saw the other day, in the hands of a certain person, several Mathematical Papers, which, he told me, were transcribed from your Manuscripts. They chiefly related to the Doctrine of Series and Fluxions, and seemed to be taken out of the Treatises you wrote on those subjects in the years 1666 and 1671....These papers, I observed, had been very incorrectly copied, so that I endeavoured all I could, to diswade the Possessour of them from getting them printed, of which nevertheless he seemed very fond....I have since met with another Person, who told me, he had likewise a Copy of your Manuscripts. But he would not let me see them, or inform me how he came by them. I imagine, when you lent any of your Friends your papers, the person they got to transcribe them, took a double Copy, which is a frequent practice, in order to make profit by it, so that they are in different hands' (King's College, Cambridge. Keynes MS 143.1, reproduced in David Brewster's *Memoirs of the Life, Writings and Discoveries of Sir Isaac Newton*, **2** (Edinburgh, 1855): Appendix v.1: 440–3). A month later, 'having since been permitted to transcribe some...Copies of several of your Manuscripts..., I take the Liberty to send them to you, that you may compare them with

source of information on several occasions, notably in 1676 when he came to compile his two celebrated mathematical epistles for Leibniz,(27) and was not averse to revealing its content privately to an interested party. Collins himself would appear never to have seen the tract—at least, we have no positive evidence that he did—but John Craige in 1684 and William Jones about 1710 were allowed to make copies of it, and by the last years of Newton's life secondary transcripts made (not without error and distortion) from these were circulating in London and elsewhere.(28) With the passing years, of course, the creation of more sophisticated calculus procedures increasingly rendered Newton's early methods obsolete and substantial rewriting would have been necessary to bring the 1671 treatise up to date. He is known to have actively contemplated making such revision on only one occasion—'in [autumn] 1691'—when, as Joseph Raphson later recorded,

Mr. Professor *Halley*, and I, had in our hands at *Cambridge*, in order to bring it up [to London] to be printed...that Treatise..., which in the Year 1671. [Sir Isaac] had prepar'd for the Press...and which was then very much worn by having been lent out; but he at that time desir'd it back again for some further Revisal, and never could be perswaded to print it afterwards, till...at length in the Year 1704. he was pleas'd to publish (at least Part of) that Treatise reviv'd.(29)

---

the Originals, to see after what manner they have been copied....I have been likewise told by One that he had a Copy of a Manuscript of yours entitled *Geometria Analytica*, which he highly prized, but this I never saw' (Keynes MS 143.2 = Brewster, *ibid.*: Appendix v. 2: 443–4). The reluctant owner of this 'Copy' of Jones' transcript of the 1671 tract was no doubt Thomas Pellet, for Wilson elsewhere remarks that 'Mr. Jones also gave...Dr. Pellet a copy of Sir Isaac Newton's Treatise of series and fluxions, written in 1671. This was deficient in several places; for Mr. Jones was wont to curtail or otherwise disguise the papers, he communicated to his scholars, that none might make out a compleat book' (*Mathematical Tracts of the late Benjamin Robins, Esq;...*, 2 (London, 1761): Appendix: 357–8). Newton's natural reaction to this flood of unauthorized transcripts of his unpublished early calculus papers was to suppress them unmercifully. Thus, after Wilson sent him on 21 January 1720/1 his own handwritten copy (now ULC. Add. 3960.4) of a manuscript of Jones containing '3 Problems, which I take to be the 2[d], 3[d] and 4th of your Treatise wrote in 1671', he steadfastly ignored Wilson's expressed desire that 'when you have perused [them], you would be pleased to seal them up, and to leave them with your sevants, that I may have them again, upon calling for them some time or other'.

(29) Joseph Raphson, *The History of Fluxions, Shewing in a compendious Manner the first Rise of, and various Improvements made in that Incomparable Method* (London, 1715): 2–3. To clarify Raphson's meaning we have transposed two phrases. In the Latin version which appeared the same year this passage—intendedly so or not—reads considerably differently: 'Tandem... *Newtonus* Anno 1704, Tractatulum circa Annum 1676, ex Tractatu antiquiore descripserat, redivivum publici juris fecerit; quemq; circiter Ann[u]m 1691. Vir Clariss. *Halleus* & ego *Cantabrigiæ* in manibus habuissemus, tum quidem prælo paratum, & perlegendo (ab iis quibus eundem mutuo dederat) obtritum, quemq; postea revisendum repostulaverit, & ad annum usq; prædictum ejusdem publicationem distulerit' (*Historia Fluxionum, Sive Tractatus Originem & Progressum Peregregiæ Istius Methodi Brevissimo Compendio (Et quasi Synopticè) Exhibens. Per Josephum*

The evidence of Newton's extant later calculus manuscripts (which we shall reproduce in our sixth and seventh volumes) allows us to clarify Raphson's statement. In fact, the intervention of David Gregory a few weeks afterwards incited Newton to prepare a new tract on quadratures (known publicly to his contemporaries in revised, truncated form as the *Tractatus de Quadratura Curvarum*, first published in 1704 in appendix to his *Opticks*), a vastly different, severely analytical treatise narrowly in line with current presentations of the subject.

Thereafter, though its content was not wholly outdated, the 1671 treatise was essentially but of historical importance as a primary record of Newton's early achievements in calculus. Subsequent efforts, on the part of William Jones and others, to publish the tract in the early eighteenth century after the onset of the fluxion priority squabble were in a spirit of veneration for a pristine document, to make Newton's 'Right to these Inventions evident to all, even the least knowing in these matters, and to put an end for ever to all Disputes...Whereby we should have an exact and adequate Notion of Fluxions and their Uses, which cannot be had from what has been delivered by others'.(30) Henry

---

Raphsonum (London, 1715: 2). We might well suspect that Newton (who later added a thirty-page appendix to Raphson's *History* after the latter's death) had deliberately contrived the misleading Latin version of the present paragraph so as to add credence to his attempt to persuade the world at large that his 'Book of Quadratures was written in the year 1676 being for the most part extracted out of a Tract w[ch] I wrote in the year 1671 but did not finish & out of some other older Papers' (ULC. Add. 3968.29: 428[r], part of an introduction to an intended, ultimately abortive edition of his *De Quadratura Curvarum* written at this time). Certainly, he quoted it many times thereafter for that purpose in a falsification—perhaps due to a lapse in his memory—of the historical sequence of his calculus writings. For instance, in a passage intended to be added at 'Pag 207 lin penult. After *393 & 396*' of his anonymous review of his edition of the 'Royal Society's' *Commercium Epistolicum* ('An Account of the Book entituled *Commercium Epistolicum Collinij & aliorum, De Analysi promota*; published by order of the *Royal-Society* in relation to the Dispute between Mr. *Leibnitz* and Dr. *Keill*, about the Right of Invention of the Method of *Fluxions*, by some call'd the *Differential Method*', *Philosophical Transactions*, **29**: No. 342 [for January/February 1714/5]: 173–224), Newton claimed that 'in the year [16]91 D[r] Halley & M[r] Ralphson borrowed this Book of Quadratures at Cambridge & carried it with them to London as D[r] Halley hath declared before the R. Society & M[r] Ralphson hath affirmed in his Treatise of [the History of Fluxions]. And therefore the Book was in MS in those days' (ULC. Add. 3968.13: 173[v]). Halley's memory, too, was evidently not of the best, for it was on his return to London from a visit to Cambridge in the summer of 1695 that, on 7 September, he begged Newton's pardon that 'I have not yett returned you your Quadratures of Curves, having not yet transcribed them, but no one has seen them, nor shall, but by your directions' (E. F. MacPike, *Correspondence and Papers of Edmond Halley* (London, 1937): 91). The inference is that Halley had only just been shown the manuscript (probably the truncated 1693 version) of the *De Quadratura*.

(30) James Wilson to Newton, 15 December 1720 (King's College. Keynes MS 143.1 = Brewster's *Memoirs* (note (28)), **2**: 441).

(31) *A View of Sir Isaac Newton's Philosophy* (London, 1728): Preface: [a2][v]. James Wilson later contributed the revealing remark that 'He would have published Newton's Treatise on

Pemberton, the last to propose such an edition in Newton's lifetime, came nearest to achieving it. Having, as he wrote a year after the latter's death,

prevailed on him to let that piece go abroad[,] I had examined all the calculations, and prepared part of the figures; but as the latter part of the treatise had never been finished, he was about letting me have other papers, in order to supply what was wanting. But his death put a stop to that design.(31)

The treatise finally appeared in print in 1736 in John Colson's English translation(32) (made not from the autograph manuscript but from Jones' copy, the self-styled 'Geometria Analytica')(33) and soon after in Buffon's French version(34) and a retranslation back into Latin by Castiglione in 1744,(35) but the original Latin text was published only in 1779 by Samuel Horsley.(36)

Below we reproduce all known autograph manuscripts which relate, in our judgement, to the tract on infinite series and fluxions which Newton composed, essentially by joining his October 1666 fluxional tract(37) to his *De Analysi*,(38) in

Fluxions, but the owners of the copy [the claimants to Newton's estate; see I: xvii–xxi] asked more money, than the booksellers cared to advance' (*A Course of Chemistry...formerly given by the late learned Doctor Henry Pemberton...Now first published...* (London, 1771): Preface: xvi). Wilson eventually managed to see Newton's original manuscript 'many years ago...when it was in my friend Dr. Pemberton's custody' (*Mathematical Tracts* (note (28)), **2**: Appendix: 358).

(32) John Colson, *The Method of Fluxions and Infinite Series; with its Application to the Geometry of Curve-lines. By the Inventor Sir Isaac Newton, Kt'* (London, 1736): 1–140. A slightly variant English rendering which appeared the following year (reissued in 1738), of virtually the same title, is reprinted in photocopy in *The Mathematical Works of Isaac Newton*, **1** (New York, 1964): 29–136.

(33) 'The translation Mr. Colson has published of this treatise was from Mr. Jones's own copy; which, I believe, was very perfect, as far as Sir Isaac Newton had at first composed it' (Wilson, *Mathematical Tracts* (note (28)), **2**: Appendix: 358). This copy (in Jones' own hand) is entitled by him 'Artis Analyticæ Specimina sive Geometria Analytica, Auctore Isaaco Newtono, Equite Aurato'.

(34) George-Louis Leclerc, Comte de Buffon, *La Methode des Fluxions et des Suites Infinies. Par M. le Chevalier Newton* (Paris, 1740; reprinted in photocopy, Paris, 1966). Buffon's 'Preface' (pages iii–xxx) is a sincere, intendedly impartial account of the discovery of infinitesimal calculus which firmly asserts Newton's priority over Leibniz and is critical of Berkeley's attack (in *The Analyst*, London, 1734) on the naïve logical basis of Newtonian fluxional theory. This inspired James Wilson's historically valuable critique which he appended to the second volume of his edition of Benjamin Robins' *Mathematical Tracts* (note (28)), **2**: 308–76 in 1761.

(35) 'Methodus Fluxionum et Serierum Infinitarum, Cum ejusdem Applicatione ad Curvarum Geometriam' (*Isaaci Newtoni Opuscula Mathematica, Philosophica et Philologica*, **1** (Lausanne and Geneva, 1744): 29–199). Colson's 1736 English version was the unique source for this Latin version (made in ignorance of the location of any copy of Newton's original text).

(36) Under Jones' title of 'Artis Analyticæ Specimina, sive Geometria Analytica' (*Isaaci Newtoni Opera quæ exstant Omnia*, **1** (London, 1779) [reprinted in photocopy, Stuttgart–Bad Cannstatt, 1964]: 389–518).

(37) See I, **2**, 7.

(38) See II, **2**, 3.

the winters of the years 1670 and 1671. These comprehend, no doubt, the 'papers...to supply what was wanting' which Newton promised Pemberton a little before his death.

The opening document, a loose slip of paper now to be found tucked between the leaves of the manuscript of Newton's 'Problemes of Curves',(39) bears a preliminary list of problems for an extensive Latin work on the model of the 1666 tract. Most of these listed propositions correspond narrowly to equivalent problems in the fluxional portion of Newton's 1671 tract as written, but it will be evident that no considerations of the centres of gravity of the areas and arc-lengths of curves or of 'solids' appear in the work subsequently written (although both topics were several times included by Collins in his contemporary summaries of its intended content).(40)

The autograph text of the 1671 treatise itself, consisting of the main 'Geometria Analytica' section printed by Horsley in 1779(41) and an unpublished supplementary chapter on geometrical fluxions, is reproduced in Section 2. Lacking, however, the former's initial leaf (numbered 1 and 2 in Newton's pagination), we have been forced to restore it from existing secondary transcripts, particularly that (William Jones') used by Horsley for the same purpose(42) and a later summary of the tract, the self-styled 'Tractatus de Seriebus infinitis et Convergentibus', which is here repeated in appendix. Objections to our positive identification of the title-less autograph manuscript reproduced with Newton's treatise may perhaps be forestalled if we quote an 'exact copy' of the opening of the latter's Problem 1 made by Newton himself about 1718 for his abortive *Historia brevis methodi serierum ex Monumentis antiquis desumpta*:

Fluentes designabo finalibus literis v, x, y, z et celeritates quibus singulæ a motu generante fluunt et augentur designabo literis l, m, n, r respective.

(39) Reproduced in II, 2, 1, §2.

(40) See, for example, Sherburne's 1675 advertisement (note (25) above) and Collins' letters to Borelli and Francis Vernon in December 1671 (quoted in the following appendix). As late as the spring of 1673 Collins communicated a similar account (the draft of which exists in private possession) to Oldenburg, who duly transmitted it to Leibniz, probably in a lost appendix to his letter of 6 April. (Compare J. E. Hofmann, *Die Entwicklungsgeschichte der Leibnizschen Mathematik während des Aufenthaltes in Paris (1672–1676)* (Munich, 1949): 21.) Among extracts from Oldenburg's letter Leibniz noted that 'Pro Geometria solida & Curvilinea Neuton invenit...methodum generalem...pro quadratura (saltem per approximationes) figurarum curvilinearum, rectificatione curvarum, & inventione centrorum gravitatis & soliditatis, solidorum rotundorum et secundorum segmentorum in ipsis...' (*Correspondence of Isaac Newton*, 2, 1960: 236). A right-hand column in which Leibniz hurriedly jotted down explanatory snippets from his conversation with Collins in October 1676 is empty at this point.

(41) See note (36).

(42) Or so we conjecture. Compare 2, §1: note (2) below.

(43) ULC. Add. 3968.19: 264v: '[Apographum]...Ex Tractatu de methodis serierum et fluxionum anno 1671 composito.' An immediately following sentence quotes the title of

Prob. 1. Relatione quantitatum fluentium inte[r] se data, fluxionum relationem determinare.

Solutio. Equationem qua data relatio exprimitur dispone secundum dimensiones alicujus fluentis quantitatis puta x, ac terminos ejus multiplica per quamlibet Arithmeticam Progressionem ac deinde per $\frac{m}{x}$. Et hoc opus in qualibet fluente quantitate seorsim institue. Dein omnium factorum summam pone nihilo æqualem et habebis æquationem desideratam.(43)

When this certified quotation from the 1671 treatise is compared with page 19 of the tract here reproduced, there can be little room to doubt their identity. If, further, we take into consideration the many implicit references to passages in his fluxional treatise which besprinkle Newton's contemporary correspondence, all doubts should be stilled. Thus, in communicating to Collins on 10 December 1672 his analytical method for constructing tangents to curves defined in oblique Cartesian coordinates, Newton paraphrased the first 'mode' of Problem 4, 'Curvarum Tangentes ducere', in the present tract in the instance of the curve $x^3-2x^2y+bx^2-b^2x+by^2-y^3=0$, and then added that

This...is one particular, or rather a Corollary of a Generall Method w[ch] extends it selfe w[th]out any troublesome calculation, not onely to the drawing tangents to all curve lines whether Geometrick or mechanick or how ever related to streight lines or to other curve lines but also to the resolving other abstruser kinds of Problemes about the crookedness, areas, lengths, centers of gravity &c. Nor is it (as Huddens method *de maximis et minimis* & consequently Slusius his new method of Tangents as I presume)(44) limited to æquations wch are free from surd quantities. This method I have interwoven wth that other of working in æquations by reducing them to infinite series.(45)

'Prob. 2' as 'Exposita æquatione fluxiones quantitatum involvente invenire relationem quantitatum inter se': compare Newton's page 23 of the manuscript reproduced below as 2, §2. On Newton's use of literal (and not dot) fluxions see 2, §2: note (86). In an allied draft on Add. 3968.7: 115[r] Newton entered, and then cancelled, a variant 'Solutio' to Problem 1 which begins 'Fluentium v, x, y, z fluxionibus dictis l, m, n, r respectivè, Æquationem qua data relatio exprimitur dispone secundum dimensiones alicujus fluentis quantitatis puta x....'

(44) Newton is here replying to a (lost) letter from Collins a few days earlier which no doubt asked him his opinion of Sluse's recent claim to have invented a 'methodus facillima' for constructing tangents to curves. A month earlier Collins had written to James Gregory that 'there is yet another method better and much more Compendious...knowne long ago to Slusius Hudden and Roberval in which you are only to looke for the Æquation which expresses the nature of the Line, from which there may be presently and without any trouble derived another Æquation that gives the Construction of the Tangent' (*Correspondence of Isaac Newton*, **1**: 244). The topic is discussed at length by J. E. Hofmann in his *Studien zur Vorgeschichte des Prioritätstreites zwischen Leibniz und Newton um die Entdeckung der höheren Analysis. I. Materialien zur ersten mathematischen Schaffensperiode Newtons (1665–1675)* [=*Abhandlungen der Preussischen Akademie der Wissenschaften, Jahrgang 1943. Math.-naturw. Klasse*, **2** (Berlin, 1943)]: 70–6, 81–94: 'Auseinandersetzungen um die Tangentenmethode'.

(45) *Correspondence*, **1**: 247–8.

Six months later, having studied Huygens' *Horologium*(46) in some detail, he informed Oldenburg that

The rectifying curve lines by that way wch M. Hugens calls Evolution, I have been sometimes considering also, & have met wth a way of resolving it wch seems more ready & free from ye trouble of calculation then that of M. Hugens. If he please I will send it him. The Problem also is capable of being improved by being propounded thus more generally.

Curvas invenire quotascunq; quarum longitudines cum propositæ alicujus Curvæ longitudine, vel cum area ejus ad datam lineam applicata, comparari possunt.(47)

There is no need to insist on the identity of this enunciation with that of Newton's present Problem 11 (on his page 103) or how accurately he described the latter's theme to Oldenburg. Above all, in his letters of 13 June and 24 October 1676 to Oldenburg for Leibniz(48) Newton drew extensively upon his five-year-old treatise for mathematical material with which to dazzle his continental contemporary. The salient points of this pillage, too numerous to detail here, are dealt with at some length in appropriate footnotes to the passages in question.

Structurally, the 1671 tract breaks down into an introductory section (Newton's pages 1–16): an augmented version, in fact, of his *De Analysi*(49) of two years before dealing with the expansion of quantities as 'converging' infinite series; two fundamental problems (pages 17–40) treating generally and in some detail the converse operations of differentiation and integration (in Newton's

(46) Christiaan Huygens, *Horologium Oscillatorium sive De Motu Pendulorum ad Horologia aptato Demonstrationes Geometricæ* (Paris, 1673). Newton refers to its 'Pars Tertia. De linearum curvarum evolutione & dimensione' (*Horologium*: 59–90) where Huygens develops a construction for the radius of curvature, at a general point on a curve defined in perpendicular Cartesian coordinates, which is effectively identical with that developed by Newton himself in late 1664. See I, **2**, 4: passim, but particularly §2: note (75).

(47) Newton to Oldenburg, 23 June 1673 (*Correspondence*, **1**: 290).

(48) *Correspondence*, **2**, 1960: 20–32 and 110–29 respectively.

(49) Especially its first half (II: 210–32).

(50) Reproduced on I: 416–41.

(51) Newton himself compiled the following short description for an 'Enarratio plenior' of the 'fluxions' scholium (that to Lemma II of the second book) in an intended but ultimately abortive edition of his *Principia* about 1716: 'In Tractatu quem anno 1671 conscripsi, primum docui reductionem quantitatum in series convergentes per divisiones & extractiones radicum tam affectarum quam simplicium. Et his præmissis, methodum fluxionum exposui docendo solutionem plurium Problematum, quorum duo prima erant hæc.

'Prob. 1. Relatione quantitatum fluentium inter se data, fluxionum relationem invenire.

'Prob. 2. Exposita æquatione fluxiones quantitatum involvente, invenire relationem quantitatum inter se' (ULC. Add. 3968.6: 46^{r}).

'Horum Problematum solutiones in hoc Manuscripto pluribus [*sc.* verbis] prosecutus sum ut fundamentum methodi meæ generalis ex methodo serierum et methodo fluxionum compositæ; deinde per hanc methodum docui solutiones aliorum Problematum' (ULC. Add. 3968.7: 115^{r}).

terminology, constructing the 'fluxions' of given 'fluent' quantities, and vice versa); and, finally, ten 'Problemata particularia' (pages 41–114: in large part a revise of Problems 1–13 of the October 1666 fluxional tract[(50)]) which occupy themselves with questions of tangency, inflexion and curvature of a curve and, conversely, its quadrature and arc-length.[(51)] On technical points of detail we may refer the reader to our footnotes to the texts reproduced (which are, in part, a running commentary upon it)[(52)] but here we should insist on the novelty of Newton's present reformulation of the calculus of continuous increase. Hitherto, this had been expounded by him in terms of the loosely justified conceptual model of the 'velocity' of a 'moveing body', but in the present treatise Newton was led, under Barrow's guidance,[(53)] to postulate a basic, uniformly 'fluent' variable of 'time' as a measure of the 'fluxions' (instantaneous 'speeds' of flow) of a set of dependent variables which continuously alter their magnitude. As he wrote some forty years later:

I consider time as flowing or increasing by continual flux & other quantities as increasing continually in time & from y[e] fluxion of time I give the name of fluxions to the velocitys wth w[ch] all other quantities increase. Also from the moments of time I give the name of moments to the parts of any other quantities generated in moments of time. I expose time by any quantity flowing uniformly & represent its fluxion by an unit, & the fluxions of other quantities I represent by any other fit symbols, & the fluxions of their fluxions by other fit symbols & the fluxions of those fluxions by others, & their moments generated by those fluxions I represent by the symbols of the fluxions drawn into y[e] letter o & its powers o^2, o^3 &c: viz[t] their first moments by their first fluxions drawn into the letter o, their second moments by their second fluxions into o^2, & so on. And when I am investigating a truth or the solution of a Probleme I use all sorts of approximations & neglect to write down the letter o, but when I am demonstrating a Proposition I always write down the letter o & proceed exactly by the rules of Geometry without admitting any approximations.[(54)] And I found the method not upon summs & differences, but upon the solution of this probleme: *By knowing the Quantities generated in time to find their fluxions.* And this is done by finding not *prima momenta* but *primas momentorum nascentium rationes*.[(55)]...This Method is derived immediately from Nature her self,

(52) Consult also the selective critique given by J. E. Hofmann in his *Studien* (note (44)): 49–69. Hofmann takes as his text Horsley's rather over-edited 1779 Latin version (note (36)), but this should cause no serious difficulty.

(53) Compare 2, §2: notes (80), (81), (82) and (84).

(54) Newton might have added that it was his secondary purpose to present his results geometrically rather than in the algebraic form in which they were discovered. On page 91 of his present treatise (2, §2 below) he urged, for instance, that 'Postquam Curvæ alicujus area sic inventa fuerit, de constructionis demonstratione consulendum est, quacum sine Computo Algebraico quantùm liceat contexta ornetur Theorema ut evadat publicæ notitiæ dignum. Estq; demonstrandi methodus generalis quam sequentibus exemplis illustrare conabar'.

(55) Compare 2, §3: note (16) below. The 'probleme' is, of course, a later English paraphrase of the enunciation of the 1671 tract's 'Prob. 1' (on Newton's page 18 of 2, §2 below).

that of indivisibles, Leibnitian differences or infinitely small quantities not so. For there are no *quantitates primæ nascentes* or *ultimæ evanescentes*, there are only *rationes primæ quantitatum nascentium* or *ultimæ evanescentium*.(56)

The ideal is more attractive than the factual, but we should not allow the cumbrousness of much of Newton's expression of his fluxional theories and the inadequacy of his notations to blind us to the conceptual elegance and profundity of this approach to continuously varying magnitude.

In the third section, lastly, are reproduced some allied manuscripts which discuss the quadrature of curves defined (in Cartesian coordinates) by a multinomial equation, giving the reduction to an explicit quadrature in the case of an arbitrary algebraic trinomial. Convincing evidence that they are to be dated to this period is afforded by their handwriting, while a firm post-date of autumn 1676 is given by an unmistakable reference to them in a letter which Newton addressed to Collins on 8 November of that year:

As for ye method of Transmutations in general, I presume [M. Leibnitz] has made further improvements then others have done, but I dare say all that can be done by it may be done better w^thout it, by ye simple consideration of ye ordinatim applicatæ.... The advantage of ye way I follow you may guess by the conclusions drawn from it wch I have set down in my answer to M. Leibnitz: though I have not said all there.(57) For there is no curve line exprest by any æquation of three terms, though the unknown quantities affect one another in it, or y^e indices of their dignities be surd quantities (suppose $ax^{\lambda}+bx^{\mu}y^{\sigma}+cy^{\tau}=0$, where x signifies y^e base, y y^e ordinate, λ, μ, σ, τ, y^e indices of y^e dignities of x & y, & a, b, c known quantities with their signes + or −) I say there is no such curve line but I can in less than halfe a quarter of an hower tell whether it may be squared or what are y^e simplest figures it may be compared w^th, be those figures Conic sections or others. And then by a direct & short way (I dare say ye shortest ye nature of ye thing admits of for a general one) I can compare them. And so if any two figures exprest by such æquations be propounded I can by ye same rule compare them if they may be compared. This may seem a bold assertion because it's hard to say a figure may or may not be squared or compared wth another, but it's plain

(56) Quoted from an early draft (ULC. Add. 3968.41: 83^r, late 1714?) of his review of the 'Royal Society's' *Commercium Epistolicum D. Johannis Collins* (London, 1712) which later appeared anonymously in the *Philosophical Transactions*. Compare this latter 'Account of the Book' (note (29)): 207 following.

(57) Namely, in his *epistola posterior* of 24 October 1676 for Leibniz. At one point in that letter (*Correspondence*, **2**: 117) Newton had hinted guardedly that 'Pro trinomijs etiam et alijs quibusdam Regulas quasdem concinnavi'.

(58) *Correspondence*, **2**: 179–80. An immediately preceding 'Manuscript by Newton on Quadratures [? 1676]' (*ibid.*, **2**: 171–4), found by H. W. Turnbull 'in a confused set of [autograph] papers' in private possession, is in fact a stray from Newton's first version (late 1691) of his *De Quadratura Curvarum* and will be reproduced, together with other sheets from the same draft sequence, in its chronological place in our sixth volume.

to me by ye fountain I draw it from, though I will not undertake to prove it to others. The same method extends to æquations of four terms & others also but not so generally.(58)

We shall not attempt to improve upon this description, but would warn the unwary that this passage, first published by William Jones in 1711,(59) was used by Newton himself on several occasions during his dispute with the Leibnizians over calculus priority as though it referred to an early draft of his *De Quadratura Curvarum*.(60)

In all these papers—above all the 1671 treatise—Newton's breadth of vision and his complementary mastery of technical detail are as impressive as ever, and they will remain always of fundamental importance to any understanding of his methods of fluxions and infinite series. It is tragic that he himself could never feel an urgent necessity to publish them—if he had done so, scientific history might well subsequently have taken a different course. But let us not waste words when the reader can form his own opinion from a study of the texts themselves, as they are here reproduced.

(59) In his Newtonian compendium, *Analysis Per Quantitatum Series, Fluxiones, ac Differentias: Cum Enumeratione Linearum Tertii Ordinis* (London, 1711): 'Excerpta Ex Epistolis D. *Newtoni* Ad Methodum Fluxionum, et Serierum Infinitarum Spectantibus': 38.

(60) Compare note (29) above. So, in preface to a proposed re-edition about 1716 of his *De Quadratura Curvarum*, Newton professed that he 'wrote [the Book of Quadratures] in the year 1676, except the Introduction & Conclusion, extracting most of it out of old Papers. And when I had finished it & the 7th 8th 9th & 10th Propositions wth their Corollaries were fresh in memory, I wrote upon them to Mr Collins that Letter wch was dated 8 Novem. 1676 & published by Mr Jones' (ULC. Add. 3968.29: $458^r/454^r$). Newton's recollection of the events preceding his letter to Collins was decidedly poor, for the first draft of the *De Quadratura* was not to be penned by him till fifteen years afterwards!

APPENDIX

NEWTON'S REFERENCES TO THE 1671 TRACT IN HIS *COMMERCIUM*: AN EXTRACT FROM A DRAFT PREFACE

[Late 1712].(1)

From the autograph in the University Library, Cambridge(2)

The *Analysis* quoted in the beginning of this Collection(3) shews that [Mr Newton] had...a general method [of solving Problemes by reducing them to equations finite or infinite whether those equations include moments (the exponents of fluxions) or do not include them,(4) & by deducing fluents & their moments from one another by means of those equations] in the year 1669. And by the Letters & Papers wch follow, it appears that in the year 1671, at the desire of his friends he composed a larger Treatise upon this same method (p. 27. l. 10 [C], 27 [D] & p. 71. l. 4 [G_1], 26 [B]), that it was very general & easy without sticking at surds or mechanical curves & extended to the finding tangents areas lengths centers of gravity & curvatures of Curves &c (p. 27 [C and D], 30 [E], 85 [G_5], that in Problemes reducible to quadratures it proceeded by the Propositions since printed in the book of Quadratures,(5) wch Propositions are there founded upon the method of fluents (p. 72 [G_2], 74 [G_3], 76 [G_4]), that it extended to the extracting of fluents out of æquations involving their fluxions & proceeded in difficulter cases by assuming the terms of a series & determining them by the conditions of the Probleme (p. 86 [G_5]), that it determined the curve by the length thereof (p. 24 [A]) & extended to inverse

(1) It is surely no longer necessary to justify the historical fact that the *Commercium Epistolicum D. Johannis Collins et Aliorum de Analysi promota* (London, 1712) was, though issued 'Jussu Societatis Regiæ', entirely Newton's own inspiration and work, and that the Royal Society committee (originally composed of Halley, William Jones, Machin, Arbuthnot, Hill and Burnet, though later several others, including Aston, de Moivre and Brook Taylor, were coopted) which officially sponsored it was but a blind front. Newton's autograph drafts of the book's *Ad Lectorem*, the 'Committee's' concluding judgement, and many of the footnotes and textual interpolations still exist—even the fine detail of the organization of the book is his, as numerous preliminary English and Latin draft schemes testify. In the present appendix we reproduce an extract from an unpublished English draft preface to the volume in which Newton directs the reader's attention to excerpts from letters quoted in the book which relate to his 1671 calculus tract. In line with the demands of modern scholarship we do not give the *Commercium*'s Latin versions of the *pièces justificatives* but reproduce corresponding extracts from the original documents (of which B, C and D below are not yet published).

(2) Add. 3968.39: 583^r^, a revised version of Add. 3968.37: 551^r^.

(3) The *De Analysi*, printed on pages 3–20 of the *Commercium* (note (1)).

(4) That is, whether the equation is simply algebraic or is a 'fluxional' (differential) one.

Problems of tangents & others more difficult, & was so general as to reach almost all Problemes except numeral ones like those of Diophantus (p. 55 [F], 85, 86 [G_5])....

[A=John Collins to James Gregory, 24 December 1670.][(6)]

...When M[r] Dary had published his *Miscellanies* he sent one of them to M[r] Newton, who sent M[r] Dary the same Series for the area of a Zone of a Circle (which doubtlesse is well derived and excellent) that I sent you,[(7)] that M[r] Dary might thereby correct and examine an error in an approach in his *Miscellanies*. And by D[r] Barrows meanes I have since had some few Series more out of Newtons generall method, and by discourse gather they are Analytically derivable from the given properties of any figure....[(8)]

...He hath as I understand wrote a treatise of this method, and made the length of the portions or intire Elliptick line, and of the quadrat[rix] Curve *DV* above instance[d]....

[B=John Collins to Jean Bertet, 21 February 1670/1.][(9)]

...To compose an intire Systeme of Algebra is an excellent Designe, and deserves much to be encouraged, and aboundantly the more in regard there is here about 4 yeares since invented by M[r] Isaac Newton a generall Analyticall method for the quadrature of all Curvilinear Spaces as well geometrick Curves as Mechanick that have any common Propertie. Hence all that depends upon Quadratures is likewise performed, as the streightening of all Curves, the finding of their Tangents & Centers of gravitie[,] Round solids and their Second Segments, Surfaces, (but not the Surfaces of Solids whose Axes incline as of

(5) '...all the ten first Propositions of the Book of Quadratures except the first and sixt are in the Tract w[ch] I wrote in the year 1671 tho not in the same words & some not in words but in equations' (ULC. Add. 3968.29: 454[r], quoted from an intended preface of Newton's to an abortive edition of his *De Quadratura Curvarum* about 1716). The 'Tables at the end of the tenth Proposition for squaring of some Curves & comparing others with the Conic Sections' are, of course, an unaltered borrowing from the 1671 tract (see 2, §2: note (549) below).

(6) Quoted from the original in Royal Society MS LXXXI, No. 19 (reproduced in *The Correspondence of Isaac Newton*, **1**, 1959: 54–5). A Latin version ('Quum D. *Dary* Miscellanea sua in lucem edidit....Tractatum hac de re scripsit...nec non Areæ supradictæ') is given in the *Commercium* (note (1)): 24.

(7) In Collins' letter to Gregory of 24 March 1669/70, the relevant portion of which is quoted at *Correspondence*, **1**: 28.

(8) The remainder of Collins' paragraph has been quoted *in extenso* on page 4 above.

(9) Extracted from the unpublished original draft in Royal Society MS LXXXI, No. 21 (dated in Collins' hand '21 Febr. 1670' on its cover). A Latin version is printed in the *Commercium* (note (1)): 26–7.

Parabolicall Conoids &c, that remaines as an insuperable difficulty to posterity.) And all this performed infinitely true, without any extraction of rootes and that by ayd of an infinite Rationall Series, whereof there may be many variously applyed to one and the same figure, as to the Circle, one to find the Area of the whole or any part others to the inscripts or adscripts, so that giving any Sine, Tangent, or Secant, you may find the length of the Arch and the Converse by other Series for that purpose, so that now it is become more easy to calculate the arch to any Sine and the converse then out of a Sine given, to find the Sine of the double arch.(10) And all this is no other then the method used by Mercator in his *Logarithmotechnia* for the quadrature of the Hyperbola rendred generall....(11)

[C=John Collins to G. A. Borelli, ? December 1671.](12)

Reverend Pere

...(13) The Introduction which [Gerrard Kinckhuysen] calls *Stelkonst* is translated and fitted for the presse by Mr Isaac Newton, who is now Mathematicall Professor at Cambridge. To which he addes his generall Method of Analyticall quadratures, by which he computes the Area of all regular Curvilinear figures that have any common Property[,] the streightening of the said Curved Lines[,] the finding of their Centres of gravity, their round Solids and the Surfaces thereof if begot by rotation, and the second Segmts of those Solids. Yea any Logcall Sine Tangent or Secant in the Canon being given to find the correspondt arch without finding the naturall Sine tangent or Secant and the Converse and this generally without any Extraction of rootes. A Specimen hereof I subjoyne in relation to the Circle....(14)

---

(10) Presumably by using the identity $\sin 2\theta = 2\sin\theta\sqrt{[1-\sin^2\theta]}$.

(11) Collins added, in the unpublished following sentence, that 'I hope Mr Newton who hath taken paines to write Annotations upon Kinckhuysens Introduction to Algebra which is translated out of belgick into Latin and fitted for the Presse, will enrich the Commonwealth of Learning with this Doctrine'.

(12) An extract from Collins' unpublished original draft in Royal Society MS LXXXI, No. 22 (which bears 'December 1671' on its cover). A Latin version is published in the *Commercium* (note (1)): 27.

(13) Collins began this paragraph: 'There are many good Bookes of Algebra in low Dutch, to wit Stampioen, Smyter, Martin van Wilkins *Officina Algebræ*, Frans van der Huyps, Wouter Verstap of figurate Arithmetick, J Jacob Fergusons *Labyrinthus Algebræ* and three Bookes of Gerrard Kinckhuysen to wit his Introduc[t]io[n], his Analyticall Conicks and his Bookes of Geometricall Problemes the which if translated into Latin would be of more use to Students than the works of Des Chartes with Schootens Commentaries.'

(14) In the original a following blank is left for the insertion of this 'Specimen' series.

[D=John Collins to Francis Vernon (and Richard Towneley?), 26 December 1671.][(15)]

M[r] Vernon

S[r]...[(16)] D[r] Barrow tells me that M[r] Newton hath almost compleated for the presse (which I am to take care of here) Kinckhuysens Introduction to Algebra with Additions of his owne, to which he subjoynes his owne generall Analyticall method of infinite Series,[(17)] whereby all Curvilinear Spaces (as well Geometricall as Mechanicall in Des Chartes sense the figures being deter[d] by some one or more common properties) have their Areas[,] the Lengths of their Curved Lines, Centers of gravitie, round Solids and Surfaces if begott by rotation computed. Hence Multitudes of Series for the Circle it selfe[:] yea giving any Number as a Log[call] Sine Tang[t] or Secant you may by an easy Calculation without any Extraction of rootes [&] without any tables find the Correspond[t] Arch and the Converse infinitely true, and that without finding the naturall Sine tang Sec[t]. So great and much more is the benefitt of this Excellent doctrine. I speake no more than I know. And with these he likewise will send up 20 Optick Lectures, which D[r] Barrow reckons one of the greatest performances of Ingenuity this age hath affoarded. I replyed twas fitt to hasten the Impression in regard Hugens intended a treatise of Dioptricks *et de Evolutione Curvarum* and he wished the Newes hereof might the rather Hasten then hinder Hugenius supposing it altogeather improbable that their Hypotheses or Deductions can be the same....[(18)]

---

(15) An extract from the unpublished original draft in Royal Society MS LXXXI, No. 23. A Latin version is given in the *Commercium* (note (1)): 27–8. Collins has noted on the cover of his draft 'To M[r] Vernon 26 December 1671 and M[r] Tounley'.

(16) A cancelled first version of the following phrase reads 'D[r] Barrow informes me that the said M[r] Newton hath prepared for the Presse...'. (In the previous paragraph Collins had expounded for Vernon his impressions of Newton's new 'Tellescope', finding that 'it is somewhat difficult to place it upon an Object, and some object that it is not so lightsome as an ordinary Tube'.)

(17) In the *Commercium*'s printed Latin version (note (15)) Newton inserted a footnote at this point: '*N.B.* Hic Tractatus unus idemq; est ac ille, cujus mentionem fecerat D. *Newtonus* in Epistola Octob. 24. 1676. data [see G below], per D. *Oldenburgum* D. *Leibnitio* communicata; & in quo methodi Serierum infinitarum & Fluxionum simul explicabantur, ut ubi loci memorat'.

(18) Whatever the truth of the rumour Collins had picked up, when Huygens' *Horologium Oscillatorium* appeared in 1673 it contained no section on dioptrics. Compare note (46) of the preceding introduction.

[E = Newton to John Collins, 10 December 1672.][19]

...This S[r] is one particular, or rather a Corollary of a Generall Method w[ch] extends it selfe w[th]out any troublesome calculation, not onely to the drawing tangents to all curve lines whether Geometrick or mechanick or how ever related to streight lines or to other curve lines but also to the resolving other abstruser kinds of Problems about the crookedness, areas, lengths, centers of gravity of curves &c....This method I have interwoven wth that other of working in æquations by reducing them to infinite series....[20]

[F = Newton to Oldenburg for Leibniz, 13 June 1676][21]

...Ex his videre est quantum fines Analyseos per hujusmodi infinitas æquationes ampliantur: quippe quæ earum beneficio, ad omnia, pene dixerim, problemata (si numeralia Diophanti et similia excipias) sese extendit. Non tamen omninò universalis evadit, nisi per ulteriores quasdem methodos eliciendi series infinitas. Sunt enim quædam Problemata in quibus non liceat ad series infinitas per divisionem vel extractionem radicum simplicium affectarumve pervenire: sed quomodo in istis casibus procedendum sit jam non vacat dicere, ut neque alia quædam tradere quæ circa reductionem infinitarum serierum in finitas, ubi rei natura tulerit, excogitavi....

[G = Newton to Oldenburg for Leibniz, 24 October 1676][22]

[G_1] ...ante annos quinque cum, suadentibus amicis, consilium cœperam edendi Tractatum de refractione Lucis et coloribus quem tunc in promptu habebam; cœpi de his seriebus iterum cogitare, & tractatum de ijs etiam conscripsi ut utrumq simul ederem....Ipse autem tractatum meum non

(19) Quoted from an autograph extract in Royal Society MS LXXXI, No. 26, of which a Latin version is printed in the *Commercium* (note (1)): 30. The original letter as sent to Collins (now in the possession of the Earl of Macclesfield) is reproduced in full in *The Correspondence of Isaac Newton*, **1**, 1959: 247–52.

(20) A fuller version of this passage has been quoted on page 15 above.

(21) The celebrated *epistola prior* to Leibniz in 1676, given in full in the *Commercium* (note (1)): 49-57, which at one time formed manuscript No. 48 in Royal Society MS LXXXI or so we presume. (For the history of MS LXXXI, Halliwell's numbering in his catalogue of the Royal Society's manuscripts in 1840, see the *Report on Leibnitz–Newton MSS. in the Possession of the Royal Society of London* (London, 1880): 21. We presume that the originals of this and other letters published in the *Commercium* were lost some time after 25 October 1714 when the Council 'Ordered, That the Originals of the Papers used in the Comercium Epist., lately printed by the Society, be sealed up and put into the Iron Chest to be ready to be produced upon occasion'.) The following extract (*Commercium*: 54–6) is reproduced from the autograph copy returned to Newton (ULC. Add. 3977.2, published in *Correspondence*, **2**, 1960: 20–31, especially 29).

penitus absolveram ubi destiti a proposito, neqꝫ in hunc usqꝫ diem mens rediit ad reliqua adjicienda. Deerat quippe pars illa qua decreveram explicare modum solvendi Problemata quæ ad Quadraturas reduci nequeunt, licet aliquid de fundamento ejus posuissem.

[G_2] Cæterùm in tractatu isto series infinitæ non magnam partem obtinebant: Alia haud pauca congessi, inter quæ erat methodus ducendi tangentes quam solertissimus Slusius ante annos duos tresve tibi communicavit, de qua tu, suggerente Collinsio, rescripsisti eandem mihi etiam innotuisse.(23) Diversa ratione in eam incidimus. Nam res non eget demonstratione prout ego operor. Habito meo fundamento nemo potuit tangentes aliter ducere, nisi volens de recta via deviaret. Quinetiam non hic hæretur ad æquationes radicalibus unam vel utramqꝫ indefinitam quantitatem involventibus utcunqꝫ affectas, sed absqꝫ aliqua talium æquationum reductione (quæ opus plerumqꝫ redderet immensum) tangens confestim ducitur. Et eodem modo se res habet in quæstionibus de Maximis et Minimis, alijsqꝫ quibusdam de quibus jam non loquor. Fundamentum harum operationum, satis obvium quidem, quoniam jam non possum explicationem ejus prosequi sic potius celavi....(24) Hoc fundamento conatus sum etiam reddere speculationes de Quadratura curvarum simpliciores, perveniqꝫ ad Theoremata quædam generalia....

[G_3] ...At quando hujusmodi Curva aliqua non potest geometricè quadrari, sunt ad manus alia Theoremata pro comparatione ejus cum Conicis Sectionibus, vel saltem cum alijs figuris simplicissimis quibuscum potest comparari; ad quod sufficit etiam hoc ipsum unicum jam descriptum Theorema, si debitè concinnetur. Pro trinomijs etiam et alijs quibusdam Regulas quasdem concinnavi. Sed in simplicioribus vulgoqꝫ celebratis figuris vix aliquid relatu dignum reperi quod evasit aliorum conatus nisi fortè longitudo Cissoidis ejusmodi censeatur. Ea sic construitur....(25)

(22) Newton's *epistola posterior* to Leibniz in 1676, given in full in the *Commercium* (note (1)): 67–86. The following extracts (*Commercium*: 71, 71–2, 74, 76 and 85–6 respectively) are made from the autograph original as sent (British Museum. Add. 4294.1, reproduced in *Correspondence*, **2**: 110–29): either this or the copy (ULC. Add. 3977.4) retained by Newton in 1676 presumably once formed manuscript No. 55 in Royal Society MS LXXXI.

(23) See note (44) of the preceding introduction.

(24) As Newton observed in a footnote at this point in the *Commercium* (note (1)): 72, the following anagram (here omitted) may be rearranged to read '*Data Æquatione quotcunque fluentes quantitates involvente, Fluxiones invenire; & vice versa.* Prior pars Problematis solvitur per Regulam Binomii ...Posterior pars...solvitur regrediendo a momentis ad fluentes; quod ubi hæretur fieri solet quadrando figuras; & ubi ad quadraturas hæretur, extrahendo fluentes per Regulas quatuor, quarum duas *Newtonus* in Epistola priore [F] explicuit, duas alias sub finem hujus Epistolæ literis transpositis occultavit'. See note (30) below, and compare the 1676 Newtonian 'Memorandum' reproduced in II, **2**, 1, §3: note (25).

(25) Newton proceeds to quote the rectification of the general cissoidal arc entered in Problem 11 of his 1671 tract; see **2**, §2: note (733) below.

[G_4] ...Seriei[26] a D. Leibnitio pro quadratura Conicarum Sectionum propositæ affinia sunt Theoremata quædam quæ pro comparatione Curvarum cum Conicis sectionibus in Catalogum dudum retuli. Possum utiq; cum Conicis sectionibus geometricè comparare curvas omnes numero infinitè [multas][27] quarum ordinatim applicatæ sunt....[28]

[G_5] ...Ubi dixi omnia pene Problemata solu[t]ilia[29] existere, volui de ijs præsertim intelligi circa quæ Mathematici se hactenus occuparunt, vel saltem in quibus ratiocinia mathematica locum aliquem obtinere possunt. Nam alia sanè adeò perplexis conditionibus implicata excogitare liceat ut non satis comprehendere valeamus et multò minùs tantarum computationum onus sustinere quod ista requirerent. Attamen ne nimium dixisse videar, inversa de tangentibus Problemata sunt in potestate, aliaq; illis difficiliora: ad quæ solvenda usus sum duplici methodo, una concinniori, altera generaliori: Utramq; visum est impræsentia literis transpositis consignare, ne propter alios idem obtinentes, institutum in aliquibus mutare cogerer....[30] Inversum hoc Problema de tangentibus quando tangens inter punctum contactus et axem

(26) The inverse-tangent series, developed by a method which had been exactly anticipated by Nilakaṇṭha, though this was not then known in Europe (see page 34, note (5) below).

(27) Corrected from the original's 'infinitè infinitas' in accordance with Newton's expressed wish to Oldenburg on 14 November 1676 (*Correspondence*, **2**, 1960: 181). However, in 1712 Newton preferred the phrase 'infinities infinita' (*Commercium* (note (1)): 76).

(28) The significance of this passage (and its continuation here omitted) is discussed in 2, §2: note (570) below.

(29) Amended from 'solubilia'. See Newton to Oldenburg, 26 October 1676 (*Correspondence*, **2**: 162). However, 'solubilia' is retained in the *Commercium* (note (1)): 85.

(30) As Newton noted in 1712 (*Commercium* (note (1)): 86), when suitably transposed the following anagram becomes '*Una Methodus consistit in extractione fluentis quantitatis ex æquatione simul involvente fluxionem ejus: altera tantum in assumptione Seriei pro quantitate qualibet incognita ex qua cætera commode derivari possunt, & in collatione terminorum homologorum æquationis resultantis, ad eruendos terminos assumptæ Seriei*'. Compare his 1676 'Memorandum' quoted in II, **2**, 1, §3: note (25). Newton went on to assert in 1712 that 'Analysin per Fluentes & earum Momenta in æquationibus tam infinitis quam finitis, *Newtonus* in his Epistolis ad Regulas quatuor reduxit. Per primam extrahitur Fluens ex Binomiis, adeoque ex æquationibus quibuscunque non affectis in Serie infinita, & Momentum fluentis simul prodit, quo evanescente Series in Æquationem finitam redit. Per secundam extrahitur Fluens ex æquationibus affectis Fluxionem non involventibus. Per tertiam extrahitur Fluens ex æquationibus affectis Fluxionem simul involventibus. Per quartam eruitur Fluens ex conditionibus Problematis'. Compare note (24).

(31) The tractrix of constant 'tangens' 1 is defined by the Cartesian equation

$$y = \log[(1+\surd[1-x^2])/x]-\surd[1-x^2].$$

The curve is not discussed elsewhere in Newton's known papers and its systematic exploration was undertaken only some fifteen years later by Huygens (see his *Œuvres complètes*, **10**: 407, 418; and compare Basnage de Beauval's *Histoire des Ouvrages des Sçavans*, 1693: 244).

figuræ est datæ longitudinis, non indiget his methodis. Est tamen Curva illa Mechanica, cujus determinatio pendet ab area Hyperbolæ.(31) Ejusdem generis est etiam Problema quando pars axis inter tangentem et ordinatim applicatam datur.... quando in triangulo rectangulo quod ab illa axis parte & tangente ac ordinatim applicata constituitur, relatio duorum quorumlibet laterum per æquationem quamlibet definitur, Problema solvi potest absq; mea methodo generali, sed ubi pars axis ad punctum aliquod positione datum terminata ingreditur vinculum tunc res aliter se habere solet.(32)

---

(32) This passage is examined at length in Christoph J. Scriba's 'The Inverse Method of Tangents: A Dialogue between Leibniz and Newton (1675–1677)', *Archive for History of Exact Sciences*, **2**, 1964: 113–37, especially 125–30: 'Comments on Newton's Epistola Posterior'. Newton suggests that, where a curve is defined with respect to perpendicular Cartesian coordinates x and y, his general 'analyticall' methods can deal with any sort of fluxional relationship of the type $f(y, s) = 0, f(y, t) = 0$ or $f(s, t) = 0$, in which s is the 'subtangens' $y\,dx/dy$ and t the corresponding 'tangens'. As Newton observes, difficulties increase enormously when the abscissa x enters the given relationship.

1

A PRELIMINARY SCHEME FOR A TRACT ON FLUXIONS

[1670?][1]

From the original in private possession

Prob [1. Relatione quantitatum inter se datâ, fluxionum relationem determinare.

2. Expositâ æquatione fluxiones quantitatum involvente, invenire relationem quantitatum inter se.][2]

3. De Max. et Min.[3]

4. Tangentes ducere.[4]

5. Curvarum curvaturam ad datum punctum cognoscere.[5]

6., 7., 8. Invenire curvas pro arbitrio multas quarum {areæ / Longit / Centr: grav} ad rectilineas comparari possunt.[6]

9., 10., 11. Invenire curvas pro arbitrio multas quarum {areæ / longit / centr grav} ad propositas curvilineas comparari possunt.[7]

12., 13., 14. Propositæ alicujus curvæ {aream / longit / cen. grav} determinare.[8]

16., 17., 18. Propositis duabus curvis {areas / longit / cent. grav} conferre.[9]

---

(1) We suggest this rough date of composition on the basis of our assessment of the style and quality of handwriting of this autograph draft outline. This Latin scheme for a tract on fluxional methods and infinite series was evidently composed some little time after the short English tract on 'Problems of curves' (II, **2**, 1, §1), of whose appended list of problems on the 'curvity', 'Areas', 'lengths' and 'centers of gravity' of 'lines' it is strongly reminiscent. There can be little doubt that the lengthy following tract (2, §§1/2 below) is its detailed amplification.

Translation

Problem [1. Given the relation of quantities to one another, to determine the relation of their fluxions.

2. When an equation involving the fluxions of quantities has been expounded, to find the relation of the quantities to one another.](2)

3. On maxima and minima.(3)

4. To draw tangents.(4)

5. To ascertain the curvature of curves at a given point.(5)

6./7./8. To find arbitrarily many curves whose areas/lengths/ centres of gravity may be compared with rectilinear figures.(6)

9./10./11. To find arbitrarily many curves whose areas/lengths/ centres of gravity may be compared with proposed curvilinear figures.(7)

12./13./14. To determine the area/length/centre of gravity of any proposed curve.(8)

16./17./18. In two proposed curves to contrast their areas/lengths/ centres of gravity.(9)

---

(2) These first two problems, which are the theoretical foundation for the following fluxional applications, are lacking in Newton's text but are here restored for completeness' sake from their revised equivalents in 2, §2.

(3) Compare the following tract's 'Prob 3. Determinare maximas et minimas' (Newton's pp. 41–2).

(4) Amplified as 'Prob 4. Curvarum Tangentes ducere' in the sequel (Newton's pp. 43–52).

(5) Compare 'Prob: 5. Curvæ alicujus ad datum punctum curvaturam invenire' and 'Prob 6. Curvaturæ...qualitatem determinare' (Newton's pp. 53–66/67–70) in 2, §2.

(6) The first two of these may be loosely correlated with Problems 7 and 10 of the following tract (Newton's pp. 71/97–102), viz. 'Curvas pro arbitrio multas invenire quarum areæ (longitudines) per finitas æquationes designari possunt'.

(7) The two former are amplified in the following Problems 8 and 11 (Newton's pp. 72–6/ 103–7): 'Curvas pro arbitrio multas invenire quarum areæ ad aream datæ alicujus Curvæ relationem habent per finitas æquationes designabilem' and '...quarum longitudines cum propositæ alicujus curvæ longitudine...comparari possunt'.

(8) The first two are treated at some length in the following tract's 'Prob: 9. Propositæ alicujus Curvæ aream determinare' (Newton's pp. 75*a*–75*l*/77–96) and 'Prob 12. Curvarum Longitudines determinare' (Newton's pp. 109–114). No 'Prob 15' is listed.

(9) None of these are discussed as such in the completed portion of the following tract, though the first two are partly subsumed into its Problems 7–12.

19. Ex data $\left\{\begin{array}{l}\text{area}\\ \text{longit}\\ \text{cen grav}\end{array}\right\}$ ordinatim applicatas determinare.(10)

20. De reductione Mechanicarū curvarū ad formam Geometricam.

21. De solidis.

22. Curvas de datis proprietatibus investigare.

23. De infinitis æquationibus in finitas convertendis.(11)

---

(10) This Problem is not treated separately in the 1671 fluxional tract as such, though it is evidently the inverse of the preceding Problems 12–14 and its first two components are tacitly subsumed into that tract's Problems 7–12.

(11) These last four listed problems are not explicitly discussed in the following tract but we may suppose that they, together with the preceding Problems 8, 11, 14 and 18, were to be the subject of the concluding portion which Newton never found opportunity to write. Their titles are not sufficiently precise to make an accurate assessment of their intended content possible. We may perhaps identify Problem 20 with the area-preserving 'convolution' of a 'spiral' curve defined in a polar coordinate system into an equivalent 'parabola' defined with respect to corresponding Cartesian coordinates: a widely known mid-seventeenth-century transformation which Newton outlines on page 75 of his 1671 tract (see 2, §2: note (444)). Problem 21 would perhaps concern itself very briefly with measuring the volumes and surface-areas of solids of revolution, given the equations of their generating curves. The 'proprietates datæ' of Problem 22 were evidently to be geometrical equivalents of simple (linear?) differential equations relating the tangent slope—and, just possibly, curvature—at a general point on a curve defined in some standard coordinate system, presumably Cartesian or polar. The curve which they define would then have to be constructed, under given initial conditions, to correspond with the analytical resolution of those 'fluxional' equations. Problem 23, finally, is least tangible of all. We may guess that it would have involved identification of the structure of an infinite sequence, then establishing a simple finite form for its derivative and lastly evaluating the integral of this in terms of known algebraic, circular and logarithmic functions.

19. From the area/length/centre of gravity given, to determine the ordinates.(10)
20. On the reduction of mechanical curves to geometrical form.
21. On solids.
22. To investigate curves from their given properties.
23. On turning infinite equations into finite ones.(11)

The integral so obtained will, if its convergence be establishable, be the finite equal of the given 'æquatio infinita'.

The mensuration of simple 'spindles' of revolution, long the province of practical gaugers with their innumerable practical rules until successfully invaded by Kepler in his *Nova Stereometria Doliorum Vinariorum, in primis Austriaci, figuræ omnium aptissima* (Linz, 1615 [= *Gesammelte Werke*, **9** (Munich, 1960): 7–133]), had since been an area of intensive research. Most recently, James Gregory in Propositions 19–54 of his *Geometriæ Pars Universalis, Inserviens Quantitatum Curvarum transmutationi & mensuræ* (Padua, 1668: 42–102) had codified contemporary geometrical knowledge of the topic, and we may be sure that Newton read this portion of Gregory's book with both interest and discernment. (Compare A. Prag's analysis of the *Geometria* in the *James Gregory Tercententenary Memorial Volume* (London, 1939): 487–509, especially 498–501. Newton's personal copy of the work is now Trinity College. NQ. 9. 48².) In August 1672, indeed, in response to Gregory's communication of an analytical expression for the 'content of ye second segments of an ellipsoid' Newton passed on his own rule for assessing the volume of a 'Parabolick spindle exactly enough for practice' (*Correspondence*, **1**: 229–30; see pages 568–9 below) and this we may fairly cite as an instance of what he intended to insert in his present Problem 21. No extended discussion by him of the geometry of solids exists, though a version of Gregory's series for the second segment of a 'sphæroid' was presented to Leibniz in his *epistola prior* in June 1676 (*Correspondence*, **2**: 28–9) and he was in the mid-1680's to communicate to Hunt a rule for evaluating the content of a paraboloidal second segment (*ibid.*, **2**: 478–80) which makes elegant use of a simple integral transformation—essentially $\int r^{-2}y^3 . dz = \int y . dz - \frac{1}{8}\int v . dx$, where $x = 2z^2/r$ and $v^2 + (x-r)^2 = y^2 + z^2 = r^2$—akin to those employed in the 1671 fluxional tract which follows.

2

THE TRACT 'DE METHODIS SERIERUM ET FLUXIONUM'

[Winter 1670–1671][(1)]

§1. THE MISSING FIRST LEAF OF NEWTON'S AUTOGRAPH ORIGINAL

Collated from contemporary copies by Gregory and Jones[(2)]

[*TRACTATUS DE METHODIS SERIERUM ET FLUXIONUM.*][(3)]

Animadvertenti plerosqꝫ Geometras, posthabitâ fere Veterum syntheticâ methodo, Analyticæ excolendæ plurimum incumbere, et ejus ope tot tantasqꝫ difficultates superasse ut pene omnia extra curvarum quadraturas et similia quædam nondum penitùs enodata videantur exhausisse: placuit sequentia quibus campi analytici terminos expandere juxta ac curvarum doctrinam promovere possem in gratiam discentium breviter compingere.

Cùm in numeris et speciebus operationes computandi persimiles sint, neqꝫ differre videantur nisi in characteribus quibus quantitates in istis definitè, in his indefinitè designantur: demiror quòd doctrinam de numeris decimalibus nuper inventam (si quadraturam Hyperbolæ per N. Mercatorem demas)[(4)]

---

(1) 'The last winter...I began to new methodise ye discourse of infinite series, designing to illustrate it wth such problems as may (some of y[m] perhaps) bee more acceptable then the invention of working by such series. But being suddenly diverted...I have not yet had leisure to return to those thoughts, & I feare I shall not renew them before winter' (ULC. Add. 3977.9, an autograph draft of Newton's letter to Collins on 20 July 1671). Other evidence for this firm dating is reviewed in the introduction to the present Part 1.

(2) The autograph of the present tract (ULC. Add. 3960.14, reproduced as §2 following) lacks its first leaf, evidently numbered '1' and '2' in Newton's pagination. Already in 1691 when Raphson, together with Halley, examined it he found it 'very much worn by having been lent out' (*History of Fluxions* (London, 1715): 3) though he made no mention of any deficiency in it. We may suppose that the first leaf became detached from the body of the text some time before 1777 when Horsley scrutinized the autograph manuscript for publication, for in his printed version of the 1671 tract, the *Geometria Analytica*, he chose to collate Newton's text with a not very accurate copy (now in private possession) made by William Jones about 1710 and a defective secondary transcript from Jones' copy made by James Wilson about 1720. (See Horsley's *Isaaci Newtoni Opera Omnia*, **1** (London, 1779): 390. The Wilson secondary copy is no longer traceable.) Since neither the punctuation nor the capitalization of Jones' transcript are Newtonian in style, in the present version we have standardized both to accord with that of the remaining autograph text, taking into especial account the textual variants present in

Translation

[*A TREATISE OF THE METHODS OF SERIES AND FLUXIONS*][(3)]

Observing that the majority of geometers, with an almost complete neglect of the ancients' synthetical method, now for the most part apply themselves to the cultivation of analysis and with its aid have overcome so many formidable difficulties that they seem to have exhausted virtually everything apart from the squaring of curves and certain topics of like nature not yet fully elucidated: I found it not amiss, for the satisfaction of learners, to draw up the following short tract in which I might at once widen the boundaries of the field of analysis and advance the doctrine of curves.

Since the operations of computing in numbers and with variables are closely similar—indeed there appears to be no difference between them except in the characters by which quantities are denoted, definitely in the one case, indefinitely so in the latter—, I am amazed that it has occurred to no one (if you except N. Mercator with his quadrature of the hyperbola)[(4)] to fit the doctrine

---

the equivalent opening passages of David Gregory's *Tractatus de Seriebus infinitis et Convergentibus* (reproduced in the Appendix below).

(3) Lacking the opening page of Newton's autograph (see note (2)) we cannot know his preferred title for the present tract. Very probably, indeed, in its original form it was untitled (except perhaps for a date), for in retrospect Newton invariably referred to it in oblique terms as 'that larger Tract w[ch] I wrote in the year 1671' (ULC. Add. 3968.31: 448[r], a draft for the 'Observations' on Leibniz' letter to Conti of 9 April 1716, which he attached to the 1718 reissue of Raphson's *History of Fluxions*) or equivalently as 'Tractatus quem anno 1671 conscripsi' (ULC. Add. 3968.6: 46[r], a draft composed about 1715 for an unpublished revised version of the *Principia*'s 'fluxions scholium'). The title we here conjecture is adapted from an 'Apographum Schediasmatis' (ULC. Add. 3968.19: 263[r]) for his unpublished *Historia brevis methodi serierum ex Monumentis antiquis desumpta*, in which he makes quotation of the present tract's Problems 1 and 2 'Ex Tractatu de methodis serierum et fluxionum anno 1671 composito'. William Jones gave his copy (note (2)) the heading of *Artis Analyticæ Specimina sive Geometria Analytica. Auctore Isaaco Newtono, Equite Aurato* and Samuel Horsley borrowed its first phrase for the title of his first printed version of the tract in its original Latin in 1779. Jones' copy was likewise the common source for Colson's first English translation of the piece (London, 1736) and the pirated version which appeared a year later, which both bore the title of *A Treatise of the Method of Fluxions and Infinite Series, With its Application to the Geometry of Curve Lines.*

(4) Nicolaus Mercator, in his *Logarithmotechnia: sive Methodus construendi Logarithmos nova, accurata, & facilis...: cui...accedit Vera Quadratura Hyperbolæ* (London, 1668: 31–3: Propositio XVII. *Quadrare Hyperbolam*), had used a Wallisian indivisibles summation coupled with a 'reductio per divisionem' in evaluating $\log(1+x) = \lim_{n\to\infty} \sum_{0\leqslant i\leqslant n} (-1)^i i^{-1} x^i$ as

$$\int_0^x (1-x+x^2-x^3 \ldots).dx$$

('hoc est, = numero terminorum...minus summâ eorundem terminorum, plus summâ quadratorum ab iisdem, minus summâ cuborum... &c'), the area under the hyperbola

nemini in mentem venerit speciebus itidem accommodare præsertim cùm ad præclariora viam aperiat.(5) Hujus autem de speciebus doctrinæ cum eodem modo ad Algebram relata sit ac doctrina decimalium numerorum ad vulgarem Arithmeticam: operationes Additio, Subtractio, Multiplicatio, Divisio et extractio Radicum exinde addisci possunt modo lector utriusq̧, et Arithmeticæ et Algebræ vulgaris, peritus fuerit et noverit correspondentiam inter decimales numeros ac terminos Algebraicos in infinitum continuatos; scilicet quòd singulis numerorum locis proportione decimali dextrorsum perpetuò decrescentibus correspondent singuli specierum termini secundum seriem dimensionum numeratorum vel denominatorum uniformi progressione in infinitum continuatam (prout factum in sequentibus) ordinati. Et quemadmodum commoditas decimalium in eo consistit ut fractiones omnes et radicales in eos reductæ quodammodo naturam integrorum induant: sic etiam infinitarum specierum commoditas est quòd per eas abstrusiorum terminorum genera (quales sunt fractiones a compositis quantitatibus denominatæ, compositarum radices, et radices affectarum æquationum) possunt ad simplicium genus reduci, ad infinitas nempe fractionum series numeratores ac denominatores simplices habentium, in quibus nullæ sunt aliorum difficultates propemodum insuperabiles.(6) Imprimis itaq̧ reductiones aliarum quantitatum ad hujusmodi terminos et methodos computandi minus obvias ostendam, dein hanc Analysin ad solutiones problematum applicabo.

Reductiones per divisionem et extractionem radicum è sequentibus exemplis cum similibus operandi modis in Arithmeticâ decimali et speciosa collatis elucesce[n]t.

$(1+x)y = 1$. See also J. E. Hofmann, 'Nicolaus Mercators Logarithmotechnia (1668)', *Deutsche Mathematik*, **3**, 1938: 446–66; and compare his comprehensive monograph 'Nicolaus Mercator (Kaufmann), sein Leben und Wirken, vorzugsweise als Mathematiker', *Akademie der Wissenschaften und der Literatur* [*in Mainz*]*: Abhandlungen der Math.-Naturw. Klasse, Jahrgang 1950, Nr. 3* (Wiesbaden, 1950): 63–5.

(5) Here Newton betrays his relative ignorance of the contemporary mathematical scene. In point of fact Johann Hudde on at least two occasions (compare C. I. Gerhardt, *Der Briefwechsel von Gottfried Wilhelm Leibniz mit Mathematikern* **1** (Berlin, 1899): 228; and *The Correspondence of Isaac Newton* **1**, 1959: 38 n. 4) claimed priority in discovering the 'Mercator' expansion of the logarithmic function, which of course Newton had found for himself in the spring of 1665 in his Wallisian researches (see I, **1**, 3, §3.3). We may forgive him for not knowing that a century and a half before the Hindu mathematician Nilakaṇṭha had derived the inverse-tangent series by a 'reductio per divisionem', exactly as James Gregory and Leibniz were to do in January 1671 and late 1673 respectively: no hint of this reached Europe, it would appear, till 1835. (See A. R. A. Iyer's and R. Tampurān's edition of the 1639 Malayālam compendium *Yuktibhāṣā*, Trichur, 1948: 113 ff.; also C. T. Rajagopal and T. V. V. Aiyar, 'On the Hindu Proof of Gregory's Series' and 'A Hindu Approximation to π', *Scripta Mathematica*, **17**, 1951: 65–74; **18**, 1952: 25–30; and K. M. Marar and C. T. Rajagopal, 'On the Hindu Quadrature of the Circle', *Journal of the Bombay Branch of the Royal Asiatic Society*, **20**, 1944: 65–82.) Both Gregory and Leibniz, of course, drew their inspiration from Mercator's quadrature of the

recently established for decimal numbers in similar fashion to variables, especially since the way is then open to more striking consequences.(5) For since this doctrine in species has the same relationship to Algebra that the doctrine in decimal numbers has to common Arithmetic, its operations of Addition, Subtraction, Multiplication, Division and Root-extraction may easily be learnt from the latter's provided the reader be skilled in each, both Arithmetic and Algebra, and appreciate the correspondence between decimal numbers and algebraic terms continued to infinity: namely, that to each single place in a decimal sequence decreasing continually to the right there corresponds a unique term in a variable array ordered according to the sequence of the dimensions of numerators or denominators continued in uniform progression to infinity (as you will see done in the sequel). And just as the advantage of decimals consists in this, that when all fractions and roots have been reduced to them they take on in a certain measure the nature of integers; so it is the advantage of infinite variable-sequences that classes of more complicated terms (such as fractions whose denominators are complex quantities, the roots of complex quantities and the roots of affected equations) may be reduced to the class of simple ones: that is, to infinite series of fractions having simple numerators and denominators and without the all but insuperable encumbrances which beset the others.(6) I will first, consequently, show how to reduce other quantities to terms of this sort, revealing less familiar methods of doing so, and then I will apply this Analysis to the resolution of problems.

Reductions by division and root extraction will become clear from the following examples when the similar ways of operating in decimal and specious arithmetic are compared.

---

hyperbola (see H. W. Turnbull's *Gregory Tercentenary Memorial Volume*, London, 1939: 173, 357; and J. E. Hofmann's *Entwicklungsgeschichte der Leibnizschen Mathematik während des Aufenthaltes in Paris (1672–1676)*, Munich, 1949: 32–6).

(6) A somewhat over-optimistic pronouncement, echoed five years later in his letter of 13 June 1676 to Oldenburg for Leibniz: 'Ex his videre est quantum fines Analyseos per hujusmodi infinitas æquationes ampliantur: quippe quæ earum beneficio, ad omnia, pene dixerim, problemata (si numeralia Diophanti et similia excipias) sese extendit' (*Correspondence*, **2**, 1960: 29). In his reply on 11 June 1677 (*Correspondence*, **2**: 215) Leibniz pointed, among other things, to the difficulty of defining the complex roots of equations by infinite series: 'si in...generali illa æquationis affectæ indefinitæ extractione...[Neutonus] idem præstari posset ut scilicet inter extrahendum radices ex æquationibus vel binomiis invenire liceret radices rationales finitas, quando eæ insunt, vel etiam irrationales tunc dicerem methodum serierum infinitarum ad summam perfectionem esse productam. Opus esset tamen præterea discerni posse varias æquationis ejusdem radices item necesse esset ope serierum discerni æquationes possibles ab impossibilibus. Quod si hæc nobis obtinuerit vir in his studiis maximus...tunc in methodo serierum infinitarum quæ divisione atque extractione inveniuntur vix quicquam amplius optandum restabit.'

EXEMPLA REDUCTIONUM PER DIVISIONEM(7)

Proposito $\frac{aa}{b+x}$, divide aa per $b+x$ ad hunc modum.

$$
\begin{array}{l}
b+x)aa+0\left(\frac{aa}{b}-\frac{aax}{bb}+\frac{aaxx}{b^3}-\frac{aax^3}{b^4}+\frac{aax^4}{b^5}\quad \&c:\right. \\
\quad aa+\frac{aax}{b} \\
\hline
\quad 0-\frac{aax}{b}+0 \\
\qquad -\frac{aax}{b}-\frac{aaxx}{bb} \\
\hline
\qquad 0+\frac{aaxx}{bb}+0 \\
\qquad\quad +\frac{aaxx}{bb}+\frac{aax^3}{b^3} \\
\hline
\qquad\quad 0-\frac{aax^3}{b^3}+0 \\
\qquad\qquad -\frac{aax^3}{b^3}-\frac{aax^4}{b^4} \\
\hline
\qquad\qquad 0+\frac{aax^4}{b^4}\quad \&c.
\end{array}
$$

Et prodit $\frac{aa}{b}-\frac{aax}{bb}+\frac{aaxx}{b^3}-\frac{aax^3}{b^4}$ &c. quæ series in infinitum continuata tantùm valet ac $\frac{aa}{b+x}$. Vel posito x primo(8) [divisoris termino hoc modo...].

---

(7) A revision of the equivalent section in the *De Analysi* (II: 212).

EXAMPLES OF REDUCTION BY DIVISION.(7)

If the reduction of $\frac{a^2}{b+x}$ be proposed, divide a^2 by $b+x$ in this way:

$$
\begin{array}{l}
b+x)\,a^2+0\left(\dfrac{a^2}{b}-\dfrac{a^2x}{b^2}+\dfrac{a^2x^2}{b^3}-\dfrac{a^2x^3}{b^4}+\dfrac{a^2x^4}{b^5}\ldots\right.\\
\qquad a^2+\dfrac{a^2x}{b}\\
\qquad\quad -\dfrac{a^2x}{b}+0\\
\qquad\quad -\dfrac{a^2x}{b}-\dfrac{a^2x^2}{b^2}\\
\qquad\qquad +\dfrac{a^2x^2}{b^2}+0\\
\qquad\qquad +\dfrac{a^2x^2}{b^2}+\dfrac{a^2x^3}{b^3}\\
\qquad\qquad\quad -\dfrac{a^2x^3}{b^3}+0\\
\qquad\qquad\quad -\dfrac{a^2x^3}{b^3}-\dfrac{a^2x^4}{b^4}\\
\qquad\qquad\qquad +\dfrac{a^2x^4}{b^4}\ldots
\end{array}
$$

And there comes out $\frac{a^2}{b}-\frac{a^2x}{b^2}+\frac{a^2x^2}{b^3}-\frac{a^2x^3}{b^4}\ldots$, a series which when continued to infinity is equivalent to $\frac{a^2}{b+x}$. Or on setting x as the first(8) [term of the divisor in this manner...]

(8) Here begins the first page ('3' in Newton's pagination) of the autograph text reproduced as §2 following.

§2. THE MAIN AUTOGRAPH TEXT

From the original in the University Library, Cambridge(1)

||[3] ||divisoris termino hoc modo

$$x+b)aa\ (\text{prodibit}\ \frac{aa}{x}-\frac{aab}{xx}+\frac{aabb}{x^3}-\frac{aab^3}{x^4}\quad \&\text{c.}$$

Ad eundem modum fractio $\frac{1}{1+xx}$ reducitur ad $1-xx+x^4-x^6+x^8$ &c vel ad $x^{-2}-x^{-4}+x^{-6}-x^{-8}$ &c.

Et fractio $\frac{2x^{\frac{1}{2}}-x^{\frac{3}{2}}}{1+x^{\frac{1}{2}}-3x}$ ad $2x^{\frac{1}{2}}-2x+7x^{\frac{3}{2}}-13x^2+34x^{\frac{5}{2}}$ &c.

Ubi obiter notandum est quòd usurpo x^{-1}, x^{-2}, x^{-3}, x^{-4}, &c: pro $\frac{1}{x}$, $\frac{1}{xx}$, $\frac{1}{x^3}$, $\frac{1}{x^4}$; & $x^{\frac{1}{2}}$, $x^{\frac{3}{2}}$, $x^{\frac{5}{2}}$, $x^{\frac{1}{3}}$, $x^{\frac{2}{3}}$, &c: pro $\surd x$, $\surd x^3$, $\surd x^5$, $\surd c{:}x$, $\surd c{:}xx$; & $x^{-\frac{1}{2}}$, $x^{-\frac{2}{3}}$, $x^{-\frac{1}{4}}$, &c pro $\frac{1}{\surd x}$, $\frac{1}{\surd c{:}xx}$, $\frac{1}{\surd 4{:}x}$. Idq; ob analogiam rei, quæ deprehendi potest ex hujusmodi geometricis progressionibus x^3, $x^{\frac{5}{2}}$, xx, $x^{\frac{3}{2}}$, x, $x^{\frac{1}{2}}$, x^0 (sive 1,) $x^{-\frac{1}{2}}$, x^{-1}, $x^{-\frac{3}{2}}$, x^{-2} &c.

Ad hunc modum pro $\frac{aa}{x}-\frac{aab}{xx}+\frac{aabb}{x^3}$ &c scribi potest $aax^{-1}-aabx^{-2}+aabbx^{-3}$ &c. Et sic vice $\sqrt{aa-xx}$ scribi potest $\overline{aa-xx}|^{\frac{1}{2}}$; et $\overline{aa-xx}|^{2}$ vice quadrati ex $aa-xx$; et $\left.\frac{abb-y^3}{by+yy}\right|^{\frac{1}{3}}$ vice $\surd C{:}\frac{abb-y^3}{by+yy}$. Et sic in alijs. Unde meritò potestates distingui possunt in affirmativas et negativas, integras, et fractas.(2)

---

(1) Add. 3960.14, a thick gathering of some fifty folio sheets folded and slit into fours together with several further inserted sheets. The tract's two states and its essential incompleteness are evident: in the body of the text several blank pages have been left for the intended completion of a chapter, while the 132 pages which have been written upon are followed by some sixty virgin sides. A first extended version of the piece is entered continuously on sides paginated by Newton 3–[114] and doubtless represents his hard work during the winter of 1670–71 to 'new methodise ye discourse of infinite series' as it had appeared in the *De Analysi* (II, 2, 2). In later revision (about the autumn of 1671?) pages 51–2 were cancelled, being replaced by new pages [51 bis/52 bis] while their content was inserted at an earlier place on new pages [45a/45b]; in a similar way new pages [59a] and [85a] intervened after pages 59 and 85, while the tabulation of integrals on pages 75–6 was cancelled in favour of the dozen inserted pages [75a–75l]. A modern pencilled pagination (of *c.* 1940) which omits pages 51–2 from its pagination blurs these differences by numbering the sheets continuously from 3 to 132.

Translation

...term of the divisor in this manner

$$x+b)a^2 \text{ (will produce } \frac{a^2}{x}-\frac{a^2b}{x^2}+\frac{a^2b^2}{x^3}-\frac{a^2b^3}{x^4}\ldots.$$

In much the same way the fraction $\frac{1}{1+x^2}$ is reduced to

$$1-x^2+x^4-x^6+x^8\ldots \quad \text{or to} \quad x^{-2}-x^{-4}+x^{-6}-x^{-8}\ldots.$$

And the fraction $\frac{2x^{\frac{1}{2}}-x^{\frac{3}{2}}}{1+x^{\frac{1}{2}}-3x}$ to $2x^{\frac{1}{2}}-2x+7x^{\frac{3}{2}}-13x^2+34x^{\frac{5}{2}}\ldots$.

Here by the way it must be noted that I employ $x^{-1}, x^{-2}, x^{-3}, x^{-4}, \ldots$ in place of $\frac{1}{x}, \frac{1}{x^2}, \frac{1}{x^3}, \frac{1}{x^4}, \ldots$; and $x^{\frac{1}{2}}, x^{\frac{3}{2}}, x^{\frac{5}{2}}, x^{\frac{1}{3}}, x^{\frac{2}{3}}, \ldots$ in place of $\sqrt{x}, \sqrt{x^3}, \sqrt{x^5}, \sqrt[3]{x}, \sqrt[3]{x^2}, \ldots$; and $x^{-\frac{1}{2}}, x^{-\frac{2}{3}}, x^{-\frac{1}{4}}, \ldots$ in place of $\frac{1}{\sqrt{x}}, \frac{1}{\sqrt[3]{x^2}}, \frac{1}{\sqrt[4]{x}}, \ldots$. And that on account of the similarity of form which may be detected in geometrical progressions of this sort, $x^3, x^{\frac{5}{2}}, x^2, x^{\frac{3}{2}}, x, x^{\frac{1}{2}}, x^0$ (or 1), $x^{-\frac{1}{2}}, x^{-1}, x^{-\frac{3}{2}}, x^{-2}, \ldots$.

In this way for $\frac{a^2}{x}-\frac{a^2b}{x^2}+\frac{a^2b^2}{x^3}\ldots$ may be written $a^2x^{-1}-a^2bx^{-2}+a^2b^2x^{-3}\ldots$. And thus in place of $\sqrt{a^2-x^2}$ may be written $(a^2-x^2)^{\frac{1}{2}}$; and $(a^2-x^2)^2$ in place of the square of a^2-x^2; and $\left(\frac{ab^2-y^3}{by+y^2}\right)^{\frac{1}{3}}$ in place of $\sqrt[3]{\frac{ab^2-y^3}{by+y^2}}$. And so in other cases. In consequence we may justifiably distinguish powers as positive, negative, integral and fractional.(2)

(2) That is, all rational numbers are allowable indices. Whether or not Newton here intentionally excluded 'surdæ quantitates' (the set of non-rational reals) it is clear that he did not do so in general: in his *epistola posterior* of 24 October 1676 to Oldenburg for Leibniz, for example, he challenged the latter to resolve the 'affected' equation $(x^{\sqrt{2}}+x^{\sqrt{7}})^{3\sqrt{\frac{2}{3}}} = y$ (*Correspondence of Isaac Newton*, **2**, 1960: 129). Newton's present insistence on spelling out the meaning of negative and fractional powers may appear strange if we recall that the logarithmic correspondence between x and y defined by $a^x = y$ had been universally accepted by mathematicians for half a century. In fact, however, contemporary algebraists were still wildly inconsistent in their notations for indices and literal nomenclature for general powers was first systematically introduced only in 1657 in Wallis' *Mathesis Universalis* (I, **1**, Introduction: Appendix 2, note (20)). See F. Cajori's *History of Mathematical Notations*, **1** (Chicago, 1928): 335–56.

EXEMPLA REDUCTIONUM PER EXTRACTIONEM RADICUM[(3)]

Proposito $aa+xx$, radicem ejus ut sequitur extrahes,

$$aa+xx\left(a+\frac{xx}{2a}-\frac{x^4}{8a^3}+\frac{x^6}{16a^5}-\frac{5x^8}{128a^7}+\frac{7x^{10}}{256a^9}-\frac{21x^{12}}{1024a^{11}}\right. \&c$$

$$aa$$

$$\overline{0\ +xx}$$

$$+xx+\frac{x^4}{4aa}$$

$$-\frac{x^4}{4aa}$$

$$-\frac{x^4}{4aa}-\frac{x^6}{8a^4}+\frac{x^8}{64a^6}$$

$$\frac{x^6}{8a^4}-\frac{x^8}{64a^6}$$

$$\frac{x^6}{8a^4}+\frac{x^8}{16a^6}-\frac{x^{10}}{64a^8}+\frac{x^{12}}{256a^{10}}$$

$$-\frac{5x^8}{64a^6}+\frac{x^{10}}{64a^8}-\frac{x^{12}}{256a^{10}}$$

$$-\frac{5x^8}{64a^6}-\frac{5x^{10}}{128a^8}+\frac{5x^{12}}{512a^{10}}\quad [\&c]$$

$$\frac{7x^{10}}{128a^8}-\frac{7x^{12}}{512a^{10}}\quad [\&c]$$

$$\frac{7x^{10}}{128a^8}+\frac{7x^{12}}{256a^{10}}\quad [\&c]$$

$$-\frac{21x^{12}}{512a^{10}}\quad [\&c]$$

||[4] || et prodit $a+\frac{xx}{2a}-\frac{x^4}{8a^3}$ &c.[(4)] Ubi notandum[(5)] quòd circa finem operis eos omnes terminos negligo quorum dimensiones transcenderent dimensiones ultimi termini ad quem cupio quotientem solummodò produci, puta $\frac{x^{12}}{a^{11}}$. Potest etiam ordo terminorum inverti ad hunc modum $xx+aa$, et radix erit $x+\frac{aa}{2x}-\frac{a^4}{8a^3}+\frac{a^6}{16x^5}$ &c.

Sic ex $aa-xx$ radix est $a-\frac{xx}{2a}-\frac{x^4}{8a^3}-\frac{x^6}{16a^5}$ &c.

Et ex $x-xx$ est $x^{\frac{1}{2}}-\frac{1}{2}x^{\frac{3}{2}}-\frac{1}{8}x^{\frac{5}{2}}-\frac{1}{16}x^{\frac{7}{2}}$ &c.

(3) Taken with but slight verbal revision from the equivalent passage in *De Analysi* (II: 214–18).

EXAMPLES OF REDUCTIONS BY ROOT EXTRACTION[3]

When the quantity a^2+x^2 is proposed, you will extract its root as follows:

$$a^2+x^2\left(a+\frac{x^2}{2a}-\frac{x^4}{8a^3}+\frac{x^6}{16a^5}-\frac{5x^8}{128a^7}+\frac{7x^{10}}{256a^9}-\frac{21x^{12}}{1024a^{11}}\dots\right.$$

$$\begin{array}{l}
\underline{a^2} \\
\quad +x^2 \\
\quad \underline{x^2+\frac{x^4}{4a^2}} \\
\qquad -\frac{x^4}{4a^2} \\
\qquad \underline{-\frac{x^4}{4a^2}-\frac{x^6}{8a^4}+\frac{x^8}{64a^6}} \\
\qquad\qquad \frac{x^6}{8a^4}-\frac{x^8}{64a^6} \\
\qquad\qquad \underline{\frac{x^6}{8a^4}+\frac{x^8}{16a^6}-\frac{x^{10}}{64a^8}+\frac{x^{12}}{256a^{10}}} \\
\qquad\qquad\quad -\frac{5x^8}{64a^6}+\frac{x^{10}}{64a^8}-\frac{x^{12}}{256a^{10}} \\
\qquad\qquad\quad \underline{-\frac{5x^8}{64a^6}-\frac{5x^{10}}{128a^8}+\frac{5x^{12}}{512a^{10}}\dots} \\
\qquad\qquad\qquad\quad \frac{7x^{10}}{128a^8}-\frac{7x^{12}}{512a^{10}}\dots \\
\qquad\qquad\qquad\quad \underline{\frac{7x^{10}}{128a^8}+\frac{7x^{12}}{256a^{10}}\dots} \\
\qquad\qquad\qquad\qquad -\frac{21x^{12}}{512a^{10}}\dots
\end{array}$$

and there comes out $a+\frac{x^2}{2a}-\frac{x^4}{8a^3}\dots$.[4] Here note that towards the end of the operation I neglect all the terms whose dimensions would surpass the dimensions of the final term, say $\frac{x^{12}}{a^{11}}$, to which alone I desire to extend the quotient. The order of the terms may be inverted also in this way, x^2+a^2, and the root will then be $x+\frac{a^2}{2x}-\frac{a^4}{8a^3}+\frac{a^6}{16x^5}\dots$.

So of a^2-x^2 the root is $a-\frac{x^2}{2a}-\frac{x^4}{8a^3}-\frac{x^6}{16a^5}\dots$.

Of $x-x^2$ it is $x^{\frac{1}{2}}-\frac{1}{2}x^{\frac{3}{2}}-\frac{1}{8}x^{\frac{5}{2}}-\frac{1}{16}x^{\frac{7}{2}}\dots$.

(4) For convergence, of course, $|x/a|$ should be less than unity.

(5) An enclitic 'venit' is here cancelled. As a gloss on Newton's text, we have indicated those lines shortened in the previous root-extraction scheme by inserting '&c' at appropriate points.

Et ex $aa+bx-xx$ est $a+\frac{bx}{2a}-\frac{xx}{2a}-\frac{bbxx}{8a^3}$ &c.[(6)]

Et ex $\frac{1+axx}{1-bxx}$ est $\frac{1+\frac{1}{2}ax^2-\frac{1}{8}aax^4+\frac{1}{16}a^3x^6}{1-\frac{1}{2}bx^2-\frac{1}{8}bbx^4-\frac{1}{16}b^3x^6}$ &c. factâq; insuper divisione, fit

$$1\begin{array}{l}+\frac{1}{2}b\\+\frac{1}{2}a\end{array}x^2\begin{array}{l}+\frac{3}{8}bb\\+\frac{1}{4}ab\\-\frac{1}{8}aa\end{array}x^4\begin{array}{l}+\frac{5}{16}b^3\\+\frac{3}{16}abb\\-\frac{1}{16}aab\\+\frac{1}{16}a^3\end{array}x^6\quad\&c.$$

Operationes verò per debitam[(7)] præparationē non rarò abbreviari possunt; Ut in allato exemplo ad extrahendam $\sqrt{\frac{1+axx}{1-bxx}}$, si non eadem fuisset numeratoris ac denominatoris forma, utrumq; multiplicassem per $\sqrt{1-bxx}$ & sic prodijsset $\frac{\sqrt{:1\begin{array}{l}+a\\-b\end{array}xx-abx^4:}}{1-bxx}$, et reliquum opus perficeretur extrahendo radicem numeratoris tantùm ac dividendo per denominatorem.

Ex hisce credo manifestum est quo pacto radices aliæ possunt extrahi et quælibet compositæ quantitates (quibuscunq; radicibus vel denominatoribus perplexæ, ut hic videre est $x^3+\frac{\sqrt{x-\sqrt{1-xx}}}{\sqrt{3:axx+x^3}}-\frac{\sqrt{5:x^3+2x^5-x^{\frac{3}{2}}}}{\sqrt{3:x+xx-\sqrt{2x-x^{\frac{2}{3}}}}}$) in series infinitas simplicium terminorum reduci.

De affectarum æquationum reductione[(8)]

Propositis verò affectis æquationibus, modus quo radices earum ad hujusmodi series reduci possint obnixiùs explicari debet, idq; cùm earum doctrina quam hactenus in numeris exposuerunt Mathematici,[(9)] per ambages (superfluis
‖[5] etiam operationibus adhibitis)[(10)] tradatur, ut in specimen ‖ operis in speciebus non debeat adhiberi. Imprimis itaq; numerosam affectarum æquationum resolutionem compendiosè tradam, dein speciosam similiter explicabo.

Proponatur æquatio $y^3-2y-5=0$ resolvenda, et sit 2 numerus utcunq;[(11)] inventus qui minùs quàm decimâ sui parte differt a radice quæsitâ. Tum

(6) This sentence is a late addition in the margin. In the last term for 'xx' read 'x^3'.

(7) Newton here copied 'æquationis' (of the equation) from the *De Analysi* (II: 216) and then cancelled it.

(8) This section is considerably amplified from the equivalent passage in the *De Analysi* (II: 218–32). Excerpts were communicated to Leibniz five years later in the *epistola prior* of 13 June 1676 (*Correspondence*, **2**, 1960: 23–5).

(9) Notably Viète, whose somewhat laborious method for solving numerical equations Newton had mastered as an undergraduate. See I, **1**, 2, §1 and compare II, **2**, **3**: note (45).

(10) Here 'obscurè' (obscurely) is justly cancelled. However cumbrous and inefficient

Of a^2+bx-x^2 it is $a+\frac{bx}{2a}-\frac{x^2}{2a}-\frac{b^2x^{[3]}}{8a^3}\ldots$.[(6)]

And of $\frac{1+ax^2}{1-bx^2}$ it is $\frac{1+\frac{1}{2}ax^2-\frac{1}{8}a^2x^4+\frac{1}{16}a^3x^6\ldots}{1-\frac{1}{2}bx^2-\frac{1}{8}b^2x^4-\frac{1}{16}b^3x^6\ldots}$; this, moreover, after the division has been carried out, becomes

$$1+(\tfrac{1}{2}b+\tfrac{1}{2}a)\,x^2+(\tfrac{3}{8}b^2+\tfrac{1}{4}ab-\tfrac{1}{8}a^2)\,x^4+(\tfrac{5}{16}b^3+\tfrac{3}{16}ab^2-\tfrac{1}{16}a^2b+\tfrac{1}{16}a^3)\,x^6\ldots.$$

But in practice these operations may not infrequently be shortened by appropriate preparation.[(7)] As in the example advanced, to extract $\sqrt{\frac{1+ax^2}{1-bx^2}}$, if its numerator and denominator had not been of the same form, I should have multiplied both by $\sqrt{1-bx^2}$, producing $\frac{\sqrt{1+(a-b)\,x^2-abx^4}}{1-bx^2}$, and the rest of the work might then be accomplished by merely extracting the root of the numerator and dividing by the denominator.

From these remarks it is, I believe, clear how other roots may be extracted and how any compound quantities (however they be entangled with roots and denominators, such as here

$$x^3+\frac{\sqrt{x-\sqrt{1-x^2}}}{\sqrt[3]{ax^2+x^3}}-\frac{\sqrt[5]{x^3+2x^5-x^{\frac{3}{2}}}}{\sqrt[3]{x+x^2-\sqrt{2x-x^{\frac{2}{3}}}}}\Bigg)$$

may be reduced to an infinite series of simple terms.

The reduction of affected equations.[(8)]

When, however, affected equations are proposed, the manner in which their roots might be reduced to this sort of series should be more closely explained, the more so since their doctrine, as hitherto expounded by mathematicians[(9)] in numerical cases, is delivered[(10)] in a roundabout way (and indeed with the introduction of superfluous operations) and in consequence ought not to be brought in to illustrate the procedure in species. In the first place, then, I will discuss the numerical resolution of affected equations briefly but comprehensively, and subsequently explain the algebraical equivalent in similar fashion.

Let the equation $y^3-2y-5=0$ be proposed for solution and let the number 2 be found, one way or another,[(11)] which differs from the required root by less

---

existing numerical techniques were, it would be unfair to say that all were presented in an obscure manner.

(11) The simplest way, one systematized by Stevin in 1594 and summarized by Kinckhuysen in the *Algebra* to which Newton had recently devoted so much of his time (see II, **3**, 1, §1: note (108)), would be to observe, where $f(y)\equiv y^3-2y-5$, that $f(2{\cdot}1)>0>f(2{\cdot}0)$ and so conclude by continuity that $f(2+\epsilon)=0$, $0<\epsilon<0{\cdot}1$.

pono $2+p=y$, et pro y substituo[12] $2+p$ in æquationem, et inde nova prodit

$$p^3+6pp+10p-1=0.$$

cujus radix p exquirenda est ut quotienti addatur. Nempe (neglectis p^3+6pp ob parvitatem) $10p-1=0$ sive $p=0,1$ ad veritatem proxime accedit. Scribo itaq; 0,1 in quotiente & suppono $0,1+q=p$, et hunc ejus fictitiū valorem ut ante substituo, et prodit $q^3+6,3qq+11,23q+0,061=0$. Et cùm $11,23q+0,061=0$ veritatem prope appropinquet sive ferè sit $q=-0,0054$ (dividendo nempe 0,061 per 11,23 donec tot eliciantur figuræ quot loca primis figuris hujus et principalis quotientis exclusivè intercedunt, quemadmodum hic duo sunt inter 2, & 0,005) scribo $-0,0054$ in inferiori parte quotientis siquidem negativa sit, et supponens $-0,0054+r=q$, hunc ut priùs substituo. Et sic operationem ad placitum produco, pro more subjecti diagrammatis.

$$\begin{cases} +2,10000000 \\ -0,00544852 \end{cases}$$

2,09455148[$=y$]

$2+p=y.$	y^3	$+8$	$+12p$	$+6pp$	$+p^3$
	$-2y$	-4	$-2p$		
	-5	-5			
	Summa	-1	$+10p$	$+6pp$	$+p^3$
$0,1+q=p.$	$+p^3$	$+0,001$	$+0,03q$	$+0,3qq$	$+q^3$
	$+6pp$	$+0,06$	$+1,2$	$+6,$	
	$+10p$	$+1,$	$+10,$		
	-1	-1			
	Summa	0,061	$+11,23q$	$+6,3qq$	$+q^3$
$-0,0054+r=q.$	$+q^3$	−0,0000001~~57464~~	+0,000~~08748~~*r*	−~~0~~,~~0162~~*rr*	+~~1~~*r*³
	$+6,3qq$	+0,0001837~~08~~	−0,068~~04~~	+~~6~~,~~3~~	
	$+11,23q$	$-0,060642$	$+11,23$		
	$+0,061$	$+0,061$			
	Summa	$+0,0005416$	$+11,162r$		
$-0,00004852+s=r.$					

(13)

‖[6] ‖ Opus verò sub fine (præsertim in æquationibus plurium dimensionum) hac methodo multùm abbreviabitur. Determinato quousq; velis radicem extrahi, tot loca post primā figuram coefficientis penultimi termini æquationum in dextra parte diagrammatis resultantium adnumera, quot supersunt loca in quotiente complenda, et subsequentes decimales neglige. In ultimo verò termino decimales post tot plura loca neglige quot in quotiente complentur

(12) 'hunc sibi valorem' (its present value) is cancelled.

(13) This is little different from the equivalent table in the *De Analysi* (II: 218). Note, however (compare II, **2**, **3**: note (47)) that Newton has reinserted the term $q^3 \approx -0{\cdot}00000016$ in the last stage, though he has not bothered to round it off to 0·0000002 in his subsequent cancellations there to $O(10^{-8})$. Since his resulting root $y = 2{\cdot}09455148$ is correct to $8D$ we may here perhaps convict Newton yet again of over-eagerness to achieve a result whose

than its tenth part. I then set $2+p = y$, and in place of y in the equation I substitute $2+p$. From this there arises the new equation $p^3+6p^2+10p-1 = 0$, whose root p is to be sought for addition to the quotient. Specifically, (when p^3+6p^2 is neglected because of its smallness) we have $10p-1 = 0$, or $p = 0{\cdot}1$ narrowly approximates the truth. Accordingly, I write 0·1 in the quotient and, supposing $0{\cdot}1+q = p$, I substitute this fictitious value for it as before. There results $q^3+6{\cdot}3q^2+11{\cdot}23q+0{\cdot}061 = 0$. And since $11{\cdot}23q+0{\cdot}061 = 0$ closely approaches the truth, in other words very nearly $q = -0{\cdot}0054$ (by dividing 0·061 by 11·23, that is, until there are obtained as many figures as places which, excluding the bounding ones, lie between the first figures of this quotient and of the principal one—here, for instance, there are two between 2 and 0·005), I write $-0{\cdot}0054$ in the lower part of the quotient seeing that it is negative and then, supposing $-0{\cdot}0054+r$ equal to q, I substitute this value as previously. And in this way I extend the operation at pleasure after the manner of the diagram appended.

$$\begin{cases} +2{\cdot}10000000 \\ -0{\cdot}00544852 \end{cases}$$

$$2{\cdot}09455148\ [=y]$$

$2+p = y.$	y^3	$+8$	$+12p$	$+6p^2$	$+p^3$
	$-2y$	-4	$-2p$		
	-5	-5			
	Total	-1	$+10p$	$+6p^2$	$+p^3$
$0{\cdot}1+q = p.$	$+p^3$	$+0{\cdot}001$	$+0{\cdot}03q$	$+0{\cdot}3q^2$	$+q^3$
	$+6p^2$	$+0{\cdot}06$	$+1{\cdot}2$	$+6$	
	$+10p$	$+1$	$+10$		
	-1	-1			
	Total	$0{\cdot}061$	$+11{\cdot}23q$	$+6{\cdot}3q^2$	$+q^3$
$-0{\cdot}0054+r = q.$	$+q^3$	$-0{\cdot}0000001\not{5}\not{7}\not{4}\not{6}\not{4}$	$+0{\cdot}000\not{0}\not{8}\not{7}\not{4}\not{8}r$	$-\not{0}{\cdot}\not{0}\not{1}\not{6}\not{2}r^2$	$+\not{1}r^3$
	$+6{\cdot}3q^2$	$+0{\cdot}0001837\not{0}\not{8}$	$-0{\cdot}068\not{0}\not{4}$	$+\not{6}{\cdot}\not{3}$	
	$+11{\cdot}23q$	$-0{\cdot}060642$	$+11{\cdot}23$		
	$+0{\cdot}061$	$+0{\cdot}061$			
	Total	$+0{\cdot}0005416$	$+11{\cdot}162r$		
$-0{\cdot}00004852+s = r.$					

(13)

Near the end, however, (especially in equations of several dimensions) the work will be much shortened by this method. When you have decided how far you wish the root to be extracted, count off as many places from the first figure of the coefficient of the last term but one in the equations resulting on the right side of the diagram as there remain places to be filled up in the quotient, and neglect the decimals which follow after. But in the final term neglect the decimals after as many more places as there are decimal places filled up in the

accuracy is greater than his technique in fact will allow. The numbers shown as cancelled by faint oblique strokes were first eliminated in the manuscript by superscript cancellation dots not here reproduced. (Compare I, 2, 2, §1: note (3).)

loca decimalia. Inq; antepenultimo termino neglige omnes post tot pauciora loca. Et sic deinceps, Arithmeticè progrediendo per intervallū istud locorum, sive quod perinde est, tot figuras passim elidendo quot in penultimo termino, modò depressissima earum loca sint in Arithmeticâ progressione juxta seriem terminorum, aut circulis compleri subintelligantur ubi res aliter eveniat.(14) Sic in exemplo jam posito, si cupiam ut quotiens ad octavum tantùm decimalium locum compleatur; inter substituendum $0,0054+r$ pro q, ubi quatuor loca decimalia in quotiente complentur ac totidem supersunt complenda, potui figuras in inferioribus quinq; locis omisisse quas eapropter lineolâ transversim notavi; imò primum terminum r^3, etsi coefficientē 99999 habuisset, potui tamen penitus omisisse. Expunctis itaq; figuris istis, pro subsequente operatione prodit summa $0,0005416+11,162r$, quæ per divisionē ad usq; præscriptum terminum peractum dat $-0,00004852$ pro r, quod quotientem ad optatum periodum complet. Deniq; negativam partem quotientis ab affirmativâ subduco, et oritur 2,09455148 quotiens absoluta.

Præterea notandum est quòd sub initio operis si dubitarem an $0,1=p$ ad veritatem satis accederet, vice $10p-1=0$ finxissem $6pp+10p-1=0$, et ejus radicis nihilo propioris primam figuram in quotiente scripsissem. Et hoc modo secundam vel etiam tertiam quotientis figuram explorare convenit ubi in
||[7] æquatione secundaria circa quam || versaris, quadratum coefficientis penultimi termini non sit decies major quàm factus ex ultimo termino ducto in coefficientem termini antepenultimi.(15) Quinimò laborem plerumq; minues, præsertim in æquationibus plurimarum dimensionum, si figuras omnes quotienti addendas hoc modo (id est extrahendo minorem radicum ex tribus ultimis terminis æquationis(16) ejus secundariæ) quæras. Sic enim figuras duplo plures in quotiente quâlibet vice lucraberis.

His in numeris sic ostensis,(17) consimiles operationes in speciebus explicandæ restant, de quibus convenit(18) sequentia prænoscere.

(14) A clumsily expressed passage whose meaning is not wholly clarified in the example given. What Newton intends may be stated in general terms thus: if at any stage the subsidiary equation is $\sum_i (k_i s^i) = 0$ and if before the first significant figure there are α zeros in the decimal expansion of k_i and β zeros correspondingly in that of s, then, supposing that we are interested only in the first γ decimal places of the final quotient, we may safely neglect those terms $k_i s^i$ for which $\alpha_i + i\beta > \gamma$ and in the rest we need consider only the first $[\delta]$ figures in the decimal expansion of s, where $\gamma = \alpha_i + i(\beta+\delta)$.

(15) This rule, copied from the corresponding passage in the *De Analysi* (II: 220), is probably empirical but, when all terms in the secondary equation which involve powers of p higher than the second are wholly neglected, it may be supported as follows: since (where A, B and C are positive quantities)

$$\frac{C}{B} - \frac{-B+\sqrt{[B^2+4AC]}}{2A} \leqslant \frac{C}{10B} \quad \text{for} \quad B^2 \geqslant 8{\cdot}1AC,$$

a fortiori when $B^2 > 10AC$ the positive root of $Ap^2+Bp-C = 0$ differs from $p = C/B$ by less

quotient, and in the last term but two neglect all after as many fewer. And so on, progressing arithmetically by that interval of places or, what is the same, cancelling everywhere as many figures as there are in the last but one term provided their lowest places be in arithmetical progression in accord with the series of terms, but alternatively these are to be understood to be filled up with zeros when the circumstances prove otherwise.(14) Thus in the example now propounded, should I desire to complete the quotient to the eighth place of decimals only, while substituting $0{\cdot}0054+r$ for q (at which stage four decimal places in the quotient are entered and the same number remain to be filled in) I could have omitted figures in the five lower places: these I have on that account scored with a small oblique stroke—indeed, even though the first term r^3 had the coefficient 99999, I could have omitted it entirely. Consequently, when those figures are expunged, for the following operation the total comes to $0{\cdot}0005416+11{\cdot}162r$, and this, upon performing division as far as the prescribed term, yields $-0{\cdot}00004852$ for r, so completing the quotient to the desired period. Finally I subtract the negative portion of the quotient from the positive one, and there arises $2{\cdot}09455148$ for the finished quotient.

It should be noted, furthermore, that were I to doubt at the beginning of the operation whether $0{\cdot}1=p$ were a close enough approach to the truth, instead of $10p-1=0$ I should have supposed $6p^2+10p-1=0$ and written the first figure of its numerically smallest root in the quotient. And in this way it proves convenient to work out the second or even third figure of the quotient when in the secondary equation with which you are concerned the square of the coefficient of the last but one term is less than ten times the product of the last term multiplied into the coefficient of the last but two term.(15) Indeed, you will for the most part lighten your labour, particularly in equations of several dimensions, should you search out all the figures to be added to the quotient in this way (namely, by extracting the lesser of the roots out of the three last terms of the secondary equation).(16) For thus at any stage you will gain twice as many figures in the quotient.

Now that I have thus shown how to proceed in numerical cases,(17) the closely similar operations in species remain to be explained. To this end it is appropriate(18) first to establish the following:

than one-tenth of the latter. In the present example $A=6$, $B=10$ and $C=1$, so that $B^2=16\frac{2}{3}AC$; accordingly, to $O(10^{-2})$ it is sufficient to assume $p=1/10$.

(16) Following the *De Analysi* version Newton first continued 'novissimè resultantis' (most recently resulting) before defining this equation as 'secundariæ'.

(17) Having in his account of the resolution of numerical equations done little more than revise his previous version in the *De Analysi*, Newton here inserts an extended preliminary before passing on to an augmented version of the 'Literalis æquationum affectarum resolutio' (II: 222–32).

(18) 'juvabit' (it will be helpful) is cancelled.

1 Quod e speciebus coefficientibus[19] aliqua præ reliquis (si sint plures) insignienda sit, ea nempe quæ est, aut fingi potest esse omnium longè minima vel maxima vel datæ quantitati vicinissima.[20] Cujus rei causa est, ut ob ejus dimensiones in numeratoribus vel denominatoribus terminorum quotientis perpetim auctas, illi termini continuò minores et inde quotiens radici propinquior evadat, sicut ante de specie x in exemplis reductionum per divisionem et extractionem radicum manifestum esse potest. Pro isthâc verò specie in sequentibus ut plurimùm usurpabo etiam x vel z, quemadmodum et y, p, q, r, s &c pro specie radicali extrahenda.

2 Siquando fractiones complexæ[21] vel surdæ quantitates in æquatione propositâ vel post in operatione occurrant, tolli debent per methodos Analystis satis notas. Quemadmodum si habeatur $y^3+\frac{bb}{b-x}yy-x^3=0$, multiplico per $b-x$ et ex facto $by^3-xy^3+bbyy-bx^3+x^4=0$ valorem y elicio.[22] Vel possum fingere $y\times\overline{b-x}=v$, et sic scribendo $\frac{v}{b-x}$ pro y, orietur

$$v^3+bbyy-bbx^3+2bx^4-x^5=0.^{(23)}$$

dein extractâ radice v, divido quotientem per $b-x$ ut obtineatur valor y. Item si proponatur $y^3-xy^{\frac{1}{2}}+x^{\frac{4}{3}}=0$ fingo $y^{\frac{1}{2}}=v$, et $x^{\frac{1}{3}}=z$, et sic scribendo vv pro y et z^3 pro x, oritur $v^6-z^3v+z^4=0$; qua æquatione resolutâ restituo y et x. Scilicet
||[8] radix invenietur $v=z+z^3+6z^5$ &c, et restitutis y et x orietur || $y^{\frac{1}{2}}=x^{\frac{1}{3}}+x+6x^{\frac{5}{3}}$ &c; et quadrando, $y=x^{\frac{2}{3}}+2x^{\frac{4}{3}}+13xx$ &c. Ad eundem modum siquæ sint negativæ dimensiones ipsorū x et y, tollo[24] multiplicando per easdem x et y. Sic habito $x^3+3xxy^{-1}-2x^{-1}-16y^{-3}=0$, multiplico per x et y^3, oriturq; $x^4y^3+3x^3yy-2y^3-16[x]=0$. Et habito $x=\frac{aa}{y}-\frac{2a^3}{yy}+\frac{3a^4}{y^3}$ duco in y^3 et oritur $xy^3=yy-2y+3$.[25] Et sic de cæteris

3 Æquatione sic præparatâ, opus ab inventione primi termini quotientis initium sumit, de quâ ut et consimili subsequentium terminorum inventione hæc esto regula generalis cùm species indefinita (x vel z) parva esse fingitur,[26]

(19) In the root meaning of 'jointly acting'.

(20) A first cancelled version of the preceding reads 'e speciebus radicem definientibus aliquam...insignio, eam nempe quam scio, aut nescius possum fingere esse omnium longè minimam vel maximam sive indefinitè parvam vel magnam datæve quantitati vicinam' (from the variables defining the root I mark out some one...: that one, namely, which I know—or, if I do not, I may suppose—to be by far the least or greatest of all, or indefinitely small or large, or close to a given quantity).

(21) Newton has written this in over 'intricatæ' (intricate), implicitly thereby cancelling it.

(22) Newton first wrote less explicitly 'resolvo' (I resolve).

(23) Read '$v^3+bbvv-b^3x^3+3bbx^4-3bx^5+x^6=0$'.

(24) Understand 'eas'. The equivalent 'facio nullas esse' is cancelled.

(25) Read '$xy^3=aayy-2a^3y+3a^4$'. Understand Newton's 'a' to be a unit.

1. Of the coefficient[19] variables some one must be marked out from the rest (if there are several): that, namely, which is—or may be conceived to be—by far the least or greatest of all or nearest to a given quantity.[20] The reason for this choice is that, through the constant increase of its dimensions in the numerators or denominators of the terms of the quotient, those terms may continually grow less and so the quotient come to be ever nearer to the root (as may appear from what was said before concerning the variable x in the examples of reductions by division and root extraction). In what follows for that variable I shall, in fact, for the most part employ x or z, just as I shall use y, p, q, r, s and so on for the radical variable to be extracted.

2. If any complex[21] fractions or surd quantities should occur at any time in the proposed equation or at a later stage in the operation, they ought to be removed by methods sufficiently well known to algebraists. For instance, if there be had $y^3+b^2y^2/(b-x)-x^3 = 0$, I multiply by $b-x$ and out of the product $by^3-xy^3+b^2y^2-bx^3+x^4 = 0$ I elicit the value of y.[22] Or I may suppose

$$y(b-x) = v$$

and thus, on writing $v/(b-x)$ for y, there will arise

$$v^3+b^2y^2-b^2x^3+2bx^4-x^5 = 0;\text{[23]}$$

then, when I have extracted the value of v, I divide the quotient by $b-x$ to obtain the value of y. Likewise, if $y^3-xy^{\frac{1}{2}}+x^{\frac{4}{3}} = 0$ be propounded, I suppose $y^{\frac{1}{2}} = v$ and $x^{\frac{1}{3}} = z$, and thus, on writing v^2 for y and z^3 for x, there arises

$$v^6-z^3v+z^4 = 0;$$

when this equation has been resolved, I restore y and x. Specifically, the root will be found to be $v = z+z^3+6z^5 \ldots$ and, after y and x are restored, there will arise $y^{\frac{1}{2}} = x^{\frac{1}{3}}+x+6x^{\frac{5}{3}}$, and on squaring $y = x^{\frac{2}{3}}+2x^{\frac{4}{3}}+13x^2 \ldots$. In the same way if there be any negative powers of x and y, these I remove by multiplying through by the same powers of x and y. Thus when there is had

$$x^3+3x^2y^{-1}-2x^{-1}-16y^{-3} = 0,$$

I multiply by xy^3 and there arises $x^4y^3+3x^3y^2-2y^3-16x = 0$. And when there is had $x = \dfrac{a^2}{y}-\dfrac{2a^3}{y^2}+\dfrac{3a^4}{y^3}$, I multiply it into y^3 and there arises $xy^3 = y^2-2y+3$.[25] And so for other cases.

3. With the equation thus prepared, the work takes as its first step the determination of the first term of the quotient. On this point, as also the closely allied determination of subsequent terms, let this be the general rule when the unbounded variable (x or z) is supposed to be small,[26] a case to which the two

(26) The following phrase is a late addition in the margin. The 'other two cases', as the sequel makes clear, occur when the variable is supposed large or near to a given quantity. Compare note (20).

ad quem casum cæteri duo casus sunt reducibiles. E terminis in quibus species radicalis (y, p, q vel r &c) non reperitur selige depressissimum respectu dimensionum indefinitæ speciei (x vel z &c), dein alium terminum in quo sit illa species radicalis selige, talem nempe ut progressio dimensionum utriusq; præfatæ speciei a termino priùs assumpto ad hunc terminum continuata, quàm maximè potest descendat vel minimè ascendat.(27) Et siqui sint alij termini quorum dimensiones cum hâc progressione ad arbitrium continuatâ conveniant, eos etiam selige. Deniq; ex his selectis terminis tanquam nihilo æqualibus quære valorem dictæ speciei radicalis,(28) et quotienti appone.

Cæterùm ut hæc regula magis elucescat, placuit insuper ope sequentis diagrammatis(29) exponere. Descripto angulo recto BAC, latera ejus BA, AC divido in partes æquales, et inde normales erigo distribuentes angulare spatium in æqualia quadrata vel parallelogramma, quæ concipio denominata esse a dimensionibus specierum x et y, prout vides in fig 1 inscriptas.(30) Deinde cùm æquatio aliqua

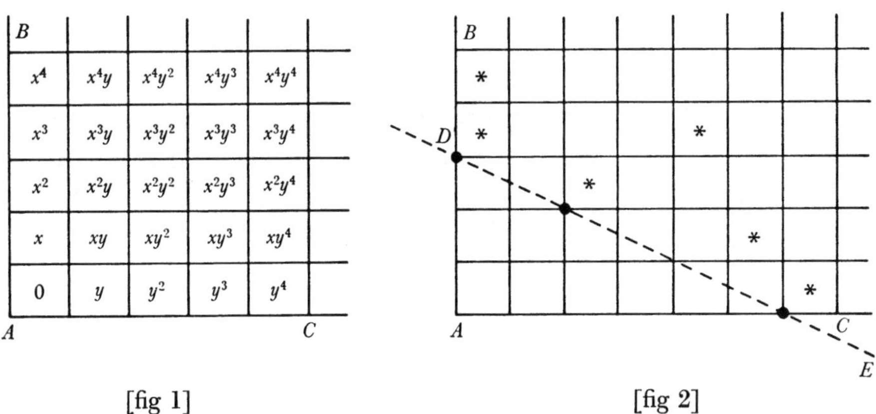

[fig 1] [fig 2]

proponitur, parallelogramma singulis ejus terminis correspondentia insignio notâ aliquâ, et Regulâ ad duo vel forte plura ex insignitis parallelogrammis

(27) In Newton's 'parallelogram' below this is the condition that the 'ruler' DE which bounds the term-entries on their lower side shall be least inclined to the horizontal. That the first term be free from y (or $p, q, r, \ldots$) is unnecessary.

(28) Newton later inserted here 'vel primum terminum ejus' (or its first term) but at once cancelled the phrase.

(29) The following passage was reproduced *verbatim* by Newton in his *epistola posterior* of 24 October 1676 to Oldenburg for Leibniz (*Correspondence*, **2**, 1960: 126–7). It was subsequently published a decade later by John Wallis (from a transcript lent him by Collins), together with the gist of the preceding resolution of literal equations, in his *Treatise of Algebra, both Historical and Practical* (London, 1685): ch. 94.

In Newton's 'parallelogram' (*sc.* rectangular array) the 'ruler' which joins the bottom left corners of the squares denominated by x^p and y^q (and hence passes through the bottom left

others are reducible. Out of the terms in which the radical variable (y, p, q or r, and so on) is not found choose the lowest with regard to the dimensions of the unbounded variable (x or z, and so on), then choose another term in which that radical variable is present, one namely such that the sequence of powers of each of the above variables continued from the term first assumed to this latter one descends most abruptly or rises the least possible.(27) And if there be other terms whose powers coincide with this sequence when continued at will, choose those also. Finally, out of these chosen terms as though set equal to nothing seek the value of the said radical variable(28) and put it in the quotient.

However, to make this rule still more evident, I thought it fitting to expound it in addition with the aid of the following diagram.(29) Describing the right angle $B\hat{A}C$, I divide its sides BA, AC into equal segments and from these raise normals distributing the space between the angle into equal squares or rectangles: these I conceive to be denominated by the powers of the variables x and y, as you see them entered in figure 1.(30) Next, when some equation is proposed, I mark the rectangles corresponding to each of its terms with some sign and apply a ruler to two or maybe several of the rectangles so marked, one of which

---

corner of the squares denominated by $x^{p(1-\lambda)}y^{q\lambda}$ for varying λ) expresses in geometrical form the dimensional equivalence of x^p, y^q and all terms $x^{p(1-\lambda)}y^{q\lambda}$ when $y = \alpha x^{p/q}$. Any given equation, when mapped term by term onto the 'parallelogram' will determine, at the bottom left corners of the corresponding squares, an equal number of points and hence a least bounding convex polygon, each of whose vertices will be one of these points. Evidently each of the lower left sides of this polygon will determine (by its 'slope' p/q) the least dimension

$$p = p(1-\lambda)+q\lambda(p/q)$$

which arises on substituting $\alpha x^{p/q}$ for y in the given equation. (In Newton's first example below, for instance, the side DE determines least $p=3$ corresponding to $y = \alpha x^{\frac{3}{6}}$, where $\alpha^2 = a$ or $2a$.) In consequence this quantity $\alpha x^{p/q}$ will serve as first term in an advancing series expansion for the root y of the given literal equation. In a similar way upper right sides of the polygon serve to determine decreasing series expansions for the same root y. Where the given equation defines, in some Cartesian coordinate system, an algebraic curve, these two types of series expansion serve respectively to determine multiple points and infinite branches of the curve. As we shall see in the fourth and seventh volumes Newton was well aware of this, but no systematic exposition of this approach appeared in print till Jean-Paul de Gua de Malves' *Usages de l'Analyse de Descartes pour découvrir, sans le secours du Calcul Différentiel, les Proprietés, ou Affections principales des Lignes Géométriques de tous les ordres* (Paris, 1740): Section seconde: 24–349). (De Gua, however, rotates the 'parallelogram' anticlockwise through 45° and renames it 'Triangle Algebrique'.) In the winter of 1691/2 Newton, in researches which will appear in the sixth volume, extended the method to extracting, as a series in x, the 'root' y of a general fluxional equation. An excerpt from these researches, communicated privately by Newton to Wallis in midsummer 1692, was immediately published by the latter in his *Opera Mathematica*, **2** (Oxford, 1693): 391–6. (Compare the contemporary English version reproduced in the *Correspondence of Isaac Newton*, **3** (Cambridge, 1961): 222–8.)

(30) The entry '0' in the bottom left-hand corner of figure 1 may seem a little curious but we should understand it to denominate the dimension of $x^0 = 1$.

applicatâ, quorũ unum sit humillimum in columnâ sinistra juxta AB,(31) et alia
||[9] ad regulam dextrorsum sita, cæteraqꝫ omnia non contingentia || regulam supra eam jaceant: seligo terminos æquationis per parallelogramma contingentia regulam designatos et inde quæro quantitatem quotienti addendam.

Sic ad extrahendam radicem, y ex $y^6-5xy^5+\frac{x^3}{a}y^4-7aaxxyy+6a^3x^3+bbx^4=0$; parallelogramma hujus terminis respondentia signo nota aliqua $*$ ut vides in schem. 2. Dein applico regulam DE ad inferiorem e locis signatis in sinistra columna, eam̃qꝫ ab inferioribus ad superiora dextrorsum gyrare facio donec alium similiter vel fortè plura e reliquis signatis locis cœperit attingere,(32) videóqꝫ loca sic attacta esse x^3, x^2y^2, & y^6. E terminis itáqꝫ $y^6-7aaxxyy+6a^3x^3$ tanquam nihilo æqualibus (et insuper si placet reductis ad $v^6-7vv+6=0$(33) ponendo $y=v\times\surd ax$) quæro valorem y, et invenio quadruplicem $+\surd ax$, $-\surd ax$, $+\surd 2ax$ & $-\surd 2ax$, quorum quemlibet pro initio quotientis accipere liceat prout e radicibus quampiam extrahere decretum est.(34)

Sic ex $y^5-byy+9bxx-x^3=0$ seligo $-byy+9bxx$, et inde obtineo $+3x$ pro initiali termino quotientis.

Et ex $y^3+axy+aay-x^3-2a^3=0$ seligo $y^3+aay-2a^3$, et radicem ejus $+a$ scribo in quotiente.

Et ex $xxy^5-3c^4xyy-c^5xx+c^7=0$ seligo xxy^5+c^7, quod exhibet $[-]\surd 5:\frac{c^7}{xx}$(35) pro initio quotientis. Et sic de cæteris.

Cæterùm invento hoc termino, si is contingat esse negativæ potestatis, æquationem per eandem indefinitæ speciei potestatem deprimo, eo ut non opus sit inter solvendum deprimere, et insuper ut regula de superfluis terminis elidendis mox tradenda aptè possit adhiberi. Sic proposito

$$8z^6y^3+az^6yy-27a^9=0,$$

cujus quotiens exordiri debet a $\frac{3a^3}{2zz}$, deprimo per zz, ut fiat

$$8z^4y^3+az^4yy-27a^9z^{-2}=0,$$

antequam solutionem ineo.

Subsequentes quotientum termini eâdem methodo ex æquationibus secundarijs inter operandum prodeuntibus eruuntur, sed ut plurimum leviori curâ.

(31) This is, of course, an unnecessary restriction.

(32) The equivalent 'quàm primùm attingat' is cancelled.

(33) The next dozen lines (up to the end of the paragraph 'Cæterùm...ineo') are written on a slip of paper glued in by Newton. When the manuscript is placed in ultra-violet light the original draft beneath can be read without too much difficulty, but proves to be almost identical with the redraft pasted in over it apart from some messy cancellations and a few transpositions and minor insertions. The second equation following, for example, was originally set as

is to be the lowest in the left-hand column alongside AB,(31) a second to the right touching the ruler, and all the rest not in contact with the ruler should lie above it. I then choose the terms of the equation which are marked out by the rectangles in contact with the ruler and thence seek the quantity to be added to the quotient.

So to extract the root y from

$$y^6-5xy^5+(x^3/a)\,y^4-7a^2x^2y^2+6a^3x^3+b^2x^4=0,$$

I mark the rectangles answering to its terms with some sign $*$, as you see done in the second illustration. I then apply the ruler DE to the lower corner of the places marked out in the left-hand column and make it swing to the right from bottom to top until in like fashion it begins to touch a second or maybe several together of the other marked places. Those so touched I see to be x^3, x^2y^2 and y^6. Hence from the terms $y^6-7a^2x^2y^2+6a^3x^3$ as though set equal to nothing (and in addition, if it pleases, reduced to $v^6-7v^2+6=0$(33) by supposing $y=v\sqrt{ax}$) I seek the value of y and find it to be fourfold, $+\sqrt{ax}$, $-\sqrt{ax}$, $+\sqrt{2ax}$ and $-\sqrt{2ax}$. Any of these may be acceptable as an initial term in the quotient depending on whether the decision is made to extract one or other of the roots.(34)

So out of $y^5-by^2+9bx^2-x^3=0$ I choose $-by^2+9bx^2$, and thence obtain $+3x$ for the initial term of the quotient.

And out of $y^3+axy+a^2y-x^3-2a^3=0$ I choose $y^3+a^2y-2a^3$ and write its root $+a$ in the quotient.

And out of $x^2y^5-3c^4xy^2-c^5x^2+c^7=0$ I choose $x^2y^5+c^7$, and this produces $-\sqrt[5]{(c^7/x^2)}$ for the opening term of the quotient. And so for other cases.

When, however, this term is found, should its power chance to be negative I depress the equation by the same power of the unbounded variable so that there be no need to depress it while it is being resolved and, in addition, so that the rule for eliminating superfluous quantities soon to be delivered may suitably be applied. So when there is proposed $8z^6y^3+az^6y^2-27a^9=0$, the quotient of which should start with $3a^3/2z^2$, I depress the equation by z^2 to make it

$$8z^4y^3+az^4y^2-27a^9z^{-2}=0$$

before I enter on its solution.

Subsequent terms of the quotients are derived by the same method from the secondary equations arising during the course of the working, but with less

---

'$y^3+aay+axy-2a^3-x^3=0$', while the paragraph beginning 'Cæterùm...' (in which the phrases 'indefinitæ speciei', 'insuper' and 'mox tradenda' are insertions) is wholly cancelled—perhaps for intended reinsertion after the next paragraph (compare note (36)).

(34) The complex roots $y=\pm\sqrt{-3ax}$ (which do not yield a real series expansion) are ignored.

(35) The manuscript lacks the necessary minus sign.

Res enim peragi solet dividendo depressissimum e terminis cum indefinitè parva specie (x, xx, x^3 &c) absq; specie radicali (p, q, r &c) affectis, per quantitatem quâcum species illa radicalis unius tantùm dimensionis abq; alterâ indefinitâ specie afficitur, et exitum scribendo in quotiente. Sic in exemplo sequente termini quotientis $\frac{x}{4}$, $\frac{xx}{64a}$, $\frac{131x^3}{512aa}$ &c eliciuntur dividendo aax, $\frac{1}{16}axx$, $\frac{131}{128}x^3$ &c per $4aa$.[36]

||[10] || His præmissis restat ut praxin resolutionis exhibeam.[37] Sit itaq;

$$y^3+aay+axy-2a^3-x^3=0$$

æquatio resolvenda, et ex ejus terminis[38] $y^3+aay-2a^3=0$ æquatione fictitiâ, juxta tertium e præmissis elicio $y-a=0$, & scribo $+a$ in quotiente. Deinde cùm $+a$ non accurate valeat y, pono $a+p=y$, et pro y in terminis æquationis in margine scriptis[39] substituo $a+p$, terminosq; resultantes ($p^3+3app+axp$ &c) rursum scribo in margine,[39] ex quibus iterum juxta tertium e præmissis excerpo terminos $+4aap+aax=0$ pro æquatione fictitiâ, quæ cùm exhibeat $p=-\frac{1}{4}x$, scribo $-\frac{1}{4}x$ in quotiente. Præterea cùm $-\frac{1}{4}x$ non accurate valeat p, pono $-\frac{1}{4}x+q=p$, & pro p in terminis marginalibus substituo $-\frac{1}{4}x+q$, terminosq; resultantes ($q^3-\frac{3}{4}xqq+3aqq$ &c) iterum scribo in margine, ex quibus denuò juxta regulam præfatam seligo terminos $4aaq-\frac{1}{16}axx=0$ pro æquatione fictitiâ quæ cum exhibeat $q=\frac{xx}{64a}$ scribo $+\frac{xx}{64a}$ in quotiente. Porrò cùm $\frac{xx}{64a}$ non accuratè valeat q, pono $\frac{xx}{64a}+r=q$ & pro q in terminis marginalibus substituo $\frac{xx}{64a}+r$, & sic opus ad placitum produco prout indicat subjectum diagramma.

		$\left(a-\frac{x}{4}+\frac{xx}{64a}+\frac{131x^3}{512aa}+\frac{509x^4}{16384a^3}\right.$ &c $[=y]$
$+a+p=y.$	$+y^3$	$+a^3+3aap+3app+p^3$
	$+axy$	$+aax\ \ +axp$
	$+aay$	$+a^3\ \ +aap$
	$-x^3$	$-x^3$
	$-2a^3$	$-2a^3$

(36) The first word 'Cæterùm' of an intended following paragraph is cancelled: see note (33).

(37) Newton once more takes up the *De Analysi* (compare note (17)).

(38) Those not involving x, that is: in the corresponding 'parallelogram' the convex polygon of points has its only lower side coincident with the base. In the *De Analysi*'s version (II: 222) Newton does not justify his selecting as 'fictive' equation that which results 'cùm x sit nulla'.

trouble for the most part. For the matter is usually performed by dividing the lowest of the terms affected with the indefinitely small variable x, x^2, x^3 and so on but without a radical variable (p, q, r and so on) by a quantity affected with a single power only of that radical variable but not with the other unbounded variable, and then writing the result in the quotient. So in the following example the quotient's terms $\frac{x}{4}$, $\frac{x^2}{64a}$, $\frac{131x^3}{512a^2}$, ... are elicited from the division of a^2x, $\frac{1}{16}ax^2$, $\frac{131}{128}x^3$, ... by $4a^2$.[36]

After these premisses it remains to illustrate the practical solution.[37] Let, therefore, $y^3+a^2y+axy-2a^3-x^3=0$ be the equation to be resolved. From its terms[38] $y^3+a^2y-2a^3=0$ as fictitious equation by the third of the premisses I elicit $y-a=0$ and write $+a$ in the quotient. Next since $+a$ is not accurately the value of y, I set $a+p=y$ and in place of y in the terms of the equation written in the margin I substitute $a+p$, writing the resulting terms $(p^3+3ap^2+axp\ldots)$ again in the margin. From these, once more by the third of the premisses, I select the terms $4a^2p+a^2x=0$ for a fictitious equation and, since it yields $p\equiv-\frac{1}{4}x$, I write $-\frac{1}{4}x$ in the quotient. Further, since $-\frac{1}{4}x$ is not accurately the value of p, I set $-\frac{1}{4}x+q=p$ and in place of p in the marginal terms I substitute $-\frac{1}{4}x+q$, writing the resulting terms $(q^3-\frac{3}{4}xq^2+3aq^2\ldots)$ once more in the margin. From these in turn by the above-stated rule I choose the terms

$$4a^2q-\tfrac{1}{16}ax^2=0$$

for a fictitious equation and, since this yields $q=\frac{x^2}{64a}$, I write $+\frac{x^2}{64a}$ in the quotient. Furthermore, since $\frac{x^2}{64a}$ is not accurately the value of q, I set $\frac{x^2}{64a}+r=q$ and in place of q in the marginal terms I substitute $\frac{x^2}{64a}+r$. And so I repeat the operation at pleasure, as the appended diagram accordingly indicates.

		$a-\frac{x}{4}+\frac{x^2}{64a}+\frac{131x^3}{512a^2}+\frac{509x^4}{16384a^3}\ldots$ $[=y]$
$a+p=y.$	y^3	$a^3+3a^2p+3ap^2+p^3$
	$+axy$	$+a^2x\ +axp$
	$+a^2y$	$+a^3\ +a^2p$
	$-x^3$	$-x^3$
	$-2a^3$	$-2a^3$

(39) In variation on the *De Analysi*'s 'margini appono' Newton here first wrote equivalently 'margini adscriptis' and then 'adscribo margini'.

$-\frac{1}{4}x+q=p.$	$+p^3$	$-\frac{1}{64}x^3+\frac{3}{16}xxq*-\frac{3}{4}xqq+q^3$
	$+3ap^2$	$+\frac{3}{16}axx\quad-\frac{3}{2}axq\quad+3aqq$
	$+axp$	$-\frac{1}{4}axx\quad+axq$
	$+4aap$	$-aax\quad+4aaq$
	$+aax$	$+aax$
	$-x^3$	$-x^3$
$+\frac{xx}{64a}+r=q.$	$+q^3$	$*$
	$-\frac{3}{4}xqq$	$*$
	$+3aqq$	$+\frac{3x^4}{4096a}\quad*+\frac{3}{32}xxr+3arr$
	$+\frac{3}{16}xxq$	$+\frac{3x^4}{1024a}\quad*+\frac{3}{16}xxr$
	$-\frac{1}{2}axq$	$-\frac{1}{128}x^3\quad-\frac{1}{2}axr$
	$+4aaq$	$+\frac{1}{16}axx\quad+4aar$
	$-\frac{65}{64}x^3$	$-\frac{65}{64}x^3$
	$-\frac{1}{16}axx$	$-\frac{1}{16}axx$

$$\left.+4aa-\tfrac{1}{2}ax\right)+\frac{131}{128}x^3-\frac{15x^4}{4096a}\left(+\frac{131x^3}{512aa}+\frac{509x^4}{16384a^3}\right.\text{.}$$ (40)

||[11] || Quod si Quotientem ad certam usq periodum produci(41) cupiam, ut x nempe in ultimo ejus termino ultra datum dimensionum numerum non ascendat; terminos(42) inter substituendum semper omitto quos nulli deinceps usui fore prævideam. Cujus rei regula esto, quòd post primum terminum ex qualibet quantitate in margine collaterali resultantem non addantur plures dextrorsum, quàm istius primò resultantis termini dimensio a periodica sive maxima dimensione quotientis deficit gradibus.(43) Ut in hoc exemplo si cupiam ut quotiens (sive x in quotiente) ad quatuor tantùm dimensiones ascendat, omitto omnes terminos post x^4, & post x^3 pono unicum tantùm. Terminos itaq post notam $*$ delendos esse concipe: et opere sic continuato donec ultimò ad terminos $\left(\frac{15x^4}{4096a}-\frac{131}{128}x^3+4aar-\frac{1}{2}axr\right)$ deveniatur in quibus (p, q, r vel s &c) residuum radicis extrahendæ sit unicæ tantùm dimensionis; tot terminos $\left(+\frac{131x^3}{512aa}+\frac{509x^4}{16384a^3}\right)$ per divisionem elicies, quot ad complendum quotientem deesse videbis. Atq ita tandem obtinebitur

$$y=a-\tfrac{1}{4}x+\frac{xx}{64a}+\frac{131x^3}{512a^2}+\frac{509x^4}{16384a^3}\quad[\&c].$$

(40) As Newton makes clear in the next paragraph those terms following the asterisk $*$ are to be omitted. In the revised version of this scheme which he introduced into his *epistola prior* of 13 June 1676 all such terms are replaced by '&c', the entries for '$+q^3$' and '$-\frac{3}{4}xqq$' at the third stage being entirely omitted. (See *Correspondence of Isaac Newton*, **2**, 1960: 24.)

$-\frac{1}{4}x+q=p.$	p^3	$-\frac{1}{64}x^3+\frac{3}{16}x^2q*-\frac{3}{4}xq^2+q^3$
	$+3ap^2$	$+\frac{3}{16}ax^2\ -\frac{3}{2}axq+3aq^2$
	$+axp$	$-\frac{1}{4}ax^2\ +axq$
	$+4a^2p$	$-a^2x+4a^2q$
	$+a^2x$	$+a^2x$
	$-x^3$	$-x^3$
$+\dfrac{x^2}{64a}+r=q.$	q^3	$*$
	$-\frac{3}{4}xq^2$	$*$
	$+3aq^2$	$+\dfrac{3x^4}{4096a}\ *\ +\frac{3}{32}x^2r+3ar^2$
	$+\frac{3}{16}x^2q$	$+\dfrac{3x^4}{1024a}\ *\ +\frac{3}{16}x^2r$
	$-\frac{1}{2}axq$	$-\frac{1}{128}x^3\quad -\frac{1}{2}axr$
	$+4a^2q$	$+\frac{1}{16}ax^2\quad +4a^2r$
	$-\frac{65}{64}x^3$	$-\frac{65}{64}x^3$
	$-\frac{1}{16}ax^2$	$-\frac{1}{16}ax^2$

$$4a^2-\tfrac{1}{2}ax\Big)\ \frac{131}{128}x^3-\frac{15x^4}{4096a}\left(\frac{131x^3}{512a^2}+\frac{509x^4}{16384a^3}\ldots\right.$$ [40]

But should I desire the quotient to be produced only to a period of defined length, so that x in the final term is not to rise beyond a given dimension, during the process of substitution I omit always the[42] terms which, I foresee, subsequently will be of no use. On this point let the rule be that, after the first term resulting from any quantity in the margin alongside, there are not to be added more on its right than the difference in degree[43] between that term first resulting and the periodic or greatest power of the quotient. As in this example should I desire that the quotient (or rather x in the quotient) rise only to the fourth power, I omit all terms after x^4 and after x^3 set one only. Accordingly, conceive the terms after the mark $*$ to be deleted and continue the work till you come finally to terms, $\dfrac{15x^4}{4096a}-\dfrac{131}{128}x^3+(4a^2-\frac{1}{2}ax)\,r$, in which the residue of the root to be extracted (p, q, r or s, and so on) is of a single dimension alone: you will then derive as many terms $\left(\dfrac{131x^3}{512a^2}+\dfrac{509x^4}{16384a^3}\right)$ by division as you see wanting to complete the quotient, And so at length will be obtained

$$y=a-\tfrac{1}{4}x+\frac{x^2}{64a}+\frac{131x^3}{512a^2}+\frac{509x^4}{16384a^3}\ldots.$$

(41) Newton first wrote, equivalently, 'continuari'.

(42) An enclitic 'istos' is cancelled.

(43) 'distat unitatibus' (distance in unities) is cancelled—rightly so since the 'quotient' series does not necessarily advance by unit powers.

Plenioris illustrationis gratia dedi aliud exemplum resolvendo

$$\tfrac{1}{5}y^5-\tfrac{1}{4}y^4+\tfrac{1}{3}y^3-\tfrac{1}{2}yy+y-z=0,^{(44)}$$

ubi proponitur inventio quotientis ad quintam tantùm dimensionem, terminiq; superflui post notam (&c) negliguntur.

		$(z+\frac{1}{2}zz+\frac{1}{6}z^3+\frac{1}{24}z^4+\frac{1}{120}z^5$ &c $[=y]$
$z+p=y.$	$+\frac{1}{5}y^5$	$+\frac{1}{5}z^5$ &c
	$-\frac{1}{4}y^4$	$-\frac{1}{4}z^4-z^3p$ &c
	$+\frac{1}{3}y^3$	$+\frac{1}{3}z^3+z^2p+zpp$ &c
	$-\frac{1}{2}y^2$	$-\frac{1}{2}z^2\ -zp-\frac{1}{2}pp.$
	$+y$	$+z\ \ +p.$
	$-z$	$-z.$
$\frac{1}{2}zz+q=p.$	$+zp^2$	$+\frac{1}{4}z^5$ &c
	$-\frac{1}{2}p^2$	$-\frac{1}{8}z^4-\frac{1}{2}z^2q$ &c
	$-z^3p$	$-\frac{1}{2}z^5$ &c
	$+z^2p$	$+\frac{1}{2}z^4\ +z^2q.$
	$-zp$	$-\frac{1}{2}z^3\ -zq.$
	$+p$	$+\frac{1}{2}z^2\ \ +q.$
	$+\frac{1}{5}z^5$	$+\frac{1}{5}z^5.$
	$-\frac{1}{4}z^4$	$-\frac{1}{4}z^4.$
	$+\frac{1}{3}z^3$	$+\frac{1}{3}z^3.$
	$-\frac{1}{2}z^2$	$-\frac{1}{2}z^2.$

$$1-z+\tfrac{1}{2}zz)+\tfrac{1}{6}z^3-\tfrac{1}{8}z^4+\tfrac{1}{20}z^5(+\tfrac{1}{6}z^3+\tfrac{1}{24}z^4+\tfrac{1}{120}z^5.$$

Atq; ita si cupiam æquationem

||[12] $$\tfrac{63}{2816}y^{11}+\tfrac{35}{1152}y^9+\tfrac{5}{112}y^7+\|\ \tfrac{3}{40}y^5+\tfrac{1}{6}y^3+y-z=0^{(45)}$$

ad usq; nonam tantùm dimensionē quotientis resolvi, ante opus initum negligo terminum $\frac{63}{2816}y^{11}$, deinde inter operandum negligo etiam omnes terminos post z^9, post z^7 pono unicum, ac duos tantum post z^5, eò quòd percipio quotientem ubiq; per gradus binarum unitatum (hoc modo z, z^3, z^5 &c) debere ascendere. Tandemq; prodit $y=z-\frac{1}{6}z^3+\frac{1}{120}z^5-\frac{1}{5040}z^7+\frac{1}{362880}z^9$.

Et hinc patet artificium quo æquationes in infinitum affectæ, vel utcunq; multis numeróve infinitis terminis constantes possunt solvi. Scilicet omnes termini ante opus initum debent negligi in quibus dimensio speciei indefinitè parvæ non affectæ cum radicali specie transcendit maximam dimensionem in quotiente desideratam vel ex quibus, substituendo pro radicali specie primum

(44) This example ($z=\log(1+y)$ to $O(y^6)$ so that $y=e^z-1$ to $O(z^6)$) is taken from the *De Analysi*'s 'inventio Basis ex area data' (II: 234). It reappears again in the *epistola posterior* of 24 October 1676 (*Correspondence*, **2**: 127–9) as an example of reversal of series.

For the sake of fuller illustration, I have given another example for solution, $\frac{1}{5}y^5-\frac{1}{4}y^4+\frac{1}{3}y^3-\frac{1}{2}y^2+y-z=0$,(44) where it is proposed to determine the quotient merely to the fifth power and superfluous terms after the symbol ... are neglected.

		$z+\frac{1}{2}z^2+\frac{1}{6}z^3+\frac{1}{24}z^4+\frac{1}{120}z^5\ldots\quad[=y]$
$z+p=y.$	$+\frac{1}{5}y^5$	$+\frac{1}{5}z^5\ldots$
	$-\frac{1}{4}y^4$	$-\frac{1}{4}z^4-z^3p\ldots$
	$+\frac{1}{3}y^3$	$+\frac{1}{3}z^3+z^2p+zp^2\ldots$
	$-\frac{1}{2}y^2$	$-\frac{1}{2}z^2\ -zp-\frac{1}{2}p^2.$
	$+y$	$+z\ \ +p.$
	$-z$	$-z.$
$\frac{1}{2}z^2+q=p.$	$+zp^2$	$+\frac{1}{4}z^5\ldots$
	$-\frac{1}{2}p^2$	$-\frac{1}{8}z^4-\frac{1}{2}z^2q\ldots$
	$-z^3p$	$-\frac{1}{2}z^5\ldots$
	$+z^2p$	$+\frac{1}{2}z^4\ \ +z^2q.$
	$-zp$	$-\frac{1}{2}z^3\ \ -zq.$
	$+p$	$-\frac{1}{2}z^2\ \ \ +q.$
	$+\frac{1}{5}z^5$	$+\frac{1}{5}z^5.$
	$-\frac{1}{4}z^4$	$-\frac{1}{4}z^4.$
	$+\frac{1}{3}z^3$	$+\frac{1}{3}z^3.$
	$-\frac{1}{2}z^2$	$-\frac{1}{2}z^2.$

$$1-z+\tfrac{1}{2}z^2)\tfrac{1}{6}z^3-\tfrac{1}{8}z^4+\tfrac{1}{20}z^5(\tfrac{1}{6}z^3+\tfrac{1}{24}z^4+\tfrac{1}{120}z^5\ldots.$$

And so, should I desire the equation

$$\tfrac{63}{2816}y^{11}+\tfrac{35}{1152}y^9+\tfrac{5}{112}y^7+\tfrac{3}{40}y^5+\tfrac{1}{6}y^3+y-z=0^{(45)}$$

to be resolved only up to the ninth power of the quotient, before beginning work I lay aside the term $\frac{63}{2816}y^{11}$, and then during the course of the working lay aside also all terms after z^9, while after z^7 I set a single one and two only after z^5, because I perceive that the quotient should advance everywhere by two degrees at a time (in this way z, z^3, z^5, ...). At length there comes

$$y=z-\tfrac{1}{6}z^3+\tfrac{1}{120}z^5-\tfrac{1}{5040}z^7+\tfrac{1}{362880}z^9\ldots.$$

Hence we have an evident expedient by which equations infinitely affected, or consisting of an arbitrarily large or infinite number of terms, may be resolved. Specifically, before work commences all terms ought to be neglected, in which the power of the indefinitely small variable not affected with the radical variable exceeds the greatest power desired in the quotient, or from which, on replacement of the radical variable by the first term of the quotient as found with the

(45) That is, $z=\sin^{-1}y$ to $O(y^{13})$ yielding $y=\sin z$ to $O(z^{11})$ below.

terminum quotientis ope tessellatæ tabulæ inventum, non nisi ejusmodi transcendentes termini possunt emergere. Sic in exemplo novissimo terminos omnes supra y^9, quamvis infinitè progrederentur, omisissem. Et sic in hâc æquatione

$$0=\begin{cases}-8+zz-4z^4+9z^6-16z^8\ \&c\\ +y \text{ in } zz-2z^4+3z^6-4z^8\ \&c\\ -yy \text{ in } zz-z^4+z^6-z^8\ \&c\\ +y^3 \text{ in } zz-\frac{1}{2}z^4+\frac{1}{3}z^6-\frac{1}{4}z^8\ \&c^{(46)}\end{cases}$$

ut radix cubica(47) ad quatuor tantùm dimensiones ipsius z extrahatur, mitto omnes in infinitum terminos post $+y^3$ in $zz-\frac{1}{2}z^4+\frac{1}{3}z^6$, et post $-yy$ in $zz-z^4+z^6$, et post $+y$ in $zz-2z^4$, et post $-8+zz-4z^4$. Et hanc tantùm æquationem

$$\tfrac{1}{3}z^6y^3-\tfrac{1}{2}z^4y^3+zzy^3-z^6yy+z^4yy-zzyy-2z^4y+zzy-4z^4+zz-8=0$$

resolvendam sumo, siquidem $2z^{-\frac{2}{3}}$ (primus nempe quotientis terminus,) pro y in reliquâ æquatione per $z^{\frac{2}{3}}$ depressâ substitutus, dat plures ubiq; quàm quatuor dimensiones.

Quæ de altioribus æquationibus dixi, ad quadraticas etiam applicari possunt. Quemadmodum si hujus

$$0=\begin{cases}yy.\\ -y \text{ in } a+x+\dfrac{xx}{a}+\dfrac{x^3}{aa}+\dfrac{x^4}{a^3}\ \&c\\ +\dfrac{x^4}{4aa}^{(48)}\end{cases}$$

radicem ad usq; periodum x^6 desiderem, mitto terminos in infinitum post $-y$ in
||[13] $a+x+\dfrac{xx}{a}$, et isthanc tantùm || $yy-ay-xy-\dfrac{xx}{a}y+\dfrac{x^4}{4aa}=0$, sive id fiat hâc lege $y=\frac{1}{2}a+\frac{1}{2}x+\dfrac{xx}{2a}-\sqrt{\tfrac{1}{4}aa+\tfrac{1}{2}ax+\tfrac{3}{4}xx+\dfrac{x^3}{2a}}$, ut solet, sive expeditiùs per methodum de affectis æquationibus jam traditam, resolvo, et exit $y=\dfrac{x^4}{4a^3}-\dfrac{x^5}{4a^4}*$, ultimo desiderato termino existente nullo.(49)

Postquam verò radices ad convenientem periodum extractæ sunt, possunt aliquando,(50) ex analogiâ seriei observatâ, in placitum produci. Sic hanc $z+\frac{1}{2}zz+\frac{1}{6}z^3+\frac{1}{24}z^4+\frac{1}{120}z^5$ &c (radicem æquationis infinitæ

$$x=y+\tfrac{1}{2}yy+\tfrac{1}{3}y^3\ \&c)^{(51)}$$

(46) To $O(z^{10})$ this may be written (when $|z| < 1$)

$$-8+z^2(1-z^2)(1+z^2)^{-3}+y(1+z^2)^{-2}-y^2(1+z^2)^{-1}+y^3\log(1+z^2) = 0.$$

(47) Namely, of the previous equation viewed as a cubic in y.

(48) To $O(x^5)$, $y^2-a^2y/(a-x)+\frac{1}{4}x^4/a^2 = 0$.

(49) Whether, in fact, the resolution of this quadratic (in y) 'per methodum de affectis æquationibus' is more expedite is dubious, since it is not evident (without constructing a

help of the squared-off table, none but excessive terms of this sort can emerge. Thus in the most recent example I would have omitted all terms above y^9 even though these formed an infinite progression. And so in this equation

$$0 = -8+z^2-4z^4+9z^6-16z^8 \ldots +y(z^2-2z^4+3z^6-4z^8 \ldots)$$
$$-y^2(z^2-z^4+z^6-z^8 \ldots)+y^3(z^2-\tfrac{1}{2}z^4+\tfrac{1}{3}z^6-\tfrac{1}{4}z^8 \ldots)^{(46)}$$

to extract the cubic root(47) only to the fourth power of z, I lay aside all terms indefinitely after $y^3(z^2-\frac{1}{2}z^4+\frac{1}{3}z^6)$, after $-y^2(z^2-z^4+z^6)$, after $y(z^2-2z^4)$ and after $-8+z^2-4z^4$, assuming that there is to be resolved only this equation

$$(\tfrac{1}{3}z^6-\tfrac{1}{2}z^4+z^2)\,y^3-(z^6-z^4+z^2)\,y^2+(-2z^4+z^2)\,y-4z^4+z^2-8 = 0.$$

For $2z^{-\frac{2}{3}}$ (the first term, that is, of the quotient), when substituted for y in the residue of the equation divided by $z^{\frac{2}{3}}$, yields everywhere more than the fourth power of z.

What I have said concerning equations of higher degree may also be applied to quadratics. For instance, were I to desire the root of this

$$0 = y^2-y(a+x+x^2/a+x^3/a^2+x^4/a^3 \ldots)+x^4/4a^{2(48)}$$

as far as the period x^6, I lay aside all terms indefinitely after $-y(a+x+x^2/a)$ and this alone, $y^2-y(a+x+x^2/a)+x^4/4a^2 = 0$, I resolve, whether in this usual form $y = \frac{1}{2}(a+x+x^2/a)-\sqrt{[\frac{1}{4}a^2+\frac{1}{2}ax+\frac{3}{4}x^2+x^3/2a]}$ or more readily by the method for affected equations just now delivered. There results $y = x^4/4a^3-x^5/4a^4$ with the final term desired turning out to be nil.(49)

Now after the roots have been extracted to a suitable period, they may sometimes(50) be extended at pleasure by observing the analogy of the series. Thus you will perpetually extend this series $z+\frac{1}{2}z^2+\frac{1}{6}z^3+\frac{1}{24}z^4+\frac{1}{120}z^5 \ldots$, the root of the infinite equation $[z] = y[-]\frac{1}{2}y^2+\frac{1}{3}y^3 \ldots$, by dividing the last term by these

---

corresponding 'parallelogram') that $-ay+x^4/4a^2 = 0$ is an admissible 'fictive' equation, so that we may then at once round off subsequent calculation to $O(x^7)$ simply by considering the approximating equation $-y(a+x+x^2/a)+x^4/4a^2 = 0$, which yields to $O(x^7)$

$$y = x^4(a-x)/4a(a^3-x^3).$$

From a logical viewpoint the elementary approach by completing the quadratic's square seems preferable, for that way we are led immediately to an exact solution

$$y = \tfrac{1}{2}(a+x+x^2/a)\,(1\mp\sqrt{[1-x^4/(a^2+ax+x^2)^2]}).$$

Newton's result follows by taking the radical sign negative, but note that (corresponding to the alternative fictitious equation $y^2-ay = 0$) the second root

$$y = a+x+x^2/a-x^4/4a^3+x^5/4a^4 \ldots$$

is equally valid. Note that Newton's '$*$' stands for the zero term '$0x^6$'.

(50) Newton has wisely cancelled 'plerumq;' (mostly) here.

(51) A slip for '$z=y-\frac{1}{2}yy+\frac{1}{3}y^3$ &c' (compare note (44)).

perpetuò produces dividendo ultimum terminum per hos ordine numeros 2, 3, 4, 5, 6, 7 &c. Et hanc $z-\frac{1}{6}z^3+\frac{1}{120}z^5-\frac{1}{5040}z^7+\frac{1}{362880}z^9$ &c dividendo per hos $2\times 3.\ 4\times 5.\ 6\times 7.\ 8\times 9$ &c.(52) Et sic in alijs.

Cæterùm in inventione primi termini quotientis et nonnunquam secundi tertijve difficultas etiamnum enodanda superest.(53) Potest enim valor ejus secundum præcedentia quæsitus esse surda sive inextricabilis radix æquationis multipliciter affectæ. Quod cùm accidit, modò non sit insuper impossibilis, illum literâ aliquâ designabis, dein operabere tanquam si cognitum haberes. Quemadmodū in exemplo $y^3+axy+aay-x^3-2a^3=0$,(54) si radix hujus $y^3+aay-2a^3=0$ fuisset surda vel ignota finxissem quamlibet (b) pro ea ponendam esse, et resolutionem (puta ad tertiam dimensionem quotientis) ut sequitur perfecissem.

$$
\begin{array}{r|l}
 & \left(b-\dfrac{abx}{cc}+\dfrac{a^4bxx}{c^6}+\dfrac{x^3}{cc}+\dfrac{a^3b^3x^3}{c^8}-\dfrac{a^5bx^3}{c^8}+\dfrac{6a^5b^3x^3}{c^{10}}\right. \quad [\&c]. \\
\hline
b+p=y. \quad +y^3 & +b^3+3bbp+3bpp+p^3 \\
+axy & +abx \ +axp \\
+aay & +aab \ +aap \\
-x^3 & -x^3 \\
-2a^3 & -2a^3 \\
\hline
\dfrac{-abx}{cc}+q=p. \quad +p^3 & -\dfrac{a^3b^3x^3}{c^6} \quad \&c \\
+3bpp & +\dfrac{3aab^3xx}{c^4}-\dfrac{6abbx}{cc}q \quad \&c \\
+axp & -\dfrac{aabxx}{cc} \quad +axq \\
+ccp & -abx \quad +ccq \\
-x^3 & -x^3 \\
+abx & +abx \\
\hline
\end{array}
$$

$$
cc+ax-\frac{6abbx}{cc}\Bigg)\ \frac{a^4bxx}{c^4}+x^3+\frac{a^3b^3x^3}{c^6}\ \Bigg(\frac{a^4bxx}{c^6}+\frac{x^3}{cc}+\frac{a^3b^3x^3}{c^8} \quad \&c.^{(55)}
$$

(52) A little variant reproduction of the *De Analysi*'s paragraph 'De serie progressionum continuanda' (II: 236–8). A following sentence 'Et hanc $a+\frac{xx}{a}-\frac{x^4}{8a^3}+\frac{x^6}{16a^5}-\frac{5x^8}{128a^7}$ &c multiplicando per hos $\frac{1}{2}\cdot\frac{-1}{4}\cdot\frac{-3}{6}\cdot\frac{-5}{8}\cdot\frac{-7}{10}$ &c.', copied from the same source, has been cancelled even though (on his page 3) Newton does not explain how the binomial coefficients in the series expansion of $\sqrt{[a^2+x^2]}$ are generated. In his 1676 letters to Leibniz, the *epistolæ prior et posterior* of 13 June and 24 October, Newton was more lavish in explanation of this and similar points. Leibniz, in his 1676 transcript of this present passage from the *De Analysi*, was to be somewhat confused by Newton's elliptical statement of the progression of coefficients in the series expansion of $\sin^{-1}x$. (See the *Correspondence of Isaac Newton*, **2**: 21, 25, 26, 29, 111–12, 122–4; and II, **2**, **3**, Appendix **1**: note (34).)

numbers in order 2, 3, 4, 5, 6, 7, ... And this series

$$z-\tfrac{1}{6}z^3+\tfrac{1}{120}z^5-\tfrac{1}{5040}z^7+\tfrac{1}{362880}z^9\ldots$$

you will produce by dividing by these numbers 2×3, 4×5, 6×7, 8×9, ...(52) And so in other cases.

In finding the first term of the quotient, however, and sometimes the second or third too, there still remains a difficulty to be smoothed out.(53) For its value, when sought according to the preceding instructions, may prove to be surd or the inextricable root of a multiply affected equation. When this happens (and provided it is not, moreover, an impossible quantity) you will designate it by some letter and then work through as though you knew its value. For instance, in the example $y^3+axy+a^2y-x^3-2a^3=0$,(54) if the root of this,

$$y^3+a^2y-2a^3=0,$$

had been surd or unknown, I should have supposed any letter b to be put for it and then have effected the solution (say to the third power of the quotient) as follows.

		$b-\frac{abx}{c^2}+\frac{a^4bx^2}{c^6}+\frac{x^3}{c^2}+\frac{a^3b^3x^3}{c^8}-\frac{a^5bx^3}{c^8}+\frac{6a^5b^3x^3}{c^{10}}\ldots$ [$=y$]
$b+p=y.$	y^3	$b^3+3b^2p+3bp^2+p^3$
	$+axy$	$+abx\ +axp$
	$+a^2y$	$+a^2b\ +a^2p$
	$-x^3$	$-x^3$
	$-2a^3$	$-2a^3$
$-\frac{abx}{c^2}+q=p.$	p^3	$-\frac{a^3b^3x^3}{c^6}\ldots$
	$+3bp^2$	$+\frac{3a^2b^3x^2}{c^4}-\frac{6ab^2x}{c^2}q\ldots$
	$+axp$	$-\frac{a^2bx^2}{c^2}\quad +axq$
	$+c^2p$	$-abx\quad +c^2q$
	$-x^3$	$-x^3$
	$+abx$	$+abx$

$$c^2+ax-\frac{6ab^2x}{c^2}\Big)\ \frac{a^4bx^2}{c^4}+x^3+\frac{a^3b^3x^3}{c^6}\ \Big(\frac{a^4bx^2}{c^6}+\frac{x^3}{c^2}+\frac{a^3b^3x^3}{c^8}\ldots^{(55)}$$

(53) Newton returns yet again, in revision, to the *De Analysi*'s solution of literal equations (II: 228).

(54) On Newton's page 10 above.

(55) At this second stage the dividend is in fact $(-3a^2b^3c^{-4}+a^2bc^{-2})\,x^2+(1+a^3b^3c^{-6})\,x^3+O(x^4)$, yielding the final quotient $(a^2bc^{-4}-3a^2b^3c^{-6})\,x^2+(c^{-2}-a^3bc^{-6}+10a^3b^3c^{-8}-18a^3b^5c^{-10})\,x^3$ to $O(x^4)$. but Newton has made appropriate substitution of $a^2+3b^2=c^2$. Compare II, **2**, 3: note (88).

||[14] || Scribens b in quotiente suppono $b+p=y$ & pro y substituo ut vides: unde prodit p^3+3bpp &c, rejectis terminis $b^3+aab-2a^3$, qui nihilo sunt æquales propterea quod b supponitur radix hujus $y^3+aay-2a^3=0$. Deinde termini $3bbp+aap+abx$ dant $\frac{-abx}{3bb+aa}$ quotienti apponendum & $\frac{-abx}{3bb+aa}+q$ substituendum pro p. Brevitatis autem gratiâ scribo cc pro $3bb+aa$, cavendo tamen ut $3bb+aa$ restituatur ubi terminos sic abbreviari posse percipiam. Completo opere assumo numerum aliquem pro a, et hanc $y^3+aay-2a^3=0$ (sicut de numerali æquatione ostensum supra) resolvo, et quamlibet ejus radicem (modo tres[56] haberet) pro b substituo. Vel potiùs hujusmodi æquationes a speciebus, ut possum, libero, præsertim ab indefinitâ; idq; pro more quem volui innuere pag 9 lin 14:[57] et pro cæteris tantùm (siquæ supersint indelebiles) pono numeros. Sic $y^3+aay-2a^3=0$ liberabitur ab a dividendo radicem per a, fietq; $y^3+y-2=0$, cujus inventa radix ducta in a substitui debet pro b.

Hactenus indefinitam speciem suposui parvam esse. Quod si datæ quantitati vicina supponatur, pro indefinitè parva differentiâ pono speciem aliquam, et hâc substitutâ, solvo ut ante. Quemadmodum in

$$\tfrac{1}{5}y^5-\tfrac{1}{4}y^4+\tfrac{1}{3}y^3-\tfrac{1}{2}y^2+y[+]a-x=0,$$

cognito vel ficto x esse ejusdem prope quantitatis ac a, pono z differentiam, & scribendo $a+z$, vel $a-z$ pro x, orietur $\frac{1}{5}y^5-\frac{1}{4}y^4+\frac{1}{3}y^3-\frac{1}{2}y^2+y-\text{vel}+z=0$ solvendum ut in præcedentibus.[58]

Sin autem species illa supponatur indefinite magna, pro reciproco ejus indefinitè parvo pono speciem aliquam, quâ substitutâ solvo ut ante. Sic habito $y^3+y^2+y-x^3=0$,[59] ubi x cognoscitur vel fingitur esse valde magnum, pro

||[15] || reciprocè parvo $\frac{1}{x}$ pono z, et substituto $\frac{1}{z}$ pro x, orietur $y^3+yy+y-\frac{1}{z^3}=0$, cujus radix est $\frac{1}{z}-\frac{1}{3}-\frac{2}{9}z+\frac{7}{81}zz+\frac{5}{81}z^3$ &c.[60] et x si placet restituto fit

$$y=x-\frac{1}{3}-\frac{2}{9x}+\frac{7}{81xx}+\frac{5}{81x^3}\quad \&c.^{(60)}$$

(56) That is, 'reales' (real ones).

(57) There, of course, a homogeneous equation in the two variables x and y was reduced to one in the single variable v by the substitution $y=v\sqrt{[ax]}$.

(58) On Newton's page 11.

(59) In a preceding cancelled first version of the present paragraph Newton, making like use of the substitution $z=x^{-1}$ again chose as his example $\frac{1}{5}y^5-\frac{1}{4}y^4+\frac{1}{3}y^3-\frac{1}{2}y^2+y-z=0$, 'cujus radix secundum præcedentia extrahitur fitq; [radix y] $\sqrt{5}:\frac{5}{x}+\frac{1}{4}-\frac{1}{8}\sqrt{5}:x-\frac{1}{12}\sqrt{5}:5x$ &c'. The correct root should be $\alpha+\frac{1}{4}-\frac{5}{24}\alpha^{-1}\ldots$, where $\alpha^5=5/x$.

Writing b in the quotient I suppose $b+p=y$ and in place of y substitute as you see. Hence there comes $p^3+3bp^2\ldots$, when the terms $b^3+a^2b-2a^3$ are rejected as being equal to nothing because b is supposed to be a root of

$$y^3+a^2y-2a^3=0.$$

Thereafter the terms $3b^2p+a^2p+abx$ give $-\dfrac{abx}{3b^2+a^2}$ to be set in the quotient and $-\dfrac{abx}{3b^2+a^2}+q$ to be substituted in place of p. For brevity's sake, however, I write c^2 for $3b^2+a^2$, but taking care to restore $3b^2+a^2$ whenever I perceive that the terms may thereby be shortened. When the working is complete, I assume some number in place of a, resolve this equation $y^3+a^2y-2a^3=0$ (as shown above [in the passage] dealing with the numerical equation) and substitute one or other of its roots (should it have three)(56) for b. Or better, as far as I can I free equations of this sort from variables, particularly from the unbounded one, and that after the manner I wished to make known on page 9, line 14,(57) while for the rest only (if there are any which cannot be removed) do I put numbers. So $y^3+a^2y-2a^3=0$ will be freed from a by dividing its roots by a, when it becomes $y^3+y-2=0$: the root of this, when found, should be multiplied by a and then substituted in place of b.

So far I have taken the unbounded variable to be small. But if it be supposed to be in the vicinity of a given quantity, for the indefinitely small difference I put some variable and, after substitution, resolve as before. For instance, in the case of $\frac{1}{5}y^5-\frac{1}{4}y^4+\frac{1}{3}y^3-\frac{1}{2}y^2+y+a-x$, if x is known or conjectured to be very nearly of the same magnitude as a, I put z for the difference and, on writing $a+z$ or $a-z$ in place of x, there will arise $\frac{1}{5}y^5-\frac{1}{4}y^4+\frac{1}{3}y^3-\frac{1}{2}y^2+y\mp z=0$ to be resolved as in the preceding.(58)

But should, however, that species be supposed indefinitely great, for its indefinitely small reciprocal I put some variable and then, after substitution, resolve as before. So when there is had $y^3+y^2+y-x^3=0$,(59) in which x is known or thought to be extremely great, in place of its reciprocal, the small quantity x^{-1}, I put z and, after substituting z^{-1} in place of x, there will arise

$$y^3+y^2+y-z^{-3}=0.$$

Its root is $z^{-1}-\frac{1}{3}-\frac{2}{9}z+\frac{7}{81}z^2+\frac{5}{81}z^3\ldots$,(60) which on restoring x (if you so wish) becomes $y=x-\frac{1}{3}-\frac{2}{9}x^{-1}+\frac{7}{81}x^{-2}+\frac{5}{81}x^{-3}\ldots$.(60)

(60) The coefficient of the term in z^3 $(=x^{-3})$ should be zero. Horsley correctly states the root to be $z^{-1}-\frac{1}{3}-\frac{2}{9}z+\frac{7}{81}z^2+\frac{14}{729}z^4\ldots$ (*Isaaci Newtoni Opera Omnia*, **1** (London, 1779): 404).

Siquando ex aliquâ harum trium suppositionum res non omninò aut non commodè succedat, ad aliam recurri potest. Sic in

$$y^4 - xxyy + xyy + 2yy - 2y + 1 = 0,$$

cùm primus terminus obtineri deberet fingendo $y^4 + 2yy - 2y + 1 = 0$,[61] quæ tamen nullam admittit possibilem radicem, Tento quid fiet aliter: quemadmodum si fingam x parùm differre a $+2$ sive esse $2 + z = x$, substituendo $2 + z$ vice x prodibit $y^4 - zzyy - 2zyy$[62]$- 2y + 1 = 0$, et quotiens exordietur ab $+1$.[63] Vel si fingam x indefinitè magnam esse, sive $\frac{1}{x} = z$, obtinebitur

$$y^4 - \frac{yy}{zz} + \frac{yy}{z} + 2yy - 2y + 1 = 0,$$

& $+z$ pro initio quotientis.[64] Et hac ratione secundum varias Hypotheses procedendo, licebit varijs modis extrahere ac designare radices.

Quod si cupias explorare quot modis id potest fieri, tentabis quænam quantitates pro indefinitâ specie in æquationem propositam substitutæ, efficient divisibilem per $y +$ vel $-$ aliquâ quantitate vel per y solum. Id quod verbi gratia in æquatione $y^3 + axy + aay - x^3 - 2a^3 = 0$ eveniet substituendo $+a$ vel $-a$, vel $-2a$, vel $\sqrt{C}\colon -2a^3$, &c pro y.[65] Atq ita possis commodè supponere quantitatem x parùm ab $+a$, vel $-a$, vel $-2a$, vel $\sqrt{C}\colon -2a^3$ differre, et inde radicem propositæ æquationis tot modis extrahere. Imò et fortasse tot alijs modis fingendo differentias istas esse indefinitè magnas.[66] Quinetiam si aliam atq aliam e speciebus radicē definientibus pro indefinitâ adhibeas, possis alijs adhuc fortasse modis propositum consequi; et etiamnum alijs substituendo valores quâcunq ratione fictos $\left(\text{quales sunt } az + bzz,\ \frac{a}{b+z},\ \frac{a+cz}{b+z}\ \&\text{c}\right)$ pro indefinitâ specie & in æquatione resultante operando sicut in præcedentibus.

Cæterùm ut conclusionum veritas constet, quotientes nempe sic extractos, dum producuntur, ita propiùs ad radicem accedere, ut minùs tandem quâvis datâ quantitate differant, adeoq in infinitum productos non omninò differre:
||[16] perpende || quod quantitates in sinistrâ columnâ dextræ partis diagrammatum,[67] sint ultimi termini æquationum quarum p, q, r, s &c existunt radices

(61) That is, $y^4 + y^2 + (y-1)^2$ which is greater than zero for all real values of y.

(62) Read '$-3zyy$'.

(63) The fictitious equation $y^4 - 2y + 1 = (y-1)(y^3 + y^2 + y - 1) = 0$ has, in fact, a second real root $y = 0{\cdot}55\ldots$.

(64) Strictly, the first term in the quotient could be $\pm z$ since the fictitious equation now becomes $-y^2/z^2 + 1 = 0$.

(65) If $y - k$ is a factor of $y^3 + a(a+x)y - x^3 - 2a^3$, then $x(x^2 - ak) = (k-a)(k^2 + ak + 2a^2)$. The cases $k = a$, $x = \pm a$; $k = x = -2a$; $k = 0$, $x = -a\sqrt[3]{2}$ yield the factorizations

$$(y-a)(y^2 + ay + [2 \pm 1]a^2); \quad (y+2a)(y^2 - 2ay + 3a^2); \quad \text{and} \quad y(y^2 + [1 - \sqrt[3]{2}]a^2)$$

respectively. Another simple factorization is determined by $k = -\frac{1}{4}a$, $x = -\frac{5}{4}a$.

If at any time the outcome should, according to one or other of these expedients, not be wholly successful or profitable, recourse may be had to another. So in $y^4-x^2y^2+xy^2+2y^2-2y+1=0$, the first term ought to be obtained by supposing $y^4+2y^2-2y+1=0$,(61) but this admits of no possible root and so I try what else is to be done. For instance, I might suppose x to be little different from 2 or that $2+z=x$: on substituting $2+z$ instead of x there comes

$$y^4-z^2y^2-[3]zy^2-2y+1=0$$

and the quotient will start with 1.(63) Or I might suppose x to be indefinitely large and take $x^{-1}=z$, when there will be obtained

$$y^4-z^{-2}y^2+z^{-1}y^2+2y^2-2y+1=0$$

and z for the beginning of the quotient.(64) And by proceeding in this manner according to various hypotheses, you will be at liberty to extract and specify the roots in various ways.

But should you desire to work out in how many ways that may be done, you will find by trial which quantities, when substituted in place of the unbound variable in the proposed equation, make it divisible by $y\pm$ some quantity or by y alone. In the equation $y^3+axy+a^2y-x^3-2a^3=0$, for example, this will occur on substitution of $+a$, $-a$, $-2a$, $\sqrt[3]{-2a^3}$, and so on for y.(65) And thus you might conveniently assume the quantity x to be little different from a or $-a$ or $-2a$ or $\sqrt[3]{-2a^3}$ and thence extract the root of the proposed equation in as many ways—indeed, perhaps also in as many further ways by supposing those differences to be indefinitely large.(66) Even more, should you employ for the unbounded one one or other of the variables defining the root, you might perhaps achieve the purpose in yet further ways; and in still more yet by substituting values arbitrarily contrived (such as $az+bz^2$, $a/(b+z)$, $(a+cz)/(b+z)$ and so on) for the unbounded species and operating on the resulting equation as in the preceding.

But to concur in the truth of these conclusions, that is, to agree that the quotients thus extracted, as they are extended, ever more closely approach the root till finally they differ from it by less than any given quantity and so, when they are infinitely extended, differ from it not at all: consider that the quantities in the left-hand column on the right side of the symbolic arrays(67) are the last terms of equations which have p, q, r, s, ... for their roots and in consequence

(66) This conjecture is apparently hazarded only for algebraic functions.

(67) The arrays in which the roots of literal equations are extracted by successively substituting $y=Y+p$, $p=\alpha_1x+q$, $q=\alpha_2x+r$, ... If, for any $s=\alpha_ix^i+t$, $s=\alpha_ix^i$ renders the coefficient of each power of x in the corresponding equation $\phi(x,s)=0$ at any stage identical to zero, then clearly $t=0$ and $y=Y+\alpha_1x+\alpha_2x^2+\ldots+\alpha_ix^i$ exactly.

et inde quòd ipsis evanescentibus, illæ p, q, r, s, id est differentiæ inter quotientē & quæsitā radicē, simul evanescunt. Adeoq; quotiens tunc non differt a radice. Quamobrem sub initio operis si terminos in dictâ columnâ sese omnes destruere videas, conclude quotientem eatinus[(68)] extractam, esse justam radicem. Sin aliter, videbis tamen terminos in quibus indefinite parva species est pauciorum dimensionum, id est longè maximos,[(69)] e columna ista perpetuò tolli, ut tandem non restent nisi datâ quâvis quantitate minores, et proinde non majores nihilo cùm opus infinitè producitur. Quare quotiens infinitè extracta fiet etiam justa radix.[(70)]

Etsi deniq; species, quas hactenus perspicuitatis grā[(71)] supposui indefinitè parvas esse, quantumvis magnæ supponātur, tamen veræ erunt quotientes, ut ut minùs citò ad justam radicem convergant[(72)] quemadmodum ex analogiâ rei constet. Sed hic radicum termini, maximæq; et minimæ quantitates spectandæ veniunt: Nam infinitarum cum finitis æquationibus communia sunt hujusmodi symptomata. Radix autem in his maxima fit vel minima quando maxima vel minima est differentia summæ affirmativorum terminorum a summâ negativorum, ac terminatur cùm indefinita quantitas (quam ideò parvam esse non immeritò finxi) non potest major sumi quin magnitudo radicis in infinitum prosiliet, hoc est fiet impossibilis.[(73)] Verbi gratiâ posito ACD semicirculo super diametro AD descripto, et BC ordinatim applicatâ: dic $AB=x$ $BC=y$ $AD=a$ et erit $y=(\sqrt{ax-xx}=)$ $\sqrt{ax}-\frac{x}{2a}\sqrt{ax}-\frac{xx}{8aa}\sqrt{ax}-\frac{x^3}{16a^3}\sqrt{ax}$ &c ut supra.[(74)] Fit ergo BC sive y maxima cum

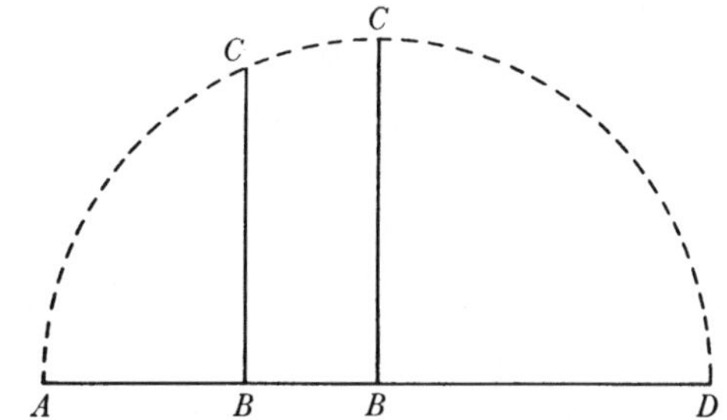

(68) Newton's variant on the usual 'eatenus'.

(69) Where x is very small $\alpha_i x^i$ will, for any finite α_i, tend rapidly to zero as i becomes large.

(70) Compare the equivalent passage in the *De Analysi* (II: 244–6). Newton is badly in need here of suitable criteria for the convergence of an infinite sum-sequence. His naïve criterion that a sum-sequence converges if all terms after an assigned one are less than any given finite quantity, however small, is—needless to say—invalid. Pietro Mengoli's proof of the divergence of the harmonic series (already known to Oresme in the fourteenth century) had been published twenty years earlier in his *Novæ Quadraturæ Arithmeticæ, seu de Additione Fractionum* (Bologna, 1650), but it is clear that neither this nor any other counter-example is known to Newton. (See A. Agostini, 'L'Opera Matematica di Pietro Mengoli', *Archives internationales d'Histoire des Sciences*, **3** (1950): 817–34, especially 822. Compare also A. P. Yushkevich, *Geschichte der Mathematik im Mittelalter* (Leipzig, 1964): 411.)

(71) Read 'gratiâ'.

(72) A first, cancelled version of the preceding reads equivalently '...quam hactenus... supposui indefinitè parvam esse, quantumvis magna supponatur quotientium utut minùs citò ad justam radicem convergentium veritas'. The phraseology of a 'series convergens' had been invented only three years before by James Gregory to denote (no doubt on the analogy of two geometrical lines 'convergent' on the point which is their common meet) a double

when those last terms vanish, the quantities $p, q, r, s, \ldots$ (the differences, namely, between the quotient and the root sought) vanish with them, so that the quotient is then not distinct from the root. For this reason should you at the beginning of the working see all the terms in the said column destroy one another, conclude that the quotient thus far extracted is the exact root. If not, you will none the less observe that the terms in which the indefinitely small variable is of comparatively low dimensions, that is, the largest(69) by far, are perpetually erased from that column till at length there remain only ones less than any assigned quantity and accordingly not distinct from zero when the work is indefinitely prolonged. Hence here too the quotient, when infinitely extracted, will become the exact root.(70)

Lastly, even though the variables, hitherto for clarity's sake supposed indefinitely small, are now taken to be as large as you wish, the quotients will still be true however reduced the swiftness with which they converge(72) to the exact root. This should be manifest by analogy. But here limits to the roots and greatest and least quantities have to be taken into consideration, for these kinds of properties are common to both finite and infinite equations. The root in these becomes greatest or least when the sums of positive and negative terms have a greatest or least difference, and is limited when the indefinite quantity (which I have on that account not without reason supposed to be small) may not be taken greater without the magnitude of the root becoming infinite, and so impossible,(73) at a bound. For example, taking ACD to be a semicircle described on the diameter AD and BC an ordinate to it, call $AB = x$, $BC = y$, $AD = a$ and then will y be $\surd(ax-x^2) = \surd ax(1-x/2a-x^2/8a^2-x^3/16a^3\ldots)$ as above.(74) In

sequence of 'termini' (bounds), one monotonically increasing, the other monotonically decreasing over a finite or infinite number of stages to a common 'limit'. (See Gregory's *Vera Circuli et Hyperbolæ Quadratura* (Padua, 1667): *Definitiones* 9/10. Newton's library copy of this book is now in Trinity College, Cambridge, NQ. 9.48[1].) The modern looser sense of a 'series' of 'terms' which 'converge' arbitrarily to a common 'limit' is seemingly Newton's generalization. Already a few months earlier, in his opening letter to Collins in January 1670, Newton had shown how to sum a given harmonic progression as an infinite 'progression... converging much more towards the truth then the former' (*Correspondence*, **1** (1959): 18). The present instance is the first known occurrence in a Newtonian mathematical text of the phrase 'series convergens'.

(73) Newton's attempted analogy between the convergence of a sum-sequence for small values of the variable x and that of the same sequence for large values of the same variable does not unfortunately hold true in general. The counter-example of $\tan^{-1}x$ and $x-\frac{1}{3}x^3+\frac{1}{5}x^5\ldots$, the latter of which converges to the former when $0 \leqslant x \leqslant 1$ but diverges to infinity for $1 < x < \infty$, serves to show that a divergent series need not necessarily 'represent' a non-real quantity. Whether Newton at this time yet knew of this 'series Gregoriana–Leibnitiana' for the inverse-tangent is highly doubtful (compare page 34, note (5) above) but the comparable counter-instance of $\log(1\pm x)$ and the 'Mercator' series $x\mp\frac{1}{2}x^2+\frac{1}{3}x^3\mp\ldots$ should have been in the forefront of his mind at this point.

(74) On Newton's page 4 above (in the case where $a = 1$).

$\surd ax$ maximè superat omnes $\frac{x}{2a}\surd ax+\frac{xx}{8aa}\surd ax+\frac{x^3}{16a^3}\surd ax$ &c, id est cùm sit $x=\frac{1}{2}a$:(75) terminabitur autem cùm sit $x=a$ quia si sumas $x \sqsubset$(76) a, summa omnium terminorum $\frac{-x}{2a}\surd ax-\frac{xx}{8aa}\surd ax-\frac{x^3}{16a^3}\surd ax$ &c erit infinita.(77) Est et alius terminus cùm ponitur $x=0$, propter impossibilitatem radicalis $\surd -ax$;(78) Quibus terminis correspondent semicirculi limites A et D.(79)

||[17] || Hactenus de modis computandi quorum pos[t]hac frequens erit usus. Jam restat ut in illustrationem hujus Artis Analyticæ tradam aliquot Problematum specimina qualia præsertim natura curvarum ministrabit. Sed imprimis observandum venit quod hujusmodi difficultates possunt omnes ad hæc duo tantùm problemata reduci quæ circa spatium motu locali(80) utcunq; accelerato vel retardato descriptum proponere licebit.

1. Spatij longitudine continuò (sive ad omne tempus)(81) data, celeritatem motûs ad tempus propositum invenire.

2. Celeritate motûs continuò datâ longitudinem descripti spatij ad tempus propositum invenire.(82)

(75) Presumably Newton would prove this formally by the Fermatian adequation method expounded in Proposition 3 below (on Newton's page 41) but it is at once obvious from the diagram that BC is greatest when $AB = \frac{1}{2}AD$.

(76) Barrow's symbol for greater inequality (compare II, 2, 3: note (145)).

(77) Once again (compare note (73)) Newton appears to believe that the divergence of a sum-series *necessarily* implies that the corresponding value of the function which it represents be non-real. Here, however, it is true that for $x > a$ the function $\surd[ax-x^2]$ is not real, so that $x = a$ is a (Newtonian) 'limes'.

(78) Read '$\sqrt{ax-xx}$ cùm x sit minor quàm 0' ($\surd[ax-x^2]$ when $x < 0$) or some equivalent.

(79) The opening words of a further sentence, 'Et hic differentia inter infinitas[...]' (And here the difference between infinite...) have been heavily cancelled: the discarded following sheet which presumably bears their continuation would appear to be irretrievably lost. The present paragraph, cancelled in the manuscript by surrounding square parentheses, has nevertheless been retained by Newton's subsequent eighteenth-century editors—evidently because of its interesting revelation of his ideas, at this time still unformed, on series-convergence—without acknowledgement of Newton's intention to suppress it. (See Samuel Horsley, *Opera* (note (60)): **1**: 405–6; and John Colson, *The Method of Fluxions and Infinite Series with its Application to the Geometry of Curve-Lines* (London, 1736): 18–19.)

(80) 'Locomotion'; that is, change of place in the Aristotelian sense, refined by the medieval scholastic physicists (M. Clagett, *The Science of Mechanics in the Middle Ages* (Oxford, 1959): 422 ff.), of movement in time through geometrical space. In the first volume (I, 2, 5, §4: note (4)) we reinforced J. E. Hofmann's conjecture that Isaac Barrow's 1664 series of Lucasian lectures may have been the inspiration for Newton's first fluxional notions in mid-1665. Only a few weeks before Newton penned the present section (in the winter of 1670–71) Barrow had published a new series of 'geometrical lectures' (*Lectiones Geometricæ: In quibus (præsertim) Generalia Curvarum Linearum Symptomata declarantur* (London, 1670)), the tenth of which, at least, Newton had read in proof. (See I: xv, note (1). Newton's unmarked library copy of Barrow's work, bound with a reissue of his *Lectiones* XVIII... of the previous year, is now in

consequence, BC or y becomes greatest when $\sqrt{ax}$ most exceeds all the other terms $\sqrt{ax}(x/2a+x^2/8a^2+x^3/16a^3\ldots)$, that is, when $x=\frac{1}{2}a$;(75) but will be limited when $x=a$ because if you take $x>a$, the sum of all the terms

$$-\sqrt{ax}(x/2a+x^2/8a^2+x^3/16a^3\ldots)$$

will be infinite.(77) There is also a second limit when x is set equal to 0 because of the impossibility of the root $\sqrt{-ax}$.(78) To these limits correspond the semicircle bounds A and D.(79)

So much for computational methods of which in the sequel I shall make frequent use. It now remains, in illustration of this analytical art, to deliver some typical problems and such especially as the nature of curves will present. But first of all I would observe that difficulties of this sort may all be reduced to these two problems alone, which I may be permitted to propose with regard to the space traversed by any local motion(80) however accelerated or retarded:

1. Given the length of the space continuously (that is, at every [instant of] time),(81) to find the speed of motion at any time proposed.

2. Given the speed of motion continuously, to find the length of the space described at any time proposed.(82)

Trinity College, Cambridge (NQ. 16. 181): its flyleaf bears the dedication in Barrow's hand, 'Isaaco Newtono Reverendus Author hunc dono dedit July 7th 1670'.) In his first three lectures, particularly the first, Barrow deals at some length with 'local' motion of 'crescent' and 'decrescent' magnitudes and their 'flux' in time. Terminology apart, however, it is tempting to pinpoint the source of Newton's present refined exposition of his fluxional concept in his renewed reading of Barrow's lectures. We shall draw further parallels in following notes.

(81) Newton has here cancelled the more precise Greek equivalent 'τὸ νῦν'. This Aristotelian phrase would hardly come naturally to his mind and we are puzzled why he did not replace it by the more accurate 'ad omnem instantiam temporis'. There may well be an oblique reference here to Barrow's assertion in the first lecture of his 1665 mathematical series that 'inter anni, mensis, diei, horæ, τὰ νῦν primum ac ultimum duratio quædam intercedit, juxta quam tempus extenditur. Ac inter motûs initium & finem medium quiddam decurso spatio respondens jacet' (*Lectiones Mathematicæ XXIII*; *In quibus Principia Matheseôs generalia exponuntur. Habitæ Cantabrigiæ A.D. 1664, 1665, 1666* (London, 1683): ${}_2$10–11; compare John Kirkby, *The Usefulness of Mathematical Learning explained and demonstrated...By Isaac Barrow* (London, 1734): 147–8). Barrow's 'τὰ νῦν primum ac ultimum' is strongly evocative of Newton's phrases 'primum momentum fluxionis' and 'ultimum momentum defluxionis' in the next section (see §3: note (16) below).

(82) Compare Barrow, *Lectiones Geometricæ* (note (80)): 7–8: 'Tempus...per rectam lineam semper designabimus...cujus partes proportionalibus temporis partibus, & puncta temporis instantibus respectivis justè respondebunt....Omni temporis instanti, seu indefinitè parvæ temporis particulæ...competit velocitatis aliquis gradus, quem mobile tunc habere concipiendum est; cui gradui respondet aliqua decursi spatii longitudo'. These two fundamental, complementary problems (elaborated as Problems 1 and 2 below) require, in the geometrical model of a line-segment traversed continuously in time, the finding of the 'celerity' or 'fluxional' speed of a variable quantity as its derivative and, conversely, the determination of that 'fluent' quantity as the integral of the fluxional speed, where in either case 'time' is the independent variable.

Sic in æquatione $xx = y$ si y designat spatij longitudinem ad quodlibet tempus quod aliud spatium x uniformi celeritate increscendo mensurat et exhibet descriptam: tunc $2mx$ designabit celeritatem(83) qua spatium y ad idem temporis momentum describi pergit; et contra. Et hinc est quod in sequentibus considerem quantitates quasi generatæ essent per incrementum continuum ad modum spatij quod mobile percurrendo describit.

Cùm autem temporis nullam habeamus æstimationem nisi quatenus id per æquabilem motum localem exponitur et mensuratur,(84) et præterea cùm quantitates ejusdem tantùm generis inter se conferri possint et earum incrementi et decrementi celeritates inter se, eapropter ad tempus formaliter spectatum in sequentibus haud respiciam, sed e propositis quantitatibus quæ sunt ejusdem generis aliquam æquabili fluxione augeri fingam(85) cui cæteræ tanquam tempori referantur, adeoq; cui nomen temporis analogicè tribui mereatur. Siquando itaq; vocabulum temporis in sequentibus occurrat (quemadmodum perspicuitatis et distinctionis gratia nonnũquàm intertexui) eo nomine non tempus formaliter spectatum subintelligi debet sed illa alia quantitas cujus æquabili incremento sive fluxione tempus exponitur et mensuratur.

Quantitates autem quas ut sensim crescentes indefinitè considero, quo distinguam ab alijs quantitatibus quæ in æquationibus quibuscunq; pro
||[18] determinatis et cognitis habendæ sunt ac || initialibus literis a, b, c, &c designantur, posthac denominabo fluentes, ac designabo finalibus literis v, x, y, et z. Et celeritates quibus singulæ a motu generante fluunt et augentur (quas possim fluxiones vel simpliciter celeritates vocitare) designabo literis l, m, n et r. Nempe pro celeritate quantitatis v ponam l et sic pro celeritatibus aliarum quantitatum x, y, et z ponam m, n, et r respectivè.(86) His præmissis, e vestigio rem aggredior, imprimis duorum jam modo propositorũ problematum solutionem exhibiturus.

---

(83) Namely, $m = \dot{x}$ and $n = \dot{y}$ respectively. Compare note (86).

(84) Compare Barrow, *Lectiones Geometricæ* (note (80)): 3: 'per se tempus quantum est, etsi quo temporis quantitas a nobis dignoscatur, advocandum sit motûs subsidium, ceu mensuræ quâ temporis quantitas æstimemus, & inter se conferamus; adéoque tempus ut mensurabile motum connotat, nec enim, si res omnes immotæ perstarent, ullo pacto quantum effluxisset temporis possemus internoscere'; *ibid.*: 6: 'Neque quisquam objiciat tempus communiter haberi pro mensura motûs, & consequenter ad hoc motûs differentias (velocioris, tardioris, accelerati, retardati) adsumendo tempus ut præcognitum definiri; nec ideo temporis quantitatem è motu, sed motûs quantitatem è tempore determinari.'

(85) A first phrasing 'aliquam...adgeneratam debemus fingere' (we ought to suppose some one generated) is cancelled.

(86) The final sentence following is a late insertion. In our English 'translation' we have dared—anachronistically—to render Newton's literal fluxions l, m, n, r (of variables v, x, y, z) by their more familiar and effective 'dotted' equivalents $\dot{v}$, $\dot{x}$, $\dot{y}$, $\dot{z}$. In the sequel we shall similarly render the fluxions k, p, q (of variables h or ϕ, s, t respectively) by $\dot{h}$ or $\dot{\phi}$, $\dot{s}$, $\dot{t}$ in the English version. As will be seen in the sixth volume Newton himself did not introduce this

So in the equation $x^2 = y$, if y designates the length of the space described in any time which is measured and represented by a second space x as it increases with uniform speed: then $2\dot{x}x$ will designate the speed[83] with which the space y at the same moment of time proceeds to be described. And hence it is that in the sequel I consider quantities as though they were generated by continuous increase in the manner of a space which a moving object describes in its course.

We can, however, have no estimate of time except in so far as it is expounded and measured by an equable local motion,[84] and furthermore quantities of the same kind alone, and so also their speeds of increase and decrease, may be compared one with another. For these reasons I shall, in what follows, have no regard to time, formally so considered, but from quantities propounded which are of the same kind shall suppose some one to increase[85] with an equable flow: to this all the others may be referred as though it were time, and so by analogy the name of 'time' may not improperly be conferred upon it. And so whenever in the following you meet with the word 'time' (as I have, for clarity's and distinction's sake, on occasion woven it into my text), by that name should be understood not time formally considered but that other quantity through whose equable increase or flow time is expounded and measured.

But to distinguish the quantities which I consider as just perceptibly but indefinitely growing from others which in any equations are to be looked on as known and determined and are designated by the initial letters a, b, c and so on, I will hereafter call them fluents and designate them by the final letters v, x, y and z. And the speeds with which they each flow and are increased by their generating motion (which I might more readily call fluxions or simply speeds) I will designate by the letters $\dot{v}$, $\dot{x}$, $\dot{y}$ and $\dot{z}$: namely, for the speed of the quantity v I shall put $\dot{v}$, and so for the speeds of the other quantities I shall put $\dot{x}$, $\dot{y}$ and $\dot{z}$ respectively.[86] With these premisses made, I now attack my objective forthwith. First, I shall display the resolution of the two problems just now proposed.

---

standard 'Newtonian' dot-notation till late 1691, and it should be clearly understood that we adopt the substitution merely as an aid to comprehension of involved arguments not easy to follow in their original dress. This convention is in fact of considerable antiquity, for at least as early as 1710 William Jones adhered to it in the transcript he made of the 1671 tract from Newton's autograph manuscript (compare §1: note (1)). All its eighteenth-century editors—notably Colson (who based his 1736 English translation (note (79)) on Jones' copy), de Buffon (who based his French version, *La Methode des Fluxions, et des Suites Infinies. Par M. le Chevalier Newton* (Paris, 1740), uniquely on Colson) and Horsley (who founded his 1779 *princeps* edition of the Latin text (note (60)) squarely on the Jones transcript even though he corrected it in proof from the autograph original)—sheepishly followed Jones' lead without any mention of Newton's original literal notation for fluxions. The practice has had its effect in seriously distorting, for example, all subsequent assessment of the relative merits of Newton's and Leibniz' early calculus symbolisms.

PROB: 1.

RELATIONE QUANTITATUM FLUENTIUM INTER SE DATÂ, FLUXIONUM RELATIONEM DETERMINARE.

SOLUTIO.

Æquationem qua data relatio exprimitur dispone secundum dimensiones alicujus fluentis quantitatis puta x, ac terminos ejus multiplica per quamlibet Arithmeticam progressionem ac deinde per $\frac{m}{x}$. Et hoc opus in qualibet fluenti quantitate seorsim institue. Dein omnium factorum summam pone nihilo æqualem, et habes æquationem desideratam.(87)

EXEMP: 1. Si quantitatum x et y relatio sit $x^3-axx+axy-y^3=0$, terminos primò secundum x ac deinde secundum y dispositos multiplico ad hunc modum.

Mult: $x^3-axx+axy-y^3$.	Mult $-y^3+axy\begin{matrix}-axx\\+x^3\end{matrix}$.
per $\frac{3m}{x}$. $\frac{2m}{x}$. $\frac{m}{x}$. 0.	per $\frac{3n}{y}$. $\frac{n}{y}$. 0.
fit $3mxx-2max+may$. $*$.	fit $-3nyy+anx$. $*$.

Et factorum summa est $3mxx-2amx+amy-3nyy+anx=0$. æquatio quæ dat relationem inter fluxiones m et n. Nempe si assumas x ad arbitrium, æquatio $x^3-axx+axy-y^3=0$ dabit y. Quibus determinatis erit

$$m.n::3yy-ax.\ 3xx-2ax+ay.$$

EXEMPL 2. Si quantitatum x, y et z relatio sit $2y^3+xxy-2cyz+3yzz-z^3=0$,

Mult $2y^3\begin{matrix}+xx\\-2cz\\+3zz\end{matrix}y-z^3$.	Mult $x[x]y\begin{matrix}+2y^3\\-2cyz\\+3yzz\\-z^3\end{matrix}$.	Mult $-z^3+3yzz-2cyz\begin{matrix}+xxy\\+2y^3\end{matrix}$.
per $\frac{2n}{y}$. 0. $-\frac{n}{y}$.	per $\frac{2m}{x}$. 0.	per $\frac{3r}{z}$. $\frac{2r}{z}$. $\frac{r}{z}$. 0.
fit $4nyy$ $*$ $+\frac{nz^3}{y}$.	fit $2mxy$ $*$.	fit $-3rzz+6ryz-2cry$ $*$.

Quare fluendi celeritatum m n et r relatio est

$$4nyy+\frac{nz^3}{y}+2mxy-3rzz+6ryz-2cry=0.$$

Cæterùm cùm tres sint hic fluentes quantitates x, y, et z, deberet alia insuper æquatio dari qua relatio inter ipsas ut et inter earum fluxiones penitiùs determinetur. Quaemadmodum si ponitur $x+y-z=0$. exinde fluxionum alia

(87) A Latin version of the equivalent differential algorithm in the 1666 fluxional tract (I: 402).

Problem 1

Given the relation of the flowing quantities to one another, to determine the relation of the fluxions.

Solution

Arrange the equation by which the given relation is expressed according to the dimensions of some fluent quantity, say x, and multiply its terms by any arithmetical progression and then by $\dot{x}/x$. Carry out this operation separately for each one of the fluent quantities and then put the sum of all the products equal to nothing, and you have the desired equation.(87)

Example 1. If the relation of the quantities x and y be $x^3-ax^2+axy-y^3=0$, I multiply the terms arranged first according to x and then to y in this way.

Multiply	$x^3 \quad -ax^2+axy-y^3$		Mult.	$-y^3+axy\begin{matrix}-ax^2\\+x^3\end{matrix}$	
by	$\dfrac{3\dot{x}}{x}.\quad \dfrac{2\dot{x}}{x}.\quad \dfrac{\dot{x}}{x}.$	$0.$	by	$\dfrac{3\dot{y}}{y}.\quad \dfrac{\dot{y}}{y}.$	$0.$
there comes	$3\dot{x}x^2-2\dot{x}ax+\dot{x}ay$	$*.$	comes	$-3\dot{y}y^2+a\dot{y}x$	$*.$

And the sum of the products is $3\dot{x}x^2-2a\dot{x}x+a\dot{x}y-3\dot{y}y^2+a\dot{y}x=0$, an equation which gives the relation between the fluxions $\dot{x}$ and $\dot{y}$. Precisely, should you assume x arbitrarily the equation $x^3-ax^2+axy-y^3=0$ will give y, and with these determined it will be $\dot{x}:\dot{y}=(3y^2-ax):(3x^2-2ax+ay)$.

Example 2. If the relation of the quantities x, y and z be

$$2y^3+x^2y-2cyz+3yz^2-z^3=0,$$

Mult. $2y^3\begin{matrix}+x^2y\\-2czy\\+3z^2y\end{matrix}\quad -z^3$	Mult. $x^2y\begin{matrix}+2y^3\\-2cyz\\+3yz^2\\-z^3\end{matrix}$	Mult. $-z^3+3yz^2-2cyz\begin{matrix}+x^2y\\+2y^3\end{matrix}$
by $\dfrac{2\dot{y}}{y}.\quad 0.\quad -\dfrac{\dot{y}}{y}$	by $\dfrac{2\dot{x}}{x}.\quad 0.$	by $\dfrac{3\dot{z}}{z}.\quad \dfrac{2\dot{z}}{z}.\quad \dfrac{\dot{z}}{z}.\quad 0.$
comes $4\dot{y}y^2\quad *\quad +\dfrac{\dot{y}z^3}{y}.$	comes $2\dot{x}xy\quad *.$	comes $-3\dot{z}z^2+6\dot{z}yz-2c\dot{z}y\quad *.$

Hence the relation of the speeds $\dot{x}$, $\dot{y}$ and $\dot{z}$ of flow is

$$4\dot{y}y^2+\dot{y}z^3/y+2\dot{x}xy-3\dot{z}z^2+6\dot{z}yz-2c\dot{z}y=0.$$

But since there are here three fluent quantities x, y and z, another equation ought also to have been given so that the relation between these and their fluxions might more thoroughly be determined. For instance, if there be set $x+y-z=0$, by which according to the rule a second relation between the

||[19] relatio juxta Regulam erit $m+n-r=0$. Confer || jam hasce cum præcedentibus æquationibus, eliminando quamlibet e tribus quantitatibus et quamlibet etiam e tribus earum fluxionibus,(88) et reliquorum relationes penitiùs determinatas obtinebis.

Siquando in æquatione propositâ insint fractiones complexæ(89) aut surdæ quantitates, pro singulis pono totidem literas, easq fingens designare quantitates fluentes,(90) operor ut ante. Dein supprimo et extermino literas ascriptitias, ut hic videre est.(91)

EXEMPL: 3. Si quantitatum x et y relatio sit $yy-aa-x\sqrt{aa-xx}$: pro $x\sqrt{aa-xx}$ scribo z et inde habeo duas æquationes $yy-aa-z=0$, et $aaxx-x^4-zz=0$, quarum prior ut ante dabit $2ny-r=0$ pro relatione celeritatum n et r, et posterior dabit $2aamx-4mx^3-2rz=0$ sive $\frac{aamx-2mx^3}{z}=r$ pro relatione celeritatum m et r. Jam r suppresso fiet $2ny\frac{-aamx+2mx^3}{z}=0$, dein restituto $x\sqrt{aa-xx}$ pro z habebitur $2ny\frac{-aam+2mxx}{\sqrt{aa-xx}}=0$ relatio inter m et n quæ quærebatur.

EXEMPL: 4. Si $x^3-ayy+\frac{by^3}{a+y}-xx\sqrt{ay+xx}=0$ designat relationem inter x et y: pono z pro $\frac{by^3}{a+y}$, et v pro $xx\sqrt{ay+xx}$ et inde nactus sum tres æquationes

$$x^3-ayy+z-v=0, \quad az+yz-by^3=0, \quad \& \quad ax^4y+x^6-vv=0.$$

Prima dat $3mxx-2any+r-l=0$, secunda dat $ar+ry+nz-3bnyy=0$, et tertia dat $4amx^3y+6mx^5+anx^4-2lv=0$ pro relationibus celeritatum l, m, n et r. Ipsorum verò r et l valores per secundam ac tertiam inventos $\left(\text{nempe } \frac{3bnyy-nz}{a+y}\right.$ pro r, et $\frac{4amx^3y+6mx^5+anx^4}{2v}$ pro $\left. l\right)$ substituo in primam et oritur

$$3mxx-2any\frac{+3bnyy-nz}{a+y}\frac{-4amx^3y-6mx^5-anx^4}{2v}=0.$$

Et vice z & v restitutis valoribus $\frac{by^3}{a+y}$ et $xx\sqrt{ay+xx}$, prodit æquatio quæsita $3mxx-2any\frac{+3abnyy+2bny^3}{aa+2ay+yy}\frac{-4amxy-6mx^3-anxx}{2\sqrt{ay+xx}}=0$ quâ relatio celeritatum m et n designatur.

(88) 'momentis' (moments) is cancelled, and Newton has made a large number of similar substitutions of 'fluxio' for 'momentum' in the next few paragraphs. In standard Newtonian usage (and in the following 'Demonstratio' in particular) the 'moment' of a quantity is the product of its fluxion and the indefinitely small increment of the base variable of 'time'.

fluxions will be $\dot{x}+\dot{y}-\dot{z}=0$. Now compare these with the preceding equations, by eliminating any one of the three quantities and also any one of their three fluxions,[88] and you will obtain a more thorough determination of the relation between the rest.

Whenever complex[89] fractions or surd quantities are present in the proposed equation, in place of each I put a corresponding letter and, supposing these to designate fluent[90] quantities, I work as before. Then I suppress and exterminate the letters ascribed, as is to be seen here.[91]

EXAMPLE 3. If the relation of the quantities x and y be

$$y^2-a^2-x\surd(a^2-x^2)=0,$$

in place of $x\surd(a^2-x^2)$ I write z and thence have two equations, $y^2-a^2-z=0$ and $a^2x^2-x^4-z^2=0$. The former of these as before will give $2\dot{y}y-\dot{z}=0$ for the relation of the speeds $\dot{y}$ and $\dot{z}$, and the latter $2a^2\dot{x}x-4\dot{x}x^3-2\dot{z}z=0$ or

$$(a^2\dot{x}x-2\dot{x}x^3)/z=\dot{z}$$

for the relation of the speeds $\dot{x}$ and $\dot{z}$. There will now, on suppression of $\dot{z}$, come $2y\dot{y}+(-a^2\dot{x}x+2\dot{x}x^3)/z=0$ and then, on restoring $x\surd(a^2-x^2)$ for z, we shall have $2\dot{y}y+(-a^2\dot{x}+2\dot{x}x^2)/\surd(a^2-x^2)=0$ as the relation which was sought between $\dot{x}$ and $\dot{y}$.

EXAMPLE 4. If $x^3-ay^2+by^3/(a+y)-x^2\surd(ay+x^2)=0$ expresses the relation between x and y, I put z for $by^3/(a+y)$ and v for $x^2\surd(ay+x^2)$ and thence obtain the three equations $x^3-ay^2+z-v=0$, $az+yz-by^3=0$ and $ax^4y+x^6-v^2=0$. The first gives $3\dot{x}x^2-2a\dot{y}y+\dot{z}-\dot{v}=0$, the second $a\dot{z}+\dot{z}y+\dot{y}z-3b\dot{y}y^2=0$ and the third $4a\dot{x}x^3y+6\dot{x}x^5+a\dot{y}x^4-2\dot{v}v=0$ for the relations of the speeds $\dot{v}$, $\dot{x}$, $\dot{y}$ and $\dot{z}$. Of these, indeed, the values of $\dot{z}$ and $\dot{v}$ found by the second and third (namely, $(3b\dot{y}y^2-\dot{y}z)/(a+y)$ in place of $\dot{z}$ and $(4a\dot{x}x^3y+6\dot{x}x^5+a\dot{y}x^4)/2v$ in place of $\dot{v}$) I substitute in the first and there arises

$$3\dot{x}x^2-2a\dot{y}y+(3b\dot{y}y^2-\dot{y}z)/(a+y)-(4a\dot{x}x^3y+6\dot{x}x^5+a\dot{y}x^4)/2v=0.$$

And when the values $by^3/(a+y)$ and $x^2\surd(ay+x^2)$ are restored instead of z and v there comes the equation sought

$$3\dot{x}x^2-2a\dot{y}y+\frac{3ab\dot{y}y^2+2b\dot{y}y^3}{a^2+2ay+y^2}-\frac{4a\dot{x}xy+6\dot{x}x^3+a\dot{y}x^2}{2\surd(ay+x^2)}=0$$

by which the relation of the speeds $\dot{x}$ and $\dot{y}$ is designated.

(89) This replaces 'intricatæ' (intricate). Compare note (21).

(90) Newton first wrote 'mutabiles' (changeable). An equivalent change is made below: see note (93).

(91) A revision of the corresponding paragraph (I: 411) in the 1666 fluxional tract.

Quo pacto in alijs casibus operandum est, quemadmodū cùm in æquatione propositâ reperiuntur surdi denominatores, radicales cubicæ, radicales intra
||[20] radicales ut $\sqrt{ax+\sqrt{aa-xx}}$ || aut alij ejusmodi perplexi termini, ex his credo manifestum esse. Quinimò si in æquatione quantitates involvantur quæ nullâ ratione geometricâ determinari et exprimi possunt, quales sunt areæ vel longitudines curvarum: tamen relationes fluxionum haud secus investigantur, prout in exemplo sequente constabit.[92]

Præparatio in Exemplum 5.

Pone BD ordinatam esse in angulo recto ad AB et quod ADH sit curva quæ per relationem inter AB et BD æquatione qualibet exhibitam definitur. AB verò dicatur x et curvæ area ADB ad unitatem applicata dicatur z. Dein erige perpendiculum AC æquale unitati et per C duc CE parallelam AB et occurrentem BD in E, et concipiendo has duas superficies ADB et $ACEB$ genitas esse per motum rectæ BED, manifestum erit quòd earum fluxiones (hoc est fluxiones quantitatum $1\times z$ et $1\times x$, sive quantitatum z et x) sunt inter se ut BD & BE lineæ generantes. Est ergo $r.m::BD.BE$ sive 1, adeoq; $r=m\times BD$. Et hinc fit quod z in æquatione quâlibet designante relationem inter x et aliam quamvis fluentem[93] quantitatem y involvi potest, et tamen fluxionum m et n relatio nihil[o] minùs inveniri.

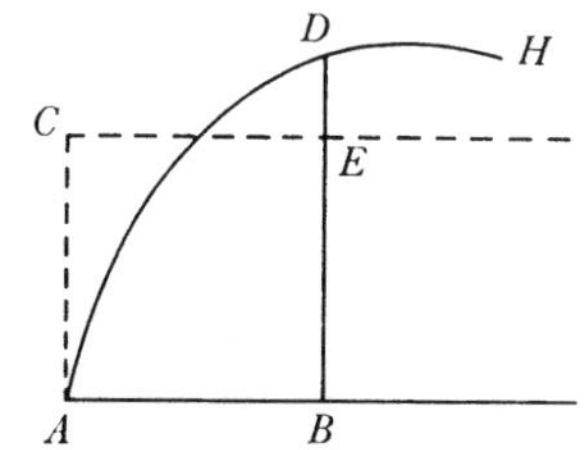

Exemplum 5. Quemadmodum si proponitur $zz+axz-y^4=0$ pro designanda relatione inter x et y, ut et $\sqrt{ax-xx}=BD$ pro curvâ determinandâ, quæ proin erit circulus: æquatio $zz+axz-y^4=0$ sicut in præcedentibus dabit $2rz+arx+amz-4ny^3=0$, pro[94] relatione celeritatum m, n, et r. Et præterea cùm sit $r=m\times BD$ sive $=m\sqrt{ax-xx}$, pro eo substitue hunc valorē, et orietur $\overline{2mz+amx\sqrt{ax-xx}}+amz-4ny^3=0$ æquatio definiens relationem celeritatum m et n.

Demonstratio.

[95]Fluentium quantitatum momenta (i.e. earum partes indefinitè parvæ, quarum additamento per singula temporis indefinita parva spatia augentur) sunt ut fluendi celeritates. Quare si cujusvis ut x momentum per factum ex ejus

(92) Compare the equivalent passages in the 1666 tract (I: 412) and the *De Analysi* (II: 242–4).

(93) This replaces 'mutabilem' (changeable). Compare note (90).

(94) In revision Newton here omitted the unnecessary gerundive 'determinanda' (determining).

How in other cases you must work (for example when there are found in the proposed equation surd denominators, radicals within radicals, such as $\surd[ax+\surd(a^2-x^2)]$, or other involved terms of the kind) is, I believe, clear from these. To be sure, even if quantities be involved in an equation which cannot be determined and expressed by any geometrical technique, such as the areas and lengths of curves, the relations of the fluxions are still to be investigated the same way, as the following example will establish.(92)

PREPARATION FOR EXAMPLE 5

Take BD an ordinate at right angles to AB and ADH to be a curve which is defined by a relationship between AB and BD displayed by any equation you please. Let AB, in fact, be called x and the area ADB under the curve divided by unity be called z. Then raise the perpendicular AC equal to unity and through C draw CE parallel to AB and meeting BD in E. By conceiving that these two surfaces ADB and $ACEB$ are generated by the motion of the straight line BED it will be evident that their fluxions (the fluxions of the quantities $1\times z$ and $1\times x$, that is, and so of the quantities z and x) are to one another as the generating lines BD and BE. Therefore it is $\dot{z}:\dot{x} = BD:BE$ (or 1), so that $\dot{z} = \dot{x}\times BD$. Hence it comes that z may be involved in any equation expressing the relation between x and any other fluent(93) quantity y, and yet the relation of the fluxions $\dot{x}$ and $\dot{y}$ may none the less be found.

EXAMPLE 5. If, for instance, $z^2+axz-y^4 = 0$ is put forward as the expression of the relationship between x and y, while $\surd[ax-x^2] = BD$ is to determine the curve (which will consequently be a circle): the equation $z^2+axz-y^4 = 0$ will, as in the preceding, yield $2\dot{z}z+a\dot{z}x+a\dot{x}z-4\dot{y}y^3 = 0$ for (94) the relation of the speeds $\dot{x}$, $\dot{y}$ and $\dot{z}$. Further, since $\dot{z}$ (or $\dot{x}\times BD$) $= \dot{x}\surd[ax-x^2]$, in its place I substitute the latter value and there will arise

$$(2\dot{x}z+a\dot{x}x)\surd[ax-x^2]+a\dot{x}z-4\dot{y}y^3 = 0$$

as the equation defining the relationship of the speeds $\dot{x}$ and $\dot{y}$.

DEMONSTRATION

(95)The moments of the fluent quantities (that is, their indefinitely small parts, by addition of which they increase during each infinitely small period of time) are as their speeds of flow. Wherefore if the moment of any particular one, say x,

(95) The whole of the following paragraph is a late addition, inserted no doubt when the widespread substitution of 'fluxio' for 'momentum' was made.

celeritate m et infinitè[96] parva quantitate o (i.e. per mo) designetur, cæterorum
||[21] v, y, z momenta per lo, no, ro designabuntur, siquidem lo, mo, no, et ro sunt || inter se ut l, m, n et r.

Jam cùm quantitatum fluentium (ut x et y) momenta (ut mo et no) sint additamenta infinitè parva quibus illæ quantitates per singula temporis infinite parva intervalla augentur, sequitur quod quantitates illæ x & y post quodlibet infinite parvum temporis intervallum futuræ sunt $x+mo$ et $y+no$. Et inde æquatio quæ relationem quantitatum fluentium ad omne tempus indifferenter designat, æque designabit relationem inter $x+mo$ et $y+no$, ac inter x et y: adeò ut $x+mo$ et $y+no$ pro quantitatibus istis vice x et y in dictam æquationem substitui possint.

Detur itaqꝫ quælibet æquatio $x^3-axx+axy-y^3=0$, et substitue $x+mo$ pro x et $y+no$ pro y et emerget

$$\left.\begin{array}{l} x^3+3moxx+3mmoox+m^3o^3 \\ -axx-2amox-ammoo \\ +axy+amoy+anox+amnoo \\ -y^3-3noyy-3nnooy-n^3o^3 \end{array}\right\}=0.$$

Jam ex Hypothesi sunt $x^3-axx+axy-y^3=0$, quibus deletis et reliquis terminis per o divisis restabunt

$$3mxx+3mmox+m^3oo-2amx-ammo+amy+anx+amno-3nyy-3nnoy-n^3oo=0.$$

Et insuper cùm o supponitur esse infinitè parvum, eo ut momenta quantitatum designare possit, termini per illud multiplicati respectu cæterorum nihil valebunt. Rejicio itaqꝫ et restat $3mxx-2amx+amy+anx-3nyy=0$, ut supra in Exempl: 1.

Hinc observare est[97] quòd termini non multiplicati per o semper evanescent, ut et illi multiplicati per o plusquam unius dimensionis: et quòd reliquorū terminorū per o divisorum ea semper erit forma quam juxta Regulam[98] habere debent. Id quod volui ostendere.

Ex hoc monstrato cætera quæ Regula involvit facilè consequentur; quemadmodum quòd in æquatione propositâ plures fluentes quantitates involvi possunt, et quòd termini non modò per numerum dimensionum quantitatum fluentium sed per quaslibet alias Arithmeticas progressiones multiplicari possunt dummodò in operatione juxta quamlibet fluentem quantitatem sit eadem terminorum differentia, et progressio secundum eundem cujusqꝫ

(96) The less precise equivalent 'indefinitè' (indefinitely) is cancelled. If we take the base variable of 'time' to be t, then (in anachronistic differential form) $o = dt$, while $l(=\dot{v})$, $m(=\dot{x})$, $n(=\dot{y})$ and $r(=\dot{z})$ are the respective instantaneous 'speeds' dv/dt, dx/dt, dy/dt and dz/dt. It follows that the 'moments' lo, mo, no, ro (or $\dot{v}o$, $\dot{x}o$, $\dot{y}o$, $\dot{z}o$ in dot-notation) are the

Plate I. Differentiation of a product: proof by geometrical fluxions (**I**, 2, §2).

be expressed by the product of its speed $\dot{x}$ and an infinitely[96] small quantity o (that is, by $\dot{x}o$), then the moments of the others, v, y, z, [...], will be expressed by $\dot{v}o$, $\dot{y}o$, $\dot{z}o$, [...] seeing that $\dot{v}o$, $\dot{x}o$, $\dot{y}o$ and $\dot{z}o$ are to one another as $\dot{v}$, $\dot{x}$, $\dot{y}$ and $\dot{z}$.

Now, since the moments (say, $\dot{x}o$ and $\dot{y}o$) of fluent quantities (x and y, say) are the infinitely small additions by which those quantities increase during each infinitely small interval of time, it follows that those quantities x and y after any infinitely small interval of time will become $x+\dot{x}o$ and $y+\dot{y}o$. Consequently, an equation which expresses a relationship of fluent quantities without variance at all times will express that relationship equally between $x+\dot{x}o$ and $y+\dot{y}o$ as between x and y; and so $x+\dot{x}o$ and $y+\dot{y}o$ may be substituted in place of the latter quantities, x and y, in the said equation.

Let there be given, accordingly, any equation $x^3-ax^2+axy-y^3=0$ and substitute $x+\dot{x}o$ in place of x and $y+\dot{y}o$ in place of y: there will emerge

$$(x^3+3\dot{x}ox^2+3\dot{x}^2o^2x+\dot{x}^3o^3)-(ax^2+2a\dot{x}ox+a\dot{x}^2o^2)$$
$$+(axy+a\dot{x}oy+a\dot{y}ox+a\dot{x}\dot{y}o^2)-(y^3+3\dot{y}oy^2+3\dot{y}^2o^2y+\dot{y}^3o^3)=0.$$

Now by hypothesis $x^3-ax^2+axy-y^3=0$, and when these terms are erased and the rest divided by o there will remain

$$3\dot{x}x^2+3\dot{x}^2ox+\dot{x}^3o^2-2a\dot{x}x-a\dot{x}^2o+a\dot{x}y+a\dot{y}x+a\dot{x}\dot{y}o-3\dot{y}y^2-3\dot{y}^2oy-\dot{y}^3o^2=0.$$

But further, since o is supposed to be infinitely small so that it be able to express the moments of quantities, terms which have it as a factor will be equivalent to nothing in respect of the others. I therefore cast them out and there remains $3\dot{x}x^2-2a\dot{x}x+a\dot{x}y+a\dot{y}x-3\dot{y}y^2=0$, as in Example 1 above.

It is accordingly to be observed[97] that terms not multiplied by o will always vanish, as also those multiplied by o of more than one dimension; and that the remaining terms after division by o will always take on the form they should have according to the rule.[98] This is what I wanted to show.

From this demonstration the other implications of the rule easily ensue—for instance, that several fluent quantities may be involved in the equation propounded, and that its terms may be multiplied not only by the number of dimensions of the fluent quantities but also by any other arithmetical progressions at all so long as throughout the operation the same difference of terms be kept for each fluent quantity and the progression is arranged in the same

corresponding 'infinitely small' increments dv, dx, dy, dz of the variables v, x, y, z. Of course, for example, $lo/mo(=\dot{v}o/\dot{x}o)=dv/dx$.

(97) 'primò' (first) is deleted. The whole paragraph is a little variant Latin translation of the corresponding passage in the 1666 tract (I: 415).

(98) This replaces 'præcedentia' (preceding).

dimensionum ordinem disponatur.(99) Et his concessis quæ præterea in exemplis 3, 4, et 5 docentur, per se manifesta sunt.

||[22] ||Prob 2

Exposita quantitate fluente ad cujus momenta relatio momentorum alterius alicujus fluentis quantitatis datur, quantitatum inter se relationem invenire.

Rationis momentorum(100) quæsitæ quantitatis ad momenta quantitatis expositæ valorem (si liber sit ab assymmetria et non afficitur denominatore aliquo plurium terminorum)(101) multiplica per expositam quantitatem, dein terminum unumquemqꝫ sigillatim divide per proprium numerum dimensionum ejusdem quantitatis; et quod oritur valebit quantitate quæsitâ.(102)

Quemadmodum si exponatur x, et y quæratur: Rationis $\frac{n}{m}$ valorem in datâ quâlibet æquatione exhibitum duc in x, et unumquemqꝫ terminum divide per numerum dimensionum ejus. Dein pone $=y$.

Exemplum 1. Si detur $\frac{n}{m}=\frac{x}{a}$, duco $\frac{x}{a}$ in x et fit $\frac{xx}{a}$, ubi cùm x sit duarum dimensionum divido per 2 et fit $\frac{xx}{2a}$ quod pono æquale y.(103)

||[23] ||Prob 2

Exposita æquatione fluxiones quantitatum involvente, invenire relationem quantitatum inter se.

Solutio Particularis

Cum hoc Problema sit præcedentis conversum,(104) contrario modo solvi debet: Utpote terminos per m multiplicatos disponendo secundum dimensiones ipsius x, dividendoqꝫ per $\frac{m}{x}$ ac deinde per numerum dimensionum aut fortasse per aliam arithmeticam progressionem, Atqꝫ idem opus in terminis per l, n, vel r multiplicatis instituendo, et resultantium summam,(105) rejectis terminis redundantibus, ponendo æqualem nihilo.(106)

---

(99) Compare the 1666 tract's 'Generall Theorem for drawing Tang^nts to crooked lines' (I: 417).

(100) 'ut $\frac{n}{m}$' (as $\dot{y}/\dot{x}$) is cancelled.

(101) Newton first wrote 'dimensionum' (dimensions) but justly cancelled it: only the equation $n/m(=\dot{y}/\dot{x}) = a/x$ is, of all equations with a single power of x in the denominator, not resolved by the present algorithm, while even that is immediately integrable.

(102) Compare the corresponding passage in the 1666 tract (I: 403–4).

sequence as the succession of dimensions in each.(99) These being granted, the further directions in Examples 3, 4 and 5 are evident in themselves.

Problem 2

When a fluent quantity is exhibited, the relationship of whose moments to those of some other fluent quantity is given, to find the relation of the quantities to one another.

Multiply the value of the ratio of the moments(100) of the quantity sought to the moments of the exhibited quantity (so long as it is free from irrationals and not affected with some denominator of several terms)(101) by the exhibited quantity, then divide each term individually by its own number of dimensions in this same quantity: what results will be the value of the quantity sought.(102)

For instance, if x be exhibited and y sought, multiply the value of the ratio $\dot{y}/\dot{x}$, displayed in any given equation, by x and then divide each term by the number of its dimensions, setting the result equal to y.

Example 1. If there be given $\dot{y}/\dot{x} = x/a$, I multiply x/a by x and there comes x^2/a. Here since x is of two dimensions I divide by 2 and there comes $x^2/2a$, which I set equal to y.(103)

Problem 2

When an equation involving the fluxions of quantities is exhibited, to determine the relation of the quantities one to another.

Particular solution

Since this problem is the converse of the preceding, it ought to be resolved the contrary way: namely, by arranging the terms multiplied by $\dot{x}$ according to the dimensions of x and dividing by $\dot{x}/x$ and then by the number of dimensions or perhaps another arithmetic progression, by carrying out the same operation in the terms multiplied by $\dot{v}$, $\dot{y}$ or $\dot{z}$, and, with redundant terms rejected, setting the total(105) of the resulting terms equal to nothing.(106)

(103) This first, cancelled draft of Problem 2 is reproduced for its intrinsic interest. The integration rule given corresponds to the 'Solutio particularis' in the revised version following: as yet there is no indication of any intended 'Solutio generalis' of the problem.

(104) The equivalent 'præcedenti contrarietur' is cancelled.

(105) A curious usage of 'summa' in the sense of the union of the partial sets of products. Newton, it will be noticed, attempts no explanation why 'redundant' terms in the sum of the products are to be omitted, but merely justifies their deletion by observing that the original fluxional equation will, 'si rectè operatus est', be obtained by differentiation 'juxta Problema 1'.

(106) In this 'particular solution' the fundamental presupposition is made that the fluxional equation is in quadrable form. Note too that no allowance is made for the additive constant of integration.

EXEMPL. Sic expositâ æquatione $3mxx-2amx+amy-3nyy+anx=0$; operor ad hunc modum,

Divido	$3mxx-2amx+amy$	Div:	$-3nyy+anx$
per $\frac{m}{x}$, & fit	$3x^3-2axx\ +axy$	per $\frac{n}{y}$, fit	$-3y^3\ +axy$
Dein divido per	3. 2. 1.	Div: per	3. 1.
et fit	$x^3-axx\ +axy.$	fit	$-y^3+axy.$

Et summa $x^3-axx+axy-y^3=0$ erit relatio desiderata quantitatum x et y. Ubi observandum venit quod etsi terminus axy bis resultavit, tamen non pono bis in hac summâ $x^3-axx+axy-y^3$ sed redundantem terminum rejicio. Et sic ubicunq; terminus aliquis bis resultat (aut sæpius si de pluribus fluentibus quantitatibus agitur,) semel tantùm in summâ terminorum scribo.

Sunt et aliæ circumstantiæ quas Artificis ingenio pro re nata observandas remitto;(107) nam supervacaneum esset his multa verba impendere, siquidem Problema non semper potest hoc artificio solvi. Addo tamen quod postquam Artifex relationem fluentium quantitatum hac methodo adeptus est, si juxta Prob: 1 potest regredi ad expositam æquationem fluxiones involventem, rectè operatus est; sin secùs, vitiosè. Sic in exemplo proposito, ubi æquationem

$$x^3-axx+axy-y^3=0$$

||[24] adeptus sum, si relatio inter m et x(108) ope primi Problematis || vicissim inde requiratur, obtinebitur æquatio exposita $3mxx-2amx+amy-3nyy+anx=0$. Unde constat æquationem $x^3-axx+axy-y^3=0$. rectè inventam fuisse. At si æquatio $m[x]-my+na=0$(109) exponeretur, et inde præscripta methodo elicerem $\frac{1}{2}xx-xy+ay=0$, pro relatione inter x et y, vitiosa foret operatio siquidem

(107) As will be shown in the sixth volume Newton elaborated some of these 'circumstances' in unpublished later chapters of his *De Quadratura Curvarum*. In particular, taking up a method privately communicated to him by Fatio de Duillier, he examined how a given first-order fluxional equation might be reduced to quadrable form by multiplying it into some term $\alpha x^i y^j$.

(108) Read 'n'.

(109) The first occurrence in a Newtonian context of Florimond de Beaune's celebrated differential equation. This first authentic inverse tangent problem had been communicated by its author to Descartes in late 1638 and the latter had, by February of the following year, reduced it (presumably by a simple geometrical argument rather than by an analytical variable change) to the condition that the curve defined by it in a rectangular Cartesian co-ordinate system should mark out a constant subtangent on the asymptote $x = y-a$. In further argument Descartes succeeded in establishing Mengolian bounds which, in effect, determine an analytical solution of the fluxional equation in the form

$$x/a = \log(x-y+a)-\log a,$$

and the corresponding curve to be logarithmic. (See I, 2, 6, §1: note (47). Compare also P. Tannery, *Mémoires Scientifiques*, **6** (Paris/Toulouse, 1926): 457–77: 'Pour l'histoire du

EXAMPLE. Thus when the equation $3\dot{x}x^2-2a\dot{x}x+a\dot{x}y-3\dot{y}y^2+a\dot{y}x=0$ is exhibited, I operate in this manner:

I divide	$3\dot{x}x^2-2a\dot{x}x+a\dot{x}y$	I divide	$-3\dot{y}y^2+a\dot{y}x$
by $\frac{\dot{x}}{x}$, making	$3x^3-2ax^2+axy$.	by $\frac{\dot{y}}{y}$, making	$-3y^3+axy$.
Then I divide by	3. 2. 1.	Then by	3. 1.
making	$x^3 \quad -ax^2+axy$	making	$-y^3+axy$.

The total $x^3-ax^2+axy-y^3=0$ will be the desired relationship of the quantities x and y. Here it should be noticed that, even though the term axy arose twice, I do not, however, put it twice into this total $x^3-ax^2+axy-y^3$ but lay aside one term as redundant. And so universally where some term arises twice (or more often if several fluent quantities are in question) I write it but once in the total of the terms.

There are other circumstances, too, which I leave to the notice of the skilled practitioner,(107) for it would be superfluous to expend a lot of words on this topic since the Problem cannot always be resolved by this practice. I may add, however, that if, after the practised mathematician has obtained a relation between the fluent quantities by this method, regress may be had by Problem 1 to the exhibited equation which involves their fluxions, the procedure has been correctly carried out, but otherwise faultily so. So in the example proposed, where I obtained the equation $x^3-ax^2+axy-y^3=0$, if in turn the relation between $\dot{x}$ and [$\dot{y}$] be thence required with the help of the first Problem, the exhibited equation $3\dot{x}x^2-2a\dot{x}x+a\dot{x}y-3\dot{y}y^2+a\dot{y}x=0$ will be got. This would consequently establish that the equation $x^3-ax^2+axy-y^3=0$ had been correctly found. But if the equation $\dot{x}x-\dot{x}y+\dot{y}a=0$(109) were to be exhibited and thence by the method prescribed I were to elicit $\frac{1}{2}x^2-xy+ay=0$ for the relation between x and y, the procedure would be faulty since therefrom in turn

problème inverse des tangentes'; and D. T. Whiteside, 'Patterns of Mathematical Thought in the later Seventeenth Century', *Archive for History of Exact Sciences*, **1** (1961): 368, note *.) Descartes' solution, contained in a letter of 20 February 1639 to de Beaune, was first published in 1667 by Clerselier in his *Lettres de Mr Descartes. Où il répond à plusieurs difficultez, qui luy ont esté proposées sur la Dioptrique, la Geometrie, & sur plusieurs autres sujets*, **3** (Paris, 1667): Lettre LXXI: 409–16. With the Latin edition of this work, *Epistolæ, partim ab Auctore Latino sermone conscriptæ, partim ex Gallico translatæ*... (which was published jointly in Amsterdam and London the following year), Newton was certainly familiar by about 1671–2 and it is tempting to suppose that he was aware of the Cartesian prehistory of his present fluxional equation

$$m/n(=\dot{x}/\dot{y}) = a/(y-x).$$

He nowhere, in fact, specifies his solution to it but in the context of the present tract he would no doubt have preferred to develop $y(=x+a+ke^{x/a})$ as an infinite series in x, perhaps with Descartes' initial condition $x=y=0$ (which determines $k=-a$).

exinde per Prob 1 vicissim produceretur $mx-my-nx+na=0$, quæ differt ab æquatione primo exposita.

Hæc itaq; perfunctoriè notata prætermittens, solutionem generalem aggredior.

||[25] ||PRÆPARATIO IN GENERALEM SOLUTIONEM.(110)

Et(111) imprimis observandum est quod in exposita æquatione symbola Fluxionum (cum sint quantitates diversi generis a quantitatibus quarum sunt fluxiones) in singulis terminis debent ad æque-multas dimensiones ascendere.(112) Et ubi res aliter se habet, alia alicujus fluentis quantitatis fluxio subintellegi debet esse unitas per quam termini depressiores toties multiplicantur ut in omnibus symbola fluxionum ad eundem dimensionum gradum ascendat. Quemadmodum si exponitur æquatio $m+mnx-axx=0$, tertiæ alicujus fluentis quantitatis ut z fluxio r subintelligi debet esse unitas per quam primus terminus m semel et ultimus axx bis multiplicetur ut fluxiones inibi ad æque-multas dimensiones ac in secundo termino mnx ascendant quasi exposita æquatio ex hac $mr+mnx-[a]rrxx=0$ derivatæ fuisset ponendo $r=1$. Et sic in æquatione $nx=yy$ debes imaginari m esse unitatem per quam terminus yy multiplicatur.

Æquationes autem in quibus duæ tantum sunt fluentes quantitates quæ ad æque multas dimensiones passim ascendunt, semper possunt ad talem formam reduci(113) ut ex una parte habeatur ratio fluxionum (velut $\frac{n}{m}$ vel $\frac{m}{n}$ vel $\frac{r}{m}$ &c)(114) et ex altera parte valor ejus rationis simplicibus terminis Algebraicis designatus; sicut hic videre est $\frac{n}{m}=2+2x-y$. Et ubi æquationibus præcedens particularis solutio non satisfacit, requiritur ut ad hanc formam reducas.

Quamobrem cum in illius rationis valore terminus aliquis a composita quantitate denominetur vel sit radicalis vel si ratio illa(115) sit æquationis radix affecta: reductio vel per divisionem, vel extractionem radicis, vel æquationis affectæ resolutionem institui debet,(116) prout in superioribus ostensum est.

(110) Significant variants in a cancelled first draft of the following 'preparation' are indicated below in appropriate footnotes.

(111) In first draft Newton here added 'in hunc finem' (to this end).

(112) For, since fluxions are derivatives with respect to an 'independent' variable of 'time', only their ratios are mathematically significant. Accordingly, when a given fluxional equation is not homogeneous with regard to the dimensions of the fluxions occurring in its individual terms, the variable of time is not independent but must be one or other of the component variables. For first-order fluxional equations the choice of 'time' variable is not restricted (as Newton remarks in the following sentence) but to give unique sense to equations of higher order it must be specified—hence the conventions in continental mathematics in the period from 1690 of 'suppositis elementis...*ds* æqualibus' (Jakob Bernoulli, 'Curvæ Dia-Causticæ ...aliaque nova his affinia', *Acta Eruditorum* (June 1693): 254) or 'en supposant *ds* constant'

by Problem 1 there would be produced $\dot{x}x-\dot{x}y-\dot{y}x+\dot{y}a=0$, an equation different from that first exhibited.

Accordingly, passing over these perfunctory remarks, I attack the general solution.

PREPARATION FOR THE GENERAL SOLUTION(110)

And(111) it must first be observed that in the exhibited equation the symbols for fluxions (since they are quantities different in kind [only] from those whose fluxions they are) ought, in each term, to rise to an equal number of dimensions.(112) And where this is not so, a further fluxion of some fluent quantity ought to be understood to be unity and the lower terms are to be multiplied by it as many times as needed to raise the symbols of fluxions everywhere to the same dimensional level. For instance, if the equation $\dot{x}+\dot{x}\dot{y}x-ax^2=0$ is exhibited, the fluxion $\dot{z}$ of some third fluent quantity such as z ought to be understood as a unit and the first term $\dot{x}$ should be multiplied by it once and the last one ax^2 twice so as to raise the fluxions in it to the same number of dimensions as in the second term $\dot{x}\dot{y}x$—as though, in effect, the exhibited equation had been derived from $\dot{x}\dot{z}+\dot{x}\dot{y}x-a\dot{z}^2x^2=0$ by setting $\dot{z}=1$. And so in the equation $\dot{y}x=y^2$ you ought to imagine that $\dot{x}$ is a unit by which the term y^2 is multiplied.

Now equations in which there are but two fluent quantities rising everywhere to an equal number of dimensions may always be reduced to such a form(113) that on one side there is had the ratio of the fluxions (as $\dot{y}/\dot{x}$ or $\dot{x}/\dot{y}$ or $\dot{z}/\dot{x}$ and so on(114) and on the other the value of that ratio expressed in simple algebraic terms, as may be seen in this example $\dot{y}/\dot{x}=2+2x-y$. And when the preceding particular solution does not satisfy the equations, it is requisite that you reduce them to this form.

When, therefore, in the value of that ratio some term has a compound term for its denominator or is a root or if that ratio(115) be the affected root of an equation, reduction either by division or root-extraction or the resolution of that affected equation should be undertaken after the manner of what was shown above.

(G. de L'Hospital to Johann Bernoulli, 16 May 1693 = *Der Briefwechsel von Johann Bernoulli*, **1** (Basel, 1955): 171) to convey that s is the base variable.

(113) In first draft Newton wrote 'convenit ut exposita æquatio ad talem formam semper reducatur' (it is convenient for the exhibited equation always to be reduced to such a form).

(114) Newton began to add a following phrase 'prout [fluxiones alijs atq; alijs symbolis exprimuntur]' (according to how the fluxions are expressed in different symbols) from his draft but cancelled it after entering its first word.

(115) Specified as '$\frac{n}{m}$' $(\dot{y}/\dot{x})$ in first draft.

(116) In his first version Newton wrote equivalently 'æquatio semper debet ad præscriptam formam reduci, idq; vel dividendo per compositum denominatorem, vel extrahendo radicem, vel affectam æquationem resolvendo'.

Quemadmodum si exponitur $na-nx-ma+mx-my=0$. Hæc imprimis reductione vel fit $\frac{n}{m}=1+\frac{y}{a-x}$, vel $\frac{m}{n}=\frac{a-x}{a-x+y}$. Et in priori casu si terminum $\frac{y}{a-x}$ a composita quantitate $a-x$ denominatum reduco ad infinitam seriem simplicium terminorum $\frac{y}{a}+\frac{xy}{aa}+\frac{xxy}{a^3}+\frac{x^3y}{a^4}$ &c dividendo numeratorem y per denominatorem $a-x$, obtinebo $\frac{n}{m}=1+\frac{y}{a}+\frac{xy}{aa}+\frac{xxy}{a^3}+\frac{x^3y}{a^4}$ &c cujus ope relatio inter x et y determinanda est.(117)

Sic exposita $nn=mn+mmxx$, sive $\frac{nn}{mm}=\frac{n}{m}+xx$, et ulteriori reductione $\frac{n}{m}=\frac{1}{2}\pm\sqrt{\frac{1}{4}+xx}$: Radicem quadraticam e terminis $\frac{1}{4}+xx$ extraho et obtineo
||[26] infinitam seriem $\frac{1}{2}+xx-x^4+2x^6-5x^8+14x^{10}$ &c || quam pro $\sqrt{\frac{1}{4}+xx}$ sub[s]tituendo prodit $\frac{n}{m}=1+xx-x^4+2x^6-5x^8$ &c. vel $\frac{n}{m}=-xx+x^4-2x^6+5x^8$ &c, prout $\sqrt{\frac{1}{4}+xx}$ additur vel subducitur a $\frac{1}{2}$.(118)

Atq; ita si exponitur $n^3+axmmn+aammn-m^3[x^3]-2m^3a^3=0$. sive

$$\frac{n^3}{m^3}+ax\frac{n}{m}+aa\frac{n}{m}-x^3-2a^3=0,$$

extraho radicem cubicè affectam et prodit $\frac{n}{m}=a-\frac{x}{4}+\frac{xx}{64a}+\frac{131x^3}{512aa}+\frac{509x^4}{16384a^3}$ &c. prout videre est ad pag 10.(119)

Cæterum hic observandum venit quod terminos solummodò pro compositis habeo qui ex parte fluentium quantitatum componuntur. Terminos ubi nulla est nisi ex parte datarum quantitatum compositio pro simplicibus habeo, siquidem ad simplices reduci possunt fingendo æquales esse alijs datis. Sic quantitates $\frac{ax+bx}{c}$, $\frac{x}{a+b}$, $\frac{bcc}{ax+bx}$, $\frac{b^4}{axx+bxx}$, $\sqrt{ax+bx}$ &c pro simplicibus habeo siquidem ad simplices $\frac{ex}{c}$, $\frac{x}{e}$, $\frac{bcc}{ex}$, $\frac{b^4}{exx}$, $\sqrt{ex}$ sive $e^{\frac{1}{2}}x^{\frac{1}{2}}$ &c reduci possunt fingendo esse $a+b=e$.(120)

Præterea quo fluentes quantitates a se invicem clarius distinguantur, fluxionem quæ in Numeratore Rationis disponitur, sive Antecedentem Rationis haud impropriè Relatam Quantitatem nominare possum, et alteram ad quam ref[er]tur, Correlatam; ut et fluentes Quantitates ijsdem respectivè nominibus insignire. Et quo sequentia promptiùs intelligantur, possis imaginari Correl[a]-

(117) The series solution is given on Newton's page 31, Example 2 (see note (149)).

(118) The equation $n/m(=\dot{y}/\dot{x}) = \frac{1}{2}(1\pm\sqrt{[1+4x^2]})$ may be integrated at once, without Newtonian restriction to $|x|<1$, as $y=\frac{1}{8}(4x\pm[2x\sqrt{[1+4x^2]}+\log(2x+\sqrt{[1+4x^2]})])$.

For instance, if the equation $\dot{y}a-\dot{y}x-\dot{x}a+\dot{x}x-\dot{x}y=0$ is exhibited, in the first place this becomes on reduction either

$$\dot{y}/\dot{x}=1+y/(a-x) \quad \text{or} \quad \dot{x}/\dot{y}=(a-x)/(a-x+y).$$

In the former case if I reduce the term $y/(a-x)$ with the compound quantity $a-x$ in its denominator into the infinite series of simple terms

$$y/a+xy/a^2+x^2y/a^3+x^3y/a^4\ldots,$$

I shall obtain $\dot{y}/\dot{x}=1+y/a+xy/a^2+x^2y/a^3+x^3y/a^4\ldots$ with whose help the relation between x and y is to be determined.(117)

So when there is exhibited $\dot{y}^2=\dot{x}\dot{y}+\dot{x}^2x^2$ or $\dot{y}^2/\dot{x}^2=\dot{y}/\dot{x}+x^2$, and by further reduction $\dot{y}/\dot{x}=\frac{1}{2}\pm\sqrt{(\frac{1}{4}+x^2)}$, I extract the square root of the terms $\frac{1}{4}+x^2$, obtaining the infinite series $\frac{1}{2}+x^2-x^4+2x^6-5x^8+14x^{10}\ldots$. On substituting this in place of $\sqrt{(\frac{1}{4}+x^2)}$ there comes

$$\dot{y}/\dot{x}=1+x^2-x^4+2x^6-5x^8\ldots \quad \text{or} \quad \dot{y}/\dot{x}=-x^2+x^4-2x^6+5x^8\ldots$$

according as $\sqrt{(\frac{1}{4}+x^2)}$ is added to or subtracted from $\frac{1}{2}$.(118)

And so if there is exhibited $\dot{y}^3+ax\dot{x}^2\dot{y}+a^2\dot{x}^2\dot{y}-\dot{x}^3x^3-2\dot{x}^3a^3=0$ or

$$(\dot{y}/\dot{x})^3+ax(\dot{y}/\dot{x})+a^2(\dot{y}/\dot{x})-x^3-2a^3=0,$$

I extract the root of this affected cubic and there comes out

$$\frac{\dot{y}}{\dot{x}}=a-\frac{x}{4}+\frac{x^2}{64a}+\frac{131x^3}{512a^2}+\frac{509x^4}{16384a^3}\ldots,$$

as may be seen on page 10.(119)

But here it should be noticed that I take as compound only those terms which are compound with regard to fluent quantities. Terms in which the only compounds are of given quantities I take as simple, seeing that they may be reduced to simple form by supposing them equal to other given magnitudes. In this way I take as simple the quantities $(ax+bx)/c$, $x/(a+b)$, $bc^2/(ax+bx)$, $b^4/(ax^2+bx^2)$, $\sqrt{(ax+bx)}\ldots$ seeing that they may be reduced to the simple ones ex/c, x/e, bc^2/ex, b^4/ex^2, $\sqrt{(ex)}$ or $e^{\frac{1}{2}}x^{\frac{1}{2}}\ldots$ by supposing $a+b=e$.(120)

Furthermore, to distinguish the fluent quantities the more sharply one from another, I may name the fluxion which is arranged to be in the numerator of the ratio (or, in other words, the antecedent of the ratio) not inappropriately the 'related quantity' and the other to which it is related the 'correlate': the fluent quantities too may be characterized by the same names respectively. And to understand what follows more readily you might imagine the correlate quantity

(119) That is, when n/m $(=\dot{y}/\dot{x})$ replaces y.

(120) Horsley (*Opera*, **1** (note (60)): **416**) absent-mindedly retains Newton's Greek 'ϵ' here.

tam Quantitatem esse Tempus vel potiùs aliam quamvis æquabiliter fluentem quantitatem qua Tempus exponitur et mensuratur, et alterum sive Relatam Quantitatem esse spatium quod mobile utcunꝗ acceleratum vel retardatum in illo tempore transigit. Et quod Problematis intensio(121) est ut e celeritate motûs ad omne tempus datâ spatium in toto tempore transactum determinetur.

Cæterùm æquationes respectu hujus Problematis in tres ordines distingui convenit.

1. In quibus duæ quantitatum fluxiones et alterutra tantùm fluens quantitas involvuntur.

2. In quibus duæ involvuntur fluentes quantitates unà cum earum fluxionibus.

3. Quæ plures duabus quantitatum fluxionibus complectuntur.

Et his præmissis, Problematis confectionem secundum hosce tres casus aggrediar.

Solutionis Cas: 1.

Fluentem quantitatem, quam unicè æquatio complectitur suppone Correlatam esse, et æquatione perinde dispositâ, (hoc est faciendo ut ex una parte habeatur fluxionis alterius ad hujus fluxionem Ratio,(122) et valor ejus in simplicibus terminis ex altera) multiplica valorem Rationis Fluxionum per Correlatam Quantitatem, dein singulos ejus terminos divide per numerum dimensionum quibus illa Quantitas inibi afficitur, et quod oritur valebit altera Fluenti Quantitate.

Sic expositâ $nn=mn+mmxx$, suppono x esse Correlatam(123) Quantitatem, et æquatione perinde reductâ habebitur(124) $\frac{n}{m}=1+xx-x^4+2x^6$ &c. Jam duco valorem in x et oritur $x+x^3-x^5+2x^7$ &c quos terminos sigillatim per numerum dimensionum divido et exitum $x+\frac{1}{3}x^3-\frac{1}{5}x^5+\frac{2}{7}x^7$ &c pono $=y$. Et isthac æquatione desiderata relatio inter x et y determinatur.

Sic habitâ $\frac{n}{m}=a-\frac{x}{4}+\frac{xx}{64a}+\frac{131x^3}{512aa}$ &c,(125) prodibit $y=ax-\frac{xx}{8}+\frac{x^3}{192a}+\frac{131x^4}{2048aa}$ &c pro determinanda relatione inter x et y.

||[27] || Et sic æquatio $\frac{n}{m}=\frac{1}{x^3}-\frac{1}{xx}+\frac{a}{x^{\frac{1}{2}}}-x^{\frac{1}{2}}+x^{\frac{3}{2}}$ dat $y=-\frac{1}{2xx}+\frac{1}{x}+2ax^{\frac{1}{2}}-\frac{2}{3}x^{\frac{3}{2}}+\frac{2}{5}x^{\frac{5}{2}}$. Nam valorem $\frac{n}{m}$ duc in x, et fit $\frac{1}{xx}-\frac{1}{x}+ax^{\frac{1}{2}}-x^{\frac{3}{2}}+x^{\frac{5}{2}}$, sive $x^{-2}-x^{-1}+ax^{\frac{1}{2}}-x^{\frac{3}{2}}+x^{\frac{5}{2}}$. Quibus terminis per numerum dimensionum divisis emergit valor assignatus y.

---

(121) Newton's variant on 'intentio'. No reference to latitude of forms is intended or extended!

(122) Changed from 'Relatio' (relation).

(123) 'fluentem' (fluent) is cancelled.

to be time (or rather any other uniformly flowing quantity by which time is expressed and measured) and the other or related quantity to be the space covered in that time by a moving body howsoever accelerated or retarded; and that the intention of the Problem is, given the speed of motion at all times, to determine the space covered in the whole time.

With respect to this problem, however, it is convenient to distinguish equations into three groups:

1. Those in which two fluxions together with one only of their fluent quantities are involved.
2. Those in which two fluent quantities are involved along with their fluxions.
3. Those embracing more than two fluxions.

On these premisses I shall now set about completing the Problem in accordance with these three cases.

The Solution: Case 1

Suppose the fluent quantity uniquely comprehended by the equation to be the correlate, and when the equation has been likewise arranged (by effecting, namely, that on its one side be had the ratio(122) of the second fluxion to that of this correlate and on the other its value in simple terms) multiply the value of the ratio of the fluxions by the correlate quantity and then divide its terms singly by the number of dimensions with which that quantity is therein endowed: what arises will be the value of the other fluent quantity.

So when $\dot{y}^2 = \dot{x}\dot{y} + \dot{x}^2x^2$ is exhibited, I take x to be the correlate(123) quantity, and when the equation has been correspondingly arranged there will be had $\dot{y}/\dot{x} = 1+x^2-x^4+2x^6 \ldots$.(124) I now multiply this value by x and there arises $x+x^3-x^5+2x^7 \ldots$, then divide its terms individually by the number of their dimensions and set the result $x+\frac{1}{3}x^3-\frac{1}{5}x^5+\frac{2}{7}x^7 \ldots$ equal to y. By that equation the desired relation between x and y is determined.

So when there is had $\dfrac{\dot{y}}{\dot{x}} = a-\dfrac{x}{4}+\dfrac{x^2}{64a}+\dfrac{131x^3}{512a^2}\ldots$,(125) there comes

$$y = ax-\frac{x^2}{8}+\frac{x^3}{192a}+\frac{131x^4}{2048a^2}\ldots$$

for determining the relation between x and y.

And thus the equation $\dot{y}/\dot{x} = 1/x^3-1/x^2+a/x^{\frac{1}{2}}-x^{\frac{1}{2}}+x^{\frac{3}{2}}$ gives

$$y = -1/2x^2+1/x+2ax^{\frac{1}{2}}-\tfrac{2}{3}x^{\frac{3}{2}}+\tfrac{2}{5}x^{\frac{5}{2}}.$$

For multiply the value of $\dot{y}/\dot{x}$ by x and it becomes $1/x^2-1/x+ax^{\frac{1}{2}}-x^{\frac{3}{2}}+x^{\frac{5}{2}}$ or $x^{-2}-x^{-1}+ax^{\frac{1}{2}}-x^{\frac{3}{2}}+x^{\frac{5}{2}}$. And on division of these terms by the number of their dimensions there emerges the value of y assigned.

(124) On taking the radical sign positive, that is, $n/m(=\dot{y}/\dot{x}) = \frac{1}{2}(1+\surd[1+4x^2])$. Compare note (118).

(125) Compare note (119).

Ad eundem modum æquatio $\frac{m}{n}=\frac{2bbc}{\sqrt{ay^3}}+\frac{3yy}{a+b}+\sqrt{by+cy}$ dat

$$x=-\frac{4bbc}{\sqrt{ay}}+\frac{y^3}{a+b}+\tfrac{2}{3}\sqrt{by^3+cy^3}.$$

Nam valore $\frac{m}{n}$ ducto in y, oritur $\frac{2bbc}{\sqrt{ay}}+\frac{3y^3}{a+b}+\sqrt{by^3+cy^3}$ sive

$$2bbca^{\frac{1}{2}}y^{\frac{1}{2}}{}^{(126)}+\frac{3}{a+b}y^3+\sqrt{b+c}\times y^{\frac{3}{2}}.$$

Et inde prodit valor x, dividendo per numerum dimensionum cujusq; termini.

Atq; ita $\frac{n}{r}=z^{\frac{2}{3}}$ dat $y=\frac{3}{5}z^{\frac{5}{3}}$. Et $\frac{n}{m}=\frac{ab}{cx^{\frac{1}{3}}}$ dat $y=\frac{3abx^{\frac{2}{3}}}{2c}$. At æquatio $\frac{n}{m}=\frac{a}{x}$ dat $y=\frac{a}{0}$. Nam $\frac{a}{x}$ ductum in x fit a, quo per numerum dimensionum (qui nullus est) diviso prodit $\frac{a}{0}$ quanti[t]as infinita pro valore y.[127]

Quamobrem siquando consimilis terminus (cujus denominator involvit[128] Correlatam Quantitatem unius tantùm dimensionis) in valore $\frac{n}{m}$ reperiatur,[129] pro Correlata Quantitate substitue summam vel differentiam inter eandem et aliam quamvis datam quantitatem pro arbitrio assumptam. Nam quantitatum[130] fluentium juxta prodeuntem æquationem eadem erit inter se fluendi relatio ac juxta æquationem primò expositam; et infinita quantitas Relata[130] hoc pacto parte infinitâ diminuetur et evadet finita, sed terminis tamen numero infinitis constans.[131]

Æquatione itaq; $\frac{n}{m}=\frac{a}{x}$ expositâ, si pro x scribam $b+x$, quantitatem b pro lubitu assumens; prodibit $\frac{n}{m}=\frac{a}{b+x}$; factâq; divisione, $\frac{n}{m}=\frac{a}{b}-\frac{ax}{bb}+\frac{axx}{b^3}-\frac{ax^{[3]}}{b^4}$ &c. Et inde Regula ut in superioribus dabit $y=\frac{ax}{b}-\frac{axx}{2bb}+\frac{ax^3}{3b^3}-\frac{ax^4}{4b^4}$ &c relationem inter x et y.[132]

Sic etiam habitâ æquatione $\frac{n}{m}=\frac{2}{x}+3-xx$, si $\left(\text{propter terminum } \frac{2}{x}\right)$ scribam

(126) Read '$2bbca^{-\frac{1}{2}}y^{-\frac{1}{2}}$'.

(127) Compare the equivalent passages in the 1666 tract (I: 403) and the *De Analysi* (II: 208).

(128) The following phrase replaces 'x', its present equal.

(129) A first cancelled continuation reads 'quantitas x debet augeri vel minui per datam quamvis quantitatem. Veluti substituendo $z+$vel$-$quavis data quantitate s pro x, ut et ejus fluxionis symbol[um] r pro m' (the quantity x ought to be increased or diminished by a given quantity: for instance, by substituting $z\pm s$ in place of x, where s is a given quantity, together with the symbol $\dot{z}$ of its fluxion in place of $\dot{x}$).

In much the same way the equation $\frac{\dot{x}}{\dot{y}} = \frac{2b^2c}{\surd(ay^3)} + \frac{3y^2}{a+b} + \surd(by+cy)$ gives $x = -\frac{4b^2c}{\surd(ay)} + \frac{y^3}{a+b} + \frac{2}{3}\surd(by^3+cy^3)$. For on multiplication of the value of $\dot{x}/\dot{y}$ by y there arises $2b^2c/\surd(ay) + 3y^3/(a+b) + \surd(by^3+cy^3)$ or

$$2b^2c[a^{-\frac{1}{2}}y^{-\frac{1}{2}}] + 3y^3/(a+b) + y^{\frac{3}{2}}\surd(b+c).$$

Thence comes the value of x on dividing each term by the number of its dimensions.

And so $\dot{y}/\dot{z} = z^{\frac{2}{3}}$ gives $y = \frac{3}{5}z^{\frac{5}{3}}$. And $\dot{y}/\dot{x} = ab/cx^{\frac{1}{3}}$ gives $y = 3abx^{\frac{2}{3}}/2c$. But the equation $\dot{y}/\dot{x} = a/x$ gives $y = a/0$: for a/x multiplied by x makes a, and when this is divided by the number of its dimensions (which is zero) there is produced the infinite quantity $a/0$ for the value of y.(127)

If therefore a similar term (one whose denominator involves but one dimension of(128) the correlate quantity) should be found in the value of $\dot{y}/\dot{x}$,(129) in place of the correlate quantity substitute the sum of or difference between it and any other given quantity arbitrarily assumed. For the fluent quantities(130) will, by the resulting equation, have the same fluxional relationship to one another as by the equation first exhibited, while the infinitely great related quantity(130) will by this means be diminished by its infinite part, coming out finite though still consisting of infinitely many terms.(131)

Accordingly, when $\dot{y}/\dot{x} = a/x$ is exhibited, if in place of x I write $b+x$, assuming the quantity b at pleasure, there will come $\dot{y}/\dot{x} = a/(b+x)$ and, after division, $\dot{y}/\dot{x} = a/b - ax/b^2 + ax^2/b^3 - ax^3/b^4 \ldots$. As in the foregoing the rule will thence give $y = ax/b - ax^2/2b^2 + ax^3/3b^3 - ax^4/4b^4 \ldots$ as the relation between x and y.(132)

So also when the equation $\dot{y}/\dot{x} = 2/x + 3 - x^2$ is had, if (on account of the term

(130) Here 'x et y' (x and y) and 'y' respectively are cancelled.

(131) The improvement introduced by the variable substitution $x = z \pm s$ is, despite Newton's hopes to the contrary, no more than illusory since the zero lower-bound implicit in his integration rule is thereby converted into a non-zero one, $\mp s$. Note the extent to which Newton here goes to avoid introducing a logarithmic function, either as an analytical 'logarithmus' or a geometrical 'hyperbola-area'.

(132) Newton essentially 'avoids' the unwanted lower limit $a \log 0 = -\infty$ to

$$\int_0^x at^{-1}.dt = \int_{-b}^{x-b} a(b+z)^{-1}.dz$$

by confusing the latter with $\int_0^x a(b+z)^{-1}.dz$ (in the series expansion of which $b > z$ is, no doubt, implicitly assumed for convergence). See note (131); and compare James Gregory's geometrical discussion of the series expansion of $a\log(1+x/b)$ in his *Exercitationes Geometricæ* (London, 1668): 9–13: 'N. Mercatoris Quadratura Hyperboles Geometrice Demonstrata'. (Newton's library copy of the work is Trinity College. NQ.9.48³.)

$1+x$ pro x, emerget $\frac{n}{m}=\frac{2}{1+x}+2-2x-xx$ terminoꝗ $\frac{2}{1+x}$ in infinitam seriem
||[28] $2-2x+2xx-2x^3+2x^4$ &c || reducto erit $\frac{n}{m}=4-4x+xx-2x^3+2x^4$ &c. Adeoꝗ juxta Regulam obtinebitur $y=4x-2xx+\frac{1}{3}x^3-\frac{1}{2}x^4+\frac{2}{5}x^5$ &c.[(133)]

Atꝗ ita si proponitur $\frac{n}{m}=x^{-\frac{1}{2}}+x^{-1}-x^{\frac{1}{2}}$. Quia terminum x^{-1} $\left(\text{sive } \frac{1}{x}\right)$ inesse video, transmuto x: quemadmodum pro eo substituendo $1-x$, et oritur $\frac{n}{m}=\frac{1}{\sqrt{1-x}}+\frac{1}{1-x}-\sqrt{1-x}$. Terminus autem $\frac{1}{1-x}$ valet $1+x+xx+x^3$ &c; et $\sqrt{1-x}$ valet $1-\frac{1}{2}x-\frac{xx}{8}-\frac{x^3}{16}$ &c. adeoꝗ $\frac{1}{\sqrt{1-x}}$ sive $\frac{1}{1-\frac{1}{2}x-\frac{1}{8}xx-\frac{1}{16}x^3\ \&c}$[(134)] valet $1+\frac{1}{2}x+\frac{3}{8}xx+\frac{5}{8}x^3$ &c.[(135)] Quamobrem (valoribus hisce substitutis) erit $\frac{n}{m}=1+2x+\frac{3}{2}xx+\frac{27}{16}x^3$ &c.[(135)] Et inde per regulam fit $y=x+xx+\frac{1}{2}x^3+\frac{27}{64}x^4$ &c.[(135)] Et sic in alijs.

Hujusmodi etiam transmutatione fluentis quantitatis æquatio in alijs casibus nonnunquam commode reduci poterit. Quemadmodum si exponitur

$$\frac{n}{m}=\frac{ccx}{c^3-3ccx+3cxx-x^3}$$

pro x scribo $c-x$ et obtineo $\frac{n}{m}=\frac{c^3-ccx}{x^3}$ sive $=\frac{c^3}{x^3}-\frac{cc}{xx}$[(136)] et inde per Reg: $y=-\frac{c^3}{2xx}+\frac{cc}{x}$. At harum transmutationum usus in sequentibus magis elucescet.

In Casū 2 Præparatio.

Hæc itaꝗ de æquationibus involventibus unicam tantum fluentem quantitatem. Cum verò utraꝗ[(137)] involvitur, æquatio imprimis ad præscriptam formam redigenda est, efficiendo scilicet ut ex una parte habeatur Fluxionum ratio æqualis aggregato simplicium terminorum ex alterâ.[(138)]

---

(133) Compare note (131). Newton's series expansion, of course, represents

$$y = \int_1^{x+1} (2t^{-1}+3-t^2).dt = 2\log(1+x)+2x-x^2-\tfrac{1}{3}x^3$$

only for $|x| \leqslant 1$.

(134) It may appear a little surprising that Newton performs the square-root extraction and inversion separately but we should remember that nowhere in the present tract does he state (and much less justify) the binomial expansion of $(1-x)^{-\frac{1}{2}}$ which would allow him to achieve the same result by a single operation. In the event a trivial numerical error in the following inversion distorts his further argument.

(135) The coefficients of the last terms in these expansions should be $\frac{5}{16}$, $\frac{81}{8}$ and $\frac{11}{32}$ respectively. Compare the previous note.

$2/x$) I write $1+x$ in place of x, there will emerge $\dot{y}/\dot{x} = 2/(1+x)+2-2x-x^2$ and then, with the term $2/(1+x)$ reduced into the infinite series

$$2-2x+2x^2-2x^3+2x^4\ldots,$$

it will be $\dot{y}/\dot{x} = 4-4x+x^2-2x^3+2x^4\ldots$, and hence by the rule will be obtained $y = 4x-2x^2+\frac{1}{3}x^3-\frac{1}{2}x^4+\frac{2}{5}x^5\ldots$.(133)

And thus if $\dot{y}/\dot{x} = x^{-\frac{1}{2}}+x^{-1}-x^{\frac{1}{2}}$ is proposed, because I see that the term x^{-1} (or $1/x$) is present, I transform x: for instance, by substituting $1-x$ for it, when there arises $\dot{y}/\dot{x} = 1/\surd(1-x)+1/(1-x)-\surd(1-x)$. Now the value of the term $1/(1-x)$ is $1+x+x^2+x^3\ldots$, while that of $\surd(1-x)$ is $1-\frac{1}{2}x-\frac{1}{8}x^2-\frac{1}{16}x^3\ldots$, so that the value of $1/\surd(1-x)$ or $1/(1-\frac{1}{2}x-\frac{1}{8}x^2-\frac{1}{16}x^3\ldots)$(134) is

$$1+\tfrac{1}{2}x+\tfrac{3}{8}x^2+\tfrac{5}{8}x^3\ldots.^{(135)}$$

Consequently, when these values are substituted there will be

$$\dot{y}/\dot{x} = 1+2x+\tfrac{3}{2}x^2+\tfrac{27}{16}x^3\ldots,^{(135)}$$

from which there comes by the rule $y = x+x^2+\frac{1}{2}x^3+\frac{27}{64}x^4\ldots$.(135) And so in other instances.

An equation could advantageously be reduced by this kind of transformation of a fluent quantity in other cases also. For instance, if

$$\dot{y}/\dot{x} = c^2x/(c^3-3c^2x+3cx^2-x^3)$$

be exhibited, in place of x I write $c-x$ and obtain $\dot{y}/\dot{x} = (c^3-c^2x)/x^3$ or

$$c^3/x^3-c^2/x^{2(136)}$$

and thence by the rule $y = -c^3/2x^2+c^2/x$. But the use of these transformations will be better evident in what follows.

Case 2: Preparation

So much, accordingly, for equations involving but a single fluent quantity. When, in fact, both are involved,(137) the equation must in the first place be reduced to the prescribed form, specifically, by contriving that on one side be had the ratio of the fluxions equal to the aggregate of simple terms on the other.(138)

---

(136) Newton forgets that the substitution $x \to c-x$ changes the sign of $n/m(= \dot{y}/\dot{x})$. Here and in the following all signs on the right-hand side must be changed.

(137) An intended clarification 'exposita æquatione' (in the exhibited equation) has been cancelled.

(138) This is unnecessarily restrictive: all that is needed in preparation is to put the fluxional equation into integrable form. But Newton remains obsessed with the idea of finding the 'root' y as an (infinite) power series in x.

Quemadmodum si exponitur $na-nx-ma+mx-my=0$, eadem per debitam reductionem vel fiet $\frac{m}{n}=\frac{a-x}{a-x+y}$, vel $\frac{n}{m}=1+\frac{y}{a-x}$. Et in posteriori casu si $\frac{y}{a-x}$ ad infinitam seriem simplicium terminorum reducatur, emerget

$$\frac{n}{m}=1+\frac{y}{a}+\frac{xy}{aa}+\frac{xxy}{a^3}+\frac{x^3y}{a^4} \quad \&c.$$

cujus ope relatio inter x et y determinanda restat. Haud secus ad exemplar eorum quæ in priori casu tradita sunt possis vel radicales vel utcunꝗ affectas

||[29] || æquationes (siquando opus est) reducere.(139)

Et præterea siquæ sunt in æquationibus sic reductis fractiones quæ denominantur a fluenti quantitate, a denominatoribus istis liberari debent per transmutationem ejus fluentis quantitatis paulo ante commemoratam. Sic exposita æquatione $nax-mxy-maa=0$ sive $\frac{n}{m}=\frac{y}{a}+\frac{a}{x}$. propter terminum $\frac{a}{x}$ assumo b ad arbitrium et pro x vel scribo $b+x$ vel $b-x$ vel $x-b$. Quemadmodum si scribam $b+x$ fiet $\frac{n}{m}=\frac{y}{a}+\frac{a}{b+x}$. Adeóꝗ termino $\frac{a}{b+x}$ in infinitam seriem per divisionem redacto erit $\frac{n}{m}=\frac{y}{a}+\frac{a}{b}-\frac{ax}{bb}+\frac{axx}{b^3}-\frac{ax^3}{b^4}$ &c.(140)

Et ad eundem modum exposita æquatione $\frac{n}{m}=3y-2x+\frac{x}{y}-\frac{2y}{xx}$; si (propter terminos $\frac{x}{y}$ & $\frac{2y}{xx}$) scribam $1-y$ pro y et $1-x$ pro x, orietur

$$\frac{n}{m}=1-3y+2x+\frac{1-x}{1-y}+\frac{2y-2}{1-2x+xx}.$$

Terminus autem $\frac{1-x}{1-y}$ per infinitam divisionem dat

$$1-x+y-xy+yy-xyy+y^3-xy^3 \quad \&c$$

ac terminus $\frac{2y-2}{1-2x+xx}$ per similem divisionem(141) dat

$$2y-2+4xy-4x+6xxy-6xx+8x^3y-8x^3+10x^4y-10x^4 \quad \&c.$$

(139) This paragraph is presumably cancelled because it repeats almost exactly an earlier one on Newton's page 25. The example is taken up again as 'Ex: 2' on his page 31 below.

(140) Newton does not pursue the solution of this fluxional equation in the sequel, but (implicitly when $x < b$) he would derive the series expansion

$$y = k+ab^{-1}+\tfrac{1}{2}(ka^{-1}+b^{-1}-ab^{-2})\,x+\ldots,$$

where k is an arbitrary constant, by the procedure exemplified below. In hindsight we can appreciate the unnecessary restriction implicit in the substitution $x \to b+x$, for the original equation $n/m(=\dot{y}/\dot{x}) = y/a+a/x$ has the exact solution $y = ae^{x/a}\int x^{-1}e^{-x/a}.dx$, which may be expanded as a power series convergent for all real (non-zero) x.

Differential equations of this type were twenty years later to become the object of intense study after Johann Bernoulli resurrected the de Beaune equation $dy/dx = a/(y-x)$ (see

For instance, if there be exhibited $\dot{y}a-\dot{y}x-\dot{x}a+\dot{x}x-\dot{x}y=0$, this will become by due reduction either $\dot{x}/\dot{y}=(a-x)/(a-x+y)$ or $\dot{y}/\dot{x}=1+y/(a-x)$. And in the latter case if $y/(a-x)$ be reduced to an infinite series of simple terms, there will emerge $\dot{y}/\dot{x}=1+y/a+xy/a^2+x^2y/a^3+x^3y/a^4\ldots$ by whose help the relation between x and y remains to be determined. Closely on the style of what has been delivered in the previous case you might (whenever there is need) reduce either radicals or equations no matter how affected.(139)

Furthermore, if in the equations thus reduced there be any fractions which have a fluent quantity in their denominators, they should be freed from those denominators by the transformation of that fluent quantity mentioned a short while ago. So when the equation $\dot{y}ax-\dot{x}xy-\dot{x}a^2=0$, that is, $\dot{y}/\dot{x}=y/a+a/x$, is exhibited, because of its term a/x I assume b arbitrarily and in place of x write $b+x$, $b-x$ or $x-b$. For instance, should I write $b+x$, it will become

$$\dot{y}/\dot{x}=y/a+a/(b+x)$$

and hence, on reducing the term $a/(b+x)$ into an infinite series by division, will be $\dot{y}/\dot{x}=y/a+a/b-ax/b^2+ax^2/b^3-ax^3/b^4\ldots$.(140)

And after the same fashion if, when the equation $\dot{y}/\dot{x}=3y-2x+x/y-2y/x^2$ is exhibited, (because of the terms x/y and $2y/x^2$) I write $1-y$ instead of y and $1-x$ instead of x, there will arise $\dfrac{\dot{y}}{\dot{x}}=1-3y+2x+\dfrac{1-x}{1-y}+\dfrac{2y-2}{1-2x+x^2}$. Now the term $(1-x)/(1-y)$ by infinite division gives $1-x+y-xy+y^2-xy^2+y^3-xy^3\ldots$ and the term $(2y-2)/(1-2x+x^2)$ by a like division(141) yields

$$2y-2+4xy-4x+6x^2y-6x^2+8x^3y-8x^3+10x^4y-10x^4\ldots.$$

note (109)), solved it analytically—apparently by use of the integrating factor $e^{x/a}$—and then (about the beginning of 1692) posed the problem to his brother Jakob and to L'Hospital. The latter without delay published a solution ('Solution du Probleme que M. de Beaune proposa autrefois à M. Descartes et que l'on trouve dans la 79. de ses lettres, Tome 3. Par Mr G***', *Journal des Sçavans*, No. 34 (for 1 September 1692): 401–3) which Johann Bernoulli equally swiftly asserted to be 'mea...solutio...tecto nomine' ('Solutio Problematis Cartesio propositi a Dn. de Beaune...', *Acta Eruditorum* (May 1693): 234), thereby starting a long-lived squabble with L'Hospital whose finer details have only recently become known. (See *Der Briefwechsel von Johann Bernoulli*, **1** (Basel, 1955): its editor Otto Spiess summarizes the dispute over de Beaune's equation on pp. 141–3.) The 'generalized de Beaune' equation

$$dy/dx = py+qy^n$$

(which yields the present Newtonian instance on setting $p = 1/a$, $q = a/x$ and $n = 0$) was given its complete solution in 1695 by Jakob Bernoulli ('Problema Beaunianum universalius conceptum, sive Solutio Æquationis nupero Decembri propositæ: $ady = ypdx+by^nqdx$', *Acta Eruditorum* (July 1696): 332–4): specifically, the substitution $y^{1-n} = z$ reduces the equation to $dz/dx = (n-1)(pz+q)$ and this, by means of the integrating factor $e^{-\pi}$, is resolved as

$$z = e^{\pi}\int qe^{-\pi}.dx, \quad \text{where} \quad \pi = (n-1)\int p.dx.$$

(141) Compare note (134). Newton chooses to square and invert rather than to expand $(1-x^2)^{-2}$ directly as a binomial series.

Quare est

$$\frac{n}{m} = -3x + 3xy + yy - xyy + y^3 - xy^3 \;\&c$$
$$+6xxy - 6xx + 8x^3y - 8x^3 + 10x^4y - 10x^4 \;\&c.^{(142)}$$

REGULA.(143)

Æquatione cùm opus est sic præparata: terminos ordina juxta dimensiones fluentium quantitatum ponendo imprimis non affectos Relata Quantitate,(144) deinde affectos minima ejus dimensione, & sic deinceps. Terminos etiam in his singulis classibus juxta dimensiones alterius Correlatæ quantitatis pariter dispone, eosq; in prima classe (i.e. quos Relata Quantitas non afficit) scribe in serie collaterali dextrorsum pergente, et cæteros in seriebus descendentibus in sinistra columnâ prout indicant subsequentia Diagrammata. Opere sic instituto Primum sive depressissimum e terminis in prima classe duc in Correlatam
||[30] Quantitatem divideq; per numerum || dimensionum, et in Quotiente, pro initiali termino valoris Relatæ Quantitatis repone. Hunc deinde in æquationis terminos in sinistrâ columnâ dispositos pro Relata Quantitate substitue, et e terminis proximè depressissimis secundum Quotientis terminum eadem ratione quâ primum elicies. Et eâdem operatione sæpiùs repetitâ Quotientem ad arbitrium producere possis.(145) Sed res exemplo clariùs patebit.

Ex: 1. Exponatur æquatio $\frac{n}{m} = 1 - 3x + y + xx + xy^{(146)}$ cujus terminos $1 - 3x + xx$ non affectos Relata quantitate y vides in suprema serie collateraliter dispositos, cæterosq; y et xy in sinistrâ columna.

	$+1-3x\;+xx$
$+y$	$*\;\;+x\;\;-xx+\frac{1}{3}x^3\;\;-\frac{1}{6}x^4+\frac{1}{30}x^5$ [&c]
$+xy$	$*\;\;\;*\;\;+xx\;\;-x^3\;\;+\frac{1}{3}x^4\;\;-\frac{1}{6}x^5+\frac{1}{30}x^6$ [&c]
Summa	$1-2x\;\;+xx-\frac{2}{3}x^3\;\;+\frac{1}{6}x^4-\frac{4}{30}x^5$ [&c]

$$y = x - xx + \tfrac{1}{3}x^3 - \tfrac{1}{6}x^4 + \tfrac{1}{30}x^5 - \tfrac{1}{45}x^6 \;\&c$$

Et imprimis terminum initialem 1 duco in Correlatam quantitatem x fitq; x, quem per numerum dimensionum 1 divisum repono in subscripta Quotiente. Dein hoc termino pro y in marginalibus substituto, vice $+y$ et $+xy$ obtineo $+x$ et $+xx$, quos e regione dextrorsum scribens, ex omnibus excerpo depress[iss]imos terminos $-3x$ & $+x$ quorum aggregatum $-2x$ ductum in x

(142) This will be taken up again as 'Ex: 3' on Newton's pages 31–2 below.

(143) Namely, for expanding the 'root' of a given first-order fluxional equation as an infinite series. It will be evident that the method is little different from that used previously in determining the roots of literal equations.

(144) In the sequel Newton usually chooses, as before, x as the 'correlate' quantity and y as the 'related' variable. But see note (152).

(145) In his choice of first term as 'depressissimus' Newton does not here allow for the unavoidable degree of arbitrariness arising in indefinite integration, one which we now allow in distinguishing between particular and general solutions to a given first-order differential

Hence

$$\dot{y}/\dot{x} = -3x+3xy+y^2-xy^2+y^3-xy^3\ldots$$
$$+6x^2y-6x^2+8x^3y-8x^3+10x^4y-10x^4\ldots.^{(142)}$$

RULE(143)

After the equation has (when necessary) been thus prepared, order its terms according to the dimensions of the fluent quantities, setting down first those not affected by the related quantity,(144) then those affected with its least dimension, and so on. Likewise arrange the terms also in each of these groups according to the dimensions of the other, correlate quantity and write these in the first group (those not affected by the related quantity, that is) in a series side by side proceeding to the right, but the rest are to be entered in descending series in the column on the left, as the following diagrams indicate. When the work has thus been organized multiply the first or lowest of the terms in the first group by the correlate quantity, then divide by the number of dimensions and store the result in the quotient for the opening term of the value of the related quantity. Next, substitute this in the equation's terms arranged in the left-hand column in place of the related quantity, and from the next lowest terms you will elicit the second term of the quotient in the same way as the first. And by repeating the same operation again and again you should be able to extend the quotient as far as you wish.(145) But the procedure will be expounded more clearly by example.

EXAMPLE 1. Let the equation $\dot{y}/\dot{x} = 1-3x+y+x^2+xy$(146) be exhibited. You see those of its terms,

$$1-3x+x^2,$$

not affected by the related quantity y arranged side by side in the top series with the others, y and xy, in the column on the left.

	$1-3x+x^2$
y	$* \quad +x-x^2+\frac{1}{3}x^3-\frac{1}{6}x^4+\frac{1}{30}x^5\ldots$
$+xy$	$* \quad * +x^2 -x^3+\frac{1}{3}x^4 -\frac{1}{6}x^5+\frac{1}{30}x^6\ldots$
Total	$1-2x+x^2-\frac{2}{3}x^3+\frac{1}{6}x^4-\frac{4}{30}x^5\ldots$

$$y = x-x^2+\tfrac{1}{3}x^3-\tfrac{1}{6}x^4+\tfrac{1}{30}x^5-\tfrac{1}{45}x^6\ldots$$

In the first place I multiply the initial term 1 by the correlate quantity x, making x, and this divided by the number of dimensions 1 I store in the quotient written beneath. Then, having substituted this term for y in the marginal ones, instead of y and xy I obtain x and x^2, and, writing these in alongside on the right, from all the terms I pick out the lowest $-3x$ and x: their sum $-2x$ multiplied by x

equation. In the following examples, in fact, he invariably chooses a 'simplest' particular solution: that, namely, which lacks a constant term. The extension from the particular solution to the more general 'solutiones infinitis modis' is briefly adumbrated in a following paragraph on Newton's page 33, but the significance of this extension is not discussed nor is any attempt made to distinguish the separate features of the components of the general solution.

(146) A second instance of a generalized de Beaune equation (note (140)): here $p = 1+x$, $q = 1-3x+x^2$ and n is again zero.

fit $-2xx$, et per numerum dimensionum 2 divisum dat $-xx$ pro secundo termino valoris y in Quotiente. Hoc proinde termino ad complendum valorem y in marginalibus $+y$ et $+xy$ abscito, oriuntur præterea $-xx$ et $-x^3$ terminis priùs oriundis $+x$ et $+xx$ adnectendi. Quo facto iterum terminos proximè depressissimos $+xx$, $-xx$, et $+xx$ in unam summā xx colligo et inde ut priùs tertium terminum $\frac{1}{3}x^3$ in valore y reponendum elicio. Iterumq; $\frac{1}{3}x[^3]$ in
‖[31] ‖ marginalium terminorum valores adscito, e proxime depressissimis $\frac{1}{3}x^3$ et $-x^3$ in unum aggregatis elicio $-\frac{1}{6}x^4$ quartum terminum valoris y. Et sic in infinitum.(147)

Ex: 2. Ad eundem modum si relationem inter x et y, habita æquatione $\frac{n}{m}=1+\frac{y}{a}+\frac{xy}{aa}+\frac{xxy}{a^3}+\frac{x^3y}{a^4}$ &c cujus terminorum series infinite progredi subintelligitur, determinare oportet.(148) Pono 1 in capite reliquosq; terminos in sinistra. Et opus deinde prosequor pro more adjuncti diagrammatis. Ubi propositum est mihi elicere valorem y ad usq; sex dimensiones x, et eâ de causâ terminos omnes quos proposito nihil conducere prævideo, inter operandum missos facio, sicut innuit nota &c quam seriebus intercisis adnexui.(149)

	$+1$						
$+\frac{y}{a}$	*	$+\frac{x}{a}$	$+\frac{xx}{2aa}$	$+\frac{x^3}{2a^3}$	$+\frac{x^4}{2a^4}$	$+\frac{x^5}{2a^5}$	&c.
$+\frac{xy}{aa}$	*	*	$+\frac{xx}{aa}$	$+\frac{x^3}{2a^3}$	$+\frac{x^4}{2a^4}$	$+\frac{x^5}{2a^5}$	&c.
$+\frac{xxy}{a^3}$	*	*	*	$+\frac{x^3}{a^3}$	$+\frac{x^4}{2a^4}$	$+\frac{x^5}{2a^5}$	&c.
$+\frac{x^3y}{a^4}$	*	*	*	*	$+\frac{x^4}{a^4}$	$+\frac{x^5}{2a^5}$	&c.
$+\frac{x^4y}{a^5}$	*	*	*	*	*	$+\frac{x^5}{a^5}$	&c.
Summa	1	$+\frac{x}{a}$	$+\frac{3xx}{2aa}$	$+\frac{2x^3}{a^3}$	$+\frac{5x^4}{[2]a^4}$	$+\frac{3x^5}{a^5}$	[&c]

$$y=x+\frac{xx}{2a}+\frac{x^3}{2aa}+\frac{x^4}{2a^3}+\frac{x^5}{2a^4}+\frac{x^6}{2a^5}\ \&c.$$

(147) In exact terms Newton's present particular solution is

$$y = 4-x+e^{\pi}(6\int e^{-\pi}.dx-4), \quad \text{where} \quad \pi = \int(1+x).dx = x+\tfrac{1}{2}x^2.$$

Newton elaborates the general series solution on his page 33 below.

(148) Newton has cancelled a first continuation 'Operationem ex adjuncto diagrammate credo satis manifestam esse' (I believe the working to be sufficiently clear from the adjoining diagram). Evidently on further consideration he decided that his figure was not self-explanatory. The series expansion given is, of course, that derived on Newton's page 29 from the fluxional equation $n/m(=\dot{y}/\dot{x}) = 1+y/(a-x)$. See note (139).

makes $-2x^2$ and on division by the number of its dimensions 2 yields $-x^2$ for the second term of y's value in the quotient. Accordingly, when this term has been adopted to fill out the value of y in the marginals y and xy, there further arise $-x^2$ and $-x^3$ to be adjoined to the terms x and x^2 which previously originated. This done, I again gather the next lowest terms x^2, $-x^2$ and x^2 into a single total x^2 and thence as before elicit the third term $\frac{1}{3}x^3$ to be stored in the value of y. And once again, adopting $\frac{1}{3}x^3$ in the values of the marginal terms, from the aggregate of the next lowest terms $\frac{1}{3}x^3$ and $-x^3$ I elicit $-\frac{1}{6}x^4$ as the fourth term in the value of y. And so on indefinitely.(147)

EXAMPLE 2. In the same manner if, when the equation

$$\dot{y}/\dot{x} = 1+y/a+xy/a^2+x^2y/a^3+x^3y/a^4\ldots$$

be had (the series of its terms being understood to proceed indefinitely), the relation between x and y must be determined:(148) I set 1 at the head and the remaining terms on the left, and then I pursue the working in the manner of the adjoining diagram. Here I proposed to elicit the value of y only up to the sixth power of x, and for that reason lay aside during the operation all terms which, I foresee, are not relevant to that end: this is intimated by the symbol '...' which I have added to the series thus cut off.(149)

	1					
$\frac{y}{a}$	$*$	$+\frac{x}{a}$	$+\frac{x^2}{2a^2}$	$+\frac{x^3}{2a^3}$	$+\frac{x^4}{2a^4}$	$+\frac{x^5}{2a^5}\ldots$
$+\frac{xy}{a^2}$	$*$	$*$	$+\frac{x^2}{a^2}$	$+\frac{x^3}{2a^3}$	$+\frac{x^4}{2a^4}$	$+\frac{x^5}{2a^5}\ldots$
$+\frac{x^2y}{a^3}$	$*$	$*$	$*$	$+\frac{x^3}{a^3}$	$+\frac{x^4}{2a^4}$	$+\frac{x^5}{2a^5}\ldots$
$+\frac{x^3y}{a^4}$	$*$	$*$	$*$	$*$	$+\frac{x^4}{a^4}$	$+\frac{x^5}{2a^5}\ldots$
$+\frac{x^4y}{a^5}$	$*$	$*$	$*$	$*$	$*$	$+\frac{x^5}{a^5}\ldots$
Total	1	$+\frac{x}{a}$	$+\frac{3x^2}{2a^2}$	$+\frac{2x^3}{a^3}$	$+\frac{5x^4}{2a^4}$	$+\frac{3x^5}{a^5}\ldots$
$y =$	x	$+\frac{x^2}{2a}$	$+\frac{x^3}{2a^2}$	$+\frac{x^4}{2a^3}$	$+\frac{x^5}{2a^4}$	$+\frac{x^6}{2a^5}\ldots$

(149) An exemplary account of Newton's method of deriving the 'solutio particularis' as an infinite series, but it is unfortunate that he has not seen that his original equation

$$n/m(=\dot{y}/\dot{x}) = 1+y/(a-x)$$

can be put in the immediately quadrable form $(a-x+y)(-1+n/m)-yn/m = 0$, which yields $(a-x+y)^2-y^2 = k$ and so $y = \frac{1}{2}[k(a-x)^{-1}-a+x]$. The present series (convergent only for $|x| < a$) is the particular case of this when $k = a^2$. The general solution may also be derived by considering the fluxional equation as a generalized de Beaune equation (note (140)) for which $p = (a-x)^{-1}$, $q = 1$ and $n = 0$.

Ex: 3. Pari methodo si proponitur æquatio(150)

$$\frac{n}{[m]} = -3x + 3xy + yy - xyy + y^3 - xy^3 + y^4 - xy^4 \;\&\text{c} \; + 6xxy - 6xx + 8x^3y - 8x^3 + 10x^4y - 10x^4 + 12x^5y - 12x^5 \;\&\text{c}.$$

et valorem y ad usq; septem dimensiones x eruere institutum est, terminos, ut in adjuncto diagrammate, in ordinem redigo et operor sicut in præcedentibus hoc tantùm excepto quod cùm hic in sinistrâ columnâ y non tantùm unius sed etiam duarum ac trium dimensionum existit (vel etiam plurium prout valorem y
||[32] ultra gradum || x^7 extrahere statuam) subjicio quadratum et cubum valoris y

	$-3x$	$-6xx$	$-8x^3$	$-10x^4$	$-12x^5$	$-14x^6$ &c
$+3xy$	*	*	$-\frac{9}{2}x^3$	$-6x^4$	$-\frac{75}{8}x^5$	$-\frac{273}{20}x^6$ &c
$+6xxy$	*	*	*	$-9x^4$	$-12x^5$	$-\frac{75}{4}x^6$ &c
$+8x^3y$	*	*	*	*	$-12x^5$	$-16x^6$ &c
$+10x^4y$	*	*	*	*	*	$-15x^6$ &c
&c						
$+yy$	*	*	*	$+\frac{9}{4}x^4$	$+6x^5$	$+\frac{107}{8}x^6$ &c
$-xyy$	*	*	*	*	$-\frac{9}{4}x^5$	$-6x^6$ &c
$+y^3$	*	*	*	*	*	$-\frac{27}{8}x^6$ &c
&c						
Summa	$-3x$	$-6xx$	$-\frac{25}{2}x^3$	$-\frac{91}{4}x^4$	$-\frac{333}{8}x^5$	$-\frac{302}{5}x^6$ [&c.]

$y = -\frac{3}{2}xx - 2x^3 - \frac{25}{8}x^4 - \frac{91}{20}x^5 - \frac{111}{16}x^6 - \frac{302}{35}x^7$ &c.

$yy = +\frac{9}{4}x^4 + 6x^5 + \frac{107}{8}x^6$ &c.

$y^3 = -\frac{27}{8}x^6$ &c.

eatenus gradatim productū, ut cùm in valoribus marginaliū terminorū dextrorsum gradibus inscribuntur, termini tot dimensionum emergant quot ad sequentem operationem requiri percipio. Et hac methodo prodit tandem

$$y = -\tfrac{3}{2}xx - [2]x^3 - \tfrac{25}{8}x^4 \;\&\text{c}$$

æquatio desiderata. Qui valor cùm sit negativus, patet alterum e quantitatibus x et y decrescere dum altera increscit. Atq; idem pariter concludi debet cum fluxionum altera affirmativa est et altera negativa.(151)

Ex: 4. Haud secus cùm Relata quantitas(152) fractis dimensionibus afficitur possis valorem ejus extrahere. Veluti si proponitur

$$\frac{m}{n} = \tfrac{1}{2}y - 4yy + 2yx^{\frac{1}{2}} - \tfrac{4}{5}xx + 7y^{\frac{5}{2}} + 2y^3,$$

ubi x in termino $2yx^{\frac{1}{2}}$ (sive $2y\sqrt{x}$) fracta dimensione $\frac{1}{2}$ afficitur: Ejus $x^{\frac{1}{2}}$ valorem e valore x paulatim elicio (extrahendo nempe radicem quadraticam) sicut in inferiori parte diagrammatis videre est; eò ut in ma[r]ginalis termini $2yx^{\frac{1}{2}}$
||[33] || valorem gradatim transferri et inseri possit.

(150) Derived on Newton's page 29 by substituting $x \to 1-x$ and $y \to 1-y$ in

$$n/m (= \dot{y}/\dot{x}) = xy^{-1} - 2x + (3 - 2x^{-2})\,y.$$

EXAMPLE 3. In a similar way if there is proposed the equation[150]

$$\dot y/\dot x = -3x+3xy+y^2-xy^2+y^3-xy^3+y^4-xy^4\ldots+6x^2y-6x^2$$
$$+8x^3y-8x^3+10x^4y-10x^4+12x^5y-12x^5\ldots$$

and it is purposed to seek out the value of y up to the seventh power of x, I reduce the terms to order as in the adjoining diagram, and operate as in the preceding, except that now in the left-hand column y is not merely of one dimension but also of two or three (or even more should I accordingly decide to extract the value of y beyond the degree of x^7) and so I introduce the square and cube of the value of y extended step by step until, when they are entered by degrees in the values of the marginal terms on the right, there emerge terms of as many dimensions as I perceive to be necessary for the following operation. And by this approach there comes out at last

	$-3x$	$-6x^2$	$-8x^3$	$-10x^4$	$-12x^5$	$-14x^6\ldots$
$3xy$	$*$	$*$	$-\frac{9}{2}x^3$	$-6x^4$	$-\frac{75}{8}x^5$	$-\frac{273}{20}x^6\ldots$
$+6x^2y$	$*$	$*$	$*$	$-9x^4$	$-12x^5$	$-\frac{75}{4}x^6\ldots$
$+8x^3y$	$*$	$*$	$*$	$*$	$-12x^5$	$-16x^6\ldots$
$+10x^4y$	$*$	$*$	$*$	$*$	$*$	$-15x^6\ldots$
$\ldots$						
$+y^2$	$*$	$*$	$*$	$+\frac{9}{4}x^4$	$+6x^5$	$+\frac{107}{8}x^6\ldots$
$-xy^2$	$*$	$*$	$*$	$*$	$-\frac{9}{4}x^5$	$-6x^6\ldots$
$+y^3$	$*$	$*$	$*$	$*$	$*$	$-\frac{27}{8}x^6\ldots$
$\ldots$						
Total	$-3x$	$-6x^2$	$-\frac{25}{2}x^3$	$-\frac{91}{4}x^4$	$-\frac{333}{8}x^5$	$-\frac{302}{5}x^6\ldots$

$$y = -\tfrac{3}{2}x^2-2x^3-\tfrac{25}{8}x^4-\tfrac{91}{20}x^5-\tfrac{111}{16}x^6-\tfrac{302}{35}x^7\ldots$$
$$y^2 = +\tfrac{9}{4}x^4+6x^5+\tfrac{107}{8}x^6\ldots$$
$$y^3 = -\tfrac{27}{8}x^6\ldots$$

$$y = -\tfrac{3}{2}x^2-2x^3-\tfrac{25}{8}x^4\ldots,$$

the desired equation. Since this value is negative it is evident that one of the quantities x and y decreases while the other increases. And the same conclusion ought likewise to be drawn when one of the fluxions is positive and the other negative.[151]

EXAMPLE 4. When the related quantity[152] is affected with fractional powers you should be able to extract its value in a not dissimilar fashion. If, say, there is proposed $\dot x/\dot y = \frac{1}{2}y-4y^2+2yx^{\frac{1}{2}}-\frac{4}{5}x^2+7y^{\frac{5}{2}}+2y^3$, where in the term $2yx^{\frac{1}{2}}$ (that is, $2y\sqrt{x}$) x is affected with the fractional power $\frac{1}{2}$, from the value of x I elicit that of $x^{\frac{1}{2}}$ little by little (namely, by extracting its square root), as may be seen in the lower part of the diagram; it may consequently be transferred and inserted in the value of the marginal term $2yx^{\frac{1}{2}}$ by degrees.

(151) This last sentence is a marginal addition in the manuscript. The constant of integration (not here considered) may be large enough to disprove Newton's assertion regarding y.

(152) A following phrase 'indefinitè quæsita' (sought indefinitely) is cancelled, presumably because it is *de trop*. Note that in this example Newton chooses x as 'related' quantity and y as the 'correlate' one, so that x is to be expanded as an infinite series in powers of y.

	$+\frac{1}{2}y \quad \cdot \quad -4yy+7y^{\frac{5}{2}}+2y^3$
$2yx^{\frac{1}{2}}$	$\cdot \quad +yy \quad \cdot \quad -2y^3+4y^{\frac{7}{2}}-2y^4$ &c
$-\frac{4}{5}xx$	$\cdot \quad \cdot \quad \cdot \quad \cdot \quad \cdot \quad -\frac{1}{20}y^4$ &c
Summa	$+\frac{1}{2}y \quad \cdot \quad -3yy+7y^{\frac{5}{2}} \quad +4y^{\frac{7}{2}}-\frac{41}{20}y^4$ &c

$$x=+\tfrac{1}{4}yy \quad \cdot \quad -y^3+2y^{\frac{7}{2}} \quad +\tfrac{8}{9}y^{\frac{9}{2}}-\tfrac{41}{100}y^5 \text{ \&c.}$$
$$x^{\frac{1}{2}}=+\tfrac{1}{2}y \quad [\cdot] \quad -yy+2y^{\frac{5}{2}}-y^3 \text{ \&c.}$$
$$xx=\tfrac{1}{16}y^4 \text{ \&c.}$$

Et sic tandem adipiscor æquationem $x=\frac{1}{4}yy-y^3+2y^{\frac{7}{2}}+\frac{8}{9}y^{\frac{9}{2}}-\frac{41}{100}y^5$ &c qua x respectu y indefinitè determinatur.

Et sic in alijs quibuscunq casibus operari licet.

Cæterùm dixi hasce solutiones infinitis modis præstari posse.(153) Et hoc fiet si non tantùm initialem quantitatem supremæ seriei sed et aliam quamvis datam quantitatem pro primo termino Quotientis ad arbitrium assumas, ac deinde opereris ut in præcedentibus.(154) Sic in primo præcedentium exemplorum si pro primo termino valoris y assumas 1, et pro y in terminis marginalibus ($+y$ & $+xy$) substituas, reliquamq operationem (cujus specimen adjunxi) sicut in præcedentibus prosequaris, ipsius y alius exurget valor

$$1+2x+x^3+\tfrac{1}{4}x^4 \text{ \&c.}$$

	$+1-3x \quad +xx$
$+y$	$+1+2x \quad \cdot \quad +x^3+\frac{1}{4}x^4$ &c
$+xy$	$\cdot \quad +x+2xx \quad \cdot \quad +x^4$ &c
Summa	$+2 \quad +3xx \quad +x^3+\frac{5}{4}x^4$ &c

$$y=1+2x \quad +x^3+\tfrac{1}{4}x^4+\tfrac{1}{4}x^5 \text{ \&c.}$$

Et sic alius atq alius exurget assumendo 2, 3, vel alium quemvis numerum pro primo ejus termino. Vel si symbolum aliquod, ut a, pro illo termino indefinitè designando usurpes, eadem operandi methodo (quam hic etiam designatam habes) elicies tandem

$$y=a+x+ax-xx+axx$$
$$+\tfrac{1}{3}x^3+\tfrac{2}{3}ax^3 \text{ \&c.}^{(155)}$$

	$+1 \quad -3x \quad +xx$
$+y$	$+a \quad \begin{matrix} +x & -xx & +\frac{1}{3}x^3 \\ +ax & +axx & +\frac{2}{3}ax^3 \end{matrix}$ &c.
$+xy$	$\cdot \quad +ax \quad \begin{matrix} +xx & -x^3 \\ +axx & +ax^3 \end{matrix}$ &c.
Summa	$\begin{matrix} +1 & -2x & +xx & -\frac{2}{3}x^3 \\ +a & +2ax & +2axx & +\frac{5}{3}ax^3 \end{matrix}$ [&c]

$$y=a \begin{matrix} +x & -xx & +\frac{1}{3}x^3 & -\frac{1}{6}x^4 \\ +ax & +axx & +\frac{2}{3}ax^3 & +\frac{5}{12}ax^4 \end{matrix} \text{ \&c.}$$

Qua inventa possis pro a substitu[e]re 1, 2, 0, $\frac{1}{2}$, aut quemvis numerum, et sic relationem inter x et y modis infinitis obtinere.

(153) As a matter of fact Newton has previously nowhere stated this explicitly. In his 'regula' on page 29 he merely stated a way of systematically deriving a 'simple' particular solution without hinting that there exist an infinite number more. Compare note (145).

(154) 'æquationibus' (equations) is cancelled.

	$\frac{1}{2}y$	$\cdot$	$-4y^2$	$+7y^{\frac{5}{2}}$	$+2y^3$	$\cdot$	$\cdot$	$\ldots$
$2yx^{\frac{1}{2}}$	$\cdot$	$\cdot$	$+y^2$	$\cdot$	$-2y^3$	$+4y^{\frac{7}{2}}$	$-2y^4\ldots$	
$-\frac{4}{5}x^2$	$\cdot$	$\cdot$	$\cdot$	$\cdot$	$\cdot$	$\cdot$	$-\frac{1}{20}y^4\ldots$	
Total	$+\frac{1}{2}y$	$\cdot$	$-3y^2$	$+7y^{\frac{5}{2}}$	$\cdot$	$+4y^{\frac{7}{2}}$	$-\frac{41}{20}y^4\ldots$	

$$x = +\tfrac{1}{4}y^2 \;\cdot\; -y^3+2y^{\frac{7}{2}} \;\cdot\; +\tfrac{8}{9}y^{\frac{9}{2}}-\tfrac{41}{100}y^5\ldots$$
$$x^{\frac{1}{2}} = +\tfrac{1}{2}y \;\cdot\; -y^2+2y^{\frac{5}{2}} -y^3\ldots$$
$$x^2 = \tfrac{1}{16}y^4\ldots$$

And so at length I gain the equation $x = \frac{1}{4}y^2 - y^3 + 2y^{\frac{7}{2}} + \frac{8}{9}y^{\frac{9}{2}} - \frac{41}{100}y^5\ldots$, by which x is determined indefinitely with respect to y.

And you may work this way in any other cases.

I have said, however, that these solutions may be presented in an infinite number of ways.(153) This will come about if you assume at will not only the initial term of the top series but also any other given quantity for the first term of the quotient, and then operate as in the preceding.(154) So in the first of the preceding examples if you assume the first term of the value of y to be 1 and substitute it in place of y in the marginal terms, y and xy, and then complete the remainder of the working (a specimen of which I have adjoined) as in the preceding, there will arise therefrom a second value of y, $1+2x+x^3+\frac{1}{4}x^4\ldots$. And in this way other differing ones will arise by assuming 2, 3, or any other number you wish for its first term. Or should you employ some symbol, as a, for designating that term indefinitely, by the same method of operation (which here you also have represented) you will at length elicit

	$1-3x\ +x^2$
y	$+1+2x \quad \cdot \quad +x^3+\frac{1}{4}x^4\ldots$
$+xy$	$\cdot \quad +x+2x^2 \quad \cdot \quad +x^4\ldots$
Total	$+2x \quad \cdot \quad +3x^2+\ x^3+\frac{5}{4}x^4\ldots$

$$y = 1+2x \quad \cdot \quad +x^3+\tfrac{1}{4}x^4+\tfrac{1}{4}x^5\ldots$$

	$1 \quad -3x \quad +x^2$
y	$a+(1+a)\,x+(-1+a)\,x^2+(\frac{1}{3}+\frac{2}{3}a)\,x^3\ldots$
$+xy$	$\cdot \quad +ax+(1+a)\,x^2+(-1+a)\,x^3\ldots$
Total	$1+a+(-2+2a)\,x+(1+2a)\,x^2+(-\frac{2}{3}+\frac{5}{3}a)\,x^3\ldots$

$$y = a+(1+a)\,x+(-1+a)\,x^2+(\tfrac{1}{3}+\tfrac{2}{3}a)\,x^3+(-\tfrac{1}{6}+\tfrac{5}{12}a)\,x^4\ldots.$$

$$y = a+(1+a)\,x+(-1+a)\,x^2+(\tfrac{1}{3}+\tfrac{2}{3}a)\,x^3\ldots.$$ (155)

And when this is found you might substitute in place of a the number 1, 2, 0, $\frac{1}{2}$ or any other, and so obtain the relation between x and y in an infinity of ways.

(155) In exact terms (compare note (147)) the general solution of

$$n/m(=\dot{y}/\dot{x}) = (1+x)\,y+(1-3x+x^2)$$

is, where $\pi = x+\frac{1}{2}x^2$, $y = 4-x+e^{\pi}(6\int e^{-\pi}.dx-4+a)$. Newton's method derives the equivalent series expansion.

Et nota quod ubi quantitas elicienda afficitur fracta dimensione (ut in præcedentium exemplorum quarto vides) convenit plerumq; unitatem (vel
||[34] alium quemvis aptum || numerum) pro primo ejus termino adhibere; immò hoc necesse est ubi radix (ad fractæ illius dimensionis valorem obtinendum) propter negativum signum nequit alias extrahi, ut et ubi nulli sunt termini in prima sive capitali classe reponendi, ex quibus initialis ille terminus eliciatur.(156)

Sic tandem hoc molestissimum et omnium difficillimum Problema, ubi duæ tantùm fluentes quantitates una cum earum fluxionibus in æquatione comprehenduntur, absolvi. Sed præter generalem methodum qua omnes difficultates complexus sum sunt aliæ plerumq; contractiores quibus opus aliquando sublevari possit, et quarum aliqua specimina ex abundanti perstringere forte non erit ingratum.

1. Siquando itaq; quantitas elicienda sit alicubi negativæ dimensionis[,] non est absolutè necessarium ut æquatio propterea ad aliam formam reducatur. Sic enim expositâ æquatione $n^{(157)}=\frac{1}{y}-xx$ ubi y est unius negat[iv]æ dimensionis,

	•	•	$-xx$
$\frac{1}{y}$	1	$-x+\frac{3}{2}xx$ &c	
Summa	1	$-x+\frac{1}{2}xx$ [&c]	

$y=1+x-\frac{1}{2}x^2+\frac{1}{6}x^3$ &c

$\frac{1}{y}=1-x+\frac{3}{2}xx$ &c

possim equidem ad aliam formam reducere, veluti scribendo $1+y$ pro y, sed expeditior erit resolutio quam in annexo diagrammate habes, ubi assumpto 1 pro initio valoris y cæteros ejus terminos ut in præcedentibus extraho, et interea valorem $\frac{1}{y}$ exinde per divisionem paulatim institutam elicio et insero in valorem marginalis termini.

||[35] || 2. Neq; semper opus est ut alterius fluentis quantitatis dimensiones sint passim affirmativæ. Nam ex æquatione $n=3+2y-\frac{yy}{x}$, absq; termini $\frac{yy}{x}$ reductione præscriptâ emerget $y=3x-\frac{3}{2}xx-4x^3$ &c.(158) Et ex $n=[-]y+\frac{1}{x}-\frac{1}{xx}$

(156) To paraphrase Newton, if the given equation in Example 4 had been

$$m/n(=\dot{x}/\dot{y}) = -\tfrac{1}{2}y-4y^2+2yx^{\frac{1}{2}}\ \ldots,$$

yielding $x=-\frac{1}{4}y^2\ldots$ as the simple integral, we are advised to insert a unit constant of integration and so derive x as $1+\frac{3}{4}y^2\ \ldots$, yielding a corresponding real value for $x^{\frac{1}{2}}$.

(157) Here and in the following paragraphs the fluxion of x is taken to be unity so that $n(=\dot{y})$ becomes equivalent to the Leibnizian differential quotient dy/dx. Compare Newton's phrase on page 38 below 'quandoquidem m supponitur esse 1'.

(158) The last term should read '$+2x^3$ &c'. This would appear to be the first historical occurrence of a case of the general 'Riccati' equation $dy/dx = n/m$ (or $\dot{y}/\dot{x}$) $= A+By+Cy^2$, for the historical background to which see G. N. Watson, *A Treatise on the Theory of Bessel Functions* (Cambridge, 1922): 1–3, 85–94, and M. Cantor, *Geschichte der Mathematik*, **3** (Leipzig, 1898): 455–60. Jacopo Francesco, Count Riccati, himself merely publicly proposed for solution in

And note that when the quantity to be elicited is affected with a fractional power (as you see in the fourth of the preceding examples) it is usually convenient to make use of unity (or some other suitable number) for its first term; indeed, this is obligatory when, to obtain the value of that fractional power, a root cannot otherwise, on account of a negative sign, be extracted and also when there are no terms to be stored in the first or head group from which that initial term may be elicited.(156)

Thus at last I have disposed of this most troublesome and difficult problem of all when only two fluent quantities together with their fluxions are involved in an equation. Apart, however, from this general method, by which I have covered all the difficulties, there are other, usually shorter ones by which the working may sometimes be lightened, and it will not be undesirable perhaps to touch on some examples of these as an extra.

1. Accordingly, if the quantity to be elicited happen at some place in it to have a negative dimension, it is not absolutely necessary on that account to reduce the equation to another form. Thus, for instance, when the equation $\dot{y}^{(157)} = y^{-1} - x^2$ is exhibited, in which y is of a single negative power, I might indeed reduce it to another form (say by writing $1+y$ in place of y) but the resolution which you have in the attached diagram will be speedier. Here, assuming 1 for the opening of y's value, I extract its other terms as in the preceding and meanwhile elicit the value of y^{-1} little by little therefrom by division, inserting it in the value of the marginal term.

	$\bullet$	$\bullet \; -x^2$	
y^{-1}	1	$-x$	$+\frac{3}{2}x^2 \ldots$
Total	1	$-x$	$+\frac{1}{2}x^2 \ldots$

$$y = 1 + x - \tfrac{1}{2}x^2 + \tfrac{1}{6}x^3 \ldots$$
$$y^{-1} = 1 - x + \tfrac{3}{2}x^2 \ldots.$$

2. Nor is there always need for the dimensions of the other fluent quantity everywhere to be positive. For from the equation $\dot{y} = 3 + 2y - y^2/x$, without reducing the term y^2/x as prescribed, there will emerge $y = 3x - \frac{3}{2}x^2 - 4x^3 \ldots$.(158)

1723—but could not resolve—the simpler differential equation $dy/dx = ax^n + by^2$ (*Acta Eruditorum* (November 1723): 502–10) and to this Leonhard Euler's suggested name of 'æquatio Riccatiana' has adhered up to the present day. Daniel Bernoulli showed the following year that finite solutions exist when $n = -4m/(2m \pm 1)$, m integral (*Acta Eruditorum* (November 1725): 473–5). Johann Bernoulli, however, had already proposed a particular case of the simple Riccati equation in 1694: 'Esto proposita æquatio differentialis hæc $xxdx + yydx = aady$, quæ, an per separationem indeterminatarum construi possit, nondum tentavi' ('Modus generalis construendi omnes æquationes differentiales primi gradus', *Acta Eruditorum* (November 1694): 435–7, especially 436). His elder brother Jakob in 1702 succeeded in reducing this by the substitution $y = -u^{-1}du/dx$ to a second-order linear equation to which he was soon after able to produce a particular solution in the form of the quotient of two infinite series. (See his letters of 15 November 1702 and 3 October 1703 to Leibniz in C. I. Gerhardt, *Leibnizens Mathematische Schriften*, **3** (Halle, 1855): 65, 75.) Twenty years later in Italy the simple Riccati equation was a main topic in correspondence between Christian Goldbach and Nicholas II Bernoulli: in a letter of 11 September 1721 Goldbach independently proposed a particular series solution of the equation which is structurally identical with Newton's present method,

(opere ad modum annexi speciminis instit[ut]o) emerget $y=\frac{1}{x}$.(159)

	$-\frac{1}{xx}+\frac{1}{x}$	
$-y$	$\cdot$	$-\frac{1}{x}$
Summa	$-\frac{1}{xx}$	0

$$y=\frac{1}{x}.$$

Ubi obiter nota quod inter modos infinitos quibus quælibet æquatio resolvi potest sæpenumero contingit aliquos esse qui ad finitum valorem quantitatis eliciendæ sicut in allato specimine finiuntur, et quos haud difficile est invenire si pro primo valoris termino symbolum aliquod assumatur. Et resolutione peractâ consulatur de symboli illius quantitate qua valor elicitus evadat finitus.

3. Porro si valor y ex æquatione $n=\frac{y}{2x}+1-2x+\frac{1}{2}xx$ eliciendus sit, id sine aliqua reductione termini $\frac{y}{2x}$ non incommodè fiet fingendo (pro more Analytico) datum esse quod quæritur. Utpote pro primo termino valoris ejus effingo $2ex$ assumendo $2e$ pro numerali coefficiente quæ nondum innotescit. Et hunc $2ex$ pro y in termino marginali $\frac{y}{2x}$ substituens prodit e quem scribo ad dextram et summa $1+e$ dabit $x+ex$ pro eodem primo termino valoris y quem prius designaveram || termino $2ex$. Pono itaq; $2ex=x+ex$ et inde

	1	$-2x$	$+\frac{1}{2}xx$	
$\frac{y}{2x}$	e	$+fx$	$+gxx$	$+hx^3$
Summa	$+1$	$-2x$	$+\frac{1}{2}xx$	$+hx^3$
	$+e$	$+fx$	$+gxx$	

$$\text{Hypotheticè } y=2ex+2fxx+2gx^3+2hx^4$$
$$\text{Consequenter } y=\begin{matrix}+x & -xx & +\frac{1}{6}x^3 & \\ +ex & +\frac{1}{2}fxx & +\frac{1}{3}gx^3 & +\frac{1}{4}hx^4\end{matrix}$$

||[36] Revera $y=2x-\frac{4}{3}xx+\frac{1}{5}x^3$.

elicio $e=1$. Adeoq; valoris y primus terminus $(2ex)$ est $2x$. Ad eundem modum pro secundo termino designando effictum $2fx^2$ usurpo et inde tandem eruo $-\frac{2}{3}$ pro valore f, adeoq; $-\frac{4}{3}xx$ pro secundo termino. Et sic effictus g in tertio termino valebit $\frac{1}{10}$, at h in quarto valebit 0,(160) et proinde cum nullos præterea terminos superesse video, concludo opus finitum esse et y valere $2x-\frac{4}{3}xx+\frac{1}{5}x^3$ præcisè.(161)

while Nicholas writing back on 14 March 1722 gave his own version of Daniel Bernoulli's finite solution (P. H. Fuss, *Correspondance Mathématique et Physique de Quelques Célèbres Géomètres du XVIIIe Siècle*, **2** (St Petersburg, 1843): 106–8, 140–3). Euler would seem responsible both for propounding the generalized Riccati equation and for first attempting (with a wide degree of success) its systematic solution: in particular, in *Problema 9* of his 'De integratione æquationum differentialium' (*Novi Commentarii Academiæ Scientiarum Petropolitanæ*, **8** (1760/1 [1763]): 3–63, especially 32 = *Opera Omnia* (1) **22** (Zurich and Leipzig, 1936): 334–94, especially 364) he reduced the equation to a generalized de Beaune equation (note (140)) by the substitution $y = p+1/z$, where p is a given particular solution. In the present Newtonian case, for example, on taking $p = 3x-\frac{3}{2}x^2+2x^3\ldots$ as the particular solution we could derive the 'solutio generalis' on Eulerian lines as $y = p+e^{\pi}/[\int x^{-1}e^{\pi}.dx+k]$, where $\pi = 2\int(1-x^{-1}p).dx$.

And from $\dot{y} = -y + x^{-1} - x^{-2}$ (the work being arranged in the fashion of the attached specimen) there will emerge $y = x^{-1}$.(159)

	$-x^{-2}+x^{-1}$	
$-y$	•	$-x^{-1}$
Total	$-x^{-2}$	0

$y = x^{-1}$.

Note here by the way that among the infinity of ways in which any equation may be resolved it very often happens that there are some which terminate at a finite value of the quantity to be elicited, as in the example just adduced. These are not difficult to find if some symbol be assumed for the first term of the value and, when the resolution has been performed, you look out for a particular value of that symbol which renders the result elicited finite.

3. Further, if the value of y is to be elicited from the equation

$$\dot{y} = y/2x + 1 - 2x + \tfrac{1}{2}x^2,$$

that may be done not inconveniently without any reduction of the term $y/2x$ by (in the usual analytic manner) supposing given what is sought. As here I express the value of the first term by $2ex$, assuming $2e$ for the numerical coefficient which is not yet known. On substituting this value $2ex$ in place of y in the marginal term $y/2x$ there results e, which I write in on the right, and the sum $1+e$

		1	$-2x$	$+\frac{1}{2}x^2$	
$y/2x$		e	$+fx$	$+gx^2$	$+hx^3$
Total		$(1+e)+$	$(-2+f)\,x$	$+(\frac{1}{2}+g)\,x^2$	$+hx^3$
By hypothesis	$y =$	$2ex$	$+2fx^2$	$+2gx^3+$	$2hx^4$
By inference	$y =$	$(1+e)\,x+$	$(-1+\frac{1}{2}f)\,x^2+$	$(\frac{1}{6}+\frac{1}{3}g)\,x^3+$	$\frac{1}{4}hx^4$
So in fact	$y =$	$2x$	$-\frac{4}{3}x^2$	$+\frac{1}{5}x^3$.	

will yield $(1+e)\,x$ for the same first term of the value of y which I had previously designated by the term $2ex$. Accordingly, I set $2ex = (1+e)\,x$ and thence elicit $e = 1$, so that the first term $(2ex)$ of the value of y is $2x$. In the same way I employ the expression $2fx^2$ to designate the second term and thence at length derive $-\frac{2}{3}$ as the value of f, so that the second term is $-\frac{4}{3}x^2$. And so the expression g in the third term will have the value $\frac{1}{10}$, but in the fourth h will be zero.(160) Hence, seeing no further remaining terms, I infer that my work is finished and that the value of y is exactly $2x - \frac{4}{3}x^2 + \frac{1}{5}x^3$.(161) In

(159) The more general solution of this de Beaune equation (note (140)), where $p = -1$, $q = x^{-1} - x^{-2}$ and $n = 0$, is $y = x^{-1} + ke^{-x}$.

(160) Since by supposition $2h = \frac{1}{4}h$.

(161) This is, of course, only the particular case $k = 0$ of the general solution

$$y = kx^{\frac{1}{2}} + 2x - \tfrac{4}{3}x^2 + \tfrac{1}{5}x^2$$

of this generalized de Beaune equation (note (140)), in which $p = \frac{1}{2}x^{-1}$, $q = 1 - 2x + \frac{1}{2}x^2$, $n = 0$. The remainder of this paragraph is a late insertion.

Ad eundem ferè modum si esset $n=\frac{3y}{4x}$, effinge $y=ex^s$ ubi e ignotum coefficientem et s numerum dimensionum similiter ignotum denotet. Et ex^s pro y substituto, prodibit $n=\frac{3ex^{s-1}}{4}$. et inde rursus $y=\frac{3ex^s}{4s}$. Conferantur jam valores y, et videbis esse $\frac{3e}{4s}=e$, adeoq; $s=\frac{3}{4}$, et e indefinitum. Quare assumpto utcunq; e, erit $y=ex^{\frac{3}{4}}$.

4. Adhæc nonnunquam opus ab altissima dimensione æquabilis quantitatis inchoari potest et ad depressiores continuo pergere. Veluti si detur

$$n=\frac{y}{xx}+\frac{1}{xx}+3+2x-\frac{4}{x},$$

& ab altissimo termino $2x$ opus inchoetur, disponendo capitalem seriem in ordine præcedentibus contrario, emerget tandem

$$y=xx+4x-\frac{1}{x} \quad \&c$$

prout in appositâ operandi forma videre est.

	$+2x$	$+3$	$-\frac{4}{x}$	$+\frac{1}{xx}$.		
$+\frac{y}{xx}$		•	$+1+\frac{4}{x}$	•	$-\frac{1}{x^3}+\frac{1}{2x^4}$	&c
Summa	$+2x$	$+4$	•	$+\frac{1}{xx}$	$-\frac{1}{x^3}+\frac{1}{2x^4}$	[&c]
	$y=xx+4x$	•	$-\frac{1}{x}$	$+\frac{1}{2xx}$	$-\frac{1}{6x^3}$	&c.

Et hic in transitu notari potest quod inter operandum potuit inter terminos $4x$ et $-\frac{1}{x}$ pro intermedio deficienti termino quælibet data quantitas inseri et sic valor y modis infinitis extrahi.(162)

5. Siquæ præterea sint fractæ dimensionum Relatæ Quantitatis indices, ad integras reduci possunt fingendo Quantitatem illam sua fracta dimensione affectam esse alij cuilibet tertiæ fluenti quantitati æqualem, et substituendo tum illam quantitatem tum fluxionem ejus ab illa fictâ æquatione oriundam pro
‖[37] Relata Quantitate et ejus fluxione. Quemadmodum si exponitur ‖ æquatio $n=3xy^{\frac{2}{3}}+y$, ubi Relata Quantitas fractâ dimensionis indice $\frac{2}{3}$ afficitur, assumpta ad arbitrium fluenti quantitate z fingo esse $y^{\frac{1}{3}}=z$ sive $y=z^3$ et fluxionum relatio juxta Prob: 1. erit $n=3rzz$. Quare substituto $3rzz$ pro n ut et z^3 pro y ac zz pro $y^{\frac{2}{3}}$, emerget $3rzz=3xzz+z^3$ sive $r=x+\frac{1}{3}z$. Ubi z supplet vices Relatæ Quantitatis. Postquam vero valor z eo nomine eruitur utpote

$$z=\tfrac{1}{2}xx+\tfrac{1}{18}x^3+\tfrac{1}{216}x^4+\tfrac{1}{3240}x^5 \quad \&c,$$

pro z restitue $y^{\frac{1}{3}}$ et habebis desideratam relationem inter x et y, nempe

$$y^{\frac{1}{3}}=\tfrac{1}{2}xx+\tfrac{1}{18}x^3+\tfrac{1}{216}x^4 \quad \&c,$$

(162) In fact, the general solution of the de Beaune equation

$$n/m(=\dot{y}/\dot{x}) = x^{-2}y+x^{-2}-4x^{-1}+3+2x$$

much the same way if it had been $\dot{y} = 3y/4x$, suppose $y = ex^s$, where e is to denote the unknown coefficient and s the number of [x's] dimensions, likewise unknown. Now when ex^s is substituted in place of y there comes $\dot{y} = \frac{3}{4}ex^{s-1}$ and thence again $y = 3ex^s/4s$. Let these two values of y be compared and you will see that $3e/4s = e$, so that $s = \frac{3}{4}$ while e is indeterminate. Hence, on assuming e at pleasure there will be $y = ex^{\frac{3}{4}}$.

4. Sometimes again the work may be begun from the highest power of the equably flowing quantity and proceed continually towards its lower ones. If there be given, say, $\dot{y} = x^{-2}y + x^{-2} + 3 + 2x - 4x^{-1}$ and the work be begun from the highest term $2x$, by arranging the head series in the contrary order to those before there will at length emerge

$$y = x^2 + 4x - x^{-1}\ldots,$$

as may be seen in the adjoining scheme of operation.

	$2x$	$+3$	$-4x^{-1}$	$+x^{-2}$	
$x^{-2}y$	$\cdot$	$+1$	$+4x^{-1}$	$\cdot$	$-x^{-3}+\frac{1}{2}x^{-4}\ldots$
Total	$2x$	$+4$	$\cdot$	$+x^{-2}$	$-x^{-3}+\frac{1}{2}x^{-4}\ldots$
$y =$	x^2	$+4x$	$\cdot$	$-x^{-1}+\frac{1}{2}x^{-2}$	$-\frac{1}{6}x^{-3}\ldots.$

And it may here be noted in passing that during the operation any given quantity you please might have been inserted between the terms $4x$ and $-x^{-1}$ in place of the missing intermediate term and thus the value of y could have been extracted in an infinity of ways.(162)

5. If, besides, the related quantity have any fractional indices in its dimensions, these may be reduced to integral ones by supposing the quantity affected by a fractional dimension equal to any other third fluent quantity and then substituting both that quantity and its fluxion, as it arises from that supposed equation, in place of the related quantity and its fluxion. For instance, if there is exhibited the equation $\dot{y} = 3xy^{\frac{2}{3}} + y$, where the related quantity is affected with the fractional dimension-index $\frac{2}{3}$, assuming the fluent quantity z at will I suppose that $y^{\frac{1}{3}} = z$ or $y = z^3$, and then by Problem 1 the relation of the fluxions will be $\dot{y} = 3\dot{z}z^2$. Hence, when $3\dot{z}z^2$ is substituted for $\dot{y}$ and likewise z^3 for y and z^2 for $y^{\frac{2}{3}}$, there will emerge $3\dot{z}z^2 = 3xz^2 + z^3$ or $\dot{z} = x + \frac{1}{3}z$. Here z fills the rôle of related quantity. But after the value of z is derived under this head (as

$$z = \tfrac{1}{2}x^2 + \tfrac{1}{18}x^3 + \tfrac{1}{216}x^4 + \tfrac{1}{3240}x^5 \ldots),$$

in place of z restore $y^{\frac{1}{3}}$ and you will have the desired relationship between x and y, namely $y^{\frac{1}{3}} = \frac{1}{2}x^2 + \frac{1}{18}x^3 + \frac{1}{216}x^4 \ldots$ or, on taking the cubes of either side,

---

is (compare note (140)) $y = x^2 + 4x - 1 + ke^{-x^{-1}}$: of this Newton gives the series expansion when $k = 1$. In view of his remarks in his preceding second observation (on the previous page) it seems unfortunate that he has not detected the finite particular solution (when $k = 0$)

$$y = x^2 + 4x - 1.$$

et cubis partium utrobicꝫ positis erit $y = \frac{1}{8}x^6 + \frac{1}{24}x^7 + \frac{1}{288}x^8$ &c.[163] Pari ratione si detur $n = \sqrt{4y} + \sqrt{xy}$, sive $2y^{\frac{1}{2}} + x^{\frac{1}{2}}y^{\frac{1}{2}}$, fingo $z = y^{\frac{1}{2}}$ sive $zz = y$, et inde per Prob 1 elicio $2rz = n$ et consequenter est $2rz = 2z + x^{\frac{1}{2}}z$ sive $r = 1 + \frac{1}{2}x^{\frac{1}{2}}$. Adeocꝫ per casum priorem hujus est $z\,(y^{\frac{1}{2}}) = x + \frac{1}{3}x^{\frac{3}{2}}$ et partibus quadratis $y = xx + \frac{2}{3}x^{\frac{5}{2}} + \frac{1}{9}x^3$. Sin valorem y modis infinitis desideres fac $z = c + x + \frac{1}{3}x^{\frac{3}{2}}$ assumpto utcuncꝫ initiali termino c, et erit $y(zz) = cc + 2cx + \frac{2}{3}cx^{\frac{3}{2}} + xx + \frac{2}{3}x^{\frac{5}{2}} + \frac{1}{9}x^3$.

Ast hæc nimis officiosè tractare videor siquidem rarissimè usui esse possunt.

Casus 3.

Problematis ubi tres vel plures quantitatum fluxiones æquatio complectitur Resolutio brevi absolvitur. Scilicet[164] inter duas quaslibet istarum quantitatum relatio (ubi ex statu Quæstionis non determinatur) quælibet effingi debet, et earum fluxionum exinde quæri, eo ut alterutra unà cum ejus fluxione ex æquatione expositâ exterminari possit. Quâ de causâ si trium insunt quantitatum fluxiones unica effingenda est æquatio, ac duæ si insunt quatuor, et sic
||[38] porro, ut exposita || æquatio in aliam tandem æquationem transformetur cui non insint plures duabus; et hâc deinde ut supra resolutâ reliquarum quantitatum relationes eruentur. Sic æquatione $2m - r + nx = 0$ exposita; quo quantitatum $x\ y$ et z (quarum fluxiones $m\ n$ et r æquatio complectitur) relationes inter se obtineam, relationem inter duas quaslibet ut x et y pro lubitu effingo, puta quod sit $x = y$, vel $2y = a + z$, vel $x = yy$ &c. Sit autem $x = yy$ et inde erit $m = 2ny$, Quare scriptis $2ny$ pro m et yy pro x, exposita æquatio transformabitur in $4ny - r + nyy = 0$. et inde relatio inter y et z emerget $2yy + \frac{1}{3}y^3 = z$. Ubi si x pro yy et $x^{\frac{3}{2}}$ pro y^3 vicissim scribatur prodibit etiam $2x + \frac{1}{3}x^{\frac{3}{2}} = z$. Adeócꝫ inter modos infinitos quibus x, y, et z ad invicem referuntur unus his æquationibus $x = yy$. $2yy + \frac{1}{3}y^3 = z$ et $2x + \frac{1}{3}x^{\frac{3}{2}} = z$ designatus investigatur.

Demonstratio.

Problema tandem cōfecimus sed[165] demonstratio superest. Et in tanta rerum copiâ ne per nimias ambages e proprijs fundamentis Syntheticè derivetur, sufficiat per Analysin sic breviter indicare. Scilicet æquatione quâlibet expositâ, postquam opus ad finem perduxeris experiri est quod ex elicitâ æquatione

(163) Newton accurately selects the substitution $y = z^3$ which converts this generalized de Beaune equation (note (140)) to linear form, but gives only the particular case $k = 9$ of the series expansion of its general solution $z = -3x - 9 + ke^{\frac{1}{3}x}$. Note that the last term in his series expansion should be '$\frac{7}{864}x^8$ &c'.

(164) This 'Casus 3' was distinguished only at a late stage in the composition of the manuscript. The following was originally joined to the last paragraph of Case 2 by 'Quamobrem de resolutione Problematis ubi...complectitur haud operæ pretium aliquid disserere; sed dicam tamen breviter quód' (Accordingly it is not worth while to discuss aspects of the resolution of the problem when...comprises..., but nevertheless let me say briefly that).

$y = \frac{1}{8}x^6 + \frac{1}{24}x^7 + \frac{1}{288}x^8 \ldots$.[163] Equally, if there be given $\dot{y} = \surd(4y) + \surd(xy)$, that is, $2y^{\frac{1}{2}} + x^{\frac{1}{2}}y^{\frac{1}{2}}$, I suppose $z = y^{\frac{1}{2}}$ or $z^2 = y$ and thence by Problem 1 elicit $2\dot{z}z = \dot{y}$. In consequence, $2\dot{z}z = 2z + x^{\frac{1}{2}}z$ or $\dot{z} = 1 + \frac{1}{2}x^{\frac{1}{2}}$, so that by the first case of this problem z (or $y^{\frac{1}{2}}$) $= x + \frac{1}{3}x^{\frac{3}{2}}$ and, on squaring, $y = x^2 + \frac{2}{3}x^{\frac{5}{2}} + \frac{1}{9}x^3$. But should you desire the value of y in an infinity of ways, make $z = c + x + \frac{1}{3}x^{\frac{3}{2}}$, assuming the initial term c at will, and there will be y (or z^2) $= c^2 + 2cx + \frac{2}{3}cx^{\frac{3}{2}} + x^2 + \frac{2}{3}x^{\frac{5}{2}} + \frac{1}{9}x^3$.

But I appear to treat these matters too seriously inasmuch as they can but on the rarest of occasions be of use.

Case 3

The resolution of the problem when the equation comprises three or more fluxions of quantities is briefly disposed of. Specifically,[164] any relation you wish (when the conditions of the inquiry do not determine one) should be presupposed between any two of those quantities, and that of their fluxions sought therefrom, so that one or other together with its fluxion may be eliminated from the equation proposed. Consequently, if the fluxions of three quantities are present, a single equation has to be thought up so that the exhibited equation may be transformed at last into another in which no more than two are present, but two have to be contrived if four be present, and so on; then after this equation has been resolved as above, the relations between the remaining quantities will be established. Thus, when the equation $2\dot{x} - \dot{z} + \dot{y}x = 0$ is exhibited, to obtain the relationships of the quantities x, y and z (the fluxions $\dot{x}$, $\dot{y}$ and $\dot{z}$ of which are comprehended in the equation) one with another, I think up a relation between any two, such as x and y, at pleasure, say, that $x = y$ or $2y = a + z$ or $x = y^2$ and so on. Letting $x = y^2$ it will accordingly be $\dot{x} = 2\dot{y}y$ and therefore, on writing $2\dot{y}y$ in place of $\dot{x}$ and y^2 in place of x, the exhibited equation will be transformed into $4\dot{y}y - \dot{z} + \dot{y}y^2 = 0$ and thence the relation between y and z will emerge as $2y^2 + \frac{1}{3}y^3 = z$. Here in turn if x be written for y^2 and $x^{\frac{3}{2}}$ for y^3 there will come also $2x + \frac{1}{3}x^{\frac{3}{2}} = z$. And so among the infinity of ways in which x, y and z are related to each other one represented by these equations $x = y^2$, $2y^2 + \frac{1}{3}y^3 = z$ and $2x + \frac{1}{3}x^{\frac{3}{2}} = z$ is now found out.

Demonstration

We have at last done with the problem but its demonstration still remains. Not (in such a mass of material) to digress too much in deriving one synthetically from proper foundations, it should be sufficient to indicate it briefly by analysis in this way: namely, when any equation is exhibited and you have carried through the work to its conclusion, you may test that the exhibited equation

(165) The less explicit opening 'Jam deniqꝫ horum' (Now finally their) is cancelled.

exposita vicissim (per Prob 1) eruetur. Et proinde quantitatum relatio in elicita æquatione exigit[166] relationem fluxionum in exposita, et contra: sicut ostendendum erat. Sic æquatione $n = x$ expositâ elicietur $y = \frac{1}{2}xx$ et inde vicissim (per prob 1) $n = mx$ sive $= x$ quandoquidem m supponitur esse 1. Et sic ex
pag 30 $n = 1 - 3x + y + xx + xy$ provenit $y = x - xx + \frac{1}{3}x^3 - \frac{1}{6}x^4 + \frac{1}{30}x^5 - \frac{1}{45}x^6$ &c et inde
‖[39] vicissim per Prob 1, $n = 1 - 2x + x^2 - \frac{2}{3}x^3 + \frac{1}{6}x^4 - \frac{2}{15}x^5$ &c. Qui ‖ duo valores ipsius n conveniunt, ut patet substituendo $x - xx + \frac{1}{3}x^3 - \frac{1}{6}x^4 + \frac{1}{30}x^5$ &c pro y in priori.

Cæterùm in æquationum reductione adhibui operationem de qua præterea rationem reddere oportet: estq; transmutatio fluentis quantitatis per connexionem cum quantitate data. Sunto[167] AE et ae lineæ utrinq; infinitæ per quas

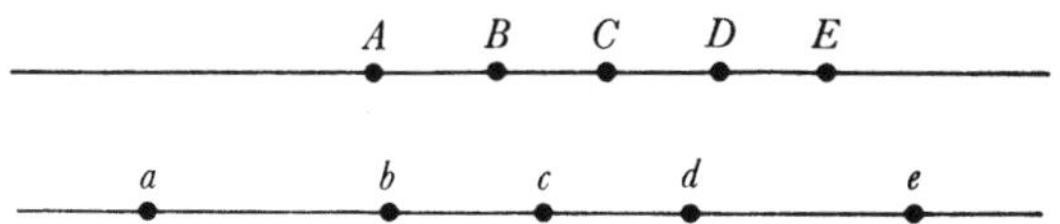

mobilia duo e longinquo trajiciantur simul attingentia locos A et a, B et b, C et c, D et d &c; et sit B punctum a cujus et rei mobilis distantiâ in AE motus æstimetur ita ut $-BA$, BC, BD, BE successive sint fluentes quantitates quando mobile sit in locis A, C, D, E. Sitq; b consimile punctum in altera linea: et erunt $-BA$ ac $-ba$ contemporaneæ fluentes quantitates, ut et BC ac bc, BD ac bd, BE ac be &c. Quod si vice punctorum B et b substituantur A et c ad quæ tanquam quiescentia motus referantur, tunc 0 & $-ca$, AB et $-cb$, AC et 0, AD et cd, AE et ce &c erunt contemporaneæ fluentes quantitates. Mutantur itaq; fluentes quantitates additione et substractione datarum AB et bc, sed non mutantur quoad motûs celeritatem et fluxionis mutuum respectum: nam ejusdem longitudinis sunt partes contemporaneæ AB et ab, BC et bc, CD et cd; DE et de in utroq; casu. Et sic in æquationibus quibus hæ quantitates designantur partes contemporaneæ quantitatum non ideo mutantur quod earum absoluta longitudo datâ aliquâ augeatur vel minuatur. Unde constat Propositum: Nam Problematis hujus scopus propriè non alius est quam contemporaneas partes sive absolutarum quantitatum (v, x, y, aut z) contemporaneas differentias data fluendi ratione descriptas determinare. Et perinde est cujusnam sint absolutæ longitudinis quantitates illæ dummodo contemporaneæ sive correspondentes earum differen-
‖[40] tiæ cum exposita ‖ fluxionum relatione conveniant.

Potest et hujus rei ratio sic Algebraicè reddi. Proponatur $n = mxy$, et finge $x = 1 + z$, eritq; (per Prob 1) $n = r$. Adeoq; pro $n = mxy$ scribi potest $r = my + mzy$. Jam cum sit $n = r$, patet quantitates x et z etsi non sint ejusdem longitudinis,

(166) We might expect the putative subjunctive 'exigat' here.

(167) Compare the equivalent use of this dynamical model in the 1666 tract (I: 385) to justify a differentiation algorithm.

will in turn (by Problem 1) result from the elicited one and, in consequence, that the relationship of the quantities in the elicited equation does entail that between the fluxions in the exhibited one, and conversely so (as was to be shown). Thus from the exhibited equation $\dot{y} = x$ will be elicited $y = \frac{1}{2}x^2$ and from this in turn (by Problem 1) $\dot{y} = \dot{x}x$, that is, x since $\dot{x}$ is taken to be unity. And thus page 30
from $\dot{y} = 1-3x+y+x^2+xy$ there is produced

$$y = x-x^2+\tfrac{1}{3}x^3-\tfrac{1}{6}x^4+\tfrac{1}{30}x^5-\tfrac{1}{45}x^6 \ldots,$$

and thence in turn, by Problem 1, $\dot{y} = 1-2x+x^2-\frac{2}{3}x^3+\frac{1}{6}x^4-\frac{2}{15}x^5 \ldots$. These two values of $\dot{y}$ agree, as is evident on substituting $x-x^2+\frac{1}{3}x^3-\frac{1}{6}x^4+\frac{1}{30}x^5 \ldots$ in place of y in the former.

For the rest, in the reduction of equations I have employed an operation of which I should give some further account: this is the transformation of a fluent quantity by adjoining a given quantity to it. Let[167] *AE* and *ae* be lines indefinitely extended either way, along which pass from a distance two moving objects reaching the places *A*, *a*; *B*, *b*; *C*, *c*; *D*, *d*; ... together. Let *B* be a point the distance of which from the moving object measures its motion in *AE* (so that $-BA$, *BC*, *BD*, *BE* are successive values of a fluent quantity when the moving object is at the places *A*, *C*, *D*, *E*) and let *b* be the equivalent point in the second line: then will $-BA$ and $-ba$, and also *BC* and *bc*, *BD* and *bd*, *BE* and *be*, and so on, be contemporaneous values of fluent quantities. Now if instead of the points *B* and *b* as 'stationary' points to which the motions are referred there be substituted *A* and *c*, then will 0 and $-ca$, *AB* and $-cb$, *AC* and 0, *AD* and *cd*, *AE* and *ce*, and so forth, be contemporaneous values of the fluent quantities. Consequently, the fluent quantities are changed [in magnitude] by the addition and subtraction of the given quantities *AB* and *bc*, but not as regards their speed of motion and fluxional relationship one to the other: for in either case the contemporaneous segments *AB* and *ab*, *BC* and *bc*, *CD* and *cd*, *DE* and *de* are of the same length. And thus in equations in which these quantities are represented contemporaneous portions of quantities are not changed accordingly when their absolute length is increased or diminished by some given quantity. Hence what was propounded is manifest: for the aim of this problem is, properly speaking, no other than to determine the contemporaneous parts or differences of absolute quantities (v, x, y or z) described at a given rate of flow, and the absolute length of those quantities is irrelevant to the agreement of their contemporaneous or correspondent differences with the exhibited relation of fluxions.

The reason for this circumstance may also be stated algebraically in this manner. Let there be proposed $\dot{x} = \dot{y}xy$ and suppose $x = 1+z$, then will there be (by Problem 1) $\dot{x} = \dot{z}$ and so in place of $\dot{x} = \dot{y}xy$ may be written $\dot{z} = \dot{y}y+\dot{y}zy$. Now since $\dot{x} = \dot{z}$, it is evident that, though the quantities x and z are not of the

pariter tamen fluere respectu ipsius y, et pares habere partes contemporaneas. Quid itaq; si ijsdem symbolis denotem quæ fluendi ratione conveniunt et ad contemporaneas differentias determinandas vice $n=mxy$ usurpem

$$n=my+mzy.^{(168)}$$

Jam deniq; quo pacto partes contemporaneæ ex æquatione quantitates involvente inveniri possint per se manifestum est. E. G. Sit $y=\frac{1}{x}+x$ æquatio. Et cum sit $x=2$ erit $y=2\frac{1}{2}$, cum verò sit $x=3$ erit $y=3\frac{1}{3}$. Ergo dum x fluit a 2 ad 3 y fluet a $2\frac{1}{2}$ ad $3\frac{1}{3}$. Adeoq; partes in hoc tempore transactæ sunt $(3-2)$ 1 et $(3\frac{1}{3}-2\frac{1}{2})$ $\frac{5}{6}$.(169)

Jactis hisce sequentium fundamentis, ad Problemata magis particularia(170) jam transeo.

||[41]

||Prob 3.

Determinare maximas et minimas.

Quantitas ubi maxima est vel minima in illo momento nec profluit nec refluit. Nam si profluit, id arguit(171) minorem fuisse et statim majorem fore quam jam est; et contra si refluit. Quamobrem fluxionem ejus per Prob: 1 quære et pone nullam esse.(172)

Exemp 1. Si maxima quantitas x in æquatione $x^3-axx+axy-y^3=0$(173) desideretur. Quantitatum x et y fluxiones quære et prodibit

$$3mxx-2amx+amy-3nyy+anx=0.$$

Positóq; $m=0$ restabit $-3nyy+anx=0$ sive $3yy=ax$. Cujus ope possis alterutrum x vel y in æquatione primariâ exterminare, et per æquationem restantem determinare alteram, et utramq; per $[-]3yy+ax=0$.

Perinde est hæc operatio ac si multiplicasses terminos propositæ æquationis per numerum dimensionum alterius fluentis quantitatis y. Unde prodit Huddeniana notissima Regula(174) quod ad obtinendum maximam aut minimam Relatam Quantitatem Æquatio juxta dimensiones Correlatæ Quantitatis disponi debet et per quamlibet Arithmeticam progressionem multiplicari. Ast cùm neq; hæc regula ad æquationes surdis quantitatibus affectas neq; ulla alia

(168) Rather confusingly, Newton momentarily switches his notation, denoting the fluxions of x and y by n and m respectively.

(169) A cancelled continuation reads 'Atq; ita cum x sit 0 erit y infinit[a] et cum x sit -1 erit $y=-2$. Ergo dum x fluit a -1 ad $+3$ y fluit ab -2 per infinitatem ad $+3\frac{1}{3}$. Adeoq; partes contemporaneæ sunt 4 et infinitum spatium' (And so, when x is zero, y will be infinite, and, when x is -1, y will equal -2. Therefore while x flows from -1 to $+3$, y flows from -2 through infinity to $+3\frac{1}{3}$, so that the contemporaneous parts are 4 and an infinite space). The idea that decreasing negative numbers 'flow' continuously to $-\infty$ and thence without

same length, they none the less have a like flow in respect of y and have their contemporaneous parts equal. Why therefore should I not denote by the same symbols those quantities which have an identical rate of flow? and why, in determining contemporaneous differences, may I not use $\dot{x} = \dot{y}y + \dot{y}zy$ instead of $\dot{x} = \dot{y}xy$?[(168)]

Now, lastly, how the contemporaneous parts may be found from an equation involving [fluent] quantities is evident of itself. For example, let $y = x^{-1} + x$ be the equation: then when $x = 2$, y will be $2\frac{1}{2}$; when indeed $x = 3$, then will $y = 3\frac{1}{3}$. In consequence as x flows from 2 to 3, y will flow from $2\frac{1}{2}$ to $3\frac{1}{3}$ and so the [corresponding] parts covered in this time are $(3-2$ or$)$ 1 and $(3\frac{1}{3} - 2\frac{1}{2}$ or$)$ $\frac{5}{6}$.[(169)]

Having laid these foundations for the following, I now pass to more specialized problems.[(170)]

Problem 3

To determine maxima and minima

When a quantity is greatest or least, at that moment its flow neither increases nor decreases: for if it increases, that proves[(171)] that it was less and will at once be greater than it now is, and conversely so if it decreases. Therefore seek its fluxion by Problem 1 and set it equal to nothing.[(172)]

Example 1. If the greatest value of x in the equation

$$x^3 - ax^2 + axy - y^3 = 0^{(173)}$$

be desired, seek the fluxions of the quantities x and y and there will come $3\dot{x}x^2 - 2a\dot{x}x + a\dot{x}y - 3\dot{y}y^2 + a\dot{y}x = 0$: then when $\dot{x}$ is set equal to zero, there will remain $-3\dot{y}y^2 + a\dot{y}x = 0$ or $3y^2 = x$. With the help of this you might eliminate one or other of x and y in the primary equation and by the resulting equation determine the second and then both by $-3y^2 + ax = 0$.

The operation is in effect the same as if you had multiplied the terms of the proposed equation by the number of dimensions of the second fluent quantity y in it. Hence comes the celebrated Rule of Hudde[(174)] that, to obtain the maxima or minima of the related quantity, the equation should be ordered according to the dimensions of the correlate one and then multiplied by an arithmetical progression. But since neither this rule nor any other I know of so far published

---

break to $+\infty$ and so forth down through the positives was a favourite idea of John Wallis'. (See his *Arithmetica Infinitorum* (I, **1**, Introduction, Appendix 2: note (20)) (Oxford, 1656): Proposition CI, Scholium: 74–5.)

(170) Namely, the following Problems 3–12 which complete the present tract.

(171) 'sequitur' (it follows) is cancelled.

(172) Or, as Newton wrote in May 1666 (I: 397), 'find ye motion...& suppose it equall to nothing'.

(173) The derivative of this equation has already been found on Newton's page 18 above.

(174) Compare I, **2**, 2, Historical note.

hactenus quod sciam evulgata absq; prævia reductione se extendat:(175) ejus rei accipe sequens exemplum.

Ex: 2. Si maxima quantitas y in æquatione $x^3-ayy+\frac{by^3}{a+y}-xx\sqrt{ay+xx}=0$ determinanda est; ipsarum x et y fluxiones quære et emerget

$$3mxx-2any\frac{+3abnyy+2bny^3}{aa+2ay+yy}\frac{-4amxy+6mx^3+anxx}{2\sqrt{ay+xx}}=0.$$

Et cum ex hypothesi sit $n=0$, neglige terminos in n ductos (id quod inter
||[42] operandum ad minuendum laborem antea fieri potuit) cæterosq; per mx || divide et restabit $3x\frac{-2ay-3xx}{\sqrt{ay+xx}}=0$, factaq; reductione exurget $4ay+3xx=0$. Cujus ope possis utramvis quantitatem x or y ex æquatione primò proposita exterminare ac deinde ex æquatione resultante (quæ cubica erit)(176) valorem alterius elicere.

Ex hoc problemate sequentium resolutio petenda est.(177)

[1.] In dato Triangulo aut Segmento cujusvis Curvæ, maximum rectangulum inscribere.(178)

[2.] Maximam vel minimam rectarum ducere quæ inter datum punctum et curvam positione datam interjacent. Sive, A dato puncto ad Curvam ducere perpendiculum.(179)

[3.] Maximam vel minimam rectarum ducere quæ per datum punctum transeuntes interjacent alijs duabus sive rectis sive curvis lineis.(180)

[4.] A puncto intra Parabolam dato rectam ducere quæ Parabolam omnium obliquissimè secabit. Et idem in alijs curvis facere.(181)

---

(175) When Newton communicated the gist of his solution of the tangent problem (see (note (191)) to Collins on 10 December 1672, he added, in an echo of the present passage, that it was 'a Corollary of a Generall Method [the fluxional algorithm developed in Problem 1] wch extends it selfe wthout any troublesome calculation...to the resolving other abstruser kinds of Problems.... Nor is it (as Huddens method *de maximis et minimis*...) limited to æquations wch are free from surd quantities' (*Correspondence of Isaac Newton*, **1** (1959): 247–8).

(176) Assuming that a positive maximum exists, that is, and so excluding the value $y = 0$. In fact, on eliminating x between $x^3-x^2\sqrt{[ay+x^2]}-ay^2+by^3/(a+y) = 0$ and $x^2 = -\frac{4}{3}ay$, there results $y^3\times[27y(a^2+(a-b)y)^2+16a^3(a+y)^2] = 0$. A similar factor $x^3 = 0$ intrudes when y is eliminated.

(177) Newton here cancelled the opening of a concluding phrase 'cujusmodi et alia permulta facilius excogitari possunt quàm (propter computandi molestiam) resolvi' when he inserted its equivalent below (note (187)).

(178) It is easily shown that a rectangle of sides ax and $b(1-x)$ may be inscribed in a triangle of base a and height b, so that the rectangle of maximum area $\frac{1}{4}ab$ will be found by maximizing the quantity $x(1-x)$. A more complex calculation will solve the problem for polygonal figures and conics, but 'inscription' will not always be uniquely defined in the case of higher curves. Where (in the case of simple closed ovals, for example) the inscription of rectangles is capable of definition the present Fermatian technique will choose rectangles whose instantaneous growth rate is zero, that is, which reach local (and not necessarily any finite) extreme values.

extends without prior reduction to equations affected with surd quantities,(175) accept the following as an example of this case.

EXAMPLE 2. If the greatest value of y in the equation

$$x^3 - ay^2 + by^3/(a+y) - x^2\surd[ay+x^2] = 0$$

is to be determined, seek the fluxions of x and y and there will emerge

$$3\dot{x}x^2 - 2a\dot{y}y + \frac{3ab\dot{y}y^2 + 2b\dot{y}y^3}{a^2+2ay+y^2} - \frac{4a\dot{x}xy + 6\dot{x}x^3 + a\dot{y}x^2}{2\surd[ay+x^2]} = 0.$$

Since, now, by hypothesis $\dot{y}$ is zero, lay aside the terms multiplied by $\dot{y}$ (to lighten the labour this could have been done beforehand during the operation) and, on dividing the remainder by $\dot{x}x$, there will be left

$$3x - (2ay+3x^2)/\surd[ay+x^2] = 0$$

and by reduction there will arise $4ay+3x^2 = 0$. With the help of this you might eliminate either of the quantities x or y from the equation first proposed and then from the resulting one (which will be a cubic)(176) elicit the value of the other.

From this problem the solution of the following is to be wrested.(177)

1. In a given triangle or segment of any curve to inscribe the greatest rectangle.(178)
2. To draw the greatest or least of the straight lines which lie between a given point and a curve given in position. In other words, to draw a normal to a curve from a given point.(179)
3. To draw the greatest or least of the straight lines passing through a given point which lie between any two other lines, straight or curved.(180)
4. From a given point inside a parabola to draw the straight line which cuts the parabola most obliquely of all. And to do the same in other curves.(181)

---

(179) This Cartesian method of constructing the normal from a point to a curve was, as we have seen (I, **2**, 2), a crucial foundation for Newton's first calculus researches.

(180) Again, of course, the concept of 'chord' is uniquely definable only in the case of polygonal figures, conics and simple convex curves. Where this is possible, the problem is to maximize $(x-X)^2+(y-Y)^2$, where the line through (x, y) and the given point meets the curve or convex figure again in the (unique real) point (X, Y).

(181) In the case of the parabola defined by $y^2 = kx$ in a perpendicular Cartesian co-ordinate system the angle between the curve at point (x, y) and the line joining this point to the given point (a, b) is

$$\tan^{-1}\frac{y-b}{x-a} - \tan^{-1}\frac{k}{2y} = \tan^{-1}\frac{y^2-2by+ak}{2(x-a)y+k(y-b)},$$

which will be a maximum or minimum when its derivative is zero. Elimination of x between this and $y^2 = kx$ yields a quartic in y, so that the problem has four solutions. (When the inclination angle is right, one of the roots is infinite: that is, three finite normals may be drawn from a given point to a given parabola, the fourth being the diameter through the point which meets the parabola in the point at infinity.) A similar solution holds in the case of higher curves.

[5.] Curvarum vertices[,] maximas aut minimas latitudines [et] puncta in quibus partes circumactæ se decussant determinare.(182)

[6.] Curvarum puncta invenire ubi maximè aut minimè curvantur.(183)

[7.] Invenire minimum angulorum in quibus rectæ ad diametros suas in data Ellipsi ordinatim applicantur.(184)

[8.] Ellipsium per data quatuor puncta transeuntium vel minimam definire vel eam quæ ad formam circularem maximè accedit.(185)

[9.] Amplitudinem sphæricæ superficiei determinare quam lux e longinquo fluens postquam ab anteriori hemisphærio refracta fuit illustrat in posteriori.(186)

Et hujusmodi alia permulta faciliùs excogitari possunt quàm (propter computandi fastidium) resolvi.(187)

||[43] ||PROB 4.

CURVARUM TANGENTES DUCERE.(188)

MODUS 1.(189)

Tangentes pro varijs relationibus curvarum ad rectas variè ducuntur.(190) Et imprimis esto *BD* recta in dato angulo ad aliam rectam *AB* tanquam basin

(182) Here, presumably, the vertices of a curve are defined as the end-points of its greatest and least diameters, while its double points (of decussation) are those where the 'girth' is zero. Yet again the concept of girth is meaningful, as Newton had already found to his cost in late 1664 (1, 2, 3, §1), only in the case of conics and simple closed curves.

(183) This is Problem 4 of the 1666 tract (I: 425–7): the topic is taken up again in Problem 6 below.

(184) This elementary problem of conic geometry is virtually resolved in Apollonius' *Conics*, **7**,7, which had recently appeared in G. A. Borelli's Latin edition from the Arabic paraphrase (*Apollonii Pergæi Conicorum Lib. V. VI. VII Paraphraste Abalphato Asphahanensi nunc primùm editi* (Florence, 1661): 281). Since conjugate diameters are parallel to a corresponding pair of supplemental chords drawn from the end-points of an axis, the problem at once reduces to finding the greatest or least inclination of a pair of supplemental chords to each other: this evidently happens when the common point of the chords is the end-point of the other axis. In analytical equivalent, if the ellipse (of eccentricity $e = \surd[1-r/q]$) is defined by $y^2 = rx-(r/q)x^2$ in a perpendicular Cartesian coordinate system, the inclination of the supplemental chords drawn between the general point (x, y) and the vertices $(0, 0)$, $(q, 0)$ is

$$\tan^{-1}\frac{y}{x}+\tan^{-1}\frac{y}{q-x} = \tan^{-1}\frac{qr}{(q-r)y}:$$

this is clearly a minimum when y attains its maximum value $\frac{1}{2}\surd[qr]$.

(185) Apparently Newton intends by a 'least' ellipse one of minimum area, while clearly one which most closely approximates a circle in shape has minimum eccentricity. Necessary tests for these are that the growth rates of the product and quotient of the main axes be zero as the ellipse drawn arbitrarily through the four given points continuously varies its shape. (Newton here implicitly invokes a Keplerian principle of continuity which can, of course, be given rigorous justification.) The computation needed to work through the problem is, in all but the simplest cases, formidable. We might, for example, determine some arbitrary point

5. To determine the vertices of curves, their greatest or least girths and the points in which their parts double back and cross one another.(182)
6. To find the points of curves where they are most or least curved.
7. To find the least of the angles in which straight lines may be ordinately applied to their diameters in a given ellipse.(184)
8. Of ellipses passing through four given points to define either the least or that which most nearly approaches the shape of a circle.(185)
9. To determine the breadth of the spherical surface which light streaming in from afar, after refraction at the front hemisphere, illuminates on the rear one.(186)

And a great many others of this kind may more easily be thought up than (because of the tediousness of computation) be resolved.(187)

Problem 4

To draw Tangents to curves(188)

Mode 1(189)

Tangents are drawn in various ways according to the various relationships of curves to straight lines.(190) And in the first place let the straight line *BD* be ordinate to another straight line *AB* as base and terminate at the curve *ED*.

$(\lambda, \alpha\lambda+\beta)$ on a given line in terms of the linear parameter λ and constants α, β, then calculate (in terms of λ) the principal axes of the conic passing through this parametric point and the four given ones; the derivative with respect to λ of the product and quotient of the resulting expressions for major and minor axes would then have to be equated to zero.

(186) In context, for 'Amplitudinem' read 'Amplitudinem maximam' (greatest breadth). This problem, developed by Newton himself in Proposition 35 of the first part of his Lucasian lectures on optics (reproduced in **3**, 1, §2 below), yields at once the radius of the primary rainbow when the latter is determined by the condition that a ray of light incident on a sphere shall exit, after suffering one internal reflection, at a maximum inclination to its incident path. For historical details see the introduction to the following Part 3.

(187) As Newton says, it is for the most part incomparably easier to define necessary conditions for the existence of an extreme value in some given range than to compute its precise position. In conceiving the preceding problems he merely creates a situation in which a variable quantity takes on the same value twice and is supposed in between times to vary continuously, evidently at some point in the interval attaining a maximum or minimum value. We need not suppose that he has in each of his stated cases (and particularly Problem 8) attempted any solution.

(188) An augmented revision of Problem 1 of the 1666 tract (I: 416–18). The two modes of representing a point in a plane for which the tangent problem was there resolved (namely, oblique Cartesian coordinates and bipolars: modes 1 and 3 below) are here extended to nine.

(189) The problem of tangents in standard oblique Cartesian coordinates. The heading was originally entered (as 'Mod: 1') in the margin opposite 'imprimis' in the second sentence following.

(190) That is, according to the system of coordinates with regard to which the curve is defined.

ordinata et ad curvam ED terminata. Et moveatur hæc ordinata per indefinitè parvum spatium ad locum bd, ita ut momento cd augeatur dum AB augetur momento Bb, cui Dc æqualis est. Jam producatur Dd donec cum AB in T conveniat et hæc tanget curvam in D vel d, eruntq; triangula dcD, DBT similia. Adeóq; $TB.BD::Dc.cd$.

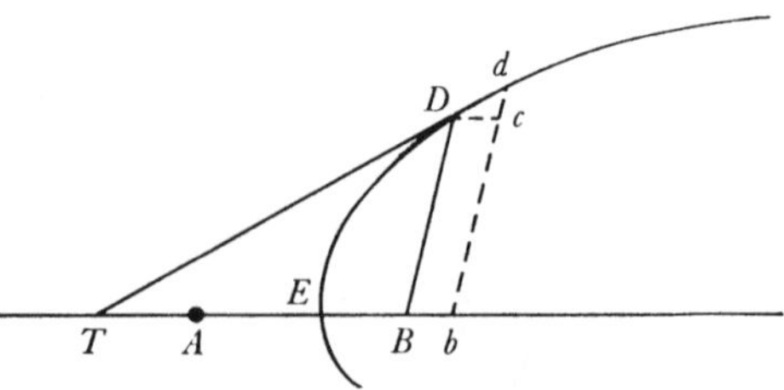

Cùm itaq; relatio BD ad AB in æquatione qualibet pro curvâ determinandâ exponitur; quære relationem fluxionum per Prob 1, et cape TB ad BD in ratione fluxionis AB ad fluxionem BD, ac TD tanget curvam in D.(191)

Exemp 1. Nominata AB x et BD y esto earum relatio

$$x^3 - axx + axy - y^3 = 0.^{(192)}$$

Et fluxionum relatio erit $3mxx - 2amx + amy - 3nyy + anx = 0$. Adeoq;

$$n.m::3xx - 2ax + ay.3yy - ax::BD(y).BT.$$

Est ergo $BT = \dfrac{3y^3 - axy}{3xx - 2ax + ay}$.(193) Dato itaq; puncto D, et inde DB et AB sive y et x; dabitur longitudo BT qua tangens TD determinatur.

Potest autem hæc operandi methodus sic concinnari. Æquationis expositæ terminos fac esse nihilo æquales; per proprium numerum dimensionum ordinatæ quantitatis multiplica, et exitum colloca in numeratore; Dein terminos ejusdem æquationis per proprium numerum dimensionum Basis multiplica et exitum per Basin divisum colloca in denominatore valoris BT.(194) Et illam BT cape ad partes adversus A si valor ejus sit affirmativus, aut versus A si sit negativus.

---

(191) Having heard from Collins on 8 December 1672 (in a letter no longer traceable) that Sluse had recently communicated to Oldenburg his own independently discovered version of the present Huddenian tangent-rule, Newton wrote back two days later, communicating his own variant of the tangent algorithm worked out for the cubic

$$x^3 - 2x^2y + bx^2 - b^2x + by^2 - y^3 = 0$$

(compare 'Exemp 1' following) but also taking care to hint at the subtle complexities of the present Problem 4: 'This...is one particular, or rather a Corollary of a Generall Method w[ch] extends it selfe w[th]out any troublesome calculation...to the drawing tangents to all curve lines whether Geometrick or mechanick or how ever related to streight lines or to other curve lines.... Nor is it (as Huddens method de maximis et minimis & consequently Slusius his new method of Tangents as I presume) limited to æquations wch are free from surd quantities. This method I have interwoven wth that other of working in æquations by reducing them to infinite series' (*Correspondence of Isaac Newton*, **1**: 247–8).

Let this ordinate move through an indefinitely small space to the position *bd* so that it increases by the moment *cd* while *AB* increases by the moment *Bb*, equal to *Dc*. Now let *Dd* be extended till it meets *AB* in *T*: this will then touch the curve in *D* or *d* and the triangles *dcD*, *DBT* will be similar, so that

$$TB:BD = Dc:cd.$$

When therefore the relationship of *BD* to *AB* is exhibited in any equation by which the curve is to be determined, seek the relation of the fluxions by Problem 1 and take *TB* to *BD* in the ratio of the fluxion of *AB* to that of *BD*: then will *TD* touch the curve at *D*.(191)

Example 1. Naming *AB* x and *BD* y, let their relation be

$$x^3 - ax^2 + axy - y^3 = 0.^{(192)}$$

The relation of the fluxions will be $3\dot{x}x^2 - 2a\dot{x}x + a\dot{x}y - 3\dot{y}y^2 + a\dot{y}x = 0$, and so $\dot{y}:\dot{x} = 3x^2 - 2ax + ay : 3y^2 - ax = BD$ (or y): BT. Therefore

$$BT = (3y^3 - axy)/(3x^2 - 2ax + ay).^{(193)}$$

Consequently, given the point *D*, and thence *DB* and *AB* or y and x, there will be given the length *BT* by which the tangent *TD* is determined.

This method of working may, however, be neatly presented in this way. Make the terms of the exhibited equation equal to nothing, multiply them by their respective number of dimensions of the ordinate quantity and place the result in the numerator; then multiply the terms of the same equation by their respective number of dimensions of the base and, having divided the result by the base, place it in the denominator of *BT*'s value.(194) Take *BT* in the direction away from *A* if its value be positive, or towards *A* if it be negative.

(192) In borrowing this example from his page 41 above Newton has apparently failed to notice that it here represents the degenerate pair of the line $x - y = 0$ and the ellipse

$$x^2 + xy + y^2 - ax = 0.$$

Since the former has unit slope, his algorithm merely finds (in unduly complicated form) the slope at a general point on the latter conic. Compare the following note.

(193) When $x \neq y$ BT will evidently be the subtangent $-y(x+2y)/(2x+y-a)$ of the ellipse $x^2 + xy + y^2 = ax$, for then $3y^2 - ax = -(x+2y)(x-y)$ and

$$3x^2 - 2ax + ay = (2x + y - a)(x - y).$$

A cancelled first opening to the next sentence reads 'Quamobrem data vel assumpta utcunq; longitudine *AB* pro x et inde per expositam [æquationem]'. (Therefore, being given or assuming the length *AB* in place of x in any manner whatever, and from this by means of the exhibited equation....)

(194) The Huddenian algorithm applied to tangents: compare the 'universal theorem for drawing tangents to crooked lines when x & y intersect at any determined angle' developed by Newton in May 1665 (I: 279–80). Note how Newton attaches a necessary negative sign to the absolute value of *BT* by ordaining in the following sentence that its geometrical sense be reversed.

$$\begin{matrix} & 0. & 0. & 1. & 3. \\ \end{matrix}$$

||[44] Sic æquatio $x^3 - axx + axy - y^3 = 0$ per superiores || numeros multiplicata dat

$$\begin{matrix} & 3. & 2. & 1. & 0. \\ \end{matrix}$$

$axy - 3y^3$ pro numeratore, et per inferiores multiplicata ac divisa per x dat $3xx - 2ax + ay$ pro denominatore valoris BT.

Geom: Cart: p 42 Sic æquatio $y^3 - byy - cdy + bcd + dxy = 0$ (quæ designat Parabolam secundi generis cujus beneficio Descartes construxit æquationes 6 dimensionum)[(195)] primâ fronte dat $\frac{3y^3 - 2byy - cdy + dxy}{dy}$, sive $\frac{3yy}{d} - \frac{2by}{d} - c + x = BT$.

Et sic $aa - \frac{r}{q}xx - yy = 0$ (quæ designat Ellipsin cujus centrum A) dat $\frac{-2yy}{-2\frac{r}{q}x}$ sive $\frac{qyy}{rx} = BT$. Et sic in alijs.

Et nota quod nihil interest cujusnam sit angulus ordinationis ABD.[(196)]

Ast hæc Regula se ad æquationes surdis quantitatibus affectas Curvasq Mechanicas non extendit. In istis casibus ad fundamentalem methodum recurrendum est.

pag 19[(197)] Exempl 2. Esto $x^3 - ayy + \frac{by^3}{a+y} - xx\sqrt{ay+xx} = 0$ æquatio designans relationem inter AB et BD, et per Prob 1 relatio fluxionum erit

$$3mxx - 2any + \frac{3abnyy + 2bny^3}{aa + 2ay + yy} \frac{-4amxy - 6mx^3 - anxx}{2\sqrt{ay+xx}} = 0.$$

Atq adeò est $3xx \frac{-4axy - 6x^3}{2\sqrt{ay+xx}} . 2ay \frac{+3abyy + 2by^3}{aa + 2ay + yy} \frac{-axx}{2\sqrt{ay+xx}} (::n.m)::BD.BT$.

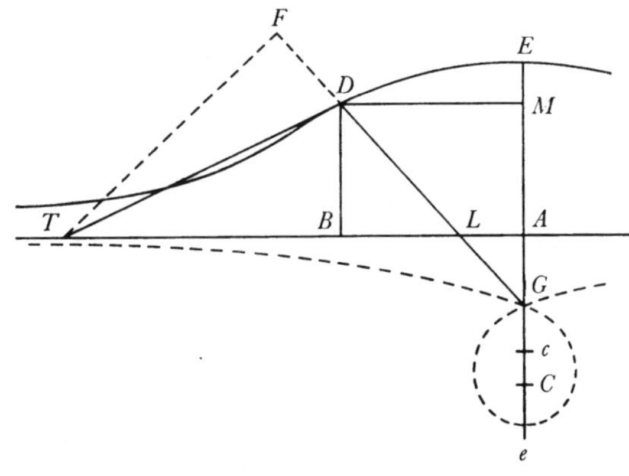

Exempl 3. Sit ED Conchoïdes Nichomedea[(198)] Polo G, Asymptoto AT et intervallo LD descripta. Sitq $GA = b$, $LD = c$, $AB = x$ et $BD = y$. Et propter similia triangula DBL et $[D]MG$ erit $\underset{\sqrt{cc-yy}}{LB} . \underset{y}{BD} :: \underset{x}{DM} . \underset{b+y}{MG}$ Adeoq $\overline{b+y}$ in $\sqrt{cc-yy} = yx$. Nactus hanc æquationem fingo $\sqrt{cc-yy} = z$

||[45] et sic duas æquationes || $bz + yz = yx$ et $zz = cc - yy$ habeo. Quarum ope

(195) Namely, the Cartesian trident $dxy = (b-y)(y^2 - cd)$ used by Descartes in the third book of his *Geometrie* to construct the real roots of a general sextic by its meets with a defined circle (*Geometria*, ($_2$1659): 97–104: 'Modus generalis construendi Problemata omnia, reducta ad Æquationem, sex dimensiones non excedentem'). Compare I: 495. Earlier, in his second book, Descartes had generated his 'second-order parabola' as the meet of a translating parabola

$$\begin{matrix} & 0. & 0. & 1. & 3. \\ \text{So the equation} & x^3 & -ax^2 & +axy & -y^3 = 0 \\ & 3. & 2. & 1. & 0. \end{matrix}$$

when multiplied by the superior numbers gives $axy-y^3$ for the numerator, and when multiplied by the lower ones and then divided by x it gives $3x^2-2ax+ay$ for the denominator of BT's value.

So the equation $y^3-by^2-cdy+bcd+dxy=0$ (which represents the second order parabola with whose aid Descartes constructed equations of six dimensions)[195] gives at first glance $(3y^3-2by^2-cdy+dxy)/dy$ or

$$3y^2/d-2by/d-c+x=BT.$$

Descartes, *Geometry:* p. 42.

And so $a^2-(r/q)\,x^2-y^2=0$ (which represents an ellipse of centre A) gives $-2y^2/-2(r/q)\,x$ or $qy^2/rx=BT$. And so in other instances.

Note, too, that the size of the angle $\widehat{ABD}$ of ordination is of no concern here.[196]

But this rule does not extend to equations affected with surd quantities and to mechanical curves. In those cases you must revert to the fundamental method.

EXAMPLE 2. Let $x^3-ay^2+by^3/(a+y)-x^2\sqrt{[ay+x^2]}=0$ be the equation representing the relationship between AB and BD. By Problem 1 the relation of the fluxions will be $3\dot{x}x^2-2a\dot{y}y+\dfrac{3ab\dot{y}y^2+2b\dot{y}y^3}{a^2+2ay+y^2}-\dfrac{4a\dot{x}xy+6\dot{x}x^3+a\dot{y}x^2}{2\sqrt{[ay+x^2]}}=0$ and

See p. 19[197]

so $3x^2-\dfrac{4axy+6x^3}{2\sqrt{[ay+x^2]}}:2ay+\dfrac{3aby^2+2by^3}{a^2+2ay+y^2}-\dfrac{ax^2}{2\sqrt{[ay+x^2]}}\;(=\dot{y}:\dot{x})=BD:BT.$

EXAMPLE 3. Let ED be a conchoid of Nichomedes[199] described with pole G, asymptote AT and distance LD, and let $GA=b$, $LD=c$, $AB=x$ and $BD=y$. Because of the similar triangles DBL and DMG there will be

$$LB(\sqrt{[c^2-y^2]}):BD(y)=DM(x):MG(b+y)$$

and so $(b+y)\sqrt{[c^2-y^2]}=yx$. Having gained this equation I suppose

$$\sqrt{[c^2-y^2]}=z$$

and thus have the two equations $bz+yz=yx$ and $z^2=c^2-y^2$. With their help

with a rotating radius vector (*Geometria*, $_2$1659: 23). Newton's marginal reference relates, somewhat illogically, to a passage in the second book (*Geometria*, $_2$1659: 42: 'Aliud Exemplum [inveniendi datarum curvarum contingentes] in Parabola secundi generis') where he begins computation of its subnormal: compare Newton's third example on I: 277.

(196) Compare I: 279–80 and the equivalent passage in the 1666 tract (I: 417).

(197) See Example 4 on (Newton's) page 19, and compare Example 2 on page 42.

(198) The Cartesian equivalent 'prima Conchoïdes Veterum' (first conchoid of the ancients) is cancelled. With the addition of the tangent DT, indeed, and apart from the insertion (in broken line) of the conchoid's lower branch and the construction lines DF, FT, Newton's figure is a close copy of Descartes' (*Geometria*, $_2$1659: Liber II: 49). The points C, c are the centres of curvature at the points E, e on the conchoid (see Newton's page 66 below). Though Newton does not make this limitation explicit, the 'ordination' angle $\widehat{ABD}$ is right.

fluxiones[199] quantitatum x, y, et z (per Prob 1) quæro et e prima prodit $br+yr+nz=nx+my$, ac e secunda $2rz=-2ny$, sive $rz+ny=0$. E quibus exterminato r, oritur $\frac{-bny}{z}-\frac{nyy}{z}+nz=nx+my$. Quâ resolutâ fit

$$y.z-\frac{by}{z}-\frac{yy}{z}-x(::n.m)::BD.BT.$$

Cùm ergo BD sit $=y$, erit $BT=z-x\frac{-by-yy}{z}$. Hoc est $-BT=AL+\frac{BD\times GM}{BL}$. Ubi signum $-$ ipsi BT præfixum denotat punctum T ad partes adversus A capiendum esse.

SCHOL: Et hinc obiter inventio puncti disterminantis concavam et convexam partem Conchoidis prodit. Nempe cùm AT sit omnium minima, erit D ejusmodi punctum.[200] Esto itaq; $AT=v$, et cùm sit $BT=-z+x\frac{+by+yy}{z}$ erit $v=-z+2x\frac{+by+yy}{z}$. Ubi ad opus abbreviandum pro x substitue $\frac{bz+yz}{y}$ valorem e superioribus erutum et fiet $\frac{2bz}{y}+z\frac{+by+yy}{z}=v$. Unde per Prob 1 fluxionibus l, n, et r quæsitis, et per Prob 3 supposita $l=0$, emerget

$$\frac{2br}{y}-\frac{2bnz}{yy}+r+\frac{bn+2yn}{z}\frac{-bry-ryy}{zz}=(l=)0.$$

In hâc deniq; substitue $\frac{-ny}{z}$ pro r et $cc-yy$ pro zz (valores r et zz e superioribus petendos) et facta reductione obtinebitur $y^3+3byy-2bcc=0$. Cujus æquationis constructione dabitur y sive AM; et per M acta MD ipsi AB parallela incidet in punctum flexûs contrarij D.[201]

[45a] [202]‖ Præterea si curva Mechanica est cujus tangentem ducere oportet, quantitatum fluxiones ut in exemplo 5 Prob 1 quærendæ sunt, cæteraq; ut in præcedentibus peragenda.

(199) Newton first wrote 'velocitates' (speeds).

(200) Newton does not prove this Huygenian criterion (see II, 2, 2, §2: note (9)) and it is not in fact universally valid. In analytical terms a necessary test for the subtangent

$$AT = t = x-y(dx/dy)$$

to attain a minimum (or, in general, an extreme value) is that

$$-dt/dx = y(dy/dx)(d^2x/dy^2) = 0.$$

Evidently the subtangent AT reaches a local maximum or minimum value at any intersection of a given curve $f(x, y) = 0$ with the base AB ($y = 0$), while it becomes infinite for any point on the curve of zero slope. (Conversely any inflexion point on the curve which lies at an intersection with AB does not correspond with any local extreme value of AT.) When, however, the given curve has an inflexion outside the base AB at a point where the curve has non-zero

I seek the fluxions(199) of the quantities x, y and z (by Problem 1), and from the first there comes $b\dot{z}+y\dot{z}+\dot{y}z = \dot{y}x+\dot{x}y$ and from the second $2\dot{z}z = -2\dot{y}y$, or $\dot{z}z+\dot{y}y = 0$. On eliminating $\dot{z}$ from these there arises

$$-b\dot{y}y/z-\dot{y}y^2/z+\dot{y}z = \dot{y}x+\dot{x}y,$$

and by resolving this there comes $y:(z-by/z-y^2/z-x)$ $(=\dot{y}:\dot{x}) = BD:BT$. In consequence, since BD is equal to y, $BT = z-x-(by+y^2)/z$, that is,

$$-BT = AL+BD\times GM/BL.$$

Here the sign $-$ prefixed to BT indicates that the point T must be taken on the side away from A.

SCHOLIUM. From this, by the way, the location of the point separating concave and convex portions of the conchoid results immediately—specifically, when AT is least of all, D will be a point of this type.(200) Let, therefore, $AT = v$ and then, since $BT = -z+x+(by+y^2)/z$, will there be

$$v = -z+2x+(by+y^2)/z.$$

Here, to shorten work, in place of x substitute $(bz+yz)/y$, a value derived from the preceding, and there will come $2bz/y+z+(by+y^2)/z = v$. Hence when, following Problem 1, the fluxions $\dot{v}$, $\dot{y}$ and $\dot{z}$ have been found and, following Problem 3, $\dot{v}$ is supposed zero, there will emerge

$$2b\dot{z}/y-2b\dot{y}z/y^2+\dot{z}+(b\dot{y}+2y\dot{y})/z-(b\dot{z}y+\dot{z}y^2)/z^2 = (\dot{v} =)\ 0.$$

In this, finally, substitute $-\dot{y}y/z$ for $\dot{z}$ and c^2-y^2 for z^2 (values of $\dot{z}$ and z^2 to be had from the preceding) and after reduction there will be obtained

$$y^3+3by^2-2bc^2 = 0.$$

The construction of this equation will give y or AM, and when MD is drawn through M parallel to AB it will meet with the point D of contrary flexure.(201)

(202)Further, if the curve whose tangent must be drawn is mechanical, the fluxions of quantities are to be sought as in Example 5 of Problem 1 and the rest performed as in the preceding.

---

slope, an extreme value of AT will correspond to the (sufficient) condition for an inflexion point $d^2x/dy^2 = 0$. Happily, in Newton's present example of the conchoid the inflexion point D obeys both these requirements. See also R.-F. de Sluse's *Miscellanea* (Liège, 1668): 117–30.

(201) See II: 200. In first draft of page 45 Newton here passed straight on to the following Mode 2. Six pages later, having (on a first version of his following pages 51–2) added three concluding examples which illustrate the application of Mode 1 to 'mechanical' (that is, transcendental) curves, Newton at once decided to insert them earlier in his scheme at their present place, to his original instances adding the two further Examples 7 and 8 below.

(202) The two following pages, not numbered by Newton himself but here paginated as 45a/45b for convenience of reference, are interposed in accordance with Newton's instruction that 'This leafe must bee inserted in the middle of pag 45'.

EXEMPL 4. Sunto AC et AD duæ curvæ quibus recta BCD ad Basin AB in dato angulo applicata occurrit in C et D. Et appelletur $AB=x$, $BD=y$, et $\frac{\text{area } ACB}{1}=z$;(203) et per Prob: 1, Præparat: ad Exempl 5, erit $r=m\times BC$.

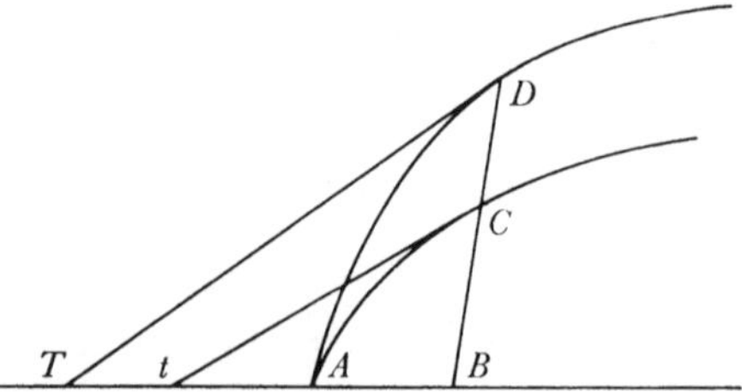

Jam sit AC circulus aut curva quævis nota et ad alteram curvam AD definiendam exponatur quævis æquatio cui z intexta est(204) veluti $zz+axz=y^4$. Et per Prob 1 erit $2rz+axr+amz=4ny^3$. Et scripto $m\times BC$ pro r, fiet $2mz\times BC+amx\times BC+amz=4ny^3$. Adeoq₃ $2z\times BC+ax\times BC+az.\ 4y^3(::n.m)::BD.BT$. Quamobrem si ex natura curvæ AC detur ordinata BC et area ACB sive z, dabitur punctum T per quod tangens DT transibit.

Ad eundem modum si $3z=2y$ sit æquatio ad curvam AD, erit

$$3r(3m\times BC)=2n.$$

Adeoq₃ $3BC.2(::n.m)::BD.BT$. Et sic in alijs.

EXEMPL: 5. Sit $AB=x$, $BD=y$ ut ante, et Curvæ cujusvis(205) AC longitudo sit z; ductâq₃ ad eam tangente Ct, erit $Bt.Ct::m.r$, sive $r=\frac{m\times Ct}{Bt}$.

Jam ad aliam curvam AD cujus tangens ducenda est, detur quælibet æquatio in qua z involvitur, puta si $z=y$, erit $r=n$. Adeoq₃ $Ct.Bt(::n.m)::BD.BT$. Invento autem T age DT tangentem.(206)

Sic posito $xz=yy$ erit $mz+rx=2ny$, et pro r scripto $\frac{m\times Ct}{Bt}$, emerget $mz+\frac{mx\times Ct}{Bt}=2ny$. Quare est $z+\frac{x\times Ct}{Bt}.2y::BD.[B]T$.

EXEMPL: 6. Sit AC circulus aut alia quævis nota curva quam tangat Ct, et sit AD alia curva cujus tangentem DT ducere oportet, et quæ definitur assu-

(203) The division by unity is made to preserve homogeneity in dimensions.

(204) In his first version of this example Newton supposed $A\hat{B}C=\frac{1}{2}\pi$ with

$$BC=\sqrt{[ax-x^2]},$$

so that the curve AC is a circle of diameter a. His present method of constructing the curve AD in accord with some given defining relation $f(x,y,z)=0$, where $AB=x$, $BD=y$ and $(ACB)=z$, is evidently a generalization of the simple areal transformation $ry=z$ applied several times by Isaac Barrow in his *Lectiones Geometricæ* (note (80)): see, for instance, Lectiones X, §11 and XI, §19 (pages 78 and 90–1 of the 1670 edition). Barrow's transform is itself closely akin to the one developed in 1658 by William Neil and generalized in 1668 by James Gregory in Proposition 6 of his *Geometriæ Pars Universalis* (Padua, 1668): 17–19, Newton's library copy of which is now in Trinity College, Cambridge (NQ. 9.48[2]). Compare D. T. Whiteside,

EXAMPLE 4. Let AC and AD be two curves met by the straight line BCD, applied to the base AB at a given angle, in C and D. Let also AB be called x, BD y and the area ACB divided by unity(203) z. Then, by Problem 1: Preparation for Example 5, will there be $\dot{z} = \dot{x} \times BC$.

Now let AC be a circle or any known curve you like and to define the other curve AD let there be exhibited any equation at will, in which z is interwoven,(204) as for instance $z^2 + axz = y^4$. By Problem 1 there will be $2\dot{z}z + ax\dot{z} + a\dot{x}z = 4\dot{y}y^3$ and, on writing $\dot{x} \times BC$ in place of $\dot{z}$, this will become

$$2\dot{x}z \times BC + a\dot{x}x \times BC + a\dot{x}z = 4\dot{y}y^3,$$

and so $(2z \times BC + ax \times BC + az) : 4y^3 (= \dot{y} : \dot{x}) = BD : BT$. Wherefore if from the nature of the curve AC there be given the ordinate BC and area ACB (or z), the point T will be given through which the tangent DT shall pass.

After the same fashion if $3z = 2y$ be the equation to the curve AD, then will it be $3\dot{z}$(or $3\dot{x} \times BC$) $= 2\dot{y}$ and so $3BC : 2 (= \dot{y} : \dot{x}) = BD : BT$. And so in other instances.

EXAMPLE 5. Let $AB = x$, $BD = y$ as before, and the length of any curve AC(205) be z: then, when Ct is drawn tangent to it, $Bt : Ct = \dot{x} : \dot{z}$, or

$$\dot{z} = \dot{x} \times Ct/Bt.$$

Now to any other curve AD whose tangent is to be drawn let there be given any equation you please involving z, say, $z = y$, when $\dot{z} = \dot{y}$ and so $Ct : Bt (= \dot{y} : \dot{x}) = BD : BT$. When T is found, draw DT tangent.(206)

So, supposing $xz = y^2$, there will be $\dot{x}z + \dot{z}x = 2\dot{y}y$ and with $\dot{x} \times Ct/Bt$ written in place of $\dot{z}$ there will emerge $\dot{x}z + \dot{x}x \times Ct/Bt = 2\dot{y}y$. Hence

$$(z + x \times Ct/Bt) : 2y = BD : BT.$$

EXAMPLE 6. Let AC be a circle or any known curve you wish to which Ct shall be tangent, and AD another curve whose tangent DT must be drawn and

---

'Patterns of Mathematical Thought in the later Seventeenth Century', *Archive for History of Exact Sciences*, **1** (1961): 179–388, especially 366–7.

(205) In first draft Newton set $BC = \sqrt{[ax + x^2]}$ as the 'æquatio ad curvam AC (quæ proinde erit Hyperbola)'.

(206) The method of constructing the curve AD by means of some given defining equation $f(x, y, z) = 0$, where $AB = x$, $BD = y$, $\widehat{AC} = z$ is Newton's generalization of a method of Fermat's (*De Linearum Curvarum cum Lineis Rectis Comparatione Dissertatio Geometrica*, Toulouse, 1660: Proposition 6 [= *Œuvres de Pierre Fermat* (ed. P. Tannery and C. Henry), **1** (Paris, 1891): 228-33]), restated more concisely by James Gregory (*Geometriæ Pars Universalis* (note (204)): Propositio 9: 24–5) and yet again by Barrow (*Lectiones Geometricæ* (note (80)): Lectio x §§5/6: 76–7). Newton's example (in which $\widehat{AC} = BD$) is, to be sure, exactly Fermat's original example but he almost certainly took as his model the equivalent accounts in Gregory and Barrow.

mendo AB = arcui AC, et (CE, ac BD in dato angulo ad AB ordinatis) referendo BD ad CE vel AE in æquatione aliqua. Dic ergo AB vel $AC = x$, $BD = y$, $AE = z$, et $CE = v$, et patet l, m, et r fluxiones ipsarum CE, AC, et AE esse inter se ut sunt CE, Ct et Et, Adeoq; $m \times \frac{CE}{Ct} = l$, et $m \times \frac{Et}{Ct} = r$.

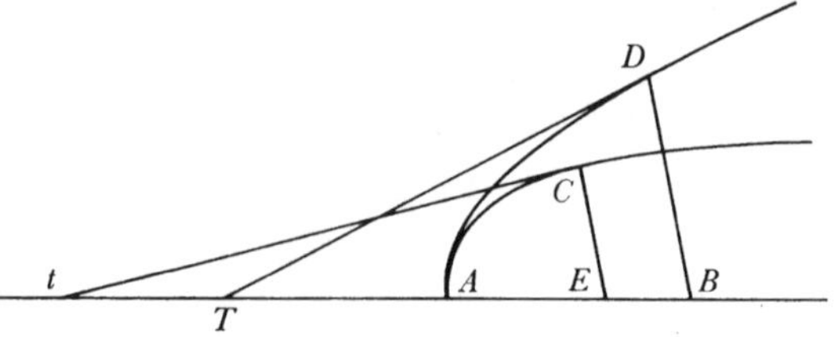

Detur jam quælibet æquatio ad definiendam Curvam AD, veluti $y = z$, et erit $n = r$, Adeoq; $Et.Ct(::n.m)::BD.BT$.

Vel detur $y = z + v - x$, et erit $n = (l + r - m =)\frac{m \times CE + Et - Ct}{Ct}$. Adeoq; $CE + Et - Ct.Ct(::n.m)::BD.BT$.

Vel deniq; detur $ayy = v^3$, et erit $2any = (3lvv =)\frac{3mvv \times CE}{Ct}$. Adeoq; $3vv \times CE.2ay \times Ct::BD.BT.$(207)

Exempl 7. Sit FC circulus quem tangat [CS], sitq; FD Curva quæ definitur assumendo quamvis relationem applicatæ DB ad FC arcum quem DA ad centrum ducta intercipit. Et demissa CE in circulo
||[45b] || applicata dic AC vel $AF = 1$, $AB = x$, $BD = y$, $AE = z$, $CE = v$, $CF = t$ et ipsius t fluxionem k, et erit $k[z]\left(= \frac{k \times CE}{CS}\right) = l$,(208) et $-kv\left(= \frac{k \times - ES}{CS}\right) = r$.(208) Ubi pono r negativè, quòd AE diminuitur dum EC augetur. Est insuper $AE.EC::AB.BD$, adeoq; $zy = vx$ et inde per Prob: 1, $ry + nz = lx + mv$. Et hæc, exterminatis l, r, et v, faciunt $nx - kyy - kxx = my$.

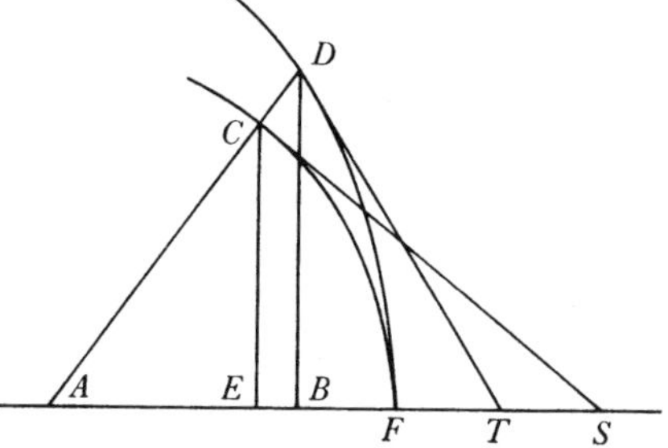

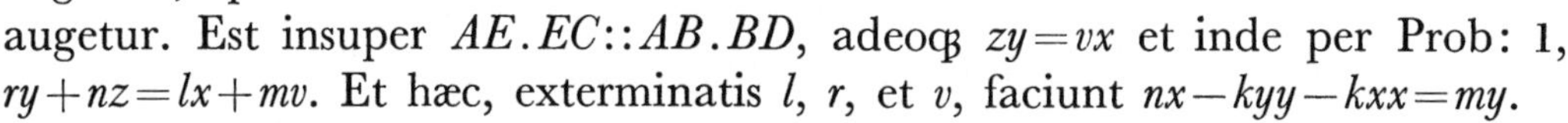

Definiatur jam curva DF æquatione quavis a qua valor k hic substituendus deduci possit: puta sit $t = y$ (æquatio ad primam Quadratricem,(209)) et per Prob 1 erit $k = n$. Adeoq; $nx - nyy - nxx = my$. Unde $y.xx + yy - x(::n.m)::DB(y).BT$. Quare $BT = xx + yy - x$. et $AT = xx + yy = \frac{AD^q}{AF}$.

(207) A further Newtonian generalization of the Fermatian structure underlying the previous example (see note (206)). Note that the defining equation $f(x, y, z, v) = 0$ of the curve AD is referred to new coordinates $AB(=\widehat{AC}) = x$, $BD = y$ inclined at the same angle as those ($AE = z$, $CE = v$) which determine the curve AC. Since

$$dx^2 = dz^2 - 2dz.dv.\cos A\hat{E}C + dv^2,$$

the variables x, z, v are not independent.

which is defined by assuming AB equal to the arc $\widehat{AC}$ and (where CE and BD are ordinate at a given angle to AB) by relating BD to CE or AE in some equation. Call therefore AB or $\widehat{AC} = x$, $BD = y$, $AE = z$ and $CE = v$. Evidently the fluxions $\dot{v}$, $\dot{x}$ and $\dot{z}$ of CE, AC and AE are to one another as CE, Ct and Et, and so

$$\dot{x}\times CE/Ct = \dot{v}, \quad \dot{x}\times Et/Ct = \dot{z}.$$

Now let any equation be given to define the curve AD, as for instance $y = z$: then will there be $\dot{y} = \dot{z}$, and so $Et:Ct(=\dot{y}:\dot{x}) = BD:BT$.

Or let there be given $y = z+v-x$: then will there be

$$\dot{y} = (\dot{v}+\dot{z}-\dot{x} =)\,\dot{x}(CE+Et-Ct)/Ct$$

and consequently $(CE+Et-Ct):Ct(=\dot{y}:\dot{x}) = BD:BT$.

Or, finally, let there be given $ay^2 = v^3$: then will there be

$$2a\dot{y}y = (3\dot{v}v^2 =)\,3\dot{x}v^2\times CE/Ct \quad \text{and so} \quad 3v^2\times CE:2ay\times Ct = BD:BT.^{(207)}$$

EXAMPLE 7. Let FC be a circle touched by CS and FD a curve defined by assuming any relationship you wish between the ordinate DB and the arc $\widehat{FC}$ cut off by drawing DA to the [circle's] centre. Letting fall the circle ordinate CE, call AC or $AF = 1$, $AB = x$, $BD = y$, $AE = z$, $CE = v$, $\widehat{CF} = t$ and its fluxion $\dot{t}$: then will there be $\dot{t}z(=\dot{t}\times CE/CS) = \dot{v}^{(208)}$ and

$$-\dot{t}v(=\dot{t}\times -ES/CS) = \dot{z}.^{(208)}$$

(I put $\dot{z}$ negative here since AE diminishes as EC increases.) Further

$$AE:EC = AB:BD$$

and hence $zy = vx$, and from this by Problem 1 $\dot{z}y+\dot{y}z = \dot{v}x+\dot{x}v$. On eliminating $\dot{v}$, $\dot{z}$ and v these make $\dot{y}x-\dot{t}y^2-\dot{t}x^2 = \dot{x}y$.

Now let the curve DF be defined by any equation, from which may be drawn the value of $\dot{t}$ here to be replaced: let, say, $t = y$ (the equation to the primary quadratrix)(209) and by Problem 1 there will be $\dot{t} = \dot{y}$, and consequently

$$\dot{y}x-\dot{y}y^2-\dot{y}x^2 = \dot{x}y.$$

Hence $y:(x^2+y^2-x)\,(=\dot{y}:\dot{x}) = DB(y):BT$. Therefore $BT = x^2+y^2-x$ and $AT = x^2+y^2 = AD^2/AF$.

(208) These equalities imply $AE:EC:AC = CE:ES:CS$, so that the triangles AEC, CES are similar: hence the angle $A\hat{E}C$ (and so $A\hat{B}D$) must be right. Correspondingly below Newton sets $AD^2 = AB^2+BD^2$. Note the unusual choice of k to denote the fluxion of t (elsewhere usually—for instance on Newton's pages 72 and 107 below—taken as q).

(209) Hence, since $t = \tan^{-1}(y/x)$, the explicit Cartesian defining equation of this classical quadratrix of Dinostratus is $x = y\cot y$. The result $AT = AD^2/AF\,(x-y\,dx/dy = x^2+y^2)$ is Barrow's (*Lectiones Geometricæ* (note (80))): Lectio x, Appendix, Example v: 84).

Ad eundem modum si sit $tt = by$, proveniet $2kt = bn$, et inde $AT = \frac{b \times ADq}{2t \times AF}$. Et sic in alijs.(210)

Exempl 8. Quod si AD sumatur æqualis arcui FC,(211) existente ADH spirali Archimedea, tum stantibus jam positis linearum nominibus, est (propter ang: ABD rect:) $xx + yy = tt$. et inde per Prob 1 $mx + ny = kt$. Est etiam $AD.AC :: DB.CE$, adeoqꝫ $tv = y$ et inde per Prob 1 $kv + tl = n$. Deniqꝫ est fluxio arcus FC ad fluxionem rectæ CE ut AC ad AE sive ut AD ad AB hoc est $k.l :: t.x$, et inde $kx = tl$. Confer jam inventas æquationes et videbis esse $kv + kx = n$, et inde $mx + ny (= kt) = \frac{nt}{v + x}$. Atqꝫ adeò (completo parallelogrammo $ABDQ$) si fiat

$$QD.QP (:: BD.BT :: n. -m) :: x.y - \frac{t}{v + x},$$

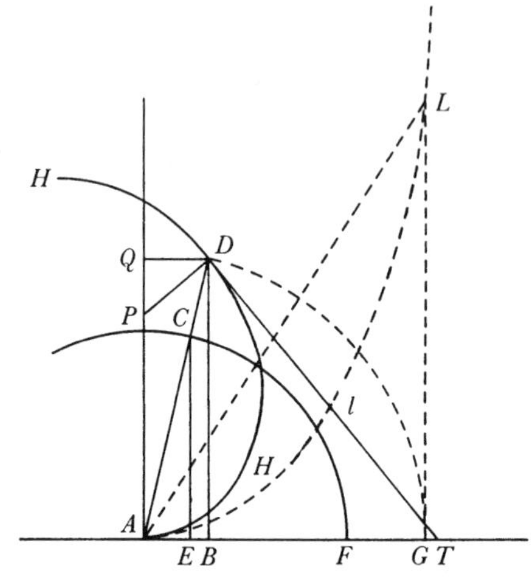

hoc est si capiatur $AP = \frac{t}{v + x}$, erit PD ad spiralem perpendicularis.(212)

Ex his opinor satis manifestum est quo pacto curvarum omnium tangentes ducendæ sunt.(213) Attamen non abs re erit si præterea confectionem Problematis ubi curvæ alijs quibuscunqꝫ modis ad rectas referuntur ostendero, ut e pluribus Methodis facillima et simplicissima(214) semper possit adhiberi.

||[45 *rursum*] || Mod: 2.(215)

Sit itaqꝫ D punctum in curva a quo subtensa DG ducitur ad datum punctum G ac DB in dato quovis angulo(216) ordinatur ad Basin AB. Punctum verò D per infinitè parvum intervallum Dd in curva fluat, inqꝫ GD sumatur GK æqualis Gd et compleatur parallelogrammum $dbBC$. Et erunt DK ac DC contemporanea momenta ipsarum GD et BD,(217) quibus nempe di-

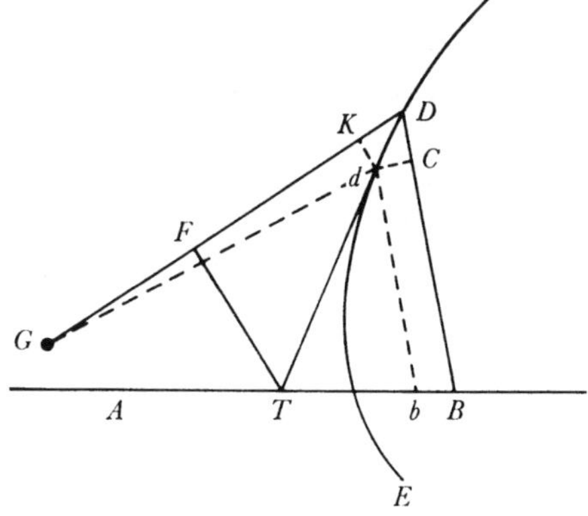

(210) In its structure this example is a Newtonian variant on the generalized quadratrix discussed by Barrow in his *Lectiones Geometricæ* (note (80)): Lectio x, §10: 77–8. Not surprisingly, Barrow's example is the 'Quadratrix communis'. Compare Mode 8 below.

(211) Newton continues to assume the curve FC to be a circle of centre A.

(212) The 'involuted' spiral equivalent of the preceding example, based presumably on Barrow's *Lectiones Geometricæ* (note (80)): Lectio x, §§7–9: 77. Barrow, it should be noted,

In much the same way if $t^2 = by$, there will come $2\dot{t}t = b\dot{y}$ and thence $AT = b \times AD^2/2t \times AF$. And so in other instances.(210)

EXAMPLE 8. But if AD be assumed equal to the arc $\widehat{FC}$,(211) when the curve ADH proves to be an Archimedean spiral, then (the denominations of the lines already settled on still standing) it will, because the angle $A\hat{B}D$ is right, be $x^2+y^2 = t^2$ and thence by Problem 1 $\dot{x}x+\dot{y}y = \dot{t}t$. Also $AD:AC = DB:CE$, so that $tv = y$ and thence, by Problem 1, $\dot{t}v+t\dot{v} = \dot{y}$. Finally, the fluxion of the arc $\widehat{FC}$ to that of the straight line CE is as AC to AE or as AD to AB; that is, $\dot{t}:\dot{v} = t:x$ and thence $\dot{t}x = t\dot{v}$. Compare these equations now found and you will see that it is $\dot{t}v+\dot{t}x = \dot{y}$ and thence $\dot{x}x+\dot{y}y(=\dot{t}t) = \dot{y}t/(v+x)$. Consequently if, on completing the rectangle $ABDQ$, there is made

$$QD:QP(=BD:BT=\dot{y}:-\dot{x}) = x:y-t/(v+x),$$

in other words, if there be taken $AP = t/(v+x)$, then PD will be normal to the spiral.(212)

From these instances it is, I believe, sufficiently clear how tangents are to be drawn to every curve.(213) Nevertheless it will not be irrelevant for me to indicate besides how the problem may be accomplished when curves are related in any other manner to straight lines—accordingly, you will then(214) be able to employ of these several methods that which is easiest and simplest.

MODE 2(215)

So let D be a point in a curve from which a subtense DG is drawn to a given point G while DB is an ordinate to the base AB at any given angle.(216) Let this point D, indeed, flow through the indefinitely small distance Dd in the curve, and in GD take GK equal to Gd, then complete the parallelogram $dbBC$. Then will DK and DC be contemporaneous moments of GD and BD,(217) those,

---

allows the curve FC to be general but at once (§9) takes as his example Archimedes' 'spiralis circularis'. The broken lines in the figure relate to Newton's page 75.

(213) In first draft of the following Newton wrote (on his page 45) 'Cæterùm ad Tangentium doctrinam revertamur earum determinationem jam aggress[ur]i cum curvæ alijs modis ad rectas referuntur' (But let us return to the doctrine of tangents and be now ready to set about their determination when curves are related in other ways to straight lines).

(214) In his revision of an earlier draft Newton has here omitted a clarifying phrase 'pro singulis curvis' (for individual curves).

(215) 'Focus-directrix' coordinates, an evident generalization of the conic case when BD and GD are in constant ratio.

(216) In this instance a trivial generalization since DB is merely a constant multiple (namely cosec $A\hat{B}D$) of the perpendicular let fall from D to AB. In a first version of the accompanying figure A was the foot of the perpendicular from G onto AB and $A\hat{B}D$ was a right angle.

(217) It follows at once that DK, DC are proportional to the fluxions of GD, BD.

||[46] minuuntur dum D transfertur ad d. Jam || Dd rectâ producatur donec cum AB conveniat in T et ab isto T ad subtensam GD demittatur perpendiculum TF et erunt Trapezia $DCdK$ ac $DBT[F]$ simila adeoq $DB.DF::DC.DK$.

Cùm itaq relatio BD ad GD in æquatione qualibet, pro curva definienda exponitur, quære relationem fluxionum et cape FD ad DB in ratione fluxionis GD ad fluxionem BD. Dein ab F erige perpendiculum FT quod cum AB concurrat in T et acta TD curvam tanget in D. Cape autem DF versus G si sit affirmativa; sin secus, cape ad contrarias partes.

Exempl 1. Dic $GD=x$ et $BD=y$ et esto earum relatio

$$x^3-axx+axy-y^3=0.^{(218)}$$

eritq fluxionum ratio $3mxx-2amx+amy+axn-3nyy=0$. Atq adeò

$$3xx-2ax+ay.3yy-ax(::m.n^{(219)})::DB(y).DF.$$

Est ergo $DF=\dfrac{3y^3-axy}{3xx-2ax+ay}$. Adeoq dato quolibet in curva puncto D, et inde BD et GD sive y et x; dabitur punctum F: Unde si normalem FT erigas; ad ejus concursum cum basi AB ducta DT curvam tanget.

Et hinc patet Regulam perinde ac in priori casu concinnari posse. Scilicet æquationis expositæ terminos omnes ad easdem partes dispone et sigillatim per dimensiones ordinatæ y multiplica et exitum colloca in Numeratore. Dein terminos ejus sigillatim per dimensiones subtensæ x multiplica, et exitum per subtensam illam x divisum colloca in Denominatore valoris DF. Illamq DF cape ad partes contra G si sit affirmativa, sin secus, cape ad easdem partes. Et nota quod nihil intersit quanto intervallo punctum G distat a Basi AB, si fortè distat, neq quinam sit angulus ordinationis ABD.

Sic æquatio superior $x^3-axx+axy-y^3=0$ prima fronte dat $axy-3y^3$ pro numeratore et $3xx-2ax+ay$ pro Denominatore valoris DF.

Sic etiam $a+\frac{b}{a}x-y=0$, (quæ æquatio est ad Conicam sectionem[(220)]) dat $-y$ pro numeratore et $\frac{b}{a}$ pro denominatore valoris DF, quæ ideo erit $=\dfrac{-ay}{b}$.[(221)]

(218) Again (compare notes (192) and (193)) this degenerate locus

$$(x-y)(x^2+xy+y^2-ax) = 0$$

is somewhat thoughtlessly used to exemplify the general method. If α is the perpendicular distance of G from AB and $k = \operatorname{cosec}A\hat{B}D$, then the present coordinate system is related to a standard Cartesian one (in which G is the origin) by $x = \sqrt{[X^2+Y^2]}$, $y = k(X+\alpha)$. Evidently $x^2+xy+y^2 = ax$ is the defining equation of a quartic curve, while $x = y$ represents the conic $X^2+Y^2 = k^2(X+\alpha)^2$ of focus G, directrix $AB(X+\alpha = 0$ or $y = 0)$ and eccentricity k. Since $DF = (dx/dy)DB$ it follows in the latter case that $DF = DB$.

namely, by which they are diminished as D passes into d. Now let Dd be produced in a straight line till it meet AB in T and from that point T let fall the perpendicular TF to the subtense GD: the quadrilaterals $DCdK$ and $DBTF$ will then be similar, so that $DB:DF = DC:DK$.

When, therefore, the relationship of BD to GD is exhibited in any equation at pleasure for determining the curve, seek the relation of the fluxions and take FD to DB in the ratio of the fluxion of GD to that of BD. Then from F erect the perpendicular FT to its meet with AB in T, and TD when drawn will touch the curve at D. Take DF, however, in the direction of G if it is positive, but otherwise take it the contrary way.

EXAMPLE 1. Call $GD = x$ and $BD = y$ and let their relationship be

$$x^3 - ax^2 + axy - y^3 = 0.^{(218)}$$

Then will the relation of the fluxions be $3\dot{x}x^2 - 2a\dot{x}x + a\dot{x}y + ax\dot{y} - 3\dot{y}y^2 = 0$, and so $3x^2 - 2ax + ay : 3y^2 - ax (= [\dot{y}:\dot{x}]) = DB(y):DF$. Therefore

$$DF = (3y^3 - axy)/(3x^2 - 2ax + ay),$$

so that, given any point D in the curve and consequently BD and GD or y and x, the point F will be given. Hence, should you erect the normal FT, DT drawn to its meet with the base AB will be tangent to the curve.

From this it is evident that a rule may be contrived exactly as in the former case, namely: arrange all terms on the same side of the exhibited equation, multiply them individually by the dimensions of the ordinate y in them and place the result in the numerator; then multiply its terms individually by the dimensions of the subtense x in them, divide the result by that subtense x and place it in the denominator of the value of DF. And take DF in the direction away from G if it is positive, but otherwise take it the same way. Note, too, that neither the distance of the point G from the base AB—if it happens to be distant from it—nor the size of the ordination angle $A\hat{B}D$ matter at all.

So the previous equation $x^3 - ax^2 + axy - y^3 = 0$ gives at first glance $axy - 3y^3$ for the numerator of DF and $3x^2 - 2ax + ay$ for the denominator of its value.

So also $a + (b/a)\,x - y = 0$ (the equation of a conic)[(220)] gives $-y$ for the numerator and b/a for the denominator of the value of DF, and this is consequently equal to $-ay/b$.[(221)]

---

(219) Read '$n.m$' and compare note (168).

(220) Much as in note (218) this conic will be of eccentricity $(a/b)\,\mathrm{cosec}\,A\hat{B}D$, having its focus at G and the line $y = a$ for directrix.

(221) In the parabolic case ($a = b \sin A\hat{B}D$) this yields the well known property that the perpendicular at G to GD meets the tangent DT on the directrix $y = a$.

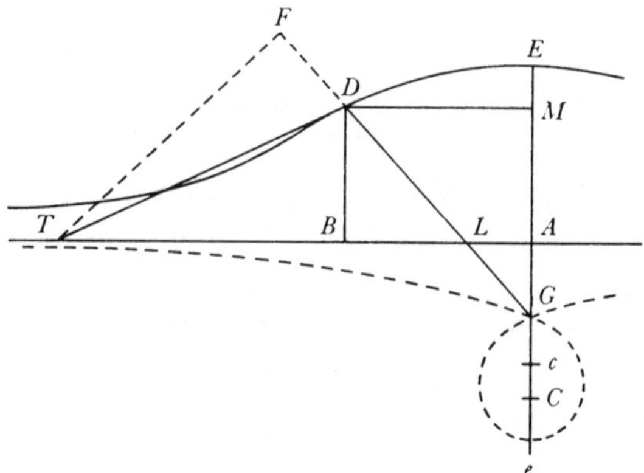

Et sic in Conchoide, (ubi res ex-
||[47] peditiùs absolvitur quàm || ante)[222] posito $GA=b$, $LD=c$, $GD=x$, et $BD=y$, erit $\underset{y}{BD}.\underset{c}{DL}::\underset{b}{GA}.\underset{x-c}{GL}.$ Adeoq $xy-cy=cb$, sive $xy-cy-cb=0$. Quæ æquatio juxta Regulam dat $\frac{xy-cy}{y}$ hoc est $x-c=DF$. Produc ergo GD ad F ut sit $DF=LG$, et ad F erige normalem FT occurrentem Asymptoto AB in T, et acta DT Conchoidem tanget.[223]

Siquando compositæ quantitates in æquatione reperiantur ad methodum generalem recurrendum est, nisi ubi malueris æquationem reducere.

Exempl: 2. Si detur æquatio $\overline{b+x}\sqrt{cc-yy}=yx$[224] pro relatione inter GD et BD determinanda, fluxionum relationem juxta Prob 1 quære. Utpote ficto $\sqrt{cc-yy}=z$, æquationes $bz+yz=yx$ et $cc-yy=zz$ habebis, et inde fluxionum m n et r relationes $br+yr+nz=nx+ym$, et $-2ny=2rz$. Et exterminatis r et z orietur $n\sqrt{cc-yy}\frac{-bny-nyy}{\sqrt{cc-yy}}-nx=my$. Est ergo

$$y.\sqrt{cc-yy}\frac{-by-yy}{\sqrt{cc-yy}}-nx^{(225)}(::n.m)::BD(y).DF.$$

Mod 3[226]

Præterea si Curva ad duas subtensas AD et BD referatur quæ a datis punctis A ac B ductæ ad Curvam conveniunt:[227] concipe punctum illud D per infinitè parvum spatium Dd in curva profluere et in AD et BD cape $AK=Ad$ et $BC=Bd$ et erunt KD et CD contemporanea momenta linearum AD et BD. Cape jam DF ad BD in ratione momenti DK ad momentum DC (i.e. in ratione fluxionis Lineæ AD ad Fluxionem lineæ BD,) et erige perpendicula BT, FT concurrentia in T eruntq trapezia $DFTB$ ac $DKdC$ similia, et proinde diagonalis DT curvam tanget.

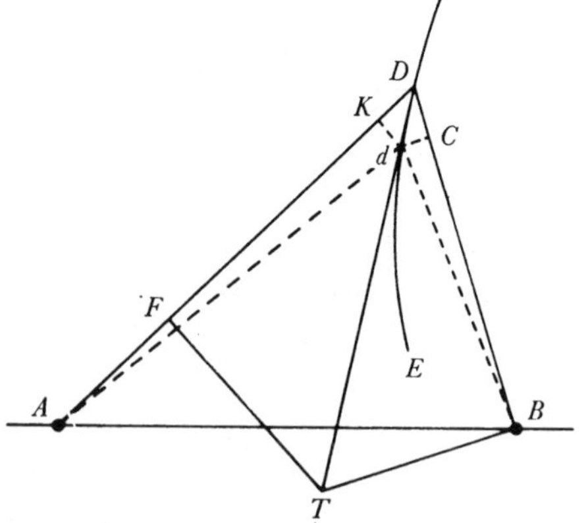

Per æquationem itaq qua relatio inter AD et BD definitur, quære relationem
||[48] fluxionum ope Prob 1, et cape || FD ad BD in eadem ratione.

So too in the conchoid (where the matter is more speedily dispatched than before)(222) it will, on putting $GA = b$, $LD = c$, $GD = x$ and $BD = y$, be

$$BD(y):DL(c) = GA(b):GL(x-c),$$

so that $xy-cy = cb$, or $xy-cy-cb = 0$. This equation by the rule gives $(xy-cy)/y$, that is, $x-c$, equal to DF. Therefore produce GD to F such that $DF = LG$ and at F erect the normal FT meeting the asymptote AB in T, and then DT when drawn will touch the conchoid.(223)

But whenever compound quantities occur in the equation, recurrence must be had to the general method unless you prefer to reduce the equation.

Example 2. If the equation $(b+x)\sqrt{[c^2-y^2]} = yx$(224) be given for determining the relationship between GD and BD, seek the relation between the fluxions by Problem 1. As for instance, supposing $\sqrt{[c^2-y^2]} = z$ you will have the equations $bz+yz = yx$ and $c^2-y^2 = z^2$, and thence between $\dot{x}$, $\dot{y}$ and $\dot{z}$ the fluxional relationships $b\dot{z}+y\dot{z}+\dot{y}z = \dot{y}x+y\dot{x}$ and $-2\dot{y}y = 2\dot{z}z$. So, on elimination of $\dot{z}$ and z, there will arise $\dot{y}\sqrt{[c^2-y^2]}-(b\dot{y}y+\dot{y}y^2)/\sqrt{[c^2-y^2]}-\dot{y}x = \dot{x}y$, and therefore it is $y:(\sqrt{[c^2-y^2]}-(by+y^2)/\sqrt{[c^2-y^2]}-x)\ (=\dot{y}:\dot{x}) = BD(y):DF.$

Mode 3(226)

If, moreover, a curve be referred to two subtenses AD and BD drawn from the given points A, B and intersecting on the curve, let that point D flow along the curve through the indefinitely small space Dd and, on taking in AD and BD $AK = Ad$ and $BC = Bd$, KD and CD will be contemporaneous moments of the lines AD and BD. Now take DF to BD in the ratio of the moment DK to the moment DC (that is, in the ratio of the fluxion of the line AD to that of the line BD) and erect the perpendiculars BT, FT meeting in T. The quadrilaterals $DFTB$ and $DKdC$ will then be similar and in consequence the diagonal DT will touch the curve.

By means, accordingly, of the equation by which the relationship between AD and BD is defined seek the relation of the fluxions with the help of Problem 1 and take FD to BD in the same ratio.

(222) See Newton's page 44 above (from where the accompanying figure is repeated).

(223) Compare I: 394 and II: 198.

(224) This standard Cartesian equation of the conchoid will, of course, no longer represent that curve in the present system of coordinates. However, much of the computational work on Newton's page 44 can be reused here, though Newton could well have chosen a more interesting second example (at the cost, admittedly, of a little extra effort).

(225) Read '$-x$' simply.

(226) Construction of the subtangent in standard bipolar coordinates. Compare the equivalent passage in the 1666 tract (I: 417–18).

(227) The less precise equivalent 'terminantur' (terminated) is cancelled.

Exempl. Posito $AD=x$ et $BD=y$ sit earum relatio $a+\frac{ex}{d}-y=0$ (quæ æquatio est ad Ellipses secundi generis quarum proprietates ad Lucem refringendam Des-Cartes in Lib 2 Geometriæ docuit [(228)]) et fluxionum relatio erit $\frac{em}{d}-n=0$. Est itaq

$$e.d(::n.m)::BD.DF.$$

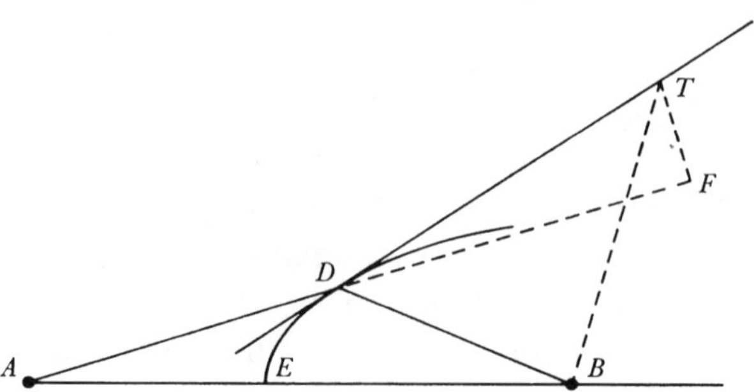

Et pari ratione si $a-\frac{ex}{d}-y=0$, erit $e.-d::BD.DF$. In priori casu cape DF versus A, et ad contrarias partes in posteriori.

Coroll: 1. Hinc si $d=e$ (quo casu curva evadit conica sectio) erit $DF=DB$. et inde triangula DFT, DBT æqualia, angulusq FDB a tangente bisecabitur.[(229)]

Coroll 2. Hinc etiam quæ Des-Cartes de his curvis circa refractiones haud absq circuitu[(230)] demonstravit, per se manifesta sunt: siquidem DF ac DB (quæ sunt in data ratione d ad e) respectu sinus totius DT sint sinus angulorum DTF ac DTB[:] id est incidentiæ radij AD in superficiem curvæ, et reflectionis vel refractionis ejus DB. Estq par ratio de refractionibus Conicarum Sectionum si modo punctorum A vel B alterutrum infinitè distare concipiatur.[(231)]

Perfacile est hanc regulam pro more præcedentium concinnare et pluribus exemplis donare. Quinimò ubi curvæ alijs quibuscunq modis ad rectas referuntur, et ad præcedentes formas haud commodè reduci possunt; perfacile est alias Regulas ad harum exemplar pro re nata excogitare.

Mod. 4

Quemadmodum si rectæ BD circa datum punctum B volventis punctum D sit ad Curvam aliquam, et C sit intersectio ejus cum rectâ AC positione datâ; habeaturq relatio inter BC et BD quacunq æquatione designata; Age BF parallelam AC, eiq occurrat DF normalis
||[49] ad || $B[D]$. Et ad DF itidem erige normalem FT, et cape in ratione ad BC

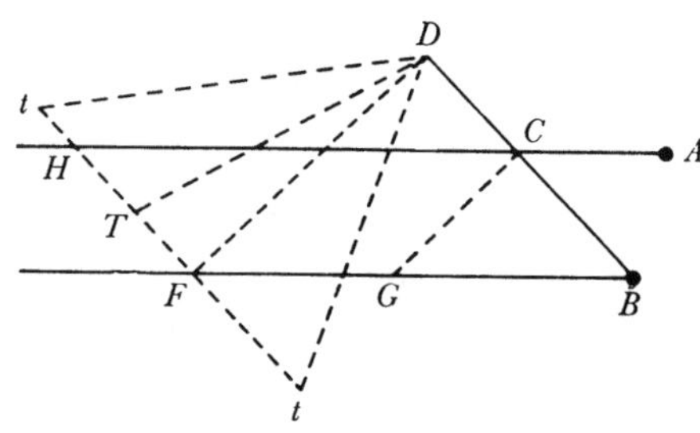

(228) Namely, the quartic refractive ovals discussed by Descartes in his *Geometrie* as an example of a curve defined by bipolars (*Geometria*, $_2$1659: Liber II: 50–60). Having explained their 'proprietates...concernentes reflexiones & refractiones' Descartes in fact computes the

Example. Take $AD = x$ and $BD = y$, and let their relation be

$$a+(e/d)x-y = 0$$

(the equation of the second-order ovals whose light-refracting properties are explained by Descartes in Book 2 of his *Geometry*).[228] The relation of the fluxions will be $(e/d)\dot{x}-\dot{y} = 0$ and accordingly $e:d(=\dot{y}:\dot{x}) = BD:DF$.

And in like manner if $a-(e/d)x-y = 0$, then $e:-d = BD:DF$. In the former case take DF towards A, but in the opposite direction in the latter.

Corollary 1. Hence if $d = e$ (in which case the curve turns out to be a conic) there will be $DF = DB$ and thence the triangles DFT, DBT congruent, so that the angle $F\widehat{D}B$ will be bisected by the tangent.[229]

Corollary 2. Hence also what Descartes demonstrated concerning these curves with regard to refraction—not without circuitousness[230]—are evident of themselves, seeing that DF and DB (which are in the given ratio of d to e) are, in respect of the whole sine DT, the sines of the angles $D\widehat{T}F$ and $D\widehat{T}B$—that is, of AD the ray of incidence at the curve interface and DB that of reflection or refraction. A like argument holds for conic refractions if only one or other of the points A, B be conceived to be infinitely distant.[231]

It is extremely easy to polish this rule after the fashion of the preceding ones and to deck it out with several examples. Indeed, when curves are related to straight lines in any other ways but cannot readily be reduced to the preceding forms, it is exceedingly easy to contrive other rules after the pattern of these, as the occasion requires.

Mode 4

For instance, if the point D in the straight line BD rotating round the given point B belong to some curve, let C be its intersection with the straight line AC given in position, and suppose the relationship between BC and BD to be expressed by any equation. Draw BF parallel to AC, till its meet with DF, normal to BD, and to DF likewise erect the normal FT, taking it to BC in the

---

subnormal (and hence the curve slope) at a general point on his 'quatuor genera novarum Ovalium Opticæ inservientium' by reducing their bipolar defining equation to its quartic equivalent in standard perpendicular Cartesian coordinates and then applying his general subnormal technique (see I, **2**, 2, Historical introduction). As we have seen (I: 551–8), Newton had already occupied himself with the differential properties of the Cartesian oval in autumn 1664 and he was soon to return to them once more in Book 1 of his Lucasian *Lectiones Opticæ* (see **3**, 1, §2: Proposition 34 below).

(229) Compare I: 394, 553–4.

(230) In the brief discussion of the Cartesian oval which he inserted in Book 1, Section 14, of his *Principia* Newton fifteen years later allowed himself a rare personal remark to the same effect. See **3**, Appendix 3: note (7) below, and compare I, **3**, Appendix 1, §2: note (24).

(231) In which case the present method reduces to Mode 2! Compare note (220) above.

quam habet fluxio ipsius BD ad fluxionem ipsius BC: Actóꝗ DT curvam tanget.(232)

MOD 5.

Sin, dato puncto A, æquatio relationem inter AC at BD designat, duc CG parallelam DF, et cape FT in ratione ad BG quam habet fluxio BD ad fluxionem AC.(233)

MOD 6.

Vel deniꝗ si æquatio relationem inter AC et CD definit: conveniant AC et FT in H, et cape HT in ratione ad BG quam habet fluxio CD ad fluxionem AC.(234) Et sic in alijs.

De Spiralibus.

MOD 7.

Haud secus absolvitur Problema(235) ubi curvæ non ad rectas sed ad alias curvas lineas (uti solent Mechanicæ)(236) referuntur. Sit BG circuli periferia in

---

(232) In proof, let Bcd (meeting CG and DF in γ and δ) be an indefinitely near position of the radius vector BCD. Then in the limit as Bcd coincides with BCD the limit-increments γc, δd of BC, BD become proportional to their fluxions, where

$$\frac{\gamma c}{\delta d} = \frac{\gamma c}{\gamma C}\times\frac{\gamma C}{\delta D}\times\frac{\delta D}{\delta d} = \frac{BD}{FD}\times\frac{BC}{BD}\times\frac{FD}{FT} = \frac{BC}{FT},$$

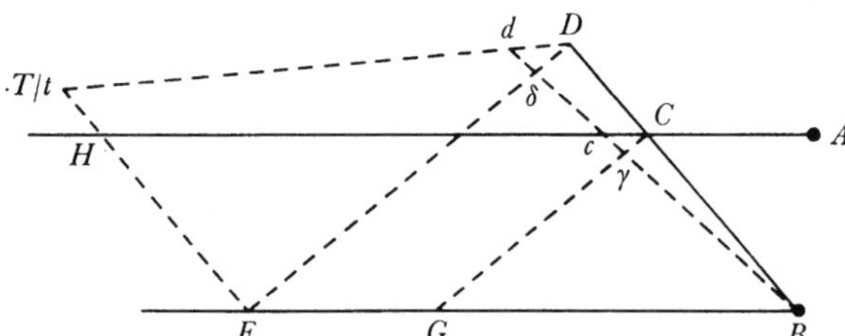

so that fluxion (BC) : fluxion $(BD) = BC:FT$. Note that any given length BC (which merely serves to determine the direction of the vector BD) determines *two* points C in AH, so that the coordinate system (as indeed those of Modes 2 and 3 preceding) can be used consistently only in the case of curves mirror-symmetrical round an axis (here the perpendicular to AC through B).

(233) Much as in note (232), in the limit as Bcd coincides with BCD there follows

$$\frac{Cc}{\delta d} = \frac{Cc}{C\gamma}\times\frac{C\gamma}{D\delta}\times\frac{D\delta}{\delta d} = \frac{BG}{CG}\times\frac{CG}{FD}\times\frac{FD}{FT} = \frac{BG}{FT},$$

where Cc, δd are the limit-increments of AC, BD and so proportional to their fluxions. This monopolar system of coordinates (like that of the preceding Mode 4) would appear to be Newton's present invention.

(234) A generalization of standard oblique Cartesian coordinates (to which the present system reduces when the point B is at infinity) already used by Newton in October 1664 (I: 173) to define the locus (D) satisfying $AC = CD$ before converting to standard perpendicular

ratio of the fluxion of BD to that of BC. When DT is drawn it will touch the curve.(232)

Mode 5

But if, given the point A, the equation should express the relationship between AC and BD, draw CG parallel to DF and take FT to BG in the ratio which the fluxion of BD has to the fluxion of AC.(233)

Mode 6

Or, finally, if the equation defines the relationship between AC and CD, let AC and FT meet in H and then take HT to BG in the ratio which the fluxion of CD has to the fluxion of AC.(234) And so in other instances.

Mode 7

Regarding spirals

The problem will be accomplished no differently(235) when curves are related not to straight lines but to other curves (as the mechanical ones(236) usually

---

Cartesian coordinates. Justification of Newton's construction is a corollary of those for Modes 4 and 5 since, in the terms of notes (232) and (233),

$$\frac{\text{fluxion}\,(AC)}{\text{fluxion}\,(CD)} = \frac{Cc}{\delta d-\gamma c} = \frac{BG}{FT-BC}$$

in the limit as Bcd coincides with BCD.

(235) Newton first wrote equivalently 'Præterea...possis tangentes nihil secus ducere' (You should, furthermore, be able to draw tangents no differently).

(236) Those not 'geometrically' representable in Cartesian coordinates. The distinction between 'geometrical' and 'mechanical' curves is, of course, that made by Descartes at the opening of the second book of his *Geometrie*, 'prenant...pour Geometrique ce qui est precis & exact, & pour Mechanique ce qui ne l'est pas' (*Discours de la Methode* (Leyden, 1637): 316 = *Geometria*, $_2$1659: 18: 'si Geometricum censeamus illud...quod omnino perfectum atque exactum est, & Mechanicum quod ejusmodi non existit'). Descartes himself rejected curve lines from geometry for the plausible (if erroneous) Aristotelian reason that 'la proportion, qui est entre les droites & les courbes n'estant pas connuë, & mesme ie croy ne le pouuant estre par les hommes, on ne pourrait rien conclure de là qui fust exact & assuré' (*Discours*: 340–1 = *Geometria*: 39: 'cum ratio, quæ inter rectas & curvas existit, non cognita sit, nec etiam ab hominibus (ut arbitror) cognosci queat; nihilque inde, quod exactum atque certum est, concludere possimus'). Since, however, all 'geometrical' curve arcs are algebraically expressible in a standard Cartesian coordinate system, his fruitful dichotomy between 'geometrical' and 'mechanical' curves (according as their standard Cartesian defining equations are, as we would say, algebraic or transcendental) is not endangered by the fallaciousness of Descartes' heuristic attempted justification of it. Newton was well aware that the problem of determining whether the general circle arc was 'geometrical' (a problem attacked, with little real success, by James Gregory in his *Vera Circuli et Hyperbolæ Quadratura*: see note (72) and compare M. Dehn and E. Hellinger, 'On James Gregory's *Vera Quadratura*' in the *James Gregory Tercentenary Memorial Volume* (ed. H. W. Turnbull) (London, 1939): 468–78) had not been solved but here—and indeed universally elsewhere—he follows informed contemporary opinion (rightly so) in classifying the circle as a 'mechanical' (transcendental) curve.

cujus semidiametro AG, dum circa centrum A convolvitur, moveatur utcunq; punctum D et spiralem ADE describat. Et concipe Dd et(237) partem curvæ infinitè parvam per quam D fluit, et in AD cape $AC = Ad$ et erunt CD ac Gg contemporanea momenta rectæ AD et periferiæ(238) BG. Duc ergo At parallelam Cd, id est perpendicularem AD, et cum ea tangens DT conveniat in T, eritq; $CD.Cd::AD.AT$. Sit insuper Gt parallela tangenti, et erit

$$Cd.Gg(::Ad \text{ vel } AD.AG)::AT.At.$$

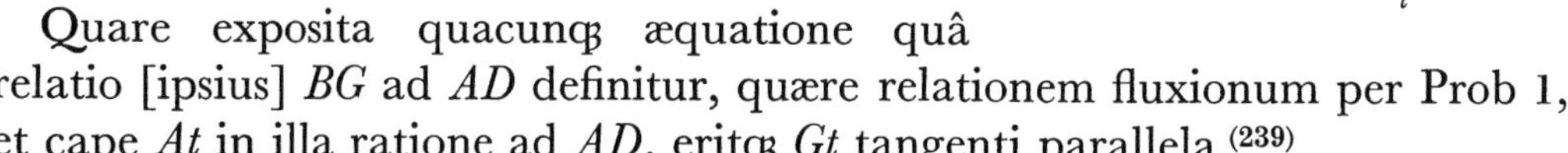

Quare exposita quacunq; æquatione quâ relatio [ipsius] BG ad AD definitur, quære relationem fluxionum per Prob 1, et cape At in illa ratione ad AD. eritq; Gt tangenti parallela.(239)

Exempl 1. Dictis $BG = x$ et $AD = y$, sit earum relatio $x^3 - axx + axy - y^3 = 0$(240) et ope Prob 1 emerget $3xx - 2ax + ay$. $3yy - ax(::n.m)::AD.At::AP.AG$.(241)
||[50] Puncto t sic invento duc || Gt eiq; parallelam DT, et illa Curvam tanget.

Exempl 2. Si sit $\frac{ax}{b} = y$ (quæ æquatio est ad Spiralem Archimedeam) erit $\frac{am}{b} = n$. Adeoq; $a.b(::n.m)::AD.At$. Unde obiter si TA producatur ad P ut sit $AP.AB::a.b$, PD ad curvam recta erit.

Exempl 3. Si $xx = by$,(242) erit $2mx = bn$. Adeoq; $2x.b::AD.At$. Et sic tangentes ad quascunq; spirales(243) nullo negotio determinari possunt.

(237) Horsley (*Opera* (note (60)), **1**: 440) corrects this to 'esse' but Newton might have intended 'ut'.

(238) 'arcus' (arc) is cancelled.

(239) This method of constructing the tangent to an arbitrary curve defined in standard polar coordinates is essentially Newton's own, though it has evident affinities with Archimedes' construction of the tangent to the simple 'spiralis Archimedea' in Proposition 20 of his work *On Spirals* (compare E. J. Dijksterhuis, *Archimedes* (Copenhagen, 1956): 268–74) and with the equivalent constructions for the class of infinite spirals $a \times \widehat{BG}^m = AD^n$ given in the seventeenth century by Fermat, Torricelli, Stefano degli Angeli and others. (Of course, Newton's construction of the polar subtangent arises naturally from his implied definition of the tangent as being coincident with the vanishingly small chord Dd and we need not look to his reading in others as the source of his inspiration.) Note the tacit introduction of the analytical polar coordinates $\widehat{BG} = x$, $AD = y$ in the following examples in order to determine a corresponding curve from a given 'polar' defining equation $f(x, y) = 0$. As we have seen (I: 374) this usage had been foreshadowed by Newton in October 1665 (where an Archimedean spiral is determined by means of the radius vector) but the present examples would seem to be the first occasion on which an analytical polar notation was freely and systematically used.

(240) One more occurrence of this degenerate locus (compare notes (192) and (218)). Here $x = y$ will represent an Archimedean spiral while $x^2 + xy + y^2 = ax$ will (when $AB \geqslant a/\pi$) be a double-looped curve with a double point at the origin A.

tend to be). Let BG be a circle circumference, in the radius AG of which, as it revolves round the centre A, the point D moves any how, describing the spiral ADE. Conceive Dd [to be] an infinitely small portion of the curve along which D flows, and in AD take $AC = Ad$: then will CD and Gg be contemporaneous moments [respectively] of the straight line AD and of the circumference(238) BG. Draw therefore At parallel to Cd, that is, perpendicular to AD, and let the tangent DT meet it in T: then will there be $CD:Cd = AD:AT$. Again, let Gt be parallel to the tangent and there will be

$$Cd:Gg(= Ad\,(\text{or } AD):AG) = AT:At.$$

Hence, when any equation defining the relation of $\widehat{BG}$ to $\widehat{AD}$ is exhibited, seek the relationship of the fluxions by Problem 1 and take At in that ratio to AD: then will Gt be parallel to the tangent.(239)

Example 1. Calling $\widehat{BG} = x$ and $AD = y$, let their relation be

$$x^3 - ax^2 + axy - y^3 = 0^{(240)}$$

and by the help of Problem 1 there will emerge

$$3x^2 - 2ax + ay : 3y^2 - ax (= \dot{y}:\dot{x}) = AD:At = AP:AG.^{(241)}$$

After the point t is thus found, draw Gt and DT parallel to it: the latter will be tangent to the curve.

Example 2. If there be $(a/b)\,x = y$ (the equation of an Archimedean spiral), then $(a/b)\,\dot{x} = \dot{y}$, so that $a:b(= \dot{y}:\dot{x}) = AD:At$. Hence by the way, if TA be produced to P such that $AP:AB = a:b$, PD will be at right angles to the curve.

Example 3. If $x^2 = by$,(242) then $2\dot{x}x = b\dot{y}$, so that $2x:b = AD:At$. And in this way tangents to any sort of spirals(243) may be determined without trouble.

(241) The following construction of the subnormal AP (an immediate corollary of the general method which is not restricted to the present example) is a late addition in the manuscript. Evidently Newton would have inserted it earlier if he had had room.

(242) A Fermatian 'infinite' spiral with an exceptionally interesting history. Notably, Fermat himself had suggested to Mersenne in 1636 that it was the Galilean fall-path to the earth's centre of a heavy body falling 'naturally' (that is, initially with the earth's uniform speed of rotation) from the earth's surface, and to show his calculus powers squared its general segment: no later than spring 1637 he was able to construct its tangent, if at least we may believe Carcavi. (See Cornélis de Waard, 'Études de Fermat sur la spirale de Galilée' in *Correspondance du P. Marin Mersenne*, **6** (Paris, 1960): 376–82.) Within a few years Roberval, Torricelli and soon Wallis, Stefano degli Angeli and James Gregory were hard at work codifying the elementary differential properties of more general 'infinite' spirals and their solids of revolution. Though Angeli published several books at Venice between 1660 and 1667 on the subject and though he had been closely familiar with Wallis' *Arithmetica Infinitorum* five years before, Newton very probably chose the present example merely as the spiral analogue of the Cartesian equation of a parabola.

(243) These general spirals need not, as the naïve student of Newton's examples and accompanying figure might suppose, pass through the origin A.

De Quadratricibus.

MOD 8.

Ad hæc si curva sit ejusmodi ut per centrum A ductâ utcunꝗ AGD quæ circulo in G, curvæꝗ in D occurrat, relatio inter arcum BG et rectam DH quæ in dato angulo(244) ad Basin AB ordinata est, æquatione quavis definiatur: Concipe punctum D per infinitè parvum intervallum ad d in curva moveri et completo parallelogrammo $dhHK$ productâꝗ Ad ad C ut sit $AC=AD$, erunt Gg arcûs BG et DK ordinatæ DH contemporanea momenta. Produc jam Dd rectà ad T ubi cum AB conveniat et demitte TF in DC perpendicularem, eruntꝗ trapezia $DKdC$, $DHTF$ similia; atꝗ adeo $DK.DC::DH.DF$. Et præterea si Gf ad AG normalis erigatur quæ cum AF concurrat in f propter parallelas DF, Gf erit $DC.Gg::DF.Gf$. Quamobrem ex æquo(245) est $DK.Gg::DH.Gf$, hoc est, ut momenta sive fluxiones linearum DH et BG.

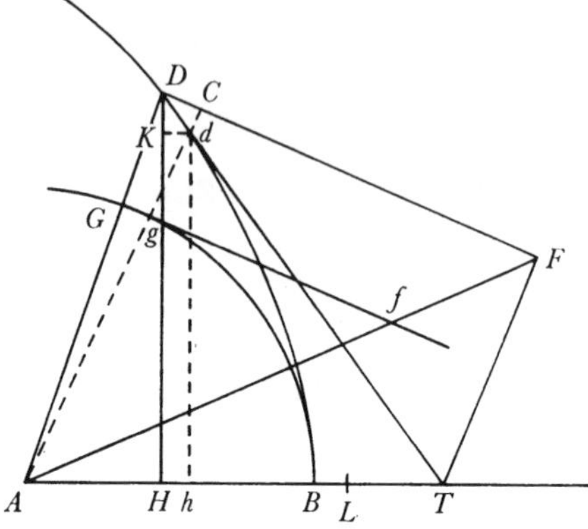

Per æquationem itaꝗ quâ relatio BG ad DH definitur quære rationem fluxionum per Prob. 1 et in ea ratione cape Gf (tangentem circuli BG) ad DH. Age DF parallelam Gf quæ cum Af producta conveniat in F et ad F erige normalem FT occurrentem AB in T et acta DT Quadraticē tanget.(246)

EXEMPL 1. Nominatis $BG=x$, ac $DH=y$, esto $xx=by$(247) et (per Prob 1) erit $2mx=bn$. Adeoꝗ $2x.b(::n.m)::DH.Gf$. Et invento f, cætera ut præscriptum est determinabis.

||[51] || Cæterum hæc Regula forte sic elegantior evadet[:] Fac $m.n::AB.AL$. Dein $AL.AD::AD.AT$, et DT curvam tanget. Nam propter æqualia triangula AFD, ATD,(248) est $AD\times DF=AT\times DH$. Adeoꝗ

$$AT.AD\left(::DF.DH,\text{ sive }\frac{n}{m}Gf\right)::AD.\frac{n}{m}AG,\text{ sive }AL.^{(249)}$$

EXEMPL 2. Esto $x=y$ (quæ æquatio est ad Veterū Quadratricem)(250) et per Prob 1 erit $m=n$. Adeoꝗ $AB.AD::AD:AT$.

(244) In the comparable Example 7 in Mode 1 above (on Newton's pages 45a–45b) the angle $A\hat{H}D$ was restricted to being right: see note (208).

(245) In context a curious phrase: we would expect 'componendo' (by compounding).

(246) Newton's generalization of his own 1665 method of constructing the quadratrix's tangent (see I: 379, 418) with the trivial variant that the circle arc through D concentric with $\widehat{BG}$ is here omitted.

MODE 8

Regarding quadratrixes.

If again the curve be of such a kind that, when AGD is drawn unrestrictedly through the centre A to meet its circle in G and the curve in D, the relationship between the arc $\widehat{BG}$ and the straight line DH (ordinate at a given angle(244) to the base AB) be defined by any equation, conceive the point D to move along the curve through an infinitely small distance to d, complete the parallelogram $dhHK$ and produce Ad to C so that $AC = AD$: then will Gg and DK be contemporaneous moments of the arc $\widehat{BG}$ and the ordinate DK. Now produce Dd in a straight line to its intersection T with AB and drop TF perpendicular to DC: then the quadrilaterals $DKdC$ and $DHTF$ will be similar, and so

$$DK:DC = DH:DF.$$

Further, if Gf be raised normally to AG till it meet AF in f, on account of the parallels DF and Gf there will be $DC:Gg = DF:Gf$. Therefore by equals(245) $DK:Gg = DH:Gf$, that is, in the ratio of the moments, and so the fluxions, of the lines DH and $\widehat{BG}$.

Accordingly, by the equation defining the relationship between $\widehat{BG}$ and DH seek the ratio of the fluxions by Problem 1 and in that ratio take Gf (tangent to the circle $\widehat{BG}$) to DH. Draw DF parallel to Gf till it meet Af produced in F, and at F raise the normal FT meeting AB in T. When DT is drawn it will touch the quadratrix.(246)

EXAMPLE 1. Naming $\widehat{BG} = x$ and $DH = y$, let $x^2 = by$(247) and (by Problem 1) there will be $2\dot{x}x = b\dot{y}$, so that $2x:b(=\dot{y}:\dot{x}) = DH:Gf$. And when f is found, you will determine the other points as prescribed.

However, this rule may perhaps be more elegantly recast in this way. Make $\dot{x}:\dot{y} = AB:AL$, then $AL:AD = AD:AT$, and DT will be tangent to the curve. For, because of the equal triangles AFD and ATD,(248) it is

$$AD \times DF = AT \times DH,$$

so that $AT:AD(= DF:DH$, that is, $(\dot{y}/\dot{x}).Gf) = AD:(\dot{y}/\dot{x}).AG$, that is, AL.(249)

EXAMPLE 2. Let $x = y$ (the equation to the ancients' quadratrix)(250) and by Problem 1 there will be $\dot{x} = \dot{y}$, so that $AB:AD = AD:AT$.

(247) Once more (compare note (242)) the Cartesian equation of a parabola is used—for want of a more rewarding example—to furnish an example in the present context of 'quadratrix' coordinates.

(248) Equal in area, that is, since the lines DA and FT (both perpendicular to DF) are parallel. Newton now assumes that DH is perpendicular to AT.

(249) For the triangles AGf, ADF are similar while the circle radii AB and AG are equal.

(250) When the ordinate angle $A\hat{H}D$ is right. Here Newton's preceding improvement yields Barrow's subtangent as an immediate corollary: see note (209).

EXEMPL 3. Esto $axx = y^3$ et erit $2amx = 3nyy$. Fac ergo

$$3yy.2ax(::m.n)::AB.AL.$$

Dein $AL.AD::AD.AT$.

Atq; ita tangentes aliarum Quadratricum utcunq; compositarum possis expeditè determinare.(251)

MOD. 9.(252)

Si deniq; ABF sit curva quævis data quam tangat recta Bt, et rectæ BC in dato angulo ad basem AC applicatæ pars BD inter hanc et aliam curvam DE intercepta relationem ad curvæ portionem AB in æquatione quacunq; definitam habeat: alterius curvæ tangentem DT duces capiendo in hujus tangente, TB in ea relatione ad BD, quam habet fluxio curvæ AB ad fluxionem rectæ BD.(253)

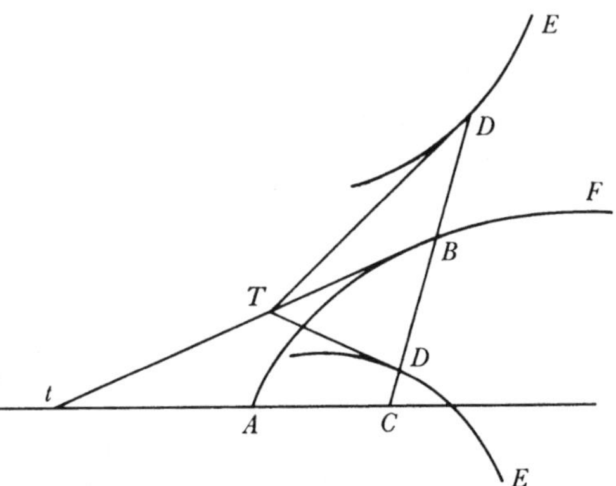

EXEMPL 1. Dictis $AB = x$ et $BD = y$, esto $ax = yy$(254) et per Prob: 1, erit $am = 2ny$, adeoq; $a.2y(::n.m)::BD.BT$.

EXEMPL 2. Sit $\frac{a}{b}x = y$ (æquatio ad Trochoidem(255) si modò ABF sit circulus) et erit $\frac{a}{b}m = n$. Adeoq; $a.b::BD.BT$.

(251) In first draft Newton added at this point—having not yet determined to insert the following final Mode 9—'et hujusmodi Regulas pro alijs quibuscunq; Mechanicarum Curvarum generibus excogitare' (and think up rules of this sort for any other kinds of mechanical curves). He then continued with a first, partially cancelled original version of Examples 4–6 on page 45a (beginning 'Siquando in æquationibus Mechanicæ quantitates involvantur, earum fluxiones ut in Exemplo 5 Prob 1 quærere oportet...' much as later) and at once, without citing any cognate problems, concluded the present Problem 4 with the words 'Et hæc in explicationem hujus methodi generalis qua curvarum omnium tangentes absq; solitâ calculi molestiâ promptè et concinnè determinantur, adduxisse sufficiat' (Let it suffice, in explanation of this general method by which tangents of all curves are readily and effectively determined without the usual heavy weight of computation, that I have adduced these points).

(252) Newton has attached no accompanying marginal subhead to this final Mode (a late insertion in the first version of the text made when his original pages 51/52 were substantially incorporated in the revised page 45a). On the analogy of those preceding it might aptly have been titled 'De Trochoidibus' (Regarding cycloids).

(253) For if cbd is an indefinitely near parallel to CBD, meeting the curves AB and DE in b and d, in the limit as it coincides with CBD the tangents Bt, DT coincide with the vanishing chords Bb, Dd. Hence, when $d\delta$ is drawn parallel to bB, $BT:BD = Bb:\delta D$, where Bb (or $\widehat{Ab} - \widehat{AB}$) and δD (or $bd - BD$) are limit-increments of $\widehat{AB}$ and BD, and therefore proportional to their fluxions. In structure, this Mode is Newton's extension (to the case where the defining relation between $\widehat{AB}$ and BD is general) of the simple trochoidal case (in which the ratio of

EXAMPLE 3. Let $ax^2 = y^3$ and there will be $2a\dot{x}x = 3\dot{y}y^2$. So make

$$3y^2 : 2ax (= \dot{x} : \dot{y}) = AB : AL,$$

then $AL : AD = AD : AT$.

And in this way you might speedily determine the tangents of other quadratrixes no matter how complex.(251)

MODE 9(252)

If, finally, ABF be any given curve touched by the straight line Bt, let BD, that part of the straight line BC (ordinate at a given angle to the base BC) which is intercepted between this and a second curve DE, have a relationship to the portion $\widehat{AB}$ of the curve which is defined in any equation. You will draw the tangent DT of the other curve by taking BT, in this [first] one's tangent, in proportion to BD as the ratio of the fluxion of the arc $\widehat{AB}$ to that of the straight line BD.(253)

EXAMPLE 1. Calling $\widehat{AB} = x$ and $BD = y$, let $ax = y^2$(254) and by Problem 1 there will be $a\dot{x} = 2\dot{y}y$, so that $a : 2y (= \dot{y} : \dot{x}) = BD : BT$.

EXAMPLE 2. Let $(a/b)\,x = y$ (the equation to a cycloid(255) provided $\widehat{ABF}$ be a circle) and there will be $(a/b)\,\dot{x} = \dot{y}$, so that $a : b = BD : BT$.

$\widehat{AB}$ to BD is constant) discussed by James Gregory in Proposition 8 of his *Geometriæ Pars Universalis* ((note (204)): 22–4) and, following him, by Barrow in his *Lectiones Geometricæ* ((note

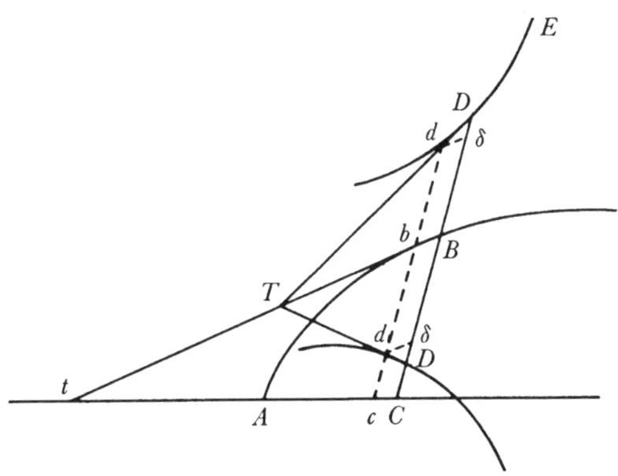

(80)): Lectio x, §§1–4: 75–6). Gregory himself borrowed heavily, both structurally and formally, from the construction of the tangent to the primary cycloid invented eight years previously by Wren. See note (255) and compare D. T. Whiteside, 'Patterns of Mathematical Thought' (note (204)): 349–50.

(254) Yet again (compare note (247)) the Cartesian defining equation of a parabola is used to give a not very rewarding example in an alien coordinate system.

(255) When, that is, $A\hat{C}B$ is a right angle. The present Example 2 is essentially Gregory's Proposition 8 (note (253)) and, when $a = b$, yields Christopher Wren's 1658 construction of the cycloid's tangent (first published in John Wallis' *Tractatus Duo. Prior, De Cycloide et corporibus inde genitis*... (I, 1, Introduction, Appendix 2: note (22)) (Oxford, 1659): 62–3: 'De rectâ Tangente Cycloidem primariam').

Et nihilo difficiliùs tangentes, ubi ipsius BD ad AC vel ad BC relatio[256] in æquatione quavis exprimitur, vel ubi curvæ alijs quibuscunq; modis ad rectas aliasve curvas referuntur, possis ducere.

Sunt etiam alia non pauca Problemata quorum solutiones ex hisce fluunt.[257] Cujusmodi sunt;

1. Invenire punctum Curvæ ubi tangens est ad Basin (vel quamvis positione datam rectam) parallela vel perpendicularis vel in alio quovis angulo inclinata.[258]

2. Invenire punctum ubi tangens maxime minimève ad Basin aut aliam
||[52] positione datam rectam inclinatur. Hoc est invenire || confinium flexûs contrarij.[259] Hujus autem specimen in Conchoide jam ante[260] exhibui.

3. A dato quovis extra curvæ perimetrum puncto rectam ducere quæ cum perimetro aut angulum contactûs aut rectum angulum, aut alium quemvis datum conficiet.[261] Hoc est tangentes vel perpendiculares vel aliter ad curvam inclinatas rectas a dato quovis puncto ducere.

4. A dato quovis intra Parabolam puncto rectam ducere quæ maximum minimumve quem potest angulum cum perimetro ejus conficiet. Et idem de alijs curvis intellige.[262]

5. Rectam ducere quæ duas positione datas curvas, vel eandem curvam (si[263] potest) in duobus punctis tangat.[264]

6. Curvam quamvis sub datis conditionibus ducere quæ aliam positione datam curvam in dato puncto tanget.[265]

(256) Compare Modes 4, 5 and 6 above.

(257) 'parturiunt' (are begotten) is cancelled.

(258) Newton discussed this problem of finding the points at which a given curve has given slope a few years earlier in his 'Problems of Curves' (II, 2, 1, §2.2: Problem 4. 'To find the points where ye curve hath a given inclination to the basis').

(259) That is, the point of 'bending back' (inflexion) at which the curve changes its convexity (and so crosses its tangent at this point). Immediately, a necessary and sufficient condition for such a point on a (smoothly continuous) curve defined by a Cartesian equation $f(x, y) = 0$ is that d^2y/dx^2 be zero. Compare I, 2, 6, §4: note (47).

(260) In the scholium to Example 3 on his page 45. There, however, Newton had found the conchoid's inflexion points by minimizing the subtangent $AT = x-y(dx/dy)$: see note (200). His earlier construction of the conchoid's inflexions in May 1666 (I: 398–9) was based directly on the present suggested procedure of maximizing the tangent slope.

(261) The condition that a line drawn through the given point (a, b) shall be inclined at the given angle α to the given curve defined (in perpendicular Cartesian coordinates) by $f(x, y) = 0$ is that

$$\tan^{-1}\frac{y-b}{x-a} = \tan^{-1}\left(-\frac{f_x}{f_y}\right) \pm \alpha:$$

this is the defining equation of a curve which will meet the given curve in the manner required. Note that the present problem is a necessary preliminary to solving Problem 3 on Newton's page 42 above.

And you might with no more difficulty draw tangents when the relation of BD to AC or BC[256] is expressed in any equation, or when curves are related in any other manner to straight lines or other curves.

There are also not a few other problems whose solutions derive from these principles, such as:

1. To find the point in a curve where the tangent is parallel to the base (or any other straight line given in position) or perpendicular to it or inclined to it at any given angle.[258]
2. To find the point where a tangent is most or least inclined to the base or to another straight line given in position—to find, in other words, the bound of contrary flexure.[259] I have already displayed an example of this above in the conchoid.[260]
3. From any given point not in the perimeter of a curve to draw a straight line which shall make with that perimeter a contact angle or a right angle, or any other given angle you wish[261]—from any given point, that is, to draw tangents or normals to the curve, or straight lines at any other inclination to it.
4. From any given point inside a parabola to draw a straight line which shall make with its perimeter the greatest or least angle possible. And understand the same problem with regard to other curves.[262]
5. To draw a straight line to touch two curves given in position, or the same curve in two points (provided this is possible).[264]
6. To draw any curve, under given conditions, which shall touch another curve given in position at a given point.[265]

---

(262) This repeats Problem 4 on Newton's page 42. It is, of course, an immediate extension of the preceding problem. Compare note (181).

(263) The strengthening adverb 'modò' is cancelled.

(264) Evidently, if the two given curves are defined in a perpendicular Cartesian coordinate system by $f(x, y) = 0$ and $F(X, Y) = 0$ respectively, the condition for the tangents at (x, y) and (X, Y) to these curves to be coincident is

$$\frac{Y-y}{X-x} = -\frac{f_x}{f_y} = -\frac{F_X}{F_Y}.$$

Impossibilities arise when (as in the case of two concentric circles) the eliminant has no real roots or, again, when the equalities demanded by the tangency condition prove inconsistent (as when the two curves are the same conic or cubic).

(265) This clearly reduces to the problem of drawing a curve *sub datis conditionibus* to touch a given line at a given point in it (tangent to the given curve at that point). Several years before Newton had resolved problems of this type where a conic is to be drawn through given points to touch given lines: compare Problems 2 and 3 of II, **1**, 3, §2.

7. Luce in quamlibet curvam superficiem incidente, cujusvis radij refractionem determinare.(266)

Horum et similium Problematum confectiones, ubi non obstat computandi tædium, non sunt ita difficiles ut ijs explicandis immorari opus sit. Et Geometris, credo, magis gratum erit sic tantùm recensuisse.

||[53] ||PROB: 5.

CURVÆ ALICUJUS AD DATUM PUNCTUM CURVATURAM INVENIRE.(267)

Problema cum primis elegans videtur et ad curvarum scientiam utile. In ejus autem constructionem generalia quædam præmittere convenit.

1. Ejusdem circuli eadem est undiqꝫ curvatura et inæqualium circulorum curvaturæ sunt reciprocè proportionales diametris. Si alicujus diameter diametro alterius duplo minor est, ejus periferiæ curvatura erit duplo major, si diameter triplo minor est curvatura erit triplo major, &c.

2. Si Circulus Curvam aliquam ad partem concavam in dato puncto tangat, sitqꝫ talis magnitudinis ut alius contingens circulus in angulis contactûs proximè punctum istud(268) interscribi nequeat, circulus ille ejusdem est curvitatis ac Curva in isto puncto contactûs.(269) Nam circulus, qui inter curvam et alium circulum juxta punctum contactus interjacet, minus deflectit a curva ejusqꝫ curvaturam magis appropinquat quam ille alius circulus; et proinde curvaturam ejus maximè appropinquat inter quem et Curvam non alius quisquam potest intercedere.

3. Itaqꝫ centrum curvaminis(270) ad aliquod Curvæ punctum est centrum tangentis circuli æqualiter incurvatæ;(271) et sic radius vel semidiameter curvaminis(270) est pars perpendiculi ad istud centrum terminata.

4. Et proportio curvaminis(270) ad diversa ejus puncta e proportione curvaminis(270) circulorum æque curvorum sive e reciproca proportione radiorum curvaminis(270) innotescit.

---

(266) Once the tangent slope at a general point on the given (plane) curve is known, the problem reduces at once to the elementary problem of finding the refraction of an incident ray inclined at a known angle to a linear interface. In his *Lectiones Opticæ* Newton, developing an approach suggested by Barrow in his own optical lectures, was already by this means able to construct the diacaustic of a point with regard to a given curved interface and a known refractive index of light point by point (see **3**, 1, §2: Propositions 32 and 33). Once the diacaustic is constructed, to find the refraction of a ray incident at any point on the given interface from the given point it remains only (by the present Problem 4) to construct the tangent from that arbitrary point to the diacaustic.

(267) An augmented Latin revise of I, **2**, 4, §2 and a corresponding passage in the 1666 tract (I: 419–21).

(268) That is, between the curve and the circle of curvature in the immediate vicinity of the contact point.

(269) In equivalent terms (compare Newton's page 62 below) this basic Newtonian

7. To determine the refraction of any ray when light falls on any curved surface you please.(266)

The accomplishing of these and similar problems is not, where computational tediousness proves no obstacle, so difficult as to require me to dwell on their explanation. Mathematicians will, I believe, take it better that I have merely reviewed them in the present manner.

PROBLEM 5

TO FIND THE CURVATURE OF ANY CURVE AT A GIVEN POINT(267)

The problem has the mark of exceptional elegance and of being pre-eminently useful in the science of curves. In preface, however, to its construction it is convenient to set down certain generalities:

1. The same circle has everywhere the same curvature, and the curvatures of unequal circles are inversely proportional to their diameters. If the diameter of one is twice as small as that of a second, the curvature of its circumference will be twice as great; if its diameter is three times as small, its curvature will be three times as great; and so on.

2. If a circle touch some curve on its concave side at a given point and be of such a size that no other tangent circle can be drawn between in the contact angles neighbouring that point,(268) that circle has the same curvature as the curve at that point of contact.(269) For a circle which lies between a curve and another circle in the vicinity of the contact point deviates less from the curve and more approximates its curvature than that second circle, and consequently most approximates its curvature when no other circle can be inserted between it and the curve.

3. Accordingly, the centre of curvature at some point of a curve is the centre of an equally curved circle; and thus the radius (or semidiameter) of curvature is that portion of the normal which ends at that centre.

4. And the ratio of curvature at its various points is known from the ratio of the curvature of equally curved circles or from the inverse ratio of the radii of curvature.

---

condition is that a circle of curvature at a point on a curve shall have 3-point contact with it there. Since only one circle can be drawn through three non-collinear points, the uniqueness of such an 'osculating' circle is guaranteed.

(270) Here (and indeed frequently below) the equivalent 'curvitatis' is cancelled. On his page 56 Newton will introduce a further equivalent 'curvedo', one favoured indeed by Leibniz and the Bernoullis when they independently rediscovered the fundamentals of an analytical theory of curvature expounded by Newton in the present problem. We might perhaps dare in translation to distinguish between 'curvitas' (curvity), 'curvatura' (curvature) and 'curvamen' or 'curvedo' (curvedness), but no such separation is made in the present English version.

(271) Newton intends 'incurvati'.

Problema itaq; ad hunc locum redijt ut radius vel centrum curvaminis inveniatur.

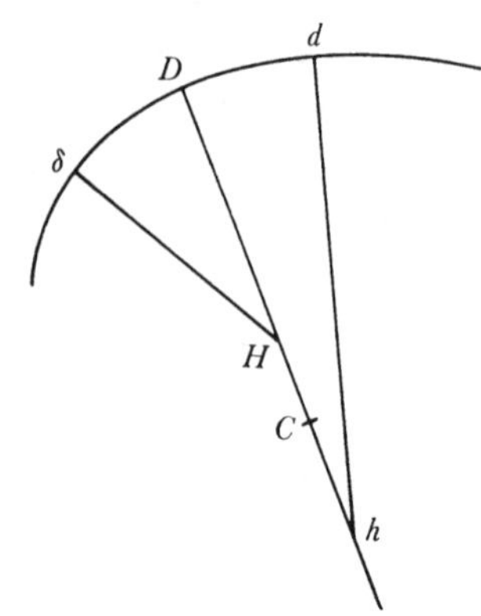

Concipe ergo quod ad tria curvæ puncta δ, D, ac d ducantur perpendicula quorum quæ sunt ad D et δ conveniant in H; et quæ ad D et d, conveniant in h. Et puncto D existente medio si major est curvitas a parte $D\delta$ quam Dd, erit DH[272] $\sqsupset dh$. Sed quo perpendicula δH ac dh propiora sunt intermedio perpendiculo, eò minùs distabunt puncta H et h. Et con-
||[54] venientibus || tandem perpendiculis, coalescent. Coalescent autem in puncto C et erit illud C centrum curvaminis ad curvæ punctum D cui perpendicula insistunt. Id quod per se manifestum est.

Hujus autem C varia sunt symptomata quæ ad ejus determinationem inservire possunt: Quemadmodum

1. Quod sit concursus perpendiculorum hinc et inde a DC infinitè parùm distantium.[273]

2. Quod perpendiculorum finitè parùm distantium intersectiones hinc et inde dirimit ac disterminat. Ita ut quæ sunt a parte curviori $D\delta$ citiùs ad H conveniant, et quæ sunt ex alterâ minùs curvâ parte Dd remotiùs conveniant ad h.[274]

3. Si DC dum curvæ perpendiculariter insistat moveri concipiatur, illud ejus punctum C (si demas motum accedendi vel recedendi a puncto insistentiæ C)[275] minimè movebitur sed centri motionis rationem habebit.[276]

4. Si centro C intervallo DC circulus describatur, non potest alius describi circulus qui juxta contactum interjacebit.[277]

(272) Read 'δH'.

(273) In historical fact all three primary discoverers, Apollonius, Huygens and Newton, of the evolute properties of curves took this as their basic defining 'symptom'. (See Apollonius, *Conics*, **5**, 51–2; Christiaan Huygens, *Œuvres complètes*, **14** (The Hague, 1920): 387–405; and I, **2**, 4: passim.) In the sequel (see note (279)) Newton qualifies the approach as 'simplicissimum' but its implications were not equally self-evident to all his contemporaries. In 1692, for example, Leibniz argued that, since normals to a curve are defined by 2-point contact with a 'normal' circle, the circle of curvature which is defined by the limit-meet of two indefinitely near normals is determined as the circle which has 4-point contact with a curve at a point, and that, more generally, an osculation of degree n is determined by a $(2n+2)$-point contact of a circle with a curve: in his own words 'Statueram ego, *contactum* continere duas intersectiones coincidentes; *osculum* continere plures contactus coincidentes, & osculum quidem primi gradus esse, quando coincidunt contactus duo, seu intersectiones quatuor; osculum secundi gradus, quando coincidunt intersectiones sex aut contactus tres &c' ('G.G.L. Generalia de Natura Linearum, anguloque contactus & osculi, provolutionibus, aliisque cognatis, & eorum usibus nonnullis', *Acta Eruditorum* (September 1692): 440–6, especially 440). In the present problem, in contrast, Newton reveals his mastery of the problem of the point-contact of a circle with a curve: in general, an osculating circle can have only 3-point contact with a curve (in which

The problem accordingly reduces to this point, that the radius or centre of curvature is to be found.

Imagine, therefore, that at the three points δ, D and d of a curve normals are drawn, and let those at D and δ meet in H, those at D and d in h, the point D being the middle one. If the curvature on the side $D\delta$ be greater than that on Dd, then DH(272) $< dh$. But the nearer the normals δH and dh are to the intermediate one, the less will be the distance between the points H and h, and when at length the normals meet they will coincide. Let them coincide in the point C: then will that point C be the centre of curvature at the point D on the curve at which they are normal. This is evident of itself.

Of this point C, however, there are various defining conditions which can serve to determine it. For instance:

1. It is the meet of normals at indefinitely small distances from DC on its either side.(273)

2. It separates and dichotomizes the intersections of normals in the finitely small neighbourhood on one side and the other, so that those on the more curved portion $D\delta$ meet more rapidly at H, while those on the other, less curved portion Dd do so more remotely at h.(274)

3. If DC while continuing to stand normal to the curve be conceived to move, that point C of it will (if you except its motion towards and away from the point C(275) at which it stands normal) not move at all but will be in the nature of a centre of motion.(276)

4. If a circle be described with centre C and radius DC, no other circle can be described which shall lie between [it and the curve] in the vicinity of the contact point.(277)

---

case the 'curvature' of the curve is defined as that of the circle) but may in addition share a fourth common point with it at a few points where the curve's curvature attains a local maximum or minimum value. The latter are the 'puncta flexûs maximi minimive' sought by him on his page 65 below: compare note (354).

(274) This somewhat useless 'symptom' does not, of course, hold true in the immediate vicinity of a point of maximum or minimum curvature.

(275) Read 'D'. This is an acute observation on Newton's part: in the comparable context of the diacaustic of a curve with respect to a point it was so little obvious to Johann Bernoulli in September 1693 that the increments CH, Ch are infinitely small in comparison to the curve increments $D\delta$, Dd that he accused L'Hospital of committing 'un terrible calcul' in his 'solution des caustiques par refraction' by making such an assumption. Bernoulli's letter to L'Hospital is lost but its gist is evident from the latter's reply on 7 October 1693 (*Briefwechsel von Johann Bernoulli* (note (140)), **1**: 191–5, especially 192) in which it is correctly asserted that, with respect to the curve increment (here Dd, say), 'la differentielle [Ch] est nulle'.

(276) The centre of curvature defined as the point on the normal to the curve at the point of contact which is instantaneously at rest. Compare I: 264.

(277) More exactly, any such circle will not be distinguishable from it to the second order of the infinitely small.

5. Deniq; si alterius alicujus tangentis circuli centrum ut H vel h paulatim ad hujus centrum C accedat donec tandem conveniat, tunc aliquod e punctis in quibus circulus ille curvam secavit simul conveniet [ad] punctum contactûs D.(278)

Et unumquodq; horum symptomatum ansam præbet diversimodè resolvendi Problema. Nos autem primum tanquam simplicissimum(279) eligemus.

Ad quodlibet Curvæ punctum D esto DT tangens, DC perpendiculum et C centrum curvaminis ut ante. Sitq; AB basis ad quam DB in angulo recto applicatur,(280) et cui DC occurrit in P. Age DG parallelam AB, et CG perpendiculum, inq; eo cape Cg cujuslibet datæ magnitudinis, et age $g\delta$ perpendiculum quod occurrat DC in δ: eritq; $Cg.g\delta(::TB.BD)::$ fluxio Basis ad fluxionem Applicatæ.(281) Concipe præterea punctum D per infinitè parvum intervallum Dd in curva promoveri, et actis dE ad DG et Cd ad curvam normalibus quarum Cd occurrit DG in F et

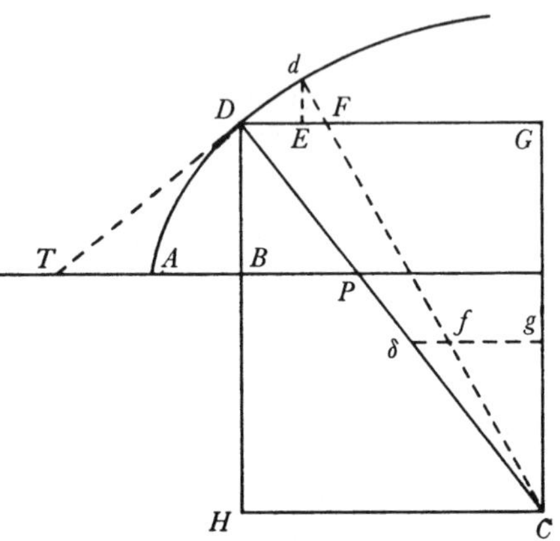

||[55] δg in f; erit DE momentum Basis, || dE momentum Applicatæ, ac δf contemporaneum momentum rectæ $g\delta$. Estq; $DF = DE + \frac{dE \times dE}{DE}$.(282) Habitis itaq; horum momentorum sive quod perinde est fluxionum generantium rationibus, habebitur ratio GC ad datam gC (quippe quæ est DF ad δf) et inde punctum C determinabitur.

Sit ergo $AB = x$, $BD = y$, $Cg = 1$, et $g\delta = z$ et erit $1.z::m.n$ seu $z = \frac{n}{m}$,(283) hujus autem z momentum δf dic $r \times o$ (factum nempe ex velocitate et infinite parva quantitate,(284)) eritq; momentum $DE = m \times o$, $dE = n \times o$, et inde $DF = mo + \frac{nno}{m}$. Est ergo $Cg(1).\,CG::(\delta f.DF::)\,ro.\,mo + \frac{nno}{m}$. Adeoq; $CG = \frac{mm+nn}{mr}$.

Cùm insuper Basis fluxioni m (ad quam tanquam(285) uniformem fluxionem

(278) Newton will elaborate this approach, on the lines of his February 1665 researches (I: 262) on his pages 62–3 below. A tangent circle shares two coincident points with the curve and the approach to coincidence at the point of contact of a third common point determines the corresponding circle as osculating and so to be of the same curvature as the curve at the 3-point contact.

(279) See note (273) above.

(280) Still a very necessary restriction on the coordinate system: indeed, as we shall see in the sixth volume, Newton could attack the problem of finding the curvature of a curve defined in an oblique Cartesian system (other than by reduction to perpendicular coordinates) only

5. If, lastly, the centre, as H or h, of some other tangent circle gradually approach the centre C of this one till at length it coincide with it, then some one of the points in which the former circle cut the curve will simultaneously coincide with the point D of contact.(278)

Each one of these defining properties provides a means of resolving the problem in a different way. We, however, shall choose the first as being the simplest.(279)

At any point D you please of the curve let DT be tangent, DC normal and C the curvature centre, as before. Again, let AB be the base to which DB is applied at right angles(280) and which is met by DC in P. Draw DG parallel to AB and CG perpendicular, taking in it Cg of any given size, and raise the perpendicular $g\delta$ till it meets DC in δ: then will there be $Cg{:}g\delta(=TB{:}BD)$ the ratio of the fluxion of the base to that of the ordinate.(281) Imagine, furthermore, that the point D advances through the infinitely small distance Dd and draw dE perpendicular to DG and Cd normal to the curve, the latter meeting DG in F and δg in f: then will DE be the moment of the base, dE that of the ordinate and δf the contemporaneous moment of the straight line $g\delta$. Also

$$DF = DE + dE^2/DE.^{(282)}$$

Accordingly, when the ratios of these moments—or, what is the same, those of the generating fluxions—are had, there will be obtained the ratio of GC to the given quantity gC (seeing that this is that of DF to δf) and by this the point C will be determined.

Let, therefore, $AB=x$, $BD=y$, $Cg=1$ and $g\delta=z$ and it will be $1{:}z=\dot{x}{:}\dot{y}$ or $z=\dot{y}/\dot{x}$,(283) calling z's moment $\delta f=\dot{z}o$ (the product, namely, of its velocity and an infinitely small quantity(284)): the moment DE will then be $\dot{x}o$, $dE=\dot{y}o$ and thence $DF=\dot{x}o+\dot{y}^2o/\dot{x}$. Therefore $Cg(1){:}CG=(\delta f{:}DF=)\,\dot{z}o{:}(\dot{x}o+\dot{y}^2o/\dot{x})$, so that $CG=(\dot{x}^2+\dot{y}^2)/\dot{x}\dot{z}$.

Since, in addition, we are free to assign any velocity at all to the fluxion $\dot{x}$ of

---

after he had acquired, in the early 1690's, a working knowledge of the Taylor expansion of a function in terms of its successive derivatives.

(281) That is, as DE to dE.

(282) In the limit as d coincides with D, so that DC may be taken to be parallel to dC.

(283) Newton first wrote somewhat abruptly 'et $g\delta$ valebit $\frac{n}{m}$' (and the value of $g\delta$ will be $\dot{y}/\dot{x}$).

(284) Understand 'temporis incrementum repræsentante' (representing the increment of time): o is the increment of the conventional variable of time and hence the 'moment' of all variable segments in the figure will be the product of o and their corresponding fluxional 'speeds'. Compare the 'Demonstratio' on Newton's pages 20–1 above.

(285) 'Correlatam et' (correlate and) is cancelled.

cæteras referre convenit) liberum sit quancunq; velocitatem tribuere, dic esse 1,(286) et erit $n=z$, et $CG=\frac{1+zz}{r}$.(287) Et inde $DG=\frac{z+z^3}{r}$, ac

$$DC=\frac{\overline{1+zz}\sqrt{1+zz}}{r}.^{(288)}$$

Expositâ itaq; quâvis æquatione quâ relatio BD ad AB pro curvâ definiendâ designetur, imprimis quære relationem inter m et n per Prob 1, et interea substitue 1 pro m et z pro n.(289) Dein ex æquatione resultante per idem Prob 1 quære relationem inter m, n, et r et interea substitue 1 pro m et z pro n ut ante. Atq; ita per priorem operationem obtinebis valorem z, et per posteriorem obtinebis valorem r; quibus habitis, produc DB ad H versus concavam partem curvæ ut sit $DH=\frac{1+zz}{r}$, et age HC parallelam AB et perpendiculo DC occurrentem in C, eritq; C centrum curvaturæ ad curvæ punctum D. Vel cùm sit $1+zz=\frac{PT}{BD}$, fac $DH=\frac{PT}{r\times BP}$,(290) vel $DC=\frac{DP^3}{r\times DB^3}$.

EXEMPL: 1. Sic exposita $ax+bxx-yy=0$, æquatione ad Hyperbolam cujus latus rectum est a ac transversum $\frac{a}{b}$;(291) emerget (per Prob 1) $a+2bx-2zy=0$, (scriptis nempe 1 pro m et z pro n in æquatione resultante, quæ secus foret $am+2bmx-2ny=0$). Et hinc denuo prodit $2b-2zz-2ry=0$ scriptis iterum 1 pro m et z pro n. Per priorem est $z=\frac{a+2bx}{2y}$, et per posteriorem $r=\frac{b-zz}{y}$.(292) Dato itaq; quovis curvæ puncto D et per consequentiam x et y, ex his dabuntur z et r, quibus cognitis fac $\frac{1+zz}{r}=GC$ vel DH, et age HC.(293)

Quemadmodum si definitè sit $a=3$, $b=1$, adeoq; $3x+xx=yy$ Hyperbolæ
||[56] conditio: et si || assumatur $x=1$, erit $y=2$, $z=\frac{5}{4}$, $r=-\frac{9}{32}$. Invento H, erige HC occurrentem perpendiculo DC priùs ducto. Vel quod perinde est fac $HD.HC(::1.z)::1.\frac{5}{4}$. et age DC curvedinis Radium.(294)

---

(286) Accordingly, the increment DE is now taken to be o.

(287) The continuation 'vel $=\frac{PT}{r\times BP}$ siquidem est $1+zz=\frac{PT}{PB}$' (that is, $PT/\dot{z}\times BP$ seeing that $PT/BP = 1+z^2$) is cancelled.

(288) Except that now the tangent slope $n/m = \dot{y}/\dot{x}$ is measured explicitly by the variable z, Newton essentially repeats his equivalent construction of the radius of curvature in 'Prob 2d' of the 1666 fluxional tract (I: 420).

(289) That is, $n/m(=\dot{y}/\dot{x})$ since $m(=\dot{x})$ is taken as unity.

(290) Newton first continued '$HC=\frac{PT}{r\times BD}$'. Compare note (287).

the base (to which fluxion, supposed[285] uniform, it is convenient to relate the others), call it unity[286] and there will be $\dot{y} = z$ and $CG = (1+z^2)/\dot{z}$,[287] and thence $DG = (z+z^3)/\dot{z}$, while $DC = (1+z^2)^{\frac{3}{2}}/\dot{z}$.[288]

Accordingly, when any equation is exhibited by which the relationship of BD to AB for defining the curve is to be expressed, in the first instance seek the relation between $\dot{x}$ and $\dot{y}$ by Problem 1, and meantime substitute 1 in place of $\dot{x}$ and z for $\dot{y}$.[289] Then from the resulting equation by the same Problem 1 seek the relationship between $\dot{x}$, $\dot{y}$ and $\dot{z}$ and at the same time substitute 1 in place of $\dot{x}$ and z for $\dot{y}$ as before. And thus by the first operation you will obtain the value of z, and by the latter that of $\dot{z}$. When you have these, produce DB to H in the direction of the concave side of the curve so that $DH = (1+z^2)/\dot{z}$ and draw HC parallel to AB and meeting the normal DC in C: C will be the centre of curvature at the point D of the curve. Alternatively, since $1+z^2 = PT/BD$, make

$$DH = PT/(\dot{z}\times BP),^{(290)} \quad \text{or} \quad DC = DP^3/(\dot{z}\times DB^3).$$

Example 1. So when $ax+bx^2-y^2 = 0$ is exhibited (an equation to a Hyperbola whose latus rectum is a and transverse diameter a/b[291]), there will emerge, by Problem 1, $a+2bx-2zy = 0$ (namely, on writing 1 in place of $\dot{x}$ and z for $\dot{y}$ in the resulting equation, which would otherwise be $a\dot{x}+2b\dot{x}x-2\dot{y}y = 0$). And hence there comes again $2b-2z^2-2\dot{z}y = 0$, on writing 1 in place of $\dot{x}$ and z for $\dot{y}$ a second time. By the former $z = (a+2bx)/y$ and by the latter

$$\dot{z} = (b-z^2)/y.^{(292)}$$

Accordingly, given any point D of the curve and in consequence x and y, from these will be given z and $\dot{z}$: then, when these are known, make

$$(1+z^2)/\dot{z} = GC \text{ or } DH$$

and draw HC.[293]

For instance if, in definite terms, $a = 3$ and $b = 1$, so that $3x+x^2 = y^2$ is the [defining] condition of the hyperbola, and if it be assumed that $x = 1$, then $y = 2$, $z = \frac{5}{4}$ and $\dot{z} = -\frac{9}{32}$. Having found H, erect HC to its meet with DC, the normal previously drawn. Or, what comes to the same, make

$$HD\!:\!HC(= 1\!:\!z) = 1\!:\!\tfrac{5}{4}$$

and draw DC, the radius of curvature.[294]

(291) As elsewhere Newton supposes his constants to be positive quantities.

(292) That is, $r(=\dot{z}) = [4by^2-(a+2bx)^2]/4y^3 = -a^2/4y^3$.

(293) A concluding phrase 'perpendiculo occurrentem in C et erit C centrum' (meeting the normal in C and C will be the centre) is cancelled.

(294) This paragraph is a late addition squashed into small spaces at the foot of page 55 and top of page 56.

Siquando computationem non admodum perplexam fore censeas, possis indefinitos valores ipsorum r et z in $\frac{1+zz}{r}$ valore CG substituere. Et sic in hoc exemplo per debitam reductionem obtinebis $DH^{(295)}=y+\frac{4y^3+4by^3}{aa}$. Cujus tamen DH valor per calculum negativus prodit sicut in exemplo numerali videre est. At hoc tantùm arguit DH ad partes versus B capiendam esse. Nam si fuisset affirmativus ad contrarias partes duxisse oporteret.

Coroll. Hinc si signum symbolo $+b$ præfixum mutetur, ut fiat $ax-bxx-yy=0$ æquatio ad Ellipsin; erit $DH=y+\frac{4y^3-4by^3}{aa}$.

At posito $b=0$ ut æquatio fiat $ax-yy=0$ ad Parabolam; erit $DH=y+\frac{4y^3}{aa}$. Indeq͛(296) $DG=\frac{1}{2}[a]+2x$.

Ex his facilè colligitur radium curvaturæ cujusvis conicæ sectionis valere $\frac{4DP^{\text{cubum}}}{aa}$ (297).

Exempl 2. Si $x^3=ayy-xyy$ (æquatio ad Cissoidem Dioclis)(298) exponatur; per Prob 1 imprimis obtinebitur $3xx=2azy-2xzy-yy$; ac deinde

$$6x=2ary+2azz-2zy-2xry-2xzz-2zy.$$

Adeoq͛ est $z=\frac{3xx+yy}{2ay-2xy}$. Et $r=\frac{3x-azz+2zy+xzz}{ay-xy}$. Dato itaq͛ quolibet Cissoidis puncto et inde x et y, dabuntur z et r: Quibus cognitis fac $\frac{1+zz}{r}=CG$.

Exempl: 3. Si detur $\overline{b+y}\sqrt{cc-yy}=xy$ æquatio ad Conchoidem, ut supra;(299) Finge $\sqrt{cc-yy}=v$, et emerget $bv+yv=xy$. Jam harum prior (viz $cc-yy=vv$) per Prob 1 dat $-2yz=2vl$ (scripto nempe z pro n,) et posterior dat $bl+yl+zv=y+xz$. Et ex his æquationibus rite dispositis determinantur l et z. Ut autem r præterea determinetur, e novissimâ æquatione extermina fluxionem l substituendo $\frac{-yz}{v}$, et emerget $-\frac{byz}{v}-\frac{yyz}{v}+zv=y+xz$.(300) æquatio quæ

(295) As Newton at once points out, read '$-DH$' accurately, since $z^2=a^2/4y^2+b$ and so $(1+z^2)/r=(a^2/4y^2+1+b)/(-a^2/4y^3)$, where the negative sign of $r(=\dot{z})=-a^2/4y^3$ determines the convexity of $\widehat{Dd}$.

(296) The continuation 'facilè colligitur esse' (it is easily gathered that) is cancelled, but the result is indeed immediate since in the parabolic case ($b=0$) $z=a/2y$ with $DG=DH\times z$.

(297) For $DP=y\sqrt{[1+z^2]}$ and so the curvature radius DC is DP^3/ry^3 where (note (292)) $ry^3=-\frac{1}{4}a^2$. This result, apparently Newton's original discovery, was found independently thirty-five years later by John Keill, who published it in his celebrated 'Epistola ad... Edmundum Halleium...de Legibus Virium Centripetarum' (*Philosophical Transactions*, **26** (1708–9) [1710]): No. 317 (for September/October 1708): 174–88, especially 177–9).

Should you at any time judge the computation not to be unduly complex, you might substitute the indefinite values of $\dot{z}$ and z in $(1+z^2)/\dot{z}$, the value of CG. And so in the present example you will obtain by appropriate reduction DH(295) $= y+(4y^3+4by^3)/a^2$. (The value of DH, however, comes out negative by calculation, as may be seen in the numerical example, but this shows only that DH must be taken towards B in direction, for if it had been positive it would have had to be drawn the opposite way.)

Corollary. Hence if the sign prefixed to the symbol $+b$ be changed, so that the equation becomes $ax-bx^2-y^2=0$, one to an ellipse, there will be

$$DH = y+(4y^3-4by^3)/a^2.$$

But when b is set equal to zero so that the equation becomes $ax-y^2=0$, one to a parabola, there will be $DH = y+4y^3/a^2$ and thence(296) $DG = \frac{1}{2}a+2x$.

From these cases it is readily gathered that the radius of curvature of any conic has the value $4DP^3/a^2$.(297)

EXAMPLE 2. If $x^3 = ay^2-xy^2$ (an equation to the cissoid of Diocles)(298) be exhibited, in the first instance you will obtain by Problem 1

$$3x^2 = 2azy-2xzy-y^2,$$

and then $6x = 2a\dot{z}y+2az^2-2zy-2x\dot{z}y-2xz^2-2zy$, so that

$$z = (3x^2+y^2)/(2ay-2xy) \quad \text{and} \quad \dot{z} = (3x-az^2+2zy+xz^2)/(ay-xy).$$

Accordingly, given any point at all of the cissoid, and thence x and y, there will be given z and $\dot{z}$. When these are known make $(1+z^2)/\dot{z} = CG$.

EXAMPLE 3. If there be given $(b+y)\sqrt{[c^2-y^2]} = xy$ (the equation to the conchoid, as above),(299) suppose $\sqrt{[c^2-y^2]} = v$ and there will emerge

$$bv+yv = xy.$$

Now the first of these (viz: $c^2-y^2=v^2$) yields by Problem 1 $-2yz = 2v\dot{v}$ (namely, when z is written in place of $\dot{y}$) while the latter gives

$$b\dot{v}+y\dot{v}+zv = y+xz,$$

and from these equations when rightly arranged are determined the values of $\dot{v}$ and z. To determine $\dot{z}$ in addition, however, eliminate the fluxion $\dot{v}$ from the most recent equation by substituting $-yz/v$ for it and there will emerge

$$-byz/v-y^2z/v+zv = y+xz,^{(300)}$$

(298) Compare II, 1, 1, §3: note (58).

(299) In Example 3 on Newton's page 44.

(300) In a cancelled continuation Newton multiplied through by v to obtain

$$\text{'}-byz-yy[z]+zvv = yv[+]x[v]z\text{'}.$$

fluentes quantitates sine aliquibus earum fluxionibus (prout exigit resolutio Prob 1^{mi}) complectitur. Hinc itaq; per Prob 1 elicies

$$-\frac{bzz}{v}-\frac{byr}{v}+\frac{byzl}{vv}-\frac{2yzz}{v}-\frac{yyr}{v}+\frac{yyzl}{vv}+rv+zl=2z+xr.$$

||[57] Qua æquatione in ordinem reductâ et concinnatâ, dabitur r. || Inventis autem z et r fac $\frac{1+zz}{r}=CG$.

Si penultimam æquationem per z divisisses, exinde postmodum per Prob 1 obtinuisses $-\frac{bz}{v}+\frac{byl}{vv}-\frac{2yz}{v}+\frac{yyl}{vv}+l=2-\frac{yr}{zz}$,(301) æquationem priore simpliciorem pro determinando r.

Dedi quidem hoc exemplum ut modus operandi in surdis æquationibus constaret. At Conchoidis curvatura sic breviùs inveniri potuit. Æquationis $\overline{b+y}\sqrt{cc-yy}=xy$ partibus quadratis et per yy divisis, exurgit

$$\frac{bbcc}{yy}+\frac{\begin{matrix}2bcc\\ \end{matrix}}{y}\begin{matrix}-bb\\+cc\end{matrix}-2by-yy=xx.$$

Et inde per Prob 1 exoritur $-\frac{2bbccz}{y^3}-\frac{2bccz}{yy}-2bz-2yz=2x$. sive $-\frac{bbcc}{y^3}-\frac{bcc}{yy}-b-y=\frac{x}{z}$. Et hinc denuo per Prob 1 exoritur

$$\frac{3bbccz}{y^4}+\frac{2bccz}{y^3}-z=\frac{1}{z}-\frac{xr}{zz}.$$

Per priorem exitum determinatur z, et per posteriorem r.

EXEMPL 4.(302) Sit $IADF$ Trochois ad circulum ALE (cujus diameter est AE) accommodata; et ordinatâ BD secante circulum in L, dic $AE=a$, $AB=x$, $BD=y$, $BL=v$, et arc[um] $AL=t$ ejusq; arcûs fluxionem dic k. Et imprimis (ducto PL semidiametro) erit fluxio Basis AB ad fluxionem arcus AL ut BL ad PL; hoc est m sive $1.k::v.\frac{1}{2}a$. Atq; adeo

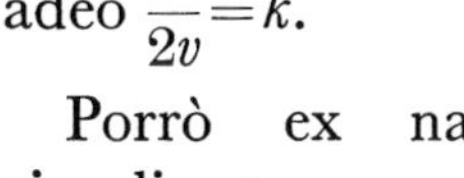

$\frac{a}{2v}=k$.

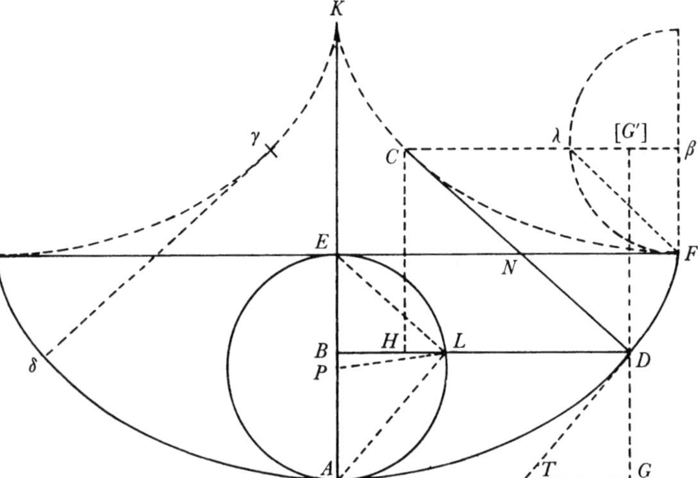

Porrò ex natura circuli est $ax-xx=vv$. Et inde per Prob: 1 $a-2x=2lv$, sive $\frac{a-2x}{2v}=l$.

an equation which comprehends the fluent quantities without any of their fluxions (as the solution of Problem 1 demands). Hence accordingly you will elicit by Problem 1 that

$$-\frac{bz^2}{v}-\frac{by\dot{z}}{v}+\frac{byz\dot{v}}{v^2}-\frac{2yz^2}{v}-\frac{y^2\dot{z}}{v}+\frac{y^2z\dot{v}}{v^2}+\dot{z}v+z\dot{v}=2z+x\dot{z}.$$

This equation, when reduced to order and tidied up, will give $\dot{z}$. Then, after z and $\dot{z}$ are found, make $(1+z^2)/\dot{z}=CG$.

If you had divided the last equation but one by z, from it you would subsequently, by Problem 1, have obtained

$$-\frac{bz}{v}+\frac{by\dot{v}}{v^2}-\frac{2yz}{v}+\frac{y^2\dot{v}}{v^2}+\dot{v}=2-\frac{y\dot{z}}{z^2},\text{(301)}$$

an equation simpler than the previous one for determining $\dot{z}$.

I have, to be sure, given this example to make the method of operation in surd equations familiar. However, the curvature of the conchoid could have been found more briefly this way. On squaring either side of the equation

$$(b+y)\surd[c^2-y^2]=xy$$

and dividing through by y^2 there arises

$$b^2c^2y^{-2}+2bc^2y^{-1}+(c^2-b^2)-2by-y^2=x^2,$$

and thence, by Problem 1, $-2b^2c^2y^{-3}z-2bc^2y^{-2}z-2bz-2yz=2x$, that is, $-b^2c^2y^{-3}-bc^2y^{-2}-b-y=xz^{-1}$, and from this again by Problem 1 there arises $3b^2c^2y^{-4}z+2bc^2y^{-3}z-z=z^{-1}-xz^{-2}\dot{z}$. By the former result is determined z, and by the latter $\dot{z}$.

EXAMPLE 4.(302) Let $IADF$ be a cycloid fitted to the circle ALE (of diameter AE). Where the ordinate BD cuts the circle in L, call $AE=a$, $AB=x$, $BD=y$, $BL=v$ and the arc $\widehat{AL}=t$ and the fluxion of this arc call $\dot{t}$. In the first place, on drawing the radius PL, the fluxion of the base AB will be to the fluxion of the arc $\widehat{AL}$ as BL to PL, that is, $\dot{x}$ (or 1)$:\dot{t}=v:\frac{1}{2}a$, so that $a/2v=\dot{t}$.

Moreover, from the nature of the circle $ax-x^2=v^2$ and therefrom by Problem 1 $a-2x=2\dot{v}v$, or $(a-2x)/2v=\dot{v}$.

(301) Alternatively, this follows from the last equation on dividing through by z and substituting for x its equal $-(by+y^2)/v+v-y/z$.

(302) Compare I: 373. The differential properties of the cycloid here developed form a necessary preliminary to Newton's subsequent discussion, probably independently of Huygens, of simple harmonic motion in a vertical cycloid. (See **2**, Introduction below, especially the appended mathematical note.)

Adhæc ex natura Trochoidis[303] est $LD=\text{arc[ui]}\ AL$; adeoq; $v+t=y$. Et inde per Prob: 1, $l+k=z$.

Deniq; pro fluxionibus l et k valores hic substituantur et emerget $\frac{a-x}{v}=z$. Unde per Prob: 1 deducitur $\frac{-al}{vv}+\frac{xl}{vv}-\frac{1}{v}=r$. Et his inventis fac $\frac{1+zz}{r}=-DH$ et erige HC.[304]

Coroll. Cæterum ex his consectatur,

1. Quod sit $DH=2BL$ et $CH=2BE$, sive quod EF in N bisecat CD radium curvaminis. Et hoc patebit substituendo valores r et z jam inventos in æquatione $\frac{1+zz}{r}=DH$ et exitum probè reducendo.[305]

2. Hinc Curva FCK in qua centrum curvaminis indefinite versatur est alia
||[58] || huic æqualis Trochois[306] cujus vertices ad I et F adjacent hujus cuspidibus. Nam circulus $F\lambda$ æqualis ALE et similiter positus describatur et agatur $C\beta$ parallela EF circuloq; occurrens in λ; et erit arc[us] $F\lambda$ ($=$arc. $EL=NF$)$=C\lambda$.

3. CD quæ recta est ad Trochoidem IAF, contingit Trochoidem IKF in C.[307]

4. Hinc (inversis Trochoidibus) si superioris Trochoidis cuspidi K pondus ad distantiam KA sive $2EA$ filo appensum innitatur, et undulante pondere filum se applicet ad Trochoidis partes KF et KI hinc inde obsistentes ne in rectum distendatur, et cogentes ut ad earum normam dum digreditur a perpendiculo paulatim desuper inflectatur, parte CD sub infimo contactûs puncto manente rectâ: pondus in inferioris Trochoidis perimetro movebitur, utpote cui filum CD semper perpendiculare est.

5. Est itaq; tota fili longitudo KA æqualis perimetro Trochoidis KCF, ejusq; pars CD æqualis parti perimetri CF.[308]

---

(303) The primary cycloid is generated by the circle ALE rolling along the base line IEF, and accordingly $EN = LD = \widehat{AL}$ measures the horizontal advance of the point D from A. Compare Newton's Schootenian definition of a general cycloid on I: 378.

(304) Namely, to meet the normal DC in C. In more familiar analytical terms, if $A\hat{P}L = \theta$ be taken as parameter, then $\widehat{AL} = LD = t = \frac{1}{2}a\theta$, $AB = x = \frac{1}{2}a(1-\cos\theta)$, $BL = v = \frac{1}{2}a\sin\theta$ and so $BD = y = t+v = \frac{1}{2}a\cos^{-1}[1-2x/a]+\sqrt{[x(a-x)]}$: hence

$$k = \dot{t} = a/2v \quad\text{and}\quad l = \dot{v} = (a-2x)/2v$$

on setting $\dot{x} = 1$, so that $z = \dot{y} = \dot{t}+\dot{v} = (a-x)/v = \sqrt{[(a-x)/x]}$ and therefore

$$r = \dot{z} = -a/2vx$$

(to which Newton's result reduces on substituting $v^2 = x(a-x)$ and $l/x = a/2vx-1/v$).

(305) Since $1+z^2 = a/x$, it follows that $DH = (1+z^2)/\dot{z} = -2v = 2LB$ and so

$$CH = HD\times z = 2(a-x) = 2EB.$$

(306) Newton has cancelled a first continuation 'quam cuspides hujus contingunt in ejus verticibus' (which is touched in its vertices by the present one's cusps), that is, at I and F.

Further, from the nature of the cycloid[303] it is $LD = \text{arc } \widehat{AL}$, so that $v+t = y$, and thence, by Problem 1, $\dot{v}+\dot{t} = (\dot{y} \text{ or})\ z$.

Finally, in place of the fluxions $\dot{v}$ and $\dot{t}$ let there here be substituted their values and there will emerge $(a-x)\,v^{-1} = z$. Hence is derived by Problem 1 $-(a-x)\,\dot{v}v^{-2}-v^{-1} = \dot{z}$. After these are found, make $(1+z^2)/\dot{z} = -DH$ and erect HC.[304]

Corollaries. For the rest, the following are immediate consectaries:

1. $DH = 2BL$ and $CH = 2BE$, in other words EF bisects the radius of curvature in N. This will be manifest on substituting the values of $\dot{z}$ and z now found in the equality $(1+z^2)/\dot{z} = DH$ and appropriately reducing the result.[305]

2. In consequence, the curve FCK in which the centre of curvature is indefinitely located is a second, congruent cycloid[306] whose vertices (at I and F) are adjacent to the first one's cusps. For let the circle $F\lambda$ equal and similarly situated to ALE be described and draw $C\beta$ parallel to EF and meeting the circle in λ: there will then be arc $\widehat{F\lambda}(= \text{arc } \widehat{EL} = NF) = C\lambda$.

3. CD, at right angles to the cycloid IAF, touches the cycloid IKF in C.[307]

4. Hence (inverting the cycloids) at the cusp K of the upper cycloid support a weight hanging by a thread at the distance KA or $2EA$. As the weight vibrates, the thread will wrap itself along the parts KF and KI of the cycliod: these on either side resist its straight-line extension, compelling it, as it departs from the vertical, gradually to mould itself to their shape in its upper portion while leaving its remaining part CD below the lowest point of contact extended in a straight line. Accordingly, the weight will move in the perimeter of the lower cycloid since the thread CD is always perpendicular to it.

5. Consequently, the total length KA of the thread is equal to the perimeter of the cycloid KCF, and its part CD to the portion CF of that perimeter.[308]

---

(307) For the limit-increment of the evolute arc $\widehat{FC}$ will coincide with the normal DC (compare note (275) above). It is an immediate corollary that the tangent at C to the cycloidal arc $\widehat{KCF}$ is parallel to the corresponding chord λF in the generating circle: a result known to Fermat in 1636 (see Roberval's 'Observations sur la Composition des Mouvemens, et sur le Moyen de trouver les Touchantes des Lignes Courbes', *Divers Ouvrages de Mathematique de M. de Roberval* [= *Memoires de l'Academie Royale des Sciences Depuis 1666 jusqu'à 1699*, **6** (Paris, ${}_2$1730): 1–478]: 'Onziéme Exemple, de la Roulette ou Trochoïde': 79–81) and implicit in Descartes' 1638 construction of the cycloid's normal (see I, **2**, 6, §2: note (1)) but first established with classical rigour by Christopher Wren in 1658 in his 'De rectâ Tangente Cycloidem primariam' (note (255)).

(308) Wren's 1658 rectification of the cycloid's arc (first published in 1659 in John Wallis' *Tractatus De Cycloide* (note (255)): 64–70: ''Εὐθυσμὸς Curvæ lineæ Cycloidis primariæ') by setting $\widehat{CF} = (CD \text{ or})\ 2CN$ is immediate. (Compare II: 192, note (9).) An equivalent result was known to Roberval about 1640 (*Divers Ouvrages* (note (307)): 419–27).

The essence of the first five of Newton's present corollaries had been discovered by Christiaan Huygens between 1 and 6 December 1659 (N.S.) by considering the centre C of curvature to be the limit-meet of normals to the cycloidal arc ADF in the immediate vicinity of D and then

6. Cum filum circa mobile punctum C tanquam centrum undulando convolvatur; superficies per quam tota CD continuò trajicitur erit ad superfic[i]em per quam pars CN supra rectam IF simul trajicitur ut CD^q ad CN^q hoc est ut 4 ad 1. Est itaqꝫ area CFN quarta pars areæ CFD, et area $KCNE$ quarta pars areæ $ACDB$.[309]

7. Quinimò cùm subtensa EL sit æqualis et parallela CN, et circa immobile centrum E perinde ac CN circa mobile centrum C circumagitur, æquales erunt superficies per quas simul trajiciuntur; nempe area CFN et circuli segmentum EL.[310] Et inde area NFD tripla erit segmenti istius, ac tota $EADF$ tripla semicirculi.

8. Deniqꝫ cùm pondus D attingit punctum F, totum filum circum Trochoidis perimetrum KCF flectetur, radio curvaminis CD manente nullo. Et proinde Trochois IAF ad ejus cuspidem F curvior est quàm quilibet circulus, et cum tangente βF productâ constituit angulum contactus infinitè majorem quàm circulus cum rectâ potest constituere.[311]

||[59] Sunt etiam anguli contactûs Trochoidalibus infinitè majores || et illis deinceps alij infinite majores et sic in infinitum, et tamen maximi sunt infinitè

deducing that the base IEF bisects the curvature radius CD. (See his *Œuvres complètes*, **17** (The Hague, 1932): 142–5.) Almost at once, however, he saw the advantage, in terms of simplicity if not of heuristic grasp of the subtleties involved, of recasting his deduction of these properties of the cycloid's evolute to follow as a particular consequence of the rectification 'per Wrennij inventum' rather than as an instance of a general curvature method (*Œuvres*, **14** (1920): 404–5; also compare the editorial *avertissement* on pp. 205–7). Yet, despite repeated entreaties from several quarters, Huygens delayed publication of his results for over a decade and they appeared in print for the first time only in his *Horologium Oscillatorium sive De Motu Pendulorum ad Horologia aptato Demonstrationes Geometricæ* (Paris, 1673: Pars Tertia, Propositio v: 66–9; compare also page 82 where he remarks that he discovered the cycloid's evolute 'primùm omnium' by his general curvature method). How many of Huygens' cycloidal results had been brought privately to Newton's attention before he began to draft the present tract in late 1670 is not known. As we saw in the first volume (I: 373) he knew the basic property that the cycloid's base bisects all radii of curvature at a time, late October 1665, when it is difficult enough to establish any personal contact on his part with Barrow in Cambridge, let alone any foreign mathematician. On the other hand the gist of Huygens' 1659 discovery that the cycloid is the isochrone and that by suitable 'cheeks' a pendulum bob can be made to traverse it, hinted at in his correspondence with Robert Moray and perhaps directly communicated by word of mouth during his 1661 visit, was already a topic of conversation in scientific circles in London by January 1662. Indeed, on 24 January of that year Moray communicated to Huygens on behalf of its author, the Royal Society's president, William Brouncker, a 'Demonstration of the Equality of Vibrations in a Cicloid-Pendulum' whose last paragraph deduces that the evolute of the cycloid is a congruent cycloid from the rectification of its general arc. Having rightly found fault with its earlier, dynamical portion Huygens in his reply a week later was not in the mood to be impressed: 'La proprieté de la Cycloide, de ce que par son evolution, il se descrit une courbe pareille n'estoit pas difficile a demonstrer apres que Monsieur Wren a decouuert la dimension de cette ligne, mais a trouuer methodiquement la dite proprietè comme j'ay fait, il y avoit plus de peine' (*Œuvres complètes*, **4** (1891): 51; compare also pp. 30–1). Newton would no doubt have agreed.

6. Since the thread revolves around the mobile point C as centre while it vibrates, the surface through which the whole line CD continuously passes will be to that through which its part CN above the straight line IF simultaneously travels as CD^2 to CN^2, that is, as 4 to 1. Accordingly, the area CFN is one-fourth the area CFD, and the area $KCNE$ one-fourth the area $ACDB$.(309)

7. Further, indeed, since the subtense EL is equal and parallel to CN and circles the stationary centre E as CN goes around the mobile centre C, the surfaces—namely, the area CFN and the circle segment EL—across which they simultaneously pass will be equal.(310) And from this the area NFD will be triple that segment, and the whole area $EADF$ triple that of the semicircle.

8. When, lastly, the weight D arrives at the point F, the thread will wholly be bent around the cycloid's perimeter KCF, the radius of curvature CD there having a zero stationary value. In consequence, the cycloid IAF is more curved at its cusp F than any circle, forming with its tangent βF produced a contact angle infinitely greater than a circle can with a straight line.(311)

There are also contact angles infinitely greater than cycloidal ones and others in turn infinitely greater than these, and so on infinitely, but still the greatest

(309) Read '$KCDA$'. This familiar way of evaluating the cycloid's area, deducible at once from its evolute properties and perhaps already known to Huygens (compare the diagrams reproduced in his *Œuvres complètes*, **17** (1932): 142–5), could scarcely have been suspected by those mathematicians of the preceding generation who first determined its area. Roberval, apparently the prime discoverer in the spring of 1638, compared the cycloid's total area with that of the generating circle with the aid of its sinusoidal 'socia', the 'compagne de la roulette' (see his 'De Trochoide ejusque Spatio', first published in his *Divers Ouvrages* [(note (307)): 361–82] in 1693; and compare his announcement of the result to Fermat the next June, first published in *Varia Opera Mathematica D. Petri de Fermat* (Toulouse, 1679): 155 [= *Œuvres de Fermat*, **2** (Paris, 1894): 147]). The methods contrived by Descartes in the summer of 1638 (communicated to Mersenne on 27 July (N.S.) in a letter first published in Clerselier's *Lettres de M. Descartes*, **3** (Paris, 1667): 363–78, especially 366–70 [= *Correspondance du P. Marin Mersenne*, **7** (Paris, 1962): 407–12]) and by Torricelli half a dozen years later (who first published the correct quadrature in his *Opera Geometrica* (Florence, 1644): 87–9) depended on even more cumbrous dissections of the cycloid.

(310) This corollary, since it depends essentially upon a tangent (rather than a curvature) property of the cycloid, might more logically have been appended to Example 2 of Mode 9 on Newton's page 51 above. To any one familiar with Wren's construction of the cycloid's tangent (note (255)) this variant evaluation of the cycloid's area would be evident, but we can nevertheless trace no precursor in the extensive literature published on the cycloid between 1659 and 1670.

(311) Newton off-handedly and without any justification introduces the radically novel idea of classifying angles of contact ('horn' angles) in terms of the difference of the curvatures of the two tangent curves (one of which is taken to be a straight line, of 'zero' curvature). All previous quantitative discussions of contact angles, from Greek times through the medieval period to Wallis' *De Angulo Contactus et Semicirculi Disquisitio Geometrica* (I, **1**, Introduction, Appendix 2: note (20); compare I: 90), were restricted to the angles formed by tangent circles at their point of contact. This measure of a contact angle, developed in following paragraphs, will be generalized in Problem 6 to yield an 'index' of curvature of a curve (see note (374) below).

minores rectilineis. Sic $xx = ay$. $x^3 = byy$. $x^4 = cy^3$, $x^5 = dy^4$ &c denotant seriem curvarum quarum quælibet posterior cum Basi constituit angulum contactus infinitè majorem quàm prior cum eadem Basi potest constituere. Estq; angulus contactus quem prima $xx = ay$ constituit, ejusdem generis cum circularibus, et ille quem secunda $x^3 = byy$ constituit, ejusdem generis cum Trochoidalibus.(312) Et quamvis subsequentium anguli angulos præcedentium perpetim infinitè superant, tamen anguli rectilinei(313) magnitudinem nunquam possunt assequi.

Ad eundem modum $x = y$. $xx = ay$. $x^3 = bby$. $x^4 = c^3y$ &c denotant seriem linearum quarum subsequentium anguli ad vertices cum basibus confecti sunt angulis præcedentiū perpetim infinitè minores.(314) Quinetiam inter angulos contactus duorum quorumlibet ex his generibus possunt alia angulorū se infinite superantium intercedentia genera in infinitum excogitari.

Angulorum verò contactus unum genus esse infinitè majus alio constat cùm unius generis curva utcunq; magna inter rectam tangentem et alterius generis curvam quantumvis parvam juxta punctum contactus non potest interjacere: sive cujus angulus contactus necessariò continet(315) alterius angulum contactûs ut partem totius. Sic curv[æ] $x^4 = cy^3$ angulus contactûs quem cum basi constituit, necessario continet angulum contactus curvæ $x^3 = byy$. Qui verò se (316)mutuò superare possunt anguli sunt ejusdem generis, uti de præfatis angulis Trochoidis et hujus curvæ $x^3 = byy$ contigit.

Ex his patet curvas in quibusdam punctis(317) posse infinitè rectiores esse vel infinitè curviores quolibet circulo et tamen formam curvarum non ideo amittere. Sed hæc in transitu.

||[59a] ||(318) EXEMPL: 5. Esto *ED* Quadratrix ad circulum centro *A* descriptum pertinens, ac *DB* ad *AE* normaliter demissâ dic $AB = x$. $BD = y$ et $AE = 1$. eritq;

(312) The radius of curvature at the general point (x, y) of the curve $x^{m+1} = \alpha y^m$ is $m^2(m+1)^{-1}\alpha^{\frac{1}{m}}x^{1-\frac{1}{m}}$, of $O(x^{1-\frac{1}{m}})$, while the corresponding radii for the circle $x^2 = y(a-y)$ and cycloid $y = \frac{1}{2}b\cos^{-1}(1-2x/b) - \sqrt{[x(b-x)]}$ are $\frac{1}{2}a$ and $2b^{\frac{1}{2}}x^{\frac{1}{2}}$ respectively.

(313) Namely, one of finite measure.

(314) The radius of curvature at a general point on the curve $x^{n+1} = \beta^n y$ is

$$[n(n+1)]^{-1}\beta^n x^{1-n}(1+(n+1)^2\beta^{-2n}x^{2n})^{\frac{3}{2}},$$

of $O(x^{1-n})$ when x is small.

(315) 'juxta punctum contactus' (in the neighbourhood of the point of contact) is again understood.

(316) Newton evidently understands 'multiplicati' (when multiplied), appealing to Euclid's definition of proportional magnitudes (*Elements*, v: Def. 5). A classification of contact angles by the difference of curvature of the tangent curves fails when these are of the same curvature: for distinguishing contact angles of this type an intrinsic measure involving variations of curvature has to be defined. As Edward Kasner has shown (compare his *précis* of 'The Recent Theory of the Horn Angle', *Scripta Mathematica*, **11** (1945): 263–7) there are difficulties also in classifying contact angles by the difference of curvature of the tangent curves, since this intrinsic measure is not invariant under a general conformal (angle-preserving) transformation. In the case where

are infinitely less than rectilinear ones. So $x^2 = ay$, $x^3 = by^2$, $x^4 = cy^3$, $x^5 = dy^4$, ... denote a series of curves, each subsequent one of which forms with its base a contact angle infinitely greater than the previous one can possibly do with the same base. Further, the contact angle formed by the first is of the same class as circular ones, and that formed by the second of the same class as cycloidal ones.(312) And though the angles of subsequent curves are perpetually infinitely larger than those of the predecessors [in sequence], nevertheless they can never attain the magnitude of a rectilinear one.(313)

In the same way $x = y$, $x^2 = ay$, $x^3 = b^2y$, $x^4 = c^3y$, ... denote a series of lines, succeeding curves of which have the angles made at their vertices with their bases perpetually infinitely less than corresponding angles of their predecessors.(314) Furthermore, between the contact angles of any two of these classes still other intermediate classes of angles infinitely in excess of one another may indefinitely be devised.

It is accepted, however, that one class is infinitely greater than another when no curve of one class, however great, can lie between a tangent straight line and any curve of the second class, however small, in the vicinity of the point of contact—in other words, a contact angle of the first necessarily contains(315) a contact angle of the second class as a part of its whole. So the contact angle which the curve $x^4 = cy^3$ forms with its base necessarily contains the corresponding contact angle of the curve $x^3 = by^2$. Of course, angles which are(316) capable of being larger than one another are of the same class, as happened in the case of the angles of the cycloid and the present curve $x^3 = by^2$.

From these considerations it is evident that curves at certain points(317) can be infinitely straighter or infinitely more curved than any circle and yet still not, on that account, lose their curvilinear shape. But this in passing.

(318)EXAMPLE 5. Let ED be a quadratrix pertaining to a circle described with centre A, and then, letting fall DB normally to AE, call $AB = x$, $BD = y$

---

the tangent curves are of differing curvatures $1/\rho_1$, $1/\rho_2$ and their contact point $(\rho_1, s_1) \equiv (\rho_2, s_2)$ is defined intrinsically in terms of their respective arc-lengths s_1, s_2, then the difference $\rho_1 - \rho_2$ of their radii of curvature and also the difference $d\rho_1/ds_1 - d\rho_2/ds_2$ of their Newtonian 'indices' of curvature are each only relatively invariant under conformal transformation, while the simplest absolute invariant is $(\rho_1 - \rho_2)^2/(d\rho_1/ds_1 - d\rho_2/ds_2)$. When $\rho_1 = \rho_2$, the simplest intrinsic measure of a contact angle which is absolutely invariant under conformal transform (which would appear a minimum condition for a satisfactory measure) is exceedingly complicated. Under affine and projective transforms the simplest invariants are $1/\rho_1 - 1/\rho_2$ and ρ_1/ρ_2. Evidently, corresponding to the 'Newtonian' k-section of a contact angle between curves of curvature radii ρ_1, ρ_2 by a tangent circle of radius $k^{-1}\rho_1 + (1-k^{-1})\rho_2$, the latter yields an angle section by a circle of radius $\rho_1^{1/k}\rho_2^{(1-1/k)}$.

(317) Namely, at inflexions and cusps.

(318) The following example is a late insertion on a separate sheet (here paginated '59a'), in whose margin Newton dictates 'This pag: must bee inserted at the end of pag 59'.

$nx-nyy-nxx=my$ ut supra.(319) quæ æquatio, scriptis 1 pro m et z pro n, fit $zx-zyy-zxx=y$; Et inde per Prob 1 elicitur

$$rx-ryy-rxx+zm-2zmx-2zny=n.$$

Factâq; reductione et scriptis iterum 1 pro m et z pro n, erit $r=\frac{2zzy+2zx}{x-xx-yy}$. Inventis autem z et r fac $\frac{1+zz}{r}=DH$, et age HC ut supra.

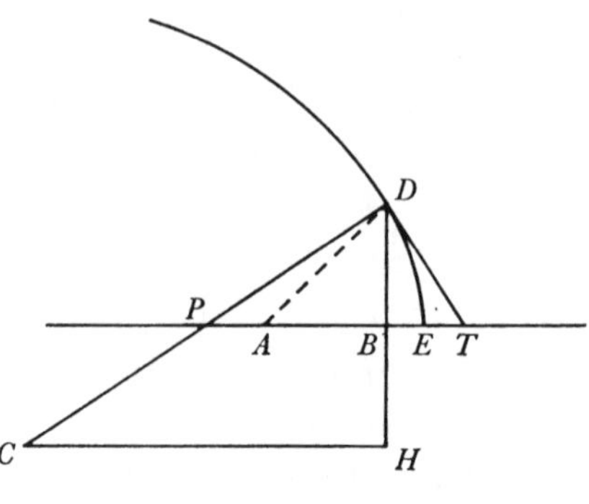

Si constructionem concinnare placet, perbrevem invenies, nempe ad DT duc normalem DP occurrentem AT in P, et fac esse $2AP.AE::PT.CH$.

Scilicet est $z=\frac{y}{x-xx-yy}=\frac{BD}{-BT}$, et $zy=\frac{BD^q}{-BT}=-BP$. et $zy+x=-AP$, et $\frac{2z}{x-xx-yy}$ in $zy+x=\frac{2BD}{AE\times BT^q}$ in $-AP=r$.(320) Præterea est $1+zz=\frac{PT}{BT}$, $\left(\text{utpote}=1+\frac{BD^q}{BT^q}=\frac{DT^q}{BT^q},\right)$ adeóq; $\frac{1+zz}{r}=\frac{PT\times AE\times BT}{-2BD\times AP}=DH$. Deniq; est $BT.BD::DH.CH=\frac{PT\times AE}{-2AP}$. Ubi valor negativus tantum arguit CH capiendam esse ad partes DH versus AB.

Eadem methodo Spiralium et aliarum quarumvis Curvarum curvatura calculo brevissimo determinari potest.

(321)Ad curvaturam præterea, cum curvæ alijs modis ad rectas referuntur, sine prævia reductione determinandam, jam potuit hæc methodus applicari perinde ut in determinando Tangentes factum est. Sed cùm omnes Geometricæ curvæ ut et Mechanicæ (præsertim ubi definientes conditiones ad infinitas æquationes uti post ostendam reducantur)(322) ad rectangulas ordinatas referri possent,
||[60] videor satis præstitisse. || Qui plura desiderat haud difficulter proprio Marte supplebit[,] præsertim si in ejus rei illustrationem ex abundanti methodum pro Spiralibus adjecero.(323)

Esto BK circulus, A centrum ejus, B punctum in circumferentia datum, ADd spiralis,(324) DC perpendiculum ejus, et C centrum curvitatis ad punctum D.

(319) See page 45b above.

(320) Here $BT = y\dot{x}/\dot{y} = y/z$. Note that Newton has added the unit length AE in the denominator to obtain the correct dimension (-1) for $r = \dot{z}$.

(321) The following paragraph revises a cancelled first version on Newton's pages 59/60.

(322) Compare Problem 12 below. Presumably the quadrature of mechanical curves by prior reduction to infinite series was to have been systematically pursued in 'that part [of the 1671 tract] w[ch] related to the solution of Problems not reducible to Quadratures' which was, in fact, never written (compare ULC. Add. 3968.9: 97[r]). In a first version of this parenthesis

and $AE = 1$: there will be $\dot{y}x-\dot{y}y^2-\dot{y}x^2 = \dot{x}y$, as above.(319) On writing 1 in place of $\dot{x}$ and z for $\dot{y}$, this equation becomes $zx-zy^2-zx^2 = y$, and from this is elicited by Problem 1 $\dot{z}x-\dot{z}y^2-\dot{z}x^2+z\dot{x}-2z\dot{x}x-2z\dot{y}y = \dot{y}$ and then, after reduction and 1 again written in place of $\dot{x}$ and z for $\dot{y}$, will it be

$$\dot{z} = (2z^2y+2zx)/(x-x^2-y^2).$$

Having found z and $\dot{z}$ make $(1+z^2)/\dot{z} = DH$ and draw HC as above.

If you desire a neat construction, you will find a very short one: namely, draw DP normally to DT meeting AT in P and make $2AP:AE = PT:CH$.

In explanation, $z = y/(x-x^2-y^2) = -BD/BT$, $zy = -BD^2/BT = -BP$, $zy+x = -AP$ and so $\dfrac{2z}{x-x^2-y^2}\times(zy+x) = \dfrac{2BD}{AE\times BT^2}\times -AP = \dot{z}$.(320) Further $1+z^2 = \dfrac{PT}{BT}\left(\text{since it equals } 1+\dfrac{BD^2}{BT^2} = \dfrac{DT^2}{BT^2}\right)$ and so

$$\frac{1+z^2}{\dot{z}} = \frac{PT\times AE\times BT}{-2BD\times AP} = DH.$$

Finally $BT:BD = DH:CH$ or $\dfrac{PT\times AE}{-2AP}$. Here the negative value shows merely that CH must be taken on the side of DH towards AB.

By the same method the curvature of spirals and any other curves you wish may be determined by a very short calculation.

(321)This method, furthermore, could now be applied to determining curvature without prior reduction when curves are related to straight lines in other ways, exactly as was done in determining tangents. But since all geometrical curves—and mechanical ones too (particularly where the defining conditions may be reduced to infinite equations, as I shall afterwards show)(322)—can be referred to right-angled ordinates, I believe I have done enough. Anyone who desires more will provide it without difficulty by his own efforts, especially if, as a bonus, in illustration of the point I add a method for spirals.(323)

Let BK be a circle, A its centre, B a given point in the circumference, ADd a spiral,(324) DC a normal to it and C the centre of curvature at the point D.

Newton wrote less definitely 'mediantibus infinitis æquationibus uti posthac ostendetur' (by means of infinite equations, as will afterwards be shown).

(323) That is, as will be obvious from the following, a method of constructing the centre of curvature at a general point on a curve defined in a system of polar coordinates.

(324) Newton has drawn his 'spiral' through the pole A in exact illustration of his three following examples, all of whose defining equations lack a constant term. In general this restriction is not intended: by a 'spiralis' (that is, 'helix' in terminology current in the earlier part of the century) Newton signifies any curve which makes one or more revolutions round a pole and which it is therefore natural to define in a system of 'polar' coordinates. Notice that, since x is the radian measure of $B\hat{A}D$ and y is the length of the corresponding radius vector

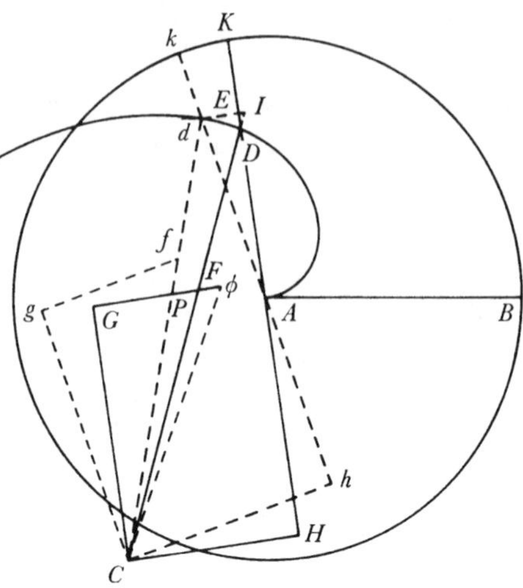

Ductâq; ADK recta, et ei parallela et æquali[(325)] CG, ut et normali GF occurrente CD in F; dic AB vel $AK=1=CG$, $BK=x$, $AD=y$, & $GF=z$. Præterea concipe punctum D per infinitè parvum spatium Dd in spirali moveri, et perinde per d agi semidiametrum Ak, eiq; parallelam et æqualem Cg, et normalem gf occurrentem Cd in f, cui etiam GF occurrit in P; Produc GF ad ϕ ut sit $G\phi=gf$, et ad AK demitte normalem dE et produc donec cum CD conveniat ad I: Et ipsarum BK, AD ac $G\phi$[(326)] contemporanea momenta erunt Kk, DE, et $F\phi$[(327)], quæ proinde dicentur $m\times o$, $n\times o$, et $r\times o$.

Jam est $AK.AE(AD)::kK.dE=oy$ ubi assumo $m=1$ ut supra.[(328)] Item $CG.GF::dE.ED=oyz$,[(329)] adeóq; $yz=n$. Præterea

$$CG.CF::dE.dD=oy\times CF::dD.dI=oy\times CF^q.$$

Adhæc propter ang $PC\phi(=\text{ang}\,GCg)=\text{ang}\,DAd$, et

$$\text{ang}\,CP\phi(=\text{ang}\,CdI=\text{ang}\,EdD+\text{rect}:)=\text{ang}\,ADd,$$

triangula $CP\phi$ et ADd sunt similia, et inde $AD.Dd::CP(CF).P\phi=o\times CF^q$, unde aufer $F\phi$ et restabit $PF=o\times CF^q-o\times r$. Deniq; demissa CH normali ad AD est $PF.dI::CG.EH$ vel $DH=\dfrac{y\times CF^q}{CF^q-r}$. Vel substituto $1+zz$ pro CF^q, erit $DH=\dfrac{y+yzz}{1+zz-r}$.[(330)]

AD, Newton's analytical coordinates are exactly our modern ones (compare note (239)). In ignorance of Newton's (still unpublished) present fluxional tract the Bernoulli brothers, Jakob and Johann, and especially L'Hospital in his 1696 *Analyse des Infiniment Petits* popularized an equivalent system of polar coordinates first introduced by Newton on his page 106 below (see note (707)) in which $AD = y$ and $x = y\times B\hat{A}D$, so that $DE = dy$ and $dE = dx$. This has the evident merit that all differential formulæ in standard Cartesian coordinates are simple corollaries of polar equivalents when y is taken to be infinitely great, but (through the hard work of such expositors of the present variant as Euler and Clairaut) fell from mathematical favour in the middle years of the eighteenth century.

(325) 'cujusvis datæ longitudinis' (of any arbitrary length) is cancelled.

(326) Read 'GF'.

(327) Since $F\phi = gf-GF$.

(328) In the comparable Cartesian case on Newton's page 55 (see note (286)). In other words, $\widehat{BK} = x$ is taken as base variable, so that henceforth $\widehat{Kk} = o$ (and $dE = oy$, as Newton at once infers).

(329) Newton has with some hesitation written in '$= on$', cancelled it, reinserted it and cancelled it again here: the insertion serves, of course, to make the transition to the conclusion $yz = n$ (or $\dot{y}$) less abrupt.

Drawing the straight lines DK and, parallel and equal to it,(325) CG as also its perpendicular GF meeting CD in F, call AB or $AK = CG$ unity, $BK = x$, $AD = y$ and $GF = z$. Further, imagine the point D to move along the spiral through the infinitely small space Dd and likewise that the radius Ak is drawn through d and also, parallel and equal to it, Cg and its perpendicular gf meeting Cd in f, while the latter is met also by GF in P. Produce GF to ϕ so that $G\phi = gf$ and to AK drop the normal dE, producing it till it meets CD at I: the contemporaneous moments of BK, AD and $G[F]$ will then be Kk, DE and $F\phi$,(327) so call these $\dot{x}o$, $\dot{y}o$ and $\dot{z}o$.

Now when I assume $\dot{x} = 1$, as above,(328) $AK:AE$ (or AD) $= kK:dE$, and so $dE = oy$. Likewise $CG:GF = dE:ED$ and $ED = oyz$,(329) so that $yz = \dot{y}$. Further $CG:CF = dE:dD$ (or $oy \times CF$) $= dD:dI$ and so $dI = oy \times CF^2$. Again, because $\widehat{PC\phi}(=\widehat{GCg}) = \widehat{DAd}$ and $\widehat{CP\phi}(=\widehat{CdI} = \widehat{EdD}+\frac{1}{2}\pi) = \widehat{ADd}$, the triangles $CP\phi$ and ADd are similar and thence $AD:Dd = CP$ (or CF)$:P\phi$, which is therefore $o \times CF^2$. Take $F\phi$ from this and there will remain $PF = o\,.\,(CF^2 - \dot{z})$. Finally, on letting fall CH normal to AD, it is $PF:dI = CG:EH$ or $DH = y \times CF^2/(CF^2 - \dot{z})$: that is, when $1+z^2$ is substituted in place of CF^2, $DH = y(1+z^2)/(1+z^2-\dot{z})$.(330)

(330) As an immediate corollary the radius of curvature at D is

$$CD = \frac{y(1+z^2)^{\frac{3}{2}}}{1+z^2-r \text{ (or } \dot{z})}.$$

The structure of Newton's present curvature method is evidently modelled on that previously used by him (on his pages 54–5 above) in the Cartesian case. Much as before the length $GF = z$ represents the slope $\dot{y}/y\dot{x}$ of the curve at D and the method depends fundamentally on evaluating its increment $F\phi = gf - GF$ corresponding to the increment $\widehat{Kk} = o$ of the base variable $\widehat{BK} = x$: here an important role is played by the pairs of triangles $CP\phi$, ADd and CPF, CdI which become similar in the limit as o vanishes. Arguably, the 'natural' choice of x as base variable is not the best. If the curve's arc-length, measured along $\widehat{AD}$, be taken as the independent variable and so its limit-increment $\widehat{Dd} = \omega$ as 'constant', on letting fall the perpendiculars AM, Am (meeting CD in μ) from A onto CD, Cd it follows at once that $DM = y^2\dot{x}$, $AM = y\dot{y}$ and $\mu m = Am - AM = d(AM) = \omega(\dot{y}^2 + y\ddot{y})$, $DE = \omega\dot{y}$, $dE = \omega y\dot{x}$ with $DE^2 + dE^2 = Dd^2$ or $y^2\dot{x}^2 + \dot{y}^2 = 1$; accordingly

$$CD = \frac{MD \times Dd}{Dd - \mu m} = \frac{y^2\dot{x}}{1-\dot{y}^2-y\ddot{y}} = \frac{1}{\dot{x}-\ddot{y}/y\dot{x}}.$$

(The equivalence of this with Newton's result follows if we note that, since $z = \dot{y}/y\dot{x}$,

$$1/(1+z^2) = y^2\dot{x}^2 = 1-\dot{y}^2 \quad \text{and therefore} \quad r = \dot{z}/\dot{x} = \dot{y}\ddot{y}/z\dot{x}(1-\dot{y}^2)^2 = y\ddot{y}(1+z^2)^2.)$$

The latter approach—when remodelled in terms of the variant polar coordinate system then in popular use on the continent (see note (324))—is, in historical fact, that communicated by L'Hospital to Johann Bernoulli in December 1694 (see the latter's *Briefwechsel* (note (112)), **1**: 253) and published two years later by him in his *Analyse des Infiniment Petits pour l'Intelligence des Lignes Courbes* (Paris, 1696): Section v, §77: 78–9. Bernoulli quickly riposted in January 1695 with an equivalent, somewhat cumbrous deduction of L'Hospital's formula by a method in which second-order increments are expressly taken (see his *Briefwechsel*, **1**: 256; and compare L'Hospital's *Analyse*: Section iv, §64: 58).

Et nota quod in hujusmodi computationibus quantitates (ut AD et AE) pro
||[61] æqualibus habeo quarum ratio a ratione æqua||litatis non nisi infinitè parùm, differt.

Ex his autem prodit hujusmodi Regula: Relatione inter x et y per quamlibet æquationem definitâ, quære relationem fluxionum m et n ope Prob 1, et substitue 1 pro m et yz pro n. Deinde ex æquatione prodeunte quære denuò per Prob 1 relationem inter m n et r et iterum substitue 1 pro m. Prior exitus per debitam reductionem dabit n et z et posterior dabit r, quibus cognitis fac $\frac{y+yzz}{1+zz-r}=DH$, et erige normalem HC spiralis perpendiculo DC priùs ducto occurrentem in C, et erit C centrum curvaminis. Vel quod eodem recidit cape $CH.HD::z.1$, et age CD.

Exempl 1. Si detur $ax=y$ æquatio ad Spiralem Archimedeam; erit per Prob 1 $am=n$ sive (scripto 1 pro m et yz pro n) $a=yz$. Et hinc denuò per Prob 1 exit $0=nz+yr$. Quare ex dato quolibet spiralis puncto D et inde longitudine AD sive y, dabuntur $z\left(=\frac{a}{y}\right)$ et $r\left(=-\frac{nz}{y}\text{ sive}=\frac{-az}{y}\right)$(331): Quibus cognitis fac $1+zz-r.1+zz::DA(y).DH$. Et $1.z::DH.CH$.

Et hinc facilè deducitur hujusmodi constructio. Produc AB ad Q ut sit $AB.\text{arc}\,BK::\text{arc}\,BK.BQ$, et fac $AB+AQ.AQ::DA.DH::a.HC$.(332)

Exempl. 2. Si $axx=y^3$ definit relationem inter BK et AD: obtinebis (per Prob 1) $2amx=3ny^2$, sive $2ax=3zy^3$, et inde rursus $2am=3ry^3+9znyy$. Est itaq; $z=\frac{2ax}{3y^3}$ et $r=\frac{2a-9zzy^3}{3y^3}$. Quibus cognitis fac $1+zz-r.\ 1+zz::DA.DH$. Vel opere concinnato, fac $9xx+6.9xx+4::DA.DH$.(333)

Exempl 3. Ad eundem modū si $axx-bxy=y^3$ determinat relationem BK ad AD, orietur $\frac{2ax-by}{bxy+3y^3}=z$, et $\frac{2a-2bzy-bzzxy-9zzy^3}{bxy+3y^3}=r$. Ex quibus DH, et inde punctum C determinatur ut ante.

(331) Since $a = n$ (or $\dot{y}$) on setting m (or $\dot{x}$) $= 1$.

(332) 'Here', as Samuel Horsley rightly urges in a remark entered in the margin of the manuscript in October 1777, 'a particular figure is required'. In his published version of the

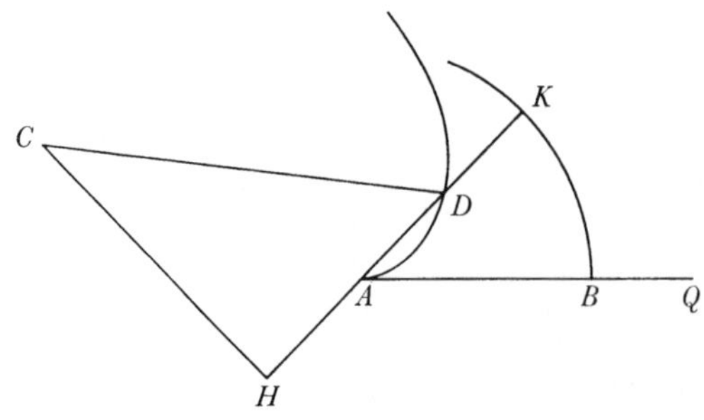

And note that in computations of this sort I consider as equal those quantities (such as AD and AE) whose ratio does not differ from one of equality except by an infinitely small amount.

There results, in consequence, a rule of this sort: When the relationship between x and y has been defined by any equation, seek the relation of the fluxions $\dot{x}$ and $\dot{y}$ with the help of Problem 1 and substitute 1 in place of $\dot{x}$ and yz for $\dot{y}$. Then from the resulting equation search out afresh by Problem 1 the relationship between $\dot{x}$, $\dot{y}$ and $\dot{z}$ and again substitute 1 in place of $\dot{x}$. By due reduction the former result will give $\dot{y}$ and z and the latter $\dot{z}$. When these are known, make $\dfrac{y(1+z^2)}{1+z^2-\dot{z}} = DH$ and raise the perpendicular HC meeting the spiral's normal DC, previously drawn, in C: then will C be the centre of curvature. Or what comes to the same take $CH:HD = z:1$ and draw CD.

EXAMPLE 1. If there be given $ax = y$, the equation to an Archimedean spiral, then by Problem 1 $a\dot{x} = \dot{y}$, that is, (on writing 1 in place of $\dot{x}$ and yz for $\dot{y}$) $a = yz$. And from this again by Problem 1 there comes $0 = \dot{y}z+y\dot{z}$. Consequently, given any point D of the spiral and thence the length of AD or y, from these will be given $z(=a/y)$ and $\dot{z}(=-\dot{y}z/y$, that is, $-az/y$(331)). Once these have been determined make $(1+z^2-\dot{z}):(1+z^2) = DA$ (or y)$:DH$ and $1:z = DH:CH$.

From this a construction of this sort is easily inferred. Produce AB to Q so that $AB:\text{arc}\,\widehat{BK} = \text{arc}\,\widehat{BK}:BQ$ and make $(AB+AQ):AQ = DA:DH = a:HC$.(332)

EXAMPLE 2. If $ax^2 = y^3$ defines the relationship between BK and AD, you will obtain (by Problem 1) $2a\dot{x}x = 3\dot{y}y^2$, that is, $2ax = 3zy^3$, and from this again $2a\dot{x} = 3\dot{z}y^3+9z\dot{y}y^2$. Accordingly, $z = 2ax/3y^3$ and $\dot{z} = (2a-9z^2y^3)/3y^3$. When these have been found out, make $(1+z^2-\dot{z}):(1+z^2) = DA:DH$; or, on putting a finishing touch, $(9x^2+[10]):(9x^2+4) = DA:DH$.(333)

EXAMPLE 3. In the same manner if $ax^2-bxy = y^3$ determines the relationship of BK to AD, there will result $(2ax-by)/(bxy+3y^3) = z$ and

$$(2a-2bzy-bz^2xy-9z^2y^3)/(bxy+3y^3) = \dot{z}.$$

From these DH and thence the point C are determined as before.

Latin text two years later (*Opera Omnia* (note (60)), **1**: 452) he contented himself with following Colson's 1736 emendation (*Method of Fluxions* (note (79)): 68) by adding the point Q to Newton's figure. In the accompanying figure AD is the Archimedean spiral defined by $AD = a\times \widehat{BK}$: hence on setting $\widehat{BK} = x$, $AD = y$ it follows (since $AB = 1$) that $AQ = x^2+1$ and so

$$DH = \frac{y(1+a^2/y^2)}{1+a^2/y^2-(-a^2/y^2)} = \frac{y(1+x^{-2})}{1+2x^{-2}} = \frac{AD\times AQ}{AB+AQ}.$$

(333) On eliminating y it follows that $z = \frac{2}{3}x^{-1}$ and r (or $\dot{z}$) $= -\frac{2}{3}x^{-2}$, so that, correctly, $DA/DH = (9x^2+10)/(9x^2+4)$.

Et sic aliarum quarumvis spiralium curvaturam nullo negotio determinabis.
||[62] Imo et ad horum exemplar Regulas pro || quibuslibet curvarum generibus[(334)] excogitare.

Absolvi tandem Problema sed cum methodum adhibuerim a vulgaribus operandi modis satis diversam, et ipsum Problema non sit ex eorum numero quorum contemplatio apud Geometras increbuit: in ablatæ solutionis illustrationem et confirmationem non gravabor aliam solutionem attingere,[(335)] magis obviam[(336)] et usitatis in ducendo tangentes methodis affinem.[(337)] Utpote si centro et intervallo quovis circulus describi concipiatur, qui curvam quamlibet in pluribus punctis secet, et circulus ille contrahetur vel dilatetur donec duo intersectionum puncta conveniant, is curvam ibidem tanget. Et præterea si centrum ejus accedere vel recedere a puncto contactûs fingatur, donec tertium intersectionis punctum cum prioribus in puncto contactûs conveniat, is æque curvus ac Curva in illo puncto contactûs evadet. Quemadmodum in ultimo quinq; symptomatum centri curvaminis supra monui,[(338)] e quorum singulis dixi Problema diversimodè confici potuisse.

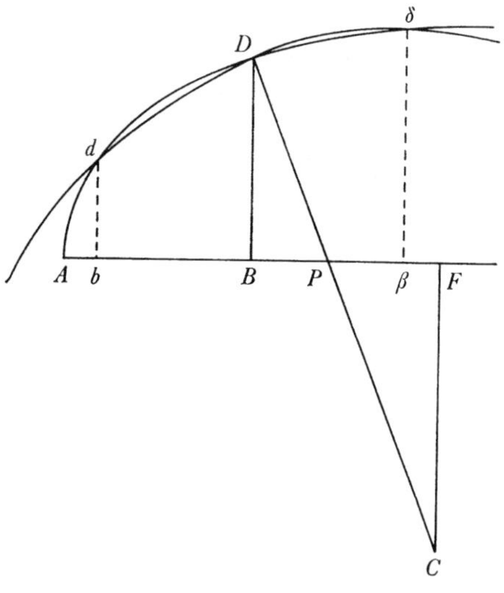

Centro itaq; C et radio CD describatur circulus secans curvam in punctis d, D, ac δ. Et demissis db, DB, $\delta\beta$, et CF ad Basin AB normalibus: dic $AB=x$, $BD=y$, $AF=v$, $FC=t$, ac $DC=s$; et erit $BF=v-x$, ac $DB+FC=y+t$; Quorum quadratorum aggregatum æquatur quadrato DC. Hoc est $vv-2vx+xx+yy+2ty+tt=ss$.[(339)] Quam [æquationem] si placet abbreviare possis fingendo $vv+tt-ss=$symbolo cuivis qq, et evadet $xx-2vx+yy+2ty+qq=0$. Postquam verò t, v, et qq inveneris si s desideres fac$=\sqrt{vv+tt-qq}$.

Proponatur jam quælibet æquatio pro Curva definienda cujus flexuræ quantitatem invenire oportet et ejus ope alterutram quantitatem x vel y extermina et emerget æquatio cujus radices (db, DB, $\delta\beta$ &c si extermines x, vel Ab, AB, $A\beta$ &c si extermines y) sunt ad intersectionum puncta (d, D, δ &c). Et

(334) For example, those defined with respect to the semi-intrinsic coordinate system introduced by Newton in Problem 10 of his 1666 tract (I: 434–6), which he will take up again in Problem 10 of the present treatise on his pages 99–100 below. In Newton's text some suitable verb, probably 'possis' (compare Horsley, *Opera Omnia* (note (60)), **1**: 453), is to be understood.

(335) Newton first continued 'in quam veri simile est Geometras statim incidisse ut incœpissent speculari' (upon which it is likely that geometers at once stumbled as soon as they began to speculate). The Greek attempts to explore the curvature of conics are probably no longer completely documented, but it is clear that Apollonius in the fifth book of his *Conics*

And in this manner you will determine the curvature of any other spiral curves without trouble. To be sure, on the model of those given you might devise rules for any kinds of curves you please.(334)

At long last I have disposed of the problem. However, the method I have employed is amply different from common ways of operation while the problem itself is not numbered among those widely considered by mathematicians, and I will therefore not hesitate, in illustration and corroboration of the solution obtained, to touch upon another(335) at once more obvious(336) and closer to the usual methods of drawing tangents.(337) In point, if a circle be conceived to be drawn with any centre and [radial] distance so as to intersect any curve in several points, then if that circle should contract or expand till two points of intersection coincide, it will touch the curve at that place. Furthermore, if its centre be imagined to approach or recede from this point of contact till a third point of intersection coincide with the former ones in the contact point, it will prove to be exactly as curved as the curve at the point of contact. This is the approach I noticed above(338) in the last of the five conditions defining a centre of curvature, following each of which I said the problem could be accomplished in a different manner.

Accordingly, with centre C and radius CD let there be described a circle cutting a curve in the points d, D and δ. And letting fall db, DB, $\delta\beta$ and CF perpendicular to the base AB, call $AB = x$, $BD = y$, $AF = v$, $FC = t$ and $DC = s$. Then $BF = v - x$ while $DB + FC = y + t$, and the sum of their squares equals the square of DC, that is, $v^2 - 2vx + x^2 + y^2 + 2ty + t^2 = s^2$. (339)You might, if you so please, shorten this equation by supposing $v^2 + t^2 - s^2$ equal to any symbol q^2, and there will then come $x^2 - 2vx + y^2 + 2ty + q^2 = 0$. If, after you have found t, v and q^2, you should want s, make it equal to $\surd[v^2 + t^2 - q^2]$.

Now let there be proposed any equation you will for defining a curve, the quantity of whose flexure must be found. By its help eliminate one or other of the quantities x and y, and there will emerge an equation whose roots (db, DB, $\delta\beta$, and so on should you eliminate x, or Ab, AB, $A\beta$, and so on should you

---

followed the same approach as that used initially by both Huygens and Newton: namely, the centre of curvature at a point is constructed as the limit-meet of normals to the curve in the immediate vicinity of the point. The present 'solutio' in which the circle of curvature is constructed as the circle which has 3-point contact at the point would appear to be Newton's undivided first discovery, though it was later to be found independently by Jakob Bernoulli. See notes (273) and (340).

(336) Newton has wisely cancelled 'magis simplicem' (simpler) as inapposite.

(337) A reference to Descartes' subnormal rule (compare I, **2**, 2), of which the present 'solutio' is an evident generalization.

(338) On his page 54 above: compare note (278).

(339) The remainder of this paragraph is a late insertion.

||[63] proinde cùm || ex istis tres evadent æquales, circulus et curvam continget et erit ejusdem curvitatis ac curva in puncto contactus. Æquales autem evadent conferendo æquationem cum alia totidem dimensionum æquatione fictitia cujus tres sunt æquales radices ut docuit Cartesius; vel expeditiùs multiplicando terminos ejus bis per Arithmeticam progressionem.(340)

EXEMPL. Sit $ax=yy$ æquatio ad Parabolam, et exterminato x (substituendo nempe in æquatione superiori(341) valorem ejus $\frac{yy}{a}$) prodibit

$$\begin{array}{llll} \frac{y^4}{aa} & -\frac{2v}{a}yy & +2ty & +qq=0. \\ & +yy & & \\ 4. & 2. & 1. & 0. \\ 3. & 1. & 0. & -1. \end{array}$$

cujus e radicibus y tres debent fieri æquales. Et in hunc finem terminos per Arithmeticam progressionem bis multiplico ut hic videre est, et exit $\frac{12y^4}{aa}-\frac{4v}{a}yy+2yy=0$. sive $v=\frac{3yy}{a}+\frac{1}{2}a$. (342)Unde facilè colligitur esse $BF=2x+\frac{1}{2}a$ ut supra.(343)

Quamobrem dato quovis Parabolæ puncto D, duc perpendiculum DP et in axe cape $PF=2AB$ et erige normalem FC occurrentem DP in C et erit C desideratum centrum curvitatis.(344)

Idem in Ellipsi et Hyperbola præstare possis sed calculo satis molesto, et in alijs curvis utplurimùm fastidiosissimo.(345)

DE QUÆSTIONIBUS QUIBUSDAM(346) COGNATIS.

Ex hujus Problematis resolutione consectantur aliorum nonnullorum confectiones. Cujusmodi sunt

(340) 'ut docuit Huddenius'; that is, 'by Huddenius his method' as Newton wrote in February 1665 (I: 262) in first describing this construction technique. The present 'circulus... ejusdem curvitatis ac curva in puncto contactus' is exactly the 'osculum primi gradus' postulated by Jakob Bernoulli when, having independently discovered it, he first published the method in 1692 ('Additamentum ad Solutionem Curvæ Causticæ... una cum Meditatione de Natura Evolutarum, & variis Osculationum Generibus', *Acta Eruditorum* (March 1692): 110–16, especially 114). Exactly as Newton, Bernoulli uses Huddenian multipliers to evaluate the condition of 3-point contact: in his own words 'Hoc [osculum] quia consistit in concursu trium intersectionum, pono nuperam æquationem pro his intersectionibus inventam habere tres radices æquales, eamque proinde bis multiplico per Progress. Arithm. aut brevius semel per productum duarum, & quod resultat, cum alia aliisve per productum 2 progr. similiter quæsitis æquationibus varie confero...: Ea enim suppeditabit lineam, in qua sumtum quodvis punctum centrum esse potest circuli alicujus curvam propositam primo gradu osculantis,

eliminate y) pertain to the points of intersection (d, D, δ, and so on). And therefore when three of those come to be equal, the circle will both touch the curve and at the point of contact be of the same curvature as it. They will, however, come to be equal when the equation is identified with a second, fictitious equation which has three equal roots, as Descartes has taught; or, more speedily when its terms are multiplied twice by an arithmetical progression.(340)

EXAMPLE. Let $ax = y^2$ be the equation to a parabola and, when x is eliminated (namely, by substituting in its place in the above equation(341) its value y^2/a), there will come

$$\begin{array}{llll} a^{-2}y^4+ & (-2a^{-1}v+1)\,y^2+ & 2ty+ & q^2 = 0, \\ 4. & 2. & 1. & 0. \\ 3. & 1. & 0. & -1. \end{array}$$

three of whose roots y should become equal. To this end I multiply the terms twice by an arithmetical progression, as may here be seen, and there results $12a^{-2}y^4 + (-4a^{-1}v+2)\,y^2 = 0$, or $v = 3y^2/a + \frac{1}{2}a$.

(342)Hence it is easily gathered that $BF = 2x + \frac{1}{2}a$, as above.(343)

In consequence, given any point D of the parabola, draw the perpendicular DP and in the axis take $PF = 2AB$: on erecting the normal FC meeting DP in C, the point C will be the desired centre of curvature.(344)

You might carry through the same procedure in the ellipse and hyperbola, though the computation would be troublesome enough, while in other curves it would almost always be repellent.(345)

CERTAIN RELATED QUESTIONS

The solutions of some other problems follow closely on the resolution of this. Such as:

cujusque cum Evoluta identitatem *Hugenium* notasse...supra diximus'. (Compare Huygens' *Horologium Oscillatorium* (Paris, 1673): Pars Tertia: 59–90.) The parallel does not end here for Bernoulli, as Newton, takes as his 'Exemplum' a parabola of parameter a and applies exactly the same double row of multipliers '4.2.1.0' and '3.1.0. − 1'. It is, of course, entirely impossible that Jakob Bernoulli could have known of Newton's still unpublished prior investigation.

(341) Namely, $x^2 - 2vx + y^2 + 2ty + q^2 = 0$.

(342) 'Hoc est $AF = 3x + \frac{1}{2}a$' (that is, $AF = 3x + \frac{1}{2}a$) is cancelled.

(343) See the *Corollarium* to Example 1 on Newton's page 56 above.

(344) Note that the subnormal BP is half the parameter (latus rectum) a.

(345) The method is, of course, restricted to curves whose Cartesian defining equation is algebraic and free of 'surds'.

(346) The enclitic phrase 'HUIC PROBLEMATI' is cancelled.

1. *Invenire punctum ubi linea datam habet curvaturam.*

Sic in Parabola $ax=yy$(347) si punctum quæratur ad quod radius curvaturæ sit datæ longitudinis f: e centro curvaturæ ut prius invento radium determinabis esse $\frac{[a]+4x}{2[a]}\sqrt{[aa]+4[a]x}$, quem pone æqualem f. Et factâ reductione emerget $x=-\frac{1}{4}a+\sqrt{C: \frac{1}{16}a[f]f}$.

2. *Invenire punctum rectitudinis.*

Punctum rectitudinis voco ad quod radius flexionis infinitus evadit, sive centrum infinitè distans; quale est ad verticem Parabolæ $ax^3=y^4$.(348) Et hoc
‖[64] idem plerumqꝫ(349) limes est flexionis contrariæ cujus ‖ determinationem supra posui. Sed et alia haud inelegans ex hoc Problemate scaturit. Nempe quo
Fig longior est radius flexionis eo minor evadit angulus DCd, et pariter momentum
[pag 54] δf, adeoqꝫ fluxio quantitatis z unà diminuitur, ita ut per ejus radij infinitatem prorsus evanescat. Quære ergo r et suppone nullam esse.(350)

Quemadmodum si limitem flexûs contrarij in Parabola secundi generis cujus ope Cartesius construxit æquationes sex dimensionum(351) determinare oportet. Ad illam Curvam æquatio est $x^3-bxx-cdx+bcd+dxy=0$. Et hinc per Prob 1 exit $3mxx-2bmx-cdm+dmy+dxn=0$. Quæ, scripto 1 pro m et z pro n, fit $3xx-2bx-cd+dy+dxz=0$: Unde rursus per Prob 1 exit $6mx-2bm+dn+dmz+dxr=0$, et hæc, scripto iterum 1 pro m, z pro n, et 0 pro r, fit $6x-2b+2dz=0$. Jam extermina z scribendo pro dz valorem $2b-3x$(352) in æquatione $3xx-2bx-cd+dy+dxz=0$, et proveniet $-cd+dy=0$, sive $y=c$. Quamobrem ad punctum A erige perpendiculum $AE=c$, et per E duc ED parallelam AB, et punctum D ubi Parabolæ partem convexo-concavam secuerit erit in confinio flexionis contrariæ.

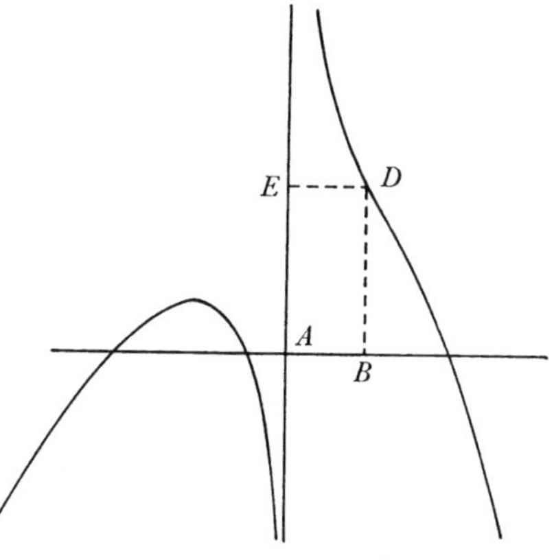

(347) Newton's first choice of defining equation was '$rx=yy$', the denotation of whose latus rectum is evidently that used by Newton in the present treatise to denote the fluxion of the tangent slope z. Uncorrected occurrences of 'r' in the sequel have been replaced by '$[a]$'.

(348) Not the best example to choose as a first illustration since, when $x=y=0$, then $z=\dot{y}=\frac{3}{4}a^{\frac{1}{4}}x^{-\frac{1}{4}}$ and $r=\dot{z}=-\frac{3}{16}a^{\frac{1}{4}}x^{-\frac{5}{4}}$ are both infinite, while

$$\lim_{x\to 0}\,(1+z^2)^{\frac{3}{2}}/r$$

is zero! Colson (*Method of Fluxions* (note (79)): 72) has the plausible emendation $a^3x=y^4$.

(349) Newton excepts the case of a point on a curve where its curvature is zero but where it does not cross its tangent. See note (353) below. Understand the figure on page 154.

1. *To find a point where a line has a given curvature*

Thus in the parabola $ax = y^2$(347) if a point be sought at which the radius of curvature is of a given length f, from the centre of curvature found as before you will determine the radius to be $\frac{1}{2}a^{-\frac{1}{2}}(a+4x)^{\frac{3}{2}}$. Set this equal to f and after reduction there will emerge $x = -\frac{1}{4}a + \sqrt[3]{[\frac{1}{16}af^2]}$.

2. *To find a point of straightness*

I call a 'point of straightness' one at which the radius of curvature comes to be infinite or where the centre becomes infinitely distant—such as at the vertex of the parabola $ax^3 = y^4$.(348) For the most part(349) this is identical with the limit of contrary flexure, a determination of which I set down above. But a second, not inelegant one, springs from the present problem: specifically, the longer the radius of curvature the less the angle $D\widehat{C}d$ comes to be, and likewise the moment δf; in consequence the fluxion of the quantity z diminishes with it and as a result, when the radius becomes infinite, it vanishes altogether. Seek $\dot{z}$, therefore, and take it to be zero.(350) Figure [on page 54]

For instance, should the limit of contrary flexure have to be determined in the second-order parabola with whose aid Descartes constructed equations of sixth degree,(351) the equation to that curve is $x^3 - bx^2 - cdx + bcd + dxy = 0$. Hence by Problem 1 there results $3\dot{x}x^2 - 2b\dot{x}x - cd\dot{x} + d\dot{x}y + dx\dot{y} = 0$ and this, on writing 1 for $\dot{x}$ and z in place of $\dot{y}$, becomes $3x^2 - 2bx - cd + dy + dxz = 0$. Hence a second time by Problem 1 there results $6\dot{x}x - 2b\dot{x} + d\dot{y} + d\dot{x}z + dx\dot{z} = 0$ and this, after again writing 1 for $\dot{x}$, z in place of $\dot{y}$ and 0 for $\dot{z}$, becomes $6x - 2b + 2dz = 0$. Now eliminate z by writing in place of dz in the equation

$$3x^2 - 2bx - cd + dy + dxz = 0$$

its value $[b] - 3x$, and there will arise $[-bx] - cd + dy = 0$ or $y = c[+(b/d)\,x]$. Therefore at the point A erect the perpendicular $AE = c[-\sqrt[3]{(b^4c/d^2)}]$ and through E draw ED parallel to AB: the point D in which it cuts the convexo-concave portion of the parabola will be at the boundary of contrary curvature.

---

(350) A sufficient test for a point on a curve to be one of 'rectitude' only when the tangent-slope z at the point is not itself infinite; that is, when the instantaneous direction of the curve at the point is not parallel to the y-axis. As we have seen (I: 424–5) Newton had already invoked the criterion that the curvature is zero at 'ye points distinguishing twixt ye concave & convex portions of crooked lines' in Problem 3 of his October 1666 fluxional tract, but L'Hospital would appear to have been the first to develop it systematically in print in his *Analyse des infiniment Petits* (note (330)): Section IV, Proposition II: 60–3.

(351) The Cartesian trident (see note (195)), whose defining equation Newton here takes as $dxy = (b-x)(x^2-cd)$.

(352) Read '$b-3x$', an error which Newton carries through the rest of his calculation. We have left the Latin text uncorrected, but have made appropriate amendment in our English translation. Neither Colson nor Horsley notice the mistake.

Similiq; methodo alia rectitudinis puncta quæ non interjacent partibus contrariè flexis determinari possunt. Veluti si $x^4-4ax^3+6aaxx-b^3y=0$ Curvam definiat, exinde per Prob 1 imprimis producetur $4x^3-12axx+12aax-b^3z=0$ et hinc denuò $+12xx-24ax+12aa-b^3r=0$, Ubi suppone $r=0$ et factâ reductione prodibit $x=a$. Quamobrem sume $AB=a$ et BD normaliter erecta curvæ in desiderato rectitudinis puncto occurret.(353)

||[65] || 3. *Invenire punctum flexûs infiniti.*(354)

Quære radium curvaminis et suppone nullum esse. Sic ad Parabolam secundi generis æquatione $x^3=ayy$ definitam, erit radius ille

$$CD=\frac{4a+9x}{6a}\sqrt{4ax+9xx},$$

qui nullus evadit cùm sit $x=0$.(355)

4. *Flexûs maximi minimive punctum determinare.*

Ad hujusmodi puncta radius curvaturæ aut maximus aut minimus evadit. Quare centrum curvaturæ ad id temporis momentum nec versus punctum contactûs neq; ad contrarias partes movetur sed penitus quiescit.(356) Quæratur itaq; fluxio Radij CD; vel expeditiùs, quæratur fluxio alterutrius rectæ BH vel AK,(357) et supponatur nulla.

(353) Evidently the condition for this second class of 'straightness' point is that the curve should have zero curvature there but the slope of the curve should increase or decrease continuously in the vicinity of the point: in other words, the tangent at the point should meet

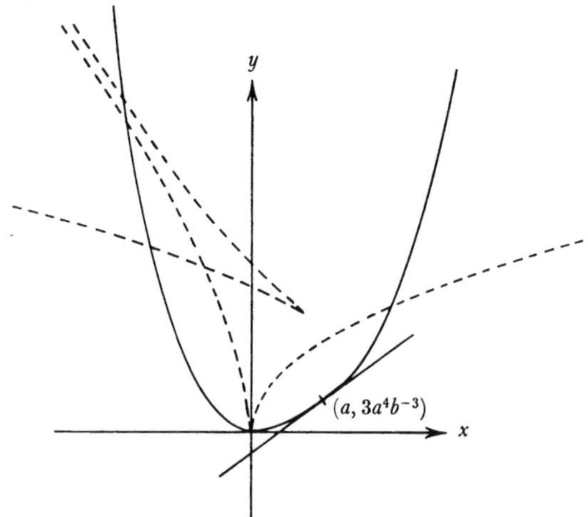

the curve there in a $2n$-ple point ($n \geqslant 2$). Here the tangent $b^3y = a^3(4x-a)$ meets the quartic $b^3y = (x-a)^4+4(x-a)a^3+3a^4$ in the (unique) quadruple point $(a, 3a^4/b^3)$ of zero curvature. This is evident visually in our sketch, where the evolute of the curve is shown as a broken line.

And by a similar method other points of straightness not lying between parts oppositely curved may be determined. As if $x^4-4ax^3+6a^2x^2-b^3y=0$ should define the curve, therefrom by Problem 1 will be produced in the first instance $4x^3-12ax^2+12a^2x-b^3z=0$ and from this once more

$$12x^2-24ax+12a^2-b^3\dot{z}=0.$$

Here take $\dot{z}=0$ and, after reduction, there will come forth $x=a$. Take, therefore, $AB=a$ and when BD is erected at right angles it will meet the curve in the desired point of straightness.(353)

3. *To find a point of infinite flexure*(354)

Seek out the radius of curvature and take it to be zero. Thus in the second-order parabola defined by the equation $x^3=ay^2$ that radius will be

$$CD=\tfrac{1}{6}a^{-1}x^{\frac{1}{2}}(4a+9x)^{\frac{3}{2}},$$

which is zero when $x=0$.(355)

4. *To find a point of greatest or least flexure*

At points of this kind the radius of curvature comes to be greatest or least. At that instant of time, in consequence, the centre of curvature neither moves towards the point of contact nor recedes the opposite way but is completely at rest.(356) Accordingly, let there be sought the fluxion of the radius CD—or, more speedily, that of one or other of the lines BH or AK(357)—and let it be taken equal to zero.

(354) The point at which a curve has infinite curvature is evidently a cusp, or 'point de rebroussement' as Johann Bernoulli named it in April 1694 (see his *Briefwechsel* (note (112)), **1**: 207). In a cancelled first draft Newton dubbed it 'punctum flexionis infinitæ' and proceeded to determine it by the equivalent argument that 'Ad ejusmodi puncta radius curvaturæ nullus est, et fluxio quantitatis z infinita. Quære ergo fluxionem ejus r, et suppone infinitam esse, hoc est denominatorem valoris ejus pone $=0$' (At points of this nature the radius of curvature is zero and the fluxion of the quantity z infinite, so seek its fluxion $\dot{z}$ and take it to be infinite—in other words, set the denominator of its value equal to nothing).

(355) In first draft Newton absentmindedly computed the projection $(1+z^2)/\dot{z}$ of the curvature radius, finding 'radius ille $CD=\dfrac{4a+9x}{6a}\sqrt{\dfrac{x}{a}}$'. In either case (compare the previous note) the condition that the 'denominator' $x^{\frac{1}{2}}$ of r (or $\dot{z}$) $=\frac{3}{4}a^{-\frac{1}{2}}/x^{\frac{1}{2}}$ be zero is that $x=0$.

(356) As a result, at the point C on the evolute there will be a cusp. In May 1665 (compare I: 268–71) Newton had failed to see that not all apparent extreme values of the radius of curvature correspond to extreme values of the analytical function denoting its length. Since he now fails to point this subtlety out, we may assume that the possibility of the curvature radius at non-real points on a curve being itself a real length had still not occurred to him. Fortunately for his present argument the simple general parabolas and hyperbolas he here considers in example are well behaved in this respect.

(357) Newton's diagram has no point 'K' but it is presumably the meet of HC with the y-axis (the vertical through A). Since the curvature centre C is supposed instantaneously at rest, so also is its distance $BH=AK$ from the x-axis AB.

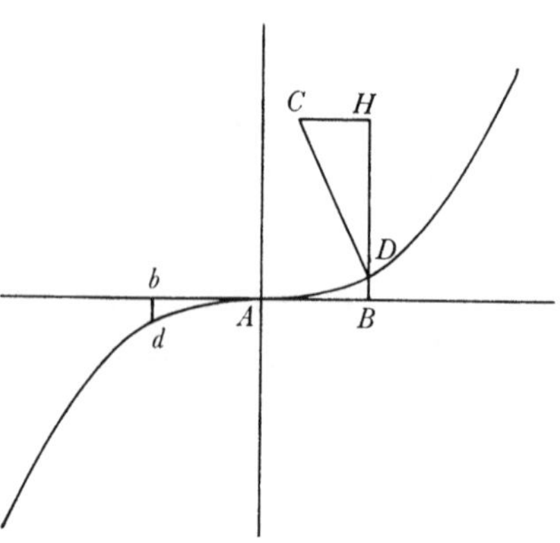

Quemadmodum si de Parabola secundi generis $x^3=aay$ quæstio proponatur: imprimis ad curvaturæ centrum determinandum invenies $DH=\dfrac{aa+9xy}{6x}$,(358) adeoq; est $BH=\dfrac{aa+15xy}{6x}$, dic autem $BH=v$, et erit $\dfrac{aa}{6x}+\dfrac{5}{2}y=v$, unde juxta Prob 1 educitur $-\dfrac{aam}{6xx}+\dfrac{5n}{2}=l$. Jam vero l ipsius BH fluxionem suppone nullam esse, et insuper cùm ex hypothesi sit $x^3=aay$, et inde per Prob 1 $3mxx=aan$ posito $m=1$ substitue $\dfrac{3xx}{aa}$ pro n et emerget $45x^4=a^4$.(359) Cape ergo $AB=\surd 4:\dfrac{a^4}{45}$. Et BD normaliter erecta occurret curvæ in puncto maximæ curvaturæ. Vel, quod perinde est fac

$$AB.BD::3\surd 5.1.^{(360)}$$

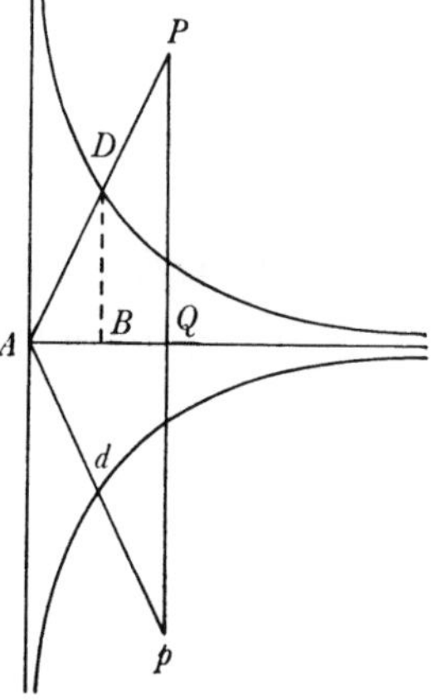

Ad eundem modum Hyperbola secundi generis per æquationem $xyy=a^3$ designata maximè flectitur in punctis D, d, quæ determinabis sumendo $AQ=1$ in Basi, et erigendo [normaliter] $QP=\surd 5$ eiq; æqualem Qp ex altera parte et agendo AP et Ap, quæ curvæ occurrent in desideratis punctis D ac d.(361)

5. *Locum centri curvaminis determinare, sive Curvam describere in quâ centrum istud perpetuo versatur.*(362)

||[66] Trochoidis centrum curvaminis in alia Trochoide || versari ostensum est.(363) Et sic Parabolæ centrum istud in alia secundi generis (quam æquatio $axx=y^3$ definit) Parabola versatur, ut inito calculo facilè constabit.(364)

(358) That is, $a^2(1+z^2)/a^2r$ where z (or $\dot{y}$) $=3a^{-2}x^2=3y/x$ and r (or $\dot{z}$) $=6a^{-2}x$.

(359) A first continuation ‘Factâq; reductione $x=\surd 4:\dfrac{a^4}{45}$’ (and on reduction $x=\sqrt[4]{[a^4/45]}$) is cancelled.

(360) For $AB:BD=x:y=a^2:x^2$. The last sentence is a late insertion.

(361) Since z (or $\dot{y}$) $=-\frac{1}{2}a^{\frac{3}{2}}x^{-\frac{3}{2}}$ and r (or $\dot{z}$) $=\frac{3}{4}a^{\frac{3}{2}}x^{-\frac{5}{2}}$, the distance $y+(1+z^2)/r$ of the curvature centre from AB is $\frac{4}{3}(a^{-\frac{3}{2}}x^{\frac{5}{2}}+a^{\frac{3}{2}}x^{-\frac{1}{2}})$, which will be least when $\frac{5}{2}a^{-\frac{3}{2}}x^{\frac{3}{2}}-\frac{1}{2}a^{\frac{3}{2}}x^{-\frac{3}{2}}=0$ or $x^3=\frac{1}{5}a^3$. The result follows equivalently by determining the stationary value of the curvature radius $(4x^3+a^3)^{\frac{3}{2}}/6a^{\frac{3}{2}}x^2$.

Should, for instance, the question be proposed for the second-order parabola $x^3 = a^2y$, to determine the centre of curvature you will find in the first place $DH = (a^2+9xy)/6x$(358) and consequently $BH = (a^2+15xy)/6x$. Call $BH = v$ and then $\frac{1}{6}a^2x^{-1}+\frac{5}{2}y = v$, from which according to Problem 1 is deduced $-\frac{1}{6}a^2\dot{x}x^{-2}+\frac{5}{2}\dot{y} = \dot{v}$. But now take the fluxion $\dot{v}$ of BH to be zero and in addition, since by hypothesis $x^3 = a^2y$ and thence by Problem 1 $3\dot{x}x^2 = a^2\dot{y}$, taking $\dot{x} = 1$ substitute $3x^2/a^2$ in place of $\dot{y}$ and there will emerge $45x^4 = a^4$.(359) So take $AB = [\pm]\, a/\sqrt[4]{45}$, and then BD when erected normally will meet the curve in a point of maximum curvature. Or what amounts to the same make

$$AB:BD = 3\surd 5:1.^{(360)}$$

After the same fashion the second-order hyperbola represented by the equation $xy^2 = a^3$ will be most curved at the points D, d which you will determine by taking AQ in the base equal to unity and raising at right angles $QP = \surd 5$ and, on the other side, its equal Qp, and then drawing AP and Ap: these will meet the curve in the desired points D and d.(361)

5. *To determine the locus of the centre of curvature, or to describe the curve in which that centre is perpetually situated*(362)

The centre of curvature of the cycloid has been shown to be situated in a second cycloid.(363) And thus for a parabola that centre is situated in another second-order parabola (one defined by an equation $ax^2 = y^3$), as will readily be agreed when the calculation is undertaken.(364)

---

(362) The defining equation $F(X, Y) = 0$ of the evolute of the curve $f(x, y) = 0$ defined in perpendicular Cartesian coordinates is obtained most generally by eliminating x and y between the latter and

$$X = x-z(1+z^2)/\dot{z},$$

$$Y = y+(1+z^2)/\dot{z}.$$

This is, in effect, the solution outlined in the October 1666 fluxional tract (I, 2, 7: Problem 9: 432–4).

(363) See Corollary 2 to Example 4 on Newton's pages 57–8 above. Analytically,

$$Y = \tfrac{1}{2}\pi a-\tfrac{1}{2}a\cos^{-1}[3-2X/a]-\surd[(X-a)\,(2a-X)]$$

is the defining equation of the evolute (a congruent cycloid) of

$$y = \tfrac{1}{2}a\cos^{-1}[1-2x/a]+\surd[x(a-x)]:$$

compare notes (304) and (305).

(364) See Newton's pages 98–9 below and compare Problem 9, Example 1 of the 1666 tract (I: 433). In analytical terms, $\frac{27}{16}\alpha Y^2 = (X-\frac{1}{2}\alpha)^3$ is the defining equation of the evolute (a semi-cubic parabola) of $y^2 = \alpha x$.

6. *Luce in quamlibet curvam incidente, invenire focum sive concursum radiorum circa quodpiam ejus punctum refractorum.*(365)

Curvaturam ad istud Curvæ punctum quære, et centro radioq; curvaturæ Circulum describe; Dein quære concursum radiorum a Circulo circa istud punctum refractorū. Nam idem erit concursus refractorum a propositâ Curvâ.

7. His addi potest *particularis inventio curvaturæ ad vertices curvarum ubi normaliter secant Bases.* Nempe punctum in quo Curvæ perpendiculum cum Basi conveniens ipsam ultimò secuerit, est centrum curvaturæ ejus.(366) Quamobrem habitâ relatione inter Basin x et rectangulum applicatum y et inde (per Prob 1) relationem inter fluxiones m et n; valor ny,(367) si in eo scribas 1 pro m et fingas $y=0$, erit radius curvaturæ.

Sic in Ellipsi $ax-\frac{a}{b}xx=yy$, est $\frac{am}{2}-\frac{amx}{b}=ny$, qui valor ny si supponas $y=0$ et consequenter $x=0$ et scribas 1 pro m evadet $\frac{1}{2}a$ radius curvaturæ. Et sic ad vertices Hyperbolæ et Parabolæ radius curvaturæ erit etiam dimidium lateris recti.(368)

Atq; ita ad Conchoiden æquatione $\frac{bbcc}{xx}+\frac{2bcc}{x}\frac{+cc}{-bb}-2bx-xx=yy$ definitam valor ny ope Prob 1 invenietur $\frac{-bbcc}{x^3}-\frac{bcc}{xx}-b-x$. Qui supponendo $y=0$, et inde $x=c$ vel $-c$ evadet $\frac{-bb}{c}-2b-c$ vel,

$$\frac{bb}{c}-2b+c$$

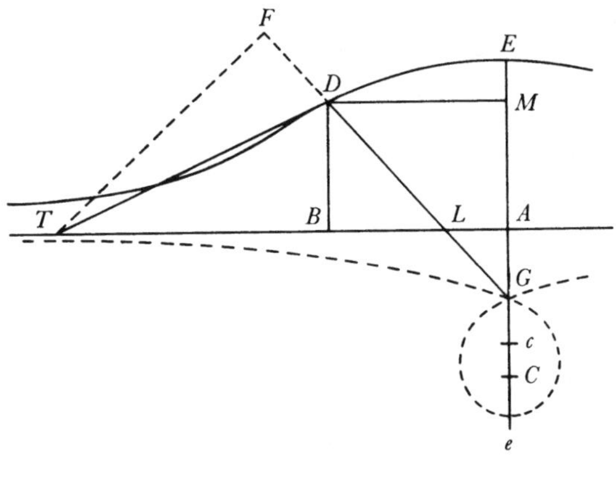

Fig [pag 44] radius curvaturæ.(369) Fac ergo $AE.EG::EG.EC$, et $Ae.eG::eG.ec$, et habes curvaturæ centra C et c ad vertices conjugatarum Conchoidum E et e.

(365) That is, to construct the diacaustic of a given point with respect to a given interface between two media of known refractive index. The solution sketched by Newton, which reduces the problem to a case already discussed by Barrow, is that outlined in Section IV, Proposition 33, of the first book of his deposited *Lectiones Opticæ* (reproduced in **3**, 1, §2 below: compare note (25) of that section). Analytically, if an incident ray through the given point meets the interface (of refractive ratio μ) after a distance x at angle $\frac{1}{2}\pi-i$, then the diacaustic point is in the refracted ray at distance y from the point of refraction defined by

$$1-[\rho/x]\cos i = \sqrt{[\mu^2\sec^2 i-\tan^2 i]}\,(1-[\rho/y]\sqrt{[1-\mu^{-2}\sin^2 i]}),$$

where ρ is the radius of curvature of the interface at the point of refraction and $\mu = \sin i/\sin r$.

6. *When light falls upon any curve, to find the focus or meet of rays refracted around any point in it*(365)

Seek the curvature at that point of the curve and with the centre and radius of curvature describe a circle. Then seek the meet of rays refracted by this circle around the point, for this will be the same as the meet of [rays] refracted by the propounded curve.

7. To these may be added *a particular way of finding curvature at the vertices of curves when they meet their base at right angles*. Specifically, the point in which the normal to the curve, meeting the base, last intersected it is the centre of its curvature.(366) In consequence, when the relationship between the base x and the ordinate y at right angles is had, and thence (by Problem 1) that between the fluxions $\dot{x}$ and $\dot{y}$, the value of $\dot{y}y$(367) derived by writing 1 in place of $\dot{x}$ in it and then supposing $y = 0$ will be the radius of curvature.

Thus, in the ellipse $ax-(a/b)\,x^2 = y^2$, it is $\frac{1}{2}a\dot{x}-(a/b)\,\dot{x}x = \dot{y}y$, and if in this value of $\dot{y}y$ you suppose $y = 0$ (and consequently $x = 0$) and write 1 for $\dot{x}$, the radius of curvature will come out as $\frac{1}{2}a$. And thus at the vertices of the hyperbola and parabola also the radius of curvature will be half the latus rectum.(368)

And so for the conchoid defined by the equation

$$b^2c^2x^{-2}+2bc^2x^{-1}+(c^2-b^2)-2bx-x^2 = y^2,$$

the value of $\dot{y}y$ will, with the aid of Problem 1, be found to be

$$-b^2c^2x^{-3}-bc^2x^{-2}-b-x.$$

By supposing $y = 0$ (and thence $x = \pm c$) in this, the radius of curvature will come out as $\mp b^2/c-2b\mp c$.(369) Therefore make $AE:EG = EG:EC$ and $Ae:eG = eG:ec$ and you have the centres, C and c, of curvature at the vertices, E and e, of the conjugate branches of the conchoid. Figure [on page 44]

(366) In analytical terms, where the coordinates x and y are at right angles, at a point indefinitely near to the vertex (0, 0) on any curve which there meets its base normally x, y and $1/n$ (or $\dot{y}^{-1}$) are each vanishingly small, while for finite curvature at the vertex r (or $\dot{n}$) must be infinitely great; hence

$$\lim_{x,\,y,\,n^{-1}\to 0}(x+yn) = \lim_{y,\,n^{-1}\to 0}\left(\frac{y}{n^{-1}}\right) = \lim_{n,\,r\to\infty}\left(\frac{n}{-rn^{-2}}\right) = \lim_{n,\,r\to\infty}\left(-\frac{(1+n^2)^{\frac{3}{2}}}{r}\right).$$

(367) The length of the subnormal measured from the foot of the ordinate (equal to that, $x+ny$, measured from the vertex in the limit as $x\to 0$).

(368) Compare I, 2, 2: passim. In the general conic $y^2 = ax\pm(a/b)\,x^2$ of latus rectum a and major axis (latus transversum) b the length of the subnormal counted from the vertex (0, 0) is $\frac{1}{2}a+(1\pm a/b)\,x$, so that the radius of curvature there is independent of the length of the major axis.

(369) That is, $\mp(b\pm c)^2/c$. For convenience the figure on page 44 is here repeated.

||[67] ||PROB 6.

CURVATURÆ AD DATUM CURVÆ ALICUJUS PUNCTUM QUALITATEM DETERMINARE.

Per qualitatem Curvaturæ intelligo formam ejus quatenus est plus vel minùs inæquabilis, sive quatenus plus vel minùs variatur in processu per diversas partes Curvæ. Sic interroganti qualis sit circuli curvatura, responderi potest quod sit uniformis, sive invariata; et interroganti qualis sit curvatura Spiralis
Fig[370] quæ describitur per motum puncti *D* cum acceleratâ celeritate *AD* in recta *AK*
[pag 49] uniformiter circa centrum *A* gyrante progredientis ab *A*, adeo ut recta *AD* ad arcum *BK* dato puncto *K* descriptum rationem habeat numeri ad Logarithmum ejus.[371] responderi potest quod sit uniformiter variata sive quod sit æquabiliter inæquabilis. Et sic aliæ curvæ in singulis earum punctis aliquales[372] pro curvaturæ variatione denominari possunt.

Quæritur itaqꝫ Curvaturæ circa aliquod Curvæ punctum inæquabilitas sive variatio. Qua de causa animadvertendum est

1. Quod ad puncta in similibus curvis similiter posita similis est inæquabilitas sive variatio curvaturæ.

2. Et quod momenta radiorum curvaturæ ad illa puncta sunt proportionalia contemporaneis momentis curvarum, et fluxiones fluxionibus.[373]

3. Atqꝫ adeò quod ubi fluxiones illæ non sunt proportionales dissimilis erit inæquabilitas curvaturæ. Utpote major erit inæquabilitas ubi major est ratio fluxionis radij curvaturæ ad fluxionem Curvæ, Adeoqꝫ fluxionum ratio illa non immeritò dici potest index inæquabilitatis sive variationis curvaturæ.[374]

(370) The figure on Newton's page 49 (and the similar one on page 60) represents a spiral in which the radius *AB* is tangent at the pole *A*. In line with the corresponding text on page 69 below we have inserted in our English text a more general figure in which the spiral meets *AB* in a point A' distinct from the pole *A*.

(371) On taking $\widehat{BK} = x$, $AD = y$, $AB = 1$ and $AA' = a$ the defining polar equation of this logarithmic spiral will be $x = \log(y/a)$, so that when the speed of rotation ($\dot{x}$) of *AD* is supposed uniform the corresponding speed of increase ($\dot{y} = \dot{x}y$) of *AD* will be accelerated in proportion as $y = \ddot{y}/\dot{x}^2$. It follows (compare note (393) below) that *AD* bisects the (right) angle between the tangent *DT* and normal *DP* at the point *D*, and, on drawing *PAT* perpendicular to *AD*, that *P* is the centre of curvature at *D* and *DT* is the length of the spiral $\widehat{AA'D}$ (counted from the pole *A* over the infinite number of its circumvolutions necessary to attain the point *D*): immediately $PD = \widehat{AA'D}$ and so their increments too are equal, so that the spiral's curvature is 'equably inequable' in Newton's sense.

(372) 'aliqualis' (perhaps a contraction of 'aliqua[nto æquabi]lis') is evidently Newton's *ad hoc* coinage here. Since no single English equivalent apparently exists we have likewise invented the word 'aliqual' in our translation. In his reproduction of Newton's manuscript in 1779 Horsley, evidently embarrassed with a word which (unlike so many other Newtonian inventions) never passed into common usage, printed 'aliquales' in italic (*Opera Omnia* (note (60)), **1**: 459) while in his 1736 English version Colson (*Method of Fluxions* (note (79)): 76) paraphrases it as 'inequably inequable'.

Problem 6

To determine the quality of curvature at a given point on any curve

By the 'quality' of curvature I understand its form in so far as it is more or less inequable, in other words, as it varies more or less in proceeding along different parts of the curve. So to anyone enquiring what the quality of curvature of a circle is, the reply can be given that it is uniform or invariant; while to any one asking the quality of curvature of a spiral described by the motion of the point *D* proceeding from *A* with accelerated speed *AD* in the straight line *AK* as it rotates uniformly around the centre *A* (so that the line *AD* bears to the arc $\widehat{BK}$ described by the given point *K* the ratio of a number to its logarithm[371]), the answer can be made that it is uniformly variant or equably inequable. And thus other curves in their individual points may be designated 'aliqual' [or inequably inequable][372] according to the variation of their curvature.

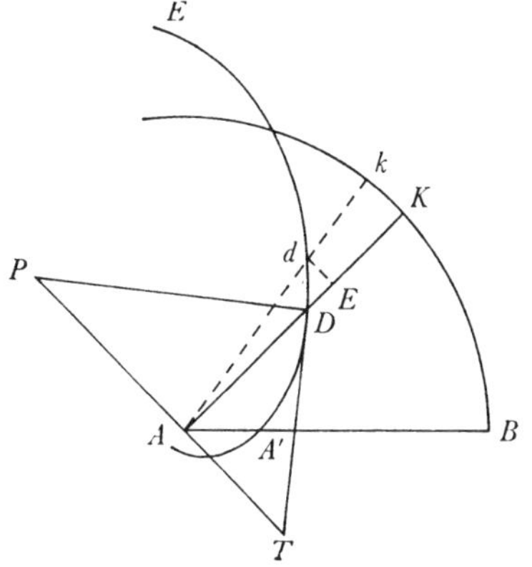

Figure[370] [on page 49]

Accordingly, the inequability or variation of curvature in the immediate neighbourhood of some point in a curve is sought. For this purpose it should be observed:

1. At points similarly located in similar curves there is a like inequability or variation of curvature.

2. The moments of the radii of curvature at those points are proportional to the contemporaneous moments of the curves, and their fluxions correspondingly so.[373]

3. Consequently, when those fluxions are not proportional their inequability of curvature will not be similar: in fact, the inequability will be greater where the ratio of the fluxion of the radius of curvature to the fluxion of the curve is greater, and therefore that fluxional ratio may not undeservedly be called an 'index' of inequability or variation of curvature.[374]

(373) Evidently in similar figures the radii of curvature at corresponding points and the lengths of arc between pairs of corresponding points will be in the same, given proportion.

(374) Where the arc-length (counted from some given point) is s and the radius of curvature is ρ, Newton suggests that $d\rho/ds$ is an acceptable 'index' of the 'shape' of the curve at the point (ρ, s). Though not (see note (316)) being invariant under conformal transformation, Newton's measure of the variation of curvature has the considerable merit of being intrinsic (and so independent of the coordinate system with regard to which any particular curve is defined) while plausibly evaluating the 'shape' of the logarithmic ('proportional') spiral as varying uniformly and—as a particular case—that of the 'perfectly shaped' circle as varying not at all.

Ad Curvæ alicujus AD puncta D ac d infinitè parùm distantia sunto radij curvaturæ DC ac dc, et existente Dd momento Curvæ erit Cc contemporaneum momentum radij curvaturæ, et $\frac{Cc}{Dd}$ index inæquabilitatis curvaturæ. Nempe tanta dicetur inæquabilitas illa, quantum esse indicat rationis illius $\frac{Cc}{Dd}$ quantitas. Sive curvatura dicetur tanto dissimilior curvaturæ circuli.

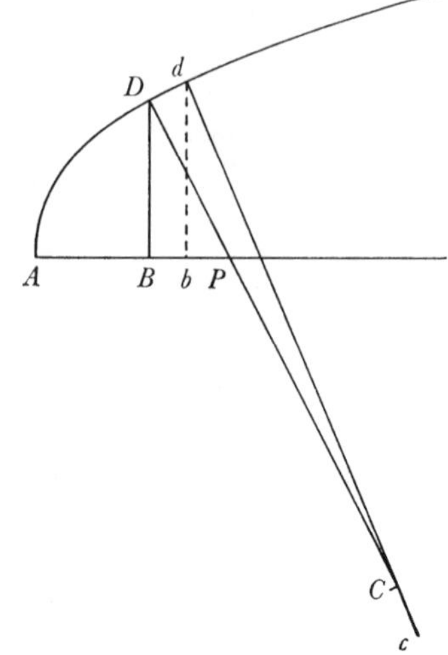

Demissis jam ad quamlibet AB occurrentem DC
||[68] in P, || rectangulis applicatis DB ac db dic $AB=x$, $BD=y$, $DP=t$, $DC=v$, et inde $Bb=m\times o$, eritq; $Cc=l\times o$, et $BD.DP::Bb.Dd=\frac{tmo}{y}$. ac $\frac{Cc}{Dd}=\frac{ly}{tm}$ sive $=\frac{ly}{t}$ supposito $m=1$. Quamobrem relatione inter x et y per quamlibet æquationem definitâ, et inde juxta Prob 4 & 5 invento perpendiculo DP sive t et radio curvaturæ v, ejusq; radij fluxione l per Prob 1; dabitur index inæquabilitatis curvaturæ $\frac{ly}{t}$.(375)

Exempl: 1. Sit $2ax=yy$ (æquatio ad Parabolam) et per Prob 4 erit $BP=a$, adeoq; $DP=\sqrt{aa+yy}=t$. Item per Prob 5 BF(376)$=a+2x$ et

$$BP.DP::BF.DC=\frac{at+2tx}{a}=v.$$

Jam æquationes $2ax=yy$ et $aa+yy=tt$,(377) et $\frac{at+2tx}{a}=v$ per Prob 1 dant $2am=2ny$, et $2ny=2kt$, et $\frac{ak+2kx+2tm}{a}=l$. Quibus ordinatis et posito $m=1$, orientur $n=\frac{a}{y}$, $k=\frac{ny}{t}$ vel $=\frac{a}{t}$ et $l=\frac{ak+2kx+2t}{a}$. Et sic inventis n, k, et l habebitur $\frac{ly}{t}$(378) index inæquabilitatis curvaturæ.

Quemadmodum si in numeris definiatur $a=1$, sive $2x=yy$, et $x=\frac{1}{2}$, erit $y(\sqrt{2x})=1$, $n\left(\frac{a}{y}\right)=1$, $t(\sqrt{aa+yy})=\sqrt{2}$, $k\left(\frac{a}{t}\right)=\sqrt{\frac{1}{2}}$, et $l\left(\frac{ak+2kx+2t}{a}\right)=3\sqrt{2}$. Adeoq; $\frac{ly}{t}=3$ indici inæquabilitatis.

Sin autem definiatur $x=2$, erit(379) $y=2$, $n=\frac{1}{2}$, $t=\sqrt{5}$, $k=\sqrt{\frac{1}{5}}$ et $l=3\sqrt{5}$, Adeoq; $\frac{ly}{t}=6$ index(380) inæquabilitatis. Quamobrem inæquabilitas Curvaturæ

At the points D and d, infinitely little apart, of some curve AD let the radii of curvature be DC and dc. Then, Dd being the moment of the curve, will Cc be the contemporaneous moment of the radius of curvature, and Cc/Dd the index of inequability of the curvature: namely, the magnitude of that inequability will be said to be exactly that shown by the quantity of the ratio Cc/Dd—in other words, the curvature will be said to be unlike the curvature of a circle by that amount.

Letting fall now, to any line AB meeting DC in P, the perpendicular ordinates DB and db, call $AB = x$, $BD = y$, $DP = t$, $DC = v$ and thence $Bb = \dot{x}o$. Then $Cc = \dot{v}o$ and, since $BD:DP = Bb:Dd$, $Dd = t\dot{x}o/y$, while $Cc/Dd = \dot{v}y/t\dot{x}$, that is, $\dot{v}y/t$ on taking $\dot{x} = 1$. In consequence, where the relation between x and y is defined by any equation, after the normal DP (or t) and radius of curvature (v) have been found from it according to Problems 4 and 5, and also the fluxion $\dot{v}$ of this radius by Problem 1, the index of inequability of curvature will be given as $\dot{v}y/t$.(375)

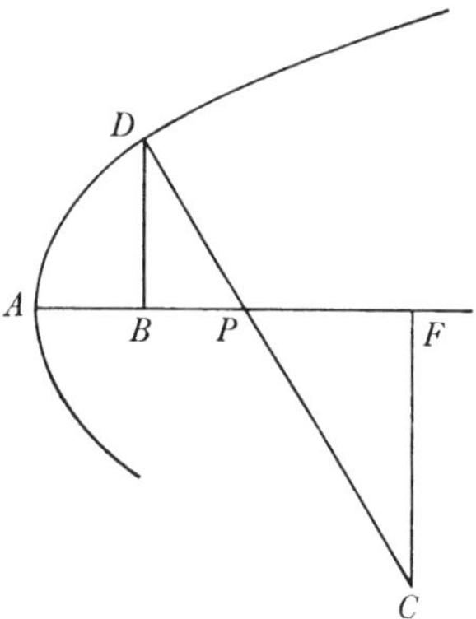

Example 1. Let $2ax = y^2$ (an equation to the parabola) and by Problem 4 there will be $BP = a$, and so $DP = \sqrt{[a^2+y^2]} = t$. Likewise by Problem 5 BF(376) $= a+2x$ and $BP:DP = BF:DC$, so that $DC = (at+2tx)/a = v$. Now by Problem 1 the equations $2ax = y^2$, $a^2+y^2 = t^2$(377) and $(at+2tx)/a = v$ yield $2a\dot{x} = 2\dot{y}y$, $2\dot{y}y = 2\dot{t}t$ and $(a\dot{t}+2\dot{t}x+2t\dot{x})/a = \dot{v}$. On ordering these and setting $\dot{x} = 1$ there will arise $\dot{y} = a/y$, $\dot{t} = \dot{y}y/t$ or a/t and $\dot{v} = (a\dot{t}+2\dot{t}x+2t)/a$. With $\dot{y}$, $\dot{t}$ and $\dot{v}$ thus found the index of inequability of curvature will be had as $\dot{v}y/t$.(378)

For instance if, in a numerical case, the values $a = 1$ (or $2x = y^2$) and $x = \frac{1}{2}$ were fixed upon, then $y(\text{or} \sqrt{[2x]}) = 1$, $\dot{y}(\text{or } a/y) = 1$, $t(\text{or } \sqrt{[a^2+y^2]}) = \sqrt{2}$, $\dot{t}$ (or a/t) $=\sqrt{\frac{1}{2}}$ and $\dot{v}$ (or $\dot{t}+2(\dot{t}x+t)/a) = 3\sqrt{2}$, so that $\dot{v}y/t = 3$ is the index of inequability.

But should it be determined that $x = 2$, then(379) $y = 2$, $\dot{y} = \frac{1}{2}$, $t = \sqrt{5}$, $\dot{t} = \sqrt{\frac{1}{5}}$ and $\dot{v} = 3\sqrt{5}$, so that $\dot{v}y/t = 6$ is the index of inequability. In consequence, the

(375) Newton assumes that the given curve is defined with respect to a system of perpendicular Cartesian coordinates. Of course (compare note (374)) the measure $d\rho/ds$ is independent of the particular choice of coordinates.

(376) The point F, not shown in Newton's figure, is the foot of the perpendicular from C onto AB. Compare Example 3 on his page 63 above.

(377) That is, $t^2 = a(a+2x)$ from k/m (or $\dot{t}/\dot{x}$) $= a/t$ at once.

(378) Namely, $y(a+2x)/(a^2+y^2)+2y/a = 3y/a$. As a result, in a parabola the Newtonian index of curvature varies as the ordinate.

(379) The semi-parameter a is taken to be unity.

(380) Read 'indici .

ad punctum Parabolæ a quo ad axin demissa ordinatim applicata æquatur lateri recto Parabolæ(381) dupla est ejus ad punctum a quo demissa ordinatim applicata æquatur dimidio ejusdem lateris recti. Hoc est curvatura in priori casu duplo dissimilior est curvaturæ circuli,(382) quàm in posteriori.

EXEMPL: 2. Sit $2ax-bxx=yy$,(383) et per Prob 4 erit $a-bx=BP$ et inde $aa-2abx+bbxx+yy=tt$, sive $aa-byy+yy=tt$. Item per Prob 5(384) erit $DH=y\dfrac{+y^3-by^3}{aa}$ ubi si [pro] $yy-byy$ substituas $tt-aa$ evadet $DH=\dfrac{tty}{a}$. Et est $BD.DP::DH.DC=\dfrac{t^3}{aa}=v$. Jam per Prob 1 æquationes $2ax-bxx=yy$ et $aa-byy+yy=tt$, et $\dfrac{t^3}{aa}=v$ dant $a-bx=ny$, et $ny-bny=tk$, et $\dfrac{3ttk}{aa}=l$. Et sic invento l, dabitur $\dfrac{ly}{t}$(385) index inæquabilitatis curvaturæ.

Sic ad Ellipsin $2x-3xx=yy$, ubi est $a=1$ et $b=3$ si supponatur $x=\frac{1}{2}$, erit $y=\frac{1}{2}$, $n=-1$, $t=\surd\frac{1}{2}$, $k=\surd 2$, $l=3\surd\frac{1}{2}$ et $\dfrac{ly}{t}=\dfrac{3}{2}$ indici inæquabilitatis curvaturæ.
||[69] Unde patet curvaturā || hujus Ellipsis ad hic definitum punctum(386) D, esse duplo minus inæquabilem (sive duplo similiorem curvaturæ circuli,(382)) quàm curvatura Parabolæ ad illud ejus punctum(386) a quo ad axin demissa ordinatim applicata æquatur dimidio ejus lateris recti.

Si conclusiones in his exemplis concinnare placet, ad Parabolam $2ax=yy$ exibit $\dfrac{ly}{t}=\dfrac{3y}{a}$(387) index inæquabilitatis et ad Ellipsin $2ax-bxx=yy$ exibit index $\dfrac{ly}{t}=\dfrac{3y-3by}{aa}\times BP$, et sic ad Hyperbolam $2ax+bxx=yy$, observata analogiâ, erit index $\dfrac{ly}{t}=\dfrac{3y+3by}{aa}\times BP$. Unde patet quod ad diversa puncta cujusvis Conicæ sectionis seorsim spectatæ curvaminis inæquabilitas est ut rectangulum $BD\times BP$. Et quod ad diversa puncta Parabolæ est ut ordinatim applicata BD.

Cæterùm cum Parabola sit simplicissima(388) linearum inæquabili curvaturâ flexarum, ejusq; curvaturæ inæquabilitas tam levi negotio determinatur (utpote cujus index sit $\dfrac{6\times\text{ordin: applic:}}{\text{lat: rect:}^{(389)}}$;) aliarum curvarum curvaturæ ad curvaturam hujus non incommodè referri possunt. Quemadmodum si quæratur qualis sit Ellipseωs $2x-3xx=yy$ curvatura ad illud ejus punctum quod definitur assumendo $x=\frac{1}{2}$: Quoniam index ejus (ut supra) sit $\frac{3}{2}$, responderi potest esse

(381) Namely at the point whose ordinate passes through the parabola's focus.

(382) Since its radius is of constant length, the 'index' of curvature in a circle is zero and so affords a useful base for comparison.

(383) An 'æquatio ad Ellipsem'.

inequability of curvature at the point of a parabola from which the ordinate let fall to the axis equals the latus rectum of the parabola(381) is twice that at its point from which the ordinate let fall equals half the same latus rectum—that is, the curvature in the former case is twice as unlike that of the circle(382) as it is in the latter.

EXAMPLE 2. Let $2ax-bx^2=y^2$,(383) and there will be by Problem 4 $a-bx=BP$ and thence $a^2-2abx+b^2x^2+y^2=t^2$, or $a^2+(1-b)\,y^2=t^2$. Likewise by Problem 5(384) $DH=y+a^{-2}(1-b)\,y^3$ and if in this you substitute t^2-a^2 in place of $(1-b)\,y^2$ there will come to be $DH=t^2y/a$. Also $BD\!:\!DP=DH\!:\!DC$, or $DC=t^3/a^2=v$. Now by Problem 1 the equations $2ax-bx^2=y^2$,

$$a^2+(1-b)\,y^2=t^2 \quad \text{and} \quad t^3/a^2=v$$

give $a-bx=\dot{y}y$, $(1-b)\,\dot{y}y=t\dot{t}$ and $3t^2\dot{t}/a^2=\dot{v}$. And thus, when $\dot{v}$ is found, will be given $\dot{v}y/t$,(385) the index of inequability of curvature.

Thus in the ellipse $2x-3x^2=y^2$, in which $a=1$ and $b=3$, if x be taken equal to $\frac{1}{2}$, then $y=\frac{1}{2}$, $\dot{y}=-1$, $t=\surd\frac{1}{2}$, $\dot{t}=\surd 2$, $\dot{v}=3\surd\frac{1}{2}$ and $\dot{v}y/t=\frac{3}{2}$, the index of inequability of curvature. It is hence evident that the curvature of this ellipse at the point D here defined(386) is twice less inequable (or twice more like the curvature of a circle(382)) than the curvature of a parabola at that point of it(386) from which the ordinate let fall to the axis is equal to half its latus rectum.

Should you desire to rework the conclusions in these examples into a unit, in the parabola $2ax=y^2$ the index of inequability $\dot{v}y/t$ will come out equal to $3y/a$,(387) in the ellipse $2ax-bx^2=y^2$ it will prove to be $3y(1-b)/a^2\times BP$, and thus, by observing analogy, in the hyperbola the index $\dot{v}y/t$ will equal $3y(1+b)/a^2\times BP$. It is consequently evident that at different points of any conic individually regarded the inequability of curvature is as the rectangle $BD\times BP$ (and in the parabola as the ordinate BD simply).

But since, however, the parabola is the simplest(388) of lines bent with an inequable curvature and its inequability of curvature is determined with such slight bother (in as much as its index is six times the ordinate divided by the latus rectum(389)), the curvatures of other curves may not inconveniently be related to that of this curve. For instance, if the quality of curvature of the ellipse $2x-3x^2=y^2$ at that of its points fixed by assuming $x=\frac{1}{2}$ be inquired after, since its index (as above) is $\frac{3}{2}$, the response may be given that it is similar to the

(384) See Example 1 on Newton's pages 55–6 above.

(385) Namely, $3a^{-2}(1-b)\,y(a-bx) = 3a^{-2}(1-b)\,y\times BP$.

(386) The points $(\frac{1}{2},\frac{1}{2})$ and, on the parabola $y^2=2ax$, $(\frac{1}{2},1)$.

(387) See note (378).

(388) Presumably because its 'index' of curvature varies linearly with its ordinate (see note (378)).

(389) That is, 2 (here).

similem curvaturæ Parabolæ $6x=yy$ ad illud ejus punctum inter quod et axin recta $=\frac{3}{2}$ ordinatim applicatur.

pag [67] & fig Sic cùm lineæ Spiralis ADE(390) jam ante descriptæ(391) fluxio sit ad fluxionem subtensæ AD in data quadam ratione, puta d ad e: versus partes concavas ejus erige ad AD normalem $AP=\frac{e}{\sqrt{dd-ee}}\times AD$, et erit P centrum curvaturæ, et $\frac{AP}{AD}$ sive $\frac{\sqrt{dd-ee}}{e}$(392) index inæquabilitatis ejus.(393) Quare Spiralis hæcce curvaturam habet ubiq3 similiter inæquabilem ac Parabola $6x=yy$ habet in illo ejus puncto a quo demittitur ad axin ordinatim applicata $=\frac{\sqrt{dd-ee}}{e}$.(394)

Et sic index inæquabilitatis ad quodvis Trochoidis punctum D (fig pag [57]) invenietur esse $\frac{AB}{BL}$.(395) Quare curvatura ejus ad idem [punctum] D tam

‖[70] inæquabilis est sive tam dissimilis curvaturæ ‖ circuli, quàm curvatura Parabolæ cujusvis $ax=yy$ ad illud ejus punctum ubi ordinatim applicata æquatur $\frac{1}{6}a\times\frac{AB}{BL}$.(396)

Ex his credo sensus Problematis satis elucescet, quo benè perspecto non difficile erit animadvertenti seriem rerum supra traditarum plura exempla de

(390) Understand '$ADdE$'. The two points E in our (collated) figure are perhaps confusing.

(391) On Newton's page 67 above (in the particular case when $d^2=2e^2$).

(392) Read '$\frac{e}{\sqrt{dd-ee}}$'.

(393) Much as in note (371), on taking $\widehat{BK}=x$, $AD=y$, $AB=1$, $AA'=a$, $\widehat{AA'D}=s$ and $e/d=\sin\alpha$, the defining condition $\dot{s}/\dot{y}=\sqrt{[1+(y\dot{x}/\dot{y})^2]}=\operatorname{cosec}\alpha$ yields $z=\dot{y}/y\dot{x}=\tan\alpha$ and hence $x\tan\alpha=\log(y/a)$ as the polar equation of the spiral. Since r (or $\dot{z}$) $=0$, the radius of curvature $PD=y\sec\alpha$ and so the curvature centre P lies in the perpendicular at A to AD; likewise, where the tangent at D meets PA in T, the arc-length

$$\widehat{AA'D}\text{ (or } s) = \int \operatorname{cosec}\alpha . dy = y\operatorname{cosec}\alpha = DT.$$

It follows that $\rho/s=\tan\alpha$; or, geometrically, $PD/TD=PA/AD=AD/AT$. Since all similar spirals have the same constant Newtonian index ($d\rho/ds=\tan\alpha$) it follows at once that they are superimposable: a defining property of the 'spira mirabilis' which tempted Jakob Bernoulli twenty years later to view the logarithmic spiral as a symbol of perpetual resurrection. As he wrote in a celebrated peroration to his paper on 'Lineæ Cycloidales, Evolutæ, Ant-Evolutæ, Causticæ, Anti-Causticæ, Peri-Causticæ. Earum usus & simplex relatio ad se invicem' (*Acta Eruditorum* (May 1692): 207–213, especially 212–13): 'Numero easdem curvas voco...quæ sibi super impositæ congruunt.... [Hæc proprietas] cum in sola nostra spira constans maneat, videtur quasi natura hoc essentiali charactere illi soli id privilegii vindicare voluisse. Cum autem ob proprietatem tam singularem tamque admirabilem mire mihi placeat spira hæc mirabilis, sic ut ejus contemplatione satiari vix queam; cogitavi, illam ad varias res symbolice repræsentandas non inconcinne adhiberi posse. Quoniam enim semper sibi similem & eandem spiram gignit, utcunque volvatur....Aut, si mavis, quia Curva nostra mirabilis in ipsa mutatione semper sibi constantissime manet similis & numero eadem, poterit esse vel

curvature of the parabola $6x = y^2$ at that of its points between which and the axis a straight line of length $\frac{3}{2}$ is ordinately applied.

Thus, because the fluxion of the spiral line $AD[d]E$ already previously described(391) is to that of the subtense AD in a certain given ratio, say d to e, towards its concave side erect $AP = AD \times e/\surd(d^2-e^2)$ normal to AD and P will be the centre of curvature with AP/AD, that is, $[e/\surd(d^2-e^2)]$, the index of its inequability.(393) Consequently, this spiral has its curvature everywhere uniformly as inequable as the parabola $6x = y^2$ has its at that point from which an ordinate of length $[e/\surd(d^2-e^2)]$ is let fall to the axis.(394)

Page 67 and figure.

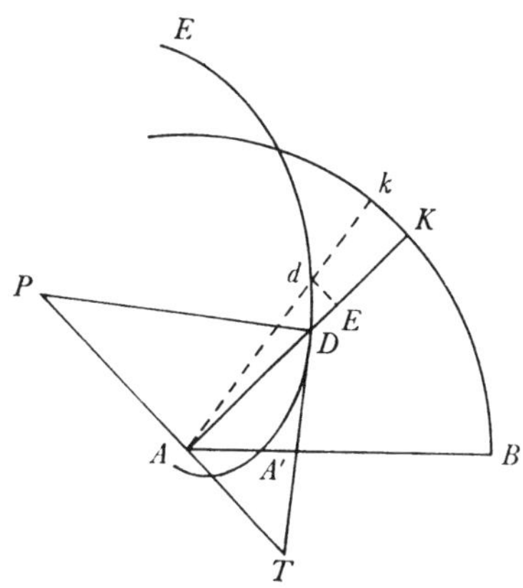

And thus the index of inequability at any point D of the cycloid (figure page 57) will be found to be AB/BL.(395) Therefore its curvature at the same point D is as inequable, or as unlike the curvature of a circle, as the curvature of any parabola at that of its points where the ordinate equals $\frac{1}{6}a \times AB/BL$.

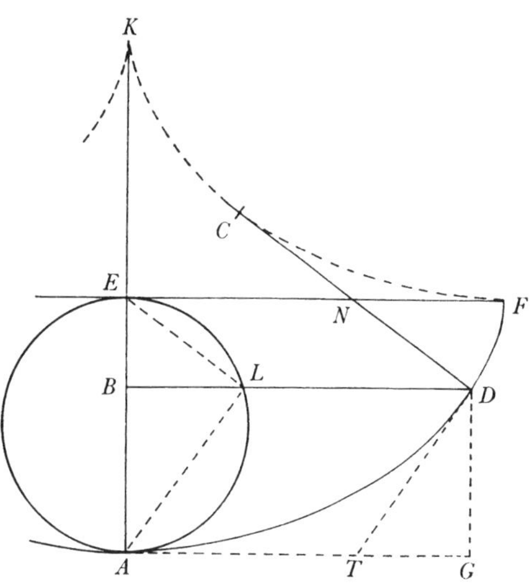

From these considerations, I believe, the sense of the problem is sufficiently clarified. When it has been closely examined it will not be difficult for anyone who devotes his attention to the sequence of arguments above presented to

fortitudinis & constantiæ in adversitatibus; vel etiam Carnis nostræ post varias alterationes, & tandem ipsam quoque mortem, ejusdem numero resurrecturæ symbolum; adeo quidem, ut si *Archimedem* imitandi hodie[r]num consuetudo obtineret, libenter Spiram hanc tumulo meo juberem incidi cum Epigraphe: *Eadem numero mutata resurget.*' For Newton the logarithmic spiral held no such revelations and his tomb in Westminster Abbey was inscribed with less esoteric, if less majestic symbols of his intellectual achievements.

(394) Much as in note (378) the Newtonian index of the parabola $ax = y^2$ is $d\rho/ds = 6y/a$ (where the curve is referred to perpendicular Cartesian coordinates). This equals the index for the present spiral $\widehat{AA'D} = (d/e) \times AD$ on setting $a = 6$. In general, computation of $d\rho/ds$ for a curve defined in polar coordinates will be formidable and the resulting index far from simple. Newton very wisely restricts himself to the simplest case in illustration.

(395) Since $\widehat{AD} = s = 2AL$ and $DC = \rho = 2EL$, the intrinsic equation of the cycloid ADF is readily seen to be $s^2+\rho^2 = 4EA^2 = KA^2$. Accordingly

$$d\rho/ds = -s/\rho = AL/EL = AB/BL.$$

(396) Compare note (394) with $y = \frac{1}{6}a \times AB/BL$.

proprio suppeditare et hujusmodi complures alias operandi methodos, prout res exiget, concinnare. Quinetiam cognata Problemata (ubi perplexa computatione non conteritur et fatigatur,) haud majori difficultate transiget: Cujusmodi sunt,

1. *Invenire punctum curvæ alicujus ubi vel nullam, vel infinitam, vel maximā aut minimam, vel datam quamvis habet inæquabilitatem curvaturæ.* Sic ad vertices Conicarum sectionum nulla est inæquabilitas curvaturæ,(397) ad cuspidem Trochoidis infinita est,(398) et ad puncta Ellipseos maxima est ubi rectangulū $BD \times BP$ fit maximum, hoc est ubi lineæ diagonales rectanguli Parallelogrammi circumscripti Ellipsin secant cujus latera tangunt illam in principalibus verticibus.(399)

2. *Curvam alicujus definitæ speciei, puta Conicam Sectionem,*(400) *determinare, cujus curvatura ad aliquod punctum & æqualis sit et similis curvaturæ alterius curvæ ad datum punctum ejus.*(401)

3. *Conicam Sectionem determinare ad cujus punctum aliquod curvatura & lineæ tangentis (respectu axis) positio sit similis curvaturæ ac tangentis positioni alterius alicujus Curvæ ad assignatum punctum ejus.* Et hujus problematis usus est ut vice Ellipsium secundi generis quarum refringendi proprietates Cartesius in Geometria demonstravit, Conicæ sectiones idem in refractionibus quàm proximè præstantes subrogari possint.(402) Atq; idem de alijs curvis intellige.

||[71] ||PROB: 7.

CURVAS PRO ARBITRIO MULTAS INVENIRE QUARUM AREÆ PER FINITAS ÆQUATIONES DESIGNARI POSSUNT.

[Fig pag 20](403) Sit AB basis curvæ, ad cujus initium A erigatur normalis $AC=1$ et agatur CE parallela AB, sit etiam DB rectangula applicata occurrens rectæ DE in E et

(397) By Example 2 above (compare note (385)) the curvature index for the general conic $y^2 = 2ax \pm bx^2$ is $3a^{-2}(1 \pm b)y(a \pm bx)$, which is zero both at the vertices of the major axis (where $y = 0$) and of the minor axis (where $bx = \mp a$).

(398) By the preceding paragraph the index of curvature at the point D in the arc $\widehat{ADF}$ is AB/BL: hence at F, where B, L coincide at E, it is $AE/0$ or infinity.

(399) As in note (385) the curvature index $d\rho/ds$ at the general point of the ellipse $y^2 = 2ax - bx^2$ is

$$3a^{-2}(1-b)y(a-bx),$$

which is greatest when $y(a-bx)$, that is $BD \times BP$, is a maximum. By equating its derivative to zero the Fermatian condition $by^2 = (a-bx)^2$ is obtained. Accordingly,

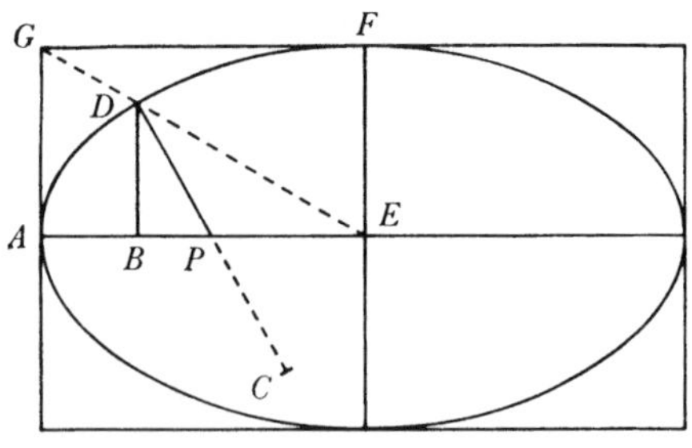

$$DB^2:BE^2 = y^2:(a/b-x)^2 = a^2/b:a^2/b^2$$
$$= (FE^2 \text{ or}) \; GA^2:AE^2,$$

so that D is the meet of the diagonal GE with the ellipse $\widehat{AF}$.

(400) 'numericè (ut loquuntur)' (numerically, as they say) is cancelled.

supply further examples on his own and, as occasion demands, formulate not a few other methods of working—indeed, (when he is not tired or worn out by intricate computation) he will deal with allied problems in no more difficult a manner. Such as:

1. *To find the point of some curve at which its inequability of curvature is zero, or infinite, or either greatest or least.* Thus at the vertices of conics there is no inequability of curvature,(397) at a cycloid's cusp it is infinite(398) and in an ellipse is greatest at those points where the rectangle $BD \times BP$ is greatest, that is, where the diagonal lines of the circumscribed rectangle cut the ellipse (its sides touching it at the principal vertices).(399)

2. *To determine*(400) *a curve of some definite species, say a conic, whose curvature at some point is both equal and similar to the curvature of another curve at some given point in it.*(401)

3. *To determine a conic, at some point of which the curvature and position of the tangent line (with regard to the axis) are similar to those of some other curve at an assigned point in it.* The use of this problem is to make it possible to choose, in place of the second-order ovals whose refraction properties Descartes demonstrated in his *Geometry*, conic sections accomplishing the same results in refraction to a high degree of approximation.(402) And understand the same for other curves.

Problem 7

To find arbitrarily many curves whose areas may be expressed by finite equations

Let AB be the base of a curve, and at its origin A let there be raised the perpendicular $AC = 1$ and let CE be drawn parallel to AB. Let also DB be an [Figure on page 20.](403)

(401) In other words, given a family of curves, to isolate the member which has given curvature and variation of curvature at a point. The family of conics would seem to be the only interesting simple case in point. The general conic $y^2 = 2ax \pm bx^2$ has curvature

$$-a^2(a^2+(1\pm b)y^2)^{-\frac{3}{2}}$$

(compare note (295)) and index of curvature variation $3a^{-2}(1\pm b)y(a\pm bx)$ at the point (x, y): evidently when these are known numerically the latus rectum $2a$ and major axis $2a/b$ of the conic may readily be computed.

(402) In the case of the conic $y^2 = 2ax \pm bx^2$ the latus rectum $2a$ and major axis $2a/b$ are to be computed from the known values of the tangent slope $(a\pm bx)/y$ and the index of curvature variation $3a^{-2}(1\pm b)y(a\pm bx)$. The application to dioptrics depends on the fact that in the immediate vicinity of a point on a given (plane) interface of known refractive ratio the 'bending' of any incident ray depends uniquely on the instantaneous direction of the interface at the point and the variation in its curvature in the immediate neighbourhood of the point. At greater distances from the point of refraction the approximation of the Cartesian oval (see note (228)) by a suitable conic will be less accurate. Since in practical dioptrics, as Newton well knew (see the concluding scholium of **3**, 1, §2 below), it is sufficient to approximate an interface by the circle which has the same curvature and tangent-direction at some mean point in it, the present optical application is entirely of theoretical interest.

(403) Newton's text has no illustrative diagram but we repeat the figure on his page 20 above, interpolating an appropriate marginal reference.

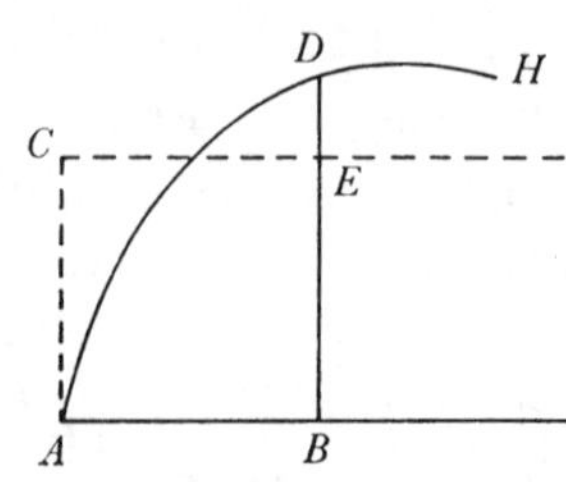

curvæ AD in D. Et concipe has areas $ACEB$ et ADB a rectis BE et BD per AB delatis generari. Et(404) earum incrementa sive fluxiones perpetim erunt ut lineæ describentes BE et BD. Quare parallelogrammum $ACEB$ sive $AB\times 1$ dic x, et curvæ aream ADB dic z: et fluxiones m et r erunt ut BE et BD, adeóq; posito $m=1$ erit $r=BD$.

Si jam ad arbitrium assumatur æquatio quævis pro definienda relatione z ad x, exinde per prob 1 elicietur r. Atq; ita duæ habebuntur æquationes quarum posterior(405) curvam definiet et prior aream ejus.

Exempla. Assumatur $xx=z$ et inde per Prob: 1 elicietur $2mx=r$, sive $2x=r$ siquidem est $m=1$.

Assumatur $\frac{x^3}{a}=z$ et inde prodibit $\frac{3xx}{a}=r$, æquatio ad Parabolā.

Assumatur $ax^3=zz$, sive $a^{\frac{1}{2}}x^{\frac{3}{2}}=z$, et emerget $\frac{3}{2}a^{\frac{1}{2}}x^{\frac{1}{2}}=r$, sive $\frac{9}{4}ax=rr$ æquatio iterum ad Parabolam.

Assumatur præterea $a^3x=zz$, sive $a^{\frac{3}{2}}x^{\frac{1}{2}}=z$ et elicietur $\frac{1}{2}a^{\frac{3}{2}}x^{-\frac{1}{2}}=r$ sive $a^3=4xrr$.(406)

Item assumatur $\frac{a^3}{x}=z$ sive $a^3x^{-1}=z$ et elicietur $-a^3x^{-2}=r$ sive $a^3+rxx=0$.(406) Ubi negativus valor ipsius r tantùm denotat BD capiendam esse ad partes contra BE.

Adhæc si assumas $ccaa+ccxx=zz$, elicies $2ccx=2zr$ et exterminato z proveniet $\frac{cx}{\sqrt{aa+xx}}=r$.(407)

Vel si assumas $\frac{aa+xx}{b}\sqrt{aa+xx}=z$, dic $\sqrt{aa+xx}=v$ et erit $\frac{v^3}{b}=z$, et inde per Prob 1 $\frac{3lvv}{b}=r$. Item æquatio $aa+xx=vv$ per Prob 1 dat $2x=2vl$ cujus ope si extermines l fiet $\frac{3vx}{b}=r=\frac{3x}{b}\sqrt{aa+xx}$.(408)

Si deniq; assumas $8-3xz+\frac{2}{5}z=zz$, elicies $-3z-3xr+\frac{2}{5}r=2rz$.(409) Quare per assumptam æquationem imprimis quære aream z, ac deinde applicatam r per elicitam.

Atq; ita ex areis qualescunq; effingas semper possis applicatas determinare.(410)

---

(404) A first, cancelled continuation reads somewhat surprisingly 'incrementum sive fluxio areæ $AC[E]B$ erit o' (the increment or fluxion of the area $ACEB$ will be o). Compare the equivalent passage in the *De Analysi* (II: 242).

(405) Namely, r (or $\dot{z}$) $= F(x)$ with respect to the perpendicular Cartesian coordinates $AB = x$, $BD = r$, where $F(x)$ is the result of eliminating z from $-f_x/f_z$ by means of the given equation $f(x, z) = 0$.

ordinate at right angles, meeting the straight line DE in E and the curve AD in D. Now conceive these areas, $ACEB$ and ADB, to be generated by the lines BE and BD as they are borne along AB;(404) their increments, and so their fluxions, will be perpetually as the describing lines BE and BD. In consequence call the rectangle $ACEB$, or $AB\times 1$, x and the area ADB under the curve z: the fluxions $\dot{x}$ and $\dot{z}$ will then be as BE and BD, so that when $\dot{x}$ is set equal to unity, $BD=\dot{z}$.

Now if any equation at will be assumed for defining the relationship of z to x, from it by Problem 1 will be elicited $\dot{z}$, and so two equations will be had, the latter of which(405) will define the curve and the former its area.

EXAMPLE. Let there be assumed $x^2=z$: thence by Problem 1 will be elicited $2\dot{x}x=\dot{z}$, that is, $2x=\dot{z}$ seeing that $\dot{x}=1$.

Let there be assumed $x^3/a=z$, and thence will be produced $3x^2/a=\dot{z}$, the equation of a parabola.

Let there be assumed $ax^3=z^2$, or $a^{\frac{1}{2}}x^{\frac{3}{2}}=z$, and there will emerge $\frac{3}{2}a^{\frac{1}{2}}x^{\frac{1}{2}}=\dot{z}$ or $\frac{9}{4}ax=\dot{z}^2$, again the equation of a parabola.

Let there be assumed, further, $a^3x=z^2$, or $a^{\frac{3}{2}}x^{\frac{1}{2}}=z$ and there will be elicited $\frac{1}{2}a^{\frac{3}{2}}x^{-\frac{1}{2}}=\dot{z}$ or $a^3=4x\dot{z}$.(406)

Likewise, let there be assumed $a^3/x=z$, or $a^3x^{-1}=z$, and there will be elicited $-a^3x^{-2}=\dot{z}$ or $a^3+\dot{z}xx=0$.(406) Here the negative value of $\dot{z}$ indicates merely that BD is to be taken in the direction opposite to BE.

Again, should you assume $c^2a^2+c^2x^2=z^2$, you will elicit $2c^2x=2z\dot{z}$ and, after z is eliminated, there will prove to be $cx/\sqrt{[a^2+x^2]}=\dot{z}$.(407)

Or should you assume $(a^2+x^2)^{\frac{3}{2}}/b=z$, call $\sqrt{[a^2+x^2]}=v$ and it will be $v^3/b=z$, and thence by Problem 1 $3\dot{v}v^2/b=\dot{z}$. Likewise, the equation $a^2+x^2=v^2$ gives, by Problem 1, $2x=2v\dot{v}$, and if by its aid you eliminate $\dot{v}$ there will come to be $3vx/b=\dot{z}=3x\sqrt{[a^2+x^2]}/b$.(408)

Finally, should you assume $8-3xz+\frac{2}{5}z=z^2$, you will elicit

$$-3z-3x\dot{z}+\tfrac{2}{5}\dot{z}=2\dot{z}z.\text{(409)}$$

Accordingly, in the first instance seek the area z by means of the assumed equation, and then the ordinate $\dot{z}$ by means of the elicited one.

And thus might you always determine the ordinates from areas, whatever kind you think up.(410)

(406) That is, 'æquatio ad Hyperbolam cubicam'.

(407) A modified kappa curve.

(408) A kappa curve whose asymptotes are at infinity.

(409) Explicitly $2r(\frac{2}{5}-3x) = (2r+3)(\frac{2}{5}-3x\pm\sqrt{[32+(\frac{2}{5}-3x)^2]})$.

(410) Having (in Problem 1) developed at length the computational procedure necessary to derive fluxions from given fluents, the present multiplication of examples scarcely seems necessary.

||[72]

||Prob: 8.

Curvas pro arbitrio multas invenire quarum areæ ad aream datæ alicujus Curvæ relationem habent per finitas æquationes designabilem.

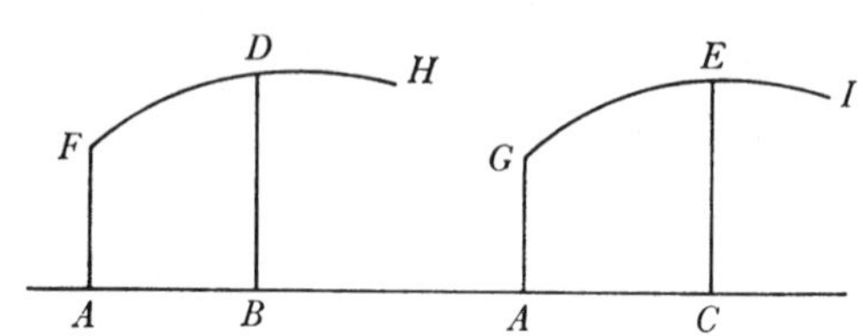

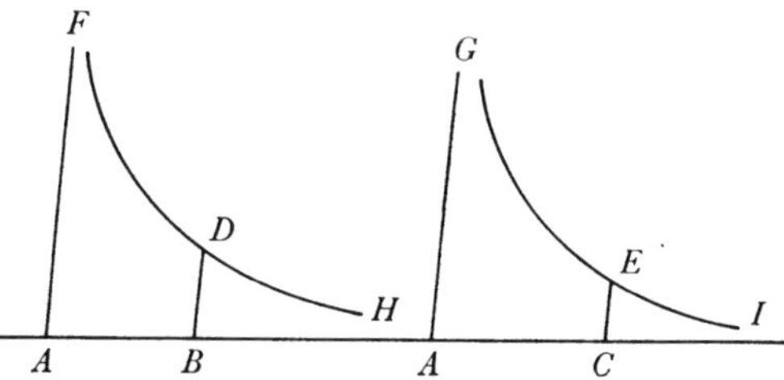

(411)Sit FDH data curva, ac GEI quæsita et earum applicatas(412) DB et EC concipe super Basibus AB et AC erectas incedere: Et arearum quas ita transigunt incrementa sive fluxiones erunt ut applicatæ illæ ductæ in earum velocitates incedendi, hoc est in fluxiones basium. Sit ergo $AB=x$, $BD=v$, $AC=z$ ac $CE=y$, area $AFDB=s$, & area $AGEC=t$,(413) ac arearum fluxiones sint p, et q, nempe p ipsius s, et q ipsius t: eritꝗ $m\times v.r\times y::p.q$. Quare si supponatur $m=1$, et $v=p$,(414) ut supra, erit $ry=q$ et inde $\frac{q}{r}=y$.

Assumantur itaꝗ duæ quævis æquationes quarum una definiat relationem arearum s ac t, et altera relationem basium x et z et inde per Prob 1 quærantur fluxiones q et r, et statuatur $\frac{q}{r}=y$.(415)

Exemp 1. Data curva AFD sit circulus æquatione $ax-xx=vv$ designatus, et quærantur aliæ curvæ quarum areæ adæquant aream ejus. Ex hypothesi ergo est $s=t$ et inde $p=q=v$. et $y=\left(\frac{q}{r}=\right)\frac{v}{r}$. Superest ut r determinetur assumendo relationem aliquam inter bases x et z.(416)

Veluti si fingas $ax=zz$ erit per Prob 1 $a=2rz$. Quare substitue $\frac{a}{2z}$ pro r et fiet $y=\left(\frac{v}{r}=\right)\frac{2vz}{a}$. Est autem $v=(\sqrt{ax-xx}=)\frac{z}{a}\sqrt{aa-zz}$, adeoꝗ $\frac{2zz}{aa}\sqrt{aa-zz}=y$, æquatio ad curvam(417) cujus area æquatur areæ circuli.

(411) A variant set of illustrative figures (in which the points A and G are coincident) has been entered at a late stage in the margin at this point. We reproduce them below immediately before Example 3.

(412) In the first instance these are taken to be inclined to their bases at an equal oblique angle, but Newton will at once (see note (414)) implicitly assume these to be perpendicular ordinates.

Problem 8

To find arbitrarily many curves whose areas bear to the area of some given curve a relationship expressible by finite equations

(411)Let FDH be the given curve, GEI that sought, and imagine their ordinates(412) DB and EC to advance erect upon the bases AB and AC: the increments, and so the fluxions, of the areas thus traversed will then be as those ordinates multiplied into their speeds of advance, that is, into the fluxions of the bases. So let $AB = x$, $BD = v$, $AC = z$ and $CE = y$ with the area $AFDB = s$, the area $AGEC = t$(413) and the fluxions of these areas $\dot{s}$ and $\dot{t}$ (namely, $\dot{s}$ that of s and $\dot{t}$ that of t): then will $\dot{x}v : \dot{z}y = \dot{s} : \dot{t}$. Hence if (as above) $\dot{x}$ be supposed equal to unity and $v = \dot{s}$,(414) then $\dot{z}y = \dot{t}$ and thence $\dot{z}/\dot{t} = y$.

Accordingly, let any two equations be assumed, one of which is to define the relation of the areas s and t, the second that of the bases x and z, and from these let there be sought the fluxions $\dot{s}$ and $\dot{t}$ and set $\dot{t}/\dot{z} = y$.(415)

Example 1. Let the given curve AFD be a circle represented by the equation $ax - x^2 = v^2$, and let other curves be sought equalling its own in area. By hypothesis, therefore, $s = t$ and thence $\dot{s} = \dot{t} = v$ and $y = (\dot{t}/\dot{z} =)\, v/\dot{z}$. It remains to determine $\dot{z}$ by assuming some relation between the bases x and z.(416)

If, for instance, you should suppose $ax = z^2$, there will, by Problem 1, be $a = 2\dot{z}z$. Substitute therefore $a/2z$ in place of $\dot{z}$ and there will come

$$y = (v/\dot{z} =)\, 2vz/a.$$

However $v = (\sqrt{[ax - x^2]} =)\, a^{-1}z\sqrt{[a^2 - z^2]}$, so that $2a^{-2}z^2\sqrt{[a^2 - z^2]} = y$ is the equation of the curve(417) whose area equals that of the circle.

---

(413) In modern terms, on setting $\sin^{-1}k = A\hat{B}D = A\hat{C}E$, it follows that

$$(AFDB) = s = k\int v.dx \quad \text{and} \quad (AGEC) = t = k\int y.dz.$$

(414) Since, in the terminology of the previous note, p/m (or $\dot{s}/\dot{x}$) $= kv$, Newton's assumption (on taking m (or $\dot{x}$) $= 1$) that $p = v$ implies that $k = 1$ and so $A\hat{B}D = A\hat{C}E = \frac{1}{2}\pi$.

(415) If generally these relationships are defined by the equations $f(s, t, v, x) = 0$ and $\phi(x, z) = 0$, then, on differentiating, $vf_s + yrf_t + lf_v + f_x = 0$ (where

$$p \text{ (or } \dot{s}) = v, \quad q \text{ (or } \dot{t}) = yr \text{ (or } y\dot{z})$$

and l is Newton's denomination of the fluxion $\dot{v}$) and also $\phi_x + r\phi_z = 0$: that is, when r is eliminated, $(vf_s + lf_v + f_x)\,\phi_z = yf_t\phi_x$, from which x may be eliminated in terms of z by means of $\phi(x, z) = 0$.

(416) Here, in terms of note (415), $f \equiv s - t$ and so $f_s = -f_t = 1$: hence

$$\sqrt{[ax - x^2]} = v = -y\phi_x/\phi_z.$$

(417) A kappa quartic, to be precise. Here $\phi \equiv ax - z^2$, so that $\phi_x = a$, $\phi_z = -2z$ and the equation results on substituting $x = z^2/a$.

Ad eundem modum si fingas $xx=z$, proveniet $2x=r$, et inde $y=\left(\frac{v}{r}=\right)\frac{v}{2x}$, et exterminato v et x fiet $y=\frac{\sqrt{az^{\frac{1}{2}}-z}}{2z^{\frac{1}{2}}}$.(418)

||[73] || Vel si fingas $cc=xz$, proveniet $0=z+xr$; et inde $\frac{-vx}{z}=y=-\frac{c^3}{z^3}\sqrt{az-cc}$.(419)

Atq; ita si fingas $ax+\frac{s}{1}$(420) $=z$, ope Prob 1 obtinebitur $a+p=r$ et inde $\frac{v}{a+p}=y=\frac{v}{a+v}$ quæ Curvam Mechanicam(421) designat.

Exempl 2. Detur iterum Circulus $ax-xx=vv$ et quærantur Curvæ quarum areæ ad aream ejus habeant aliam quamlibet assumptam relationem. Veluti si assumes $cx+s=t$, et præterea fingas $ax=zz$, mediante Prob 1 elicies $c+p=q$ et $a=2rz$. Quare est $y=\left(\frac{q}{r}=\right)\frac{2cz+2pz}{a}$, et substituto $\sqrt{ax-xx}$ pro p, et $\frac{zz}{a}$ pro x fit $y=\frac{2cz}{a}+\frac{2zz}{aa}\sqrt{aa-zz}$.

Quod si assumas $s-\frac{2v^3}{3a}=t$, et $x=z$, invenies ope Prob 1 $p-\frac{2lvv}{a}=q$ et $1=r$. Adeoq; $y=\left(\frac{q}{r}=\right)p-\frac{2lvv}{a}$ sive $=v-\frac{2lvv}{a}$. Jam vero pro exterminando l, æquatio $ax-xx=vv$ per Prob 1 dat $a-2x=2vl$ et proinde est $y=\frac{2vx}{a}$ ubi si supprimas v et x substituendo $\sqrt{ax-xx}$ et z, emerget $y=\frac{2z}{a}\sqrt{az-zz}$.

Sin assumas $ss=t$, et $x=zz$ emerget $2ps=q$, et $1=2rz$ atq; adeò $y=\left(\frac{q}{r}=\right)4psz$, et pro p et x substitutis $\sqrt{ax-xx}$ et zz fiet $y=4szz\sqrt{a-zz}$ æquatio ad Curvam Mechanicam.(422)

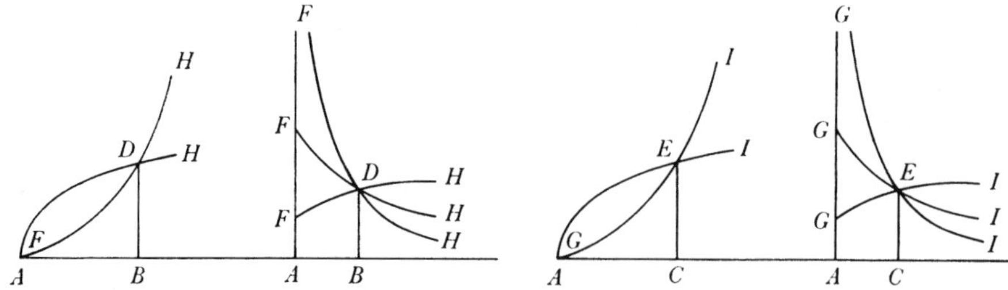

Exempl: 3. Ad eundem modum figuræ assumptam relationem ad aliam quamvis datam figuram habentes inveniuntur. Sic datâ Hyperbolâ $cc+xx=vv$,

(418) Here $\phi \equiv x^2-z$, so that $\phi_x = 2x$, $\phi_z = -1$ and the result follows on substituting $x = \sqrt{z}$.

In the same manner, should you suppose $x^2 = z$, there will come out $2x = \dot{z}$ and thence $y = (v/\dot{z} =) \, v/2x$ and, after elimination of v and x this will become $y = \sqrt{[az^{\frac{1}{2}} - z]}/2z^{\frac{1}{2}}$.(418)

Or should you suppose $c^2 = xz$, there will come out $0 = z + x\dot{z}$, and thence $-vx/z = y = -c^3z^{-3}\sqrt{[az-c^2]}$.(419)

And thus should you suppose $ax + s = z$, by the help of Problem 1 will be obtained $a + \dot{s} = \dot{z}$ and thence $v/(a+\dot{s}) = y = v/(a+v)$, which represents a mechanical curve.(421)

Example 2. Let the circle $ax - x^2 = v^2$ be again given and let there be sought curves whose areas bear to its own any other assumed relationship. Should you, for instance, assume $cx + s = t$ and suppose, moreover, that $ax = z^2$, by means of Problem 1 you will elicit $c + \dot{s} = \dot{t}$ and $a = 2\dot{z}z$. Therefore

$$y = (\dot{t}/\dot{z} =) \, 2a^{-1}z(c+\dot{s})$$

and this, when $\sqrt{[ax-x^2]}$ is substituted in place of $\dot{s}$ and z^2/a in place of x, becomes $y = 2cz/a + 2z^2\sqrt{[a^2-z^2]}/a^2$.

But if you assume $s - 2v^3/3a = t$ and $x = z$, you will find with the aid of Problem 1 that $\dot{s} - 2\dot{v}v^2/a = \dot{t}$ and $1 = \dot{z}$, so that $y = (\dot{t}/\dot{z} =) \, \dot{s} - 2\dot{v}v^2/a$, that is, $v - 2\dot{v}v^2/a$. Now, however, for eliminating $\dot{v}$ the equation $ax - x^2 = v^2$ yields, by Problem 1, $a - 2x = 2v\dot{v}$ and there is consequently $y = 2vx/a$. Here, if you suppress v and x by substituting $\sqrt{[ax-x^2]}$ and z in their place, there will emerge $y = 2z\sqrt{[az-z^2]}/a$.

But should you assume $s^2 = t$ and $x = z^2$, there will emerge $2\dot{s}s = \dot{t}$ and $1 = 2\dot{z}z$, so that $y = (\dot{t}/\dot{z} =) \, 4\dot{s}sz$. With $\sqrt{[ax-x^2]}$ and then z^2 replacing $\dot{s}$ and x respectively this will become $y = 4sz^2\sqrt{[a-z^2]}$, the equation to a mechanical curve.(422)

Example 3. In the same manner are found figures which bear an assumed relationship to any other given one. Thus, given the hyperbola $c^2 + x^2 = v^2$,

---

(419) $\phi \equiv xz - c^2$, $\phi_x = z$, $\phi_z = x$ and x is replaced by c^2/z.

(420) Newton divides by unity to preserve dimensional equivalence.

(421) Evidently so since $s = \int \sqrt{[ax-x^2]} \, . \, dx$ is the area of a segment of a circle of diameter a, and so not 'geometrical' (compare note (236)), while $y = v/(a+v)$ where

$$v^2 = a^{-2}(z-s)(a^2-z+s).$$

(422) For, much as in note (421),

$$s = \int_0^{z^2} \sqrt{[ax-x^2]} \, . \, dx = \tfrac{1}{2}z(z^2 - \tfrac{1}{2}a)\sqrt{[a-z^2]} - \tfrac{1}{8}a^2\sin^{-1}(1-2z^2/a)$$

is not 'geometrical'. The component sections of Example 2 follow in an immediate way by setting, in the notation of note (415), $f \equiv cx + s - t$, $\phi \equiv ax - z^2$; $f \equiv s - \frac{2}{3}a^{-1}v^3 - t$, $\phi \equiv x - z$; and $f \equiv s^2 - t$, $\phi \equiv x - z^2$ respectively.

si assumas $s=t$ et $xx=cz$ elicies per Prob 1 $p=q$ et $2x=cr$ et inde $y\left(=\frac{q}{r}\right)=\frac{cp}{2x}$, et substitutis $\sqrt{cc+xx}$ pro p(423) et $c^{\frac{1}{2}}z^{\frac{1}{2}}$ pro x, proveniet $y=\frac{c}{[2]z}\sqrt{cz+zz}$.

Atq; ita si assumas $xv-s=t$, et $xx=cz$, elicies $v+lx-p=q$, et $2x=cr$. Est autem $v=p$ et inde $lx=q$.(424) Quare $y\left(=\frac{q}{r}\right)=\frac{cl}{2}$. Jam vero $cc+xx=vv$ ope Prob 1 dat $x=lv$. Adeóq; est $y=\frac{cx}{2v}$ et substitutis $\sqrt{cc+xx}$ pro v et $c^{\frac{1}{2}}z^{\frac{1}{2}}$ pro x, fit $y=\frac{cz}{2\sqrt{cz+zz}}$.(425)

Exempl 4. Adhæc si detur Cissoides $\frac{xx}{\sqrt{ax-xx}}=v$(426) ad quam relatæ aliæ figuræ sunt inveniendæ, et ea de causa assumatur $\frac{x}{3}\sqrt{ax-xx}+\frac{2}{3}s=t$, finge $\frac{x}{3}\sqrt{ax-xx}=h$ ejusq; fluxionem k et erit $h+\frac{2}{3}s=t$ et inde per Prob 1 $k+\frac{2}{3}p=q$. Aequatio autem $\frac{ax^3-x^4}{9}=hh$ per Prob 1 dat $\frac{3axx-4x^3}{9}=2kh$ ubi si extermines h fiet $k=\frac{3ax-4xx}{6\sqrt{ax-xx}}$. Quare cùm præterea sit $\frac{2}{3}p=(\frac{2}{3}v=)\frac{4xx}{6\sqrt{ax-xx}}$ erit

‖[74] $\frac{ax}{2\sqrt{ax-xx}}=q$. Porro ad determinandum ‖ z et r assumatur $\sqrt{aa-ax}=z$ et ope Prob 1 emerget $-a=2rz$. Quare est

$$y\left(=\frac{q}{r}=\frac{-zx}{\sqrt{ax-xx}}=\sqrt{\frac{zzx}{a-x}}=\sqrt{ax}\right)=\sqrt{aa-zz}.^{(427)}$$

Quæ æquatio cùm sit ad circulum, habebitur relatio arearum circuli et Cissoidis.

Atq; ita si assumpsisses $\frac{2x}{3}\sqrt{ax-xx}+\frac{1}{3}s=t$ et $x=z$ prodijsset $y=\sqrt{az-zz}$ æquatio denuò ad circulum.(428)

(423) Since p (or $\dot{s}$) $=v$ where m (or $\dot{x}$) $=1$.

(424) Obviously so since $t=xv-s$ (or $\int v.dx) = \int x.dv$.

(425) Here, in the terminology of note (415), where the base hyperbola is defined by $v=\sqrt{[x^2+c^2]}$, the two component sections follow much as before by positing $f\equiv s-t$, $\phi\equiv x^2-cz$; and $f\equiv -xv+s-t$, $\phi\equiv x^2-cz$ respectively.

(426) Strictly, this is the pair of the cissoid $x^3=v^2(a-x)$ and the line $x=0$.

(427) The last three terms should be preceded by a minus sign.

(428) Here, of course, there is more than a little contrivance on Newton's part. Since the derivative of $x^{\frac{3}{2}}(a-x)^{\frac{1}{2}}$ may be expressed in either of the forms

$$\tfrac{3}{2}x^{\frac{1}{2}}(a-x)^{\frac{1}{2}}-\tfrac{1}{2}x^{\frac{3}{2}}(a-x)^{-\frac{1}{2}}=\tfrac{3}{2}ax^{\frac{1}{2}}(a-x)^{-\frac{1}{2}}-2x^{\frac{3}{2}}(a-x)^{-\frac{1}{2}},$$

the integral $s=\int x^{\frac{3}{2}}(a-x)^{-\frac{1}{2}}.dx$ may be evaluated in either of two simple ways. First, on taking

should you assume $s = t$ and $x^2 = cz$, you will elicit by Problem 1 that $\dot{s} = \dot{t}$ and $2x = c\dot{z}$, and thence $y = (\dot{t}/\dot{z} =)\ c\dot{s}/2x$: with $\sqrt{[c^2+x^2]}$ substituted in place of $\dot{s}$(423) and then $c^{\frac{1}{2}}z^{\frac{1}{2}}$ in place of x there will come out $y = c\sqrt{[cz+z^2]}/2z$.

And thus if you assume $xv - s = t$ and $x^2 = cz$, you will elicit $v + \dot{v}x - \dot{s} = \dot{t}$ and $2x = c\dot{z}$. However $v = \dot{s}$ and thence $\dot{v}x = \dot{t}$;(424) therefore $y = (\dot{t}/\dot{z} =)\ \frac{1}{2}c\dot{v}$. But now $c^2+x^2 = v^2$ yields, with the help of Problem 1, $x = \dot{v}v$, so that $y = cx/2v$, and this, with $\sqrt{[c^2+x^2]}$ substituted for v and then $c^{\frac{1}{2}}z^{\frac{1}{2}}$ replacing x, becomes

$$y = cz/2\sqrt{[cz+z^2]}.^{(425)}$$

EXAMPLE 4. Moreover, if the cissoid $x^2/\sqrt{[ax-x^2]} = v$(426) be given and other figures related to it are to be found, assuming for the purpose that

$$\tfrac{1}{3}x\sqrt{[ax-x^2]} + \tfrac{2}{3}s = t,$$

suppose $\frac{1}{3}x\sqrt{[ax-x^2]} = h$ and its fluxion to be $\dot{h}$, and there will be $h+\frac{2}{3}s = t$ and thence, by Problem 1, $\dot{h}+\frac{2}{3}\dot{s} = \dot{t}$. The equation $\frac{1}{9}(ax^3-x^4) = h^2$ gives, by Problem 1, $\frac{1}{9}(3ax^2-4x^3) = 2\dot{h}h$, and if you here eliminate h it will become $\dot{h} = (3ax-4x^2)/6\sqrt{[ax-x^2]}$. Furthermore, since $\frac{2}{3}\dot{s} = (\frac{2}{3}v =)\ 4x^2/6\sqrt{[ax-x^2]}$, there will consequently be $ax/2\sqrt{[ax-x^2]} = \dot{t}$. Moreover, to determine z and $\dot{z}$ assume $\sqrt{[a^2-ax]} = z$ and with the help of Problem 1 there will emerge $-a = 2\dot{z}z$: consequently

$$y(= \dot{t}/\dot{z} = -zx/\sqrt{[ax-x^2]} = \sqrt{[z^2x/(a-x)]} = \sqrt{[ax]}) = \sqrt{[a^2-z^2]}.^{(427)}$$

Since this equation is that of a circle, the relationship between the areas of a circle and a cissoid will be had.

And thus if you had assumed $\frac{2}{3}x\sqrt{[ax-x^2]}+\frac{1}{3}s = t$ and $x = z$, there would have been produced $y = \sqrt{[az-z^2]}$, the equation again of a circle.(428)

$h = \frac{1}{3}x^{\frac{3}{2}}(a-x)^{\frac{1}{2}}$, we may set $s = \frac{3}{2}(t-h)$ provided yr (or $y\dot{z}$) $= \frac{1}{2}ax^{\frac{1}{2}}(a-x)^{-\frac{1}{2}}$: here Newton ordains that $y^2+z^2 = a^2$, whence $z^2 = a(a-x)$. Alternatively (and more easily) we may set $s = 3(t-2h)$ provided yr (or $y\dot{z}$) $= x^{\frac{1}{2}}(a-x)^{\frac{1}{2}}$, from which at once $z = x$ on taking $y^2 = z(a-z)$. The latter quadrature of the cissoid's area is virtually that first derived by Huygens in the early spring of 1658, and by him communicated (without proof) to Sluse and Wallis later the same year; compare his *Œuvres complètes*, **2** (The Hague, 1889): 164, 170, 178. Wallis replied on 1 January 1659 (Huygens' *Œuvres*, **2**: 296), communicating a proof of the quadrature by means of the indivisibles techniques he had developed in his *Arithmetica Infinitorum* (Oxford, 1656). The same result was achieved simultaneously by Pierre de Fermat in a 'De Cissoide Fragmentum' published only in 1891 (*Œuvres de Fermat*, **1** (Paris, 1891): 285–8). Compare Josepha and J. E. Hofman, 'Die erste Quadratur der Kissoide', *Deutsche Mathematik*, **5** (1940–1): 571–84. Wallis' letter to Huygens, published by him within a few weeks as a 'Tractatus...Epistolaris, In qua agitur de Cissoide et Corporibus inde Genitis...' (*Tractatus Duo* (note (255)): 75–123, especially 81.90), was familiar to Newton from his study of Wallis' works in the winter of 1664/5. Even fresher in his mind, however, would have been Wallis' improved deduction of the cissoid's area as presented in Proposition XXIX of the latter's *Mechanicorum, sive Tractatus de Motu Corporum Pars Secunda* (London, 1670). (Newton thanked Collins on 11 July 1670 for sending him this second volume of 'Dr Wallis his Mechanicks', now Trinity College. NQ. 16.149; see his *Correspondence*, **1** (Cambridge, 1959): 16. It is un-

Haud secus si detur curva aliqua Mechanica, possunt aliæ ad eam relatæ curvæ Mechanicæ inveniri, sed ad eliciendum Geometricas convenit ut e rectis ab invicem Geometricè dependentibus aliqua(429) pro Basi adhibeatur, et ut area ad parallelogrammum complementalis quæratur(430) supponendo fluxionem ejus valere Basin ductam in fluxionem ordinatim applicatæ.

Fig [pag 57] EXEMPL 5. Sic Trochoide ADF propositâ, refero ad Basin AB et completo parallelogrammo $ABDG$ quæro complementalem superficiem ADG concipiendo descriptam esse per motum rectæ GD, et proinde fluxionem ejus valere illam GB(431) in celeritatem progrediendi ductam, hoc est $x \times l$. Jam cùm AL sit parallela tangenti DT, erit AB ad BL ut fluxio ejusdem AB ad fluxionem applicatæ BD hoc est ut 1 ad l. Quare est $l = \frac{BL}{AB}$, adeóque $x \times l = BL$, et proinde area ADG describitur fluxione BL; Atque adeo cùm area circularis ALB eadem fluxione describatur æquales erunt.(432)

Fig(433) Pari ratione si concipias ADF esse figuram arcuum sive sinuum versorum,(434) hoc est cujus applicata BD æquatur arcui AL: cùm fluxio arcus AL sit ad fluxionem Basis AB ut PL ad BL, hoc est $l.1 :: \frac{1}{2}a.\sqrt{ax-xx}$ erit $l = \frac{a}{2\sqrt{ax-xx}}$ Adeoque $l \times x$ fluxio areæ ADG erit $\frac{ax}{2\sqrt{ax-xx}}$. Quare si ad ipsius AB punctum B recta

---

likely that at the time he drafted the present Problem 8 he had yet seen the 'Demonstratio D. Hugenii' first published by Wallis in late 1671 in the third, concluding part of his 'Mechanicks' (*Mechanicorum...Pars Tertia* (London, 1671): Cap. xv, Prop. II).) The elegant geometrical derivation of Huygens' quadrature given by James Gregory in his *Exercitationes Geometricæ* (London, 1688: 23–4) was also known to him in 1670.

(429) Newton first wrote 'recta aliqua Geometricè noscibilis' (some geometrically knowable straight line).

(430) In Newton's analytical terms $vx - s$ (or $\int v.dx$), that is, $\int x.dv$; compare note (424). The device, well known to his contemporaries, is the moden technique of integration 'by parts'.

(431) Read 'GD'.

(432) Much as before (on his page 57 above) Newton defines the cycloid $\widehat{ADF}$ by $AB = x$, $AE = a$, $BL = \sqrt{[x(a-x)]}$ ($AL = \frac{1}{2}a\cos^{-1}[1-2x/a]$) and $BD = v = BL + \widehat{AL}$, so that

$$l \text{ (or } \dot{v}/\dot{x}) = BL/x.$$

At once area $AGD = \int x.dv = \int BL.dx =$ semi-circle segment (ALB). Newton here generalizes a result due to Torricelli that the area contained between the cycloidal arc $\widehat{ADF}$, the tangent at A and the tangent at F is equal to (ALE), half the generating circle. (See Torricelli's 'Appendix de Dimensione Spatij Cycloidalis' to his 'De Dimensione Parabolæ' in his *Opera Geometrica* (Florence, 1644): $_1$85–90.)

(433) The reader is apparently again referred to the figure of the cycloid on Newton's page 57 though clearly a new diagram is required.

(434) Namely, the sinusoid $2x/a = 1 - \cos v$, where (in the figure we have inserted in our English version) $AB = x$, $BD = v$ and $AE = a$. Note that P is the centre of the semicircle ALE and that $EF = \widehat{ALE} = \frac{1}{2}\pi a$. This sine-curve was first described by Roberval in the mid-1630's as the 'compagne de la Roulette' (Trochoidis Comes) and it was he who, in Proposition 4 of his tract 'De Trochoide ejusque Spatio' (*Divers Ouvrages* (note (307))): 361–3, published only

No differently, if some mechanical curve be given, other mechanical curves related to it may be found. But to elicit geometrical ones it is convenient to employ one or other of the straight lines dependent geometrically upon each other[(429)] as the base, and then to seek the complement of the area to the rectangle [of base and ordinate][(430)] by supposing the value of its fluxion to be the base multiplied into the fluxion of the ordinate.

EXAMPLE 5. Thus should the cycloid ADF be proposed, I relate it to the base AB and, having completed the rectangle $ABDG$, seek the complementary surface ADG by conceiving it to be described by the motion of the straight line GD, and consequently that the value of its fluxion is that line $G[D]$ multiplied into the speed of its onward motion, that is, $x\dot{v}$. Now, since AL is parallel to the tangent DT, AB will be to BL as the former's fluxion to that of the ordinate BD, that is, as 1 to $\dot{v}$. Hence $\dot{v} = BL/AB$, so that $x\dot{v} = BL$ and consequently the area ADG is described at the rate of flow BL. Accordingly since the circle area ALB is described at the same rate of flow, these will be equal.[(432)]

Figure on page 57.

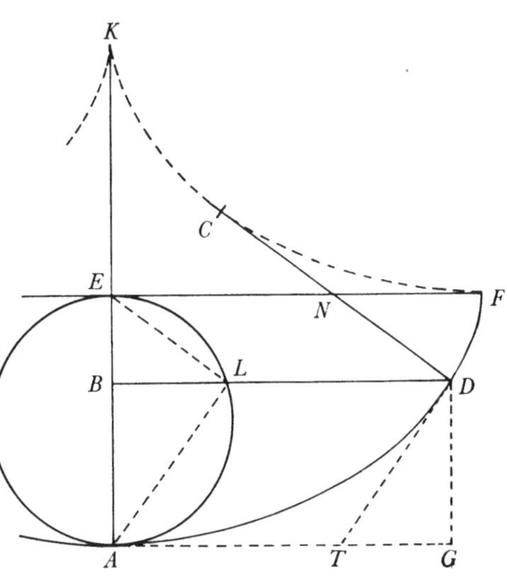

By like reasoning, if you conceive ADF to be the 'figure of arcs' or versed-sine curve[(434)] (whose ordinate BD, namely, is equal to the [circle] arc $\widehat{AL}$), the fluxion of the arc $\widehat{AL}$ is to that of the base AB as $PL:BL$, that is, $\dot{v}:1 = \frac{1}{2}a:\sqrt{[ax-x^2]}$; hence it will be $\dot{v} = \frac{1}{2}a/\sqrt{[ax-x^2]}$, so that $\dot{v}x$ (the fluxion of the area ADG) will be $\frac{1}{2}ax/\sqrt{[ax-x^2]}$. In consequence, if at the point B of AB a straight line equal to $\frac{1}{2}ax\sqrt{[ax-x^2]}$

Figure[(433)]

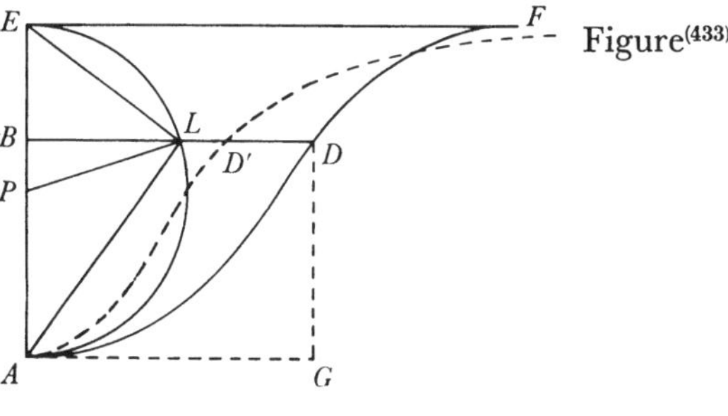

in 1693), first determined its quadrature. Independently of Roberval, John Wallis in 1659 derived the sinusoidal area $ADFE$ by 'distorting' (in Newton's preceding cycloidal figure) the 'trilineum' $ADFEL$ parallel to EF such that the semicircle $\widehat{ALE}$ collapses into its diameter ABE (so that points L and B coincide), 'estque hæc [Ellipsis expansa], ni fallor, ea curva quam *Cycloidis Comitem* vel *Sociam* (*la Compagne de la Roulette*) appellat [Pascal]'. (See his *Tractatus Duo* (note (255)): 27, 99–100. Wallis there derives the sinusoid more directly as the development, his 'Ellipsis expansa', of a plane 45° cut of a right cylinder of diameter AE.) However, Newton's present terminology of 'figura sinuum versorum' is evidently taken from Wallis' revised account of the curve in Proposition XVII of his *Mechanicorum, sive de Motu Corporum Pars Secunda* (note (428)).

æqualis $\frac{ax}{2\sqrt{ax-xx}}$ in angulo recto applicari concipiatur, illa ad curvam quandam Geometricam[(435)] terminabitur cujus area Basi AB adjacens æquatur areæ ADG.[(436)]

Et sic alijs figuris per arcuum circuli, Hyperbolæ vel cujusvis Curvæ ad arcuum istorum sinus rectos vel versos[(437)] aut alias quasvis geometricè determinabiles rectas lineas in datis angulis applicationem constitutis, æquales
‖[75] Geometricæ ‖ figuræ inveniri possunt.

Fig[(438)] [pag 45b] Circa Spiralium areas levissimum est negotium. Utpote centro convolutionis[(439)] A radio quovis AG descripto arcu DG occurrente AF in G et spirali in D; cùm arcus ille ad instar lineæ super Basi AG incedentis describat Spiralis Aream $AHDG$, ita ut ejus areæ fluxio sit ad fluxionem rectanguli $1\times AG$, ut arc[us] GD ad 1; si rectam GL arcui isti æqualem erigas illa similiter incedendo super eadem AG describet aream ALG æqualem areæ Spiralis $AHDG$, curva ElL[(440)] existente Geometricâ.[(441)] Et præterea si subtensa AL ducatur, erit triang $ALG(=\frac{1}{2}AG\times GL=\frac{1}{2}AG\times GD)=$sectori AGD, adeoq complementalia segmenta ALl et ADH erunt etiam æqualia.[(442)] Et hæc non tantum Spirali Archimedeæ (ubi AlL evadit Parabola Apolloniana)[(443)] sed et alijs quibuscunq

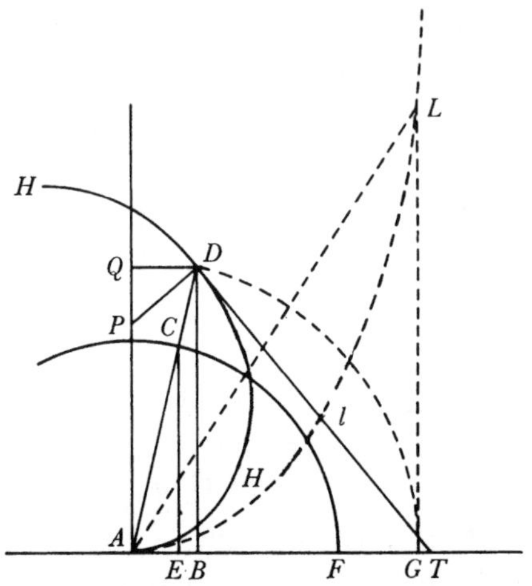

(435) We have inserted this in our figure as the broken-line curve AD', where

$$BD'(=u,\text{ say})=\tfrac{1}{2}ax^{\frac{1}{2}}(a-x)^{-\frac{1}{2}}.$$

To be precise, $u=\frac{1}{2}ax/\sqrt{[x(a-x)]}$ is the pair of the line $x=0$ and the conchoidal elliptical hyperbolism $4u^2(a-x)=a^2x$.

(436) In Newton's terms, since

$$v=\cos^{-1}(1-2x/a),\quad l\ (\text{or } \dot{v}/\dot{x})=\tfrac{1}{2}a/BL\quad (\text{where } BL^2=ax-x^2)$$

so that $(AGD)=\int x.dv=\int u.dx=(AD'B)$ on taking $BD'=u=\frac{1}{2}ax^{\frac{1}{2}}(a-x)^{-\frac{1}{2}}$. He might well have added, in equivalent geometrical form at least, the corollary that

$$\int u.dx=\tfrac{1}{2}a(v-\sqrt{[ax-x^2]});$$

that is, on taking $A\hat{P}L=\theta$ $(x=\frac{1}{2}a(1-\cos\theta),\ v=\frac{1}{2}a\theta)$, $\int\frac{1}{4}a^2(1-\cos\theta).d\theta=\frac{1}{4}a^2(\theta-\sin\theta)$.

(437) In the case of circle arcs these yield variants of the preceding sinusoid.

(438) For ease of consultation we here repeat Newton's figure. As before, note that he has drawn the spiral AHD through the pole A in 'Archimedean' style: in general, the spiral need never (even after an infinite number of revolutions) pass through the pole.

(439) The terminology of 'convoluting' a 'parabola' into a 'spiral' of equal area is evidently borrowed from Wallis' *Tractatus Duo* (note (255)): 105–8, though the latter there uses 'convolutio' to denote a Gregorian *length*-preserving 'involution'. Cavalieri in 1635 used 'commutatio' to describe the present area-preserving transformation. Compare note (443).

is conceived to be applied at right angles, it will terminate at a certain geometrical curve,(435) the area of which, bordering on the base AB, is equal to the area ADG.(436)

And, in this way, to other figures formed by applying the arcs of a circle, hyperbola or any curve at given angles to the sines or versed sines of those arcs(437) or to any other geometrically determinable straight lines, geometrical figures may be found equal.

Regarding spiral areas the bother is exceedingly slight. In point, where the arc $\widehat{DG}$ described on the convolution centre(439) A with any radius AG meets AF in G and the spiral in D, since that arc describes the spiral area $AHDG$ in the fashion of a line advancing upon the base AG, the fluxion of its area will accordingly be to the fluxion of the rectangle $1\times AG$ as the arc $\widehat{GD}$ to 1. If now you erect the straight line GL equal to that arc, this as it advances in a similar way upon the same base AG will describe the area ALG equal to the spiral area $AHDG$, the curve ElL(440) proving to be geometrical.(441) Furthermore, if the subtense AL be drawn, then triangle

Figure(438) [on page 45b.]

$$ALG(=\tfrac{1}{2}AG\times GL=\tfrac{1}{2}AG\times \widehat{GD}) = \text{sector } AGD,$$

so that the complementary segments ALl and ADH will also be equal.(442) And this procedure obtains not only for the Archimedean spiral (in which case ALl comes out to be an Apollonian parabola)(443) but to any others whatsoever, so

(440) Read 'AlL'. Newton confuses the pole A with the foot of the perpendicular from C onto AB.

(441) When, that is, the defining relation between the polar coordinates AC and DG is likewise algebraic.

(442) Where, more generally, AG meets the spiral in a point A' distinct from the pole A, and Gg is an indefinitely small increment of it, since universally $GL = \widehat{GD}$ it follows readily that the corresponding areal increments $(GLlg)$ and $(GDdg)$ are equal: hence, $(A'GL) = (A'GDH)$ and therefore $(ALA') = (ADHA')$. It will be evident that the 'spiral' $D'A'D$ and 'parabola' $L'A'L$ are tangent at A'. Note that $l\lambda(= gl-GL$, that is,

$$gd-GD = d\delta+Gg\times \widehat{GD}/AG)$$

is not equal to $d\delta$, so that the 'differential' triangles $Ll\lambda$, $Dd\delta$ are not congruent: a point not obvious to such amateur mathematicians as Thomas Hobbes, who in the early 1640's in Roberval's presence confused this area-preserving transformation with the nearly similar one which preserves the arc-length $\widehat{A'D} = \widehat{A'L}$ (see note (704) below).

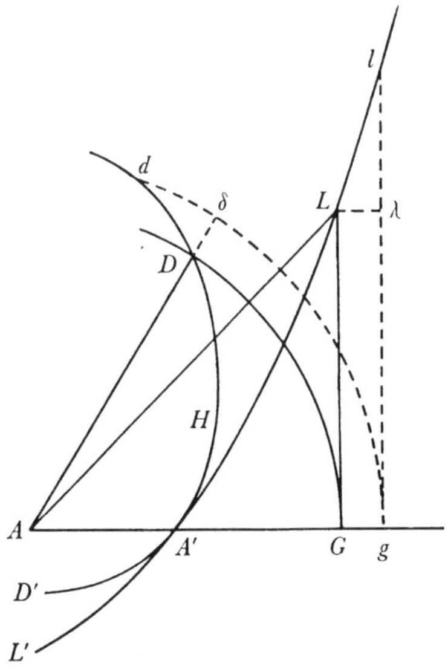

(443) In the Archimedean spiral

$$(AG =)\ AD = k\times(\widehat{GD}/AG) \quad \text{and so} \quad AG^2 = k\times\widehat{GD} = k\times GL.$$

Accordingly (since here A' coincides with the pole A), the curve AL will be a conic parabola

conveniunt, adeo ut omnes eodem negotio in æquales Geometricas converti possint.(444)

Possem plura hujus construendi Problematis specimina(445) afferre, sed hæc sufficiant(446) cùm sint adeò generalia ut quicquid hactenus circa curvarum areas inventum fuerit, vel ni fallor inveniri possit, aliquo saltem modo complectantur, et utplurimùm leviori curâ sine solitis ambagibus determinent.

Præcipuus autem hujus & præcedentis Problematis usus est, ut assumptis conicis sectionibus vel quibuslibet notæ magnitudinis curvis, aliæ curvæ quæ cum his conferri possunt, investigentur, et earum definientes æquationes in Catalogum ordinatim disponantur. Et constructo ejusmodi Catalogo, cùm curvæ alicujus area quæritur, si æquatio ejus definiens vel immediatè in Catalogo reperiatur, vel in aliam quam Catalogus complectitur transformari potest, exinde cognosces aream ejus. Quinetiam Catalogus ille determinandis Curvarum longitudinibus, centris gravitatum, solidis per convolutionem generatis, solidorum superficiebus, et cuilibet fluenti quantitati per analogam fluxionem generatæ, inservire potest.(447) Ast quomodo formandus sit et utendus in sequente Problemate patebit ubi duplicem exhibuimus.(448)

---

of parameter k. As one would expect, this canonical application of the 'convolution' transformation was the first to suggest itself historically, and is indeed latent in Propositions 23–5 of Archimedes' tract *On Spirals* (compare E. J. Dijksterhuis, *Archimedes* (Copenhagen, 1956): 274–80). The first published comparison of an Apollonian parabola with its 'convoluted' Archimedean spiral would appear to be due to Cavalieri, who in 'Liber VI, In quo de Spatiis Helicis, et Solidis inde genitis...Speculatio instituitur' of his *Geometria Indivisibilibus Continuorum nova quadam Ratione promota* (Bologna, 1635) systematically expounded the niceties of the correspondence in some seventy pages and also hinted, in a concluding *Corollarium Generale*, how the 'commutatio' might be extended to pairs of higher-order curves. A decade later Torricelli introduced the spiral convolute of the parabola (in a slightly variant form) in Proposition XVII of his 'De Dimensione Parabolæ...: In quo quadratura parabolæ XX. modis absolvitur, partim Geometricis, Mecanisque; partim ex indivisibilium Geometria deductis rationibus' (*Opera Geometrica* (Florence, 1644): $_2$1–115, especially 73–4). In 1660 James Gregory's teacher, Stefano degli Angeli, was encouraged to generalize 'analogia illa, quam Phœnix ingeniorum Torricellius consideravit inter spiralem linearem & parabolam quadraticam' in Proposition VI of his *De Infinitarum Spiralium Spatiorum Mensura, Opusculum Geometricum* (Venice, 1660): 15–16, applying it to a comparison of the higher order parabolas $LG^p = k \times AG^q$ and their 'convolute' corresponding spirals $(\widehat{DG}/AD)^p = k \times AD^{q-p}$. (The preface, signatures a4^v–b3^v, of Angeli's work contains a potted history of 'infinite' spirals up to 1660. See also Angeli's *De Infinitis Parabolis Liber Quintus* (Venice, 1663): 90–101.)

(444) Compare Problem 20 of the first section (also 1: note (11)). In analytical terms, on denoting $AA' = R$, $AD = r$, $A'\widehat{A}D = \theta$, $A'G = x$ and $GL = y$ the 'convolution' relates the 'mechanical' spiral $f(r, \theta) = 0$ and the 'geometrical' parabola $\phi(x, y) = 0$ by $r = x+R$, $\theta = y/(x+R)$: evidently the transformation preserves area since

$$\int_0^x y.dx = \left(\int_R^r r\theta.dr = \right) \tfrac{1}{2}r^2\theta - \int_0^\theta \tfrac{1}{2}r^2.d\theta$$

and, correspondingly,
$$\tfrac{1}{2}y(x+R) - \int_0^x y.dx = \int_0^\theta \tfrac{1}{2}r^2.d\theta.$$

much so that all may be converted into equivalent geometrical curves with the same [minimum of] trouble.(444)

I might have presented further examples of the construction of this problem,(445) but let these suffice(446) since they are so general as to embrace—in some way at least—whatever has till now been found out regarding the areas of curves or indeed, unless I am wrong, can be so found, and for the most part determine them with less trouble and apart from the usual circuitous deviations.

However, when conics or any other curves of known magnitude are assumed, this problem and that preceding find their outstanding use in searching out other curves which may be compared with them and in arranging their defining equations methodically in a catalogue. And when—a catalogue of this sort having been constructed—the area of some curve is required, if its defining equation shall either be located directly in the catalogue or be transformable into another which is encompassed in the catalogue, you will ascertain its area therefrom. Indeed, that catalogue can serve to determine the lengths of curves, their centres of gravity, the solids generated by their revolution and the surfaces of those solids, and any fluent quantity generated by analogous fluxion.(447) But how this is to be formed and used will be evident in the following problem where we have displayed both.(448)

(445) That is, examples of area-preserving transformations in other coordinate systems than the (perpendicular) Cartesian and polar ones previously considered.

(446) The less definite variant 'hæc sufficere possunt' (these can suffice) is cancelled.

(447) Newton wrote to Collins to the same effect on 10 December 1672, asserting of his 'Generall Method' that it 'extends it selfe wthout any troublesome calculation...to the resolving other abstruser kinds of Problems about the crookedness, areas, lengths, centers of gravity of curves &c' (*Correspondence*, **1** (1959): 247).

(448) This final sentence, reproduced neither by Horsley in his Latin edition (*Opera Omnia* (note (60)), **1**: 470) nor equivalently by Colson in his 1736 English version (*Method of Fluxions* (note (79)): 86), was added inconspicuously in a bottom margin of the manuscript when a first draft of the following 'Catalogus Curvarum aliquot ad rectilineas figuras relatarum ope Prob 7 constructa' (introduced, much as later, by the words 'In hujus rei illustrationem accipe impræsentia sequentem curvarum aliquot simpliciorum Catalogum, ubi...') was entered by Newton on his pages 75–6. This catalogue of curves, comprehending (with some minor variants) the 'Ordo primus' and first three entries in 'Ordo tertius' and 'Ordo quartus' of the first table on Newton's page 77 below, was composed in the first instance as a continuation of the present Problem 8 but it was at once revised, with extensive additional material, on pages 77–9. Subsequently, after Newton had entered his 'Ordo decimus' on page 77, he saw the advantage of appending a second catalogue of curves whose areas may be related to those of conics (that is, evaluated in terms of circular and hyperbolic functions) and decided to expound the comparison of curve areas in a separate Problem 9. The supplementary pages here marked 75a–75l (ULC. Add. 3960.14: 83–93) were then inserted by Newton after his page 75, the abrupt transition to page 77 being smoothed by a short preliminary passage on page 75l. The diagrams which illustrate the cancelled text on Newton's page 76 are set with the revised version on page 77.

||[75a] ||PROB: 9.

PROPOSITÆ ALICUJUS CURVÆ AREAM DETERMINARE.(449)

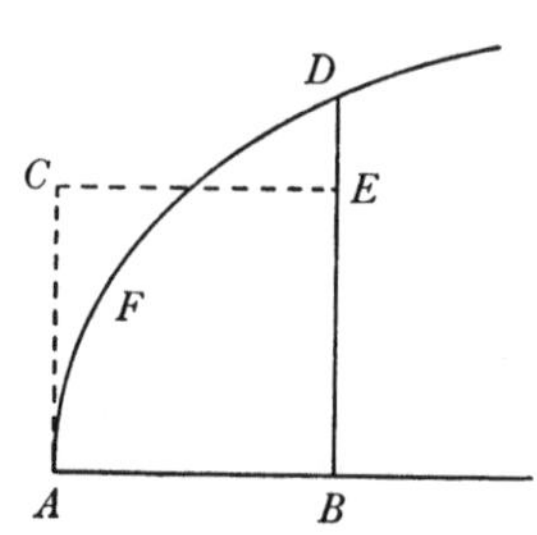

Problematis resolutio in eo fundatur ut quantitatum fluentium relatio ex relatione fluxionum (per Prob: 2) eliciatur. Et imprimis si recta BD cujus motu quæsita area $AFDB$ describitur, super basi AB positione datâ erectè incedat, concipe ut supra parallelogrammum $ABEC$ a parte ejus BE unitatem æquante interea describi. Et posita BE fluxione parallelogrammi erit BD fluxio areæ quæsitæ.(450)

Dic ergo $AB=x$, et erit etiam $ABEC(=1\times x)=x$ et $BE=m$[:] dic insuper aream $AFDB=z$, et erit $BD=r$ ut et $=\frac{r}{m}$, eo quod sit $m=1$. Et proin(451) per æquationem definientem BD simul definitur fluxionum ratio $\frac{r}{m}$, et exinde per Prob 2, Cas 1, elicietur relatio fluentium quantitatum x et z.(452)

EXEMPLA 1MA.(453) *Ubi BD sive r valet simplicem aliquam quantitatem.*

Detur $\frac{xx}{a}=r$ vel $=\frac{r}{m}$ æquatio nempe ad Parabolam, et (per Prob 2) emerget $\frac{x^3}{3a}=z$. Est ergo $\frac{x^3}{3a}$ sive $\frac{1}{3}AB\times BD=$ areæ Parabolicæ $AFDB$.

Detur $\frac{x^3}{aa}=r$ æquatio ad Parabolam secundi generis et (per Prob 2) emerget $\frac{x^4}{4aa}=z$, hoc est $\frac{1}{4}AB\times BD=$ areæ $AFDB$.

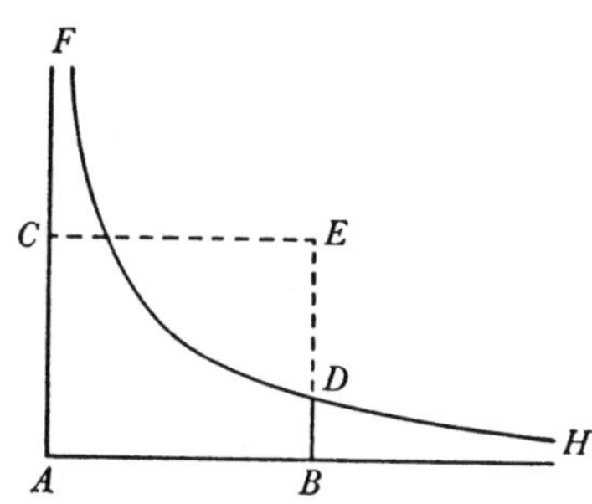

Detur $\frac{a^3}{xx}=r$ sive $a^3x^{-2}=r$ æquatio ad Hyperbolam secundi generis, et emerget $-a^3x^{-1}=z$ sive $-\frac{a^3}{x}=z$: hoc est $AB\times BD=$ areæ infinite longæ $HDBH$ ex altera parte applicatæ BD jacentis, ut innuit valor negativus.(454)

Atq; ita si detur $\frac{a^4}{x^3}=r$, emerget $-\frac{a^4}{2xx}=z$.

||[75b] || Præterea sit $ax=rr$, æquatio iterum ad Parabolam, et(455) proveniet $\frac{2}{3}a^{\frac{1}{2}}x^{\frac{3}{2}}=z$, hoc est $\frac{2}{3}AB\times BD=$ areæ $AFDB$.

Sit $\frac{a^3}{x}=rr$, et fiet $2a^{\frac{3}{2}}x^{\frac{1}{2}}=z$, sive $2AB\times BD=AFDB$.

(449) First drafts of portions of pages 75a, 75b, 75j, 75k and 75l exist on a loose preliminary worksheet (ULC. Add. 3960.3: 29–32). Where these are significantly variant from the revised version here reproduced, mention will be made in an appropriate footnote.

Problem 9

To determine the area of any proposed curve[449]

The resolution of the problem is based on that of establishing the relationship between fluent quantities from one between their fluxions (by Problem 2). And, in the first place, if the straight line *BD*, by whose motion the area *AFDB* sought is described, advance erect on the base *AB* given in position, conceive (as above) that the rectangle *ABEC* is described by that part of it, *BE*, equal to unity: on taking *BE* to be the fluxion of the rectangle, *BD* will be the fluxion of the area sought.[450]

Call $AB = x$, therefore, and there will be also $ABEC(= 1\times x) = x$ and $BE = \dot{x}$: further, call the area $AFDB = z$ and then $BD = \dot{z}$ or, equivalently, $\dot{z}/\dot{x}$, since $\dot{x} = 1$. Consequently,[451] by the equation defining *BD* is at once defined the fluxional ratio $\dot{z}/\dot{x}$, and from this (by Problem 2, Case 1) will be elicited the relationship of the fluent quantities x and z.[452]

First examples:[453] *when BD (that is, $\dot{z}$) is, in value, some simple quantity.*

Let there be given $x^2/a = \dot{z}$ or $\dot{z}/\dot{x}$, namely, the equation to a parabola, and there will (by Problem 2) emerge $x^3/3a = z$. Therefore $x^3/3a$ or $\frac{1}{3}AB\times BD$ is equal to the parabolic area *AFDB*.

Let there be given $x^3/a^2 = \dot{z}$, the equation to a second-order parabola, and there will (by Problem 2) emerge $x^4/4a^2 = z$, that is,

$$\tfrac{1}{4}AB\times BD = \text{area } AFDB.$$

Let there be given $a^3/x^2 = \dot{z}$ or $a^3x^{-2} = \dot{z}$, the equation to a second-order hyperbola, and there will emerge $-a^3x^{-1} = z$ or $-a^3/x = z$: that is, $AB\times BD$ equals the infinitely extended area *HDBH* lying on the further side of the ordinate *BD*, as its negative value conveys.[454]

And thus, should there be given $a^4/x^3 = \dot{z}$ there will emerge $-a^4/2x^2 = z$.

Furthermore, let $ax = \dot{z}^2$, the equation again to a parabola, and[455] there will come out $\frac{2}{3}a^{\frac{1}{2}}x^{\frac{3}{2}} = z$, that is, $\frac{2}{3}AB\times BD = \text{area } AFDB$.

Let $a^3/x = \dot{z}^2$ and there will come $2a^{\frac{3}{2}}x^{\frac{1}{2}} = z$, or $2AB\times BD = AFDB$.

---

(450) Compare Newton's page 20 above. The manuscript at this point lacks any referent figure, and we have accordingly repeated the diagram on page 20, omitting point H and inserting a new point F.

(451) 'data *BD*' (given *BD*) is cancelled.

(452) That is, $\int BD\,.\,dx$.

(453) Compare 'Reg: I' of the *De Analysi* (II: 206–8).

(454) As in the preceding parabolic examples Newton implicitly assumes that the area $z = \int r$ (or $\dot{z}$) $.\,dx$ has the lower bound $x = 0$. As before, the manuscript lacks a clarifying diagram (compare note (450)): that reproduced here is our restoration.

(455) 'ope Pr[ob 2]' (with the aid of Problem 2) is cancelled.

Sit $\frac{a^5}{x^3}=rr$, et fiet $-\frac{2a^{\frac{5}{2}}}{x^{\frac{1}{2}}}=z$, sive $2AB\times BD=HDBH$.
Sit $axx=r^3$, et fiet $\frac{3}{5}a^{\frac{1}{3}}x^{\frac{5}{3}}=z$, sive $\frac{3}{5}AB\times BD=AFDB$.
Et sic in alijs.(456)

EXEMPLA 2^{A}. *Ubi r valet plures ejusmodi connexas quantitates.*(457)

Sit $x+\frac{xx}{a}=r$, et fiet $\frac{xx}{2}+\frac{x^3}{3a}=z$.

Sit $a+\frac{a^3}{xx}=r$, et fiet $ax-\frac{a^3}{x}=z$.

Sit $3x^{\frac{1}{2}}-\frac{5}{xx}-\frac{2}{x^{\frac{1}{2}}}=r$ et fiet $2x^{\frac{3}{2}}+\frac{5}{x}-4x^{\frac{1}{2}}=z$.

EXEMP: 3. *Ubi prævia reductio per divisionem requiritur.*

Detur $\frac{aa}{b+x}=r$, æquatio ad Hyperbolam Apollonianam et factâ in infinitum divisione, evadet $r=\frac{aa}{b}-\frac{aax}{bb}+\frac{aaxx}{b^3}-\frac{aax^3}{b^4}$ &c. Et inde per Prob 2 (ut in secundis exemplis) obtinebitur $z=\frac{aax}{b}-\frac{aaxx}{2bb}+\frac{aax^3}{3b^3}-\frac{aax^4}{4b^4}$ &c.(458)

Detur $\frac{1}{1+xx}=r$ et per divisionem elicietur $r=1-xx+x^4-x^6$ &c vel etiam $r=\frac{1}{xx}-\frac{1}{x^4}+\frac{1}{x^6}$ &c. Indeq per Prop 2,(459) $z=x-\frac{1}{3}x^3+\frac{1}{5}x^5-\frac{1}{7}x^7$ &c $=AFDB$ vel $z=-\frac{1}{x}+\frac{1}{3x^3}-\frac{1}{5x^5}$ &c $=HDBH$.(460)

Detur $\frac{2x^{\frac{1}{2}}-x^{\frac{3}{2}}}{1+x^{\frac{1}{2}}-3x}=r$, et per divisionem evadet(461)

$$r=2x^{\frac{1}{2}}-2x+7x^{\frac{3}{2}}-13x^2+34x^{\frac{5}{2}} \quad \&c.$$

et inde per Prop 2,(459) $z=\frac{4}{3}x^{\frac{3}{2}}-xx+\frac{14}{5}x^{\frac{5}{2}}-\frac{13}{[3]}x^3+\frac{68}{7}x^{\frac{7}{2}}$ &c.

||[75c] ||EXEMPL: 4. *Ubi prævia reductio per extractionem radicum requiritur.*

Detur $r=\sqrt{aa+xx}$ æquatio nempe ad Hyperbolam et radice ad usq terminos infinitè multos extractâ, evadet $r=a+\frac{xx}{2a}-\frac{x^4}{8a^3}+\frac{x^6}{16a^5}-\frac{5x^8}{112a^7}$ &c Atq inde ut in præcedentibus $z=ax+\frac{x^3}{6a}-\frac{x^5}{40a^3}+\frac{x^7}{112a^5}-\frac{5x^9}{1008a^7}$ &c.(462)

Ad eundem modum si detur $r=\sqrt{aa-xx}$ æquatio scilicet ad circulum, obtinebitur $z=ax-\frac{x^3}{6a}-\frac{x^5}{40a^3}-\frac{x^7}{112a^5}-\frac{5x^9}{1008a^7}$ &c.

(456) Note that Newton here assumes the lower integration bound to be either $x = 0$ (parabolic cases) or $x = \infty$ (hyperbolic instances), whichever corresponds to $z = 0$.
(457) 'terminos' (terms) is cancelled.

Let $a^5/x^3 = \dot{z}^2$ and there will come $-2a^{\frac{5}{2}}x^{\frac{1}{2}} = z$, or $2AB \times BD = HDBH$.

Let $ax^2 = \dot{z}^3$ and there will come $\frac{3}{5}a^{\frac{1}{3}}x^{\frac{5}{3}} = z$, or $\frac{3}{5}AB \times BD = AFDB$.

And so in other instances.(456)

SECOND EXAMPLES: *when the value of $\dot{z}$ is compounded of several quantities of this kind.*(457)

Let $x + x^2/a = \dot{z}$ and there will become $\frac{1}{2}x^2 + \frac{1}{3}x^3/a = z$.

Let $a + a^3/x^2 = \dot{z}$ and there will become $ax - a^3/x = z$.

Let $3x^{\frac{1}{2}} - 5x^{-2} - 2x^{-\frac{1}{2}} = \dot{z}$ and there will become $2x^{\frac{3}{2}} + 5x^{-1} - 4x^{\frac{1}{2}} = z$.

THIRD EXAMPLES: *when prior reduction by division is required.*

Let there be given $a^2/(b+x) = \dot{z}$, the equation of an Apollonian hyperbola, and after division is made to infinity there will result

$$\dot{z} = a^2b^{-1} - a^2b^{-2}x + a^2b^{-3}x^2 - a^2x^{-4}b^3 \ldots.$$

From this by Problem 2 will be obtained (as in the second examples)

$$z = a^2b^{-1}x - \tfrac{1}{2}a^2b^{-2}x^2 + \tfrac{1}{3}a^2b^{-3}x^3 - \tfrac{1}{4}a^2b^{-4}x^4 \ldots.^{(458)}$$

Let there be given $1/(1+x^2) = \dot{z}$ and by division there will be elicited $\dot{z} = 1 - x^2 + x^4 - x^6 \ldots$, or alternatively $\dot{z} = x^{-2} - x^{-4} + x^{-6} \ldots$. And thence, by [Problem] 2, $z = x - \frac{1}{3}x^3 + \frac{1}{5}x^5 - \frac{1}{7}x^7 \ldots = AFDB$ or

$$z = -x^{-1} + \tfrac{1}{3}x^{-3} - \tfrac{1}{5}x^{-5} \ldots = HDBH.^{(460)}$$

Let there be given $(2x^{\frac{1}{2}} - x^{\frac{3}{2}})/(1 + x^{\frac{1}{2}} - 3x) = \dot{z}$, and by division there will result(460) $\dot{z} = 2x^{\frac{1}{2}} - 2x + 7x^{\frac{3}{2}} - 13x^2 + 34x^{\frac{5}{2}} \ldots$, and thence, by [Problem] 2,

$$z = \tfrac{4}{3}x^{\frac{3}{2}} - x^2 + \tfrac{14}{5}x^{\frac{5}{2}} - \tfrac{13}{3}x^3 + \tfrac{68}{7}x^{\frac{7}{2}} \ldots.$$

FOURTH EXAMPLES: *when prior reduction by root-extraction is required.*

Let there be given $\dot{z} = \sqrt{[a^2 + x^2]}$, namely, the equation to a Hyperbola, and after the root has been extracted to infinitely many terms there will result $\dot{z} = a + \frac{1}{2}x^2/a - \frac{1}{8}x^4/a^3 + \frac{1}{16}x^6/a^5 - \frac{5}{[128]}x^8/a^7 \ldots$, and thence, as in the preceding, $z = ax + \frac{1}{6}x^3/a - \frac{1}{40}x^5/a^3 + \frac{1}{112}x^7/a^5 - \frac{5}{[1152]}x^9/a^7 \ldots.$(462)

In the same way, if there be given $\dot{z} = \sqrt{[a^2 - x^2]}$, namely, the equation to a circle, there will be obtained $z = ax - \frac{1}{6}x^3/a - \frac{1}{40}x^5/a^3 - \frac{1}{112}x^7/a^5 - \frac{5}{[1152]}x^9/a^7 \ldots.$

---

(458) In exact terms $\int_0^x a^2/(b+x)\,.\,dx = a^2\log(1+x/b)$. (459) Read 'Prob 2'.

(460) $(AFDB) = \int_0^x (1+x^2)^{-1}.\,dx[= \tan^{-1}x]$, while

$$(DBH) = \int_x^\infty (1+x^2)^{-1}.\,dx = \int_0^{1/x} (1+y^2)^{-1}.\,dy.$$

(461) Compare Newton's page 3 above.

(462) That is, $\int_0^x \sqrt{[a^2+x^2]}\,.\,dx = \frac{1}{2}x\sqrt{[a^2+x^2]} + \frac{1}{2}a^2\log\{x/a + \sqrt{[1+x^2/a^2]}\}$. Newton has carelessly copied the series expansion of $(a^2+x^2)^{\frac{1}{2}}$ from his page 3 above.

Atq; ita si detur $r=\sqrt{x-xx}$ æquatio iterum ad circulum proveniet extrahendo radicem $r=x^{\frac{1}{2}}-\frac{1}{2}x^{\frac{3}{2}}-\frac{1}{8}x^{\frac{5}{2}}-\frac{1}{16}x^{\frac{7}{2}}$ &c adeoq; est $z=\frac{2}{3}x^{\frac{3}{2}}-\frac{1}{5}x^{\frac{5}{2}}-\frac{1}{28}x^{\frac{7}{2}}-\frac{1}{72}x^{\frac{9}{2}}$ &c.

Sic $r=\sqrt{aa+bx-xx}$ æquatio denuò ad circulum per extractionem radicis dat $r=a+\dfrac{bx}{2a}-\dfrac{xx}{2a}-\dfrac{bbxx}{8a^3}$(463) &c unde per Prop 2(459) elicitur

$$z=ax+\frac{bxx}{4a}-\frac{x^3}{6a}-\frac{bbx^3}{24a^3}\text{(463)} \quad \&\text{c.}\text{(464)}$$

Et sic $\sqrt{\dfrac{1+axx}{1-bxx}}=r$, per debitam reductionem(465) dat $r=1\begin{matrix}+\frac{1}{2}b\\+\frac{1}{2}a\\ \end{matrix}x^2\begin{matrix}+\frac{3}{8}bb\\+\frac{1}{4}ab\\-\frac{1}{8}aa\end{matrix}x^4$ &c.

Unde per Prop 2(459) fit $z=x\begin{matrix}+\frac{1}{6}b\\+\frac{1}{6}a\\ \end{matrix}x^3\begin{matrix}+\frac{3}{40}bb\\+\frac{1}{20}ab\\-\frac{1}{40}aa\end{matrix}x^5$ &c.

Sic deniq; $r=\sqrt{3:a^3+x^3}$ per extractionem radicis cubicæ(466) dat

$$r=a+\frac{x^3}{3aa}-\frac{x^6}{9a^5}+\frac{5x^9}{81a^8} \quad \&\text{c.}$$

Indeq; $z=ax+\dfrac{x^4}{12aa}-\dfrac{x^7}{63a^5}+\dfrac{x^{10}}{162a^8}$ &c$=AFDB$. vel etiam

$$r=x+\frac{a^3}{3xx}-\frac{a^6}{9x^5}+\frac{5a^9}{81x^8} \quad \&\text{c.}$$

Indeq; $z=\dfrac{xx}{2}-\dfrac{a^3}{3x}+\dfrac{a^6}{36x^4}-\dfrac{5a^9}{567x^7}$ &c$=HDBH$.

EXEMPL: 5. *Ubi prævia reductio per æquationis affectæ resolutionem requiritur.*

Si curva per æquationem $r^3+aar+axr-2a^3-x^3=0$ definiatur, extrahe radicem(467) et proveniet $r=a-\dfrac{x}{4}+\dfrac{xx}{64a}+\dfrac{131x^3}{512aa}$ &c. Unde ut in prioribus obtinebitur $z=ax-\dfrac{xx}{8}+\dfrac{x^3}{192a}+\dfrac{131x^4}{2048aa}$ &c.

Sin $r^3-crr-2xxr-ccr+2x^3+c^3=0$(468) sit æquatio ad curvam, resolutio dabit triplicem radicem(469) nempe $r=c+x-\dfrac{xx}{4c}+\dfrac{x^3}{32cc}$ &c et $r=c-x+\dfrac{3xx}{4c}-\dfrac{15x^3}{32cc}$ &c,

(463) Read '$-\dfrac{bbx^3}{8a^3}$' and so '$-\dfrac{bbx^4}{32a^3}$' respectively. Compare note (6) above.

(464) When, that is, $a^2 \neq 0$. Somewhat disappointingly the present expansion cannot, unlike its exact equivalent

$$\tfrac{1}{4}(ab-(b-2x)\sqrt{[a^2+bx-x^2]})+\tfrac{1}{8}(4a^2+b^2)\sin^{-1}2\,\frac{b\sqrt{[a^2+bx-x^2]}-a(b-2x)}{4a^2+b^2},$$

comprehend the previous instance in which $a=0$, $b=1$.

(465) Compare Newton's page 4 above.

(466) The first occurrence of a cube-root expansion in the present treatise. Newton, of course, writes down the corresponding series as a particular case of his general binomial expansion (I, 1, 3, §3.2 and especially §4) but his ignorant reader is presumably intended to

And thus if there be given $\dot{z} = \sqrt{[x-x^2]}$, again the equation of a circle, on extracting the root there will prove to be

$$\dot{z} = x^{\frac{1}{2}} - \tfrac{1}{2}x^{\frac{3}{2}} - \tfrac{1}{8}x^{\frac{5}{2}} - \tfrac{1}{16}x^{\frac{7}{2}} \ldots, \quad \text{so that} \quad z = \tfrac{2}{3}x^{\frac{3}{2}} - \tfrac{1}{5}x^{\frac{5}{2}} - \tfrac{1}{28}x^{\frac{7}{2}} - \tfrac{1}{72}x^{\frac{9}{2}} \ldots.$$

So $\dot{z} = \sqrt{[a^2+bx-x^2]}$, once more the equation of a circle, yields by extraction of the root $\dot{z} = a+\tfrac{1}{2}bx/a-\tfrac{1}{2}x^2/a-\tfrac{1}{8}b^2x^{[3]}/a^3 \ldots$, and from this by [Problem] 2 is elicited $z = ax+\tfrac{1}{4}bx^2/a-\tfrac{1}{6}x^3/a-\dfrac{1}{[32]}b^2x^{[4]}/a^3 \ldots$.(464)

And so $\sqrt{[(1+ax^2)/(1-bx^2)]} = \dot{z}$ by due reduction(465) yields

$$\dot{z} = 1+(\tfrac{1}{2}a+\tfrac{1}{2}b)\,x^2+(-\tfrac{1}{8}a^2+\tfrac{1}{4}ab+\tfrac{3}{8}b^2)\,x^4 \ldots.$$

Hence, by [Problem] 2, there comes

$$z = x+(\tfrac{1}{6}a+\tfrac{1}{6}b)\,x^3+(-\tfrac{1}{40}a^2+\tfrac{1}{20}ab+\tfrac{3}{40}b^2)\,x^5 \ldots.$$

So finally $\dot{z} = \sqrt[3]{[a^3+x^3]}$ by extraction of the cube root(466) yields

$$\dot{z} = a+\tfrac{1}{3}x^3/a^2-\tfrac{1}{9}x^6/a^5+\tfrac{5}{81}x^9/a^8 \ldots,$$

and in consequence $z = ax+\tfrac{1}{12}x^4/a^2-\tfrac{1}{63}x^7/a^5+\tfrac{1}{162}x^{10}/a^8 \ldots =$ area *AFDB*. Or alternatively $\dot{z} = x+\tfrac{1}{3}a^3/x^2-\tfrac{1}{9}a^6/x^5+\tfrac{5}{81}a^9/x^8 \ldots$, and thence

$$z = \tfrac{1}{2}x^2-\tfrac{1}{3}a^3/x+\tfrac{1}{36}a^6/x^4-\tfrac{5}{567}a^9/x^7 \ldots = \text{area } HDBH.$$

FIFTH EXAMPLES: *when prior reduction by the resolution of an affected equation is required.*

If the curve be defined by the equation $\dot{z}^3+a^2\dot{z}+ax\dot{z}-2a^3-x^3 = 0$, extract the(467) root and there will come out $\dot{z} = a-\tfrac{1}{4}x+\tfrac{1}{64}x^2/a+\tfrac{131}{512}x^3/a^2 \ldots$. Hence, as before, there will be obtained $z = ax-\tfrac{1}{8}x^2+\tfrac{1}{192}x^3/a+\tfrac{131}{2048}x^4/a^2 \ldots$.

But if $\dot{z}^3-c\dot{z}^2-2x^2\dot{z}-c^2\dot{z}+2x^3+c^3 = 0$(468) be the equation to the curve, its resolution will yield a triple root:(469) namely, $\dot{z} = c+x-\tfrac{1}{4}x^2/c+\tfrac{1}{32}x^3/c^2 \ldots$,

$$\dot{z} = c-x+\tfrac{3}{4}x^2/c-\tfrac{15}{32}x^3/c^2 \ldots,$$

---

verify it by direct computation on the lines of the square-root expansion worked on his page 3 above.

(467) 'affectam' (affected) is cancelled. For the working of this example compare his page 10 above with y replaced by r (or $\dot{z}$).

(468) Newton first wrote 'Sic æquationis $r^3+rr+r-x^3=0$ radix est' (So of the equation $\dot{z}^3+\dot{z}^2+\dot{z}-x^3 = 0$ the root is...), but at once decided to defer this example till page 75e, evidently because evaluation of z as an infinite series depends on integrating $\tfrac{2}{9}x^{-1}$ (a case he discusses separately below).

(469) The 'fictitious' equation here is $r^3-cr^2-c^2r+c^3 = (r-c)^2\,(r+c) = 0$. On substituting $r = c+\alpha x$ it follows that $2c(\alpha^2-1)+(\alpha^3-2\alpha+2)\,x = 0$, so that $\alpha = +1+\beta x+\gamma x^2+\ldots$ (from which $\beta = -\tfrac{1}{4}c^{-1}$, $\gamma = \tfrac{1}{32}c^{-2}$) or $\alpha = -1+\beta' x+\gamma' x^2+\ldots$, from which

$$\beta' = \tfrac{3}{4}c^{-1},\ \gamma' = -\tfrac{15}{32}c^{-2}.$$

Corresponding substitution of $r = -c+ax$ yields $4c^2a+2c(1-2a^2)\,x+(a^3-2a+2)\,x^2 = 0$, from which $a = -\tfrac{1}{2}c^{-1}x-\tfrac{1}{2}c^{-2}x^2+\tfrac{1}{4}c^{-4}x^4 \ldots$.

et $r = -c - \frac{xx}{2c} - \frac{x^3}{2cc} + \frac{x^5}{4c^4}$ &c. et inde trium correspondentium arearum valores $z = cx + \frac{1}{2}xx - \frac{x^3}{12c} + \frac{x^4}{128cc}$ &c, $z = cx - \frac{1}{2}x^{[2]} + \frac{x^3}{4c} - \frac{15x^4}{128cc}$ &c, ac

$$z = -cx - \frac{x^3}{6c} - \frac{x^4}{8cc} + \frac{x^6}{24c^4} \quad \&c.$$

||[75d] || De Curvis Mechanicis hic nihil adjicio, siquidem reductio ad formam Geometricarum post ostendetur.(470)

Cæterùm cum sic inventi valores z areis quandoq; ad Basis finitam partem AB, quandoq; ad partem BH infinitè versus H productam, et quandoq; ad utramq; partem sitis secundum diversos eorum terminos competant: quò debitus areæ ad quamlibet Basis portionem sitæ valor assignetur, Area illa semper ponenda est æqualis differentiæ valorum z partibus Basis ad initium et finem istius areæ terminatis competentium.

E.G. Ad curvam(471) quam æquatio $\frac{1}{1+xx} = r$ definit inventum est(472) $z = x - \frac{1}{3}x^3 + \frac{1}{5}x^5$ &c. Jam ut quantitatem areæ $bdDB$ adjacentis parti Basis bB determinem, a valore z qui fit ponendo $AB = x$ subduco valorem z qui fit ponendo $Ab = x$, et (distinctionis gratia scriptâ X majuscula pro AB et x minusculâ pro Ab) restat $X - \frac{1}{3}X^3 + \frac{1}{5}X^5$ &c $-x + \frac{1}{3}x^3 - \frac{1}{5}x^5$ &c(473) valor areæ illæ $bdDB$. Unde si Ab seu x ponatur nullum habebitur tota area

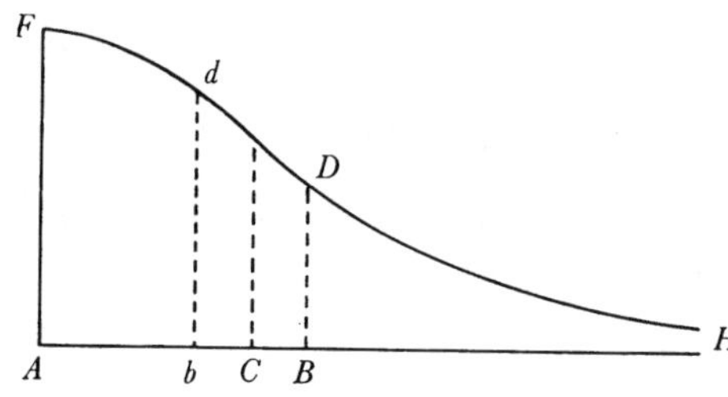

$$AFDB = X - \tfrac{1}{3}X^3 + \tfrac{1}{5}X^5 \quad \&c.^{(474)}$$

Ad eandem Curvam inventum est etiam $z = -\frac{1}{x} + \frac{1}{3x^3} - \frac{1}{5x^5}$ &c. unde rursus(475) juxta præcedentia erit area illa $bdBD = \frac{1}{x} - \frac{1}{3x^3} + \frac{1}{5x^5}$ &c $-\frac{1}{X} + \frac{1}{3X^3} - \frac{1}{5X^5}$ &c.(476) Adeoq; si AB seu X statuatur infinitum, area adjacens bdH a parte H similiter infinite longa valebit $\frac{1}{x} - \frac{1}{3x^3} + \frac{1}{5x^5}$ &c. siquidem posterior series

$$-\frac{1}{X} + \frac{1}{3X^3} - \frac{1}{5X^5} \quad \&c$$

propter infinitatem denominatorum evanescat.

Ad Curvam(477) æquatione $a + \frac{a^3}{xx} = r$ designatam, inventum est $ax - \frac{a^3}{x} = z$.(478)

(470) See Newton's page 89 below, where he effects the reduction by the area-preserving transformation introduced on his page 75 (compare note (444)). This paragraph is a late addition, presumably inserted when Newton came to write page 89.

and $\dot{z} = -c - \frac{1}{2}x^2/c - \frac{1}{2}x^3/c^2 + \frac{1}{4}x^5/c^4 \ldots$. From these come the values of three corresponding areas, $z = cx + \frac{1}{2}x^2 - \frac{1}{12}x^3/c + \frac{1}{128}x^4/c^2 \ldots$,

$z = cx - \frac{1}{2}x^2 + \frac{1}{4}x^3/c - \frac{15}{128}x^4/c^2 \ldots$, and $z = -cx - \frac{1}{6}x^3/c - \frac{1}{8}x^4/c^2 + \frac{1}{24}x^6/c^4 \ldots$.

Regarding mechanical curves I here add nothing, seeing that their reduction to geometrical form will be disclosed subsequently.(470)

However, the values of z thus found relate, in accordance with the differences in their terms, sometimes to areas situated along a finite section AB of the base, at others to ones along the portion BH infinitely extended in the direction of H, and at still others to areas situated along either section. Accordingly, to assign a due value to an area adjacent to any portion of the base, that area must always be set equal to the difference of the values of z belonging to those sections of the base which are terminated at the beginning and end of that area.

For example, in regard to the curve(471) defined by the equation $1/(1+x^2) = \dot{z}$ it has been found(472) that $z = x - \frac{1}{3}x^3 + \frac{1}{5}x^5 \ldots$. Now, to determine the magnitude of the area $bdDB$ adjacent to the portion bB of the base, from the value of z which comes by setting $AB = x$ I subtract the value of z which comes by setting $Ab = x$, and then (after, for distinction's sake, writing capital X in the case of AB and small x in that of Ab) there remains

$$X - \tfrac{1}{3}X^3 + \tfrac{1}{5}X^5 \ldots - x + \tfrac{1}{3}x^3 - \tfrac{1}{5}x^5 \ldots,^{(473)}$$

the value of the area $bdDB$. Hence if Ab, that is, x, be set zero, there will be had the whole area $AFDB = X - \frac{1}{3}X^3 + \frac{1}{5}X^5 \ldots$.(474)

For the same curve it was found also that $z = -x^{-1} + \frac{1}{3}x^{-3} - \frac{1}{5}x^{-5} \ldots$, and consequently, in line with the preceding, the area $bdDB$ will again equal $x^{-1} - \frac{1}{3}x^{-3} + \frac{1}{5}x^{-5} \ldots - X^{-1} + \frac{1}{3}X^{-3} - \frac{1}{5}X^{-5} \ldots$.(476) Accordingly, if AB, that is, X, be determined as infinity, the adjacent area bdH, likewise infinite in extent towards H, will have the value $x^{-1} - \frac{1}{3}x^{-3} + \frac{1}{5}x^{-5} \ldots$ since the latter series

$$-X^{-1} + \tfrac{1}{3}X^{-3} - \tfrac{1}{5}X^{-5} \ldots$$

(because each of its denominators is infinity) vanishes.

In regard to the curve(477) represented by the equation $a + a^3/x^2 = \dot{z}$ it was

(471) A conchoidal elliptical hyperbolism.

(472) See the second entry in Example 3 on page 75b.

(473) That is, $\tan^{-1}X - \tan^{-1}x$.

(474) Note that $AB = X$ must not be greater than $AF = 1$ if the series is to converge. The complementary case where $AB > AF$ is discussed in the next paragraph.

(475) The equivalent following phrase 'ut ante invenietur' is cancelled.

(476) Namely, $\tan^{-1}(1/x) - \tan^{-1}(1/X) = \int_{1/X}^{1/x} (1+y^2)^{-1}.dx$. Compare note (460).

(477) A parabolic hyperbola, to be precise.

(478) See the first part of Example 2 on page 75b.

Unde fit $aX-\frac{a^3}{X}-ax+\frac{a^3}{x}=$ areæ $bdDB$. Hæc autem evadit infinita sive x fingatur nulla sive X infinita et proinde utraq; area $AFDB$ et bdH infinitè magna est, ac solæ partes intermediæ (qualis $bdDB$) exhiberi possunt.(479) Id quod semper evenit ubi basis x tum in numeratoribus aliquorum tum in denominatoribus aliorum terminorum valoris z reperitur. Ubi vero x in numeratoribus solummodo, ut in primo exemplo, reperitur; valor z competit areæ sitæ ad AB cis parallelè incedentem. Et ubi in denominatoribus tantùm, ut in secundo exemplo; valor ille mutatis omnium terminorum signis, competit areæ omni ultra parallelè incedentem infinitè productæ.

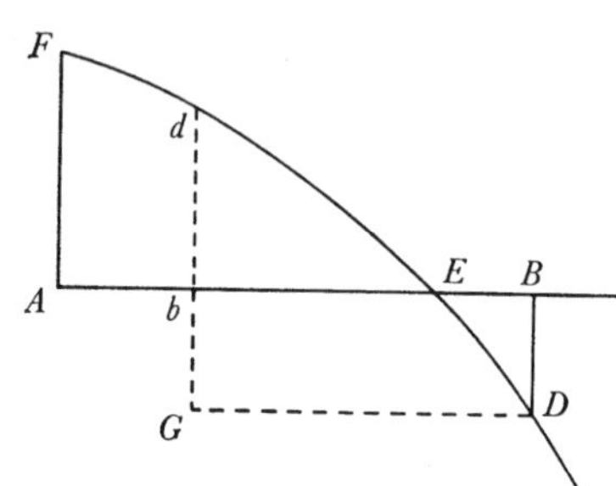

Siquando Curva linea secat Basin inter puncta b et B puta in E, vice areæ habebitur arearum ad diversas Basis partes differentia $bdE-BDE$, cui si addatur rectangulum $BDGb$ obtinebitur area $dEDG$.(480)

||[75e] || Præcipuè autem notandum est quod ubi in valore r terminus aliquis per x unius tantùm dimensionis dividitur, area illi termino correspondens pertinet ad Hyperbolam conicam et proinde per infinitam seriem(481) seorsim exhibenda est; quemadmodum in sequentibus factum.

Sit $\frac{a^3-aax}{ax+xx}=r$ æquatio ad Curvam(482) et per divisionem fiet

$$r=\frac{aa}{x}-2a+2x-\frac{2xx}{a}+\frac{2x^3}{aa} \quad \&c.$$

Indeq(483) $z=\boxed{\frac{aa}{x}}-2ax+xx-\frac{2x^3}{3a}+\frac{x^4}{2aa}$ &c. Et area

$$bdDB=\boxed{\frac{aa}{X}}-2aX+X^2-\frac{2X^3}{3a} \;\&c -\boxed{\frac{aa}{x}}+2ax-xx+\frac{2x^3}{3a} \;\&c.^{(484)}$$

Ubi per notas $\boxed{\frac{aa}{X}}$ et $\boxed{\frac{aa}{x}}$ designo areolas terminis $\frac{aa}{X}$ et $\frac{aa}{x}$ competentes. Jam ut $\boxed{\frac{aa}{X}}-\boxed{\frac{aa}{x}}$ investigetur, fingo Ab seu x definitam(485) esse et bB indefinitam seu fluentem lineam, quam itaq; si dicam y, erit $\boxed{\frac{aa}{x+y}}=$ areæ isti Hyperbolicæ(486) adjacenti bB, nempe $\boxed{\frac{aa}{X}}-\boxed{\frac{aa}{x}}$. Est autem, factâ divisione,

$$\frac{aa}{x+y}=\frac{aa}{x}-\frac{aay}{xx}+\frac{aayy}{x^3}-\frac{aay^3}{x^4} \quad \&c.$$

(479) In a finite manner, that is. The remainder of the paragraph is a late addition.

found that $ax - a^3/x = z$.[478] Hence the area $bdDB$ is equal to

$$aX - a^3/X - ax + a^3/x.$$

This, however, proves to be infinite whether x be supposed zero or X infinity and consequently both the areas $AFDB$ and bdH are infinite in magnitude, while only intermediate portions of them (such as $bdDB$) can be displayed.[479] This always happens when the base x is found in the numerators of some of the terms in z's value at the same time as in the denominators of others. When, in fact, x is found only in numerators—as in the first example—the value of z relates to the area situated along AB on the near side of one of the parallel ordinates. But when it is present only in denominators (as in the second example) that value, with the signs of all its terms changed, relates to the whole area infinitely extended beyond an ordinate.

If ever the curved line cuts its base between the points b and B, say in E, instead of the area [sought] there will be had the difference ($bdE - BDE$) of the areas on opposite sides of the base. If to this difference the rectangle $BDGb$ be added, there will be obtained the area $dEDG$.[480]

It should, however, be particularly noted that, when some term in the value of $\dot{z}$ is divided by a unit power of x alone, the area corresponding to that term belongs to a conic hyperbola and must accordingly be displayed by an infinite series[481] on its own: for instance as is done in the following.

Let $(a^3 - a^2x)/(ax + x^2) = \dot{z}$ be the equation to the curve.[482] By division will come $\dot{z} = a^2/x - 2a + 2x - 2x^2/a + 2x^3/a^2 \ldots$, from which[483]

$$z = [\textstyle\int s^2/x] - 2ax + x^2 - \frac{2}{3}x^3/a + \frac{1}{2}x^4/a^2 \ldots$$

and so area

$$bdDB = [\textstyle\int a^2/X] - 2aX + X^2 - \frac{2}{3}X^3/a \ldots - [\textstyle\int a^2/x] + 2ax - x^2 + \frac{2}{3}x^3/a \ldots.^{(484)}$$

Here by the symbols $[\int a^2/X]$ and $[\int a^2/x]$ I represent the component areas corresponding to the terms a^2/X and a^2/x. Now to find out $[\int a^2/X] - [\int a^2/x]$ I suppose Ab, that is, x, to be a definite[485] line-segment and bB an indefinite or fluent one. Consequently if I call the latter y, $[\int a^2/(x+y)]$ will equal the hyperbolic area[486] adjacent to Bb, namely $[\int a^2/X] - [\int a^2/x]$. But, when the division is made, $a^2/(x+y) = a^2/x - a^2y/x^2 + a^2y^2/x^3 - a^2y^3/x^4 \ldots$, and consequently the

(480) Precisely, $(BDGb) = [(dEDG) - (bdE)] + (BDE)$.

(481) Or, of course, a logarithmic function. Newton, however—here as elsewhere—is unwilling to introduce any symbolic equivalent to 'log' or 'exp'.

(482) A hyperbolic hyperbolism.

(483) As on II: 227/9, in our English version of the following we 'translate' Newton's 'square' integral operators by their (incomplete) Leibnizian equivalents '$\int$'.

(484) That is, $a^2 \int_x^X (a-y)/y(a+y) \, . dy = a^2 \Big[\log y - 2 \log (a+y) \Big]_x^X$.

(485) 'datam' (given) is cancelled.

(486) Namely, $\int a^2/(x+y) \, . dy$. The disadvantage of Newton's integral operator that it does not define the base variable will be apparent.

Et proinde tota area quæsita

$$bdDB = \frac{aay}{x} - \frac{aayy}{2xx} + \frac{aay^3}{3x^3} \;\&\text{c} \; -2aX + XX - \frac{2X^3}{3a} \;\&\text{c} + 2ax - xx + \frac{2x^3}{3a} \;\&\text{c.}$$

Ad eundem modum AB seu X pro definita linea adhiberi potuit et sic prodijsset $\boxed{\frac{aa}{X}} - \boxed{\frac{aa}{x}} = \frac{aay}{X} + \frac{aayy}{3XX} + \frac{aay^3}{3X^3} + \frac{aay^4}{4X^4}$ &c.(487)

Quinetiam si bisecetur bB in C et assumatur AC esse definitæ longitudinis et Cb ac CB indefinitæ. Tum dicto $AC = e$ et Cb vel $CB = y$, erit

$$bd = \frac{aa}{e-y} = \frac{aa}{e} + \frac{aay}{ee} + \frac{aayy}{e^3} + \frac{aay^3}{e^4} + \frac{aay^4}{e^5} \quad \&\text{c},$$

indeq; area Hyperbolica parti Basis bC adjacens

$$\frac{aay}{e} + \frac{aayy}{2ee} + \frac{aay^3}{3e^3} + \frac{aay^4}{4e^4} + \frac{aay^5}{5e^5} \quad \&\text{c.}$$

Erit etiam $CB = \frac{aa}{e+y} = \frac{aa}{e} - \frac{aay}{ee} + \frac{aayy}{e^3} - \frac{aay^3}{e^4} + \frac{aay^4}{e^5}$ &c et inde area alteri basis parti CB adjacens $\frac{aay}{e} - \frac{aayy}{2ee} + \frac{aay^3}{3e^3} - \frac{aay^4}{4e^4} + \frac{aay^5}{5e^5}$ &c. Et harum arearum summa $\frac{2aay}{e} + \frac{2aay^3}{3e^3} + \frac{2aay^5}{5e^5}$ &c valebit $\boxed{\frac{aa}{X}} - \boxed{\frac{aa}{x}}$.(488)

Sic æquatione $r^3 + rr + r - x^3 = 0$(489) ad Curvam(490) existente, ejus radix erit

||[75f] $r = x - \frac{1}{3} - \frac{2}{9x} + \frac{7}{81xx} + \frac{5}{81x^3}$ &c. Unde fit || $z = \frac{1}{2}xx - \frac{1}{3}x - \boxed{\frac{2}{9x}} - \frac{7}{81x} - \frac{5}{162xx}$ &c, et area $bdDB = \frac{1}{2}XX - \frac{1}{3}X - \boxed{\frac{2}{9X}} - \frac{7}{81X}$ &c $-\frac{1}{2}xx + \frac{1}{3}x + \boxed{\frac{2}{9x}} + \frac{7}{81x}$ &c, hoc est $= \frac{1}{2}XX - \frac{1}{3}X - \frac{7}{81X}$ &c $- \frac{1}{2}xx + \frac{1}{3}x + \frac{7}{81x}$ &c $-\frac{4y}{9e} - \frac{4y^3}{27e^3} - \frac{4y^5}{45e^5}$ &c.

Potest autem terminus iste Hyperbolicus utplurimùm commodè devitari mutando initium Basis, id est, augendo vel minuendo eam per datam aliquam quantitatem. Quemadmodum in exemplo priori ubi $\frac{a^3 - aax}{ax + xx} = r$ erat æquatio ad Curvam, si faciam b esse initium Basis, et fingens Ab cujuslibet esse determinatæ longitudinis puta $\frac{1}{2}a$, pro Basis residuo bB jam scribam x: Hoc est si diminuam Basem per $\frac{1}{2}a$ scribendo $x + \frac{1}{2}a$ pro x: evadet $\frac{\frac{1}{2}a^3 - aax}{\frac{3}{4}aa + 2ax + xx} = r$, et per

(487) That is, $\int_y^0 a^2/(X-y)\,.dy$.

(488) Equivalently, $(BDdb) = \int_{-y}^{y} a^2/(e+y)\,.dy = a^2 \log(X/x)$.

whole area $bdDB$ sought is

$$\frac{a^2y}{x}-\frac{a^2y^2}{2x^2}+\frac{a^2y^3}{3x^3}\ldots-2aX+X^2-\frac{2X^3}{3a}\ldots+2ax-x^2+\frac{2x^3}{3a}\ldots.$$

In the same way AB, that is, X, could have been employed for the definite line, producing $[\int a^2/X]-[\int a^2/x] = a^2y/X+\frac{1}{2}a^2y^2/X^2+\frac{1}{3}a^2y^3/X^3+\frac{1}{4}a^2y^4/X^4\ldots$.(487)

Indeed, if bB be bisected in C and AC taken to be of a definite length, while Cb and CB be indefinite, then, on calling $AC = e$ and Cb or $CB = y$, will bd equal $\frac{a^2}{e-y}=\frac{a^2}{e}+\frac{a^2y}{e^2}+\frac{a^2y^2}{e^3}+\frac{a^2y^3}{e^4}+\frac{a^2y^4}{e^5}\ldots$ and thence the hyperbolic area adjacent to the portion bC of the base will be $\frac{a^2y}{e}+\frac{a^2y^2}{2e^2}+\frac{a^2y^3}{3e^3}+\frac{a^2y^4}{4e^4}+\frac{a^2y^5}{5e^5}\ldots$. Also CB will equal $\frac{a^2}{e+y}=\frac{a^2}{e}-\frac{a^2y}{e^2}+\frac{a^2y^2}{e^3}-\frac{a^2y^3}{e^4}+\frac{a^2y^4}{e^5}\ldots$ and from this the area adjacent to the second part CB is $\frac{a^2y}{e}-\frac{a^2y^2}{2e^2}+\frac{a^2y^3}{3e^3}-\frac{a^2y^4}{4e^4}+\frac{a^2y^5}{5e^5}\ldots$. The aggregate,

$$\frac{2a^2y}{e}+\frac{2a^2y^3}{3e^3}+\frac{2a^2y^5}{5e^5}\ldots,$$

of these areas will be the value of $[\int a^2/X]-[\int a^2/x]$.(488)

So when the equation to the curve(490) is $\dot{z}^3+\dot{z}^2+\dot{z}-x^3=0$,(489) its root will be $\dot{z}=x-\frac{1}{3}-\frac{2}{9}x^{-1}+\frac{7}{81}x^{-2}+\frac{5}{81}x^{-3}\ldots$. Hence comes

$$z=\tfrac{1}{2}x^2-\tfrac{1}{3}x-[\textstyle\int\frac{2}{9}x^{-1}]-\frac{7}{81}x^{-1}-\frac{5}{162}x^{-2}\ldots$$

and the area $bdDB$ equals

$$\tfrac{1}{2}X^2-\tfrac{1}{5}X-[\textstyle\int\frac{2}{9}X^{-1}]-\frac{7}{81}X^{-1}\ldots-\frac{1}{2}x^2+\frac{1}{3}x+[\int\frac{2}{9}x^{-1}]+\frac{7}{81}x^{-1}\ldots,$$

that is, $\frac{1}{2}X^2-\frac{1}{3}X-\frac{7}{81}X^{-1}\ldots-\frac{1}{2}x^2+\frac{1}{3}x+\frac{7}{81}x^{-1}\ldots-\frac{4}{9}y/e-\frac{4}{27}y^3/e^3-\frac{4}{45}y^5/e^5\ldots$.

This hyperbolic term may, however, for the most part be bypassed conveniently by changing the origin of the base, increasing or diminishing it, that is, by some given quantity. For instance, in the former of my examples where $(a^3-a^2x)/(ax+x^2)=\dot{z}$ was the equation to the curve, I might make b the origin of the base and, supposing Ab to be of any determinate length you please, say a, now write x for the residue bB of the base—that is, I might diminish the base by $\frac{1}{2}a$ by writing $x+\frac{1}{2}a$ in place of x. There will then result

$$(\tfrac{1}{2}a^3-a^2x)/(\tfrac{3}{4}a^2+2ax+x^2)=\dot{z}$$

(489) See Newton's pages 14–15, and compare note (468). As we observed in note (60) the following series expansion of r (or $\dot{z}$) is incorrect.

(490) A serpentine elliptical hyperbolism which 'snakes' round the asymptote

$$x=r\ (\text{or}\ \dot{z})+\tfrac{1}{3},$$

having its inflexion point at the origin.

divisionem $r=\frac{2}{3}a-\frac{28}{9}x+\frac{200xx}{27a}$ &c. Unde fit $z=\frac{3}{2}ax-\frac{14}{9}xx+\frac{200x^3}{81a}$ &c $=$ areæ *bdDB*.

Et sic pro initio Basis adhibendo aliud atq; aliud ejus punctum, potest area cujusvis curvæ modis infinitis exprimi.

Potuit etiam æquatio $\frac{a^3-aax}{ax+xx}=r$ in duas series infinitas resolvi prodeunte $r=\frac{a^3}{xx}-\frac{a^4}{x^3}+\frac{a^5}{x^4}$ &c $-a+x-\frac{xx}{a}+\frac{x^3}{aa}$ &c[(491)] ubi terminus per x unius tantùm dimensionis divisus non reperitur. Sed hujusmodi series, ubi dimensiones x in unius numeratoribus et alterius denominatoribus infinitè ascendunt, minùs aptæ sunt ex quibus z per computum Arithmeticum obtineri possit, cùm in ejus valore numeri pro speciebus substituuntur.[(492)]

Instituenti computum hujusmodi numerosum, postquam valor areæ in speciebus habetur, haud aliquid difficile occurret. Tamen in præcedentem doctrinam penitùs illustrandam exemplum unum et alterum subjungere placuit.

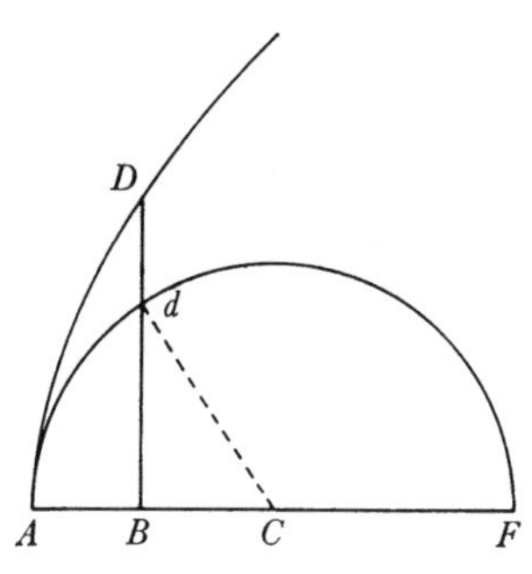

Proponatur Hyperbola *AD* quam æquatio $\sqrt{x+xx}=r$ designat, utpote cujus vertex est ad *A*, et uterq; Axis æquatur unitati. Et e præcedentibus Area ejus *ADB*[(493)] erit $\frac{2}{3}x^{\frac{3}{2}}+\frac{1}{5}x^{\frac{5}{2}}-\frac{1}{28}x^{\frac{7}{2}}+\frac{1}{72}x^{\frac{9}{2}}-\frac{5}{704}x^{\frac{11}{2}}$ &c hoc est $x^{\frac{1}{2}}$ in
||[75g] $\frac{2}{3}x+\frac{1}{5}xx-\frac{1}{28}x^3+\frac{1}{72}x^4$ || $-\frac{5}{704}x^5$ &c. Quæ series infinitè producitur multiplicando ultimum terminum continuò per succedaneos terminos hujus progressionis $\frac{1,3}{2,5}x.\ \frac{-1,5}{4,7}x.\ \frac{-3,7}{6,9}x.\ \frac{-5,9}{8,11}x.\ \frac{-7,11}{10,13}[x]$ &c. Nempe primus terminus $\frac{2}{3}x^{\frac{3}{2}}$ in $\frac{1,3}{2,5}x$ facit $\frac{1}{5}x^{\frac{5}{2}}$ secundum terminum. Hic in $\frac{-1,5}{4,7}x$ facit $-\frac{1}{28}x^{\frac{7}{2}}$ tertium terminum. Hic in $\frac{-3,7}{6,9}x$ facit $\frac{1}{72}x^{\frac{9}{2}}$ quartum terminum. Et sic in infinitum.[(494)] Sumatur jam *AB*[(495)] cujuslibet longitudinis puta $\frac{1}{4}$, et hunc numerum scribe pro x ejusq; radicem $\frac{1}{2}$ pro $x^{\frac{1}{2}}$, et primus terminus $\frac{2}{3}x^{\frac{3}{2}}$ sive $\frac{2}{3}\times\frac{1}{8}$ in decimalem fractionem reductus evadit 0,08333333 &c. Hic in $\frac{1,3}{2,5,4}$ facit 0,00625[(496)] secundam terminum. Hic in $\frac{-1,5}{4,7,4}$ facit $-0,0002790178$ &c[(496)]

(491) That is, $a^3x^{-2}/(1+a/x)-a/(1+x/a)$.

(492) A gross understatement! When one portion of the series converges (say for $|x|<1$) the other will diverge to infinity, and conversely so (when $|x|>1$).

(493) In exact terms, $\frac{1}{4}(1+2x)\sqrt{[x+x^2]}-\frac{1}{8}\log\{1+2(x+\sqrt{[x+x^2]})\}$.

and, by division, $\dot{z} = \frac{2}{3}a - \frac{28}{9}x + \frac{200}{27}x^2/a \ldots$. Hence comes

$$z = \frac{3}{2}ax - \frac{14}{9}x^2 + \frac{200}{81}x^3/a \ldots = \text{area } bdDB.$$

And by employing in this way now one, now another of its points as origin of the base, the area of any curve can be expressed in an infinity of ways.

The equation $(a^3 - a^2x)/(ax+x^2) = \dot{z}$ could also have been resolved into two infinite series, producing

$$\dot{z} = a^3/x^2 - a^4/x^3 + a^5/x^4 \ldots - a + x - x^2/a + x^3/a^2 \ldots.^{(491)}$$

Here a term which is divided by the first power of x alone is not found. But series of this sort, in which the dimensions of x rise infinitely in the numerators of one and the denominators of the second, are less suitable for our obtaining z from them by arithmetical calculation when in its value numbers are substituted for variables.(492)

After the value of the area is determined in terms of variables, any one who starts out on a numerical computation of this sort will not encounter anything that is difficult. Nevertheless, in more thorough illustration of the preceding doctrine I have thought it appropriate to add one or two examples.

Let there be proposed the hyperbola AD represented by the equation $\sqrt{[x+x^2]} = \dot{z}$. Evidently its vertex is at A and either axis is equal to unity. From what has gone before its area ADB(493) will be

$$\tfrac{2}{3}x^{\frac{3}{2}} + \tfrac{1}{5}x^{\frac{5}{2}} - \tfrac{1}{28}x^{\frac{7}{2}} + \tfrac{1}{72}x^{\frac{9}{2}} - \tfrac{5}{704}x^{\frac{11}{2}} \ldots,$$

that is, $x^{\frac{1}{2}}(\frac{2}{3}x + \frac{1}{5}x^2 - \frac{1}{28}x^3 + \frac{1}{72}x^4 - \frac{5}{704}x^5 \ldots)$. This series is produced indefinitely through continually multiplying its last term by successive terms of this progression: $\dfrac{1.3}{2.5}x, \dfrac{-1.5}{4.7}x, \dfrac{-3.7}{6.9}x, \dfrac{-5.9}{8.11}x, \dfrac{-7.11}{10.13}x, \ldots$. Namely, the first term $\frac{2}{3}x^{\frac{3}{2}}$ multiplied by $\dfrac{1.3}{2.5}x$ makes the second term $\frac{1}{5}x^{\frac{5}{2}}$; this multiplied by $\dfrac{-1.5}{4.7}x$ makes $-\frac{1}{28}x^{\frac{7}{2}}$, the third term; this by $\dfrac{-3.7}{6.9}x$ makes $\frac{1}{72}x^{\frac{9}{2}}$, the fourth term; and so on infinitely.(494) Now let AB(495) be taken of any length, say $\frac{1}{4}$, and write this number in place of x and its root $\frac{1}{2}$ in place of $x^{\frac{1}{2}}$. The first term $\frac{2}{3}x^{\frac{3}{2}}$ or $\frac{2}{3} \times \frac{1}{8}$, when reduced to a decimal fraction, then comes out as 0·08333 333 ...; this multiplied into $\dfrac{1.3}{2.5.4}$ makes 0·00625,(496) the second term; this into $\dfrac{-1.5}{4.7.4}$ makes

---

(494) $\int x^{\frac{1}{2}}(1+x)^{\frac{1}{2}}.dx = \int \sum_{0\leqslant i\leqslant\infty} \binom{\frac{1}{2}}{i} x^{i+\frac{1}{2}}.dx = \sum_{0\leqslant i\leqslant\infty} u(i).x^{i+\frac{3}{2}}$, where

$$\frac{u(i+1)}{u(i)} = \frac{3+2i}{5+2i} \times \frac{-(2i-1)}{2(i+1)}.$$

(495) ‘seu x’ (or x) is cancelled.

(496) That is, $\frac{1}{160}$ and $-\frac{1}{3584}$ respectively.

tertium terminum. Et sic in infinitum. Terminos autem quos sic gradatim elicio dispono in duas Tabulas[,] affirmativos nempe in unam et negativos in aliam, et addo, ut hic vides.

+0,08333,33333,333333.	−0,00027,90178,571429.
625,00000,000000	34679,066051
2,71267,361111	834,465027
5135,169396	26,285354
144,628917	961296
4,954581	38676
190948	1663
7963	75
352	4
16	−0,00028,25719,389575.
1	
+0,08961,09885,646618.	

Dein a summa affirmativorum aufero summam negativorum et restat 0,08932 84166 257043 quantitas areæ Hyperbolicæ ADB(497) quam quærere oportuit.

Proponatur jam circulus AdF quem æquatio $\sqrt{x-xx}=r$ designat, hoc est cujus diameter AF sit unitas, et e præcedentibus area ejus $Ad[B]$ erit $\frac{2}{3}x^{\frac{3}{2}}-\frac{1}{5}x^{\frac{5}{2}}-\frac{1}{28}x^{\frac{7}{2}}-\frac{1}{72}x^{\frac{9}{2}}$ &c.(498) In qua serie cùm termini non differant a terminis seriei supra exprimentis aream Hyperbolicam nisi in signis + et −, nihil aliud agendum restat quam ut eosdem numerales terminos cum alijs signis nectamus, subducendo nempe connexas ambarum præfatarum
‖[75h] Tabularum summas 0,08989 35605 036193 ‖ a primo termino duplicato 0,16666 66666 666666 et residuum 0,07677 31061 630473 erit areæ circularis portio AdB, posito scilicet AB quadrante diametri.(499) Atq; ita videre est quod etsi areæ circuli et Hyperbolæ non conferantur ratione geometrica, tamen utraq; eodem computo arithmetico prodit.(500)

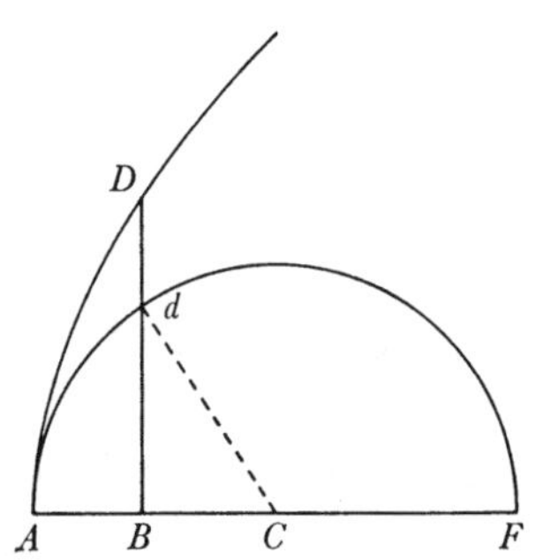

(497) Namely, $\frac{1}{8}(\frac{3}{4}\sqrt{5}-\log\{\frac{1}{2}(3+\sqrt{5})\})$. Note that Newton (in the first row of his right-hand column, for example) has rounded off his calculations to the nearest significant figure.

(498) $\frac{1}{8}\cos^{-1}(1-2x)-\frac{1}{4}(1-2x)\sqrt{[x-x^2]}$ in exact equivalent.

(499) On setting $x=\frac{1}{4}$, namely, in which case $(AdB)=\frac{1}{24}\pi-\frac{1}{32}\sqrt{3}$.

(500) There may well be an intended reference here to James Gregory's *Vera Circuli et Hyperbolæ Quadratura* (Padua, 1667), which Newton had just recently read (see notes (72) and (236)). In that work it had been Gregory's unsuccessfully accomplished aim to show that

$-0{\cdot}00027\,90178\ldots$,[496] the third term: and so on infinitely. The terms which I elicit stage by stage in this way I set down in two arrays, namely, in one the positive ones and the negative ones in the other, and then add them, as you see here:

+0·08333 33333 33333 3	−0·00027 10178 57142 9
625 00000 00000 0	34679 06605 1
2 71267 36111 1	834 46502 7
5135 16939 6	26 28535 4
144 62891 7	96129 6
4 95458 1	3867 6
19094 8	166 3
796 3	7 5
35 2	4
1 6	−0·00028 25719 38957 5.
1	
+0·08961 09885 64661 8.	

I then take the negative sum from the positive one and there remains

0·08932 84166 25704 3

as the quantity of the hyperbolic area *ADB*[497] which we were required to find.

Now let there be proposed the circle *AdF* represented by the equation $\surd[x-x^2] = \dot{z}$ (whose diameter *AF* is unity, that is) and from the preceding its area *AdB* will be $\frac{2}{3}x^{\frac{3}{2}}-\frac{1}{5}x^{\frac{5}{2}}-\frac{1}{28}x^{\frac{7}{2}}-\frac{1}{72}x^{\frac{9}{2}}\ldots$.[498] In this series, since its terms do not differ from those of the series above which expresses the area of a hyperbola except in its + and − signs, nothing else remains to be done except to connect the same numerical terms with other signs, namely, by taking away the connected sums of both the above-mentioned arrays, 0·08989 35605 03619 3, from double the first term, 0·16666 66666 66666 6, and the remainder

0·07677 31061 63047 3

will be the portion *AdB* of the circle area (on setting *AB* the fourth part of the diameter, that is).[499] And it may thus be seen that, even though the areas of a circle and a hyperbola may not be compared by a geometrical technique, nevertheless each is forthcoming from the same arithmetical computation.[500]

neither a general circle segment nor a general hyperbola segment can be compared 'geometrically' (that is, by a sequence of finite algebraic operations) with their base triangles. Compare D. T. Whiteside, 'Patterns of Mathematical Thought in the later Seventeenth Century' (note (109)): 268–70.

Inventa circuli portione AdB, exinde tota area facilè eruitur. Nempe radio dC acto duc Bd seu $\frac{1}{4}\sqrt{3}$ in BC seu $\frac{1}{4}$ et facti dimidium $\frac{1}{32}\sqrt{3}$ seu

$$0,05412\,65877\,365274$$

valebit triangulum CdB, quod adde areæ AdB et habebitur Sector

$$ACd = 0,13089\,96938\,995747,$$

cujus sextuplum 0,78539 81633 974482 est area tota.(501)

Et hinc obiter exit peripheriæ longitudo 3,14159 26535 897928, dividendo nempe aream per quadrantem diametri.(502)

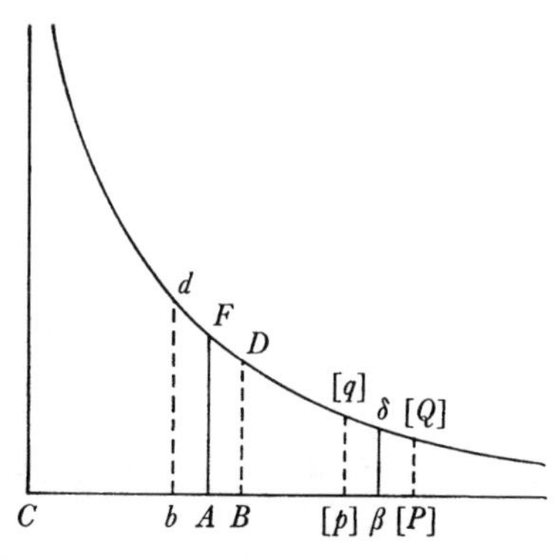

Hisce calculum areæ inter Hyperbolam dFD et ejus Asymptoton CA interjectæ subnectimus.(503) Sit C centrum Hyperbolæ et posito $CA = a$, $AF = b$, et $AB = Ab = x$; erit $\frac{ab}{a+x} = BD$, et $\frac{ab}{a-x} = bd$ et inde area $AFDB$(504) $= bx - \frac{bxx}{2a} + \frac{bx^3}{3aa} - \frac{bx^4}{4a^3}$ &c, et area $AFdb$(504) $= bx + \frac{bxx}{2a} + \frac{bx^3}{3aa} + \frac{bx^4}{4a^3}$ &c ac earum summa $bdDB$(504) $= 2bx + \frac{2bx^3}{3aa} + \frac{2bx^5}{5a^{[4]}} + \frac{2bx^7}{7a^{[6]}}$ &c. Ponamus jam $CA = AF = 1$, et Ab vel $AB = \frac{1}{10}$, existente $Cb = 0,9$ et $CB = 1,1$: et substituendo hos numeros pro a b et x, primus seriei terminus evadet 0,2, secundus 0,00066 666 &c, tertius 0,000004, et sic deinceps ut vides in hac Tabula.(505)

	0,20000,00000,000000
	66,66666,666666
	40000,000000
	285,714286
	2,222222
	18182
	154
	1
Summa	0,20067,06954,621511 = areæ $bdDB$.

(501) This is geometrically evident since $AB = BC = \frac{1}{2}AC$ and therefore $A\hat{C}d = \frac{1}{3}\pi$: hence sector $(ACd) = \frac{1}{24}\pi$ and so $(ABd) = (ACd) - \triangle BCd = \frac{1}{24}\pi - \frac{1}{32}\sqrt{3}$, as in note (499). Clearly (ACd) is one-sixth of the total circle area $(\frac{1}{4}\pi)$.

(502) For in general $\pi d = \frac{1}{4}\pi d^2/\frac{1}{4}d$, where d is the circle's diameter. Accurately, the last two places of Newton's evaluation of the constant π should read '32'. Newton incorporated a shortened version of the last four paragraphs in his *epistola posterior* of 24 October 1676 to Oldenburg for Leibniz (see his *Correspondence*, **2** (1960): 121–2).

When the portion AdB of the circle has been found, the whole area is easily deduced therefrom. To be precise, having drawn the radius dC multiply Bd (or $\frac{1}{4}\sqrt{3}$) into BC (or $\frac{1}{4}$) and half the product ($\frac{1}{32}\sqrt{3}$ or 0·05412 65877 36527 4) will be the value of the triangle CdB: add this to the area AdB and the sector ACd (0·13089 96938 99574 7) will be had, six times which

(0·78539 81633 97448 2)

is the total area.[501]

Hence, by the way, the length of the circumference comes out to be

3·14159 26535 89792 8,

namely by dividing the area by the fourth part of the diameter.[502]

To these we subjoin the calculation of the area lying between the hyperbola dFD and its asymptote CA.[503] Let C be the hyperbola's centre. On setting $CA = a$, $AF = b$ and $AB = Ab = x$, there will be

$$BD = ab/(a+x) \quad \text{and} \quad bd = ab/(a-x),$$

and thence area $AFDB$[504] $= bx - \frac{1}{2}bx^2/a + \frac{1}{3}bx^3/a^2 - \frac{1}{4}bx^4/a^3 \ldots,$

and area $AFdb$[504] $= bx + \frac{1}{2}bx^2/a + \frac{1}{3}bx^3/a^2 + \frac{1}{4}bx^4/a^3 \ldots,$

while their sum $bdDB$[504] $= 2bx + \frac{2}{3}bx^3/a^2 + \frac{2}{5}bx^5/a^4 + \frac{2}{7}bx^7/a^6 \ldots.$ Now let us set $CA = AF = 1$ and Ab or $AB = \frac{1}{10}$, so that $Cb = 0·9$ and $CB = 1·1$; on substituting these numbers for a, b and x in the series its first term will come out 0·2, its second 0·00066 666..., its third 0·00000 4, and so forth as you see in this array:[505]

	0·20000 00000 00000 0
	66 66666 66666 6
	40000 00000 0
	285 71428 6
	2 22222 2
	1818 2
	15 4
	1
Total	0·20067 06954 62151 1 = area $bdDB$.

(503) Compare I, **1**, 3, §3.3/§5 and II, **2**, 1, §3. The present modest sixteen decimal places are, however, a far cry from the 57 places of the latter. In his manuscript figure Newton drew the lines PQ, pq but omitted to insert the points P, Q, p, q (which have been pencilled in by a later reader, presumably Pemberton or Horsley).

(504) Namely, $ab\log[(a+x)/a]$, $ab\log[a/(a-x)]$ and $ab\log[(a+x)/(a-x)]$ respectively.

(505) In his following computation of $\log[11/9]$ Newton has, with the exception of the second (whose last figure should be '7'), rounded off all rows to the nearest significant figure.

Quod si areæ hujus partes *Ad* et *AD* seorsim desiderentur subduc minorem *AD*
||[75i] e majori *Ad* et restabit $\frac{bxx}{a}+\frac{bx^4}{2a^3}+\frac{bx^6}{3a^5}+\frac{bx^8}{4a^7}$ &c.(506) || Ubi si 1 scribatur pro *a* et *b*, ac $\frac{1}{10}$ pro *x*, termini in decimales redacti conficient sequentem Tabulam

	0,01000,00000,000000
	5,00000,000000
	3333,333333
	25,000000
	1667
	14
Summa	0,01005,03358,535014 $=Ad-AD$.

Jam si hæc arearum differentia addatur et auferatur summæ earum priùs inventæ, aggregati dimidium 0,10536,05106,578263 erit major area *Ad*, et residui dimidium 0,09531,01798,043248 minor *AD*.(507)

Per easdem Tabulas obtinentur etiam areæ illæ *AD* et *Ad* ubi *AB* et *Ab* ponuntur $\frac{1}{100}$ sive $CB=1{,}01$ & $Cb=0{,}99$ si modo numeri in depressiora loca debitè transferantur ut hic videre est

	0,02000,00000,000000		0,00010,00000,000000
	6666,666667(508)		50,000000
	400000		3333
	28	Sum	0,00010,00050,003333 $=Ad-AD$.
Sum̄	0,02000,06667,066695 $=bD$.		

½ Aggreg 0,01005,03358,535014 $=Ad$. ½ Resid 0,00995,03308,531681 $=AD$.(509)

Et sic positis *AB* & *Ab* $\frac{1}{1000}$ seu $CB=1{,}001$ et $Cb=0{,}999$, obtinebitur $Ad=0{,}00100{,}05003{,}335835$ et $AD=0{,}00099{,}95003{,}330835$.(510)

Ad eundem modum si stantibus *CA* et $AF=1$, ponantur *AB* et $Ab=0{,}2$ vel $=0{,}02$ vel $=0{,}002$ elicientur areæ illæ(511)

$Ad=0, 22314, 35513, 142097$, et $AD=0, 18232, 15567, 939546$.

vel $Ad=0, 02020, 27073[,] 175194$, et $AD=0, 01980, 26272, 961797$.

vel $Ad=0, 00200, 2[0026, 706731]$ et $AD=0, 001[99, 80026, 626731]$.

(506) That is, $ab\log[a^2/(a^2-x^2)]$: accordingly in the following computational array (for which $a = b = 1$, $x = 0{\cdot}1$) the 'sum' $Ad-AD$ is $\log[100/99]$.

(507) Specifically, $Ad = \log[10/9]$ and $AD = \log[11/10]$.

(508) Compare note (505). Newton now rounds off the last '6' as '7'.

(509) Here $Ad = \log[100/99]$ and $AD = \log[101/100]$.

(510) $Ad = \log[1000/999]$, $AD = \log[1001/1000]$.

But if the separate portions *Ad* and *AD* of this area be desired, take away the lesser *AD* from the greater *Ad* and there will remain

$$bx^2/a+\tfrac{1}{2}bx^4/a^3+\tfrac{1}{3}bx^6/a^5+\tfrac{1}{4}bx^8/a^7\ldots.^{(506)}$$

Here if 1 is written in place of *a* and *b* with $\frac{1}{10}$ for *x*, the terms reduced to decimals will together make up the following array:

```
        0·01000 00000 00000 0
              5 00000 00000 0
                 3333 33333 3
                   25 00000 0
                        166 7
                          1 4
Total   0·01005 03358 53501 4 = Ad−AD.
```

Now if this difference of the areas be added to and taken from their sum found previously, half the aggregate, 0·10536 05156 57826 3, will be the larger area *Ad* and half the remainder, 0·09531 01798 04324 8, the smaller one *AD*.(507)

By the same arrays will also be obtained the areas *AD* and *Ad* when *AB* and *Ab* are set as $\frac{1}{100}$ (in other words, when $CB=1{\cdot}01$ and $Cb=0{\cdot}99$) if only the numbers are appropriately transferred to lower places, as is here to be seen:

```
      0·02000 00000 00000 0
               6666 66666 7(508)
                    40000 0
                        2 8
Total 0·02000 06667 06669 5 = bD.
```

```
      0·00010 00000 00000 0
                 50 00000 0
                      333 3
Total 0·00010 00050 00333 3 = Ad−AD.
```

Semi-sum 0·01005 03358 53501 4 = *Ad*.
Semi-remainder 0·00995 03308 53168 1 = *AD*.(509)

And in this way after setting $AB=Ab=\frac{1}{1000}$ (or $CB=1{\cdot}001$ and $Cb=0{\cdot}999$) there will be obtained *Ad* = 0·00100 05003 33583 5 and

AD = 0·00099 95003 33083 5.(510)

After the same fashion (with *CA* and *AF* continuing to be unity) if *AB* and *Ab* are set equal to 0·2, or 0·02 or 0·002, these areas will be derived:(511)

Ad = 0·22314 35513 14209 7, *AD* = 0·18232 15567 93954 6;

or *Ad* = 0·02020 27073 17519 4, *AD* = 0·01980 26272 96179 7;

or *Ad* = 0·00200 2[0026 70673 1], *AD* = 0·001[99 80026 62673 1].

(511) Namely, *Ad* = log[10/8], *AD* = log[12/10]; *Ad* = log[100/98], *AD* = log[102/100]; and *Ad* = log[1000/998], *AD* = log[1002/1000] respectively. In the manuscript the last pair are entered only to six and three decimal places but we have completed their value to 16 places in square brackets.

Ex inventis hisce areis jam facile est alias per solam additionem et subductionem derivare. Utpote cum sit $\frac{1,2}{0,8}$ in $\frac{1,2}{0,9}=2$, arearum pertinentium ad rationes $\frac{1,2}{0,8}$ & $\frac{1,2}{0,9}$ (hoc est, insistentium partibus Basis $1,2-0,8$ et $1,2-0,9$) summa 0,69314 71805 599453 erit area $AF\delta\beta$, existente $C\beta=2$, ut notum est.(512) Dein cum sit $\frac{1,2}{0,8}$ in $2=3$, arearum pertinentium ad $\frac{1,2}{0,8}$ et 2 summa

1,09861 22886 681097

erit area $AF\delta\beta$, existente $C\beta=3$. Pariter cùm sit $\frac{2 \text{ in } 2}{0\cdot 8}=5$, et 2 in $5=10$, per debitam arearum additionem obtinebitur $1,60933\,79124\,341004=AF\delta\beta$, existente ‖ $C\beta=5$, et $2,30258\,50929\,940457=AF\delta\beta$ existente $C\beta=10$.(513) Atq; ita cùm sit 10 in $10=100$, et 10 in $100=1000$, et $\sqrt{5 \text{ in } 10 \text{ in } 0,98}=7$, et 10 in $1,1=11$, et $\frac{1000 \text{ in } 1,001}{7 \text{ in } 11}=13$, et $\frac{100 \text{ in } 1,02}{2 \text{ in } 3}=17$, et $\frac{1000 \text{ in } 0,999}{3 \text{ in } 3 \text{ in } 3}=37$, et(514) 100 in $1,01=101$, et $\frac{1000 \text{ in } 1,002}{2 \text{ in } 3}=167$, et $\frac{1000 \text{ in } 0,998}{2}=499$. patet aream $AF\delta\beta$ per arearum supra inventarum compositionem inveniri posse, existente $C\beta=100$; 1000; 7; aut alio quolibet e recensitis numeris, et stante $[CA=AF]=1$.(515) Id quod significare volui ut Methodus construendo Logarithmorum Canoni aptissima pateret quæ areas Hyperbolicas (ex quibus Logarithmi facilè deducuntur) tot numeris primis correspondentes, quasi per binas tantum haud molestas operationes determinat. Cæterùm cùm Canon iste ex hoc fonte præ cæteris feliciter depromi videatur, quid si Constructionem ejus coronidis loco perstringam.

‖[75j]

Imprimis itaq; assumpto 0 pro Logarithmo numeri 1, et 1 pro Logarithmo numeri 10 ut solet, investigandi sunt Logarithmi primorum numerorum 2,3,5, 7, 11, 13, 17, 37,(516) dividendo inventas areas Hyperbolicas per

2, 30258 50929 940457

(512) Having previously evaluated log 10 and log 1·024 to equivalent accuracy by computing the area of appropriate segments of the hyperbola $xy = 10^{25}$, James Gregory in the Scholium to Proposition XXXIII of his *Vera Circuli et Hyperbolæ Quadratura* (note (72)) had found $\log 2 = \frac{1}{10}(3\log 10 + \log 1\cdot 024)$ to be 0·69314 71805 59945 29141 71917, correct to exactly the sixteen places given here by Newton. Compare *Christiani Hugenii Opera Varia*, 1 (Leyden, 1724): 407–62, especially 456; and C. J. Scriba, *James Gregorys frühe Schriften zur Infinitesimalrechnung* (Giessen, 1957): 20–3.

(513) Newton computes the natural logarithms of 2, 3, 5 and 10 respectively.

(514) $\frac{\text{‘}1000 \text{ in } 1,002\text{’}}{2 \text{ in } 3 \text{ in } 3}$ $(1000 \times 1\cdot 002/2 \times 3 \times 3)$ is cancelled.

Once these areas are found it is easy to deduce others from them by addition and subtraction alone. For instance, since $\frac{1\cdot 2}{0\cdot 8}\times\frac{1\cdot 2}{0\cdot 9}=2$, the sum,

$$0\cdot 69314\ 71805\ 59945\ 3,$$

of the areas relating to the ratios $\frac{1\cdot 2}{0\cdot 8}$ and $\frac{1\cdot 2}{0\cdot 9}$ (that is, of those standing upon the portions $1\cdot 2-0\cdot 8$ and $1\cdot 2-0\cdot 9$ of the base) will be the area $AF\delta\beta$, $C\beta$ being equal to 2, as is known.(512) Next, since $\frac{1\cdot 2}{0\cdot 8}\times 2=3$, the sum,

$$1\cdot 09861\ 22886\ 68109\ 7,$$

of the areas relating to the ratios $\frac{1\cdot 2}{0\cdot 8}$ and 2 will be the area $AF\delta\beta$, $C\beta$ equalling 3. Likewise, since $\frac{2\times 2}{0\cdot 8}=5$ and $2\times 5=10$, by appropriate addition of areas there will be obtained $1\cdot 60933\ 79124\ 34100\ 4=AF\delta\beta$, $C\beta$ equalling 5, and

$$2\cdot 30258\ 50929\ 94045\ 7=AF\delta\beta,$$

$C\beta$ equalling 10.(513) And thus, since $10\times 10=100$, and $10\times 100=1000$ and $\surd[5\times 10\times 0\cdot 98]=7$, and $10\times 1\cdot 1=11$ and $100\times 1\cdot 001/7\times 11=13$, and $100\times 1\cdot 02/2\times 3=17$, and $1000\times 0\cdot 999/3\times 3\times 3=37$, and(514)

$$100\times 1\cdot 01=101,\quad \text{and}\quad 1000\times 1\cdot 002/2\times 3=167,$$

and $1000\times 0\cdot 998/2=499$, it is evident that the area $AF\delta\beta$ may be found by compounding the areas discovered above, where $C\beta$ is equal to 100, 1000, 7 or any other of the numbers specified, and both CA and AF continue to be unity.(515) I wanted to point this out so as to make obvious what is the most appropriate method for constructing a canon of logarithms, one which determines the hyperbolic areas (from which the logarithms are easily deduced) corresponding to an equal number of primes by a twin operation, as it were, and that not at all troublesome. However, since such a canon appears to issue more happily from this source than any other, why should I not, as a crowning touch, glance through its construction?

In the first place, accordingly, when (as is usual) 0 is taken as the logarithm of the number 1 and 1 for that of 10, the logarithms of the primes 2, 3, 5, 7, 11, 13, 17, 37, [...](516) are to be discovered by dividing the hyperbolic areas found

(515) Compare II: 187–8. Newton communicated an augmented version of this construction of the 'logarithmi numerorum multiplicium' to Oldenburg on 24 October 1676 (*Correspondence of Isaac Newton*, 2 (1960): 124). Newton's original text here reads '$AB=BF=1$'.

(516) Newton first continued '89, 101, 167, et 499' but then cancelled these numbers together with an immediately following phrase 'Id quod fit' (as is done).

aream nempe correspondentem numero 10, vel quod eodem recidit, multiplicando per ejus reciprocum 0,43429 44819 032518.(517) Sic enim e.g. Si 0,69314 718 &c area correspondens numero 2 multiplicetur per 0,43429 [&c] facit 0,30102 99956 639812 Logarithmum numeri 2.(518)

Deinde Logarithmi numerorum omnium in Canone qui ex horum multiplicatione fiunt indagandi sunt per additionem eorum Logarithmorum, ut solet, et loca vacua postmodum interpolanda ope hujus Theorematis.(519) Sit n numerus Logarithmo donandus, x differentia inter illum et proximos numeros hinc inde æqualiter distantes quorum logarithmi habentur,(520) ac d semissis differentiæ logarithmorum, et quæsitus logarithmus numeri n obtinebitur addendo $d+\frac{dx}{2n}+\frac{dx^3}{12n^3}$ [&c] logarithmo minoris numeri. Nam si numeri exponantur per Cp, $C\beta$ et CP. et existente rectangulo CBD vel $C\beta\delta=1$ ut supra,(521) ac erectis parallelè incedentibus pq et PQ, si n scribatur pro $C\beta$ et x pro βp vel βP, erit area $pqQP$ sive $\frac{2x}{n}+\frac{2x^3}{3n^3}+\frac{2x^5}{5n^5}$ &c ad 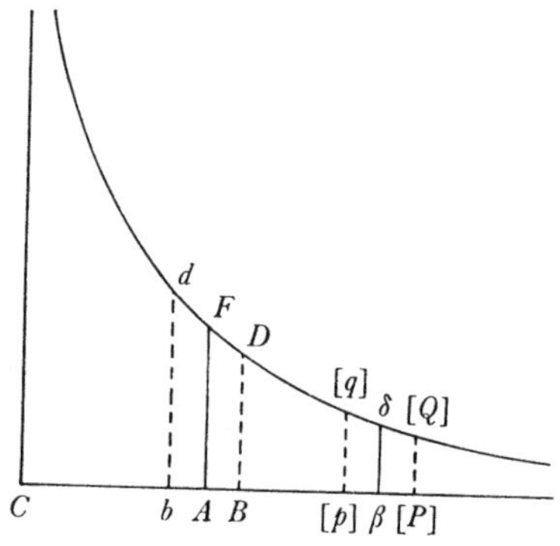
||[75k] aream $pq\delta\beta$ || sive $\frac{x}{n}+\frac{xx}{2nn}+\frac{x^3}{3n^3}$ &c, ut differentia inter logarithmos extremorum numerorum sive $2d$, ad differentiam inter logarithmos minoris et medij, quæ proinde erit

$$\frac{\frac{dx}{n}+\frac{dxx}{2nn}+\frac{dx^3}{3n^3}\ \&c}{\frac{x}{n}+\frac{x^3}{3n^{[3]}}+\frac{x^5}{5n^{[5]}}\ \&c},$$

hoc est facta divisione $d+\frac{dx}{2n}+\frac{dx^3}{12n^3}$ &c.(522)

(517) Newton's computation of this basic constant ($\log_{10}e$) is preserved on ULC. Add. 3960.3: 31. In a first, careless division he made it to be 0·43429 44819 03252 26 but then found its product by log 10 (his present value for which is perhaps taken from Gregory's Proposition XXXII) to be $1+10^{-15}$ to $O(10^{-18})$. Without further ado, he then subtracted $100/2·3... = 43$ from the incorrect reciprocal, correctly adjusting its last three places to read '1 83'.

(518) This multiplication was effected by Newton on ULC. Add. 3960.3: 31 likewise. He there found the sixteenth and seventeenth decimal places to be '18' (correctly '19[5 ...]') but the error in the last place disappears when, as here, the result is rounded off to sixteen places.

(519) In a first opening to this paragraph on ULC. Add. 3960.3: 32 Newton wrote 'Deinde Logarit[h]mi numerorum omnium intra limitem Canonis qui ex horum multiplicatione fiunt per additionem inventorum Logarithmorum eruendi sunt, ut notum est. Tum incipiendo a fine Canonis loca vacua solitaria i.e. quibus loca plena immediatè ad utramq; partem adjacent, interpolanda sunt ope Theorematis mox traditi...' (Then the logarithms of all numbers within the bounds of the canon which are formed by multiplication of these are to be rooted out by addition of their found logarithms, as is well known. Then, starting from the end of the canon, the individual empty places (those, that is, which full places immediately adjoin on either side) are to be interpolated with the aid of a theorem to be described in a moment). Note the phrase

by 2·30258 50929 94045 7 (namely, the area corresponding to the number 10) or, what comes to the same, multiplying them by its reciprocal

0·43429 44819 03251 8.[517]

Thus, for example, if the area 0·69314 718 ... corresponding to the number 2 be multiplied by 0·43429 ..., it makes 0·30102 99956 63981 2, the logarithm of the number 2.[518]

Then the logarithms of all numbers in the canon which come from multiplying these together are to be hunted down (as usual) by addition of their logarithms, and the places still vacant afterwards filled in with the aid of this theorem.[519] Let n be the number whose logarithm is to be assigned, x the difference between it and the nearest numbers equally distant from it on either side whose logarithms are known,[520] and d the half-difference of their logarithms: the required logarithm of the number n will then be obtained by adding

$$d+\tfrac{1}{2}dx/n+\tfrac{1}{12}dx^3/n^3\ldots$$

to the logarithm of the lesser number. For if the numbers are expressed by Cp, $C\beta$ and CP and if n be written in place of $C\beta$ and x instead of βp or βP (where the rectangle $CB\times BD$ or $C\beta\times\beta\delta$ continues to be unity, as above,[521] and the parallel ordinates pq and PQ have been erected), the area $pqQP$ (that is, $2x/n+\frac{2}{3}x^3/n^3+\frac{2}{5}x^5/n^5\ldots$) to the area $pq\delta\beta$ (or $x/n+\frac{1}{2}x^2/n^2+\frac{1}{3}x^3/n^3\ldots$) as the difference between the logarithms of the extreme numbers (or $2d$) to the difference between those of the lesser and middle ones, and this difference therefore will be $\dfrac{dx/n+\frac{1}{2}dx^2/n^2+\frac{1}{3}dx^3/n^3\ldots}{x/n+\frac{1}{3}x^3/n^3+\frac{1}{5}x^5/n^5\ldots}$, that is, on performing the division, $d+\frac{1}{2}dx/n+\frac{1}{12}dx^3/n^3\ldots$.[522]

'loca vacua...interpolanda': Newton later came to distinguish narrowly between 'interpolating' given 'full places' and 'intercalating' new 'empty places' between the former (compare *Correspondence of Isaac Newton*, **2** (1960): 149, note (8)), a usage which is here reversed.

(520) In a cancelled phrase at this point Newton denoted the 'logarithmus minoris numeri' $\log(n-x)$ by l (compare the first draft on ULC. Add. 3960.3: 32), finding the 'quæsitus logarithmus' $\log n$ directly as $l+d+\frac{1}{2}dx/n+\ldots$.

(521) See page 75h. The accompanying figure is here repeated for convenience.

(522) Given the two bounding logarithms $\log(n+x)$ and $\log(n-x)$ of semi-difference

$$d = \sum_{1\leqslant i\leqslant\infty}\frac{1}{2i-1}\left(\frac{x}{n}\right)^{2i-1},$$

Newton approximately bisects the interval $2x$ of the argument by equating

$$\log n-\log(n-x) = \sum_{1\leqslant i\leqslant\infty} i^{-1}\left(\frac{x}{n}\right)^i \quad\text{with}\quad d\left(1+\sum_{1\leqslant i\leqslant\infty} a_i\left(\frac{x}{n}\right)^{2i-1}\right).$$

It follows that $\log n = \log(n-x)+d(1+a_1x/n+a_2x^3/n^3+\ldots)$, where $a_1=\frac{1}{2}$, $\frac{1}{3}a_1+a_2=\frac{1}{4}$, $\frac{1}{5}a_1+\frac{1}{3}a_2+a_3=\frac{1}{6}$, ..., and accordingly $a_2=\frac{1}{12}$, $a_3=\frac{7}{180}$, $a_4=\frac{181}{7560}$, $a_5=\frac{1903}{113400}$, ...; compare Horsley, *Opera Omnia* (note (60)), **1**: 482, note θ.

Hujus autem seriei duos primos terminos $d+\frac{dx}{2n}$ pro Canone construendo sat accuratos existimo etiamsi ad usq; quatuordecim vel forte quindecim figurarum loca(523) logarithmi producerentur, si modò numerus logarithmo donandus non sit minor quam 1000. Quod sane calculum haud difficilem præbere potest siquidem x utplurimùm erit unitas vel numerus binarius. Non opus est tamen omnia loca beneficio hujus regulæ interpolare.(524) Nam logarithmi numerorum qui prodeunt e multiplicatione vel divisione numeri novissimè transacti per numeros quorum logarithmi prius habebantur obtineri possunt per additionem vel suductionem eorum logarithmorum. Quinetiam per(525) differentias logarithmorum et illarum differentiarum secundas differentias tertiasq; si opus est, loca vacua expeditiùs impleri possunt, adhibitâ tantùm prædictâ regulâ ubi ad obtinendū illas differentias continuatio aliquot locorum plenorum desideratur.(526)

Eadem methodo Regulæ pro intercalatione Logarithmorum inveniri possunt ubi e tribus numeris(527) dantur logarithmi minoris et medij, vel medij et majoris, licet numeri non sint in Arithmetica progressione.

Imò et hujus methodi vestigijs insistendo Regulæ pro construendis artificialium sinuum et Tangentium Tabulis sine adminiculo naturalium(528) haud difficulter depromi possunt. Sed hæc in transitu.

(523) That is, 'decimalia' (decimal), as Newton himself wrote in first draft on ULC. Add. 3960.3: 31.

(524) Compare note (519) for this present non-standard Newtonian sense of the word.

(525) Newton here first wrote 'observando analogiam inter' (by observing the analogy between), as in his preliminary draft on ULC. 3960.3: 31.

(526) In fact, when Newton came to summarize the present passage for Leibniz in his *epistola posterior* of 24 October 1676 he made no mention of the 'prædicta regula', urging only that the basic logarithms in the canon 'semel atq; iterum per dena intervalla interpola[ndi sunt]' (*Correspondence of Isaac Newton*, **2** (1960): 124). It is unlikely that he had at this time any grand scheme of logarithmic subtabulation: his researches, indeed, in the general theory of finite differences (to be reproduced in the next volume) commence only in 1675. In late 1670 Newton had certainly read Nicolaus Mercator's 1668 *Logarithmotechnia* (his library copy of which is now Trinity College, Cambridge. NQ. 9.48) and he was therefore broadly familiar with the 'Tabella differentiarum' described in Mercator's Proposition III (pages 11–14) and applied *ad hoc* in simple cases in Proposition VII (pages 15–18). He could scarcely yet have known of the more elaborate subtabulation algorithms developed on similar lines in an obscure chapter of Gabriel Mouton's *Observationes Diametrorum Solis et Lunæ* (Lyons, 1670): Liber III, Caput III 'De nonnullis numerorum proprietatibus': 368–95. There exists no evidence that Newton was ever aware of the elaborate 'modi inserendi duos/quatuor Logarithmos inter duos proximos datos' expounded half a century before by Henry Briggs in Chapter XIII of his *Arithmetica Logarithmica, sive Logarithmorum Chiliades Triginta* (London, 1624): 27–32.

(527) In a first version of this passage on ULC. Add. 3960.3: 30 Newton continued 'in Arithmetica progressione dantur Log[i] minoris et medij vel medij et majoris. Vel etiam ubi differentiæ nume[rorum] sunt in data quavis ratione' (in arithmetic progression the logarithms of the lesser and mean ones or of the mean and greater ones are given, or even where the

But I consider that the first two terms, $d+\frac{1}{2}dx/n$, are accurate enough for constructing a canon even though the logarithms are extended to fourteen or perhaps fifteen(523) places of figures, provided that the number whose logarithm is to be assigned is not less than 1000. And this, to be sure, can lead to no difficulties in calculation seeing that x for the most part is one or two units. Nor is there yet need to interpolate(524) all [vacant] places with the assistance of this rule. For the logarithms of numbers resulting from multiplication or division of the number most recently treated by numbers whose logarithms were previously to hand can be obtained by the addition or subtraction of their logarithms: indeed, vacant places may more speedily be filled up by(525) [considering] the differences of logarithms and the second differences of those differences—and third ones, too, if there is need—with the above-cited rule used merely where, to obtain those differences, some extension of the number of full places is desirable.(526)

By the same method rules may be found for the intercalation of logarithms when of three numbers(527) the logarithms of the lesser and the middle ones or of the middle and greater ones are given, even though the numbers should not be in arithmetical progression.

In fact, by following in the steps of this method rules for constructing tables of logarithmic sines and tangents may be derived with no difficulty at all and without the assistance of natural ones.(528) But this in passing.

---

differences of the numbers are in any given ratio). In the present manuscript he has cancelled 'in Arithmetica progressione' for 'continuè proportiona[li]bus', and this in turn for 'æqualiter differentibus': that too he cancelled when its sense was incorporated in his revised final phrase. Newton's meaning will be clear: if we wish to compute the λ-th 'mean' logarithm $\log n$ between $\log(n+\lambda x)$ and $\log(n-(2-\lambda)x)$, we may evolve a suitable Newtonian algorithm by equating $\log(n+\lambda x)-\log n$ with $d(\lambda+\sum_{1\leqslant i\leqslant\infty} b_i(x/n)^i)$, where

$$d = \tfrac{1}{2}[\log(n+\lambda x)-\log(n-(2-\lambda)x)].$$

(528) In first draft on ULC. Add. 3960.3: 30 Newton added at this point 'vel Canonis Logarithmorum' (or of a canon of logarithms). Evidently, we follow the structure of his present approach by seeking the infinite series expansion of

$$\frac{\log\sin(n+x)-\log\sin n}{\log\sin(n+x)-\log\sin(n-x)} = \frac{\log(1+X)+\log\cos x}{\log(1+X)-\log(1-X)},$$

where $X = \cot n.\tan x$, which may be expanded as a series when $|X| < 1$ and $|\tan x| < 1$ since

$$\log\cos x = -\int_0^x \tan y.dy = -\int_0^{\tan x} z/(1+z^2).dz = \sum_{1\leqslant i\leqslant\infty}(-1)^i\tan^{2i}x/2i.$$

Immediately, on making $2D = \log(1+X)-\log(1-X) = 2(X+\frac{1}{3}X^3+\ldots)$ and finding

$$f(\cot n,\ \tan x) \equiv \tfrac{1}{2}(\log(1+X)+\log\cos x)/D,$$

we may set $\log\sin n = \log\sin(n+x)-f(\cot n,\ \tan x)$ 'sine adminiculo Canonis Logarithmorum' provided only that we know the value of $\tan x$. The latter we may readily compute *ab initio* (where x is in radians) from the infinite series $x+\frac{1}{3}x^3+\frac{2}{15}x^5+\ldots$ for small x.

[75 1][(529)] ‖ Hactenus Curvarum[(530)] quæ per æquationes minùs simplices definiuntur Quadraturam mediante reductione in æquationes ex infinite multis terminis simplicibus constantes ostendimus. Cum verò ejusmodi curvæ per finitas etiam æquationes nonnunquam[(531)] quadrari possint vel saltem comparari cum alijs curvis quarum areæ quodammodo pro cognitis habeantur,[(532)] quales sunt sectiones conicæ: eapropter sequentes duos Theorematum catalogos in illum usum ope Propositionis $7^{æ}$ & $8^{æ}$[(533)] ut promisimus[(534)] constructos, jam visum est adjungere. Horum prior exhibet areas curvarum quæ quadrari possunt,[(535)] et posterior complectitur curvas quarum areas cum areis conicarum sectionum
‖[77] conferre liceat. In utrisq̧ literæ latinæ ‖[(536)] d, e, f, g et h datas quasvis quantitates, x et z bases curvarum, v et y parallelè incedentes, et s ac t areas ut supra denotant.[(537)] Græcæ autem η et θ quantitati z suffixæ denotant ejusdem z dimensionum numerum sive sit integer vel fractus, sive affirmativus aut negativus. Veluti si sit $\eta=3$ erit $z^{\eta}=z^{3}$, $z^{2\eta}=z^{6}$, $z^{-\eta}=z^{-3}$ sive $\frac{1}{z^{3}}$, $z^{\eta+1}$ vel $z^{+\overset{\eta}{1}}=z^{4}$, & $z^{\eta-1}$ vel $z^{-\overset{\eta}{1}}=zz$. Insuper in valoribus arearum abbreviandi causâ scribitur R vice radicalis illius $\sqrt{e+fz^{\eta}}$ vel $\sqrt{e+fz^{\eta}+gz^{2\eta}}$ quâ valor incedentis y afficitur.

Catalogus Curvarum aliquot ad rectilineas figuras relatarum, ope Prob 7 constructus.[(538)]

Curvarum[(539)] — *Arearum valores.*

Ordo primus.

$$dz^{\eta-1}=y. \qquad \frac{d}{\eta}z^{\eta}=t.$$

Ordo secundus.[(540)]

$$\frac{dz^{\eta-1}}{ee+2efz^{\eta}+ffz^{2\eta}}=y. \qquad \frac{dz^{\eta}}{\eta ee+\eta efz^{\eta}}=t, \text{ vel } \frac{-d}{\eta ef+\eta ffz^{\eta}}=t.$$

(529) See note (448). Newton drafted this transitional page on ULC. Add. 3960.3: 30, significant variants in which are reproduced in following footnotes.

(530) As in first draft Newton here continued 'tantùm simpliciorum areas per finitas æquationes exhibuimus, aliarum verò' (we have exhibited the areas merely of the simpler curves by means of finite equations, but [have exposed the quadrature] of others) but then cancelled the phrase.

(531) Newton was a little more optimistic in his first version, writing 'sæpenumero' (frequently).

(532) Newton first wrote simply 'faciliùs inveniuntur' (are more easily found).

(533) Read 'Problematis 7^{i} & 8^{i}'.

(534) See Newton's page 75 above.

(535) That is, 'exactè' (outright).

Hitherto we have exposed the quadrature of curves(530) defined by less simple equations by the technique of reducing them to equations consisting of infinitely many simple terms. However, curves of this kind may sometimes(531) be squared by means of finite equations also, or at least compared with other curves (such as conics) whose areas may, after a fashion, be accepted as known.(532) For this reason I have now decided to add the two following catalogues of theorems constructed, as promised,(533) for this use with the help of [Problems] 7 and 8. The former of these displays the areas of curves which can be squared,(535) while the latter embraces curves whose areas are allowably compared with the areas of conics. In either the italic letters(536) d, e, f, g and h denote, as above,(537) any given quantities, x and z the bases of curves, v and y parallel ordinates, and s and t areas; while the Greek suffixes η and θ of the quantity z denote the number of z's dimensions, whether integral or fractional, positive or negative. For instance, if $\eta = 3$, then $z^\eta = z^3$, $z^{2\eta} = z^6$, $z^{-\eta} = z^{-3}$ or $1/z^3$, $z^{\eta+1}\ldots = z^4$, and $z^{\eta-1}\ldots = z^2$. In addition, in the values of the areas to abbreviate R is written instead of the radical, $\surd[e+fz^\eta]$ or $\surd[e+fz^\eta+gz^{2\eta}]$, which affects the value of the ordinate y.

A Catalogue of some curves related to rectilinear figures, constructed with the help of Problem 7(538)

Curves(539)	*Values of the areas*
First order	
$dz^{\eta-1} = y.$	$(d/\eta)\, z^\eta = t.$
Second order(540)	
$dz^{\eta-1}/(e+fz^\eta)^2 = y.$	$(d/\eta e)\, z^\eta/(e+fz^\eta) = t$, or $-(d/\eta f)/(e+fz^\eta) = t.$

(536) Before he added the previous supplementary page 751 Newton began the present sentence 'In hujus rei illustrationem accipe imprǣsentia sequentem curvarum aliquot simpliciorum Catalogum, ubi' (In illustration of this topic accept, for the present, the following catalogue of some of the simpler curves, in which). See note (448).

(537) On Newton's page 72.

(538) Here, specifically, by the variable transform $z^\eta = x$. The reader should not anachronistically confuse Newton's constant 'd' in the following tables with a Leibnizian differential operator.

(539) In first draft on his page 76 Newton here added 'expositarum' (exhibited).

(540) On taking $A = e+fz^\eta$ it follows that $y = dz^{\eta-1}A^{-2}$ and so $t = \int y.dz = -d/\eta fA$. In his *epistola posterior* to Oldenburg for Leibniz in October 1676 Newton inserted a celebrated generalization of this result in line with the two following 'orders': namely, on taking $y = dz^\theta A^\lambda$,

$$t = \int y.dz = \frac{d}{\eta f} z^{\theta+1-\eta} A^{\lambda+1}\left(\frac{1}{s} - \frac{r-1}{s(s-1)}\cdot\frac{e}{f}z^{-\eta} + \frac{(r-1)(r-2)}{s(s-1)(s-2)}\cdot\frac{e^2}{f^2}z^{-2\eta}\ldots\right),$$

in which $r = (\theta+1)/\eta$ and $s = \lambda+r$. See the *Correspondence of Isaac Newton*, **2** (Cambridge, 1960): 115 and 153, note (26). Compare also note (574) below.

Curvarum	*Arearum valores.*
Ordo tertius.(541)	
1. $dz^{\overset{\eta}{-1}}\sqrt{e+fz^{\eta}}=y.$	$\frac{2d}{3\eta f}R^{3}=t.$
2. $dz^{\overset{2\eta}{-1}}\sqrt{e+fz^{\eta}}=y.$	$\frac{-4e+6fz^{\eta}}{15\eta ff}dR^{3}=t.$
3. $dz^{\overset{3\eta}{-1}}\sqrt{e+fz^{\eta}}=y.$	$\frac{16ee-24efz^{\eta}+30ffz^{2\eta}}{105\eta f^{3}}dR^{3}=t.$
4. $dz^{\overset{4\eta}{-1}}\sqrt{e+fz^{\eta}}=y.$	$\frac{-96e^{3}+144eefz^{\eta}-180effz^{2\eta}+210f^{3}z^{3\eta}}{945\eta f^{4}}dR^{3}=t.$
Ordo quartus.(542)	
1. $\frac{dz^{\eta-1}}{\sqrt{e+fz^{\eta}}}=y.$	$\frac{2d}{\eta f}R=t.$
2. $\frac{dz^{2\eta-1}}{\sqrt{e+fz^{\eta}}}=y.$	$\frac{-4e+2fz^{\eta}}{3\eta ff}dR=t.$
‖[78] ‖3. $\frac{dz^{3\eta-1}}{\sqrt{e+fz^{\eta}}}=y.$	$\frac{16ee-8efz^{\eta}+6ffz^{2\eta}}{15\eta f^{3}}dR=t.$
4. $\frac{dz^{4\eta-1}}{\sqrt{e+fz^{\eta}}}=y.$	$\frac{-96e^{3}+48eefz^{\eta}-36effz^{2\eta}+30f^{3}z^{3\eta}}{105\eta f^{4}}dR=t.$

His adjiciantur sequentia magis generalia Theoremata quibus via ad altiora sternitur.(543)

Ordo quintus.	*Arearum valores.*
1. $2\theta ez^{\overset{\theta}{-1}}\begin{smallmatrix}+2\theta\\+3\eta\end{smallmatrix}fz^{\overset{\theta}{\substack{+\eta\\-1}}}$ in $\frac{1}{2}\sqrt{e+fz^{\eta}}=y.$	$z^{\theta}R^{3}=t.$
2. $2\theta ez^{\overset{\theta}{-1}}\begin{smallmatrix}+2\theta\\+3\eta\end{smallmatrix}fz^{\overset{\theta}{\substack{+\eta\\-1}}}\begin{smallmatrix}+2\theta\\+6\eta\end{smallmatrix}gz^{\overset{\theta}{\substack{+2\eta\\-1}}}$ in $\frac{1}{2}\sqrt{e+fz^{\eta}+gz^{2\eta}}=y.$	$z^{\theta}R^{3}=t.$
Ordo sextus.	
1. $\frac{2\theta ez^{\overset{\theta}{-1}}\begin{smallmatrix}+2\theta\\+\eta\end{smallmatrix}fz^{\overset{\theta}{\substack{+\eta\\-1}}}}{2\sqrt{e+fz^{\eta}}}=y.$	$z^{\theta}R=t.$
2. $\frac{2\theta ez^{\overset{\theta}{-1}}\begin{smallmatrix}+2\theta\\+\eta\end{smallmatrix}fz^{\overset{\theta}{\substack{+\eta\\-1}}}\begin{smallmatrix}+2\theta\\+2\eta\end{smallmatrix}gz^{\overset{\theta}{\substack{+2\eta\\-1}}}}{2\sqrt{e+fz^{\eta}+gz^{2\eta}}}=y.$	$z^{\theta}R=t.$

Curves	Values of the areas
Third order(541)	
1. $dz^{\eta-1}\sqrt{[e+fz^{\eta}]} = y.$	$(2d/\eta f)\times\frac{1}{3}R^3 = t.$
2. $dz^{2\eta-1}\sqrt{[e+fz^{\eta}]} = y.$	$(2d/\eta f)\,(-\frac{2}{15}e/f+\frac{1}{5}z^{\eta})\,R^3 = t.$
3. $dz^{3\eta-1}\sqrt{[e+fz^{\eta}]} = y.$	$(2d/\eta f)\,(\frac{8}{105}e^2/f^2-\frac{4}{35}(e/f)z^{\eta}+\frac{1}{7}z^{2\eta})\,R^3 = t.$
4. $dz^{4\eta-1}\sqrt{[e+fz^{\eta}]} = y.$	$(2d/\eta f)\,(-\frac{48}{945}e^3/f^3+\frac{24}{315}(e^2/f^2)z^{\eta}-\frac{6}{63}(e/f)z^{2\eta}+\frac{1}{9}z^{3\eta})\,R^3 = t.$
Fourth order(542)	
1. $dz^{\eta-1}/\sqrt{[e+fz^{\eta}]} = y.$	$(2d/\eta f)\,R = t.$
2. $dz^{2\eta-1}/\sqrt{[e+fz^{\eta}]} = y.$	$(2d/\eta f)\,(-\frac{2}{3}e/f+\frac{1}{3}z^{\eta})\,R = t.$
3. $dz^{3\eta-1}/\sqrt{[e+fz^{\eta}]} = y.$	$(2d/\eta f)\,(\frac{8}{15}e^2/f^2-\frac{4}{15}(e/f)z^{\eta}+\frac{1}{5}z^{2\eta})\,R = t.$
4. $dz^{4\eta-1}/\sqrt{[e+fz^{\eta}]} = y.$	$(2d/\eta f)\,(-\frac{48}{105}e^3/f^3+\frac{24}{105}(e^2/f^2)z^{\eta}-\frac{6}{35}(e/f)z^{2\eta}+\frac{1}{7}z^{3\eta})\,R = t.$

To these let there be added the following more general theorems by means of which the path is laid to deeper considerations.(543)

	Values of the areas
Fifth order	
1. $(2\theta ez^{\theta-1}+(2\theta+3\eta)fz^{\theta+\eta-1})\times\frac{1}{2}\sqrt{[e+fz^{\eta}]} = y.$	$z^{\theta}R^3 = t.$
2. $(2\theta ez^{\theta-1}+(2\theta+3\eta)fz^{\theta+\eta-1}$ $+(2\theta+6\eta)\,gz^{\theta+2\eta-1})\times\frac{1}{2}\sqrt{[e+fz^{\eta}+gz^{2\eta}]} = y.$	$z^{\theta}R^3 = t.$
Sixth order	
1. $(2\theta ez^{\theta-1}+(2\theta+\eta)fz^{\theta+\eta-1})\times 1/2\sqrt{[e+fz^{\eta}]} = y.$	$z^{\theta}R = t.$
2. $(2\theta ez^{\theta-1}+(2\theta+\eta)fz^{\theta+\eta-1}+(2\theta+2\eta)\,gz^{\theta+2\eta-1})$ $\times 1/2\sqrt{[e+fz^{\eta}+gz^{2\eta}]} = y.$	$z^{\theta}R = t.$

(541) Newton here (and in Ordo 4 following) takes $R = \sqrt{[e+fz^{\eta}]}$, so that $y = dz^{\lambda\eta-1}R$, $\lambda = 1, 2, 3, 4$, and therefore

$$t = \int y.dz = \frac{2d}{\eta f}z^{(\lambda-1)\eta}R^3\left(\frac{1}{p}-\frac{p-3}{p(p-2)}.\frac{e}{f}z^{-\eta}+\frac{(p-3)(p-5)}{p(p-2)(p-4)}.\frac{e^2}{f^2}z^{-2\eta}...\right),$$

in which $p = 2\lambda+1$.

(542) Here, similarly, $y = dz^{\lambda\eta-1}/R$, $\lambda = 1, 2, 3, 4$, so that

$$t = \int y.dz = \frac{2d}{\eta f}z^{(\lambda-1)\eta}R\left(\frac{1}{q}-\frac{q-1}{q(q-2)}.\frac{e}{f}z^{-\eta}+\frac{(q-1)(q-3)}{q(q-2)(q-4)}.\frac{e^2}{f^2}z^{-2\eta}...\right),$$

where $q = 2\lambda-1$.

(543) The justice of this observation will be apparent in his second table following, where the integrals of complex 'curves' are evaluated by suitably compounding corresponding 'areas'. In all six remaining 'orders' the entries in the left-hand column follow at once by finding the fluxion of (that is, differentiating) each corresponding 'areæ valor' on the right.

Curvarum	*Arearum valores.*

Ordo septimus.

$$1.\quad \frac{2\theta e z^{\substack{\theta\\-1}} \begin{smallmatrix}+2\theta\\-\eta\end{smallmatrix} f z^{\substack{\theta\\+\eta\\-1}}}{e+fz^{\eta} \text{ in } 2\sqrt{e+fz^{\eta}}}=y. \qquad\qquad \frac{z^{\theta}}{R}=t.$$

$$2.\quad \frac{2\theta e z^{\substack{\theta\\-1}} \begin{smallmatrix}+2\theta\\-\eta\end{smallmatrix} f z^{\substack{\theta\\+\eta\\-1}} \begin{smallmatrix}+2\theta\\-2\eta\end{smallmatrix} g z^{\substack{\theta\\+2\eta\\-1}}}{e+fz^{\eta}+gz^{2\eta} \text{ in } 2\sqrt{e+fz^{\eta}+gz^{2\eta}}}=y. \qquad\qquad \frac{z^{\theta}}{R}=t.$$

Ordo octavus.

$$1.\quad \frac{2\theta e z^{\substack{\theta\\-1}} \begin{smallmatrix}+2\theta\\-2\eta\end{smallmatrix} f z^{\substack{\theta\\+\eta\\-1}}}{ee+2efz^{\eta}+ffz^{2\eta}}=2y. \qquad\qquad \frac{z^{\theta}}{RR}\left(\text{sive } \frac{z^{\theta}}{e+fz^{\eta}}\right)=t.$$

$$2.\quad \frac{2\theta e z^{\substack{\theta\\-1}} \begin{smallmatrix}+2\theta\\-2\eta\end{smallmatrix} f z^{\substack{\theta\\+\eta\\-1}} \begin{smallmatrix}+2\theta\\-4\eta\end{smallmatrix} g z^{\substack{\theta\\+2\eta\\-1}}}{ee+2efz^{\eta} \begin{smallmatrix}+ff\\+2eg\end{smallmatrix} z^{2\eta}+2fgz^{3\eta}+ggz^{4\eta}}=2y. \qquad\qquad \frac{z^{\theta}}{RR}\left(\text{sive } \frac{z^{\theta}}{e+fz^{\eta}+gz^{2\eta}}\right)=t.$$

Ordo nonus, ubi (ut et in decimo) pro radicali $\sqrt{h+iz^{\eta}}$ in arearum valoribus substituitur P.

$$2\theta ehz^{\substack{\theta\\-1}} \begin{smallmatrix}+2\theta\\+3\eta\end{smallmatrix} fhz^{\substack{\theta\\+\eta\\-1}} \begin{smallmatrix}+2\theta\\+4\eta\end{smallmatrix} fiz^{\substack{\theta\\+2\eta\\-1}} \text{ in } \frac{\sqrt{e+fz^{\eta}}}{2\sqrt{h+iz^{\eta}}}=y. \qquad\qquad z^{\theta}R^{3}P=t.$$
$$\begin{smallmatrix}+2\theta\\+\eta\end{smallmatrix} ei$$

||[79] ||*Ordo decimus.*

$$2\theta ehz^{\substack{\theta\\-1}} \begin{smallmatrix}+2\theta\\+3\eta\end{smallmatrix} fhz^{\substack{\theta\\+\eta\\-1}} \begin{smallmatrix}+2\theta\\+2\eta\end{smallmatrix} fiz^{\substack{\theta\\+2\eta\\-1}} \text{ in } \frac{\sqrt{e+fz^{\eta}}}{h+iz^{\eta} \text{ in } 2\sqrt{h+iz^{\eta}}}=y. \qquad\qquad \frac{z^{\theta}R^{3}}{P}=t.$$
$$\begin{smallmatrix}+2\theta\\-\eta\end{smallmatrix} ei$$

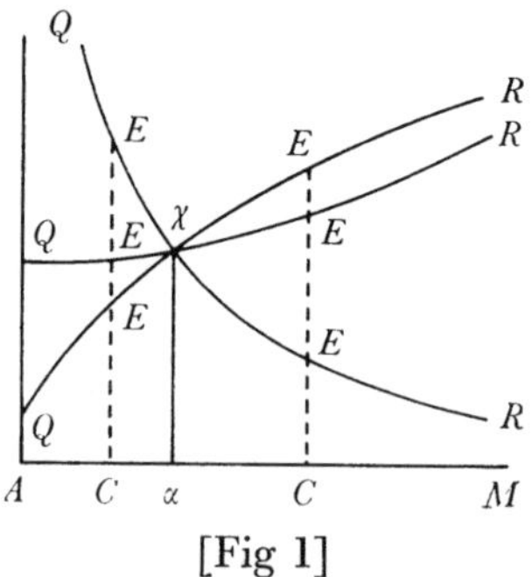

[Fig 1]

Possint et hujusmodi alia adjici,(544) sed ad alterius generis curvas quæ cum Conicis sectionibus conferri possunt jam transeo. Et in hoc Catalogo expositam Curvam linea $QE\chi R$ (fig [1]) designatam habes, cujus basis principium sit A, basis AC, parallelè incedens(545) CE, areæ principium $\alpha\chi$, et area descripta $\alpha\chi EC$. Ejus autem areæ principium sive terminus initialis (quod utplurimùm vel basis principio A insistit, vel ad infinitam distantiam recedit) invenitur quærendo basis longitudinem $A\alpha$ cùm areæ valor nullus est, et erigendo normalem $\alpha\chi$.

Curves	*Values of the areas*
Seventh order	
1. $(2\theta e z^{\theta-1} + (2\theta-\eta) f z^{\theta+\eta-1}) \times 1/2(e+fz^{\eta})^{\frac{3}{2}} = y.$	$z^{\theta}/R = t.$
2. $(2\theta e z^{\theta-1} + (2\theta-\eta) f z^{\theta+\eta-1}$ $+ (2\theta-2\eta)\, g z^{\theta+2\eta-1}) \times 1/2(e+fz^{\eta}+gz^{2\eta})^{\frac{3}{2}} = y.$	$z^{\theta}/R = t.$
Eighth order	
1. $(2\theta e z^{\theta-1} + (2\theta-2\eta) f z^{\theta+\eta-1}) \times 1/(e+fz^{\eta})^{2} = 2y.$	z^{θ}/R^{2} (or $z^{\theta}(e+fz^{\eta})^{-1}$) $= t.$
2. $(2\theta e z^{\theta-1} + (2\theta-2\eta) f z^{\theta+\eta-1}$ $+ (2\theta-4\eta)\, g z^{\theta+2\eta-1}) \times 1/(e+fz^{\eta}+gz^{2\eta})^{2} = 2y.$	z^{θ}/R^{2} (or $z^{\theta}(e+fz^{\eta}+gz^{2\eta})^{-1}$) $= t.$

Ninth order: here (and in the tenth also) P is substituted in place of the radical $\sqrt{[h+iz^{\eta}]}$ in the values of the areas.

$(2\theta e h z^{\theta-1} + [(2\theta+3\eta) fh + (2\theta+\eta)\, ei]\, z^{\theta+\eta-1}$ $+ (2\theta+4\eta) fi z^{\theta+2\eta-1}) \times \sqrt{[e+fz^{\eta}]}/2\sqrt{[h+iz^{\eta}]} = y.$	$z^{\theta}R^{3}P = t.$

Tenth order

$(2\theta e h z^{\theta-1} + [(2\theta+3\eta) fh + (2\theta-\eta)\, ei]\, z^{\theta+\eta-1}$ $+ (2\theta+2\eta) fi z^{\theta+2\eta-1}) \times \sqrt{[e+fz^{\eta}]}/2(h+iz^{\eta})^{\frac{3}{2}} = y.$	$z^{\theta}R^{3}/P = t.$

Other entries of this kind might be added also,(544) but I now pass on to curves of the second class: those which can be compared with conics. In this catalogue you have the exhibited curve represented by the line $QE\chi R$ (figure 1), whose base origin is A, its base AC, its parallel ordinate CE, the origin-line of the area $\alpha\chi$ and the area described $\alpha\chi EC$. The origin-line or initial bound of the area (which for the most part either is ordinate at the base origin A or has receded to infinity) is found by seeking the length $A\alpha$ of the base when the value of the area is zero and then erecting the normal $\alpha\chi$.

(544) For instance, the generalization of Ordo 2 given by Newton in 1676 (note (540)).
(545) Here—and also two paragraphs below—Newton first wrote equivalently 'incedens applicata'.

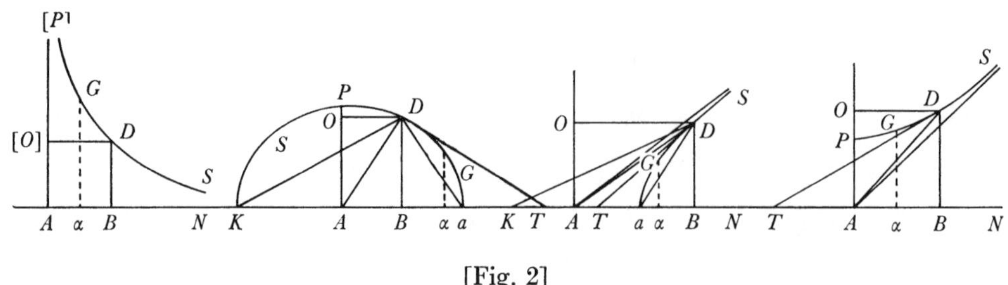

[Fig. 2]

Ad eundem modum Conicam sectionem (fig [2]) habes designatam lineâ *PDG*, cujus centrum sit *A*, vertex *a*, rectangulæ semidiametri(546) *Aa* & *AP*, basis principium *A* vel *a*, vel *α*, basis *AB* vel *aB*, vel *αB*, ordinatim applicata *BD*, tangens *DT* occurrens *AB* in *T*, subtensa *aD* et inscriptum vel ascriptum(547) rectangulum *ABDO*.

Itaq retentis jam ante definitis literis, erit $AC=z$, $CE=y$, $\alpha\chi EC=t$, AB vel $aB=x$, $BD=v$, et $ABDP$, vel $aGDB=s$, et præterea siquando ad alicujus areæ determinationem duæ Conicæ Sectiones requiruntur, posterioris area dicetur σ, basis ξ, et parallelè incedens Υ.(548)

(546) That is, semi-axes.

(547) In the archaic sense of 'adjoint'.

(548) In figure 1 α is the origin of the coordinates $\alpha C = z$, $CE = y$ and the area

$$(\alpha\chi EC) = \int_0^z y.dz = t.$$

In figure 2, where again α is set as the origin of the coordinates $\alpha B = x$, $BD = v$ (or, equivalently, $\alpha B = \xi$, $BD = \Upsilon$), the first diagram represents the rectangular hyperbola

$$d = (e+fx)v$$

referred to its asymptotes—an instance invoked in Ordo 1.1 and Ordo 3.1/2 below—while the remaining three (of an ellipse, namely, and hyperbolas with horizontal/vertical diameters) illustrate particular instances of the general central conic of defining equation $v^2 = e+fx+gx^2$. In all cases (αGDB) is $\int_0^x v.dx = s$ (or, equivalently, $\int_0^\xi \Upsilon.d\xi = \sigma$). When the conic's equation is taken to be $d = ex^2+fv^2$ (as in Ordo 2.1) or $hv^2 = df+(eh-fg)x^2$ (as in Ordo 3.1 (cancelled) or Ordo 10.1) or $v^2 = ex^2+f$ (as in Ordo 5.1*a*/2*a* and Ordo 7.1*a*/2*a*), α coincides with A, $(APDB) = \int_0^x v.dx = s$ and $\triangle ADB = \frac{1}{2}vx$, so that $(PAD) = |\frac{1}{2}xv-s|$ and $(POD) = |xv-s|$. However, when it suits his convenience Newton silently takes $s = \int_x^{Aa} v.dx$, so that here $(ADGa) = |\frac{1}{2}xv-s|$; in Ordo 5.1*a*, for example, having found $Aa = \sqrt{(-f/e)}$, $AT = -f/ex$ and so $BT = -v^2/ex$, he equates $(aGDT) = |\triangle BDT-(aBD)|$ with $|\frac{1}{2}v^3/ex-s|$. When the conic's equation is $v^2 = e+fx+gx^2$ (as in Ordo 6.2 and Ordo 8.1), then $A\alpha = -\frac{1}{2}f/g$, so that $AB = x+\frac{1}{2}f/g$ and accordingly Newton sets

$$(\alpha GDA) = |(\alpha GDB)\pm\triangle DBA| = |s-\tfrac{1}{2}xv-\tfrac{1}{4}(f/g)\,v|;$$

while in the particular case $v^2 = ex^2+fx$ the origin α coincides with a, so that $Aa = \frac{1}{2}f/e$,

$$(aDGa) = |\tfrac{1}{2}xv-s|,\quad (ADGa) = |s-\tfrac{1}{2}xv-\tfrac{1}{4}(f/e)\,v| \quad\text{and}\quad (KDGa) = |s-\tfrac{1}{2}xv-\tfrac{1}{2}(f/e)\,v|$$

In the same way you have (in figure 2) a conic section represented by the line *PDG*, whose centre is *A*, vertex *a*, rectangular semidiameters(546) *Aa* and *AP*, base origin *A*, *a* or α, base *AB*, *aB* or αB, ordinate *BD*, tangent *DT* meeting *AB* in *T*, subtense *aD* and the inscribed or ascribed(547) rectangle *ABDO*.

Accordingly, with the previously defined letters still kept, there will here be $AC = z$, $CE = y$, $\alpha\chi EC = t$, AB or $aB = x$, $BD = v$, and $ABDP$ or $aGDB = s$. Further, whenever two conics are required in the determination of some area, the area of the latter conic will be called σ, its base ξ and ordinate Υ.(548)

(as in Ordo 5.1b/2b/3 and Ordo 7.1b/2b/3). Newton's cancelled first draft of this preliminary (on his page 76 above) is insignificantly variant from the present one, except that a last paragraph has in the present manuscript been transferred to observation 6 on page 85a below.

Ingenuity aside, there would seem little to recommend these inadequately explained geometrical equivalents to the analytical expressions for the 'Arearum valores' which are listed in the following table (pp. 244–55). The more complex instances, indeed, are not amenable to being so represented and it is significant that Newton makes no attempt to translate these more involved integral equations into corresponding pictorial equivalents. For comparison-curves other than the conic it rapidly becomes impractical to develop such geometrical representations and later mathematicians wisely chose not to follow Newton's lead. See also note (594).

Catalogus Curvarum aliquot ad Conicas Sectiones relatarum ope Prob 8 constructus.(549)

	Curvarum	Sectionis Conicæ: Basis.	Sectionis Conicæ: parall: Incedens.	Arearum valores.
	Ordo primus.(550)			
	1. $\dfrac{dz^{\eta-1}}{e+fz^{\eta}}=y.$	$z^{\eta}=x.$	$\dfrac{d}{e+fx}=v.$	$\dfrac{1}{\eta}s=t=\dfrac{\alpha GDB}{\eta}.$ Fig(551)
	2. $\dfrac{dz^{2\eta-1}}{e+fz^{\eta}}=y.$	$z^{\eta}=x.$	$\dfrac{d}{e+fx}=v.$	$\dfrac{d}{\eta f}z^{\eta}-\dfrac{e}{\eta f}s=t.$
	3. $\dfrac{dz^{3\eta-1}}{e+fz^{\eta}}=y.$	$z^{\eta}=x.$	$\dfrac{d}{e+fx}=v.$	$\dfrac{d}{2\eta f}z^{2\eta}-\dfrac{de}{\eta ff}z^{\eta}+\dfrac{ee}{\eta ff}s=t.$
	Ordo secundus.(552)			
	1. $\dfrac{dz^{\frac{1}{2}\eta-1}}{e+fz^{\eta}}=y.$	$\sqrt{\dfrac{d}{e+fz^{\eta}}}=x.$	$\sqrt{\dfrac{d}{f}-\dfrac{e}{f}xx}=v.$	$\dfrac{2xv\div 4s^{(553)}}{\eta}=t=\dfrac{4}{\eta}ADGa.$
‖[80]	‖2. $\dfrac{dz^{\frac{3}{2}\eta-1}}{e+fz^{\eta}}=y.$	$\sqrt{\dfrac{d}{e+fz^{\eta}}}=x.$	$\sqrt{\dfrac{d}{f}-\dfrac{e}{f}xx}=v.$	$\dfrac{2de}{\eta f}z^{\frac{\eta}{2}}+\dfrac{4es-2exv}{\eta f}=t.$
	3. $\dfrac{dz^{\frac{5}{2}\eta-1}}{e+fz^{\eta}}=y.$	$\sqrt{\dfrac{d}{e+fz^{\eta}}}=x.$	$\sqrt{\dfrac{d}{f}-\dfrac{e}{f}xx}=v.$	$\dfrac{2de}{3\eta f}z^{\frac{3\eta}{2}}-\dfrac{2dee}{\eta ff}z^{\frac{\eta}{2}}+\dfrac{2eexv-4ees}{\eta ff}=t.$

The notes for pp. 244–55 are on pp. 256–9.

A Catalogue of some curves related to conic sections, constructed with the help of Problem 8[549]

Curves:	Conic base	Conic ordinate	Values of the areas
First order[550]			
1. $dz^{\eta-1}(e+fz^{\eta})^{-1} = y.$	$z^{\eta} = x.$	$d(e+fx)^{-1} = v.$	$\eta^{-1}s = t = \eta^{-1}(\alpha GDB)$. See figure.[551]
2. $dz^{2\eta-1}(e+fz^{\eta})^{-1} = y.$	$z^{\eta} = x.$	$d(e+fx)^{-1} = v.$	$\eta^{-1}(dz^{\eta}/f-es/f) = t.$
3. $dz^{3\eta-1}(e+fz^{\eta})^{-1} = y.$	$z^{\eta} = x.$	$d(e+fx)^{-1} = v.$	$\eta^{-1}(dz^{2\eta}/2f-dez^{\eta}/f^2+e^2s/f^2) = t.$
Second order[552]			
1. $dz^{\frac{1}{2}\eta-1}(e+fz^{\eta})^{-1} = y.$	$d^{\frac{1}{2}}(e+fz^{\eta})^{-\frac{1}{2}} = x.$	$(d-ex^2)^{\frac{1}{2}}f^{-\frac{1}{2}} = v.$	$\eta^{-1}\lvert 2xv-4s\rvert = t = 4\eta^{-1}(ADGa).$
2. $dz^{\frac{3}{2}\eta-1}(e+fz^{\eta})^{-1} = y.$	$d^{\frac{1}{2}}(e+fz^{\eta})^{-\frac{1}{2}} = x.$	$(d-ex^2)^{\frac{1}{2}}f^{-\frac{1}{2}} = v.$	$\eta^{-1}(2dz^{\frac{1}{2}\eta}+4s-2xv)\,(e/f) = t.$
3. $dz^{\frac{5}{2}\eta-1}(e+fz^{\eta})^{-1} = y.$	$d^{\frac{1}{2}}(e+fz^{\eta})^{-\frac{1}{2}} = x.$	$(d-ex^2)^{\frac{1}{2}}f^{-\frac{1}{2}} = v.$	$\eta^{-1}(\frac{2}{3}(de/f)\,z^{\frac{3}{2}\eta}-2(de^2/f^2)\,z^{\frac{1}{2}\eta}$ $+(e^2/f^2)\,(2xv-4s)) = t.$

	Curvarum	*Sectionis Conicæ* Basis.	*Sectionis Conicæ* parall: Incedens.	Arearum valores.

In sequenti tertio quartoq; ordine abbreviandi causa scribitur p pro $\sqrt{ff-4eg}$.(554)

Ordo tertius.

	Curvarum	Basis.	parall: Incedens.	Arearum valores.
1.	$\frac{dz^{\eta-1}}{e+fz^{\eta}+gz^{2\eta}}=y.$	$z^{\eta}=x.$ $z^{\eta}=\xi.$	$\frac{2dg}{f+p+gx}=v.$(555) $\frac{2dg}{f-p+g\xi}=\Upsilon.$(555)	$\frac{\sigma-s}{\eta p}=t.$
	Vel generaliùs fac	$\sqrt{\frac{d}{e+fz^{\eta}+gz^{2\eta}}}=x.$	$\sqrt{\frac{d}{g}+\frac{pp}{4gg}xx}=v.$	$\frac{xv\div 2s}{\eta}=t=\frac{2}{\eta}ADGa.$
	Vel quod perinde est	$\sqrt{\frac{dz^{2\eta}}{e+fz^{\eta}+gz^{2\eta}}}=x.$	$\sqrt{\frac{d}{e}+\frac{pp}{4ee}xx}=v.$	$\frac{2s\div xv}{\eta}=t=\frac{2}{\eta}ADGa.$
2.	$\frac{dz^{2\eta-1}}{e+fz^{\eta}+gz^{2\eta}}=y.$	$z^{\eta}=x.$ $z^{\eta}=\xi.$	$\frac{df+dp}{df+dp+gx}=v.$(556) $\frac{df-dp}{df-dp+g\xi}=\Upsilon.$(556)	$\frac{s-\sigma}{\eta p}=t.$(557)

Ordo tertius.(558)

	Curvarum	Basis.	parall: Incedens.	Arearum valores.
1.	$\frac{dz^{\eta-1}}{e+fz^{\eta}+gz^{2\eta}}=y.$	$\sqrt{\frac{d}{e+fz^{\eta}+gz^{2\eta}}}=x.$	$\sqrt{\frac{d}{g}+\frac{ff-4eg}{4gg}xx}=v.$	$\frac{xv-2s}{\eta}=t.$
	Vel sic,	$\sqrt{\frac{dz^{2\eta}}{e+fz^{\eta}+gz^{2\eta}}}=x.$	$\sqrt{\frac{d}{e}+\frac{ff-4eg}{4ee}xx}=v.$	$\frac{2s-xv}{\eta}=t.$
2.	$\frac{dz^{2\eta-1}}{e+fz^{\eta}+gz^{2\eta}}=y.$	$\sqrt{\frac{d}{e+fz^{\eta}+gz^{2\eta}}}=x.$ $fz^{\eta}+gz^{2\eta}=\xi.$	$\sqrt{\frac{d}{g}+\frac{ff-4eg}{4gg}xx}=v.$ $\frac{1}{e+\xi}=\Upsilon.$	$\frac{d\sigma+2fs-fxv}{2\eta g}=t.$

Curves:	Conic base	Conic ordinate	Values of the areas

In the third and fourth orders following, to abbreviate, p is written in place of $\sqrt{[f^2-4eg]}$.[554]

Third order

1. $dz^{\eta-1}/(e+fz^\eta+gz^{2\eta})=y.$	$z^\eta=x.$ $z^\eta=\xi.$	$2dg/(f+p+gx)=v.$[555] $2dg/(f-p+g\xi)=\Upsilon.$[555]	$\eta^{-1}(\sigma-s)/p=t.$
Or, more generally, make	$\sqrt{[d/(e+fz^\eta+gz^{2\eta})]}=x.$	$\sqrt{[dg^{-1}+\frac{1}{4}(p^2/g^2)\,x^2]}=v.$	$\eta^{-1}\,\vert 2s-xv\vert=t=2\eta^{-1}(ADGa).$
Or what is the same	$\sqrt{[dz^{2\eta}/(e+fz^\eta+gz^{2\eta})]}=x.$	$\sqrt{[de^{-1}+\frac{1}{4}(p^2/e^2)\,x^2]}=v.$	$\eta^{-1}\,\vert 2s-xv\vert=t=2\eta^{-1}(ADGa).$
2. $dz^{2\eta-1}/(e+fz^\eta+gz^{2\eta})=y.$	$z^\eta=x.$ $z^\eta=\xi.$	$d(f+p)/(df+dp+gx)=v.$[556] $d(f-p)/(df-dp+g\xi)=v.$[556]	$\eta^{-1}(s-\sigma)/p=t.$[557]

Third order[558]

1. $\dfrac{dz^{\eta-1}}{e+fz^\eta+gz^{2\eta}}=y.$	$\sqrt{[d/(e+fz^\eta+gz^{2\eta})]}=x.$	$\sqrt{[dg^{-1}+\frac{1}{4}(f^2-4eg)\,g^{-2}x^2]}=v.$	$\eta^{-1}(xv-2s)=t.$
Or thus:	$\sqrt{[d/(e+fz^\eta+gz^{2\eta})]}\,z^\eta=x.$	$\sqrt{[de^{-1}+\frac{1}{4}(f^2-4eg)\,e^{-2}x^2]}=v.$	$\eta^{-1}(2s-xv)=t.$
2. $\dfrac{dz^{2\eta-1}}{e+fz^\eta+gz^{2\eta}}=y.$	$\sqrt{[d/(e+fz^\eta+gz^{2\eta})]}=x.$ $fz^\eta+gz^{2\eta}=\xi.$	$\sqrt{[dg^{-1}+\frac{1}{4}(f^2-4eg)\,g^{-2}x^2]}=v.$ $1/(e+\xi)=\Upsilon.$	$\dfrac{d\sigma+2fs-fxv}{2\eta g}=t.$

Curvarum	Sectionis Conicæ: *Basis.*	Sectionis Conicæ: *parall: Incedens.*	*Arearum valores.*
Ordo quartus,(559) ubi abbreviandi causâ scribitur p pro $\sqrt{ff-4eg}$.			
1. $\dfrac{dz^{\frac{1}{2}\eta-1}}{e+fz^{\eta}+gz^{2\eta}}=y.$	$\sqrt{\dfrac{2dg}{f-p+2gz^{\eta}}}=x.$	$\sqrt{d\dfrac{-f+p}{2g}xx}=v.$	$\dfrac{2xv-4s-2\xi\Upsilon+4\sigma}{\eta p}=t.$
	$\sqrt{\dfrac{2dg}{f+p+2gz^{\eta}}}=\xi.$	$\sqrt{d\dfrac{-f-p}{2g}\xi\xi}=\Upsilon.$	
2. $\dfrac{dz^{\frac{3}{2}\eta-1}}{e+fz^{\eta}+gz^{2\eta}}=y.$	$\sqrt{\dfrac{2dez^{\eta}}{fz^{\eta}-pz^{\eta}+2e}}=x.$	$\sqrt{d\dfrac{-f+p}{2e}xx}=v.$	$\dfrac{4s-2xv-4\sigma+2\xi\Upsilon}{\eta p}=t.$
	$\sqrt{\dfrac{2dez^{\eta}}{fz^{\eta}+pz^{\eta}+2e}}=\xi.$	$\sqrt{d\dfrac{-f-p}{2e}[\xi\xi]}=\Upsilon.$	
‖[81] ‖*Ordo quintus*(560)			
1. $\dfrac{d}{z}\sqrt{e+fz^{\eta}}=y.$	$\dfrac{1}{z^{\eta}}=xx.$	$\sqrt{f+exx}=v.$	$\dfrac{4de}{\eta f}$ in $\dfrac{v^3}{2ex}-s=t=\dfrac{4de}{\eta f}$ in $aGDT$, vel in $APDB\div TDB$.
Vel sic,	$\dfrac{1}{z^{\eta}}=x.$	$\sqrt{fx+exx}=v.$	$\dfrac{8dee}{\eta ff}$ in: $\overline{s-\tfrac{1}{2}xv-\dfrac{fv}{4e}+\dfrac{ffv}{4eex}}=t=\dfrac{8dee}{\eta ff}$ in $aGDA+\dfrac{ffv}{4eex}$.
2. $\dfrac{d}{z^{\eta+1}}\sqrt{e+fz^{\eta}}=y.$	$\dfrac{1}{z^{\eta}}=xx.$	$\sqrt{f+exx}=v.$	$\dfrac{-2d}{\eta}s=t=\dfrac{2d}{\eta}\times APDB$, ceu(561) $\dfrac{2d}{\eta}\times aGDB$.
Vel sic,	$\dfrac{1}{z^{\eta}}=x.$	$\sqrt{fx+exx}=v.$	$\dfrac{4de}{\eta f}$ in $s-\tfrac{1}{2}xv-\dfrac{fv}{2e}=t=\dfrac{4de}{\eta f}\times aGDK$.
3. $\dfrac{d}{z^{2\eta+1}}\sqrt{e+fz^{\eta}}=y.$	$\dfrac{1}{z^{\eta}}=x.$	$\sqrt{fx+exx}=v.$	$\dfrac{-d}{\eta}s=t=\dfrac{d}{\eta}\times -aGDB$, vel $BDPK$.
4. $\dfrac{d}{z^{3\eta+1}}\sqrt{e+fz^{\eta}}=y.$	$\dfrac{1}{z^{\eta}}=x.$	$\sqrt{fx+exx}=v.$	$\dfrac{3dfs-2dv^3}{6\eta e}=t.$(562)

Curves:	*Conic base*	*Conic ordinate*	*Values of the areas*
Fourth order.[559] Here, to abbreviate, p is written in place of $\surd[f^2-4eg]$.			
1. $\dfrac{dz^{\frac{1}{2}\eta-1}}{e+fz^\eta+gz^{2\eta}}=y.$	$\surd[2dg/(f-p+2gz^\eta)]=x.$ $\surd[2dg/(f+p+2gz^\eta)]=\xi.$	$\surd[d-\frac{1}{2}(f-p)\,g^{-1}x^2]=v.$ $\surd[d-\frac{1}{2}(f+p)\,g^{-1}\xi^2]=\Upsilon.$	$\dfrac{2xv-4s-2\xi\Upsilon+4\sigma}{\eta p}=t.$
2. $\dfrac{dz^{\frac{3}{2}\eta-1}}{e+fz^\eta+gz^{2\eta}}=y.$	$\surd[2de/(f-p+2ez^{-\eta})]=x.$ $\surd[2de/(f+p+2ez^{-\eta})]=\xi.$	$\surd[d-\frac{1}{2}e^{-1}(f-p)\,x^2]=v.$ $\surd[d-\frac{1}{2}e^{-1}(f+p)\,\xi^2]=\Upsilon.$	$\dfrac{4s-2xv-4\sigma+2\xi\Upsilon}{\eta p}=t.$
Fifth order[560]			
1. $dz^{-1}\surd[e+fz^\eta]=y.$	$z^{-\eta}=x^2.$	$\surd[f+ex^2]=v.$	$(4de/\eta f)\,(v^3/2ex-s)=t=(4de/\eta f)\times(aGDT)$ or $\lvert(APDB)-(TDB)\rvert.$
	Or thus: $z^{-\eta}=x.$	$\surd[fx+ex^2]=v.$	$(8de^2/\eta f^2)\,(s-\frac{1}{2}xv-\frac{1}{4}fv/e+\frac{1}{4}f^2v/e^2x)=t$ $=(8de^2/\eta f^2)\times([aGDA]+\frac{1}{4}f^2v/e^2x).$
2. $dz^{-\eta-1}\surd[e+fz^\eta]=y.$	$z^{-\eta}=x^2.$	$\surd[f+ex^2]=v.$	$-(2d/\eta)\,s=t=(2d/\eta)\times(APDB)$ or $(2d/\eta)\times(aGDB).$
	Or thus: $z^{-\eta}=x.$	$\surd[fx+ex^2]=v.$	$(4de/\eta f)\,(s-\frac{1}{2}xv-\frac{1}{2}fv/e)=t$ $=(4de/\eta f)\times(aGDK).$
3. $dz^{-2\eta-1}\surd[e+fz^\eta]=y.$	$z^{-\eta}=x.$	$\surd[fx+ex^2]=v.$	$-(d/\eta)\,s=t=(d/\eta)\times-(aGDB)$ or $(BDPK).$
4. $dz^{-3\eta-1}\surd[e+fz^\eta]=y.$	$z^{-\eta}=x.$	$\surd[fx+ex^2]=v.$	$(d/\eta e)\,(\frac{1}{2}fs-\frac{1}{3}v^3)=t.$[562]

Curvarum	Sectionis Conicæ: Basis.	Sectionis Conicæ: parall: Incedens.	Arearum valores.
Ordo sextus.[563]			
1. $\frac{d}{z}\sqrt{e+fz^{\eta}+gz^{2\eta}}=y.$	$z^{\eta}=x.$ $\frac{1}{z^{\eta}}=\xi.$	$\sqrt{e+fx+gxx}=v.$ $\sqrt{g+f\xi+e\xi\xi}=\Upsilon.$	$\frac{4dee\xi\Upsilon+2def\Upsilon-2dffv-8dee\sigma+4dfgs}{4\eta eg-\eta ff}=t.$[564]
2. $dz^{\overset{\eta}{-1}}\sqrt{e+fz^{\eta}+gz^{2\eta}}=y.$	$z^{\eta}=x.$	$\sqrt{e+fx+gxx}=v.$	$\frac{d}{\eta}s=t=\frac{d}{\eta}$ in αGDB.
3. $dz^{\overset{2\eta}{-1}}\sqrt{e+fz^{\eta}+gz^{2\eta}}=y.$	$z^{\eta}=x.$	$\sqrt{e+fx+gxx}=v.$	$\frac{d}{3\eta g}v^3-\frac{df}{2\eta g}s=t.$
4. $dz^{\overset{3\eta}{-1}}\sqrt{e+fz^{\eta}+gz^{2\eta}}=y.$	$z^{\eta}=x.$	$\sqrt{e+fx+gxx}=v.$	$\frac{6dgx-5df}{24\eta gg}v^3\frac{+5dff-4deg}{16\eta gg}=t.$
Ordo septimus.[565]			
1. $\frac{d}{z\sqrt{e+fz^{\eta}}}=y.$	$\frac{1}{z^{\eta}}=xx.$	$\sqrt{f+exx}=v.$	$\frac{4d}{\eta f}$ in $\frac{1}{2}xv\div s=t=\frac{4d}{\eta f}$ in PAD vel in $aGDA$.
Vel sic	$\frac{1}{z^{\eta}}=x.$	$\sqrt{fx+exx}=v.$	$\frac{8de}{\eta ff}$ in $s-\frac{1}{2}xv-\frac{fv}{4e}=t=\frac{8de}{\eta ff}$ in $aGDA$.
2. $\frac{d}{z^{\overset{\eta}{+1}}\sqrt{e+fz^{\eta}}}=y.$	$\frac{1}{z^{\eta}}=xx.$	$\sqrt{f+exx}=v.$	$\frac{2d}{\eta e}$ in $s-xv=t=\frac{2d}{\eta e}$ in POD, vel in $AODGa$.
Vel sic	$\frac{1}{z^{\eta}}=x.$	$\sqrt{fx+exx}=v.$	$\frac{4d}{\eta f}$ in $\frac{1}{2}xv\div s=t=\frac{4d}{\eta f}$ in $aDGA$.
3. $\frac{d}{z^{\overset{2\eta}{+1}}\sqrt{e+fz^{\eta}}}=y.$	$\frac{1}{z^{\eta}}=x.$	$\sqrt{fx+exx}=v.$	$\frac{d}{\eta e}$ in $3s\div 2xv=t=\frac{d}{\eta e}$ in $3aDGA\div\Delta aDB$.
4. $\frac{d}{z^{\overset{3\eta}{+1}}\sqrt{e+fz^{\eta}}}=y.$	$\frac{1}{z^{\eta}}=x.$	$\sqrt{fx+exx}=v.$	$\frac{10dfxv-15dfs-2dexxv}{6\eta ee}=t.$

Curves:	Conic base	Conic ordinate	Values of the areas
Sixth order[563]			
1. $dz^{-1}\sqrt{[e+fz^{\eta}+gz^{2\eta}]} = y.$	$\begin{cases} z^{\eta} = x. \\ z^{-\eta} = \xi. \end{cases}$	$\left.\begin{matrix} \sqrt{[e+fx+gx^2]} = v. \\ \sqrt{[g+f\xi+e\xi^2]} = \Upsilon. \end{matrix}\right\}$	$\dfrac{d(4e^2\xi\Upsilon+2ef\Upsilon-2f^2v-8e^2\sigma+4fgs)}{\eta(4eg-f^2)} = t.$[564]
2. $dz^{\eta-1}\sqrt{[e+fz^{\eta}+gz^{2\eta}]} = y.$	$z^{\eta} = x.$	$\sqrt{[e+fx+gx^2]} = v.$	$(d/\eta)\,s = t = (d/\eta)\times(\alpha GDB).$
3. $dz^{2\eta-1}\sqrt{[e+fz^{\eta}+gz^{2\eta}]} = y.$	$z^{\eta} = x.$	$\sqrt{[e+fx+gx^2]} = v.$	$(d/\eta g)\,(\frac{1}{3}v^3-\frac{1}{2}fs) = t.$
4. $dz^{3\eta-1}\sqrt{[e+fz^{\eta}+gz^{2\eta}]} = y.$	$z^{\eta} = x.$	$\sqrt{[e+fx+gx^2]} = v.$	$(d/\eta g^2)\,([\frac{1}{4}gx-\frac{5}{24}f]\,v^3+[\frac{5}{16}f^2-\frac{1}{4}eg]) = t.$
Seventh order[565]			
1. $dz^{-1}/\sqrt{[e+fz^{\eta}]} = y.$	$z^{-\eta} = x^2.$	$\sqrt{[f+ex^2]} = v.$	$(4d/\eta f)\,\lvert\frac{1}{2}xv-s\rvert = t$ $= (4d/\eta f)\times(PAD)$ or $(aGDA).$
	Or thus: $z^{-\eta} = x.$	$\sqrt{[fx+ex^2]} = v.$	$(8de/\eta f^2)\,(s-\frac{1}{2}xv-\frac{1}{4}fv/e) = t$ $= (8de/\eta f^2)\times(aGDA).$
2. $dz^{-\eta-1}/\sqrt{[e+fz^{\eta}]} = y.$	$z^{-\eta} = x^2.$	$\sqrt{[f+ex^2]} = v.$	$(2d/\eta e)\,(s-xv) = t$ $= (2d/\eta e)\times(POD)$ or $(AODGa).$
	Or thus: $z^{-\eta} = x.$	$\sqrt{[fx+ex^2]} = v.$	$(4d/\eta f)\,\lvert\frac{1}{2}xv-s\rvert = t = (4d/\eta f)\times(aDGA).$
3. $dz^{-2\eta-1}/\sqrt{[e+fz^{\eta}]} = y.$	$z^{-\eta} = x.$	$\sqrt{[fx+ex^2]} = v.$	$(d/\eta e)\,\lvert 3s-2xv\rvert = t$ $= (d/\eta e)\times\lvert 3(aDGA)-\Delta aDB\rvert.$
4. $dz^{-3\eta-1}/\sqrt{[e+fz^{\eta}]} = y.$	$z^{-\eta} = x.$	$\sqrt{[fx+ex^2]} = v.$	$(d/\eta e^2)\,(\frac{5}{3}fxv-\frac{5}{2}fs-2ex^2v) = t.$

	Curvarum	Sectionis Conicæ: Basis.	Sectionis Conicæ: parall: Incedens.	Arearum valores.
‖[82]	‖*Ordo octavus.*(566)			
1.	$\dfrac{dz^{\eta-1}}{\sqrt{e+fz^{\eta}+gz^{2\eta}}}=y.$	$z^{\eta}=x.$	$\sqrt{e+fx+gxx}=v.$	$\dfrac{8dgs-4dgxv-2dfv}{4\eta eg-\eta ff}=t=\dfrac{8dg}{4\eta eg-nff}$ in $\alpha GDB\pm\Delta DBA.$
2.	$\dfrac{dz^{2\eta-1}}{\sqrt{e+fz^{\eta}+gz^{2\eta}}}=y.$	$z^{\eta}=x.$	$\sqrt{e+fx+gxx}=v.$	$\dfrac{-4dfs+2dfxv+4dev}{4\eta eg-\eta ff}=t.$
3.	$\dfrac{dz^{3\eta-1}}{\sqrt{e+fz^{\eta}+gz^{2\eta}}}=y.$	$z^{\eta}=x.$	$\sqrt{e+fx+gxx}=v.$	$\dfrac{\begin{array}{l}3dff\\-4deg\end{array}s\begin{array}{l}-2dff\\+4deg\end{array}xv-2defv}{4\eta egg-\eta ffg}=t.$
4.	$\dfrac{dz^{4\eta-1}}{\sqrt{e+fz^{\eta}+gz^{2\eta}}}=y.$	$z^{\eta}=x.$	$\sqrt{e+fx+gxx}=v.$	$\dfrac{\begin{array}{l}36defg\\-15dfff\end{array}s\begin{array}{l}+8degg\\-2dffg\end{array}xxv\begin{array}{l}-28defg\\+10dfff\end{array}xv\begin{array}{l}+10deff\\-16deeg\end{array}v}{24\eta eg^{3}-6\eta ffgg}=t.$
	Ordo nonus.(567)			
1.	$\dfrac{dz^{\overset{\eta}{-1}}\sqrt{e+fz^{\eta}}}{g+hz^{\eta}}=y.$	$\sqrt{\dfrac{d}{g+hz^{\eta}}}=x.$	$\sqrt{\dfrac{df}{h}+\dfrac{eh-fg}{h}xx}=v.$	$\dfrac{\begin{array}{l}4fg\\-4eh\end{array}s\begin{array}{l}-2fg\\+2eh\end{array}xv+2df\dfrac{v}{x}}{\eta fh}=t.$
2.	$\dfrac{dz^{\overset{2\eta}{-1}}\sqrt{e+fz^{\eta}}}{g+hz^{\eta}}=y.$	$\sqrt{\dfrac{d}{g+hz^{\eta}}}=x.$	$\sqrt{\dfrac{df}{h}+\dfrac{eh-fg}{h}xx}=v.$	$\dfrac{\begin{array}{l}4egh\\-4fgg\end{array}s\begin{array}{l}-2egh\\+2fgg\end{array}xv+\frac{2}{3}dh\dfrac{v^{3}}{x^{3}}-2dfg\dfrac{v}{x}}{\eta fhh}=t.$

Curves:	Conic base	Conic ordinate	Values of the areas
Eighth order[566]			
1. $\frac{dz^{\eta-1}}{\sqrt{[e+fz^{\eta}+gz^{2\eta}]}}=y.$	$z^{\eta}=x.$	$\sqrt{[e+fx+gx^2]}=v$	$\frac{d(8gs-4gxv-2fv)}{4\eta(eg-f^2)}=t$ $=8dg/\eta(4eg-f^2)\times[(\alpha GDB)\pm\Delta DBA].$
2. $\frac{dz^{2\eta-1}}{\sqrt{[e+fz^{\eta}+gz^{2\eta}]}}=y.$	$z^{\eta}=x.$	$\sqrt{[e+fx+gx^2]}=v.$	$\frac{d(-4fs+2fxv+4ev)}{\eta(4eg-f^2)}=t.$
3. $\frac{dz^{3\eta-1}}{\sqrt{[e+fz^{\eta}+gz^{2\eta}]}}=y.$	$z^{\eta}=x.$	$\sqrt{[e+fx+gx^2]}=v.$	$\frac{d([3f^2-4eg]\,s+[4eg-2f^2]\,xv-2efv)}{\eta g(4eg-f^2)}=t.$
4. $\frac{dz^{4\eta-1}}{\sqrt{[e+fz^{\eta}+gz^{2\eta}]}}=y.$	$z^{\eta}=x.$	$\sqrt{[e+fx+gx^2]}=v.$	

$$\frac{d([36eg-15f^2]fs+[8eg-2f^2]\,gx^2v+[10f^2-28eg]fxv+[10f^2-16eg]ev)}{6\eta g^2(4eg-f^2)}=t.$$

Curves:	Conic base	Conic ordinate	Values of the areas
Ninth order[567]			
1. $\frac{dz^{\eta-1}\sqrt{[e+fz^{\eta}]}}{g+hz^{\eta}}=y.$	$\sqrt{\frac{d}{g+hz^{\eta}}}=x.$	$\sqrt{\frac{df+(eh-fg)\,x^2}{h}}=v.$	$\frac{(eh-fg)\,(-4s+2xv)+2dfv/x}{\eta fh}=t.$
2. $\frac{dz^{2\eta-1}\sqrt{[e+fz^{\eta}]}}{g+hz^{\eta}}=y.$	$\sqrt{\frac{d}{g+hz^{\eta}}}=x.$	$\sqrt{\frac{df+(eh-fg)\,x^2}{h}}=v.$	$\frac{g(eh-fg)\,(4s-2xv)+\frac{2}{3}dhv^3/x^3-2dfgv/x}{\eta fh^2}=t.$

Curvarum	Sectionis Conicæ: Basis.	Sectionis Conicæ: parall: Incedens.	Arearum valores.
Ordo decimus.[568]			
1. $\dfrac{dz^{\eta-1}}{g+hz^{\eta}\sqrt{e+fz^{\eta}}}=y.$	$\sqrt{\dfrac{d}{g+hz^{\eta}}}=x.$	$\sqrt{\dfrac{df}{h}+\dfrac{eh-fg}{h}xx}=v.$	$\dfrac{2xv-4s}{\eta f}=t=\dfrac{4}{\eta f}ADGa.$
2. $\dfrac{dz^{2\eta-1}}{g+hz^{\eta}\sqrt{e+fz^{\eta}}}=y.$	$\sqrt{\dfrac{d}{g+hz^{\eta}}}=x.$	$\sqrt{\dfrac{df}{h}+\dfrac{eh-fg}{h}xx}=v.$	$\dfrac{4gs-2gxv+2d\frac{v}{x}}{\eta fh}=t.$
Ordo undecimus.[569]			
1. $dz^{-1}\sqrt{\dfrac{e+fz^{\eta}}{g+hz^{\eta}}}=y.$	$\left\{\begin{array}{l}\sqrt{g+hz^{\eta}}=x.\\ \sqrt{h+gz^{-\eta}}=\xi.\end{array}\right.$	$\left.\begin{array}{l}\sqrt{\dfrac{eh-fg}{h}+\dfrac{f}{h}xx}=v.\\ \sqrt{\dfrac{fg-eh}{g}+\dfrac{e}{g}\xi\xi}=\Upsilon.\end{array}\right\}$	$\dfrac{dxv^{3}z^{-\eta}-4dfs-4de\sigma}{\eta fg-\eta eh}=t.$
2. $dz^{\overset{\eta}{-1}}\sqrt{\dfrac{e+fz^{\eta}}{g+hz^{\eta}}}=y.$	$\sqrt{g+hz^{\eta}}=x.$	$\sqrt{\dfrac{eh-fg}{h}+\dfrac{f}{h}xx}=v.$	$\dfrac{2d}{\eta h}s=t.$
3. $dz^{\overset{2\eta}{-1}}\sqrt{\dfrac{e+fz^{\eta}}{g+hz^{\eta}}}=y.$	$\sqrt{g+hz^{\eta}}=x.$	$\sqrt{\dfrac{eh-fg}{h}+\dfrac{f}{h}xx}=v.$	$\dfrac{dhxv^{3}\begin{smallmatrix}-3dfg\\-deh\end{smallmatrix}s}{2\eta fhh}=t.$

Curves:	*Conic base*	*Conic ordinate*	*Values of the areas*
Tenth order[568]			
1. $\dfrac{dz^{\eta-1}}{(g+hz^\eta)\sqrt{[e+fz^\eta]}}=y.$	$\sqrt{\dfrac{d}{g+hz^\eta}}=x.$	$\sqrt{\dfrac{df+(eh-fg)\,x^2}{h}}=v.$	$\dfrac{4}{\eta f}(\tfrac{1}{2}xv-s)=t=\dfrac{4}{\eta j}\times(ADGa).$
2. $\dfrac{dz^{2\eta-1}}{(g+hz^\eta)\sqrt{[e+fz^\eta]}}=y.$	$\sqrt{\dfrac{d}{g+hz^\eta}}=x.$	$\sqrt{\dfrac{df+(eh-fg)\,x^2}{h}}=v.$	$\dfrac{4gs-2gxv+2dv/x}{\eta fh}=t.$
Eleventh order[569]			
1. $dz^{-1}\sqrt{\dfrac{e+fz^\eta}{g+hz^\eta}}=y.$	$\left\{\begin{array}{l}\sqrt{[g+hz^\eta]}=x.\\ \sqrt{[h+gz^{-\eta}]}=\xi.\end{array}\right.$	$\left.\begin{array}{l}\sqrt{[(eh-fg+fx^2)/h]}=v.\\ \sqrt{[(fg-eh+e\xi^2)/g]}=\Upsilon.\end{array}\right\}$	$\dfrac{d(xv^3z^{-\eta}-4fs-4e\sigma)}{\eta(fg-eh)}=t.$
2. $dz^{\eta-1}\sqrt{\dfrac{e+fz^\eta}{g+hz^\eta}}=y.$	$\sqrt{[g+hz^\eta]}=x.$	$\sqrt{[(eh-fg+fx^2)/h]}=v.$	$(2d/\eta h)\,s=t.$
3. $dz^{2\eta-1}\sqrt{\dfrac{e+fz^\eta}{g+hz^\eta}}=y.$	$\sqrt{[g+hz^\eta]}=x.$	$\sqrt{[(eh-fg+fx^2)/h]}=v.$	$\dfrac{d(hxv^3-[3fg+eh]\,s)}{2\eta fh^2}=t.$

NOTES TO PAGES 244–55

(549) In this 'posterior' catalogue of 'curves' and their corresponding 'areas' the area $t_\lambda = \int y.dz$ under the curve $y = f(z^\eta, \lambda)$ is converted by the substitution $x = F(z)$ into an equal area expressible in terms of x, $v = d/(e+fx)$ or $(e+fx+gx^2)^{\frac{1}{2}}$ and $s = \int v.dx$. As will be seen in the seventh volume Newton incorporated this integral table virtually unchanged in 1693 in the revised version of his *De Quadratura Curvarum* as a 'Tabula Curvarum simpliciorum quæ cum Ellipsi & Hyperbola comparari possunt'. In this latter form it first appeared in print in appendix to his *Opticks: or, a Treatise of the Reflexions, Refractions, Inflexions and Colours of Light* (London, 1704): $_2$199–204. A corrected version was published seven years later by William Jones in his *Analysis Per Quantitatum Series, Fluxiones, ac Differentias* (London, 1711) on two folding plates between pages 62 and 63, evidently from the emendations marked up by Newton in late 1709 in a copy of his first Latin *Optice* (London, 1706) which is now ULC. Adv. b.39.4.

(550) Where $y = dz^{\lambda\eta-1}/(e+fz^\eta)$, $\lambda = 1, 2, 3, \ldots$, the substitution $z^\eta = x$ yields

$$\int y.dz = (d/\eta)\int x^{\lambda-1}/(e+fx).dx = t_\lambda, \quad \text{say.}$$

Since $(d/\eta)\,x^{\lambda-1} = (\lambda-1)\,(ft_\lambda+et_{\lambda-1})$, it follows that

$$t_\lambda = \frac{d}{\eta f}\cdot\frac{x^{\lambda-1}}{\lambda-1}-\frac{e}{f}\cdot t_{\lambda-1} = \frac{d}{\eta f}x^\lambda.\sum_{1\leqslant i\leqslant(\lambda-1)}\left(-\frac{e}{f}\right)^{i-1}\cdot\frac{x^{-i}}{\lambda-i}+\left(-\frac{e}{f}\right)^{\lambda-1}\cdot\frac{s}{\eta},$$

where $s = t_0 = \int d/(e+fx).dx$, that is, $\int v.dx$.

(551) The first diagram of figure 2 above.

(552) Where $y = dz^{(\lambda-\frac{1}{2})\eta-1}/(e+fz^\eta)$, $\lambda = 1, 2, 3, \ldots$, the substitution $x^2 = d/(e+fz^\eta)$ yields $t_\lambda = \int y.dz = -(2d/\eta f)\int x^{-2(\lambda-1)}([d-ex^2]/f)^{\lambda-\frac{3}{2}}.dx$. On taking $I_\mu = \int f^{\frac{1}{2}}x^{2\mu}(d-ex^2)^{-\frac{1}{2}}.dx$, we find $s = \int v.dx = f^{-1}(dI_0-eI_1)$, $xv = f^{-\frac{1}{2}}x(d-ex^2)^{\frac{1}{2}} = f^{-1}(dI_0-2eI_1)$, $z^{\frac{1}{2}\eta} = -f^{-1}dI_{-1}$ and $z^{\frac{3}{2}\eta} = -3f^{-2}d(dI_{-2}-eI_{-1})$: Newton's (corrected) results follow on noting that

$$\eta t_1 = -2f^{-1}dI_0, \quad \eta t_2 = -2f^{-2}d(dI_{-1}-eI_0)$$

and $\eta t_3 = -2f^{-3}d(d^2I_{-2}-2deI_{-1}+e^2I_0)$, and then eliminating the I_μ, $\mu = -2, -1, 0, 1$. Note that in his tabulation of the integrals in Ordo 2.2 and Ordo 2.3 an extraneous factor e has crept into the first term of the former and the two first terms of the latter. Newton himself observed the error in late 1709 and the corrected integrals were published two years later in Jones' *Analysis* (see note (549)).

(553) By '$2xv \div 4s$' understand the absolute difference of $2xv$ and $4s$, that is $|2xv-4s|$. The notation '$\div$' is a Newtonian variant on Barrow's '$-:$', signifying the 'differentiam vel excessum; item quantitates omnes, quæ sequuntur, subtrahendas esse, signis non mutatis' (Isaac Barrow, *Euclidis Elementorum Libri XV. breviter demonstrati* (Cambridge, 1655): [A8]^v: 'Notarum explicatio').

(554) In a prior version of this following cancelled Ordo 3 every occurrence of p was written out in full as '$\sqrt{ff-4eg}$'.

(555) For '$f+p+gx$' read '$f+p+2gx$', and correspondingly for '$f-p+g\xi$' read '$f-p+2g\xi$'.

(556) For '$df+dp+gx$' read '$f+p+2gx$', and similarly for '$df-dp+g\xi$' read '$f-p+2g\xi$'.

(557) Where $y = dz^{\lambda\eta-1}/(e+fz^\eta+gz^{2\eta})$ the substitution $z^\eta = x$ yields

$$t_\lambda = (d/\eta)\int x^\lambda/(e+fx+gx^2).dx.$$

On taking $f^2-p^2 = 4eg$, $4g(e+fx+gx^2) = (f+p+2gx)(f-p+2gx)$ and immediately

$$t_1 = \frac{d}{\eta p}\int\left[\frac{2g}{f-p+2gx}-\frac{2g}{f+p+2gx}\right].dx, \quad t_2 = \frac{d}{\eta p}\int\left[\frac{f+p}{f+p+2gx}-\frac{f-p}{f-p+2gx}\right].dx.$$

The 'more general' construction of Ordo 3.1 is taken up in the revised version following (see next note).

(558) Where $y = dz^{\lambda\eta-1}/(e+fz^{\eta}+gz^{2\eta})$, on setting $p^2 = f^2-4eg$ the substitution

$$x^2 = d/(e+fz^{\eta}+gz^{2\eta})$$

yields $\quad t_\lambda = \int y.dz = -(2d/\eta)\int(2g)^{1-\lambda}(x^{-1}[4dg+p^2x^2]^{\frac{1}{2}}-f)^{\lambda-1}(4dg+p^2x^2)^{-\frac{1}{2}}.dx.$

Hence on making $I_\mu = \int 2gx^{2\mu}(4dg+p^2x^2)^{-\frac{1}{2}}.dx$ we will find that

$$s = \int v.dx = (4dgI_0+p^2I_1)/4g^2, \quad xv = (2dgI_0+p^2I_1)/2g^2$$

and lastly (since $e+\xi = dx^{-2}$) $\sigma = \int \Upsilon.d\xi = -2\int x^{-1}.dx$. Newton's primary results follow at once by observing that $t_1 = -(d/\eta g)I_0$ and $t_2 = (-d\int x^{-1}.dx+(d/g)I_0)/2\eta g$. Alternatively, on putting $J_\mu = \int 2ex^{2\mu}(4de+p^2x^2)^{-\frac{1}{2}}.dx$, the substitution $x^2 = dz^{2\eta}/(e+fz^{\eta}+gz^{2\eta})$ gives

$$t_1 = (d/\eta e)J_0,$$

which we may evaluate straightforwardly by eliminating J_1 between $4e^2s = 4deJ_0+p^2J_1$ and $2e^2xv = 2deJ_0+p^2J_1$. Since $\lambda\eta(et_\lambda+ft_{\lambda+1}+gt_{\lambda+2}) = dz^{\lambda\eta}$, all t_λ, $\lambda \geqslant 2$, may be evaluated in terms of t_1 and t_2.

(559) On setting $t_\lambda = d\int z^{(\lambda-\frac{1}{2})\eta-1}/(e+fz^{\eta}+gz^{2\eta}).dz$ the integrals

$$t_1 = p^{-1}\int[2dgz^{\frac{1}{2}\eta-1}/(f-p+2gz^{\eta})-2dgz^{\frac{1}{2}\eta-1}/(f+p+2gz^{\eta})].dz$$

and $t_2 = p^{-1}\int[2dez^{\frac{1}{2}\eta-1}/(2e+(f-p)z^{\eta})-2dez^{\frac{1}{2}\eta-1}/(2e+(f+p)z^{\eta})].dz$ are readily resolved in Newton's form by a quadruple application of the construction in Ordo 2.1 above.

(560) Where $y = dz^{-\lambda\eta-1}(e+fz^{\eta})^{\frac{1}{2}}$, on defining $I_\mu = \int x^{2\mu}(f+ex^2)^{\frac{1}{2}}.dx$ the substitution $z^{\eta} = x^{-2}$ yields $t_\lambda = \int y.dz = -(2d/\eta)I_{\lambda-1}$ and we may readily evaluate t_1, t_2 in terms of $s = \int v.dx = I_0$ and $v^3/x = 2eI_0-fI_{-1}$. Alternatively, on putting $J_\mu = \int x^{\mu}(ex^2+fx)^{-\frac{1}{2}}.dx$ the substitution $z^{\eta} = x^{-1}$ gives $t_\lambda = -(d/\eta)(eJ_\lambda+fJ_{\lambda+1})$, $\lambda = 0, 1, 2, 3$, and these integrals are at once constructible in terms of $v/x = -\frac{1}{2}fJ_{-1}$, $v = eJ_1+\frac{1}{2}fJ_0$, $s = eJ_2+fJ_1$,

$$xv = 2eJ_2+\tfrac{3}{2}fJ_1 \quad\text{and}\quad v^3 = 3e^2J_3+\tfrac{9}{2}efJ_2+\tfrac{3}{2}f^2J_1.$$

(561) Read 'seu' probably.

(562) Newton has here correctly cancelled a first, negative form '$2dv^3-3dfs$' of the present numerator.

(563) Where $y = dz^{\lambda\eta-1}(e+fz^{\eta}+gz^{2\eta})^{\frac{1}{2}}$, the substitution $z^{\eta} = x$ yields

$$t_\lambda = \int y.dz = (d/\eta)\int x^{\lambda-1}(e+fx+gx^2)^{\frac{1}{2}}.dx,$$

that is, $(d/\eta)(eI_{\lambda-1}+fI_\lambda+gI_{\lambda+1})$ on putting $I_\mu = \int x^{\mu}(e+fx+gx^2)^{-\frac{1}{2}}.dx$. We may then, as Newton, evaluate t_1, t_2, t_3 in terms of $v = \frac{1}{2}fI_0+gI_1$, $s = \int v.dx = eI_0+fI_1+gI_2$,

$$xv = eI_0+\tfrac{3}{2}fI_1+2gI_2, \quad v^3 = \tfrac{3}{2}efI_0+(\tfrac{1}{2}f^2+eg)I_1+\tfrac{3}{2}fgI_2+g^2I_3$$

and $xv^3 = e^2I_0+\frac{7}{2}efI_1+3(\frac{1}{2}f^2+eg)I_2+\frac{7}{2}fgI_3+2g^2I_4$ by eliminating I_0, I_1, I_2, I_3 and I_4. In general, since $x^{\lambda}(e+fx+gx^2)^{\frac{1}{2}} = \lambda eI_{\lambda-1}+(\lambda+\frac{1}{2})fI_\lambda+(\lambda+1)gI_{\lambda+1}$, all t_λ, $\lambda \geqslant 4$ may be expressed in terms of I_0, I_1, I_2, I_3, I_4 and so of v, s, xv, v^3 and xv^3. Similarly, on substituting z^{η} (or x) $= \xi^{-1}$, it follows that $\Upsilon = x^{-1}(e+fx+gx^2) = eI_{-2}-\frac{1}{2}fI_{-1}$, $\xi\Upsilon = -2eI_{-3}-\frac{3}{2}fI_{-2}-gI_{-1}$ and $\sigma = \int\Upsilon.d\xi = -\int x^{-3}(e+fx+gx)^{\frac{1}{2}}.dx = -eI_{-3}-fI_{-2}-gI_{-1}$; from this Newton's (corrected) construction of t_0 readily follows by eliminating I_{-3}, I_{-2}, I_{-1}, I_0, I_1 and I_2.

(564) The numerator lacks the necessary terms '$-2dfgxv+4degv$'. The error was first gently pointed out to Newton by Roger Cotes in his first known letter to him on 18 August 1709 (Joseph Edleston, *Correspondence of Sir Isaac Newton and Professor Cotes* (London, 1850): 4). As Cotes wrote, he had 'some days ago' been examining Proposition 91, Corollary 2 of Book 1 of Newton's *Principia* 'and found it to be true by ye Quadratures of ye 1st & 2^d^ Curves of ye 8th Form of ye second Table in y^r^ Treatise *De Quadrat[ura Curvarum]*'. (Compare note (549): Cotes' unpublished verification of the attraction of Newton's 'Sphæroid' is preserved in Trinity College, Cambridge, R.16.18: 24–5.) Cotes continued that 'at the same time I went over ye whole [Sixth] & Eighth Forms which agreed with my Computation excepting ye First

||[83] || Antequam Theoremata in his Curvarum classibus tradita exemplis illustrare pergam, juvabit observare.(570)

1 Quòd cùm quantitatum d, e, f, g, h et i signa omnia in æquationibus curvas definientibus affirmativa posuerim, siquando contingant esse negativa in subsequentibus Basis et incedentis lineæ Sectionis Conicæ, nec non quæsitæ Areæ valoribus mutari debent.(571)

2. Numeralium η et θ, ubi negativæ sunt, signa in arearum valoribus sunt etiam mutanda. Quinetiam ipsarum signis mutatis Theoremata novam formam induere possunt. Sic in septimo ordine posterioris Catalogi, Theorema tertium,

---

of y[e] [Sixth] & Fourth of y[e] Eighth'. The latter error was, in fact, merely a transcriptional one made by Newton in 1693 in copying the present catalogue: as such it is here of no concern. In correction of the former Cotes suggested the entry

$$\frac{4de\frac{\Upsilon^3}{\xi}-2df\frac{v^3}{x}-8dee\sigma+4dfgs}{4\eta eg-\eta ff} = t_{[0]},$$

which agrees with our emendation since $\xi = x^{-1}$ and so $\Upsilon = \xi v$, $v = x\Upsilon$.

(565) In this Ordo $y = dz^{-\lambda\eta-1}/(e+fz^{\eta})^{\frac{1}{2}}$, so that, on setting $I_\mu = \int x^{2\mu}(ex^2+f)^{-\frac{1}{2}}.dx$, substitution of $z^{\eta} = x^{-2}$ produces $t_\lambda = \int y.dz = -(2d/\eta)\,I_\lambda$ with

$$x^{2\lambda-1}(ex^2+f)^{\frac{1}{2}} = 2\lambda eI_\lambda+(2\lambda-1)\,fI_{\lambda-1}.$$

The particular integrals t_0 and t_1 may readily be evaluated in terms of

$$s = eI_1+fI_0 \quad\text{and}\quad xv = 2eI_1+fI_0$$

by eliminating I_0 and I_1, while all t_λ, $\lambda \geqslant 2$, are expressible in terms of t_0, t_1. Alternatively, on making $J_\mu = \int x^{\mu}(ex^2+fx)^{-\frac{1}{2}}.dx$, substitution of $z^{\eta} = x^{-1}$ yields $t_\lambda = -(d/\eta)\,J_\lambda$ with

$$x^{\lambda-1}(ex^2+fx)^{\frac{1}{2}} = (\lambda+1)\,eJ_\lambda+\lambda fJ_{\lambda-1}.$$

Accordingly t_0, t_1, t_2, t_3 are easily determined, as in Newton's scheme, in terms of

$$v = eJ_1+\tfrac{1}{2}fJ_0,\quad s = \int v.dx = eJ_2+fJ_1,\quad xv = 2eJ_2+\tfrac{3}{2}fJ_1 \quad\text{and}\quad x^2v = 3eJ_3+\tfrac{5}{2}fJ_2$$

by eliminating J_0, J_1, J_2, J_3.

(566) Here $y = dz^{\lambda\eta-1}/(e+fz^{\eta}+gz^{2\eta})$. On setting $I_\mu = \int x^{\mu}(e+fx+gx^2)^{-\frac{1}{2}}.dx$ the substitution $z^{\eta} = x$ yields $t_\lambda = \int y.dz = (d/\eta)\,I_{\lambda-1}$, and the particular integrals t_1, t_2, t_3 are readily computed in terms of $v = \frac{1}{2}fI_0+gI_1$, $s = \int v.dx = eI_0+fI_1+gI_2$, $xv = eI_0+\frac{3}{2}fI_1+2gI_2$ and $x^2v = 2eI_1+\frac{5}{2}fI_2+3gI_3$ by eliminating I_0, I_1, I_2, I_3.

(567) In this case, where $y = dz^{\lambda\eta-1}(e+fz^{\eta})^{\frac{1}{2}}/(g+hz^{\eta})$, substitution of $x^2 = d/(g+hz^{\eta})$ gives $t_\lambda = -(2d/\eta h^{\lambda+1})\int h^{-\frac{1}{2}}x^{-2\lambda}(d-gx^2)^{\lambda-1}(fd+(eh-fg)\,x^2)^{\frac{1}{2}}.dx$. Hence, making

$$I_\mu = \int h^{-\frac{1}{2}}x^{2\mu}(fd+(eh-fg)\,x^2)^{-\frac{1}{2}}.dx,$$

we may compute $t_1 = -(2d/\eta h^2)\,(fdI_{-1}+(eh-fg)\,I_0)$ and

$$t_2 = -(2d/\eta h^3)\,(d[fdI_{-2}+(eh-fg)\,I_{-1}]-g[fdI_{-1}+(eh-fg)\,I_0])$$

Before I proceed to give illustrative examples of the theorems presented in these classes of curves, it will be helpful to observe:[(570)]

1. Though I have taken all the signs of the quantities d, e, f, g, h and i in the defining equations of the curves to be positive, whenever they happen to be negative they should be altered in subsequent values of the conic's base and ordinate lines, and in that of the area sought also.[(571)]

2. When the numerical symbols η and θ are negative, their signs must also be altered in the values of the areas. Indeed, after their signs have been changed the theorems may take on a new form. Thus in the seventh order of the latter

---

in terms of $hs = fdI_0 + (eh - fg)I_1$, $hxv = fdI_0 + 2(eh - fg)I_1$, $hv/x = -fdI_{-1}$, and

$$h^2v^3/x^3 = -3fd(fdI_{-2} + (eh - fg)I_{-1})$$

by eliminating I_{-2}, I_{-1}, I_0 and I_1.

(568) Much as in the previous note but where now $y = dz^{\lambda\eta-1}/(e+fz^\eta)^{\frac{1}{2}}(g+hz^\eta)$, substitution of $x^2 = d/(g+hz^\eta)$ in $t_\lambda = \int y.dz$ here yields the particular integrals $t_1 = -(2d/\eta h)I_0$, $t_2 = -(2d/\eta h^2)(dI_{-1} - gI_0)$ which are at once evaluated in terms of s, xv and v/x by eliminating I_{-1}, I_0 and I_1.

(569) In this last Ordo, where $y = dz^{\lambda\eta-1}(e+fz^\eta)^{\frac{1}{2}}/(g+hz^\eta)^{\frac{1}{2}}$, on making

$$I_\mu = \int h^{-\frac{1}{2}}(x^2-g)^\mu(eh - fg + fx^2)^{\frac{1}{2}}.dx$$

substitution of $x^2 = g + hz^\eta$ yields $t_\lambda = \int y.dz = (2d/\eta h^\lambda)I_{\lambda-1}$; from which the particular integrals t_1 and t_2 are readily evaluated in terms of

$$s = \int v.dx = I_0 \quad \text{and} \quad hxv^3 = (3fg + eh)I_0 + 4fI_1.$$

Alternatively, substitution of $\xi^2 = gz^{-\eta} + h$ gives $\sigma = \int \Upsilon.d\xi = -(g/h)I_{-2}$, which may be used to eliminate I_{-2} from $hxv^3z^{-\eta} = -2egI_{-2} + (fg - eh)I_{-1} + 2fhI_0$ and so evaluate $t_0 = (2d/\eta h)I_{-1}$ in Newton's terms. On a point of mathematical style, Newton might well have chosen to replace $z^{-\eta}$ by its equivalent $h/(x^2-g) = (\xi^2 - h)/g$.

(570) These following remarks amplify a short introductory paragraph on Newton's (cancelled) page 76 preceding. When in October 1676 he referred Leibniz, in his *epistola posterior*, to 'Theoremata quædam quæ pro comparatione Curvarum cum Conicis sectionibus in Catalogum dudum retuli' (namely, in his first table above), Newton went on to say that 'Possum utiꝗ cum Conicis sectionibus geometricè comparare curvas...numero infinitè [multas]', instancing the 'ordinatim applicatæ' of curves listed in Orders 3.1/2, 4.1/2, 6.1/2, 8.1/2, 9.1/2, 10.1/2 and 11.1/2 above in support of his claim, and then observed generally: 'singula bina Theoremata sunt duo primi termini seriei in infinitum progredientis. In [Ordinibus 4.1 et 4.2] $4eg$ debet esse non majus quam ff nisi e et g sint contrarij signi: in cæteris nulla est limitatio. Horum aliqua (nempe [Ordines 3.2, 4.1, 4.2, 6.1 et 11.1]) ex areis duarum Conicarum sectionum conjunctis constant. Alia quædam, ut [Ordines 9.1, 11.1 et 11.2], sunt aliter satis composita; et omnia quidem in continuatione progressionum citò evadunt compositissima: adeò ut vix per transmutationes figurarum quibus Gregorius et alij usi sunt [see notes (442) and (704)] absꝗ ulteriori fundamento inveniri posse putem. Ego equidem haud quicquam generale in his obtinere potui antequam abstraherem a contemplatione figurarum et rem totam ad simplicem considerationem ordinatim applicatarum reducerem' (*Correspondence of Isaac Newton*, **2** (Cambridge, 1960): 119–20).

(571) A sop to those intended readers—in 1670 still the vast majority—unfamiliar with the concept of an algebraic variable having both a negative and a positive range.

signo ipsius η mutato,(572) evadit $\dfrac{d}{z^{\overset{-2\eta}{+1}}\sqrt{e+fz^{-\eta}}}=y.$ $\dfrac{1}{z^{-\eta}}=x.$ &c. hoc est $\dfrac{dz^{3\eta-1}}{\sqrt{ez^{2\eta}+fz^{\eta}}}=y.$ $z^{\eta}=x.$ $\sqrt{fx+exx}=v.$ $\dfrac{d}{\eta e}$ in $2xv-3s=t$. Et sic in alijs.(573)

3. Cujusꝗ ordinis (si secundum prioris Catalogi demas) series utrinꝗ in infinitum(574) continuari potest. Scilicet in tertij quartiꝗ ordinis seriebus prioris Catalogi, numeri coefficientes initialium terminorum (2, -4, 16, -96, 868(575) &c) generantur multiplicando numeros -2, -4, -6, -8, -10 &c in se continuò; et subsequentium terminorum coefficientes ex initialibus in tertio

||[84] || ordine derivantur multiplicando gradatim per $-\frac{3}{2}$, $-\frac{5}{4}$, $-\frac{7}{6}$, $-\frac{9}{8}$, $-\frac{11}{10}$ &c, vel in quarto ordine multiplicando per $-\frac{1}{2}$, $-\frac{3}{4}$, $-\frac{5}{6}$, $-\frac{7}{8}$, $-\frac{9}{10}$ &c. Denominatorum verò coefficientes (1, 3, 15, 105 &c) ex ductu numerorum 1, 3, 5, 7, 9 &c in se gradatim oriuntur.(576)

In secundo autem Catalogo series ordinum 1, 2, 3, 4, 9 & 10 ope solius divisionis infinitè(577) producuntur. Sic habito $\dfrac{dz^{4\eta-1}}{e+fz^{\eta}}=y$, si divisionem ad usꝗ convenientem periodum instituas, orietur e.g.

$$\frac{d}{f}z^{3\eta-1}-\frac{de}{ff}z^{2\eta-1}+\frac{dee}{f^3}z^{\eta-1}\frac{-\dfrac{de^3}{f^3}z^{\eta-1}}{[e]+fz^{\eta}}=y.$$

Priores tres termini sunt primi ordinis prioris Catalogi et quartus primæ speciei hujus ordinis:(578) unde constat aream valere $\dfrac{d}{3\eta f}z^{3\eta}-\dfrac{de}{2\eta ff}z^{2\eta}+\dfrac{dee}{\eta f^3}z^{\eta}-\dfrac{e^3}{\eta f^3}s$; positâ nempe s areâ sectionis conicæ(579) cujus basis x sit$=z^{\eta}$, et incedens applicata $v=\dfrac{d}{e+fx}$.

Quinti autem sextiꝗ ordinis series ope duarum Theorematum in quinto ordine prioris Catalogi per debitam Additionem vel subductionem infinitè(577) producuntur, ut et septimi octaviꝗ series ope Theorematum in subsequenti

(572) That is, on substituting $-\eta$ for $+\eta$.

(573) The paragraph which follows next in the manuscript, first marked '3' but subsequently renumbered '5', has been here reset in sequence on Newton's page 85 below.

(574) The 'secundum prioris Catalogi' is (compare note (540)) the *particular* case for which $\theta=\eta-1$, $\lambda=-2$ ($r=1$, $s=-1$) of the expansion of $I_\theta=d\int z^{\theta}(e+fz^{\eta})^{\lambda}.dz$ as

$$-(d/\eta fr)\,z^{\theta+1-\eta}(e+fz^{\eta})^{\lambda+1}.\sum_{0\leqslant j\leqslant k-1}\prod_{0\leqslant i\leqslant j}\left(\frac{-(r-i)e}{(s-i)fz^{\eta}}\right)+R_kI_{\theta-k\eta},$$

where $\quad r=(\theta+1)/\eta,\quad s=\lambda+r\quad$ and $\quad R_k=\prod_{0\leqslant i\leqslant k}\left(-\dfrac{(r-i-1)e}{(s-i)f}\right).$

Evidently I_θ may be expanded in the opposite direction in a similar way, yielding an 'error' term $R'_kI_{\theta+k\eta}$, though, as we shall see in the sixth volume, Newton himself did not do so till

catalogue, the third theorem comes out, when the sign of η is altered,[572] as $dz^{-(-2\eta)-1}/\surd[e+fz^{-\eta}]=y.\ z^{-(-\eta)}=x....$, that is, as $dz^{3\eta-1}/\surd[ez^{2\eta}+fz^{\eta}]=y.$ $z^{\eta}=x.\surd[fx+ex^2]=v.\ (d/\eta e)\,(2xv-3s)=t.$ And so in other cases.[573]

3. In any order (if you except the second of the former catalogue) the series may be continued to infinity in either direction.[574] Specifically, in the series of the third and fourth orders of the former catalogue, the numerical coefficients of the first terms (2, -4, 16, -96, [7]68, ...) are generated by multiplying the numbers -2, -4, -6, -8, -10, ... into one another continually, while the coefficients of subsequent terms are derived from the initial ones, in the third order, by multiplying these consecutively by $-\frac{3}{2}$, $-\frac{5}{4}$, $-\frac{7}{6}$, $-\frac{9}{8}$, $-\frac{11}{10}$, ... or, in the fourth order, consecutively by $-\frac{1}{2}$, $-\frac{3}{4}$, $-\frac{5}{6}$, $-\frac{7}{8}$, $-\frac{9}{10}$, ... The coefficients (1, 3, 15, 105, ...) of the denominators, however, arise from the step by step multiplication of the numbers 1, 3, 5, 7, 9, ... into each other.[576]

In the second catalogue the series of orders 1, 2, 3, 4, 9 and 10 are extended infinitely[577] with the aid solely of division. So, when there is had

$$dz^{4\eta-1}/(e+fz^{\eta})=y,$$

should you carry out the division up to a convenient period there will arise, for example,

$$(d/f)\,z^{3\eta-1}-(de/f^2)\,z^{2\eta-1}+(de^2/f^3)\,z^{\eta-1}-(de^3/f^3)\,z^{\eta-1}/(e+fz^{\eta})=y.$$

The three former terms belong to the first order of the former catalogue, the fourth to the first species of the present order:[578] hence it is established that the value of the area is $(d/3\eta f)\,z^{3\eta}-(de/2\eta f^2)\,z^{2\eta}+(de^2/\eta f^3)\,z^{\eta}-(e^3/\eta f^3)\,s$, where s, namely, is set as the area of the conic[579] whose base $x=z^{\eta}$ and ordinate $v=d/(e+fx)$.

But the series of the fifth and sixth orders are extended infinitely[577] with the help of the two theorems in the fifth order of the former catalogue by appropriate addition and subtraction, just as those of the seventh and eighth are with the

late in 1691. (The 'Manuscript by Newton on Quadratures [?1676]' printed by H. W. Turnbull in his edition of *The Correspondence of Isaac Newton*, **2** (1960): 171–4 belongs, in fact, to this later period.) The generalized expansions given in notes (541) and (542) above may likewise be inverted. Implicit in Newton's remark is the assumption that all such series, when continued *in infinitum*, converge—a minimal condition for which is that the error functions R_k, R'_k should become zero as $k\to\infty$.

(575) Read '768' correctly.

(576) Compare notes (541) and (542).

(577) 'indefinitè' (indefinitely) is cancelled both times.

(578) Namely, Ordo 1.1 in the latter table above.

(579) Specifically, a rectangular hyperbola: in equivalent analytical terms

$$s=\int v.dx=(d/f)\log(e/(f+x)).$$

sexto ordine; ac undecimi series ope Theorematis in decimo ordine ejusdem prioris Catalogi.[580] E.g. si præfati quinti ordinis series ultra producenda sit, finge $\theta=-4\eta$, et quinti ordinis alterius Catalogi Theorema primum evadit $-8\eta e z^{\overset{-4\eta}{-1}}-5\eta f z^{\overset{-3\eta}{-1}}$ in $\frac{1}{2}\sqrt{e+fz^{\eta}}=y$. $\frac{R^3}{z^{4\eta}}=t$. Est autem juxta quartum Theorema hujus producendæ seriei, $\left(\text{scripto } -\frac{5\eta f}{2} \text{ pro } d,\right)$ $-\frac{5\eta}{2}fz^{\overset{[-]3\eta}{-1}}\sqrt{e+fz^{\eta}}=y$. $\frac{1}{z^{\eta}}=x$. $\sqrt{fx+exx}=v$. & $\frac{10fv^3-15ffs}{12e}=t$. Quare subductis prioribus ipsarum y ac t valores restabunt $4\eta e z^{\overset{-4\eta}{-1}}\sqrt{e+fz^{\eta}}=y$, et $\frac{10fv^3-15ffs}{12e}$[581] $-\frac{R^3}{z^{4\eta}}=t$. Ipsisq; in $\frac{d}{4\eta e}$ ductis, et pro $\frac{R^3}{z^{4\eta}}$ scripto si placet xv^3, emerget quintum producendæ seriei[582] Theorema $\frac{d}{z^{\overset{4\eta}{+1}}}\sqrt{e+fz^{\eta}}=y$. $\frac{1}{z^{\eta}}=x$. $\sqrt{fx+exx}=v$, & $\frac{10dfv^3-15dffs}{48\eta ee}-\frac{dxv^3}{4\eta e}=t$.

4. Horum ordinum nonnulli ex alijs etiam possunt aliter derivari, utpote in posteriori Catalogo quintus, sextus, septimus et undecimus ab octavo, ac nonus a decimo.[583] Adeo ut omisisse potuissem, nisi quod usui esse possint, quamvis
||[85] non prorsus necessariæ. || Nonnullos tamen ordines omisi quos a primo et secundo, nec non a nono decimoq; derivasse potuissem, utpote qui denominatoribus magis compositis afficiuntur, et proinde vix ulli unquam usui esse possunt.

[584]5. Si Curvæ alicujus definiens æquatio ex pluribus æquationibus diversorum ordinum vel diversarum specierum ejusdem ordinis componatur, ejus aream ex areis correspondentibus componere oportet; cavendo tamen ut[585] signis + et − rectè connectantur. Nam parallelè incedentes parallelè incedentibus et areæ correspondentes correspondentibus areis non semper sunt simul addendæ vel simul subducendæ; sed aliquando harum summa et illarum differentia sumenda est pro nova linea incedente et area correspondente constituenda. Et hoc fieri debet cùm constituentes areæ positæ sunt ad diversam partem parallelè incedentis. Ut autem hoc incommodum cauti promptiùs

(580) Compare notes (563), (565), (566) and (569).

(581) Newton wrote '$10dfv^3-15ffs$' in the numerator but we have suppressed the stray factor d in the first term.

(582) 'dicti ordinis' (of the said order) is cancelled.

(583) In Ordo 5 the ordinate may be set as

$$(dez^{-(\lambda+1)\eta-1}+df^{-\lambda\eta-1})/(fz^{-\eta}+ez^{-2\eta})^{\frac{1}{2}};$$

in Ordo 6 as $(dez^{\lambda\eta-1}+dfz^{(\lambda+1)\eta-1}+dgz^{(\lambda+2)\eta-1})/(e+fz^{\eta}+gz^{2\eta})^{\frac{1}{2}}$; in Ordo 7 as

$$dz^{-(\lambda+1)\eta-1}/(fz^{-\eta}+ez^{-2\eta})^{\frac{1}{2}};$$

and in Ordo 11 as $(dez^{\lambda\eta-1}+dfz^{(\lambda+1)\eta-1})/(eg+(eh+fg)z^{\eta}+fhz^{2\eta})^{\frac{1}{2}}$. Similarly, in Ordo 9

$$y=(dez^{\lambda\eta-1}+dfz^{(\lambda+1)\eta-1})/(e+fz^{\eta})^{\frac{1}{2}}(g+hz^{\eta}).$$

assistance of the theorems in the immediately following sixth order, while the series of the eleventh is extended with the aid of the tenth order's theorem in the same first catalogue.[580] If, for example, the series of the above-cited fifth order has to be further extended, suppose $\theta = -4\eta$ and the first theorem of the fifth order of the other catalogue comes out as

$$(-8\eta e z^{-4\eta-1} - 5\eta f z^{-3\eta-1}) \times \tfrac{1}{2}\sqrt{[e+fz^{\eta}]} = y. \quad z^{-4\eta}R^3 = t.$$

However, according to the fourth theorem of the present series to be produced (when $-\frac{5}{2}\eta f$ is written in place of d)

$$-\tfrac{5}{2}\eta f z^{-3\eta-1}\sqrt{[e+fz^{\eta}]} = y. \quad z^{-\eta} = x. \quad \sqrt{[fx+ex^2]} = v. \quad (f/e)\,(\tfrac{5}{6}v^3 - \tfrac{5}{4}fs) = t.$$

Consequently, when the former values are taken away there will remain those of y and t: $4\eta e z^{-4\eta-1}\sqrt{[e+fz^{\eta}]} = y$ and $(f/e)\,(\frac{5}{6}v^3 - \frac{5}{4}fs) - z^{-4\eta}R^3 = t$. After these are multiplied into $d/4\eta e$ and, appropriately, xv^3 is written instead of $z^{-4\eta}R^3$, the fifth theorem of the series to be extended[582] will emerge as:

$$dz^{-4\eta-1}\sqrt{[e+fz^{\eta}]} = y. \quad z^{-\eta} = x. \quad \sqrt{[fx+ex^2]} = v.$$

$$(d/\eta e)\,([\tfrac{5}{24}f/e]\,v^3 - \tfrac{5}{16}[f^2/e]\,s - \tfrac{1}{4}xv^3) = t.$$

4. Several of these orders may also be derived from others in other ways, as for instance in the latter catalogue the fifth, sixth, seventh and eleventh from the eighth, and the ninth from the tenth.[583] Accordingly, I could have omitted them were it not that they might, although strictly superfluous, be of use. Some orders, however, which could have been derived from the first and second, and also the ninth and tenth, I have omitted seeing that they are affected with more compound denominators and therefore scarcely ever able to be of any use.

[584]5. If the defining equation of some curve be composed of several equations of different orders or different species of the same order, its area must be composed of the corresponding areas—but take care to join them[585] correctly with the signs + and −. For ordinates are not always to be added one to another or taken away from each other in exactly the way that the corresponding areas are, but sometimes the sum of the latter is to be taken when the former's difference is assumed in order to form a new ordinate line and corresponding area. This should be done when the component areas are placed on opposite sides of the ordinate. However, to enable the wary reader to avoid this inconvenience more

(584) In accordance with Newton's marginal dictate 'Hic intersere notas 5, 6, 7, 8, et 9' we here interpose paragraph 5 from page 83 above (compare note (573)) and continue with 'Notes' 6–9 as listed on an interleaved supplementary sheet (ULC. Add. 3960.14: 168^{v}) added some time after he completed the first version of his present observations (namely, in order, paragraphs 1, 2, 5, 4 and 10).

(585) 'de illarum arearum additione et subductione ut partes' (but with regard to the addition and subtraction of those areas [take care to join] their parts) is cancelled.

devitare possint, singulis arearum valoribus propria signa, (etiamsi nonnunquam negativa, ut fit in posterioris Catalogi quinto septimoqꝫ ordine,) præfixi.[586]

||[85a] Vide Exempl 1 sequ. || [587]6. De Arearum signis observandum est præterea quod $+s$ vel denotat aream Conicæ sectionis Basi adjacentem esse reliquis quantitatibus in valore t addendam, vel aream ex altera parte ordinatim applicatæ esse subducendam. Et contra $-s$ ambiguè denotat aream basi adjacentem esse subducendam, vel aream ex altera parte ordinatim applicatæ esse addendam: prout commodum videbitur. Deinde valor ipsius t si affirmativus prodierit, designat aream Curvæ propositæ adjacentem basi ejus. Et contra si fuerit negativus, designat aream ex altera parte ordinatim applicatæ.

7. Cæterùm ut Area illa certiùs definiatur,[588] prospiciendum est de limitibus ejus. Et quidem limitum ad Basin, parallelè incedentem,[589] et Curvæ perimetrum, nulla potest esse incertitudo: sed limes initialis sive principium a quo incipit descriptio ejus varias positiones obtinet. In sequentibus exemplis[590] vel est ad initiū basis vel ad infinitam distantiam, vel in concursu curvæ cum basi ejus. Sed potest alibi locari. Et ubicunqꝫ sit, invenies quærendo Basis longitudinem ad quam valor ipsius t evadit nullus, et parallelè incedentem erigendo. Nam erecta applicata illa linea erit limes quæsitus.[591]

8. Siqua pars areæ infra basin posita sit, t designabit differentiam ejus et partis supra basin.

[592]9. Siquando dimensiones terminorum in valoribus x, v et t nimis altæ vel nimis depressæ obvenerint, ad justum gradum liceat reducere dividendo vel multiplicando toties per datam quamvis quantitatem quæ vices unitatis gerere fingitur, quoties dimensiones illæ sint justo altiores vel depressiores.

||[85 *rursum*] || 10.[593] Præter præcedentes catalogos possunt etiam Catalogi Curvarum ad alias Curvas in suo genere simplicissimas (ut ad $\sqrt{e+fx^3}=v$, vel ad $x\sqrt{e+fx^3}=v$, vel ad $\sqrt{e+fx^4}=v$ &c) relatarum construi, eò ut Curvæ cujuslibet propositæ aream ex origine simplicissima possimus derivare, et cum quibus curvis affinitatem habeat cognoscere. Cæterùm præcedentes tandem exemplis aliquot illustremus.[594]

(586) A further caution to any reader without full command of the niceties of applying the general algebraic variable geometrically.

(587) Compare note (584). Paragraphs 6–9 following are a late insertion on a supplementary sheet. The present paragraph 6 lightly revises a cancelled equivalent passage on Newton's page 76.

(588) 'cautè' (warily) is cancelled.

(589) Here, as many times elsewhere, the equivalent 'ordinatim applicatam' is cancelled.

(590) Newton first wrote 'vel initio basis insistit vel ad infinitam distantiam recedit' (it is either ordinate at the origin of the base or has departed to an infinite distance).

(591) In present context, in other words, a 'limes' is taken to be that (fixed) bound to the 'area' $t = \int y.dz$ which corresponds to $t = 0$: it will be infinity, for example, when $y = kz^{-1}$.

(592) A first draft of the following paragraph is preserved on ULC. Add. 3960.3: 32, but is not significantly variant.

readily, I have prefixed to the individual values of the areas their proper signs (even when, as happens in the fifth and seventh order of the latter catalogue, these are sometimes negative).[586]

[587]6. In regard to the signs of the areas it should further be observed that $+s$ denotes either that the area of the conic adjacent to the base is to be added to the remaining quantities in t's value, or that the area on the opposite side of the ordinate is to be taken from them. Conversely, $-s$ has the double meaning either that the area adjacent to the base has to be subtracted or that the area on the opposite side of the ordinate is to be added (whichever appears the more suitable). Again, if the value of t proves to be positive, it represents the area of the curve proposed which adjoins the base. Conversely, if it be negative it represents the area on the other side of the ordinate. See Example 1 following.

7. But to determine that area more precisely, forethought is[588] to be given to its bounds. Now indeed there can be no uncertainty regarding the bounds of the curve along its base, the ordinate and its perimeter, but the initial bound or origin from which its description starts may have differing positions. In the following examples[590] it is either at the origin of the base or at an infinite distance or at the curve's meet with its base. But it can be placed elsewhere. Wherever it be, you will find it by seeking the length of the base for which the value of t comes out to be zero and raising the ordinate there. For when that ordinate is raised it will be the bound sought.[591]

8. If any portion of the area be sited below the base, t will represent the difference between it and the part above the base.

[592]9. Whenever the dimensions of the terms in the values of x, v and t chance to be too high or too low, they may allowably be reduced to an appropriate degree by dividing or multiplying as many times by any given quantity which is supposed to take on the rôle of unity as those dimensions are higher or lower than the appropriate one.

10.[593] Apart from the preceding ones, catalogues may also be constructed of curves related to others which are the simplest of their kind (to $\surd[e+fx^3] = v$, say, or $x\surd[e+fx^3] = v$, or $\surd[e+fx^4] = v$, and so on), so that we may derive the area of any curve you please from the simplest source and know to what curves it has an affinity. But now at length let us illustrate the preceding with a few examples.[594]

(593) See note (584) above. A number (10) was given to this concluding paragraph of observations on the two preceding integral tables only after paragraphs 6–9 were inserted.

(594) Newton is not known to have constructed any integral tabulations along these more general lines. Unfortunately, as time was to prove, no real advance in the construction of usable 'catalogues' of integrals was possible following the present curve-comparison approach, but had to await the creation of a general analytical theory of real functions. When the present (second) table was published in 1704 (see note (549)) it was already obsolescent and, in

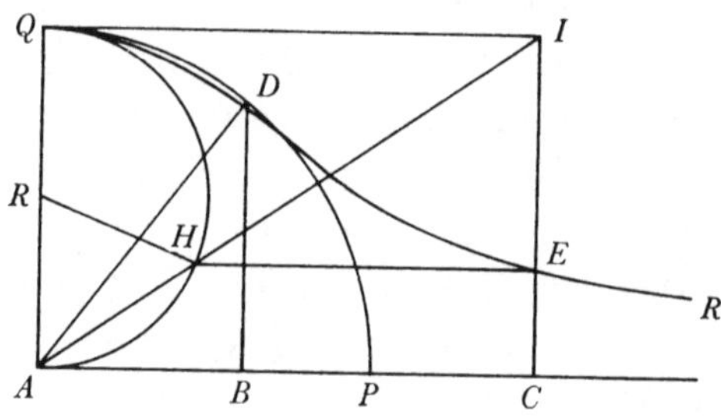

EXEMPL 1. Sit QER ejusmodi Conchoidalis(595) ut, semicirculo QHA descripto et ad diametrum AQ erecto AC perpendiculo, si compleatur parallelogrammum $QACI$, agatur diagonalis AI semicirculo occurrens in H, et ab H demittatur ad IC normalis HE, punctum E incidat in Curvam. Et quæratur area $ACEQ$. Dic itaq; $AQ=a$. $AC=z$, $CE=[y]$, et propter continuè proportionales AI, AQ, AH, EC, erit EC sive $y=\dfrac{a^3}{a^2+z^2}$.

Jam ut hæc induat formam æquationum in Catalogis, finge $\eta=2$ et pro z^2 in denominatore scribe z^η, ac $a^3z^{\frac{1}{2}\eta-1}$ pro a^3 sive a^3z^{1-1} in numeratore, et emerget $y=\dfrac{a^3z^{\frac{1}{2}\eta-1}}{a^2+z^\eta}$, æquatio primæ speciei secundi ordinis posterioris Catalogi; collatisq; terminis fit $d=a^3$, $e=a^2$ et $f=1$; adeoq; $\sqrt{\dfrac{a^3}{a^2+z^2}}=x$, $\sqrt{a^3-a^2x^2}=v$, et $xv-2s=t$.(596)

Ut autem inventi valores x et v ad justum dimensionum numerum reducantur(597) selige datam quamlibet quantitatem, velut a, per quam tanquam unitatem semel multiplicetur a^3 in valore x, et in valore v dividatur a^3 semel et a^2x^2 bis. Et hoc pacto obtinebis $\sqrt{\dfrac{a^4}{a^2+z^2}}=x$, $\sqrt{a^2-x^2}=v$, et $xv-2s=t$.(598) Quorum constructio est ejusmodi.

||[86] ||Centro A intervallo AQ describe quadrantem circuli QDP, in AC cape $AB=AH$, erige normalem BD quadranti occurrentem in D, et age AD. Et sectoris ADP duplum æquabitur areæ quæsitæ $ACEQ$. Est enim

$$\sqrt{\frac{a^4}{a^2+z^2}}\,(=\surd AQ\times EC=AH)=AB \text{ sive } x,$$

historical fact, was quickly absorbed by Roger Cotes into a general 'harmony of [circular and hyperbolic] measures' between 1709 and his death in 1716. (See his cousin Robert Smith's edition of Cotes' *Harmonia Mensurarum, sive Analysis & Synthesis per Rationum & Angulorum Mensuras promotæ* (Cambridge, 1722): Pars Secunda. 'Theoremata tum Logometrica tum Trigonometrica quæ datarum Fluxionum Fluentes exhibent per Mensuras': 43–76; and compare Smith's own compendium 'Theoremata tum Logometrica tum Trigonometrica datarum Fluxionum Fluentes exhibentia per Methodum Mensurarum ulterius extensam', *ibid.*: 111–249.)

(595) The *versiera* or (Agnesi's) 'witch', as we know it, though the name was given only in 1718 by Guido Grandi. The curve was apparently first proposed by Pierre de Fermat in the 1650's in a paper 'De Æquationum Localium Transmutatione et Emendatione ad multi-

EXAMPLE 1. Let QER be a conchoidal curve(595) defined in this way: if, after the semicircle QHA is described and AC raised perpendicular to the diameter AQ, the rectangle $QACI$ is completed, its diagonal AI drawn to meet the semicircle in H and from H is let fall the normal HE onto IC, the point E shall lie on the curve. And let the area $ACEQ$ be sought. So call $AQ = a$, $AC = z$, $CE = y$ and then, because of the continued proportionals AI, AQ, AH, EC, will EC or $y = a^3/(a^2+z^2)$.

Now to arrange that this takes on the form of the equations in the catalogues, suppose $\eta = 2$ and instead of z^2 in the denominator write z^η with $a^3z^{\frac{1}{2}\eta-1}$ in place of a^3 (that is, a^3z^{1-1}) in the numerator, and there will emerge

$$y = a^3z^{\frac{1}{2}\eta-1}/(a^2+z^\eta),$$

an equation of the first species of the second order in the latter catalogue. On comparing terms there comes $d = a^3$, $e = a^2$ and $f = 1$, so that

$$\sqrt{[a^3/(a^2+z^2)]} = x, \quad \sqrt{[a^3-a^2x^2]} = v \quad \text{and} \quad xv-2s = t.^{(596)}$$

But to reduce these values of x and v now found to an appropriate number of dimensions,(597) choose some given quantity, say a, as it were unity, and by it multiply a^3 in the value of x once, while in v's value divide a^3 once and a^2x^2 twice by it. By this means you will obtain $\sqrt{[a^4/(a^2+z^2)]} = x$, $\sqrt{[a^2-x^2]} = v$ and $xv-2s = t$.(598) This is a way to construct them.

With centre A and radius AQ describe the circle quadrant QDP, in AC take $AB = AH$, erect the normal BD meeting the quadrant in D, and draw AD. Then twice the sector ADP will equal the area $ACEQ$ sought. For

$$\sqrt{[a^4/(a^2+z^2)]}\ (= \sqrt{[AQ\times EC]} = AH) = AB \text{ or } x$$

---

modam curvilineorum inter se, vel cum rectilineis comparationem' published only twenty years afterwards (*Varia Opera* (Toulouse, 1679): 44–57, especially 54 [= *Œuvres*, **1** (Paris, 1891): 255–85, especially 279]), where, almost exactly as Newton in the present tract, he 'squared' its area by the substitution $ay = x^2$: at once $z = ax^{-1}\sqrt{[a^2-x^2]}$ and so

$$\int y.dz = a^{-1}\int x^2.dz = a^{-1}x^2z-2a^{-1}\int xz.dx,$$

that is, $x\sqrt{[a^2-x^2]}-2\int\sqrt{[a^2-x^2]}.dx$. In 1668 James Gregory introduced the 'witch' as an auxiliary curve in his quadrature of the cissoid (*Exercitationes Geometricæ* (London, 1668): 'Analogia inter Lineam Meridianam Planispheri Nautici...', Proposition VI: 23–4) and this is evidently the source for Newton's knowledge of the curve. The figure has two points R.

(596) That is, where $t = \int y.dz$ and $s = \int v.dx$, as previously.

(597) A first, cancelled draft of the following reads 'multiplico vel divido per datam quamlibet quantitatem' (I multiply or divide by a given quantity). Newton's procedure is equivalent to the transform $a^{\frac{1}{2}}x \to x$.

(598) That is, $x\sqrt{[a^2-x^2]}-2\int\sqrt{[a^2-x^2]}.dx = a^2\cos^{-1}(x/a)$.

et $\sqrt{a^2-x^2}(=\sqrt{AD^q-AB^q})=BD$ sive v; et $xv-2s=2\triangle ADB-2ABDQ$ vel etiam $=2\triangle ADB+2BDP$, hoc est vel $=-2QAD$ vel $=2DAP$: quorum valorum affirmativus $2DAP$ competit areæ $ACEQ$ adjacenti AC citra EC, et negativus $2QAD$ competit areæ $RECR$ ultra EC in infinitum protensæ.(599)

Solutiones Problematum sic inventæ nonnunquam concinnari possunt. Sic in hoc casu(600) actâ RH circuli QHA semidiametro, propter arcus QH, DP æquales, erit sector QRH dimidium sectoris DAP,(601) atq; adeò pars quarta superficiei $ACEQ$.

EXEMPL 2. Sit AGE curva(602) quam normæ(603) AEF punctum angulare E describit dum crurum alterum AE interminatum continuò transit per datum punctum A, et alterum EF datæ longitudinis super recta AF positione data prolabitur. Demitte EH ad AF normalem, et comple parallelogrammum $AHEC$, ac dictis $AC=z$, $CE=y$, et $EF=a$, propter HF, HE, HA continuè proportionales erit HA sive $y=\frac{z^2}{\sqrt{a^2-z^2}}$.

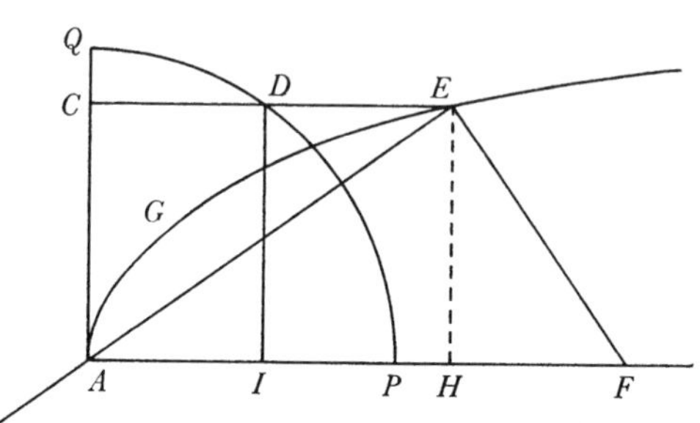

Jam ut innotescat area $AGEC$, finge $z^2=z^\eta$, sive $2=\eta$ et inde fiet $\frac{z^{\frac{3}{2}\eta-1}}{\sqrt{a^2-z^\eta}}=y$. (604)Ubi cum z sit fractæ dimensionis in numeratore, deprime valorem y dividendo per $z^{\frac{1}{2}\eta}$ et fiet $\frac{z^{\eta-1}}{\sqrt{a^2z^{-\eta}-1}}=y$, æquatio secundæ speciei septimi ordinis posterioris Catalogi. Ac terminis collatis evadet $d=1$, $e=-1$, et $f=a^2$. Adeoq; $z^2\left(=\frac{1}{z^{-\eta}}\right)=x^2.\sqrt{a^2-x^2}=v$, & $s-xv=t$. Cùm itaq; x et z æquentur, et sit $\sqrt{a^2-x^2}=v$ æquatio ad circulum cujus diameter(605) est a: centro A intervallo a sive EF describatur circulus PDQ cui occurrat CE in D, et

(599) In the figure $AC = z$, $CE = y$, $AB = x$, $BD = v$ and Newton constructs

$$(ACEQ) = \int_0^z y.dz = x\sqrt{[a^2-x^2]}+2\int_0^a \sqrt{[a^2-x^2]}.dx,$$

that is, $2\triangle ABD+2(DBP) = 2(DAP)$ or $a^2\cos^{-1}(x/a)$: compare note (598). The line AC is the cubic's real asymptote.

(600) Newton first began this paragraph simply 'Hic obiter notetur quod [actâ]' (Here note by the way that...).

(601) For $\triangle QAH = \triangle DAB$ so that $Q\hat{R}H = (2Q\hat{A}H$ or) $2D\hat{A}B$, while $QR = \frac{1}{2}DA$.

(602) The kappa curve, as we know it today following Aubry. This quartic, suggested to Sluse in 1662 by Gutschoven, was squared by Huygens the same year by a limit-increment argument which essentially mirrors Newton's fluxional proof below (note (641)). See his letter to Sluse of 25 September 1662 (N.S.) (*Œuvres complètes*, **4**, 1891): 238; compare also **14**, 1920:

and $\surd[a^2-x^2]$ $(=\surd[AD^2-AB^2])$ $= BD$ or v, while

$$xv-2s = 2\triangle ADB-2 \text{ area } ABDQ$$

or, alternatively, $2\triangle ADB+2(BDP)$, that is, either $-2(QAD)$ or $2(DAP)$: of these values the positive one $2(DAP)$ relates to the area $ACEQ$ lying along AC on the near side of EC, the negative one $2(QAD)$ to the area $RECR$ infinitely extended beyond EC.(599)

Solutions of problems found in this way may not infrequently be presented more neatly. So in this case,(600) after the radius RH of the circle QHA has been drawn, because the arcs $\widehat{QH}$, $\widehat{DP}$ are equal, the sector QRH will be half the sector DAP(601) and so one-fourth the surface $ACEQ$.

EXAMPLE 2. Let AGE be the curve(602) described by the corner point E of a right-angled rule(603) AEF while one of the legs AE, not bounded [in length], continuously passes through the given point A and the second EF, of a given length, glides forward above the straight line AF given in position. Drop EH normal to AF and complete the parallelogram $AHEC$, calling $AC = z$, $CE = y$ and $EF = a$: then, because HF, HE and HA are continual proportionals, will HA or $y = z^2/\surd[a^2-z^2]$. Now to find out the area $AGEC$ suppose $z^2 = z^\eta$, that is, $2 = \eta$, and from this there will come $z^{\frac{3}{2}\eta-1}/\surd[a^2-z^\eta] = y$.(604) Here, since z has a fractional dimension in the numerator, depress the value of y by dividing through by $z^{\frac{1}{2}\eta}$ and there will come $z^{\eta-1}/\surd[a^2z^{-\eta}-1] = y$, an equation of the second species of the seventh order in the latter catalogue. On comparing terms there will prove to be $d = 1$, $e = -1$ and $f = a^2$, so that $z^2(=z^{-(-\eta)}) = x^2$, $\surd[a^2-x^2] = v$ and $s-xv = t$. Since, therefore, x and z are equal and $\surd[a^2-x^2] = v$ is the equation of a circle whose [radius] is a, with centre A and interval a (or EF) describe the circle PDQ, to be met by CE in D, and complete the rectangle

500–1). Some time later Barrow learnt of this 'courbe de Gutschoven' by way of Collins and in 1670 inserted it as 'Exemp. I' of the analytical tangent method which he appended to the tenth lecture of his *Lectiones Geometricæ* (note (80)): 81–2. Where, however, Sluse constructed the curve by the method Newton here describes, Barrow defined it somewhat differently: namely, in terms of the present figure, draw the perpendicular at Q to AQ (which will be an asymptote) and through A any radius vector meeting this perpendicular in e, then take the point E in Ae such that $AE = Qe$. (The equivalence of the two constructions follows readily from the congruency of the triangles AEF and eQA.) We do not know if Newton was aware of Sluse's definition of Gutschoven's quartic but he would have no difficulty in discovering it independently from Barrow's variant determination. The point A is a tacnode.

(603) Or 'sector' as Newton will name it in August 1672 in a letter to Collins (see II: 156). Here and elsewhere below he borrows extensively from the terminology established by him several years before to describe his method of generating curves 'organically' (compare II, 1, 3: passim).

(604) Newton first wrote in continuation at this point 'Sed hujus formæ nulla occurrit æquatio in catalogis' (But no equation of this form occurs in the catalogues).

(605) Read 'semidiameter' or 'intervallum'.

compleatur parallelogrammum $ACDI$, eritq $AC=z$, $CD=v$, et area quæsita $AGEC(=s-xv=ACDP-ACDI)=IDP$.(606)

||[87] Exempl: 3. Sit AGE Cissois ad circulum ADQ diametro ||AQ descriptum pertinens.(607) Agatur DCE diametro normalis et curvis occurrens in D et E. Et nominatis $AC=z$, $CE=y$ et $AQ=a$, propter CD, CA, CE continuè proportionales erit CE sive $y=\dfrac{z^2}{\sqrt{az-z^2}}$, ac dividendo per z, fit $y=\dfrac{z}{\sqrt{az^{-1}-1}}$. Est itaq $z^{-1}=z^{[\eta]}$ sive $-1=[\eta]$ et inde

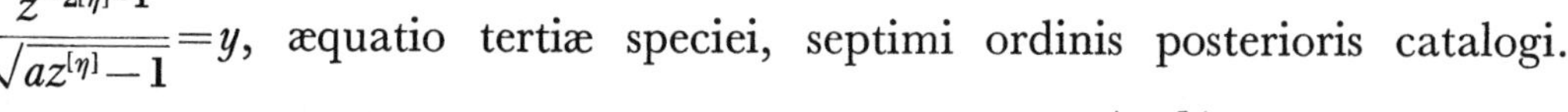

$\dfrac{z^{-2[\eta]-1}}{\sqrt{az^{[\eta]}-1}}=y$, æquatio tertiæ speciei, septimi ordinis posterioris catalogi. Collatisq terminis fit $d=1$. $e=-1$. et $f=a$. Adeoq $z\left(=\dfrac{1}{z^\eta}\right)=x$. $\sqrt{ax-x^2}=v$. et $3s-2xv=t$. Quare est $AC=x$, $CD=v$, et inde $ACDH=s$, adeoq $3ACDH-4\triangle ADC(=3s-2xv=t)=$areæ Cissoidali $ACEGA$. Vel quod perinde est 3 segment: $ADHA=$areæ $ADEGA$. sive 4 seg $ADHA=$areæ $AHDEGA$.

Exempl: 4. Esto PE prima Conchoides Veterum centro G,(608) Asymptoto AL et intervallo LE descripta. Age GAP axin ejus ac demitte EC ordinatim applicatam, dictisq $AC=z$, $CE=y$, $GA=b$, et $AP=c$, propter proportionales

$$AC.CE-AL::GC.CE,$$

erit CE sive $y=\dfrac{b+z}{z}\sqrt{c^2-z^2}$.(609)

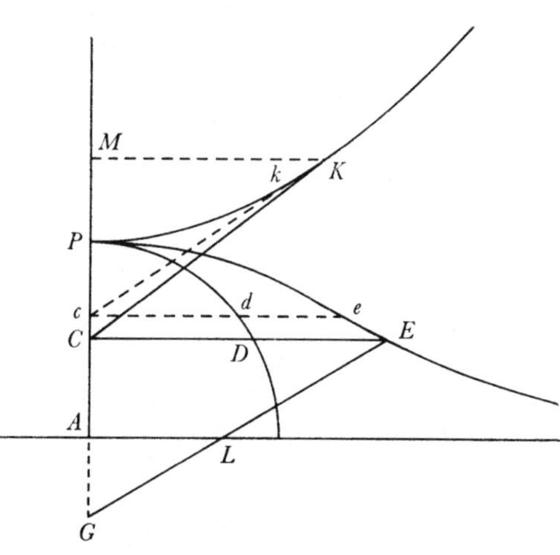

Jam ut ejus area exhinc inveniatur, partes applicatæ CE seorsim considerandæ sunt. Et quidem si illa CE ita dividatur in D ut sit $CD=\sqrt{c^2-z^2}$ ac $DE=\dfrac{b}{z}\sqrt{c^2-z^2}$ erit CD ordinatim applicata circuli centro A intervallo AP

(606) More directly, since $AB=AC$ or $x=z$, it follows that

$$(AGEC)=\int_0^z y.dz=\int_0^x x^2/\sqrt{[a^2-x^2]}.dx=\int_0^x\sqrt{[a^2-x^2]}.dx-x\sqrt{[a^2-x^2]},$$

that is, $(ACDP)-(ACDI)=(DPI)$ or $\frac{1}{2}a^2\sin^{-1}(x/a)-\frac{1}{2}x\sqrt{[a^2-x^2]}$.

(607) Compare Example 4 on Newton's pages 73–4 above. The present evaluation of the

ACDI. Then $AC = z$, $CD = v$ and the area sought $AGEC(= s - xv$, that is, $ACDP - ACDI) = IDP$.(606)

Example 3. Let *AGE* be a cissoid pertaining to the circle *ADQ* described on the diameter *AQ*.(607) Let the normal to the diameter *DCE* be drawn, meeting the curves in *D* and *E*. Then on naming $AC = z$, $CE = y$ and $AQ = a$, because *CD*, *CA*, *CE* are continued proportionals, will *CE* or $y = z^2/\sqrt{[az - z^2]}$ and, on division by z, there will come $y = z/\sqrt{[az^{-1} - 1]}$. Accordingly $z^{-1} = z^\eta$, that is, $-1 = \eta$, and thence $z^{-2\eta-1}/\sqrt{[az^\eta - 1]} = y$, an equation of the third species of the seventh order in the latter catalogue. On comparing terms there comes $d = 1$, $e = -1$ and $f = a$, so that $z(= z^{-\eta}) = x$, $\sqrt{[ax - x^2]} = v$ and $3s - 2xv = t$. Therefore $AC = x$, $CD = v$ and consequently $ACDH = s$, so that $3(ACDH) - 4\triangle ADC(= 3s - 2xv = t) =$ cissoidal area *ADEGA*. Or what is the same, $3 \times$ segment $ADHA =$ area *ADEGA*, that is, $4 \times$ segment $ADHA =$ area *AHDEGA*.

Example 4. Let *PE* be a 'first' conchoid of the ancients described with centre *G*,(608) asymptote *AL* and interval *LE*. Draw its axis *GAP* and let fall *EC* ordinate to it. On calling $AC = z$, $CE = y$, $GA = b$ and $AP = c$ there will, because of the proportionals $AC:(CE - AL) = GC:CE$, be

$$CE \quad \text{or} \quad y = (bz^{-1} + 1)\sqrt{[c^2 - z^2]}.^{(609)}$$

Now to find its area from this, portions of the ordinate *CE* must be considered separately. In fact, if that line *CE* be divided at *D* such that $CD = \sqrt{[c^2 - z^2]}$ and $DE = bz^{-1}\sqrt{[c^2 - z^2]}$, *CD* will be the ordinate of a circle described with

cissoid's area repeats the second of the two previous quadratures, yielding

$$\int_0^x x^{\frac{3}{2}}(a-x)^{-\frac{1}{2}}.dx = 3\int_0^x \sqrt{[ax - x^2]}.dx - 2x\sqrt{[ax - x^2]},$$

that is, $(AGEC) = 3(ACDH) - 4\triangle ACD$. For the broken lines in the figure see pp. 280–1.

(608) 'vertice *P*' (with vertex *P*) is cancelled. Newton's present quadrature of the (upper) segment generalizes Gregory's result for the particular conchoid in which $GA = AP$ (*Exercitationes Geometricæ* (note (595)): 21–3); earlier both Wallis (*Tractatus Duo* (note (255)): 122) and Huygens (in letters to Schooten of 5 September 1658 and 1 January 1659 published in his *Œuvres complètes*, **2** (1889): 212, 298–9) had observed that the total area comprehended between the conchoidal 'shell' *PeE* and its asymptote *AL* is infinite. Newton's quadrature was rediscovered independently by Johann Bernoulli in 1691 (see his 'Lectiones Mathematicæ, de Methodo Integralium, Aliisque conscriptæ in usum...Hospitalii...Parisiis...Annis 1691 & 1692': Lectio III. 'Variarum Curvarum Quadratura', §II [= *Opera Omnia*, **3** (Lausanne and Geneva, 1742): 400–1]) and by Roger Cotes in about 1710, the latter of whom first published the result in his 'Logometria[, sive de Mensura Rationis]' (*Philosophical Transactions of the Royal Society*, **29** (1714–16): No. 338 [for January–March, 1714]: 5–45, especially 28–9 [= *Harmonia Mensurarum* (note (594)): 25]). Both Bernoulli and Cotes stated their quadrature in the form $(APEL) = bc\log(\sec\phi + \tan\phi) + \frac{1}{2}c^2\phi$, where $P\hat{G}E = \phi$.

(609) That is, on calling $P\hat{G}E = \phi$, $y = b\tan\phi + c\sin\phi$ and so $CE = AL + CD$, or $DE = AL$. This result, invoked implicitly in Newton's working and explicitly on his pages 91–2 below, was claimed by Wallis as a 'discovery': 'Est autem & magna, inter Conchoidem & Cycloidem

descripti: adeoq; pars areæ PDC innotescet, et restabit pars altera $DPED$ invenienda. Cùm itaq; DE (pars applicatæ quacum describitur) valeat $\frac{b}{z}\sqrt{c^2-z^2}$, suppone $2=\eta$, et evadet $\frac{b}{z}\sqrt{c^2-z^\eta}=DE$, æquatio primæ speciei quinti ordinis posterioris Catalogi. Collatisq; terminis, fiet $d=b$, $e=c^2$ et $f=-1$; atq; adeò $\frac{1}{z}\left(=\sqrt{\frac{1}{z^\eta}}\right)=x$. $\sqrt{-1+ccxx}=v$. et $2bccs-\frac{bv^3}{x}=t$.

||[88] ||His inventis redige ad justum dimensionum numerum multiplicando terminos nimis depressos ac dividendo nimis altos per datam quamvis quantitatem. Id quod si fiet per c, prodibit $\frac{c^2}{z}=x$. $\sqrt{-c^2+x^2}=v$, & $\frac{2bs}{c}-\frac{bv^3}{cx}=t$.(610) Et horum constructio est ejusmodi.

Centro A, vertice principali P, et parametro $2AP$(611) Hyperbolam PK describe. Deinde a puncto C age rectam CK quæ tangat Hyperbolam in K: et erit ut AP ad $2AG$ ita area $CKPC$ ad aream quæsitam $DPED$.

EXEMPL: 5. Norma GFE ita circa polum G rotante ut ejus punctum angulare F super recta AF positione data continuò prolabitur: concipe curvam(612) PE a puncto quolibet E in crure EF sito describi. Jam ut inveniatur hujus area, demitte GA et EH ad rectam AF perpendiculares et completo parallelogrammo $AHEC$, dic $AC=z$, $CE=y$, $AG=b$ et $EF=c$, et propter proportionales $HF.EH::AG.AF$, erit $AF=\frac{bz}{\sqrt{cc-z^2}}$. Adeoq; CE sive $y=\frac{bz}{\sqrt{c^2-z^2}}-\sqrt{c^2-z^2}$. Cùm autem $\sqrt{cc-z^2}$

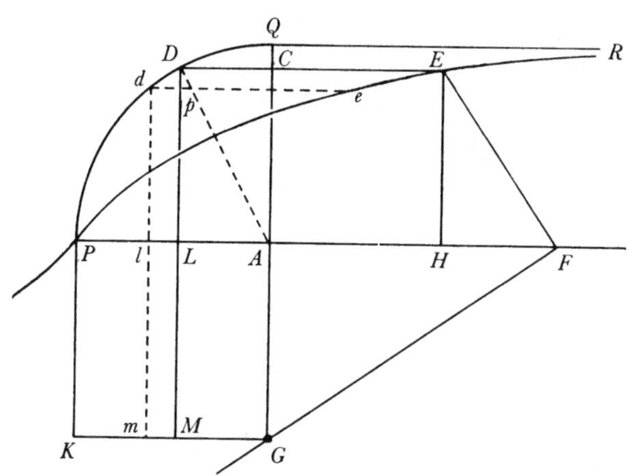

convenientia....Ut Ordinatim-applicata in Cycloide est Sinûs et Arcûs ejusdem anguli Aggregatum,...Sic in Conchoide, Ordinatim-applicata est Aggregatum Sinûs & Tangentis ejusdem anguli' (*Tractatus Duo* (note (255)): 122: 'Addendum ad Conchoidem').

(610) In explicit terms Newton determines (PDE) or $\int_z^c bz^{-1}\sqrt{[c^2-z^2]}.dz$ to be

$$(2b/c)\int_{c^2/z}^{c}\sqrt{[x^2-c^2]}.dx+bz^{-2}(c^2-z^2)^{\frac{3}{2}},$$

that is, $bc\log\frac{c+\sqrt{[c^2-z^2]}}{z}-b\sqrt{[c^2-z^2]}$. Immediately, upon adding

$$(PCD)=\int_z^c\sqrt{[c^2-z^2]}.dz=\tfrac{1}{2}c^2\cos^{-1}(z/c)-\tfrac{1}{2}z\sqrt{[c^2-z^2]}\quad\text{and}\quad(ACEL)=(b+\tfrac{1}{2}z)\sqrt{[c^2-z^2]},$$

centre A and radius AP: in consequence, the portion PDC of the area will be known and it will remain to find the other portion $DPED$. Since therefore the value of DE (the part of the ordinate by which it is described) is $bz^{-1}\surd[c^2-z^2]$, assume $2=\eta$ and there will come out $bz^{-1}\surd[c^2-z^\eta]=DE$, an equation of the first species of the fifth order in the latter catalogue. On collating terms there will come $d=b$, $e=c^2$ and $f=-1$, and accordingly $z^{-1}(=z^{-\frac{1}{2}\eta})=x$,

$$\surd[-1+c^2x^2]=v \quad \text{and} \quad 2bc^2s-bv^3/x=t.$$

Reduce these results, when found, to an appropriate number of dimensions by multiplying terms which are too depressed but dividing those too high by any given quantity. If this is done by c, there will result $c^2/z=x$, $\surd[-c^2+x^2]=v$ and $2(b/c)\,(s-\frac{1}{2}v^3/x)=t$.(610) And the manner of their construction is this.

With centre A, principal vertex P and parameter $2AP$(611) describe the hyperbola PK. Next from the point C draw the straight line CK to touch the hyperbola in K: then will AP be to $2AG$ as the area $CKPC$ to the area $DPED$ sought.

EXAMPLE 5. As the right-angled rule GFE rotates around the pole G so that its corner point F continuously glides forward along the straight line AF given in position, imagine that the curve PE(612) is described by any point E situated in the leg EF. Now to find its area, let fall GA and EH perpendicular to the line AF and, having completed the rectangle $AHEC$, call $AC=z$, $CE=y$, $AG=b$ and $EF=c$: then, because of the proportionals $HF:EH=AG:AF$, will

$$AF=bz/\surd[c^2-z^2],$$

so that CE or $y=bz/\surd[c^2-z^2]-\surd[c^2-z^2]$. But since $\surd[c^2-z^2]$ is the ordinate

---

it follows that
$$(APEL)=bc\log\frac{c+\surd[c^2-z^2]}{z}+\tfrac{1}{2}c^2\cos^{-1}(z/c):$$
a corollary due to Johann Bernoulli and Roger Cotes (note (608)).

(611) Equal, that is, to its corresponding main diameter (*latus transversum*), so that the hyperbola will be equilateral. Note Newton's ingenious use of the conic's pole-polar property to construct $AM=AP^2/AC$, that is, $x=c^2/z$. Accordingly

$$MK=\surd[x^2-c^2]=cz^{-1}\surd[c^2-z^2] \quad \text{and} \quad CM=x-z=(c^2-z^2)/z,$$

so that $\triangle CMK-(PMK)$, or the hyperbolic sector (CPK), is equal to

$$\tfrac{1}{2}cz^{-2}(c^2-z^2)^{\frac{3}{2}}-\int_c^{c^2/z}\surd[x^2-c^2]\,.\,dx.$$

Newton's construction, $(PDE)=2b/c\times(CPK)$, is an immediate deduction.

(612) An unnamed quartic, of Cartesian defining equation $y^2(c^2-z^2)=(z^2+bz-c^2)^2$, akin to a group of curves discussed two centuries later by Steiner. It is evidently unicursal, having the parallel asymptotes $z=\pm c$. Newton has obviously contrived this curve *ad hoc* as a composite of the Gutschoven quartic and the conchoid (Examples 2 and 4 preceding), having run out of suitable known curves to whose quadrature he might apply the integrals set out in his 'Catalogus posterior'. (To avoid difficulties inherent in applying the concept of a unique 'area under a curve' his choice is evidently confined to those higher curves whose defining equation may be thrown into the form $y=\pm f(z)$, that is, largely to unicursal curves and ones symmetrical round the diameter $y=0$.)

sit ordinatim applicata circuli semidiametro c descripti: circa centrum A describe talem circulum PDQ, eicꝫ $C[E]$ producta occurat in D, et erit $DE=\frac{bz}{\sqrt{c^2-z^2}}$: cujus æquationis ope restat area $PDEP$ vel $DERQ$ determinanda.(613) Supponatur ergo $\eta=2$ et evadet $DE=\frac{bz^{\eta-1}}{\sqrt{cc-z^\eta}}$ æquatio primæ speciei quarti ordinis prioris catalogi. Et collatis terminis fiet $b=d$, $cc=e$, et $-1=f$; adeocꝫ $-b\sqrt{cc-zz}(=-bR)=t$. Jam cum valor t negativus existat, et inde area per t designata jaceat ultra lineam DE; ut ejus limes initialis inveniatur quære illam ipsius z longitudinem qua t evadit nulla et invenies esse c. Quare produc AC ad Q ut sit $AQ=c$, et erige applicatam QR et erit
‖[89] $DQRED$ area illa cujus ‖valor jam inventus est $-b\sqrt{cc-zz}$.

Quod si quantitatem areæ PDE juxta basin AC positæ et cum ea coextensæ desideres, possis ignoto limite QR sic determinare. A valore quem t ad basis longitudinem AC sortita est subduc valorem ejus ad initium basis. hoc est a $-b\sqrt{cc-zz}$ subduc $-bc$ et proveniet quantitas $bc-b\sqrt{cc-zz}$ quam quæris. Comple ergo parallelogrammum $PAGK$ et ad AP demitte normalem DM quæ cum GK occurrat in M et erit parallelogrammum $P[L]MK$ æquale areæ PDE.(614)

Siquando æquatio curvam aliquam definiens non reperiatur in Catalogis, necꝫ ad simpliciores terminos ope divisionis vel alio pacto reduci possit: transformanda est in alias affinium Curvarum æquationes pro more in Prob 8 ostenso, donec tandem obvenerit aliqua cujus area ex Catalogis innotescat. Et conatibus omnimodo institutis, si nulla talis obveniat, certum est Curvam propositam necꝫ cum figuris rectilineis necꝫ cum Conicis Sectionibus comparari posse.

Ad eundem modum cùm de Curvis Mechanicis agitur illæ imprimis transformandæ sunt in æquales Geometricas prout in eodem Prob 8 ostensum fuerit,(615) ac deinde Geometricarum areæ ex Catalogis eliciendæ. Cujus rei accipe sequens exemplum.(616)

(613) Namely $\int_0^z bz/\sqrt{[c^2-z^2]}.dz$ and $\int_z^c bz/\sqrt{[c^2-z^2]}.dz$ respectively.

(614) Having evaluated the indefinite integral $\int bz/\sqrt{[c^2-z^2]}.dz = -b\sqrt{[c^2-z^2]}$, Newton finds the definite 'areas'

$$(DQRE) = \int_z^c DE.dz = b\sqrt{[c^2-z^2]} \quad \text{and} \quad (PDE) = \int_0^z DE.dz = b(c-\sqrt{[c^2-z^2]}),$$

where $AG = b$, $AL = CD = \sqrt{[c^2-z^2]}$ and $AP = AQ = c$. Note that R is at infinity.

(615) That is, by the area-preserving 'convolution' of a 'spiral' defined by a finite algebraic polar equation into a corresponding 'geometrical' Cartesian 'parabola', described on his page 75 above, or (in suitable cases) by an integration by parts, as sketched on his page 74.

of a circle described with radius c, round the centre A describe such a circle PDQ and let CE produced meet it in D: then will $DE = bz/\sqrt{[c^2-z^2]}$ and with the help of this equation there remains to be determined only the area $PDEP$ or $DERQ$.(613) Suppose therefore that $\eta = 2$ and there will result

$$DE = bz^{\eta-1}/\sqrt{[c^2-z^\eta]},$$

an equation of the first species of the fourth order in the former catalogue. On collating terms there will come $b = d$, $c^2 = e$ and $-1 = f$, so that $-b\sqrt{[c^2-z^2]}\ (= -bR) = t$. Now the value of t proves to be negative and in consequence the area represented by t lies beyond the line DE. Hence to find its initial bound seek the length of z which makes t come out zero and you will find it to be c. Therefore produce AC to Q such that $AQ = c$ and erect the ordinate QR: $DQRED$ will then be the area whose value was just now found to be $-b\sqrt{[c^2-z^2]}$.

But should you desire the quantity of the area PDE set close by the base AC and coextensive with it, you could determine it this way without knowing the bound QR. From the value of t obtaining for the base length AC subtract its value at the base's origin: in other words, from $-b\sqrt{[c^2-z^2]}$ subtract $-bc$ and there will result the quantity $bc-b\sqrt{[c^2-z^2]}$ which you seek. So complete the rectangle $PAGK$ and to AD let fall the normal DM which should meet GK in M: the rectangle $PLMK$ will then be equal to the area PDE.(614)

Should ever the equation defining some curve not be located in the catalogues nor be reducible to simpler terms with the aid of division or by some other procedure, it must be transformed into other equations of related curves after the manner shown in Problem 8 till at last one is forthcoming whose area may be found out from the catalogues. And if, after all manner of attempts have been made, no such curve should be forthcoming, it is certain that the proposed curve may be compared neither with rectilinear figures nor with conics.

In the same way when mechanical curves are at issue these must first be transformed into equivalent geometrical ones, after the manner shown in the same Problem 8,(615) and then the areas of these geometrical curves elicited from the catalogues. Take the following as an example of this point.(616)

---

(616) As an instance of integration by parts the following 'Exemplum' develops the transform $\int_0^z s.dz = sz - \int_0^z z\sqrt{[1+(dy/dz)^2]}.dz$, where $AC = z$, $CD = y$ are the coordinates of the curve (D) of arc-length $\widehat{QD} = s$, in the particular case where (D) is the central conic

$$y^2 = (b/a)z^2 + \tfrac{1}{4}b^2.$$

It will be clear that this type of transformation yields a complementary area listed in Newton's catalogues only for those 'mechanical' curves whose slope is algebraic.

EXEMPL: 6. Proponatur figura arcuum cujusvis Conicæ Sectionis ad sinus rectos applicatorum determinanda.(617) Utpote sit A centrum Conicæ Sectionis, AQ & AR semiaxes, CD ordinatim applicata ad axin AR, et PD perpendiculum ad punctum D. Sit etiam AE dicta figura Mechanica occurrens CD in E, et ex ejus natura præfinita erit CE æqualis arcui QD. Quæritur itaq; area AEC, vel parallelogrammo $ACEF$ completo, quæritur excessus AEF. In quem finem sit a latus rectum Conicæ Sectionis, et
||[90] ||b latus transversum sive $2AQ$. Sit etiam $AC=z$, et $CD=y$, eritq; $\sqrt{\frac{1}{4}bb+\frac{b}{a}zz}=y$

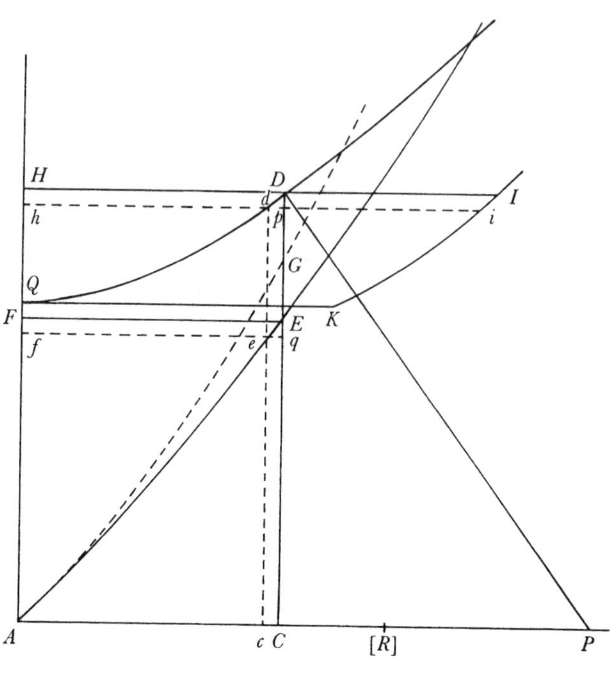

æquatio ad conicam sectionem,(618) ut notum est. Erit etiam PC(619)$=\frac{b}{a}z$ et inde $PD=\sqrt{\frac{1}{4}bb+\frac{bb+ab}{aa}zz}$. Atq; adeò cùm sit fluxio arcus QD ad fluxionem Basis AC ut PD ad CD, si fluxio basis(620) supponatur 1 erit arcus illius QD, sive applicatæ CE fluxio $\sqrt{\dfrac{\frac{1}{4}bb+\frac{bb+ab}{aa}zz}{\frac{1}{4}bb+\frac{b}{a}zz}}$. Hanc duc in FE sive z et proveniet $z\sqrt{\dfrac{\frac{1}{4}bb+\frac{bb+ab}{aa}zz}{\frac{1}{4}bb+\frac{b}{a}zz}}$ fluxio areæ AEF adeoq; si in applicata CD capias $CG=z\sqrt{\dfrac{\frac{1}{4}bb+\frac{bb+ab}{aa}zz}{\frac{1}{4}bb+\frac{b}{a}zz}}$, area AGC quam illa CG super AC incedens describet, æquabitur areæ AEF, et erit AG curva geometrica. Quæritur itaq; area AGC. Et in hunc finem substituatur z^{η} pro zz in æquatione novissima et evadet $z^{\frac{\eta}{1}-1}\sqrt{\dfrac{\frac{1}{4}bb+\frac{bb+ab}{aa}z^{\eta}}{\frac{1}{4}bb+\frac{b}{a}z^{\eta}}}=CG$, æquatio secundæ speciei

EXAMPLE 6. Let it be proposed to determine [the area of] the figure of the arcs of any conic applied to their right sines.(617) Let, for instance, A be the conic's centre, AQ and AR its semiaxes, CD ordinate to the axis AR and PD the normal at the point D. Let also AE be the said mechanical figure meeting CD in E, and from its nature as previously defined CE will be equal to the arc $\widehat{QD}$. Consequently, the area ACE is sought or alternatively, the rectangle $ACEF$ being completed, the excess AEF is required. To this end let a be the conic's latus rectum and b its transverse axis (that is, $2AQ$). Let also $AC = z$ and $CD = y$, and $\sqrt{[\frac{1}{4}b^2 + (b/a)\,z^2]} = y$ will be the equation to the conic,(618) as is known. Again, PC(619) $= (b/a)\,z$ and thence $PD = \sqrt{[\frac{1}{4}b^2 + a^{-2}(b^2+ab)\,z^2]}$. Consequently, since the fluxion of the arc $\widehat{QD}$ to that of the base AC is as PD to CD, if the fluxion of the base(620) be taken as unity, then the fluxion of that arc $\widehat{QD}$, and so of the ordinate CE, will be $\sqrt{\dfrac{\frac{1}{4}b^2 + a^{-2}(b^2+ab)\,z^2}{\frac{1}{4}b^2 + (b/a)\,z}}$. Multiply this by FE or z and there will come out $z\sqrt{\dfrac{\frac{1}{4}b^2 + a^{-2}(b^2+ab)\,z^2}{\frac{1}{4}b^2 + (b/a)\,z}}$ as the fluxion of the area AEF, and in consequence, if in the ordinate CD you take

$$CG = z\sqrt{\frac{\frac{1}{4}b^2 + a^{-2}(b^2+ab)\,z^2}{\frac{1}{4}b^2 + z^2}},$$

the area AGC which that line CG describes as it advances upon AC will equal the area AEF while AG will be a geometrical curve. The area AGC is accordingly required. To this end let z^η be substituted in place of z^2 in the most recent equation and there will result $z^{\eta-1}\sqrt{\dfrac{\frac{1}{4}b^2 + a^{-2}(b^2+ab)\,z^\eta}{\frac{1}{4}b^2 + (b/a)\,z^\eta}} = CG$, an equation of

(617) By analogy with the particular case of the circle (for which $b = -a$ in the following) the 'sinus rectus' in a central conic is the half-chord (here HD) which is subtended by the corresponding 'arcus' $\widehat{QD}$, where QH is the conic's main diameter, and the corresponding 'figura arcuum...ad sinus rectos applicatorum' is the plane curve AE defined by the Cartesian coordinates $AC = HD$ and $CE = \widehat{QD}$. Newton's assessment of this 'conic-arc' curve as 'mechanical' in the next sentence but one is, of course, exact although (compare note (236)) the proof of its transcendence lay well beyond the limits of seventeenth-century mathematical techniques.

(618) Specifically, a hyperbola in the case illustrated by Newton's figure but the present argument evidently holds good for the complementary elliptical case on taking $b \rightarrow -b$.

(619) That is, the conic subnormal $y\,dy/dz$.

(620) Namely, $AC = z$. Note that the 'differential' triangle Ddp is similar to the finite triangle PDC. Newton determines the general arc-length of a central conic $y^2 = a^2 \pm bz^2$ as an infinite series in z on his page 113 below (Example 8 of Problem 12). Compare also notes (742) and (744).

undecimi ordinis posterioris Catalogi. Et collatis utrobiꝗ terminis fit $d=1$. $e=\frac{1}{4}bb=g$. $f=\frac{bb+ab}{aa}$, et $h=\frac{b}{a}$, adeoꝗ $\sqrt{\frac{1}{4}bb+\frac{b}{a}zz}=x$, $\sqrt{-\frac{b^3}{4a}+\frac{a+b}{a}xx}=v$ & $\frac{a}{b}s=t$. Hoc est $CD=x$. $DP=v$ et $\frac{a}{b}s=t$.(621) Et inventorum talis est constructio. Ad Q erige QK perpendicularem et æqualem QA et huic parallelam[,] æqualem vero DP age HI per punctum D. Et linea KI in quam HI terminatur erit Sectio Conica(622) areaꝗ comprehensa $HIKQ$ ad aream quæsitam AEF ut b ad a, sive ut PC ad AC.

Nota, si mutes signum b, sectio Conica cujus arcui recta CG æquatur, evadet Ellipsis; et præterea si fiat $b=-a$ Ellipsis evadet circulus: In quo casu linea KI fit recta parallela AQ.(623)

||[91] ||Postquam Curvæ alicujus area sic inventa fuerit; de constructionis demonstratione consulendum est, quacum sine Computo Algebraico quantùm liceat contexta ornetur Theorema ut evadat publicæ notitiæ dignum. Estꝗ demonstrandi methodus generalis(624) quam sequentibus exemplis illustrare conabor.(625)

Demonstratio Constructionis in Exempl 5. In arcu PQ sume punctum d proximum ad D et age de ac dm parallelas DE ac DM et occurrentes DM et AP in p et l: et erit $DEed$ momentum areæ $PDEP$ et $LMml$ momentum areæ $LMKP$. Age semidiametrum AD, et concipe indefinite exiguum arcum Dd esse instar rectæ et triangula Dpd et ALD erunt similia,(626) adeoꝗ $Dp.pd::AL.LD$. Est autem $HF.EH::AG.AF$, hoc est $AL.LD::ML.DE$. et proinde $Dp.pd::ML.DE$.

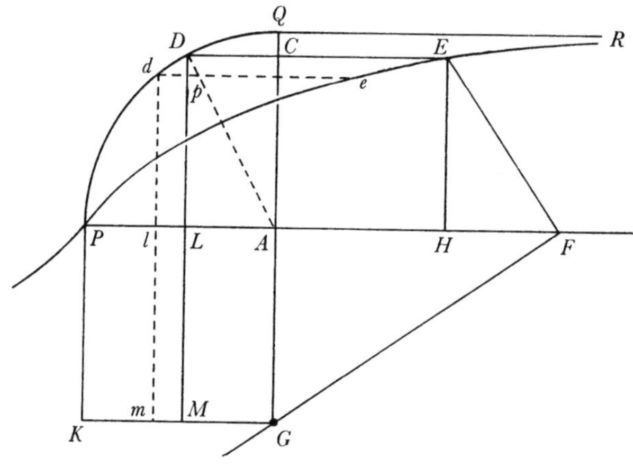

(621) Evidently the substitution $x^2 = (b/a)z^2+\frac{1}{4}b^2$ (or y^2) gives

$$(AEF) = t = \int_0^z z(ds/dz).dz = (a/b)\int_{\frac{1}{2}b}^y v.dx$$

on setting $v = \sqrt{[(1+b/a)x^2-\frac{1}{4}b^3/a]} = \sqrt{[x^2+(bz/a)^2]}$ or $\sqrt{[CD^2+CP^2]} = DP$. Note that the curves AE and AG are touched by the common tangent $y = z$ at their point A of intersection.

(622) Namely, of Cartesian equation $av^2 = (a+b)x^2-\frac{1}{4}b^3$, where $AH = (y =)\ x$ and $HI = (DP =)\ v$. The conic will be an ellipse or hyperbola according as b is less or greater than $-a$, and is obviously through $K(\frac{1}{2}b, \frac{1}{2}b)$ in all cases.

(623) More exactly the pair of lines $v = \pm\frac{1}{2}b$ parallel to AQ: of these only $v = \frac{1}{2}b$ passes through K. In this case $\widehat{QD} = CE = s = \frac{1}{2}b\cos^{-1}(2z/b)$, so that $ds/dz = -\frac{1}{2}b/\sqrt{[\frac{1}{4}b^2-z^2]}$ and therefore $(AEF) = \int_0^z -\frac{1}{2}bz/\sqrt{[\frac{1}{4}b^2-z^2]}.dz = \int_{\frac{1}{2}b}^y \frac{1}{2}b.dx = -\frac{1}{2}b(\frac{1}{2}b-y)$.

the second species of the eleventh order in the latter catalogue. On comparing related terms in each there comes $d = 1$, $e = \frac{1}{4}b^2 = g$, $f = a^{-2}(b^2+ab)$ and $h = b/a$, so that $\surd[\frac{1}{4}b^2+(b/a)z^2] = x$, $\surd[-\frac{1}{4}a^{-1}b^3+a^{-1}(a+b)x^2] = v$ and

$$(a/b)s = t.$$

That is, $CD = x$, $DP = v$ and $(a/b)s = t$.(621) And the following is the construction of these findings. At Q erect QK perpendicular and equal to QA, and parallel to this but equal to DP draw HI through the point D. Then the line KI in which HI terminates will be a conic,(622) while the contained area $HIKQ$ will be to the area AEF sought as b to a, that is, as PC to AC.

Note, should you change the sign of b, the conic to whose arc the straight line CG is equal will prove to be an ellipse; and, further, if b is made equal to $-a$ the ellipse will come out to be a circle, in which case the curve KI becomes a straight line parallel to AQ.(623)

After the area of some curve has thus been found, careful consideration should be given to fabricating a proof of the construction which as far as permissible has no algebraic calculation, so that the theorem embellished with it may turn out worthy of public utterance. A general method of proof(624) exists, indeed, and this I shall attempt to illustrate by the following examples.

PROOF OF THE CONSTRUCTION IN EXAMPLE 5. In the arc PQ take a point d very near to D and draw de and dm parallel to DE and DM, meeting DM and AP in p and l: $DEed$ will then be the moment of the area $PDEP$ and $LMml$ that of the area $LMKP$. Draw the radius AD and imagine that the indefinitely little arc $\widehat{Dd}$ is the equivalent of a straight line. The triangles Dpd and ALD will then be similar,(626) so that $Dp:pd = AL:LD$. But $HF:EH = AG:AF$, that is,

$$AL:LD = ML:DE$$

(624) Namely, the method of limit-increments of geometrical variables (whose ratios are equal to those of the corresponding Newtonian fluxions). Book 1 of Newton's *Philosophiæ Naturalis Principia Mathematica* has several classic instances (notably Propositions 41, 79–81 and 91) of an essentially algebraic argument being presented to public view in the synthetic form of an equivalent limit-increment argument. We will return to the point in our sixth volume. Newton's present rejection, where possible, of all 'computus Algebraicus' from a formally presented 'demonstratio' of a mathematical theorem is reminiscent of a passage in one of Barrow's 1664 lectures: '[Algebra vel] Analysis (eatenus intellecta, quatenus a Geometriæ vel Arithmeticæ pronunciatis et regulis distincti quid innuit)...est...duntaxat pars quædam aut species Logicæ, seu modus quidam utendi ratione circa quæstionum solutionem, inventionemque vel probationem conclusionum, qualis in aliis omnibus scientiis exercetur haud raro. Quare non est pars, aut species, sed instrumentum potius Mathematicæ subministrans; ut neque Synthesis, quæ modus est theoremata demonstrandi analysi contradistinctus inversusque' (*Lectiones Mathematicæ* (note (81)): Lectio II [1664]: 31–2).

(625) For convenience, with each of the following examples we repeat the appropriate figure from Newton's pages 85–90.

(626) Since their corresponding sides are perpendicular to each other.

Quare $Dp \times DE = pd \times ML$. Hoc est[627] momentum $DEed$ æquale momento $LMml$. Et cùm hoc de quibuslibet contemporaneis momentis indeterminatè demonstretur, patet singula[628] momenta areæ $PDEP$ esse singulis[628] contemporaneis momentis areæ[629] $PLMK$ æqualia, adeoq; totas areas[629] ex istis momentis compositas æquari. Q.E.D.

Demonstratio Constructionis in Exemplo 3. Esto $DEed$ momentum superficiei $AHDE$ ac $AdDA$ contemporaneum momentum segmenti ADH. age semidiametrum DK, et de occurrat AQ in c, estq; $Cc.Dd::DC.DK$.[630] Præterea est $DC.QA(2DK)::AC.DE$. Adeoq; $Cc.2Dd(::DC.2DK)::AC.DE$. et $Cc \times DE = 2Dd \times AC$. Jam ad periferiæ momentum Dd rectà productum (i.e. ad tangentem circuli)[631] demitte normalem AI et erit AI æqualis AC, adeoq; $2Dd \times AC (= 2Dd \times AI) = 4$ triangulis ADd. Quare

$$4 \text{ triang } ADd = Cc \times DE = \text{momento } DEed.$$

Spatij ergo $AHDE$ singula momenta sunt quadrupla momentorum contemporaneorum segmenti ADH et proinde totum illud spatium quadruplum totius segmenti. Q.E.D.

Demonstratio constructionis in Exemplo 4. Parallelam CE age indefinitè parùm distantem ce, et Hyperbolæ tangentem Ck ac demitte KM rectam ad AP: Et ex Hyperbolæ natura erit $AC.AP::AP.AM$ Adeoq; $AG^q.GL^q(::AC^q.LE^q$ sive $AP^q)::AP^q.AM^q$ ac divisim

$$AG^q.AL^q(DE^q)^{(632)}::$$
$$AP^q.AM^q - AP^q(MK^q).$$

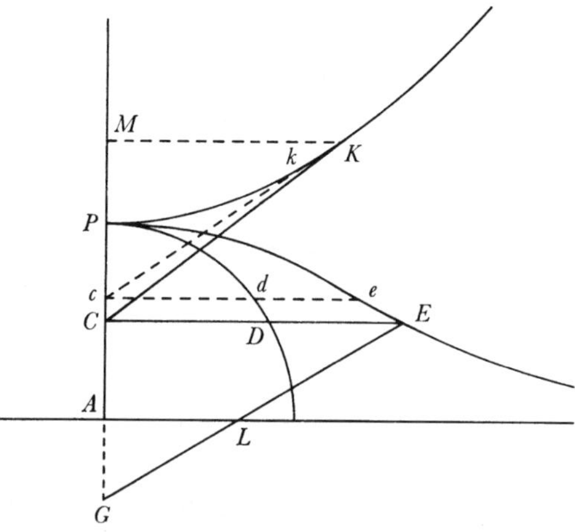

et inversè[633] $AG.AP::DE.MK$. Est
||[92] autem ||areola $DEed$ ad triangulum CKc ut altitudo DE ad semissem altitudinis KM. Hoc est, ut AG ad $\frac{1}{2}AP$. Quare omnia spatij PDE momenta ad omnia contemporanea momenta spatij PKC sunt ut AG ad $\frac{1}{2}AP$. Et proinde tota illa spatia sunt in eadem ratione. Q.E.D

(627) Understand 'neglecto infinitè parvo triangulo Dpd' (on neglecting the infinitely small triangle Dpd). Newton, indeed, at this point began an unfinished next phrase 'Dp'.

(628) Newton first wrote here somewhat less precisely 'omnia' and 'omnibus' respectively.

(629) The equivalent 'spatij' and 'spatia' are here cancelled.

(630) In the limit, that is, as the points c, d pass into coincidence with C, D.

(631) Namely, at D. The congruency of the right triangles IDA, CDA is an immediate inference since $I\hat{D}A = D\hat{Q}A = C\hat{D}A$.

and consequently $Dp:pd = ML:DE$. Hence $Dp \times DE = pd \times ML$, that is,[627] the moment $DEed$ is equal to the moment $LMml$. And since this may be proved for any contemporaneous moments without restriction, it is evident that each individual[628] moment of the area $PDEP$ is equal to a corresponding contemporaneous one of the area $PLMK$, so that the total areas composed of those moments are equal. As was to be proved.

PROOF OF THE CONSTRUCTION IN EXAMPLE 3. Let $DEed$ be a moment of the surface $AHDE$ with $AdDA$ the contemporaneous moment of the segment ADH. Draw the radius DK and let de meet AQ in c, and there is

$$Cc:Dd = DC:DK.^{(630)}$$

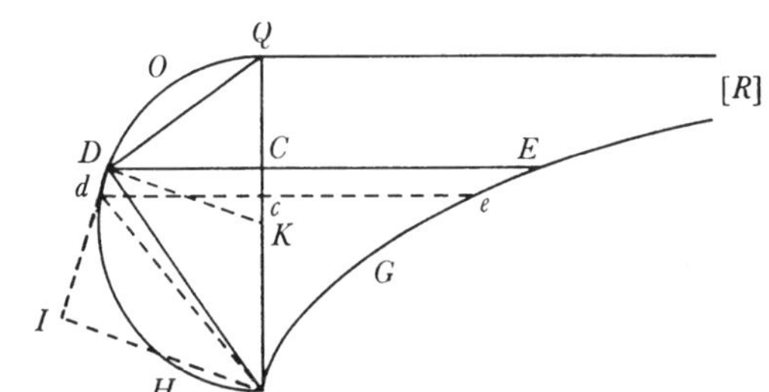

Further, $DC:QA$ (or $2DK$) $= AC:DE$, so that

$$Cc:2Dd (= DC:2DK) = AC:DE,$$

and $Cc \times DE = 2Dd \times AC$. Now to the moment Dd of the circumference extended in a straight line (that is, to the circle tangent)[631] let fall the perpendicular AI and AI will then be equal to AC, so that

$$2Dd \times AC (= 2Dd \times AI) = 4\triangle ADd.$$

Hence $4\triangle ADd = Cc \times DE =$ moment $DEed$. Therefore each moment of the space $AHDE$ is four times the contemporaneous one of the segment ADH, and consequently the total space is four times the total segment. As was to be proved.

PROOF OF THE CONSTRUCTION IN EXAMPLE 4. Parallel to CE at an indefinitely little distance away draw ce, also the hyperbola's tangent Ck, and let fall KM normal to AP. Then by the nature of the hyperbola it will be $AC:AP = AP:AM$, so that

$$AG^2:GL^2 (= AC^2:LE^2, \text{ that is, } AP^2) = AP^2:AM^2$$

and by dividing the ratios

$$AG^2:AL^2 \text{ (or } DE^2)^{(632)} = AP^2: (AM^2 - AP^2 \text{ or) } MK^2,$$

and so, inversely,[633] $AG:AP = DE:MK$. But the element of area $DEed$ is to the triangle CKc as the altitude DE to half the altitude KM, that is, as AG to $\frac{1}{2}AP$. Consequently every moment of the space PDE is to every contemporaneous moment of the space PKC as AG to $\frac{1}{2}AP$, and therefore the whole of those spaces are in the same ratio. As was to be proved.

(632) See note (609) above.

(633) Read 'permutando et radices extrahendo' (by permutation and extracting roots).

DEMONSTRATIO CONSTRUCTIONIS IN EXEMPLO 6. Parallelam et proximam CD age cd et occurrentem curvæ AE in e [necnon] age hi & fe occurrentes DC in p et q. Et erit ex Hypothesi(634) $Dd=Eq$ et ex similitudine triangulorum Dpd, DCP erit $Dp\,.\,Dd(Eq) :: CP\,.\,PD(HI)$. Adeoqȝ $Dp\times HI=Eq\times CP$, et inde $Dp\times HI$ (moment $HIih$). $Eq\times AC$ (moment $EFfe$) :: $Eq\times CP\,.\,Eq\times AC :: CP\,.\,AC$. Quare cum PC et AC sint in data ratione lateris transversi ad latus rectum Conicæ Sectionis QD,(635) et arearū $HIKQ$ et AEF momenta $HIih$ & $EFfe$ in illâ ratione, erunt ipsæ areæ in eâdem ratione. Q.E.D.

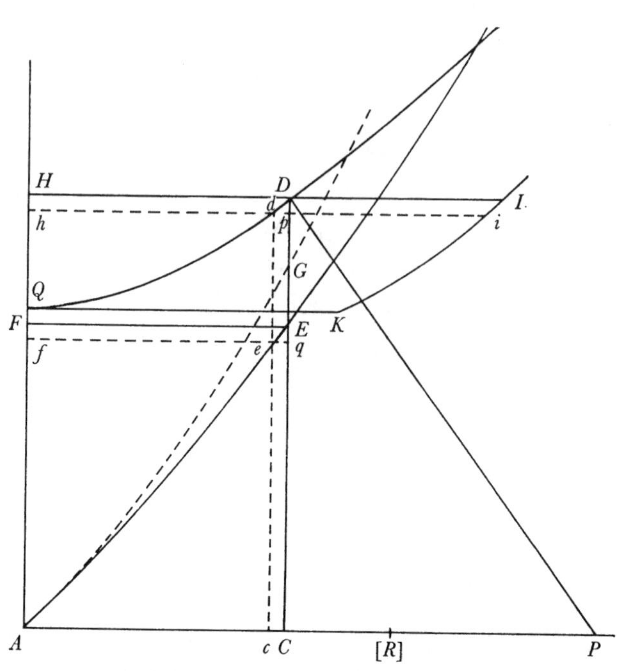

In hujusmodi demonstrationibus observandum est quod quantitates pro æqualibus habeo quarum ratio est æqualitatis. Et ratio æqualitatis censenda est quæ minùs differt ab æqualitate quàm quælibet inæqualis ratio potest assignari. Sic in postremâ demonstratione(636) posui rectangulum $Eq\times AC$, sive $FEqf$ æquale spatio $FEef$ quia (propter differentiam Eqe infinite minorem ipsis sive respectu ipsarum nullam) non habent rationem inæqualitatis.(637) Et eadem de causa posui $Dp\times HI=HIih$, & sic in alijs.

(638)Hac methodo probandi curvas per æqualitatem vel datam rationem momentorum æquales esse vel datam rationem habere hic usus sum quòd cum methodis in his rebus usitatis(639) affinitatem habeat; sed(640) magis naturalis videtur quæ genesi superficierum ex fluendi motu innititur. Sic si constructio in Exemplo 2 demonstranda sit; Ex natura circuli est fluxio rectæ ID ad fluxionem rectæ IP, ut AI ad ID: estqȝ AI ad ID ut ID ad CE ex natura Curvæ AGE: et proinde $CE\times$flux: $ID=$fluxioni areæ $ACEG$, et $ID\times$flux $IP=$fluxioni areæ

(634) Namely, that $\widehat{QD} = CE$.

(635) For $(b/a)\,z:z = (2AQ$ or$)\ b:a$.

(636) That of the construction of Example 6 immediately preceding. Understand '[quæ] potest assignari' in the previous phrase, probably.

(637) In the limit in which the increment Eq vanishes, that is.

(638) The brief diversion into the equivalent theory of geometrical fluxions which forms the next two paragraphs was, in revision, expanded and systematized into a short treatise (reproduced as §3 following) evidently intended to replace it in the final version of the 1671 tract. The revised scheme, omitted by Jones in his 1710 transcript(and, following him, by Colson

PROOF OF THE CONSTRUCTION IN EXAMPLE 6. Parallel and very close to CD draw cd, meeting the curve AE in e: draw also hi and fe, meeting DC in p and q. Then by hypothesis(634) $Dd = Eq$ while from the similarity of the triangles Dpd, DCP there will be $Dp:Dd$ (or Eq) $= CP:PD$ (or HI), so that

$$Dp \times HI = Eq \times CP,$$

and thence $Dp \times HI$ (or the moment $HIih$): $Eq \times AC$ (or the moment $EFfe$)

$$= Eq \times CP : Eq \times AC = CP:AC.$$

Hence, since PC and AC are in the given ratio of the transverse axis of the conic QD to its latus rectum(635) and the moments $HIih$ and $EFfe$ of the areas $HIKQ$ and AEF are also in that ratio, the areas themselves will be in the same ratio. As was to be proved.

In demonstrations of this sort it should be observed that I take quantities as equal whose ratio is one of equality. And a ratio of equality is to be regarded as one which differs less from equality than any ratio of inequality [which] can possibly be assigned. Thus, in the last proof(636) I set the rectangle $Eq \times AC$, that is, $FEqf$, equal to the space $FEef$ since (because their difference Eqe is infinitely less than they and so, in regard to them, zero) they have no ratio of inequality.(637) For the same reason I set $Dp \times HI = HIih$, and so in other instances.

(638)I have here used this method of proving that curves are equal or have a given ratio by means of the equality or given ratio of their moments since it has an affinity to the ones usually employed in these cases.(639) However, that based on the genesis of surfaces by their motion of flow appears(640) a more natural approach. So if the construction in Example 2 has to be demonstrated, from the nature of the circle the fluxion of the straight line ID is to that of the line IP as AI to ID, while AI is to ID as ID to CE by the nature of the curve AGE; furthermore, $CE \times$ fluxion of $ID =$ fluxion of the area $ACEG$, and $ID \times$ fluxion of

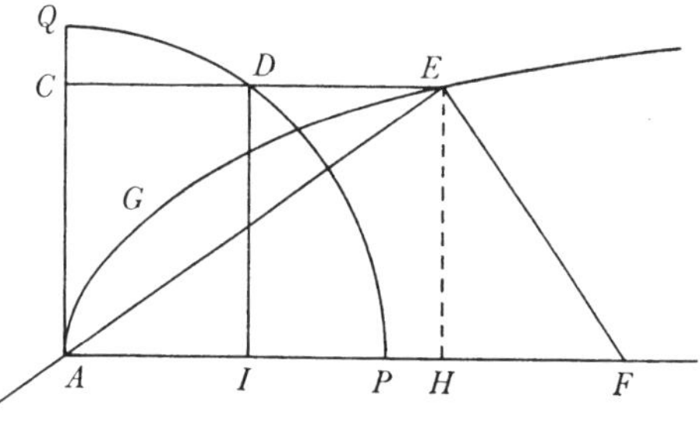

in his 1736 English version) and ignored by Horsley in his 1779 Latin version, has never been published. See §3: note (3) below.

(639) Newton refers apparently to James Gregory's extension of the Archimedean method of exhaustion as applied in the latter's *Geometriæ Pars Universalis* of 1668 (see note (204)) and also to the less rigorous limit-increment methods expounded by Isaac Barrow in the later lectures of his 1670 *Lectiones Geometricæ* (note (80)). Gregory and Barrow, of course, for the most part codified and generalized three preceding decades of research into geometrical integration by Fermat, Roberval, Torricelli, Huygens and many others. See D. T. Whiteside, 'Patterns of Mathematical Thought in the later Seventeenth Century' (note (109)): Chapter IX: 331–48.

(640) 'promptior aliquantò et' (a somewhat readier and) is cancelled.

PDI. Et propterea areæ illæ æqualiter fluendo genitæ(641) æquales erunt. Q.E.D.

Plenioris illustrationis gratia adjiciam demonstrationem Constructionis qua Cissoidis area in Exemplo 3 determi-
||[93] natur. ||Lineæ punctim(642) notatæ in schemate deleantur, et agatur DQ et Cissoidis Asymptoton QR: Et ex natura circuli est $DQ^q = AQ \times CQ$, et inde per Prob 1 $2DQ \times$ flux: ipsius $DQ =$

$$AQ \times \text{flux: } CQ.$$

Adeoq $AQ . DQ :: 2 \text{ flux}: DQ . \text{flux}: CQ$. Est et ex natura Cissoidis

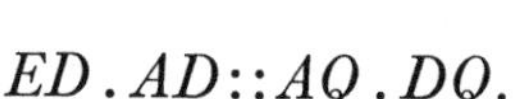

$$ED . AD :: AQ . DQ.$$

Quare $ED . AD :: 2 \text{ flux}: DQ . \text{flux}: CQ$. et

$$ED \times \text{flux } CQ = AD \times 2 \text{ flux: } DQ = 4 \times \tfrac{1}{2}AD \times \text{flux: } DQ.$$

Jam cùm DQ perpendicularis sit ad terminum ipsius AD circa A gyrantis, est $\frac{1}{2}AD \times$ flux $DQ =$ fluxioni generanti aream $ADOQ$. Est et ejus quadruplum $ED \times$ flux $CQ =$ fluxioni generanti Cissoidalem aream $QREDO$. Et proinde area illa infinitè longa $QREDO$ generatur quadrupla alterius $ADOQ$. Q.E.D.

SCHOLIUM.(643)

Per præcedentes catalogos non tantùm areæ curvarum sed et aliæ cujuscunq generis quantitates analoga fluendi ratione generatæ, e fluxionibus derivari possunt. Idq mediante hoc Theoremate, Quod quantitas cujuscunq generis sit ad unitatem congeneram ut area Curvæ ad unitatem superficialem, si modò fluxio quantitatem illam generans sit ad unitatem sui generis ut fluxio generans aream ad unitatem sui generis, hoc est ut linea super Basi normaliter incidens qua area illa describitur, ad unitatem linearem.(644) Et proinde si fluxio qualiscunq exponatur per ejusmodi lineam incedentem[,] quantitas ab illa fluxione generata exponetur per aream ab illa incedente descriptam. Vel si fluxio per eosdem terminos Algebraicos cum incedente linea exponatur, quantitas generata exponetur per eosdem cum area descripta. Æquatio itaq quæ
||[94] fluxionem ||cujuscunq generis exhibet quærenda est in prima collumna Catalogorum, et valor t in ultima collumna indicabit quantitatem generatam.

(641) Newton has cancelled the equivalent phrase 'fluxionibus æqualibus progenitæ'.

(642) Literally 'by points'. The accompanying 'scheme' is repeated from Newton's page 87. This variant quadrature of the kappa quartic had been found by Huygens in 1662; see note (602).

IP = fluxion of the area PDI. Accordingly, those areas, being generated by an equal flow, will themselves be equal. As was to be proved.

For the sake of fuller illustration, I shall add a proof of the construction by which the cissoid's area is determined in Example 3. Let the lines marked as broken(642) in the diagram be deleted, and draw DQ and the cissoid's asymptote QR. Then from the nature of the circle $DQ^2 = AQ \times CQ$, and from this by Problem 1 $2DQ \times$ fluxion $(DQ) = AQ \times$ fluxion (CQ), so that

$$AQ:DQ = 2 \text{ fluxion } (DQ):\text{fluxion } (CQ).$$

Also from the nature of the cissoid $ED:AD = AQ:DQ$. Hence

$$ED:AD = 2 \text{ fluxion } (DQ):\text{fluxion } (CQ)$$

and so $ED \times$ fluxion $(CQ) = AD \times 2$ fluxion $(DQ) = 4 \times \frac{1}{2}AD \times$ fluxion (DQ). Now since DQ is perpendicular to AD at its end-point as it rotates round A, $\frac{1}{2}AD \times$ fluxion (DQ) = fluxion generating the area $ADOQ$. Also its quadruple $ED \times$ fluxion (CQ) = fluxion generating the cissoidal area $QREDO$. Consequently that infinitely long area $QREDO$ is generated four times as big as the other $ADOQ$. As was to be proved.

SCHOLIUM(643)

By means of the preceding catalogues not merely the areas of curves but also other quantities of any kind generated at an analogous rate of flow may be derived from their fluxions, and that through the medium of this theorem: A quantity of any kind is to the unity of its own class as the area of a curve to the surface unity if only the fluxion generating that quantity shall be to the unity of its own kind as the fluxion generating the area to its own unity—that is, as the line, moving normally upon the base, by which that area is described to the linear unity.(644) In consequence, if a fluxion of whatever kind be expressed by an ordinate line of this sort, the quantity generated by that fluxion will be expressed by the area described by that ordinate. Or if the fluxion be expressed by the same algebraic terms as the ordinate line, the generated quantity will be expressed by the same ones as the area described. Accordingly, the equation which exhibits a fluxion of any kind should be sought in the first column of the catalogues, and the value of t in the last column will then disclose the quantity generated.

(643) Newton develops the point that the 'area' under a 'curve' is a standard model for all (simple Riemannian) integrals.

(644) This, of course, to preserve correct analytical dimensions. A cancelled continuation reads 'Vel breviùs quod quantitates sint analogæ quæ ex analogis fluxionibus generantur' (Or more briefly: those quantities are analogous which are generated from analogous fluxions).

Quemadmodum si $\sqrt{1+\frac{9z}{4a}}$ fluxionem cujuscunq; generis exhibeat, pone æqualem y, et ut ad formam æquationum in Catalogis reducatur substitue z^{η} pro z, sic enim evadet $z^{\overset{\eta}{-1}}\sqrt{1+\frac{9}{4a}z^{\eta}}=y$, æquatio primæ speciei tertij ordinis prioris Catalogi et collatis terminis fiet $d=1$, $e=1$, $f=\frac{9}{4a}$, et inde

$$\frac{8a+18z}{27}\sqrt{1+\frac{9z}{4a}}\left(=\frac{2d}{[3]\eta f}R^3\right)=t.$$

Est itaq; $\frac{8a+18z}{27}\sqrt{1+\frac{9z}{4a}}$ quantitas quæ generatur fluxione $\sqrt{1+\frac{9z}{4a}}$.(645)

Atq; ita si $\sqrt{1+\frac{16z^{\frac{2}{3}}}{9a^{\frac{2}{3}}}}$ designet fluxionem, per debitam reductionem (extrahendo $z^{\frac{2}{3}}$ e radicali, et scribendo z^{η} pro $z^{-\frac{2}{3}}$) habebitur $\frac{1}{z^{\overset{\eta}{+1}}}\sqrt{z^{\eta}+\frac{16}{9a^{\frac{2}{3}}}}$(646) $=y$, æquatio secundæ speciei quinti ordinis posterioris Catalogi, et collatis terminis fit $d=1$, $e=\frac{16}{9a^{\frac{2}{3}}}$, et $f=1$, Adeoq; $z^{\frac{2}{3}}\left(=\frac{1}{z^{\eta}}\right)=xx$, $\sqrt{1+\frac{16xx}{9a^{\frac{2}{3}}}}=v$, et

$$\tfrac{3}{2}s\left(=-\frac{2d}{\eta}s\right)=t.^{(647)}$$

Quibus inventis, quantitas per fluxionem $\sqrt{1+\frac{16z^{\frac{2}{3}}}{9a^{\frac{2}{3}}}}$ generata innotescet ponendo esse ad unitatem sui generis ut area $\frac{3}{2}s$ ad unitatem superficialem. Vel quod eodem recidit, ponendo quantitatem t non amplius superficiem(648) significare, sed alterius generis quantitatem quæ est ad unitatem ejusdem generis ut superficies illa ad unitatem superficialem.(649) Sic posito quod $\sqrt{1+\frac{16z^{\frac{2}{3}}}{9a^{\frac{2}{3}}}}$ designet fluxionem linearem,(650) imaginor t non ampliùs superficiem

(645) More directly, the substitution $x = 1+9z/4a$ yields

$$\int_0^z \sqrt{[1+9z/4a]}\,.\,dz = \tfrac{4}{9}a\int_1^{(1+9z/4a)^{\frac{1}{2}}} x^{\frac{1}{2}}\,.\,dx = \tfrac{8}{27}a(1+9z/4a)^{\frac{3}{2}}.$$

Compare Example 3 on Newton's page 110 below.

(646) Read '$\frac{1}{z^{\overset{2\eta}{+1}}}\sqrt{z^{\eta}+\frac{16}{9a^{\frac{2}{3}}}}$'. The error is carried through the next few lines, and thence into Example 4 on Newton's pages 110–11 below. It was not corrected either by Horsley in his *Opera Omnia*, **1** (note (60)): 501 or by Colson in his *Method of Fluxions* (note (79)): 120–1, though the former for consistency's sake alters the next stage in argument to read 'Adeoq;...$3s$ (...) $=t$'.

For instance, if $\surd[1+9z/4a]$ should exhibit a fluxion of any kind, set it equal to y and, to reduce it to the form of the equations in the catalogues, substitute z^{η} in place of z. For thus there will result $z^{\eta-1}\surd[1+\frac{9}{4}a^{-1}z^{\eta}] = y$, an equation of the first species of the third order in the former catalogue, and, when terms are compared, there will come $d = 1$, $e = 1$, $f = \frac{9}{4}a^{-1}$ and in consequence

$$\tfrac{8}{27}a(1+9z/4a)^{\frac{3}{2}}\,(= [2d/3\eta f]\,R^3) = t.$$

Accordingly $\frac{8}{27}a(1+9z/4a)^{\frac{3}{2}}$ is the quantity generated by the fluxion

$$\surd[1+9z/4a].^{(645)}$$

And thus if $\surd[1+\frac{16}{9}a^{-\frac{2}{3}}z^{\frac{2}{3}}]$ represents the fluxion, by appropriate reduction (taking $z^{\frac{2}{3}}$ outside the root and writing z^{η} in place of $z^{-\frac{2}{3}}$) there will be had $z^{-\eta-1}\surd[z^{\eta}+\frac{16}{9}a^{-\frac{2}{3}}]^{(646)} = y$, an equation of the second species of the fifth order in the latter catalogue. When terms are compared there comes $d = 1$, $e = \frac{16}{9}a^{-\frac{2}{3}}$ and $f = 1$, so that

$$z^{\frac{2}{3}}(= z^{-\eta}) = x^2, \quad \surd[1+\tfrac{16}{9}a^{-\frac{2}{3}}x^2] = v \quad \text{and} \quad \tfrac{3}{2}s(= -[2d/\eta]\,s) = t.^{(647)}$$

When these have been found, the quantity generated by the fluxion

$$\surd[1+\tfrac{16}{9}a^{-\frac{2}{3}}z^{\frac{2}{3}}]$$

will become known by taking it to the unity of its own class as the area $\frac{3}{2}s$ is to the surface unity—or, what comes to the same, by taking the quantity t to signify no longer a surface[(648)] but a quantity of another class which is to the unity of the same class as that surface is to the surface unity.[(649)] Thus, supposing that $\surd[1+\frac{16}{9}a^{-\frac{2}{3}}z^{\frac{2}{3}}]$ represents a linear fluxion,[(650)] I conceive that t now no

---

(647) On making correct comparison with an 'æquatio tertiæ speciei quinti ordinis posterioris Catalogi' Newton should have written 'Adeoq; $z^{\frac{2}{3}}\left(=\frac{1}{z^{\eta}}\right)=x$, $\sqrt{x+\frac{16xx}{9a^{\frac{2}{3}}}}=v$, et $\frac{3}{2}s\left(=-\frac{d}{\eta}s\right)=t$'. The reader should correct our English version correspondingly. More directly, $\int_0^z \surd[1+\frac{16}{9}z^{\frac{2}{3}}/a^{\frac{2}{3}}]\,.\,dz = \frac{3}{2}\int_0^{z^{\frac{2}{3}}} \surd[x+\frac{16}{9}x^2/a^{\frac{2}{3}}]\,.\,dx$, that is,

$$a(Z+\tfrac{9}{32}Z^{\frac{1}{3}})\surd[Z^{\frac{2}{3}}+\tfrac{9}{16}]-\tfrac{81}{1024}a\log(1+\tfrac{32}{9}Z^{\frac{2}{3}}+\tfrac{32}{9}Z^{\frac{1}{3}}\surd[Z^{\frac{2}{3}}+\tfrac{9}{16}]),$$

where $Z = z/a$.

(648) Newton has cancelled a first phrase 'aream Conicæ Sectionis' (the area of a conic). The area t is, to be sure, not restricted to being that of a conic but only expressible in terms of the quadrature of 'figuræ rectilineæ' and of conics. Two lines below, correspondingly, 'superficies illa' replaces 'area Conicæ Sectionis'.

(649) A first continuation 'Et hoc modo $\frac{3}{2}s$ erit quantitas per propositam fluxionem generata' (And in this manner $\frac{3}{2}s$ will be the quantity generated by the fluxion propounded) is cancelled.

(650) 'ut longitudo generata innotescat' (to make known the generated length) is cancelled.

sed lineam jam significare, eam nempe quæ ad unitatem linearem est ut area(651) quam t iuxta Catalogos designat ad unitatem superficialem, hoc est eam quæ producitur applicando aream illam ad linearem unitatem.(652) Qua ratione si linearis unitas statuatur e[,] longitudo per præfatam fluxionem generata erit $\frac{3s}{2e}$. Et hoc fundamento Catalogi illi ad longitudines curvarum, contenta solidorum & alias quascunq̧ quantitates æque ac areas curvarum determinandas applicari possunt.(653)

||[95] ||De Quæstionibus cognatis.

1. Curvarum areas per Mechanicam approximare.(654)

Methodus est ut duarum pluriumve rectilinearum figurarum valores ita componantur inter se ut valorem areæ curvæ quamproximè constituant. Sic ad circulum AFD quem æquatio(655) $x-xx=rr$ designat[,] postquam inventus est areæ $AFDB$ valor $\frac{2}{3}x^{\frac{3}{2}}-\frac{1}{5}x^{\frac{5}{2}}-\frac{1}{28}x^{\frac{7}{2}}-\frac{1}{72}x^{\frac{9}{2}}$ &c. quærendi sunt aliquot rectangulorum valores, quales sunt ipsius $BD\times AB$ valor $x\sqrt{x-xx}$ sive $x^{\frac{3}{2}}-\frac{1}{2}x^{\frac{5}{2}}-\frac{1}{8}x^{\frac{7}{2}}-\frac{1}{16}x^{\frac{9}{2}}$ &c. ac ipsius $AD\times AB$ valor $x\sqrt{x}$ sive $x^{\frac{3}{2}}$.(656) Dein hi valores per literas quaslibet diversas (quæ numeros indefinitè designent) multiplicandi sunt et addendi summæq̧ termini cum correspondentibus terminis valoris areæ $AFDB$ comparandi, ut quantum liceat evadant æquales. Quemadmodum si per e et f multiplicentur, fiet summa $\begin{matrix}e\\+f\end{matrix}x^{\frac{3}{2}}-\frac{e}{2}x^{\frac{5}{2}}-\frac{e}{8}x^{\frac{7}{2}}$ &c. cujus terminis cum terminis

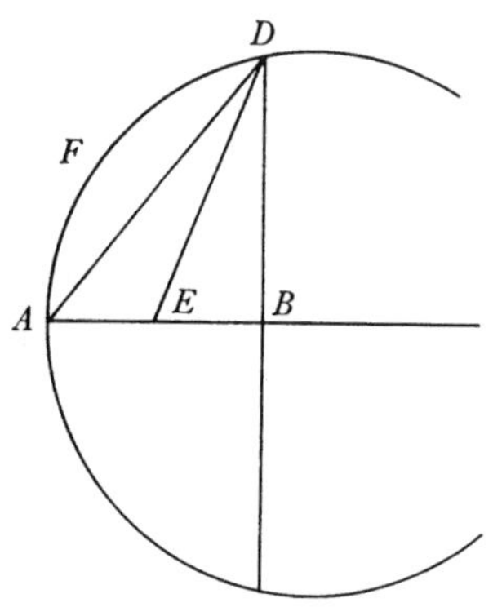

(651) Newton first wrote more fully 'area Conicæ sectionis': compare note (648).

(652) The two final sentences following are a late addition. Newton originally concluded this section with the words 'Et hoc subintellecto, longitudo quæsita erit $\frac{3}{2}s$. Quod idem de valore t in similibus casibus posthac intelligendum est' (On this assumption the required length will be $\frac{3}{2}s$. The same is always to be understood hereafter in similar cases concerning the value of t).

(653) Compare note (447).

(654) In generalization of the technique used previously to determine a given integral as a finite combination of known ones, Newton now seeks to determine the value of an integral which cannot be so expressed 'per Mechanicam': namely, he finds convergent series expansions for the given integral and comparable known functions, and then determines a finite combination of the latter which duplicates the former up to a predetermined number of terms. It seems likely that Newton here elaborates an idea lightly suggested by James Gregory in 1668 in his 'Appendicula ad Veram Circuli & Hyperbolæ Quadraturam' (*Exercitationes Geometricæ* (note (595)): 1–8, especially 6). Having discussed the properties of two 'series convergentes', one monotonically increasing, one monotonically decreasing, in generalization of his account the previous year in his *Vera Quadratura* (see note (72)), Gregory went on to observe: 'Hinc

longer signifies a surface but a line: that, namely, which is to the linear unity as the area(651) represented by t (according to the catalogues) to the surface unity—in other words, that which is produced on dividing that area by the linear unity.(652) For this reason, if the linear unity be taken as e, the length generated by the aforesaid fluxion will be $\frac{3}{2}s/e$. And on this foundation those catalogues may be applied to determining the lengths of curves, the contents of solids and any other quantities whatever equally as well as the areas of curves.(653)

Cognate Questions

1. To approximate the areas of curves by a mechanical technique(654)

The method is to compound the values of two or more rectilinear figures one with another in such a way that together they make up as nearly as possible the value of a curve's area. Thus, in the case of the circle AFD represented by the equation(655) $x-x^2=\dot{z}^2$, after the value of the area $AFDB$ has been found to be $\frac{2}{3}x^{\frac{3}{2}}-\frac{2}{5}x^{\frac{5}{2}}-\frac{1}{28}x^{\frac{7}{2}}-\frac{1}{72}x^{\frac{9}{2}}\ldots$, the values of some rectangles should be sought: such as, for instance, that of $BD\times AB=x\surd[x-x^2]$ or $x^{\frac{3}{2}}-\frac{1}{2}x^{\frac{5}{2}}-\frac{1}{8}x^{\frac{7}{2}}-\frac{1}{16}x^{\frac{9}{2}}\ldots$ and that of $AD\times BD=x\surd x$ or $x^{\frac{3}{2}}$.(656) These values are then to be multiplied by any distinct letters you please (representing numbers not determined) and added, and the terms of the aggregate compared with corresponding terms in the value of the area $AFDB$ so that they may as far as permissible equal them. For instance, if the multipliers are e and f (here), the aggregate will become

$$(e+f)\,x^{\frac{3}{2}}-\tfrac{1}{2}ex^{\frac{5}{3}}-\tfrac{1}{8}ex^{\frac{7}{2}}\ldots.$$

---

patet campus vastissimus inveniendi approximationes non solum in Circuli & Hyperbolæ mensura, sed etiam in omnium aliarum serierum convergentium terminationibus: nos tamen unam particularem methodum eligimus, quam existimamus esse reliquis faciliorem & minus prolixam, nempe ex inventorum terminorum combinatione; in hac enim prolixæ operationes Arithmeticæ evitantur'. The 'particular method' he then exemplified at length (on his pages 6–8) for the case of a general circle sector, although 'ob diversas rationes mihi satis perspectas' he chose not to expound his method for finding the inequalities he states; nevertheless 'paratus...sum...etiam approximationes exhibere, quæ veras notas octuplucent, nonuplicent, decuplicent, &c. in infinitum etiam in aliis seriebus convergentibus ex sola hac terminorum convergentium combinatione'. Newton must at once have seen that Gregory's 'approximationes' resulted from combining a number of inscribed and circumscribed 'polygona complicata' $i_n=2^n\sin 2^{-n}\theta$ and $I_n=2^n\tan 2^{-n}\theta$ so as to yield an aggregate $\theta+O(\theta^{2k+1})$, in which case the approximation is deemed by Gregory to 'k-plicate the true notes'. Compare H. W. Turnbull, *James Gregory Tercentenary Memorial Volume* (London, 1939): 459–62; and J. E. Hofmann, 'Über Gregorys systematische Näherungen für den Sektor eines Mittelpunktkegelschnittes', *Centaurus*, **1** (1950): 24–37.

(655) Forgetting that r (or $\dot{z}$) is here the fluxion of the area $AFDB=z=\int r.dx$, Horsley inconsistently retains the original form of equation '$x-xx=rr$' in his otherwise anachronistic text (*Opera Omnia*, **1** (note (60)): 501).

(656) For, since the circle's diameter is unity, $AD^2=AB$.

hisce $\frac{2}{3}x^{\frac{3}{2}}-\frac{1}{5}x^{\frac{5}{2}}-\frac{1}{28}x^{\frac{7}{2}}$ &c collatis, prodit $e+f=\frac{2}{3}$, et $-\frac{e}{2}=-\frac{1}{5}$; sive $e=\frac{2}{5}$ et $f(=\frac{2}{3}-e)=\frac{4}{15}$. Adeoq; est $\frac{2}{5}BD\times AB+\frac{4}{15}AD\times AB=$ areæ $AFDB$ proximè. Scilicet $\frac{2}{5}BD\times AB+\frac{4}{15}AD\times AB$ valet $\frac{2}{3}x^{\frac{3}{2}}-\frac{1}{5}x^{\frac{5}{2}}-\frac{1}{20}x^{\frac{7}{2}}-\frac{1}{40}x^{\frac{9}{2}}$ &c quod ab area $AFDB$ subductum relinquit solummodò errorem $\frac{1}{70}x^{\frac{7}{2}}+\frac{1}{90}x^{\frac{9}{2}}$ &c.(657)

Sic bisectâ AB in E, rectanguli $AB\times DE$ valor erit $x\sqrt{x-\frac{3}{4}xx}$(658) sive $x^{\frac{3}{2}}-\frac{3}{8}x^{\frac{5}{2}}-\frac{9}{128}x^{\frac{7}{2}}-\frac{27}{1024}x^{\frac{9}{2}}$ &[c]. et hoc collatum cum rectangulo $AD\times AB$ dat $\frac{8DE+2AD}{15}$ in $AB=$ areæ $AFDB$, errore tantùm existente $\frac{1}{560}x^{\frac{7}{2}}+\frac{1}{5760}x^{\frac{9}{2}}$ &c qui semper minor est quam $\frac{1}{1500}$ totius areæ, etiamsi $AFDB$ ponatur quadrans circuli.(659) Hoc autem Theorema sic enunciari potest. Ut 3 ad 2 ita rectangulum AB in DE plus quinta parte differentiæ inter AD ac DE ad aream $AFDB$ proxime.(660)

Atq; ita conferendo duo rectangula $AB\times ED$ et $AB\times BD$, vel omnia tria
||[96] rectangula inter se, vel adhibendo adhuc alia rectangula ||possunt aliæ regulæ excogitari, eæq; tanto exactiores quo plura rectangula adhibentur. Et idem de area Hyperbolæ ac aliarum curvarum intelligendum est.(661) Imò et per unicum tantùm rectangulum area plerumq; commode exhiberi potest, ut in prædicto circulo si capiatur AE(662) ad AB ut $\sqrt{10}$ ad 5, rectangulum $AB\times ED$ erit ad aream $AFDB$ ut 3 ad 2, errore tantùm existente $\frac{1}{175}x^{\frac{7}{2}}+\frac{11}{2250}x^{\frac{9}{2}}$ &c.(663)

2. Ex datâ areâ, Basem et incedentem lineam determinare.

Ubi area per finitam æquationem exhibetur nihil occurrit difficultatis. Ubi verò per infinitam exhibetur, affecta radix extrahenda est quæ Basem designat.

(657) As a check on the accuracy of the approximation, when $x=\frac{1}{2}$ the area

$$AFDB \text{ (or } \tfrac{1}{16}\pi) \approx \tfrac{1}{30}(3+2\sqrt{2})$$

correct to $2D$. James Gregory would have said 'hæc approximatio veras notas triplicat' (see note (654)).

(658) Since $DE^2=\frac{1}{4}AB^2+(AB-AB^2)$.

(659) $\frac{1}{15}(8DE+2AD)=\frac{2}{3}x^{\frac{1}{2}}-\frac{1}{5}x^{\frac{3}{2}}-\frac{3}{80}x^{\frac{5}{2}}-\frac{9}{640}x^{\frac{7}{2}}\ldots$. The proportion of the 'error' to the total area $AFDB$ is $(\frac{1}{560}x^{\frac{7}{2}}+\frac{1}{5760}x^{\frac{9}{2}}\ldots)/(\frac{2}{3}x^{\frac{3}{2}}-\frac{1}{5}x^{\frac{5}{2}}\ldots)=\frac{3}{1120}x^2\ldots$ for $x\leqslant 1$. When $x=\frac{1}{2}$, $AFDB=\frac{1}{16}\pi$ and so the proportional error is $\frac{\sqrt{2}}{\pi}(\frac{1}{560}+\frac{1}{11520}\ldots)\approx\frac{1}{1500}$.

(660) That is, area $AFDB\approx\frac{2}{3}AB(DE+\frac{1}{5}[AD-DE])$ or $\frac{1}{15}AB(8DE+2AD)$ as before. Newton communicated this 'constructio mechanica' to Leibniz in his *epistola prior* of 13 June 1676 (*Correspondence of Isaac Newton*, **2** (1960): 30–1), ostensibly as a gloss on Huygens' *De Circuli Magnitudine Inventa* (Leyden, 1654): Theorem VII: 9–11.

(661) The extension to the case of an elliptical segment $AFDB$ on the same diameter as the present circle is immediate. The case of a hyperbolic segment $AFDB$ of diameter AB reduces likewise to that of the equilateral hyperbola $x+x^2=r^2$ (or $\dot{z}^2$), where $AB=x$,

$$BD=r=x^{\frac{1}{2}}+\tfrac{1}{2}x^{\frac{3}{2}}-\tfrac{1}{8}x^{\frac{5}{2}}\ldots,\quad AD=\sqrt{[x+2x^2]}=x^{\frac{1}{2}}+x^{\frac{3}{2}}-\tfrac{1}{2}x^{\frac{5}{2}}\ldots,$$

On comparing its terms with these $\frac{2}{3}x^{\frac{3}{2}}-\frac{1}{5}x^{\frac{5}{2}}-\frac{1}{28}x^{\frac{7}{2}}\ldots$ there results $e+f=\frac{2}{3}$ and $-\frac{1}{2}e=-\frac{1}{5}$, or $e=\frac{2}{5}$ and $f(=\frac{2}{3}-e)=\frac{4}{15}$, so that

$$\tfrac{2}{5}BD\times AB+\tfrac{4}{15}AD\times AB = \text{area } AFDB,$$

approximately. In fact, the value of $\frac{2}{5}BD\times AB+\frac{4}{15}AD\times AB$ is

$$\tfrac{2}{3}x^{\frac{3}{2}}-\tfrac{1}{5}x^{\frac{5}{2}}-\tfrac{1}{20}x^{\frac{7}{2}}-\tfrac{1}{40}x^{\frac{9}{2}}\ldots,$$

and this taken from the area $AFDB$ leaves an error of only $\frac{1}{70}x^{\frac{7}{2}}+\frac{1}{90}x^{\frac{9}{2}}\ldots$.(657) So, when AB is bisected in E, the value of the rectangle $AB\times DE$ will be $x\sqrt{[x-\frac{3}{4}x^2]}$(658) or $x^{\frac{3}{2}}-\frac{3}{8}x^{\frac{5}{2}}-\frac{9}{128}x^{\frac{7}{2}}-\frac{27}{1024}x^{\frac{9}{2}}\ldots$, and this compared with the rectangle $AD\times AB$ yields $\frac{2}{15}(4DE+AD)\times AB=$ area $AFDB$, with an error merely of $\frac{1}{560}x^{\frac{7}{2}}+\frac{1}{5760}x^{\frac{9}{2}}\ldots$, which is always less than $\frac{1}{1500}$ of the whole area even if $AFDB$ is taken to be a quadrant of the circle.(659) This theorem may be expressed in this form. As 3 is to 2, so is the rectangle $AB\times DE$ plus one-fifth of the difference between AD and DE to the area $AFDB$, approximately.(660)

And thus, by comparing the two rectangles $AB\times ED$ and $AB\times BD$—or, indeed, all three rectangles—one with the other or by employing still other rectangles, other rules may be contrived, and these will be the more exact the greater the number of rectangles employed. The same should be understood in regard to the hyperbola's area and that of other curves.(661) Above all, the area may conveniently be exhibited in most cases by but a single rectangle, as for instance in the aforesaid circle, if [EB] be taken to AB as $\sqrt{10}$ to 5, the rectangle $AB\times ED$ will be to the area $AFDB$ as 3 to 2, the error being merely

$$\tfrac{1}{175}x^{\frac{7}{2}}+\tfrac{11}{2250}x^{\frac{9}{2}}\ldots.\text{(663)}$$

2. From the area given, to determine the base and ordinate line

When the area is exhibited by a finite equation no difficulty occurs. But when it is exhibited by an infinite one, the affected root denoting the base has to be

---

$DE=\sqrt{[x+\frac{5}{4}x^2]}=x^{\frac{1}{2}}+\frac{5}{8}x^{\frac{3}{2}}-\frac{25}{128}x^{\frac{5}{2}}\ldots$ and accordingly the area

$$AFDB=\int_0^x\sqrt{[x+x^2]}.dx=\tfrac{2}{3}x^{\frac{3}{2}}+\tfrac{1}{5}x^{\frac{5}{2}}-\tfrac{1}{28}x^{\frac{7}{2}}\ldots$$

may be approximated 'mechanically' by some suitable combination

$$\tfrac{2}{3}AB(\lambda.AD+\mu.BD+\nu.DE).$$

(662) Read 'EB', the complement of AE to AB.

(663) On taking $EB=\lambda x$ and $\mu=1-\lambda^2$, $ED=x^{\frac{1}{2}}(1-\mu x)^{\frac{1}{2}}$ and so

$$\tfrac{2}{3}AB\times ED-(AFDB)=(\tfrac{1}{5}-\tfrac{1}{3}\mu)x^{\frac{5}{2}}+\tfrac{1}{4}(\tfrac{1}{7}-\tfrac{1}{3}\mu^2)x^{\frac{7}{2}}+\tfrac{1}{24}(\tfrac{1}{3}-\mu^3)x^{\frac{9}{2}}\ldots.$$

The best approximation of area $AFDB$ by $\frac{2}{3}AB\times ED$ is accordingly had when $\mu=\frac{3}{5}$, in which case the 'error' is $\frac{1}{175}x^{\frac{7}{2}}+\frac{11}{2250}x^{\frac{9}{2}}+O(x^{\frac{11}{2}})$.

Sic ad Hyperbolam quam æquatio $\frac{ab}{a+x}=r$ designat postquam inventum est $z=bx-\frac{bxx}{2a}+\frac{bx^3}{3aa}$ &c; ut ex data area z vicissim innotescat Basis x, extrahe radicem affectam et proveniet $x=\frac{z}{b}+\frac{zz}{2abb}+\frac{z^3}{6aab^3}+\frac{z^4}{24a^3b^4}+\frac{z^5}{96a^4b^5}$ &c.[664] Et præterea si incedens r desideretur divide ab per $a+x$ hoc est per $a+\frac{z}{b}+\frac{zz}{2abb}+\frac{z^3}{6aab^3}$ &c et emerget $r=b-\frac{z}{a}+\frac{zz}{2aab}-\frac{z^3}{6a^3bb}+\frac{z^4}{24a^4b^3}$ &c.[665]

Sic ad Ellipsin quam æquatio $ax-\frac{a}{c}xx=rr$ designat, postquam inventa fuerit area $z=\frac{2}{3}a^{\frac{1}{2}}x^{\frac{3}{2}}-\frac{a^{\frac{1}{2}}x^{\frac{5}{2}}}{5c}-\frac{a^{\frac{1}{2}}x^{\frac{7}{2}}}{28cc}-\frac{a^{\frac{1}{2}}x^{\frac{9}{2}}}{72c^3}$ &c.[666] scribe v^3 pro $\frac{3z}{2a^{\frac{1}{2}}}$ ac t pro $x^{\frac{1}{2}}$, et evadet $v^3=t^3-\frac{3t^5}{10c}-\frac{3t^7}{56cc}-\frac{t^9}{48c^3}$ &c, et extracta radice $t=v+\frac{v^3}{10c}+\frac{81v^5}{1400cc}+\frac{1171v^7}{25200c^3}$ &c. Cujus quadratum $vv+\frac{v^4}{5c}+\frac{22v^6}{175cc}+\frac{823v^8}{7875c^3}$ &c valet x. Et hoc valore pro x in æquatione $ax-\frac{a}{c}xx=rr$ substituto, et extracta radice, proveniet

$$r=a^{\frac{1}{2}}v-\frac{2a^{\frac{1}{2}}v^3}{5c}-\frac{38a^{\frac{1}{2}}v^5}{175cc}-\frac{407a^{\frac{1}{2}}v^7}{2250c^3} \quad \&c^{(667)}$$

Adeoq; ex data area z et inde v sive $\sqrt{\text{cub}}: \frac{3z}{2a^{\frac{1}{2}}}$, dabitur Basis x et Incedens r. Quæ omnia ad Hyperbolam etiam accommodantur si modo signum quantitatis c ubiq; mutetur ubi existit imparium dimensionum.[668]

||[97] ||PROB 10. CURVAS PRO ARBITRIO MULTAS INVENIRE QUARUM LONGITUDINES PER FINITAS ÆQUATIONES DESIGNARI POSSUNT.

Ad hujus resolutionem via per sequentes positiones[669] sternitur.

1. Si recta DC in curvam quamvis AD perpendiculariter insistens moveri concipiatur, singula ejus puncta G, k,[670] r &c describent alias æquidistantes sibiq; perpendiculares curvas GK, gk, rs &c.[671]

(664) The last term should be $\frac{1}{120}z^5/a^4b^5$. Newton inverts the series expansion of

$$z = ab\log(1+x/a)$$

to find that of $x = a(e^{z/ab}-1)$.

(665) That is, $ab/(a+x) = be^{-z/ab}$.

(666) The series expansion of $\int_0^x \sqrt{[ax-(a/c)x^2]}.dx$. A following phrase 'Ut ex data z vicissim deter[minatur x]' (Inversely, given z to determine x) is cancelled.

extracted. So in the case of the hyperbola denoted by the equation $ab/(a+x) = \dot{z}$, after there has been found to be $z = bx - \frac{1}{2}bx^2/a + \frac{1}{3}bx^3/a^2 \ldots$, to determine in turn the base x given the area z, extract the affected root and there will come out $x = z/b + \frac{1}{2}z^2/ab^2 + \frac{1}{6}z^3/a^2b^3 + \frac{1}{24}z^4/a^3b^4 + \frac{1}{96}z^5/a^4b^5 \ldots$.(664) Further, should the ordinate $\dot{z}$ be desired, divide ab by $a+x$ (that is, by

$$a + z/b + \tfrac{1}{2}z^2/ab^2 + \tfrac{1}{6}z^3/a^2b^3 \ldots)$$

and there will emerge $\dot{z} = b - z/a + \frac{1}{2}z^2/a^2b - \frac{1}{6}z^3/a^3b^2 + \frac{1}{24}z^4/a^4b^3 \ldots$.(665)

So in the case of the ellipse represented by the equation $ax - (a/c)\,x^2 = \dot{z}^2$, after the area has been found to be

$$z = a^{\frac{1}{2}}(\tfrac{2}{3}x^{\frac{3}{2}} - \tfrac{1}{5}x^{\frac{5}{2}}/c - \tfrac{1}{28}x^{\frac{7}{2}}/c^2 - \tfrac{1}{72}x^{\frac{9}{2}}/c^3 \ldots),^{(666)}$$

write v^3 in place of $\frac{3}{2}z/a^{\frac{1}{2}}$ and t in place of $x^{\frac{1}{2}}$ and there will result

$$v^3 = t^3 - \tfrac{3}{10}t^5/c - \tfrac{3}{56}t^7/c^2 - \tfrac{1}{48}t^9/c^3 \ldots,$$

and on extracting the root $t = v + \frac{1}{10}v^3/c + \frac{81}{1400}v^5/c^2 + \frac{1171}{25200}v^7/c^3 \ldots$. The square of this, $v^2 + \frac{1}{5}v^4/c + \frac{22}{175}v^6/c^2 + \frac{823}{7875}v^8/c^3 \ldots$, is the value of x. And when this value is substituted for x in the equation $ax - (a/c)\,x^2 = \dot{z}^2$ and the root extracted, there will come out $\dot{z} = a^{\frac{1}{2}}(v - \frac{2}{5}v^3/c - \frac{38}{175}v^5/c^2 - \frac{407}{2250}v^7/c^3 \ldots)$,(667) so that from the area z and thence v, that is $\sqrt[3]{[\frac{3}{2}a^{-\frac{1}{2}}z]}$, given there will be given the base x and ordinate $\dot{z}$. All these results may be adapted to the hyperbola also, provided that wherever the quantity c exists as an odd power its sign be changed.(668)

PROBLEM 10. TO FIND AS MANY CURVES AS YOU WISH WHOSE LENGTHS MAY BE EXPRESSED BY MEANS OF FINITE EQUATIONS

The way is prepared to resolving this problem by means of the following postulates:(669)

1. If the straight line DC be imagined to move, remaining perpendicularly ordinate to the curve AD, each of its points G, [g], r and so on will describe other, equidistant curves Gk, gk, rs, and so on perpendicular to it.(671)

---

(667) The coefficient of the last term should be $\frac{284}{1575}$.

(668) The equivalent expansion for the hyperbola r (or $\dot{z}$) $= \sqrt{[ax + (a/c)\,x^2]}$ follows more straightforwardly by replacing c by $-c$ everywhere. This last sentence is a late addition.

(669) Compare Lemmas 1–3 of Problem 9 of the October 1666 tract (I: 432–3).

(670) A careless slip for 'g'.

(671) Newton first wrote, much as in Lemma 2 of his 1666 tract '..., singula ejus puncta r, ${}_2r$, ${}_3r$, ${}_4r$ &c describent alias æquidistantes subiꝗ parallelas curvas rs, ${}_2r_2s$, ${}_3r_3s$ &c quibus itidem perpendicularis erit' (each of its points r, ${}_2r$, ${}_3r$, ${}_4r$, ... will describe other, equidistant curves rs, ${}_2r_2s$, ${}_3r_3s$, ... parallel to each other, and to these it will likewise be perpendicular).

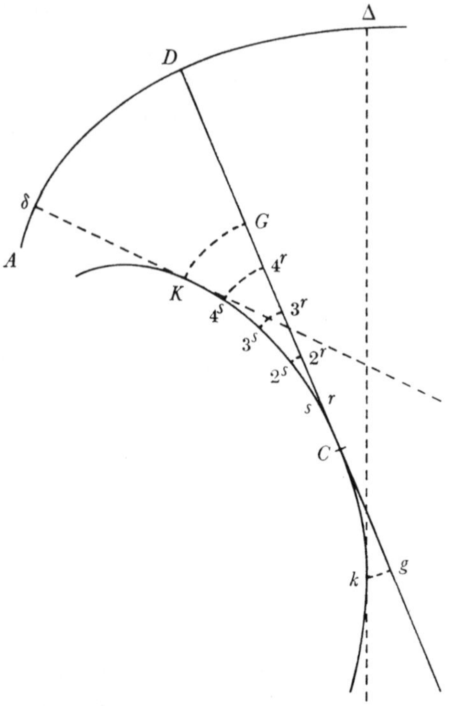

2. Si recta illa hinc inde indefinitè producatur ejus extremitates movebuntur ad contrarias plagas, et punctum quod distinguit inter contrarios motus, quodq; ideo dici potest centrum motionis, idem est cum centro curvaturæ quam curva AD habet ad punctum D, ut supra(672) diximus. Istud autem punctū esto C.

3. Si lineam AD non circularem esse sed difformiter incurvatam supponamus puta magis curvam in δ et minùs in Δ, illud centrum continuò mutabitur propiùs(673) accedens ad partes magis curvas ut in K et longiùs recedens a partibus minùs curvis ut in k, eoq; pacto lineam aliquam qualis KCk describet.

4. Hanc a centro curvaturæ descriptam lineam recta DC continuò tanget. Nam si rectæ illius punctum D moveat versus δ, ejus punctum G quod interea transit ad K et situm est ad eandem partem centri C movebit versus eandem plagam (per Positionem 2dam). Deinde si idem D moveat versus Δ punctum g quod interea transit ad k et situm est ad contrariam partem centri C movebit ad contrariam plagam[,] hoc est ad eandem plagam ad quam G in priori casu
||[98] movebat ||dum transit ad K. Et proinde K et k jacent ad eandem partem rectæ DC. Quare cum K et k indeterminatè pro quibuslibet punctis sumantur, patet totam illam curvam jacere ad eandem partem rectæ DC, proindeq; ab illa non secari sed tangi tantùm.(674)

Hic supponitur lineam $\delta D\Delta$ magis curvam esse a parte δ continuò et minùs a parte Δ. Quod si maxima minimáve curvatura fuerit ad ipsum D, tunc recta DC secabit curvam KC, sed in angulo tamen qui sit quovis rectilineo minor.(675) Quod perinde est ac si tangere dicatur. Imo punctum C in hoc casu termi[nus] est instar cuspidis, ad quem partes curvæ obliquissimo concursu desinentes se mutuò contingunt, proindeq; a recta DC quæ angulum illum contactûs dividit rectiùs dicatur tangi quàm secari.(676)

5. Recta CG æquatur curvæ CK. Nam concipe rectæ illius singula puncta r, ${}_2r$, ${}_3r$, ${}_4r$ &c describere curvarum arcus rs, ${}_2r\,{}_2s$, ${}_3r\,{}_3s$ &c interea dum per motum rectæ illius accedant ad curvam CK; et arcus illi, cùm (per Positionem primam) sint perpendiculares ad rectas quæ (per Posit 4) tangunt curvam CK, erunt

(672) In the third 'symptom' on his page 54 preceding.

2. If that line be produced indefinitely in either direction, its extremities will move opposite ways and the point which distinguishes between these opposite motions—and which on that account may be called the centre of motion—is identical with the centre of curvature possessed by the curve AD at the point D, as we have said above.(672) Let that point be C.

3. If we may suppose the line AD not to be circular but non-uniformly curved, say, more curved in δ and less so in Δ, that centre will continuously change [in position], approaching more closely to the more curved parts (as in K) and receding more distantly from the less curved portions (as in k), and for this reason will describe some such line as KCk.

4. To this line described by the centre of curvature the straight line DC will be continuously tangent. For if the point D in that straight line should move towards δ, its point G (which during the same time passes to K and is situated on the same side of the centre C) will move in the same direction (by Postulate 2). Next, if the same point D should move towards Δ, the point g (which during that time passes to k and is situated on the opposite side of the centre C) will move in the opposite direction, that is, in the same direction as G moved in the previous case while passing to K. In consequence, K and k lie on the same side of the straight line DC. Hence, since K and k are taken unrestrictedly as any points, it is evident that the whole of the curve lies on the same side of the straight line DC and so is not intersected but only touched by it.(674)

Here it is supposed that the line $\delta D\Delta$ is continuously more curved on the side δ and continuously less so on the side Δ. But should the curvature be greatest or least at D itself, then the straight line DC will intersect the curve KC, but in an angle, however, which is less than any rectilinear one.(675) But that is equivalent to saying that it touches it. In this case, indeed, the terminal point C has the form of a cusp, at which the parts of the curve end in the obliquest of meets and are tangent one to the other, and therefore it may more correctly be said to be touched rather than intersected by the straight line DC which divides that contact angle.(676)

5. The straight line CG is equal to the curve $\widehat{CK}$. For conceive each point r, ${}_2r$, ${}_3r$, ${}_4r$, ... of that line to describe the curve arcs rs, ${}_2r\,{}_2s$, ${}_3r\,{}_3s$, ... during the time in which, through the motion of that straight line, they approach the curve CK. Then since those arcs are (by Postulate 1) perpendicular to straight lines which (by Postulate 4) touch the curve CK, they will also be perpendicular to that

(673) By a slip of his pen Newton wrote 'propriùs' here.

(674) Newton assumes that a simply convex curve arc has a smoothly continuous evolute, except (as he hastens to make clear in the following paragraph) where the radius of curvature attains a local maximum or minimum at some point.

(675) In a contact angle, in other words. Compare Newton's page 59 above.

(676) This last sentence is a late addition entered on ULC. Add. 3960.14: 154.

etiam perpendiculares ad curvam illam. Quare partes istius CK inter arcus illos interjectæ quæ propter infinitam parvitatem pro rectis haberi possint æquantur intervallis eorundem arcuum, hoc est (per Posit: 1) totidem partibus rectæ CG. Et additis utrinq; æqualibus, tota CK æquabitur toti CG.

Idem constare potest imaginando singulas partes rectæ CG inter movendum successivè applicari ad singulas partes curvæ CK, easq; mensurare, perinde ut rotæ super planum per gyros promoventis circumferentia distantiam metitur quam punctum contactûs transigit.

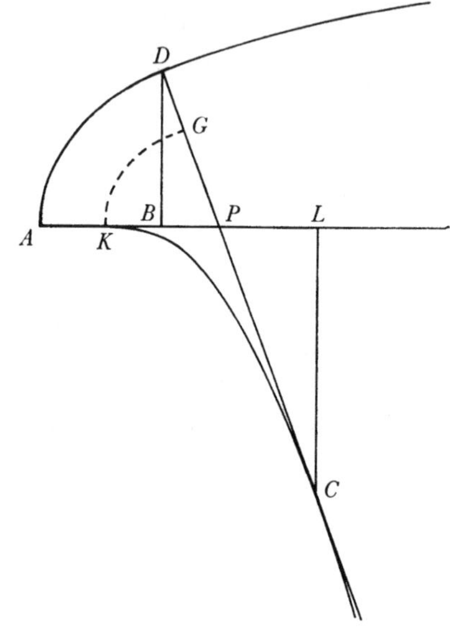

Ex his pateat Problema resolvi posse assumendo pro lubitu curvam quamvis $A\delta D\Delta$ et inde determinando alteram curvam KCk in qua assumptæ(677) centrum curvaturæ versatur. Ad rectam itaq; quamvis positione datam AB demissis perpendiculis DB, CL et in AB sumpto quovis puncto A dictisq; $AB=x$ et $BD=y$, pro curva AD definienda assumatur relatio quævis inter x et y et inde per Prob 5 elicietur punctum C quo et curva KC et ejus longitudo GC determinatur.

EXEMPL.(678) Sit $ax=yy$ æquatio ad curvam AD, Parabolam ||nempe Apollonianam.
||[99] Et per Prob: 5 invenientur $AL=\frac{1}{2}a+3x$, $CL=\frac{4y^3}{aa}$, ac $DC=\frac{a+4x}{a}\sqrt{\frac{1}{4}aa+ax}$.(679) Quibus habitis, curva KC determinatur per AL et LC et longitudo ejus per DC. Utpote cùm liberum sit ubivis in curva KC assumere puncta K et C, supponamus K esse centrum curvaturæ Parabolæ ad verticem, et positis perinde AB et BD seu x et y nullis evadet $DC=\frac{1}{2}a$,(680) estq; hæc longitudo AK vel DG quæ subducta a superiori indefinito valore DC relinquit GC seu $KC=\frac{a+4x}{a}\sqrt{\frac{1}{4}aa+ax}-\frac{1}{2}a$.

Jam si qualis sit hæc curva quantaq; ejus longitudo, non ampliùs habita relatione ad Parabolam scire desideretur; Dic $KL=z$ et $LC=v$, et erit $z(=AL-\frac{1}{2}a)=3x$ seu $\frac{1}{3}z=x$, et $\frac{az}{3}(=ax)=yy$, adeoq; $4\sqrt{\frac{z^3}{27a}}\left(=\frac{4y^3}{aa}=CL\right)=v$. sive $\frac{16z^3}{27a}=vv$. Quod indicat curvam KC esse Parabolam secundi generis. Et pro ejus longitudine prodit $\frac{3a+4z}{3a}\sqrt{\frac{1}{4}aa+\frac{1}{3}az}-\frac{1}{2}a$, scribendo $\frac{1}{3}z$ pro x in valore CG.(681)

(677) Understand 'curvæ'.

(678) Compare Example 1 of the October 1666 tract's Problem 9 (I: 433).

curve. Hence the parts of CK interposed between those arcs, which because of their infinite smallness may be considered as straight lines, are equal to the distances of those same arcs, that is, (by Postulate 1) to the corresponding number of portions of the straight line CG. When equal portions in each are added together, the whole length $\widehat{CK}$ will equal the whole, CG.

The same result can be established by imagining that during its motion individual parts of the straight line CG apply themselves one after another to corresponding portions of the curve $\widehat{CK}$ and measure them, just as the circumference of a wheel rolling forward along a plane surface measures the distance covered by its point of contact.

From these remarks it should be evident that the problem may be resolved by assuming any curve $A\delta D\Delta$ at pleasure and thence determining a second curve KCk in which is located the centre of curvature of the [curve] assumed. Accordingly, having let fall to any straight line AB given in position the perpendiculars DB and CL, take any point A in AB and call $AB = x$ and $BD = y$: then, to define the curve AD, assume any relation you wish between x and y and from it, by Problem 5, elicit the point C which determines both the curve KC and its length GC.

EXAMPLE.(678) Let $ax = y^2$ be the equation of the curve AD, namely, an Apollonian parabola. By Problem 5 there will be found $AL = \frac{1}{2}a + 3x$, $CL = 4y^3/a^2$ and $DC = \frac{1}{2}(a+4x)^{\frac{3}{2}}/a^{\frac{1}{2}}$.(679) When these are had, the curve KC is determined by means of AL and LC, and its length by DC. But since we are free to assume the points K and C anywhere in the curve KC, let us suppose K to be the parabola's centre of curvature at the vertex. Accordingly, on setting AB and BD (that is, x and y) zero, DC will prove to be $\frac{1}{2}a$,(680) and this is the length of AK or DG which, when taken from the above indefinite value of DC, leaves GC or $KC = \frac{1}{2}(a+4x)^{\frac{3}{2}}/a^{\frac{1}{2}} - \frac{1}{2}a$.

Now should you desire to know the nature of this curve and the length of its arc without reference to the parabola, call $KL = z$ and $LC = v$ and there will be $z(= AL - \frac{1}{2}a) = 3x$, that is, $\frac{1}{3}z = x$, and $\frac{1}{3}az(= ax) = y^2$, so that

$$4\surd[\tfrac{1}{27}z^3/a]\ (= 4y^3/a^2 = CL) = v \quad \text{or} \quad \tfrac{16}{27}z^3/a = v^2.$$

This reveals that the curve KC is a second-order parabola, and for its length there results $\frac{1}{2}(a+\frac{4}{3}z)^{\frac{3}{2}} - \frac{1}{2}a$ on writing $\frac{1}{3}z$ in place of x in the value of CG.(681)

(679) See the corollary to Example 1 on Newton's page 56 above.

(680) Newton first wrote 'evadet $AL = \frac{1}{2}a$' (AL will prove to be $\frac{1}{2}a$). To be sure, as C passes into the cusp K the line AK is the common limit of both DC and AL.

(681) An 'indirect' method for evaluating the arc-length $\widehat{KC} = \int_0^z \surd[1+4z/3a].dz$ of the semicubical parabola $16z^3 = 27av^2$ as the difference of the curvature radii

$$DC - AK = \tfrac{1}{2}a[(1+4z/3a)^{\frac{3}{2}} - 1]$$

at corresponding points on the parabolic involute $ax = y^2$.

Potest etiam Problema resolvi per assumptionem æquationis quæ relationem inter AP et PD (posita nempe P intersectione Basis et Perpendiculi) definiat.(682) Nam dictis $AP=x$, et $PD=y$, concipe CPD per spatium quàm minimum moveri puta ad locum Cpd, inq; CD et Cd sumpto $C\Delta$ et $C\delta$ ejusdem cujusvis datæ longitudinis puta 1, et ad CL demissis Δg, $\delta\gamma$ perpendiculis quorum Δg (quod dic z) occurrat Cd in f, et completo parallelogrammo $g\gamma\delta e$, positisq; m, n, et r fluxionibus quantitatum x y et z ut supra; erit

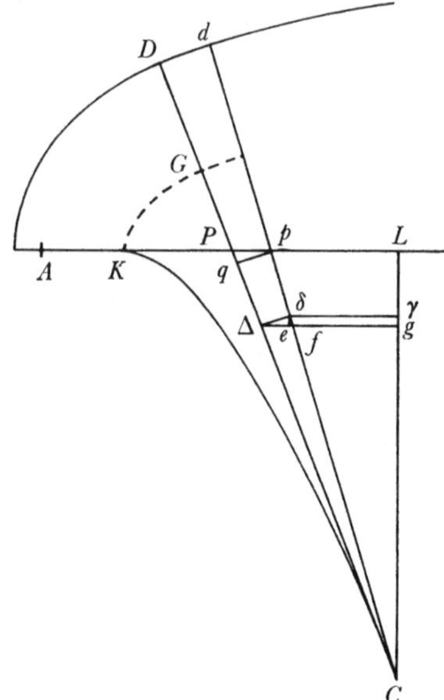

$$\Delta e\,.\,\Delta f(::\Delta e\times\Delta e\,.\,\Delta\delta\times\Delta\delta::\ Cg\times Cg\,.\,C\Delta\times C\Delta)::\frac{Cg\times Cg}{C\Delta}\,.\,C\Delta.$$

Et $\Delta f\,.\,Pp::C\Delta\,.\,CP$. Et ex æquo

$$\Delta e\,.\,Pp::\frac{Cg\times Cg}{C\Delta}\,.\,CP.$$

||[100] Est autem Pp ||momentum Basis AP cujus additamento evadit Ap, ac Δe contemporaneum momentum perpendiculi Δg cujus ablatione evadit $\delta\gamma$. Adeoq; Δe et Pp sunt ut fluxiones linearum $\Delta g(z)$ et $AP(x)$, hoc est ut r et m. Quare $r\,.\,m::\frac{Cg^{\text{quadr}}}{C\Delta}\,.\,CP$. Et proinde cùm sit $Cg^{\text{quadr}}(=C\Delta^{\text{quadr}}-\Delta g^{\text{quad}})=1-zz$, et $C\Delta=1$, erit $CP=\frac{m-mzz}{r}$. Et insuper cùm e tribus m, n, et r(683) quamlibet pro uniformi fluxione ad quam cæteræ referantur habere liceat, si ista ponatur m ejusq; quantitas unitas, evadet $CP=\frac{1-zz}{r}$.(684)

Præterea est $C\Delta(1)\,.\,\Delta g(z)::CP\,.\,PL$, et $C\Delta(1)$. $Cg(\sqrt{1-zz})::CP\,.\,CL$, Adeoq; fit $PL=\frac{z-z^3}{r}$ et $CL=\frac{1-zz}{r}\sqrt{1-zz}$. Ac deniq; acta pq parallela arcui infinitè parvo Dd seu perpendiculari DC erit Pq momentum ipsius DP cujus additamento evadit dp simul ac AP evadit Ap. Et idcirco Pp et Pq sunt ut fluxiones ipsarum $AP(x)$ et $PD(y)$, hoc est ut 1 et n. Atq; adeò cùm propter similia triangula Ppq & $C\Delta g$, $C\Delta$ ac Δg seu 1 et z sint in eadem ratione erit $n=z$. Unde talis evadit Problematis resolutio.

E proposita æquatione quæ relationem inter x et y designet(685) quære relationem fluxionum m et n per Prob 1. Et posito $m=1$ habebitur valor n cui z

(682) See Problem 10 of the October 1666 tract (I: 434–6). Newton proceeds to revise his earlier discussion of the curvature of curves defined in this semi-intrinsic coordinate system along the lines of his treatment of the Cartesian and polar cases on his pages 54–5 and 59–60 above. Compare note (334).

The problem may also be resolved by assuming an equation which shall define the relationship between AP and PD (where P, namely, is taken as the intersection of the base and the normal).(682) For, calling $AP = x$ and $PD = y$, imagine that CPD moves through the smallest space possible to the position Cpd, say, and in CD and Cd take $C\Delta$ and $C\delta$ of the same arbitrary given length, say unity, letting fall to CL the perpendiculars Δg, $\delta\gamma$. Of these let Δg (call it z) meet Cd in f and complete the rectangle $g\gamma\delta e$. Then (taking $\dot{x}$, $\dot{y}$ and $\dot{z}$ for the fluxions of the quantities x, y and z as above) will there be

$$\Delta e:\Delta f(=\Delta e^2:\Delta\delta^2 = Cg^2:C\Delta^2) = Cg^2/C\Delta:C\Delta \quad \text{and} \quad \Delta f:Pp = C\Delta:CP,$$

and so *ex æquo* $\Delta e:Pp = Cg^2/C\Delta:CP$. But Pp is the moment of the base AP by whose addition it comes to be Ap, while Δe is the contemporaneous moment of the perpendicular Δg by whose removal it comes to be $\delta\gamma$. Consequently Δe and Pp are as the fluxions of the lines $\Delta g(z)$ and $AP(x)$, that is, as $\dot{z}$ and $\dot{x}$. Hence $\dot{z}:\dot{x} = Cg^2/C\Delta:CP$, and therefore, since $Cg^2(=C\Delta^2-\Delta g^2) = 1-z^2$ and $C\Delta = 1$, then $CP = \dot{x}(1-z^2)/\dot{z}$. Moreover, since we are permitted to consider any of the three fluxions(683) $\dot{x}$, $\dot{y}$ and $\dot{z}$ as the uniform one to which the others are to be referred, if that is taken to be $\dot{x}$ and its quantity unity, there will result

$$CP = (1-z^2)/\dot{z}.^{(684)}$$

Furthermore, $C\Delta(1):\Delta g(z) = CP:PL$ while $C\Delta(1):Cg(\surd[1-z^2]) = CP:CL$, so that $PL = (z-z^3)/\dot{z}$ and $CL = (1-z^2)^{\frac{3}{2}}/\dot{z}$. Lastly, when pq is drawn parallel to the infinitely small arc $\widehat{Dd}$, that is, perpendicular to DC, Pq will be the moment of DP by whose addition it comes to be dp in just the same time that AP becomes Ap. For that reason Pp and Pq are as the fluxions of $AP(x)$ and $PD(y)$, that is, as 1 and $\dot{y}$, and in consequence, since (because of the similar triangles Ppq and $C\Delta g$) $C\Delta$ and Δg, that is, 1 and z, are in the same ratio, it will be $\dot{y} = z$. Hence the resolution of the problem comes out to be this.

From the proposed equation which expresses the relationship between x and y(685) seek the relation of the fluxions $\dot{x}$ and $\dot{y}$ by Problem 1, and when $\dot{x}$ is set equal to 1 the value of $\dot{y}$, to which z is equal, will be obtained. Then, having

(683) 'literam' is here cancelled. In our anachronistic English rendering we should have rendered it as '[dotted] letter[s]'.

(684) In modern Leibnizian form, $CP = [1-(dy/dx)^2]/(d^2y/dx^2)$. Correspondingly below, $PL = CP.(dy/dx)$ and so $CL = [1-(dy/dx)^2]^{\frac{3}{2}}/(d^2y/dx^2)$. Compare Newton's equivalent forms in his 1666 tract (I: 436).

(685) Here, and indeed at the corresponding point in Problem 10 of his 1666 tract, Newton presumes that the equation $y = f(x)$ connecting the coordinates $AP = x$, $PD = y$ defines a unique curve and thereby implicitly excludes the infinite family of circles of centre $(p, 0)$ and radius 'of curvature' $f(p)$, where p is an arbitrary parameter. In fact, $y = f(x)$ is in Cartesian terms a first-order differential equation whose particular solution is the 'unique' defined curve postulated by Newton, while the family of circles $x = p$, $y = f(p)$ are its general solutions.

æquatur. Dein substituto z pro n(686) ope æquationis novissimæ quære relationes fluxionum m n et r per idem Prob 1, et iterum substituto 1 pro m [et z pro n](687) obtinebitur valor r. Quibus habitis fac $\frac{1-nn}{r}=CP$, $z\times CP=PL$ et $CP\times\sqrt{1-nn}=CL$; et erit C ad curvam(688) cujus pars quævis KC æquatur rectæ CG differentiæ nempe tangentium ductarum a punctis C et K perpendiculariter ad curvam Dd.

Exempl. Sit $ax=yy$(689) æquatio quæ relationem inter AP et PD designet
‖[101] et per Prob: 1 primò erit $am=2yn$ seu ‖$a=2yz$. Deinde $0=2nz+2yr$ seu $\frac{-zz}{y}=r$. Indeqȝ fit $CP=\left(\frac{1-nn}{r}=\right)y-\frac{4y^3}{aa}$, $PL=(z\times CP=)\,\frac{1}{2}a-\frac{2yy}{a}$, et $CL=\frac{aa-4yy}{2aa}\sqrt{4yy-aa}$. Et a CP ac PL ablatis y et x restat $CD=-\frac{4y^3}{aa}$(690) et $AL=\frac{1}{2}a-\frac{3yy}{a}$. Aufero autem y et x quòd CP et PL ubi valores habent affirmativos cadant ad partes puncti P versus D et A, et tunc diminui debent auferendo affirmativas quantitates PD et AD. Ubi verò negativos valores obtinent, cadent ad contrarias partes puncti P et tunc augeri debent, id quod etiam fit auferendo affirmativas quantitates PD et AD.

Jam ut curvæ in qua punctum C locatur longitudo inter duo quævis puncta K et C noscatur, quæro longitudinem tangentis ad(691) punctum K et aufero a CD. Quemadmodum si K sit punctum ad quod tangens terminatur ubi $C\Delta$ et Δg seu 1 et z ponuntur æquales(692) quodqȝ proinde in ipsa basi AP situm est, scribe 1 pro z in æquatione $a=2yz$ et prodit $a=2y$. Quare pro y scribe $\frac{1}{2}a$ in valore CD nempe in $-\frac{4y^3}{aa}$, et oritur $-\frac{1}{2}a$. Estqȝ hæc longitudo tangentis ad punctum K, sive ipsius DG inter quam et superiorem indefinitum valorem CD differentia $\frac{4y^3}{aa}-\frac{1}{2}a$ est GC(693) cui curvæ pars KC æquatur.

Ut insuper pateat qualis sit hæc curva, ab AL (mutato prius signo ut evadat

(686) The equivalent phrase 'imaginando z pro n substitui' is cancelled.

(687) This necessary phrase, omitted by Newton, has been inserted in line with a first version of the preceding 'scribendo iterum 1 pro m et z pro n'.

(688) Newton first continued 'KC quæ æquatur DC auctæ vel diminutæ data aliqua quantitate DG' (KC, which is equal to DC plus or minus some given quantity DG).

(689) This repeats Example 1 of Problem 10 of the 1666 tract (I: 436–7). If the general point D is referred to the perpendicular Cartesian coordinates $AB = X$, $BD = Y$, then $x = X+YZ$, $y = Y\sqrt{[1+Z^2]}$ where $Z = dY/dX$ (compare I, 2, 7: note (123)). Accordingly, in the present instance the Cartesian defining equation of the locus is, in differential form, $a(X+YZ) = Y^2(1+Z^2)$, so that $Y^2 = a(X+\frac{1}{4}a)+W$ where $W=-\frac{1}{4}(dW/dX)^2$. Newton's Apollonian parabola $Y^2 = a(X+\frac{1}{4}a)$ is the particular solution $W = 0$; but more generally

substituted z for $\dot{y}$, with the aid of the latest equation seek the relationship of the fluxions $\dot{x}$, $\dot{y}$ and $\dot{z}$ by the same Problem 1, and when 1 is substituted in place of $\dot{x}$ and also z for $\dot{y}$[687] the value of $\dot{z}$ will again be obtained. When you have these, make $(1-\dot{y}^2)/\dot{z} = CP$, $z \times CP = PL$ and $CP \times \sqrt{[1-\dot{y}^2]} = CL$, and C will then be on the curve,[688] any part $\widehat{KC}$ of which is equal to the straight line CG (the difference, namely, of the tangents drawn from the points C and K perpendicular to the curve Dd).

EXAMPLE. Let $ax = y^2$[689] be the equation expressing the relationship between AP and PD, and in the first place, by Problem 1, there will be $a\dot{x} = 2y\dot{y}$ or $a = 2yz$. Next $0 = 2\dot{y}z + 2y\dot{z}$ or $-z^2/y = \dot{z}$. Thence comes

$$CP = ([1-\dot{y}^2]/\dot{z} =)\, y - 4y^3/a^2, \quad PL = (z \times CP =)\, \tfrac{1}{2}a - 2y^2/a$$

and $CL = -\frac{1}{2}(4y^2-a^2)^{\frac{3}{2}}/a^2$, and when y and x are taken from CP and PL there remain $CD = -4y^3/a^2$,[690] $AL = \frac{1}{2}a - 3y^2/a$. I take away y and x because when CP and PL have positive values they should fall on the side of the point P towards D and A and ought then to be diminished by taking away the positive quantities PD and AD. But when they acquire negative values they will fall on the opposite side of the point P and ought then to be increased, which again happens by taking away the positive quantities PD and AD.

Now to know the length of the curve in which the point C is located, taken between any two points K and C, I seek the length of the tangent at the[691] point K and take it from CD. For instance, should K be the point at which the tangent terminates when $C\Delta$ and Δg or 1 and z are set equal[692] and which is consequently situated in the base AP itself, write 1 in place of z in the equation $a = 2yz$ and there results $a = 2y$. Hence instead of y write $\frac{1}{2}a$ in the value of CD (in $-4y^3/a^2$, namely) and there arises $-\frac{1}{2}a$. This is the length of the tangent at the point K, or that of DG, the difference $(4y^3/a^2 - \frac{1}{2}a)$ between which and the above indefinite value CD is GC,[693] to which the portion $\widehat{KC}$ of the curve is equal.

To reveal the nature of this curve, moreover, from AL (first changing its sign

---

$W = -(X-p+\frac{1}{2}a)^2$, where p is an arbitrary parameter, and the locus is the infinite family of circles $(X-p)^2+Y^2 = ap$ corresponding to $x = p$, $y = (ap)^{\frac{1}{2}}$: compare note (685). In the parabolic case $x = X+\frac{1}{2}a$, $y = \sqrt{[Y^2+\frac{1}{4}a^2]}$.

(690) That is, the radius of curvature in the Cartesian parabola $Y^2 = a(X+\frac{1}{4}a)$ at the point $D(X, Y)$ is $4DP^3/a^2$, where a is its latus rectum. The result is (see note (297)) true for a general conic.

(691) 'datum quodpiam' (arbitrarily given) is cancelled.

(692) Again Newton restricts himself implicitly to the unique non-circular curve (a parabola) which is defined by the locus $ax = y^2$, for the differential ratio $C\Delta/\Delta g = dy/dx$ is indeterminate in the case of the family of circles $x = p$, $y = (ap)^{\frac{1}{2}}$ (see note (689)).

(693) Newton first wrote 'quæ subducta a superiori indefinito valore CD relinquit $\frac{1}{2}a - \frac{4y^3}{aa}$ pro GC' (which taken from the above indefinite value of CD leaves $\frac{1}{2}a - 4y^3/a^2$ for GC).

affirmativa) aufer AK quæ erit $\frac{1}{4}a$(694) et restabit $KL=\frac{3yy}{a}-\frac{3}{4}a$ quam dic t et in valore lineæ CL quam dic v scribe $\frac{4at}{3}$ pro $4yy-aa$ et prodibit $\frac{2t}{3a}\sqrt{\frac{4}{3}at}=v$. seu $\frac{16t^3}{27a}=vv$ æquatio ad Parabolam secundi generis ut supra.(695)

Siquando relatio inter t et v minùs commodè ad æquationem redigi possit, sufficit investigasse tantùm longitudines PC et PL. Quemadmodum si pro relatione inter AP et PD assumatur æquatio $3aax+3aay-y^3=0$.(696) Inde per
||[102] Prob 1 primò prodit $aa+aaz-yyz=0$, deinde $aar-2yzz-yyr=0$. ||Atq adeo est $z=\frac{aa}{yy-aa}$, & $r=\frac{2yzz}{aa-yy}$. Unde dantur $PC=\frac{1-nn}{r}$ & $PL=z\times PC$, quibus punctum C quod ad curvam situm est determinatur. Et longitudo curvæ inter duo ejusmodi puncta e differentia correspondentium duarum tangentium DC sive $PC-y$ innotescit.

Ex gr. Si ponatur $a=1$ et ad determinandum aliquod curvæ punctum C umatur $y=2$; evadet AP seu $x\left(=\frac{y^3-3aay}{3aa}\right)=\frac{2}{3}$. $z=\frac{1}{3}$. $r=-\frac{4}{9}$. $PC=-2$. & $PL=-\frac{2}{3}$. Deinde ad aliud punctum C determinandum si sumatur $y=3$ evadet $AP=6$. $z=\frac{1}{8}$. $r=\frac{-3}{256}$. $PC=-84$ et $PL=-10\frac{1}{2}$. Quibus habitis si auferatur y a PC restabit -4 in priori casu et -87 in secundo casu pro longitudinibus DC quarum differentia 83 est longitudo curvæ inter inventa duo puncta C et C.

Hæc ita intelligenda sunt ubi curva(697) inter puncta duo C et C vel K et C continuatur sine termino quem cuspidi assimilavimus. Sed ubi unus vel plures ejusmodi termini interjacent istis punctis (qui termini inveniuntur per determinationem maximæ aut minimæ PC(698) vel DC) longitudines singularum partium Curvæ inter illos(699) et puncta C vel K seorsim investigari debent et addi.

(694) $AP = x = y^2/a$ coincides with AK when $dy/dx = 1$ and so $y = \frac{1}{2}a$.

(695) Newton's indirect proof that the (non-circular) curve $ax = y^2$ is an Apollonian parabola $Y^2 = a(X+\frac{1}{4}a)$, since its evolute is $16t^3 = 27av^2$, where $X = t+\frac{1}{4}a$ and

$$Y = \pm a\sqrt{[\tfrac{1}{2}+3(\tfrac{1}{4}v/a)^{\frac{2}{3}}]}.$$

Compare Newton's 'exemplum' on his pages 98–9 above.

(696) On converting to perpendicular Cartesian coordinates $AB = X$, $BD = Y$ by setting $x = X+YZ$, $y = Y\sqrt{[1+Z^2]}$, where $Z = dY/dX$ (see note (689)), the corresponding differential defining condition is $3a^2(X+YZ) = Y(Y^2+Y^2Z^2-3a^2)\sqrt{[1+Z^2]}$. It would appear that, unlike in the preceding parabolic case, there is here no simple algebraic particular solution, though evidently the general solution (namely an infinite family of circles) is given by $y = p$, $x = p(\frac{1}{3}p^2/a^2-1)$, where p is an arbitrary parameter.

(697) That is, the evolute curve KC.

(698) This necessary criterion for C to be a cusp on the evolute KC is not, unfortunately, sufficient. Retaining Newton's perpendicular coordinates $AL = t$, $LC = v$, we may cite the

to make it come positive) take AK, which will be $\frac{1}{4}a$,(694) and there will remain $KL = 3y^3/a - \frac{3}{4}a$. Call this t, and in the value of the line CL (call it v) write $\frac{4}{3}at$ in place of $4y^2 - a^2$ and there will result $\frac{1}{2}(\frac{4}{3}at)^{\frac{3}{2}}/a^2 = v$ or $16t^3/27a = v^2$, the equation to a second-order parabola as above.(695)

Whenever it is less conveniently possible to reduce the relationship between t and v to an equation, it is sufficient merely to have found out the lengths PC and PL. For instance, if for the relation between AP and PD there be assumed the equation $3a^2x + 3a^2y - y^3 = 0$,(696) from this by means of Problem 1 there results first $a^2 + a^2z - y^2z = 0$, then $a^2\dot{z} - 2yz^2 - y^2\dot{z} = 0$. Consequently

$$z = a^2/(y^2 - a^2) \quad \text{and} \quad \dot{z} = 2yz^2/(a^2 - y^2),$$

from which are given $PC = (1 - \dot{y}^2)/\dot{z}$ and $PL = z \times PC$. By these the point C situated on the curve is determined, while the length of the curve between two such points is known from the difference of the two corresponding tangents DC, that is, $PC - y$.

For example, if a is set equal to 1, and, to determine some point C of the curve, y be taken equal to 2, there will prove to be AP or $x(= y^3/3a^2 - y) = \frac{2}{3}$, $z = \frac{1}{3}$, $\dot{z} = -\frac{4}{9}$, $PC = -2$ and $PL = -\frac{2}{3}$. Then if, to determine a second point C, y be taken equal to 3, there will come out $AP = 6$, $z = \frac{1}{8}$, $\dot{z} = -\frac{3}{256}$, $PC = -84$ and $PL = -10\frac{1}{2}$. When these are obtained, if y be taken from PC there will remain -4 in the first case and -87 in the second for the lengths DC, and their difference 83 is the length of the curve between the two points C found.

So we should understand the matter when the curve(697) is continuous between two points C or the points K and C without terminating in what we have likened to a cusp. But when one or more terminations of this kind—which are found by determining the greatest or least values of PC(698) or DC—lie between those points, the lengths of individual portions of the curve between them and the points C or K should be determined separately and then added.

---

circle $t^2 + (v-a)^2 = r^2$, $a > r$, as a simple counter-example to Newton's assertion: though by definition of the circle locus no point on it can be a cusp, nevertheless, on taking

$$ds = \sqrt{[dt^2 + dv^2]}$$

to be the element of arc-length, the tangent $PC = v\,ds/dv$, that is rv/t, attains the minimum length $\sqrt{[a^2 - r^2]}$ at the two points $(\pm r\sqrt{[1 - r^2/a^2]},\ a[1 - r^2/a^2])$. Such cases may be eliminated by noting that only at a cusp is either v or t simultaneously a maximum or minimum, but since the tangent slope at such *points de rebroussement* is not in general zero, extreme values of this type cannot be located by a simple Fermatian test. However, such a test for PC to attain a local extreme value yields at once $\frac{d}{ds}\left(v\frac{ds}{dv}\right) = 1 - \frac{v}{\rho} \times \frac{dt}{dv} \cdot \frac{ds}{dv} = 0$, where ρ denotes the radius of curvature $\left|\left(\frac{ds}{dv}\right)^3 \Big/ \frac{d^2t}{dv^2}\right|$. Since PC trivially attains a zero minimum value at all intersections of the curve (C) with the base AP $(v = 0)$ and since the abscissa $AL = t$ attains an extreme value when C is an inflexion point where LC is tangent to the curve, this Newtonian criterion reduces

||[103] ||PROB: 11. CURVAS INVENIRE QUOTCUNQUE QUARUM LONGITUDINES CUM PROPOSITÆ ALICUJUS CURVÆ LONGITUDINE, VEL CUM AREA EJUS AD DATAM LINEAM APPLICATÂ, OPE FINITARUM ÆQUATIONUM COMPARARI POSSUNT.

Peragitur involvendo longitudinem areamve propositæ Curvæ in æquatione quæ in præcedente Problemate assumitur ad determinandam relationem inter *AP* et *PD*. Sed ut z et r inde per Prob 1 eliciantur, fluxio longitudinis vel areæ illius priùs investigari debet.

Fluxio longitudinis ejus determinatur ponendo æqualem radici quadraticæ summæ quadratorum a fluxionibus Basis et perpendiculariter incedentis. Sit enim *RN* linea perpendiculariter incedens super Basi *MN*, et *QR* curva proposita ad quam *RN* terminatur. Dictisq; $MN=s$, $NR=t$, et $QR=v$, et earum fluxionibus p, q, et l respectivè; concipe lineam *NR* ad locum quam proximum *nr* promoveri, et demisso ad *nr* perpendiculo *Rs*, erunt *Rs*, *sr*, et *Rr* contemporanea momenta linearum *MN*, *NR*, et *QR* quorum additamentis evadunt *Mn*, *nr*, et *Qr*. Et cùm hæc sint inter se ut earundem linearum fluxiones, ac propter angulum rectum *Rsr* sit $\sqrt{Rs\times Rs+sr\times sr}=Rr$, erit $\sqrt{pp+qq}=l$.(700)

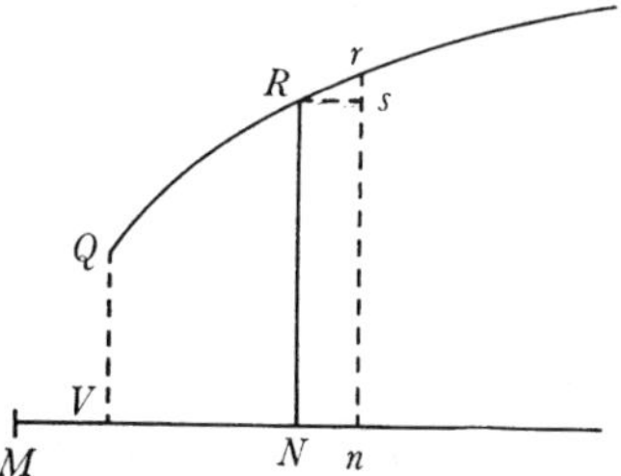

Ad determinandas autem fluxiones p et q duæ requiruntur æquationes una quæ definiat relationem inter *MN* et *NR* seu s et t, unde relatio inter fluxiones p et q eruenda est, et alia quæ definiat relationem inter *MN* vel *NR* ad datam figuram et *AP* seu x ad quæsitam, unde relatio fluxionis p vel q ad fluxionem m seu 1 innotescit.

||[104] ||Invento l, fluxiones n et r per assumptam tertiam æquationem qua longitudo *PD* sive y definitur investigandæ sunt, et capienda $PC=\frac{1-nn}{r}$, $PL=n\times PC$, ac $DC=PC-y$, ut in præcedente Problemate.(701)

to the condition $\rho = 0$ for a cusp only if simultaneously the measure s of arc-length attains a local maximum or minimum value: thus the curve $v = t^{\frac{2}{3}}+t^{\frac{4}{5}}$ will have a cusp at the origin. Newton's following criterion that C is a cusp when DC instantaneously attains a local extreme value is unexceptionable in the case of smoothly continuous curves (D) having continuously varying curvature. Compare note (354).

(699) Newton has here cancelled the superfluous noun 'terminos'.

(700) A standard exposition of rectification procedure for a curve defined with regard to the perpendicular Cartesian coordinates $MN = s$, $NR = t$ of arc-length $\widehat{QR} = v$. Much as before p, q and l (denoted by $\dot{s}$, $\dot{t}$ and $\dot{v}$ in our anachronistic English version) are the Newtonian fluxions of s, t and v. Compare Problem 13 of the October 1666 tract (I: 441).

PROBLEM 11. TO FIND ANY NUMBER OF CURVES WHOSE LENGTHS MAY BE COMPARED WITH THE LENGTH OF SOME PROPOSED CURVE, OR WITH ITS AREA DIVIDED BY A GIVEN LINE-LENGTH, BY THE AID OF FINITE EQUATIONS

Accomplishing this requires involving the length or area of the proposed curve in the equation which, in the preceding problem, is assumed in order to determine the relationship between AP and PD. But to elicit z and $\dot{z}$ from this by Problem 1, the fluxion of the length or area ought previously to be found out.

The fluxion of the length is determined by setting it equal to the square root of the sum of the squares of the fluxions of the base and the perpendicular ordinate. For let RN be the line perpendicularly ordinate upon the base MN, and QR the proposed curve at which RN terminates. Calling $MN = s$, $NR = t$ and $QR = v$, and their respective fluxions $\dot{s}$, $\dot{t}$ and $\dot{v}$, imagine the line NR to move forward to the next closest possible place nr and then, when the perpendicular Rs is let fall to nr, will Rs, sr and Rr be contemporaneous moments of the lines MN, NR and QR, by whose addition they come to be Mn, nr and Qr. Since these are to one another as the fluxions of the same lines and, because of the right angle $\widehat{Rsr}$, it is $\sqrt{[Rs^2+sr^2]} = Rr$, therefore $\sqrt{[\dot{s}^2+\dot{t}^2]} = \dot{v}$.(700)

To determine the fluxions $\dot{s}$ and $\dot{t}$, however, two equations are needed: one to define the relationship between MN and NR, that is, s and t—from this the relation between the fluxions $\dot{s}$ and $\dot{t}$ should be derived—and the second to define the relationship between MN or NR in the given figure and AP (or x) in the one sought, from which the relationship of the fluxion $\dot{s}$ or $\dot{t}$ to the fluxion $\dot{x}$ or 1 may be established.

After $\dot{v}$ is found, the fluxions $\dot{y}$ and $\dot{z}$ should be discovered by means of a third assumed equation by which the length of PD or y is defined, and then there should be taken $PC = (1-\dot{y}^2)/\dot{z}$, $PL = \dot{y}\times PC$ and $DC = PC-y$, as in the preceding problem.(701)

(701) Given the curve $\widehat{QR} = v$ defined by some equation, say $f(s, t) = 0$, with respect to the perpendicular coordinates $MN = s$, $NR = t$ (so that $dv^2 = ds^2+dt^2$), in the first three examples following Newton converts to the semi-intrinsic coordinate system introduced on his page 99 above (in which $PD = y$ is the normal to the curve (D) and $AP = x$ is the corresponding abscissa measured from some given origin A) by some stated transformation $x = \alpha(s)$, $y = \beta(s, v)$. By this means evaluation of v in terms of s and t is reduced to the equivalent problem of finding the radius of curvature

$$DC = |y-[1-(dy/dx)^2]/(d^2y/dx^2)|$$

at the corresponding point D on the constructed curve. His examples, we may note, are the circle $as-s^2 = t^2$ (which he transforms by $\alpha \equiv (as)^{\frac{1}{2}}$, $\beta \equiv \frac{2}{3}v$ and again by $\alpha \equiv s$, $\beta \equiv v^2/4a-s$) and the rectangular hyperbola $st = a^2$ (which he transforms by $\alpha \equiv a+3s$, $\beta \equiv a+3s+v$), in both of which cases he readily constructs the general radius of curvature at the point D in terms of s, t, v and their derivatives p (or $\dot{s}$), q (or $\dot{y}$) and l (or $\dot{v}$), where $p^2+q^2 = l^2$. He does not,

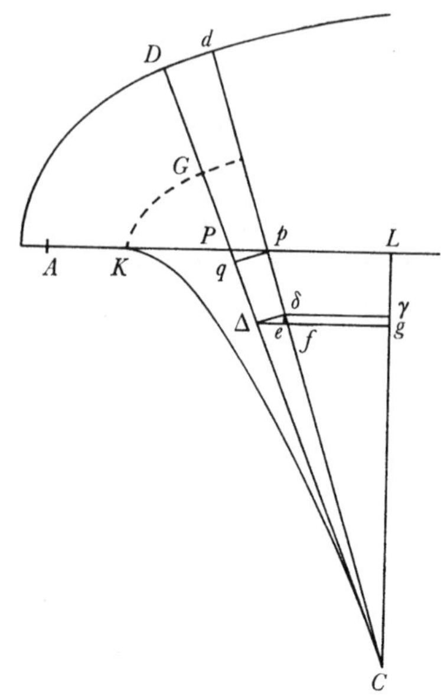

EXEMPL: 1. Sit $as-ss=tt$ æquatio ad datam curvam QR, utpote circuluum, $xx=as$ relatio inter lineas AP et MN, et $\frac{2}{3}v=y$ relatio inter longitudinem datæ curvæ QR et rectæ PD. Per primam fit $ap-2sp=2tq$ seu $\frac{a-2s}{2t}p=q$. Et inde $\frac{ap}{2t}(=\sqrt{pp+qq})=l$. Per secundam fit $2x=ap$ adeoq; est $\frac{x}{t}=l$. Et per tertiam fit $\frac{2}{3}l=n$ hoc est $\frac{2x}{3t}=z$, dein hinc fit $\frac{2}{3t}-\frac{2xq}{3tt}=r$. Quibus inventis capienda sunt $PC=\frac{1-nn}{r}$, $PL=n\times PC$, ac $DC=PC-y$ sive $PC-\frac{2}{3}QR$. Ubi patet longitudinem datæ curvæ QR inveniri non posse quin simul innotescat longitudo rectæ DC, indeq; longitudo curvæ ad quam punctum C cadit. Et contra.

EXEMPL: 2. Stante $as-ss=tt$, ponatur $x=s$ et $vv-4ax=4ay$. Perq; primam invenietur $\frac{ap}{2t}=l$ ut supra. Per secundam verò $1=p$, atq; adeo $\frac{a}{2t}=l$. Et per tertiam $2lv-4a=4an$, seu (eliminato l) $\frac{v}{4t}-1=z$, dein hinc $\frac{l}{4t}-\frac{vq}{4tt}=r$.

EXEMPL. 3. Ponantur tres æquationes $aa=st$, $a+3s=x$ et $x+v=y$. Et per primam (quæ Hyperbolam denotat) evadit $0=pt+qs$, seu $-\frac{pt}{s}=q$, et inde $\frac{p}{s}\sqrt{ss+tt}(=\sqrt{pp+qq})=l$. Per secundam evadit $3p=1$, adeoq; est $\frac{1}{3s}\sqrt{ss+tt}=l$. Et per tertiam fit $1+l=n$ sive $1+\frac{1}{3s}\sqrt{ss+tt}=z$, dein hinc fit $k=r$, posita scilicet k fluxione radicalis $\frac{1}{3s}\sqrt{ss+tt}$, quæ si fingatur æqualis ψ sive $\frac{1}{9}+\frac{tt}{9ss}=\psi\psi$, proveniet inde $\frac{2tq}{9ss}-\frac{2ttp}{9s^3}=2\psi k$. Et substituto imprimis $-\frac{pt}{s}$ pro q, deinde $\frac{1}{3}$ pro p, factaq; divisione per 2ψ, habebitur $\frac{-2tt}{27\psi s^3}=(k=)\,r$. Inventis n et r cætera peraguntur ut in exemplo primo.

||[105] ||Quod si a quovis curvæ puncto Q perpendiculum QV ad MN demittatur & curva invenienda sit cujus longitudo ex longitudine quæ oritur applicando aream $QRNV$ ad datam aliquam lineam innotescat: ponatur illa data linea e, longitudo $\frac{QRNM}{e}$ quæ ex applicatione oritur v, et ipsius v fluxio l. Et cùm fluxio

Example 1. Let $as-s^2=t^2$ be the equation to the given curve QR (a circle to be precise), $x^2=as$ the relation between the lines AP and MN, with $\frac{2}{3}v=y$ that between the lengths of the given curve QR and of PD. By the first there comes $a\dot{s}-2s\dot{s}=2t\dot{t}$ or $(a-2s)\,\dot{s}/2t=\dot{t}$ and thence $a\dot{s}/2t(=\surd[\dot{s}^2+\dot{t}^2])=\dot{v}$; by the second $2x=a\dot{s}$, so that $x/t=\dot{v}$; while by the third $\frac{2}{3}\dot{v}=\dot{y}$, that is, $2x/3t=z$, and then from this $2/3t-2x\dot{t}/3t^2=\dot{z}$. When these have been found, you must take $PC=(1-\dot{y}^2)/\dot{z}$, $PL=\dot{y}\times PC$ and $DC=PC-y$, that is, $PC-\frac{2}{3}QR$. It is evident here that the length of the given curve QR cannot be found without the length of the straight line DC—and consequently the length of the curve in which the point C falls—at once being known. And conversely so.

Example 2. With $as-s^2=t^2$ still standing, let there be put $x=s$ and $v^2-4ax=4ay$. By the first will be found $a\dot{s}/2t=\dot{v}$ as above; but by the second $1=\dot{s}$ and so $a/2t=\dot{v}$; and by the third $2\dot{v}v-4a=4a\dot{y}$ or (on eliminating $\dot{v}$) $v/4t-1=z$, and then from this $\dot{v}/4t-v\dot{t}/4t^2=\dot{z}$.

Example 3. Take the three equations $a^2=st$, $a+3s=x$ and $x+v=y$. Then by the first (which represents a hyperbola) there results $0=\dot{s}t+\dot{t}s$, or $-\dot{s}t/s=\dot{t}$ and thence $\dot{s}\surd[s^2+t^2]/s(=\surd[\dot{s}^2+\dot{t}^2])=\dot{v}$; by the second results $3\dot{s}=1$, so that $\frac{1}{3}\surd[s^2+t^2]/s=\dot{v}$; while by the third comes $1+\dot{v}=\dot{y}$, that is, $1+\frac{1}{3}\surd[s^2+t^2]/s=z$. From the last then comes $\dot{\psi}=\dot{z}$, where, of course, $\dot{\psi}$ is put for the fluxion of the radical $\frac{1}{3}\surd[s^2+t^2]/s$. The supposition that the latter equals ψ, that is,

$$\tfrac{1}{9}+\tfrac{1}{9}t^2/s^2=\psi^2,$$

yields in consequence $\frac{2}{9}t\dot{t}/s^2-\frac{2}{9}t^2\dot{s}/s^3=2\psi\dot{\psi}$, and, when in the first instance $-\dot{s}t/s$ is substituted in place of $\dot{t}$ and subsequently $\frac{1}{3}$ replaces $\dot{s}$, on carrying out division by 2ψ there will be obtained $-\frac{2}{27}t^2/\psi s^3=(\dot{\psi}=)\,\dot{z}$. After $\dot{y}$ and $\dot{z}$ are found the rest will be accomplished as in Example 1.

But if the perpendicular QV be let fall to MN from any point Q of the curve and a curve has to be found whose length is to be known from the length originating from dividing the area $QRNV$ by some given line, take this given line to be e, the length $(QRNM)/e$ originating from the division v, and the

however, broach the tricky problem of constructing the corresponding curve (D) which alone would give real point to the exercise. In a similar way, using the analogous results developed on his pages 98–9 for the evolute to a Cartesian curve, Newton might have transformed by $X=\alpha'(s)$, $Y=\beta'(s,v)$ to a curve (D') defined with respect to perpendicular coordinates $AB=X$, $BD'=Y$, and then constructed $D'C=|[1+(dY/dX)^2]^{\frac{3}{2}}/(d^2Y/dX^2)|$, but then, even in the simplest conic cases, the computations needed would already be formidable.

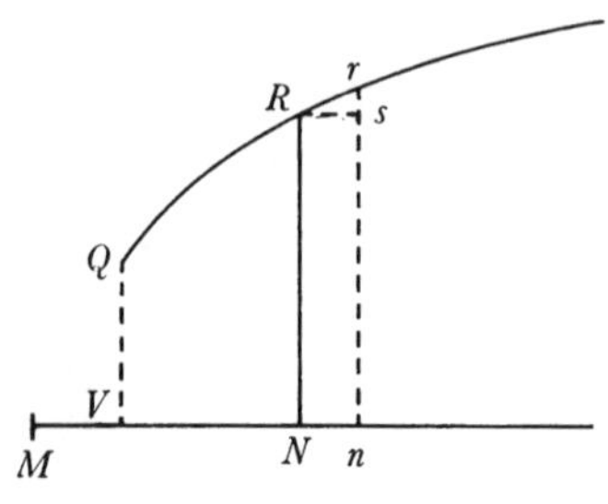

areæ $QRNV$ sit ad fluxionem areæ parallelogrammi rectanguli super MN ad altitudinem e constituti ut incedens linea NR seu t qua hæc describitur ad incedentem lineam qua illud eodem tempore describitur; et longitudinum quæ oriuntur applicando areas illas ad datam e, hoc est linearum v et MN seu s fluxiones l et p sint in eadem ratione, erit $l=\frac{pt}{e}$. Per hanc itaqꝫ regulam valor l inquirendus est, cæteraqꝫ ut in præcedentibus exemplis peragenda.(702)

Exempl: 4. Sit QR Hyperbola quam æquatio $aa+\frac{ass}{c}=tt$(703) definit, et inde juxta Prob 1 evadet $\frac{asp}{c}=tq$ sive $\frac{asp}{ct}=q$. Dein si pro alijs duabus æquationibus assumantur $x=s$, et $y=v$; prior dabit $1=p$, unde fit $l\left(=\frac{pt}{e}\right)=\frac{t}{e}$; et posterior dabit imprimis $n=l$, sive $z=\frac{t}{e}$, dein hinc $r=\frac{q}{e}$, et substituto $\frac{asp}{ct}$ sive $\frac{as}{ct}$ pro q evadet $r=\frac{as}{cet}$. Inventis n et r fac $\frac{1-nn}{r}=CP$ et $n\times CP=PL$ ut in præcedentibus, et inde punctum C adeoqꝫ curva in quam omnia ejusmodi puncta cadunt determinabitur, cujus curvæ longitudo ex longitudine DC quæ valet $CP-v$ innotescet, uti satis ostendimus.

Est et alia Methodus qua Problema resolvitur; quærendo nempe Curvas quarum fluxiones vel æquentur fluxioni Curvæ propositæ, vel ex illius et aliarum linearum fluxionibus componantur. Et hæc aliquando usui esse potest præsertim in convertendo Mechanicas curvas in æquales geometricas. Cujus rei
|| [106] insigne est Exemplum in || Spiralibus.(704)

(702) Much as before, but where now $ev = \int t.ds$, so that l/p (or $\dot{v}/\dot{s}$) $= t/e$, Newton connects (the area under) the curve QR with the curve (D) by a transformation $x = \alpha(s)$, $y = \beta(s, v)$. In his following example of the hyperbola $a^2+(a/c)s^2 = t^2$ (of main diameter $2a$ and latus rectum $2c$) he, in fact, chooses the simplest case in which $\alpha \equiv s$, $\beta \equiv v$.

(703) A first equivalent example '$aa+\frac{att}{c}=ss$' is cancelled.

(704) The development of this length-preserving transformation in the three decades preceding 1670 is a fascinating case-history in human insight and preconception which has never been systematically explored in the monograph needed to do it full justice. Along with the creation in the early seventeenth century of new analytical techniques for the mathematical treatment of curves, a fresh interest in the traditional problem of rectification was born. A first break-through in this direction came about 1641 when Roberval, anticipating Wren by some fifteen years (see note (308)), first 'straightened' the cycloid's general arc, though his 'De Longitudine Trochoidis Propositio'—essentially different in structure from Wren's—was to be published only in 1693 in his *Divers Ouvrages* (= *Memoires*, **6** (note (307)): 419–23). Unfortunately, other simple curves then known, particularly the Apollonian parabola and

fluxion of v to be $\dot{v}$. Now the fluxion of the area $QRNV$ is to that of the area of the rectangle constructed on MN to the height e as the ordinate line NR (or t) by which the former is described to the ordinate line by which the latter is in the same time: accordingly, the fluxions of the areas which arise from dividing those areas by the given quantity e, that is, the fluxions $\dot{v}$ and $\dot{s}$ of the lines v and (MN or) s, are in the same ratio and so $\dot{v} = \dot{s}t/e$. By this rule, therefore, the value of $\dot{v}$ should be sought out and the rest then accomplished as in the preceding examples.(702)

EXAMPLE 4. Let QR be a hyperbola defined by the equation

$$a^2 + (a/c)\, s^2 = t^2 \text{(703)}$$

and from it will result, according to Problem 1, $(a/c)\, s\dot{s} = t\dot{t}$ or $as\dot{s}/ct = \dot{t}$. Then, if $x = s$ and $y = v$ be assumed for the other two equations, the former will yield $1 = \dot{s}$ and hence $\dot{v} (= \dot{s}t/e) = t/e$, while the latter will give in the first place $\dot{y} = \dot{v}$, or t/e, and then as a consequence $\dot{z} = \dot{t}/e$ and so, when $as\dot{s}/ct$ (that is, as/ct) is substituted for $\dot{t}$, there will result $\dot{z} = as/cet$. After $\dot{y}$ and $\dot{z}$ are found make $(1-\dot{y}^2)/\dot{z} = CP$ and $\dot{y} \times CP = PL$, as in the preceding, and the point C—and so the curve in which all such points fall—will thereby be determined. As we have shown well enough, the length of this curve will be known from the length of DC, which is equal in value to $CP - v$.

There is also another method by which the problem is resolved: namely, by seeking curves whose fluxions either are equal to that of the proposed curve or may be compounded of it and the fluxions of other lines. This may sometimes be of use, particularly when mechanical curves are to be converted into equal geometrical ones. There is an oustanding example of this in the case of spirals.(704)

Archimedean spiral, still proved intractable to any approach aiming at their finite, algebraic rectification. It was apparently Thomas Hobbes who first suggested that the general arc-lengths of the latter curves might be compared in a way analogous to that by which Cavalieri had connected their general areas (see note (443)). In his *Examinatio & Emendatio Mathematicæ Hodiernæ, Qualis explicatur in libris Johannis Wallisii, Distributa in sex Dialogos* ((London, 1660): 'Dialogus Quintus': 122 [= W. Molesworth, *Thomæ Hobbes...Opera Philosophica*, 4 (London, 1845): 189]) Hobbes has left us a vivid (if edited) account of a visit he paid in company with Roberval to Mersenne's Minimite convent in Paris some time about the winter of 1642/3, during which, 'creta designans figuram in pariete' in support of his argument, he threw out the conjecture: 'videtur linea spiralis æqualis esse rectæ quæ subtendit Semiparabolam, cujusque quidem Axis sit æqualis Semiperimetro [?Hobbes' (false) argument here requires 'perimetro' to be consistent] circuli spiralem continentis; Basis autem ejusdem circuli Radio....Quoniam in Axe parabolæ, motus quo parabola generatur augetur juxta rationem temporum duplicatam, motus autem in Base est uniformis; item quia motus quo generatur spiralis, in circulo augetur in ratione temporum duplicata, & in Radio est uniformis; videtur similis esse generatio unius generationi alterius; & proinde...rectam illam, ut quæ eandem habet generationem, æqualem esse oportere Spirali'. Roberval at once saw the fallacy in Hobbes' reasoning (see note (707)) and accordingly 'abjecta creta errorem agnovit *Hobbius*. At *Robervallus* postridie eandem propositionem ad *Mersennum* demonstratam attulit' (Roberval's

Sit AB recta positione data, BD arcus[705] super AB tanquam Basi incedens ac interea retinens A pro centro, ADd Spiralis ad quam arcus ille perpetim terminatur, bd arcus quam proximus sive locus in quem arcus BD dum incedit proximè movetur, DC perpendicularis ad arcum bd, dG differentia arcuum, AH alia curva spirali AD æqualis,[706], BH recta super AB normaliter incedens ac terminata ad curvam AH, bh locus quam proximus in quam recta illa incedit,

---

correct argument is discussed, from an unpublished French autograph in the Bibliothèque Nationale at Paris, in Léon Auger's *Un savant méconnu: Gilles Personne de Roberval (1605–1672)* (Paris, 1962): 73–4.) Mersenne lost little time in spreading knowledge of Roberval's result among his circle of correspondents. Fermat, in his reply on 16 February 1643 (N.S.), was somewhat crushing: 'Pour les lignes courbes auxquelles vous m'écrivez que M. de Roberval a trouvé d'autres lignes égales, sur lequel vous m'alléguez l'hélice, j'appréhende qu'il y aura de l'équivoque. Il semble d'abord par les raisons des inscrits et des circonscrits que l'hélice d'Archimède est la moitié de la circonférence du cercle qui sert à la décrire, et c'était une pensée que j'avois eue il y a fort longtemps, mais je me détrompai d'abord. Si c'est celle de M. de Roberval, je m'assure qu'il ne sera pas longtemps de même avis et qu'il n'aura besoin que d'une seconde réflexion pour se dédire' (*Œuvres de Pierre Fermat*, **2** (Paris, 1891): 252). A month later Descartes was even more abrupt: 'Ie n'ay pas le loisir d'examiner ce que uous me mandez de l'helice et de la parabole' (see his letter to Mersenne of 23 March 1643 (N.S.) in his *Œuvres*, **3** (Paris, 1901): 642). The next year in January when Mersenne announced Roberval's discovery to Torricelli—'Nescio an ad te scripserim lineam parabolicam inventam æqualem alteri lineæ curvæ, nempe helici Archimedeæ..., a Robervallo...geometricè demonstratam' (*Opere di Evangelista Torricelli*, **3** (Faenza, 1919): 161, and compare 182)—the latter was more receptive, quickly formulating his own fluxional proof of the Robervallian comparison (see his *Opere*, **1**. 2 (1919): 361–2, 387, 391).

In modern terms, the fallacious rectification of the Archimedean spiral which Fermat warned against essentially confuses $\int_{\phi=0}^{\phi=2\pi} ds$ with $\int_0^{2\pi} r.d\phi = \pi R$, where the spiral has polar defining equation $r = (R/2\pi)\,\phi$ and $ds = \sqrt{[dr^2+r^2d\phi^2]}$ is the element of arc-length. (For large r the approximation is indeed very good.) This plausible 'proof' that 'The Spirall Line is equall to half the Circle of the first Revolution' occurred independently a decade afterwards to John Wallis, who, in ignorance of its basic fallacy, published it in 'indivisible' form as Proposition 5 of his *Arithmetica Infinitorum* (I, **1**, Appendix 2: note (20)) in 1656. Within a few weeks of publication Hobbes, quick to seize a god-given chance of denigrating an old enemy, laid bare the *non-sequitur* that 'the Aggregate of that infinite Number of indefinitely little Arches is...the Spirall Line made by your construction' in his forthright *Six Lessons to the Professors of the Mathematiques...in the University of Oxford* (London, 1656): Lesson v: 47, while a few years later, in his *De Infinitis Parabolis Liber Quintus, Opusculum Geometricum* ((Venice, 1663): 91–101: 'De Infinitis Spiralibus Wallisii'), Stefano degli Angeli was, if kindlier, equally firm in his denunciation of Wallis' error. (Not untypically, in the revised version of the *Arithmetica* he later incorporated in his *Opera Mathematica*, **1** (Oxford, 1695): 367 Wallis tried to bluff his way out with the interpolated remark: 'Per Spiralem, hic intelligo, non ipsam Curvam *Archimedeam*, sed, Aggregatum Arcuum...Spirali *Archimedeæ* continue inscriptorum... quem [*Figuram*] *Spuriam* dicimus'.) None the less, having spent twenty pages in attempted justification of the thesis 'Geometriam Indivisibilium esse fallacem [et] Fundamentia ipsius esse debilia', Thomas White illogically concluded his *Exercitatio Geometrica. De Geometria Indivisibilium, & Proportione Spiralis ad Circulum* (London, 1658) with an indivisibles proof in Wallisian style that 'Spiralem Archimedeam esse æqualem semiperipheriæ circuli primæ

Let AB be a straight line given in position, BD an arc[(705)] moving along on the base AB but all the while keeping A as its centre, ADd a spiral curve at which that arc perpetually terminates, bd the next possible arc (in other words, the place to which the arc BD next passes as it moves along), DC perpendicular to the arc bd, dG the difference of the arcs, AH a second curve equal[(706)] to the spiral AD, BH a straight line ordinate perpendicularly upon AB and terminating at the curve AH, bh the next possible place into which that line advances, and HK

---

revolutionis'. Unfortunately for White, the equality of the general arc of an Archimedean spiral to that of Roberval's parabola (which, unlike the arc of a circle, can be measured in terms of the logarithmic function) was finally, about December of the same year (1658), given an unchallengeably rigorous Archimedean proof by Pascal in his 'Demonstration à la maniere des Anciens de l'Egalité des Lignes Spirale & Parabolique', the terminal tract of his *Lettre[s] de A. Dettonville* ((Paris, 1659); reproduced in L. Brunschvicg, P. Boutroux and F. Gazier, *Œuvres de Blaise Pascal*, **8** (Paris, 1914): 255–82). The more general connection between corresponding higher spirals $r^n = k\phi$ and parabolas $(1+1/n)x^n = ky$ was made by Fermat, who, after reading Pascal's definitive study of the case $n = 1$, in 1660 put out the challenge 'Videat subtilis ille Geometra qui nuper æqualitatem helicis et paraboles demonstravit, an potuerit universalius concipi theorema et helices infinitæ cum infinitis parabolicis eleganter comparari' in conclusion to a series of theorems announced in an anonymous appendix to Antoine de Lalouvère's *Veterum Geometria promota in Septem de Cycloide Libris* ((Toulouse, 1660): Appendix Secunda, Pars Prima: 391–5, particularly 395 = *Œuvres de Fermat*, **1** (1891): 206–9). Pascal, however, died soon after with the challenge unanswered.

Newton could have been aware of little of this complex pattern of events in 1670. He would, however, have been familiar with John Wallis' first generalized discussion of (in his terminology) the length-preserving 'convolution' of 'parabolas' into corresponding 'spirals' in the closing pages of the latter's 1659 *Tractatus Duo* (note (255)): 105–7, where, apart from (now correctly) asserting that 'Spiralis Archimedea, non aliud est quam Parabola convoluta... Unde linearum Spiralis & Parabolicæ æqualitatem manifestam esse constat', Wallis for the first time in print introduced the logarithmic spiral as the 'convolution' of a straight line, so rectifying it. (J. Lohne has recently found that Harriot, in still unpublished manuscript dating from about 1595, was the first to light upon this 'obvious straightening' of the latter spiral but, as yet, has published no account of his discovery. Torricelli, in researches likewise unpublished till the present century, gave a more rigorous Archimedean account of this 'straight line convolute' fifty years later: compare E. Carruccio, *Evangelista Torricelli: De Infinitis Spiralibus* (Pisa, 1955).) More importantly, it will be evident that Newton had already mastered James Gregory's definitive geometrical study of the general transform (which connects, in his phrasing, equal arc-lengths of corresponding 'figuræ involutæ et evolutæ') presented with full Archimedean rigour on pages 29–41 of his *Geometriæ Pars Universalis, inserviens Quantitatum Curvarum Transmutationi & Mensuræ* (note (204)) in 1668. Though it is possible that Newton may have derived some small added inspiration from complementary passages in Isaac Barrow's *Lectiones Geometricæ* (note (80)): Lectio XII, Appendicula III: 126–7, Gregory's synthetic systematization must surely be the main source for his present analytical codification of the transform. Compare also A. Prag, 'On James Gregory's *Geometriæ Pars Universalis*' (in H. W. Turnbull's *James Gregory Tercentenary Memorial Volume* (London, 1939): 487–505): §4: 493–7.

(705) Of a circle of centre A, that is.

(706) Namely, in arc-length ($\widehat{AD} = \widehat{AH}$). Newton, for some reason not clear, first wrote 'simili[s]' (similar).

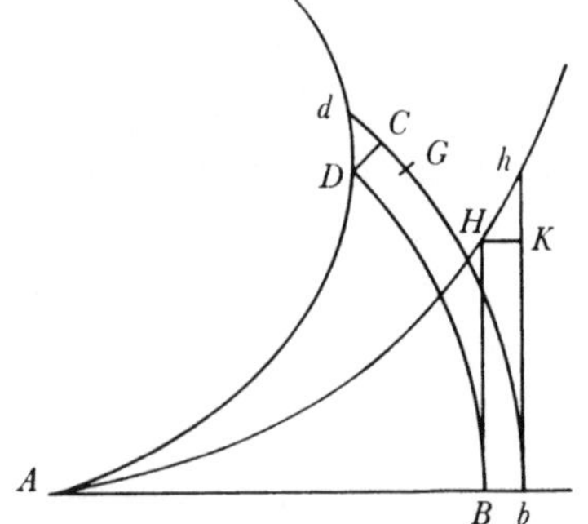

et HK perpendicularis ad bh. Et in triangulis infinitè parvis DCd ac HKh, cùm DC et HK æqualia sint eidem tertio Bb, indeq; sibi mutuo æqualia, ac Dd et Hh ex Hypothesi sint correspondentes partes æqualium curvarum et inde etiam æqualia, nec non anguli ad C et K recti, tertia etiam latera dC et hK æqualia erunt. Quare cùm insuper sit $AB \,.\, BD :: Ab \,.\, bC :: Ab - AB(Bb) \,.\, bC - BD(CG)$. Adeoq; $\frac{BD \times Bb}{AB} = CG$, si hoc auferatur a dG restabit $dG - \frac{BD \times Bb}{AB} (= dC) = hK$.

Dic itaq; $AB = z$, $BD = v$, & $BH = y$, et earum fluxiones r, l, et n respectivè; et cùm Bb, dG et hK sint earundem contemporanea momenta quorum additamentis evadunt Ab, bd, et bh, et proinde inter se sint ut fluxiones, ideo pro momentis in æquatione novissima substituantur fluxiones, juxta et notæ pro lineis et emerget $l - \frac{vr}{z} = n$. Ubi si e fluxionibus r pro æquabili habeatur et supponatur unitas esse ad quam cæteræ referantur evadet $l - \frac{v}{z} = n$.(707)

Quamobrem data per æquationem aliquam relatione inter AB et BD (sive [z](708) et v) qua Spiralis definiatur, dabitur (per Prob 1) fluxio l, et inde etiam

(707) This last sentence is a late insertion. No doubt to stress the ensuing close similarity between the respective defining equations of the 'spiral' (D) and the 'parabola' (H) of equal general arc-length, Newton has chosen as 'polar' coordinates of the former the lengths $AB = z$ and $\widehat{BD} = v = z \times B\hat{A}D$ (compare note (324)), and we should remember throughout that $AB = z$ and $B\hat{A}D = v/z$ correspond to our modern conventional coordinates: thus, on taking $B\hat{A}D = \phi$, the increment Gd $(= \widehat{bd} - \widehat{BD})$ of $\widehat{BD} = \phi z$ is seen to be compounded of $GC = (\phi.dz$ or$)$ $v.dz/z$ and $Cd = (z.d\phi$ or$)$ $z.d[v/z] = dv - v.dz/z$. The length-preserving transform of the 'spiral' (D) into the parabola (H) defined by Cartesian coordinates $A'B = x$, $BH = y$ is then denoted by ($AB = AA' + A'B$ or) $z = R + x$, and so $Bb = dz = dx$, together with $Dd = Hh$ and so ($Cd = KH$ or) $dv - v.dz/z = dy$. (Newton, in fact, illustrates the particular case of the transform in which $AA' = R$ vanishes when $y = 0$: compare notes (324) and (438).) At once, on making appropriate substitution of $R + z$ for z, there results

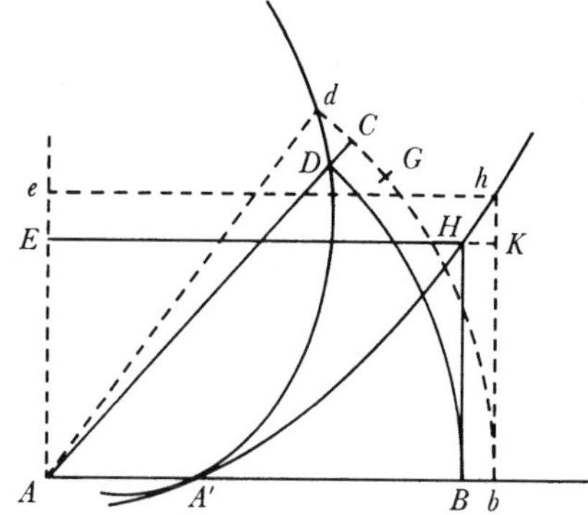

$$y = \int_0^x (dv/dx - v/[R+x]) \,.\, dx$$

perpendicular to bh. Then in the infinitely small triangles DCd and HKh, since DC and HK are equal to the same third length Bb and so to each other, while Dd and Hh by hypothesis are corresponding portions of equal curves and so also equal, and in addition the angles at C and K are right, the third sides dC and hK will also be equal. Hence, since moreover

$$AB:\widehat{BD} = Ab:\widehat{bC} = (Ab-AB \text{ or}) \; Bb:(\widehat{bC}-\widehat{BD} \text{ or}) \; \widehat{CG},$$

so that $\widehat{BD}\times Bb/AB = \widehat{CG}$, if the latter be taken from $\widehat{dG}$ there will remain $\widehat{dG}-\widehat{BD}\times Bb/AB(=\widehat{dC}) = hK$.

Accordingly, call $AB = z$, $BD = v$ and $BH = y$ and their respective fluxions $\dot{z}$, $\dot{v}$ and $\dot{y}$. Then, since Bb, dG and hK are contemporaneous moments of those quantities (by whose addition these come to be Ab, bd and bh) and are therefore to one another as their fluxions, on that account replace the moments in the most recent equation by fluxions and also line-lengths by their symbols and there will emerge $\dot{v}-v\dot{z}/z = \dot{y}$. Here if, of the fluxions, $\dot{z}$ be considered as uniform and supposed to be the unit to which the others are to be referred, there will result $\dot{v}-v/z = \dot{y}$.(707)

In consequence, when the relationship between AB and BD (that is, z(708) and v) by which the spiral is to be defined is determined by means of some equation, the fluxion $\dot{v}$ will (by Problem 1) be given and thence also the fluxion

---

as the defining equation of the curve $A'H$ when that, say $f(z, v) = f(R+x, v) = 0$, of the 'spiral' AD is given. Newton's first two examples below concern the general 'Fermatian' spiral (see note (242)) $z^p = a^{p-q}v^q$, which yields, on taking $R = 0$, $(1-q/p)^q x^p = a^{p-q}y^q$ ($p = 2$, $q = 1$ and $p = 3$, $q = 2$ respectively); while the third, $z^2(z+a) = cv^2$, gives

$$y = \int_0^x c^{-\frac{1}{2}}z(z+a)^{-\frac{1}{2}}.dz = \tfrac{1}{3}c^{-\frac{1}{2}}[(x-2a)(x+a)^{\frac{1}{2}}+2a^{\frac{3}{2}}].$$

In the former case $dv/dz = pv/qz$ and so $Cd = (p/q).GC$; in particular, when the curve AD is an Archimedean spiral ($p = 2$, $q = 1$) as in Newton's Example 1, $Cd = GC$. Historically this caused some considerable confusion to people like Hobbes (see note (704)), who in the winter of 1642/3 thought that the arcs $\widehat{AD}$ and $\widehat{AH}$ would be everywhere equal by equating the increment $GC = bd-BD$ to the increment Kh: as a result, since the increment Kh is taken to be twice its correct size (namely, $Gd = 2Cd$), when the curve AD is an Archimedean spiral, AH will be an Apollonian parabola with a parameter twice that of the parabola constructed by the present 'Gregorian' transform. In general, equation of the increments Gd and Hk defines a correspondence between the curves (D) and (H) which is evidently identical with the 'Cavalierian' area-preserving transformation discussed by Newton on his page 75 above (compare notes (442) to (444)). To confuse matters, it may, as Wallis conjectured in 1659 in his *Tractatus Duo* and Gregory proved rigorously nine years later (see note (704)), easily be shown that the Gregorian transform also conserves areas: in fact, on drawing HE parallel to AB and AE perpendicular to it, for all corresponding points D and H on the two curves there results $(AA'DA) = \frac{1}{2}\times(AA'HE)$.

(708) The manuscript reads 'x', but we have standardized this to read 'z': of course, the identity of $AB = z$ and $AB = x$ is basic in the present transformation in which $D(z, v)$ corresponds with $H(x, y)$. Compare note (707).

fluxio n ponendo æqualem $l-\frac{v}{z}$. Atqȝ hæc per Prob 2[709] dabit lineam y sive BH cujus est fluxio.

EXEMPL 1. Si detur $\frac{zz}{a}=v$, æquatio nempe ad Spiralem Archimedeam, inde per Prob. 1 elicietur $\frac{2z}{a}=l$. A quo aufer $\frac{v}{z}$ sive $\frac{z}{a}$ et restabit $\frac{z}{a}=n$, et inde per Prob 2 fit $\frac{zz}{2a}=y$. Quod indicat curvam AH cui hæc spiralis AD æquatur esse Parabolam Apollonianam cujus latus rectum existit $2a$; sive cujus incedens BH perpetuò æquatur semissi arcus BD.[710]

EXEMPL: 2. Si proponatur Spiralis quam æquatio $z^3=avv$ sive $\frac{z^{\frac{3}{2}}}{a^{\frac{1}{2}}}=v$ definit,
||[107] emerget per Prob 1 $\frac{3z^{\frac{1}{2}}}{2a^{\frac{1}{2}}}=l$, A quo si auferatur $\frac{v}{z}$ sive $\frac{z^{\frac{1}{2}}}{a^{\frac{1}{2}}}$ restabit ||$\frac{z^{\frac{1}{2}}}{2a^{\frac{1}{2}}}=n$, et inde per Prob 2 producetur $\frac{z^{\frac{3}{2}}}{3a^{\frac{1}{2}}}=y$. Hoc est $\frac{1}{3}BD=BH$, existente AH Parabola secundi generis.

EXEMPL: 3. Si ad Spiralem sit $z\sqrt{\frac{a+z}{c}}=v$. exinde per Prob 1 elicietur $\frac{2a+3z}{2\sqrt{ac+cz}}=l$, A quo si auferatur $\frac{v}{z}$ sive $\sqrt{\frac{a+z}{c}}$ restabit $\frac{z}{2\sqrt{ac+cz}}=n$. Jam cum quantitas hac fluxione n generata nequeat inveniri per ea quæ in Prob 2 habentur, nisi[711] fiat resolutio in infinitam seriem; juxta tenorem Scholij Prob 9[712] reduco ad formam æquationum in prima collumna Catalogorum substituendo z^{η} pro z, et evadit $\frac{z^{2\eta-1}}{2\sqrt{ac+cz^{\eta}}}=n$,[713] æquatio nempe secundæ speciei quarti ordinis prioris Catalogi. Et conferendo terminos fit $d=\frac{1}{2}$, $e=ac$, et $f=c$, adeoqȝ $\frac{z-2a}{3c}\sqrt{ac+cz}(=t)=y$.[714] Quæ æquatio est ad curvam geometricam AH cui spiralis AD æquatur.[715]

||[109] ||PROB 12. CURVARUM LONGITUDINES DETERMINARE.

Fluxionem curvæ lineæ in superiore Problemate ostendimus æqualem esse radici quadraticæ summæ quadratorum a fluxionibus Basis et perpendiculariter Incedentis. Et proinde si Basis fluxionem pro uniformi ac determinata mensura,

(709) Understand 'by integration' in a more general sense than the methods of deriving fluents from fluxions in Problem 2. It will be evident that Newton proposes Example 3 below precisely as an instance of a fluent which 'nequ[i]t inveniri per ea quæ in Prob 2 habentur, nisi fiat resolutio in infinitam seriem'.

(710) The complex previous history of this apparently innocent example is outlined in note (704).

$\dot{y}$, by setting it equal to $\dot{v}-v/z$. By Problem 2[(709)] this will yield the line y, that is, BH, whose fluxion it is.

EXAMPLE 1. If there be given $z^2/a=v$ (the equation, namely, to an Archimedean spiral), therefrom by Problem 1 will be derived $2z/a=\dot{v}$. From this take v/z, that is, z/a, and there will remain $z/a=\dot{y}$: from which there comes $z^2/2a=y$ by Problem 2. This indicates that the curve AH to which this spiral AD is equal is an Apollonian parabola whose latus rectum has the value $2a$—in other words, whose ordinate BH is everywhere equal to half the arc $\widehat{BD}$.[(710)]

EXAMPLE 2. Should a spiral be proposed which is defined by the equation $z^3=av^2$, that is, $z^{\frac{3}{2}}/a^{\frac{1}{2}}=v$, by Problem 1 there will emerge $\frac{3}{2}z^{\frac{1}{2}}/a^{\frac{1}{2}}=\dot{v}$. If v/z or $z^{\frac{1}{2}}/a^{\frac{1}{2}}$ be taken from this there will remain $\frac{1}{2}z^{\frac{1}{2}}/a^{\frac{1}{2}}=\dot{y}$ and thence by Problem 2 will be forthcoming $\frac{1}{3}z^{\frac{3}{2}}/a^{\frac{1}{2}}=y$, that is, $\frac{1}{3}\widehat{BD}=BH$, the parabola AH being of second order.

EXAMPLE 3. If [the equation] to the spiral be $z\surd[a+z]/\surd c=v$, therefrom by Problem 1 will be derived $\frac{1}{2}(2a+3z)/\surd[c(a+z)]=\dot{v}$. If v/z or $\surd[a+z]/\surd c$ be taken from this, there will remain $\frac{1}{2}z/\surd[c(a+z)]=\dot{y}$. Now since the quantity y generated by this fluxion $\dot{y}$ cannot be found by means of the techniques of Problem 2 (except by resolving it[(711)] into an infinite series), following the drift of the Scholium to Problem 9[(712)] I reduce it to the form of the equations in the first column of the catalogues by substituting z^η in place of z, when there results $\frac{1}{2}z^{2\eta-1}/\surd[ac+cz^\eta]=\dot{y}$,[(713)] an equation, namely, of the second species of the fourth order in the former catalogue. On comparing terms there comes $d=\frac{1}{2}$, $e=ac$ and $f=c$, so that $\frac{1}{3}(z-2a)\surd[c(a+z)]/c(=t)=y$.[(714)] This is the equation to the geometrical curve AH to which the spiral AD is equal.[(715)]

PROBLEM 12. TO DETERMINE THE LENGTHS OF CURVES

In the previous problem we showed that the fluxion of a curve line is equal to the square root of the sum of the squares of the fluxions of the base and the perpendicular ordinate. Consequently if we consider the fluxion of the base as

(711) Newton has here cancelled 'prius' (beforehand).

(712) On Newton's pages 93–4 above.

(713) Where r (or $\dot{z}$) is taken to be unity.

(714) Read '$y-\frac{2}{3}a^{\frac{3}{2}}c^{-\frac{1}{2}}$' in order to obey Newton's implicit initial condition that $AB=z$ and $BH=y$ shall be simultaneously zero (at A). On setting $\alpha=\frac{2}{3}a^{\frac{3}{2}}c^{-\frac{1}{2}}$ the resulting cubic, $(y-\alpha)^2=\frac{1}{4}a^{-3}\alpha^2(z-2a)^2(z+a)$, is easily seen to be a divergent parabola of diameter $y=\alpha$, nodate at the point $(2a,\alpha)$ and tangent to the z-axis at the origin (as indeed is the corresponding Fermatian spiral $cv^2=z^3+az^2$).

(715) The remainder of Newton's page 107 and the whole of the following page 108 has been left blank, presumably with the intention that further examples (and perhaps 'cognate questions'?) should be inserted. What their detail might have been is anyone's guess.

nimirum unitate, ad quam cæteræ fluxiones referantur, habeamus, et insuper per æquationem quæ curvam definit quæramus fluxionem Incedentis, habebitur fluxio Curvæ lineæ a qua longitudo ejus per Prob 2 elicienda est.[(716)]

Exempl 1. Proponatur Curva *FDH* quam æquatio $\frac{z^3}{aa}+\frac{aa}{12z}=y$[(717)] definit, posito scilicet $z=$Basi AB, ac $y=$incedenti DB: et ex æquatione illa per Prob 1 elicietur $\frac{3zz}{aa}-\frac{aa}{12zz}=n$, existente nimirum 1 pro fluxione ipsius z et n fluxione y. Dein additis fluxionum quadratis fit summa $\frac{9z^4}{a^4}+\frac{1}{2}+\frac{a^4}{144z^4}=qq$, et extracta radice $\frac{3zz}{aa}+\frac{aa}{12zz}=q$, indeq per Prob 2, $\frac{z^3}{aa}-\frac{aa}{12z}=t$, ubi q fluxionem Curvæ ac t longitudinem designat.

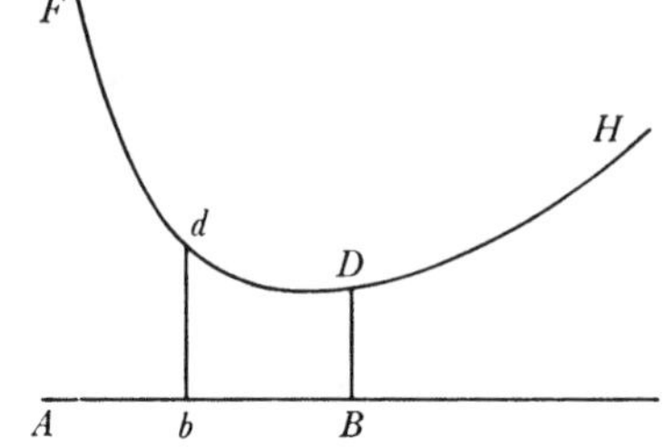

Itaq si cujusvis portionis Curvæ hujus puta *dD* longitudo desideretur[,] a punctis *d* ac *D* demitte ad *AB* perpendicula *db* ac *DB* et in valore *t* substitue quantitates *Ab* et *AB* seorsim pro *z*, ac differentia productorum erit longitudo quæsita. Quemadmodum si sit $Ab=\frac{1}{2}a$ et $AB=a$, scripto $\frac{1}{2}a$ pro z evadet $t=-\frac{a}{24}$, dein scripto a pro z evadet $t=\frac{11a}{12}$, a quo si prior valor auferatur restabit $\frac{23a}{24}$ pro longitudine *dD*. Vel si *Ab* tantùm definiatur esse $\frac{1}{2}a$ et *AB* spectetur indefinitè, restabit $\frac{z^3}{aa}-\frac{aa}{12z}+\frac{a}{24}=dD$.

Quod si cupias noscere portionem Curvæ quam t designat, finge valorem t

||[110] æquari nihilo, et evadet $z^4=\frac{a^4}{12}$, sive ||$z=\frac{a}{\sqrt{④12}}$. Adeoq si sumatur $Ab=\frac{a}{\sqrt{④12}}$, et erigatur *bd*, longitudo arcus *dD* erit t sive $\frac{z^3}{aa}-\frac{aa}{12z}$. Et hæc de alijs curvis generaliter intelligenda sunt.

Ad eundem modum quo hujus longitudinem determinavimus si pro alia Curva definienda proponatur æquatio $\frac{z^4}{a^3}+\frac{a^3}{32zz}=y$ proveniet $\frac{z^4}{a^3}-\frac{a^3}{32zz}=t$, vel si proponatur $\frac{z^{\frac{3}{2}}}{a^{\frac{1}{2}}}-\frac{1}{3}a^{\frac{1}{2}}z^{\frac{1}{2}}=y$, proveniet $\frac{z^{\frac{3}{2}}}{a^{\frac{1}{2}}}+\frac{1}{3}a^{\frac{1}{2}}z^{\frac{1}{2}}=t$. Vel generaliter si sit

(716) Newton sketches the standard procedure for rectifying a curve: namely, the arc-length t is the 'fluent' (taken between appropriate integration bounds) of the fluxion

$$\dot{t} = \sqrt{[\dot{y}^2+\dot{z}^2]},$$

a uniform, determinate measure (namely unity) to which the other fluxions shall be referred, and on top of this seek out the fluxion of the ordinate by means of the equation defining the curve, the fluxion of the curve line will be had and from that its length should be elicited by Problem 2.(716)

EXAMPLE 1. Let the curve *FDH* defined by the equation

$$z^3/a^2+a^2/12z = y^{(717)}$$

be proposed, where, of course, z = base *AB* and y = ordinate *DB*. From that equation by means of Problem 1 will be derived $3z^2/a^2-a^2/12z^2 = \dot{y}$, where to be sure the fluxion of z is taken as 1 and $\dot{y}$ is the fluxion of y. Then, on squaring these fluxions and adding, the sum comes to be $9z^4/a^4+\frac{1}{2}+a^4/144z^4 = \dot{t}^2$ and, when the root is extracted, $3z^2/a^2+a^2/12z^2 = \dot{t}$, and from this, by Problem 2, $z^3/a^2-a^2/12z = t$. Here $\dot{t}$ denotes the fluxion of the curve while t is its length.

In consequence, if the length of any portion of this curve, say *dD*, should be desired, from the points *d* and *D* let fall to *AB* the perpendiculars *db* and *DB*, and then in the value of t substitute the quantities *Ab* and *AB* separately in place of z: the difference of the results will be the length required. For instance, if $Ab = \frac{1}{2}a$ and $AB = a$, with $\frac{1}{2}a$ written for z there will come $t = -\frac{1}{24}a$, and then with a written in place of z will come $t = \frac{11}{12}a$, from which, on taking away the previous value, there will remain $\frac{23}{24}a$ for the length of *dD*. Or if *Ab* be merely defined to be $\frac{1}{2}a$, *AB* being regarded indefinitely, there will remain

$$z^3/a^2-\tfrac{1}{12}a^2/z+\tfrac{1}{24}a = dD.$$

But should you desire to know the portion of the curve which t denotes, suppose the value of t equal to zero and there will result $z^4 = \frac{1}{12}a^4$, and so $z = a/\sqrt[4]{12}$. Accordingly, if *Ab* be taken equal to $a/\sqrt[4]{12}$ and *bd* be erected, the length of the arc *dD* will be t, that is, $z^3/a^2-a^2/12z$. This is to be understood to hold for other curves in general.

In the same way as we determined the length of this one, if to define another curve the equation $z^4/a^3+/32z^2 = y$ be proposed, there will come out

$$z^4/a^3-a^3/32z^2 = t;$$

or if $z^{\frac{3}{2}}/a^{\frac{1}{2}}-\frac{1}{3}a^{\frac{1}{2}}z^{\frac{1}{2}} = y$ be proposed, there will come out $z^{\frac{3}{2}}/a^{\frac{1}{2}}+\frac{1}{3}a^{\frac{1}{2}}z^{\frac{1}{2}} = t$. Or in

where the curve is defined with regard to perpendicular Cartesian coordinates $AB = z$, $BD = y$. The analogous procedure for other types of coordinate system is not discussed in any of the nine following examples, but we may conjecture that the simple polar case at least would have been dealt with if Newton had ever completed his present tract.

(717) That is, $12a^2yz = 12z^4+a^4$. This quartic curve, which has the vertical through A ($z = 0$) for linear asymptote and is approximated otherwise in its infinite branches by the Wallis cubic $a^2y = z^3$, has apparently achieved no special name during three following centuries of mathematical history. Observe that Newton has drawn only the portion of the curve in the first quadrant: a congruent branch, symmetrical with regard to it round the centre A, also exists but the uniqueness of the 'length' $t = \int_\beta^\alpha \sqrt{[1+(dy/dz)^2]}\,.dz$ will be evident in this case.

$cz^{\theta}+\frac{z^{2-\theta}}{4\theta\theta c-8\theta c}=y$, ubi θ pro quolibet numero sive integro sive fracto designando adhibetur erit $cz^{\theta}-\frac{z^{2-\theta}}{4\theta\theta c-8\theta c}=t$.(718)

EXEMPL 2. Proponatur curva quam æquatio $\frac{2aa+2zz}{3aa}\sqrt{aa+zz}=y$ definit, et per Prob 1 obtinetur(719) $n=\frac{4a^4[z]+8aa[z]^3+4[z]^5}{3a^4y}$ sive, exterminato y, $n=\frac{2z}{aa}\sqrt{aa+zz}$ cujus quadrato adde 1, et summa erit $1+\frac{4zz}{aa}+\frac{4z^4}{a^4}$, ejusqꝫ radix $1+\frac{2zz}{aa}=q$. Unde per Prob 2 obtinetur $z+\frac{2z^3}{3aa}=t$.(720)

EXEMPL: 3. Proponatur Parabola secundi generis ad quam æquatio est $z^3=ayy$ seu $\frac{z^{\frac{3}{2}}}{a^{\frac{1}{2}}}=y$(721) et inde per Prob 1 elicietur $\frac{3z^{\frac{1}{2}}}{2a^{\frac{1}{2}}}=n$, adeoqꝫ est $\sqrt{1+\frac{9z}{4a}}\,(=\sqrt{1+nn})=q$. Jam cùm longitudo per fluxionem q generata nequeat inveniri per Prob 2 absqꝫ reductione in infinitam seriem simplicium terminorum,(722) consulo Catalogos ad Prob 9 et juxta ea quæ in Scholio(723) ejus habentur prodit $t=\frac{8a+18z}{27}\sqrt{1+\frac{9z}{4a}}$.

Et sic Parabolarum $z^5=ay^4$, $z^7=ay^6$, $z^9=ay^8$ &c longitudines inveniri possunt.(724)

EXEMPL 4. Proponatur Parabola ad quam æquatio est $z^4=ay^3$, sive $\frac{z^{\frac{4}{3}}}{a^{\frac{1}{3}}}=y$, et inde per Prob 1 orietur $\frac{4z^{\frac{1}{3}}}{3a^{\frac{1}{3}}}=n$. Adeoqꝫ $\sqrt{1+\frac{16z^{\frac{2}{3}}}{9a^{\frac{2}{3}}}}\,(=\sqrt{nn+1})=q$. Quo invento(725) consulo Catalogos juxta Scholium prædictum et facta collatione
||[111] cum secundo Theoremate quinti ordinis posterioris ||Catalogi, prodit $z^{\frac{1}{3}}=x$, $\sqrt{1+\frac{16xx}{9a^{\frac{2}{3}}}}=v$, et $\frac{3}{2}s=t$.(726) Ubi x designat basem[,] y ordinatim applicatam et s aream Hyperbolæ(727) atqꝫ t longitudinem quæ oritur applicando aream $\frac{3}{2}s$ ad unitatem linearem.

(718) The general condition for the curve $y = cz^{\theta}+dz^{\phi}$ to have a rational arc-length t is that $1+n^2$ (or $\dot{y}^2$) be an exact square, namely, q^2 (or $\dot{t}^2$). On substituting $n = c\theta z^{\theta-1}+d\phi z^{\phi-1}$ this condition is easily seen to be $1+4cd\theta\phi z^{\theta+\phi-2} = 0$, which yields $q = c\theta z^{\theta-1}-d\phi z^{\phi-1}$ and so $t = cz^{\theta}-dz^{\phi}$ at once. Evidently $\phi = 2-\theta = -1/4cd\theta$ or $d = 1/4c\theta(\theta-2)$.

(719) Namely, by squaring and finding the fluxion of the resulting equation

$$\tfrac{3}{2}a^4y^2 = \tfrac{2}{3}(a^2+z^2)^3.$$

(720) Some artifice has gone into framing this 'simple' example. Newton's evident starting point is to set q (or $\dot{t}$) $= 1+z^2/c^2$, from which n (or $\dot{y}$) $= \frac{z}{c}\sqrt{\left[2+\frac{z^2}{c^2}\right]}$, and so $t = z+\frac{1}{3}z^3/c^2$ while $y = \frac{1}{2}c\int\sqrt{[2+z^2/c^2]}\,.\,d(z^2/c^2) = \frac{1}{3}c(2+z^2/c^2)^{\frac{3}{2}}$. For simplicity c^2 has been taken equal to $\frac{1}{2}a^2$.

general if $cz^\theta + z^{2-\theta}/4\theta c(\theta-2) = y$, where θ is employed to denote any number you please, integral or fractional, then $cz^\theta - z^{2-\theta}/4\theta c(\theta-2) = t$.(718)

EXAMPLE 2. Let a curve defined by the equation $\frac{2}{3}(a^2+z^2)^{\frac{3}{2}}/a^2 = y$ be proposed, and there will be obtained by Problem 1(719) $\dot{y} = \frac{4}{3}z(a^2+z^2)^2/a^4y$ or, on eliminating y, $\dot{y} = 2z\sqrt{[a^2+z^2]}/a^2$. To the square of the latter add 1 and the sum will be $1+4z^2/a^2+4z^4/a^4$ and its root $1+2z^2/a^2$. Hence by Problem 2 will be obtained $z+\frac{2}{3}z^3/a^2 = t$.(720)

EXAMPLE 3. Let there be proposed the second-order parabola to which the equation is $z^3 = ay^2$ or $z^{\frac{3}{2}}/a^{\frac{1}{2}} = y$(721) and thence by Problem 1 will be elicited $\frac{3}{2}z^{\frac{1}{2}}/a^{\frac{1}{2}} = \dot{y}$, so that $\sqrt{[1+9z/4a]}\ (=\sqrt{[1+\dot{y}^2]}) = \dot{t}$. Now since the length generated by the fluxion $\dot{t}$ cannot be found by means of Problem 2 without reducing it to an infinite series of simple terms,(722) I consult the catalogues in Problem 9 and according to what was stated in its Scholium(723) there results

$$\tfrac{8}{27}a(1+9z/4a)^{\frac{3}{2}} = t.$$

In this way the lengths of the parabolas $z^5 = ay^4$, $z^7 = ay^6$, $z^9 = ay^8$ and so on may be found.(724)

EXAMPLE 4. Let there be proposed the parabola which has for its equation $z^4 = ay^3$ or $z^{\frac{4}{3}}/a^{\frac{1}{3}} = y$. From this by Problem 1 will arise $\frac{4}{3}z^{\frac{1}{3}}/a^{\frac{1}{3}} = \dot{y}$, so that $\sqrt{[1+\frac{16}{9}z^{\frac{2}{3}}/a^{\frac{2}{3}}]}\ (=\sqrt{[1+\dot{y}^2]}) = \dot{t}$. When this is found I(725) consult the catalogues in accordance with the aforesaid Scholium, and after comparison with the second theorem of the fifth order in the latter catalogue, there results $z^{\frac{1}{3}} = x$, $\sqrt{[1+\frac{16}{9}x^2/a^{\frac{2}{3}}]} = v$, and $\frac{3}{2}s = t$.(726) Here x denotes the base, y the ordinate and s the area of a hyperbola(727) while t is the length which arises on dividing the area $\frac{3}{2}s$ by the linear unity.

---

(721) Namely, Neil's semicubical parabola: see II, **1**, 1, §3: note (88). On his page 99 above Newton has already effected the rectification by defining this cubic as a parabolic evolute.

(722) Newton strains too hard after effect here since the substitution $1+9z/4a = x$ yields an immediate reduction to $\frac{4}{9}ax^{\frac{1}{2}}m = q$ (or $\dot{y}$), where $m = \dot{x}$ is the fluxion of x, and so at once $\frac{8}{27}x^{\frac{3}{2}} = y$ by Problem 2.

(723) On Newton's page 94 above: compare note (645).

(724) In general, where $z^{2p+1} = ay^{2p}$, $p = 1, 2, 3, \ldots$, then n (or $\dot{y}$) $= (1+1/2p)\,a^{-1/2p}z^{1/2p}$ and accordingly q (or $\dot{t}$) $= \sqrt{[1+fz^{1/p}]}$ on making $(1+1/2p)^2a^{-1/p} = f$. Hence, on substituting $z = x^p$, $t = \int px^{p-1}\sqrt{[1+fx]}\,.dx$ and the integration is readily effected for given p by Ordo 3 of the 'Catalogus prior' on Newton's page 77 by taking $z \to x$ and making $d = p$, $e = \eta = 1$. Compare note (541).

(725) Newton has cancelled 'iterum' (again).

(726) See note (647). Newton follows through the error he committed on his page 94 above. Correctly, 'facta collatione cum tertio Theoremate quinti ordinis posterioris Catalogi, prodit $z^{\frac{2}{3}} = x$, $\sqrt{x+\dfrac{16xx}{9a^{\frac{2}{3}}}} = v$, et $\frac{3}{2}s = t$' (after comparison with the third theorem of the fifth order in the latter catalogue there results...).

(727) Implicitly understanding the constant a to be positive, Newton has cancelled 'Conic[æ sectionis]' (conic).

Eadem methodo Parabolarum $z^6=ay^5$, $z^8=ay^7$, $z^{10}=ay^9$ &c longitudines etiam per aream Hyperbolæ determinantur.(728)

EXEMPL: 5. Proponatur Cissois Veterum, et existente ad eam æquatione $\frac{aa-2az+zz}{\sqrt{az-zz}}=y$,(729) inde per Prob 1 elicietur $\frac{-a-2z}{2zz}\sqrt{az-zz}=n$, et consequenter $\frac{a}{2z}\sqrt{\frac{a+3z}{z}}\,(=\sqrt{nn+1})=q$. Quæ scribendo z^η pro $\frac{1}{z}$ seu z^{-1}(730) evadit $\frac{a}{2z}\sqrt{az^\eta+3}=q$ æquatio primæ speciei quinti ordinis posterioris Catalogi et collatis terminis fiunt $\frac{a}{2}=d$, $3=e$, et $a=f$; adeoq₃ $z\left(=\frac{1}{z^\eta}\right)=xx$. $\sqrt{a+3xx}=v$, et $6s-\frac{2v^3}{x}$(731) $\left(=\frac{4de}{\eta f}\text{ in }\frac{v^3}{2ex}-s\right)=t$. Et adhibita a pro unitate per cujus multiplicationem vel divisionem hæ quantitates ad justum dimensionum numerum reducantur, evadunt $az=xx$, $\sqrt{aa+3xx}=v$, et $\frac{6s}{a}-\frac{2v^3}{ax}$(732) $=t$. Quorum hæc est constructio.

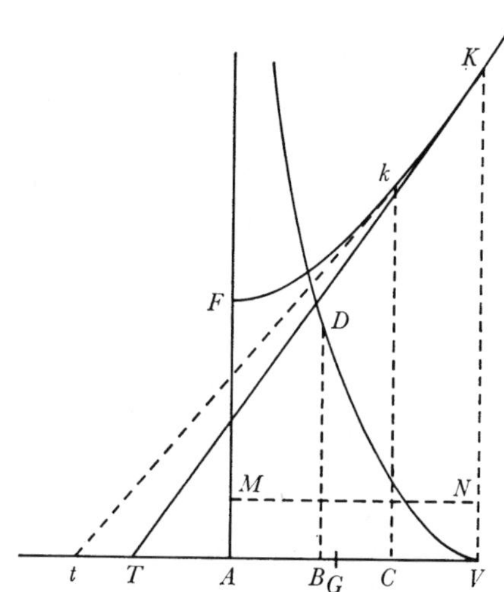

Existente VD Cissoide, AV diametro circuli ad quem aptatur, AF asymptoto ejus, ac DB perpendiculari ad AV; cum semiaxe $AF=AV$, et semiparametro $AG=\frac{1}{3}AV$ describatur Hyperbola FkK, et inter AB et AV sumpta AC media proportionali, erigantur ad C et V perpendicula Ck et VK, et agantur kt et KT rectæ tangentes Hyperbolam in k et K et occurrentes AV in t ac T, et ad AV constituatur rectangulum $AVNM$ æquale spatio $TKkt$; et Cissoidis VD longitudo erit sextupla altitudinis VN.(733)

(728) In general, where $z^{2p+2} = ay^{2p+1}$, $p = 1, 2, 3, \ldots$, then

$$n\ (\text{or } \dot{y}) = (1+1/[2p+1])\,a^{-1/[2p+1]}z^{1/[2p+1]}$$

and accordingly q (or $\dot{t}$) $= \sqrt{[1+ez^{2/[2p+1]}]}$ on taking $(1+1/[2p+1])^2a^{-1/[2p+1]} = e$. It follows, on substituting $z^2 = x^{2p+1}$, that $t = \int(p+\frac{1}{2})x^{p-1}\sqrt{[x+ex^2]}\,.dx$ and the integration is readily effected (much as in Ordo 5 of the 'Catalogus posterior' on Newton's page 81 above by taking $d = p+\frac{1}{2}$, $f = \eta = 1$) in terms of the area under the hyperbola $v^2 = ex^2+x$. Compare note (560).

(729) To be precise, this is the pair of the cissoid $y^2z = (a-z)^3$ and the line $z = a$. Compare note (426). Since $z = a$ is a bound to the present rectification integral, the added factor causes no inconvenience.

(730) On taking $\eta = -1$, that is.

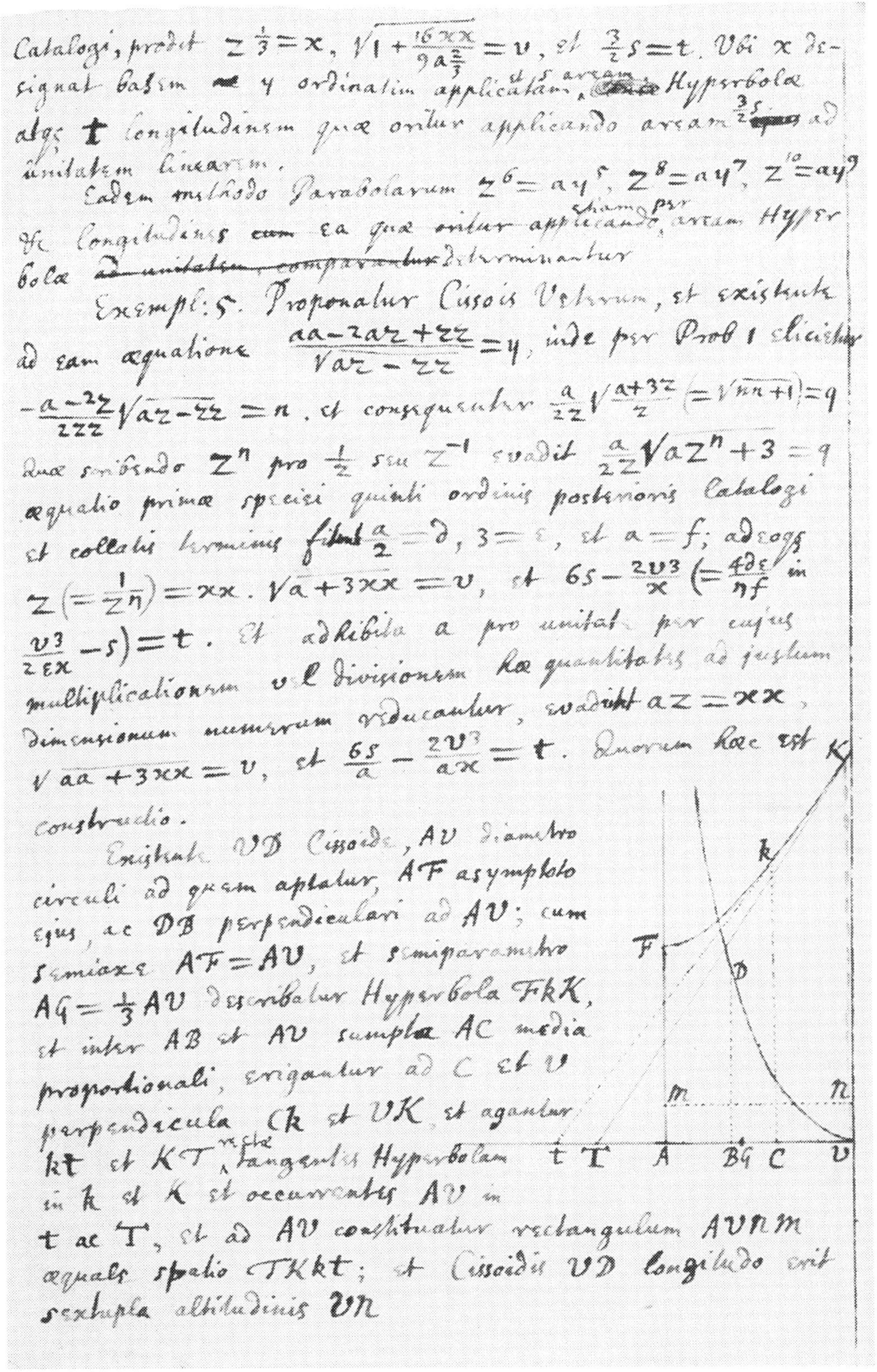

Catalogi, prodit $z^{\frac{1}{3}}=x$, $\sqrt{1+\frac{16xx}{9a^{\frac{2}{3}}}}=v$, et $\frac{3}{2}s=t$. Ubi x designat basem et y ordinatim applicatam, et s aream Hyperbolæ atque t longitudinem quæ oritur applicando aream $\frac{3}{2}s$ ad unitatem linearem.

Eadem methodo Parabolarum $z^6=ay^5$, $z^8=ay^7$, $z^{10}=ay^9$ &c longitudines ~~cum~~ ea quæ oritur applicando etiam per aream Hyperbolæ ~~ad unitatem comparantur~~ determinantur

Exempl: 5. Proponatur Cissois Veterum, et existente ad eam æquatione $\frac{aa-2az+zz}{\sqrt{az-zz}}=y$, inde per Prob 1 elicietur $-\frac{a-2z}{2zz}\sqrt{az-zz}=n$, et consequenter $\frac{a}{zz}\sqrt{\frac{a+3z}{z}}\left(=\sqrt{nn+1}\right)=q$ Quæ scribendo z^{η} pro $\frac{1}{z}$ seu z^{-1} evadit $\frac{a}{zz}\sqrt{az^{\eta}+3}=q$ æquatio primæ speciei quinti ordinis posterioris Catalogi et collatis terminis fit $\frac{a}{2}=d$, $3=e$, et $a=f$; adeoque $z\left(=\frac{1}{z^{\eta}}\right)=xx$. $\sqrt{a+3xx}=v$, et $6s-\frac{2v^3}{x}\left(=\frac{4de}{\eta f}\right.$ in $\left.\frac{v^3}{2ex}-s\right)=t$. Et adhibita a pro unitate per cujus multiplicationem vel divisionem hæ quantitates ad justum dimensionum numerum reducantur, evadit $az=xx$, $\sqrt{aa+3xx}=v$, et $\frac{6s}{a}-\frac{2v^3}{ax}=t$. Quorum hæc est constructio.

Existente VD Cissoide, AV diametro circuli ad quem aptatur, AF asymptoto ejus, ac DB perpendiculari ad AV; cum semiaxe $AF=AV$, et semiparametro $AG=\frac{1}{3}AV$ describatur Hyperbola FkK, et inter AB et AV sumpta AC media proportionali, erigantur ad C et V perpendicula Ck et VK, et agantur kt et KT rectæ tangentes Hyperbolam in k et K et occurrentes AV in t ac T, et ad AV constituatur rectangulum AVnm æquale spatio TKkt; et Cissoidis VD longitudo erit sextupla altitudinis Vn

Plate II. Rectification of the cissoid by means of a hyperbola (1, 2, §2).

By the same method the lengths of the parabolas $z^6 = ay^5$, $z^8 = ay^7$, $z^{10} = ay^9$ and so on are also determined by means of the area of a hyperbola.(728)

EXAMPLE 5. Let the cissoid of the ancients be proposed, the equation to it being $(a-z)^2/\sqrt{[z(a-z)]} = y$.(729) From this by means of Problem 1 will be deduced $-\frac{1}{2}(a+2z)\sqrt{[z(a-z)]}/z^2 = \dot{y}$ and consequently

$$\tfrac{1}{2}az^{-\frac{3}{2}}\sqrt{[a+3z]}\ (=\sqrt{[\dot{y}^2+1]}) = \dot{t}.$$

On writing z^η for $1/z$ or z^{-1}(730) there results $\frac{1}{2}az^{-1}\sqrt{[az^\eta+3]} = q$, an equation of the first species of the fifth order in the latter catalogue, and when terms are compared there comes $\frac{1}{2}a = d$, $3 = e$ and $a = f$, so that $z(=z^{-\eta}) = x^2$,

$$\sqrt{[a+3x^2]} = v \quad \text{and} \quad 6s-2v^3/x^{(731)}(=[4de/\eta f]\times[v^3/2ex-s]) = t.$$

When a is employed as a unit, through multiplication or division by which these quantities are to be reduced to a correct number of dimensions, there results $az = x^2$, $\sqrt{[a^2+3x^2]} = v$ and $6s/a-2v^3/ax$(732) $= t$. This is their construction.

Where VD is the cissoid, AV the diameter of the circle to which it is fitted, AF its asymptote and DB perpendicular to AV, let the hyperbola FkK be described with semi-axis $AF = AV$ and semi-parameter $AG = \frac{1}{3}AV$. Between AB and AV take the mean proportional AC, raise the perpendiculars Ck and VK to C and V, draw the straight lines kt and KT touching the hyperbola at k and K and meeting AV in t and T, and on AV construct the rectangle $AVNM$ equal to the area $TKkt$. The length VD of the cissoid will be six times the length of VN.(733)

---

(731) Read '$6s-\frac{v^3}{x}$'. The error is repeated by Horsley (*Opera Omnia*, **1** (note (60)): 515) and by Colson in his English version (*Method of Fluxions* (note (79)): 136–7). Correspondingly, in the next sentence read '$\frac{6s}{a}-\frac{v^3}{ax}=t$'. The mistake evidently arose when this fifth example was copied from a worksheet (now lost) since the extraneous factor '2' is not carried over into the following construction.

(732) See the previous note.

(733) In Leibnizian equivalent, Newton evaluates the general arc-length $\widehat{VD} = t$ of the (upper branch of the) cissoid $y^2z = (a-z)^3$ measured between the vertex $V(a, 0)$ and its general point $D(z, y)$, where $AB = z$, $BD = y$ are its perpendicular Cartesian coordinates, as $\int_z^a \sqrt{[1+(dy/dz)^2]}.dz = \int_z^a \frac{1}{2}az^{-\frac{3}{2}}(a+3z)^{\frac{1}{2}}.dz$. By means of the substitution $z = x^2/a$ (on the lines of his reduction of Ordo 5 in the 'Catalogus posterior' on his page 81 above) this 'fluent' is then evaluated in terms of the area $(VKkC) = s = \int_x^a v.dx$ under the hyperbola FkK defined with respect to perpendicular coordinates $AC = x$, $Ck = v$ by the equation $v^2 = a^2+3x^2$: specifically, $at = \int_x^a a^2x^{-2}\sqrt{[a^2+3x^2]}.dx = 6s-\left[\frac{v^3}{x}\right]_{x=x}^{x=a}$. Newton's ingenious construction depends on noticing that the hyperbolic subtangent $Ct = \frac{1}{3}v^2/x$, so that $\triangle Ckt = \frac{1}{6}v^3/x$, and therefore $AV\times\widehat{VD} = 6\times[(VKkC)-\triangle VKT+\triangle Ckt] = 6\times(TKkt)$. In more direct analytical terms $\widehat{VD} = t = a\left[\sqrt{3}\log(x+\sqrt{[x^2+\frac{1}{3}a^2]})-\sqrt{[3+\frac{a^2}{x^2}]}\right]_{x=\sqrt{[az]}}^{x=a}$. The construction, introduced

||[112] Fig [pag 75g] ||Exempl 6. Existente Ad ellipsi quam æquatio $\sqrt{az-2zz}=y$ definit: proponatur curva Mechanica(734) AD talis ut si Bd seu y producatur donec huic curvæ ad D occurrat, sit BD æqualis arcui Ellipticæ Ad. Jam quo hujus longitudo determinetur æquatio $\sqrt{az-2zz}=y$ dabit $\frac{a-4z}{2\sqrt{az-2zz}}=n$. Cujus quadrato si 1 addatur prodit $\frac{aa-4az+8zz}{4az-8zz}$ quadratum fluxionis arcûs Ad, et huic si iterum addatur 1 provenit $\frac{aa}{4az-8zz}$ cujus radix $\frac{a}{2\sqrt{az-2zz}}$ est fluxio curvæ lineæ AD.

Ubi si e radicali extrahatur z et pro z^{-1} scribatur z^{η},(735) habebitur

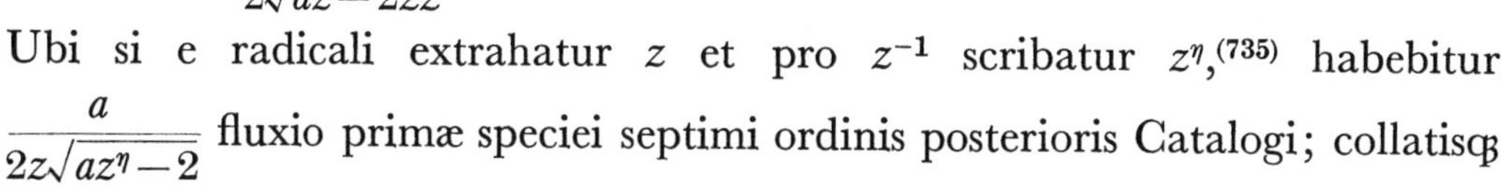

$\frac{a}{2z\sqrt{az^{\eta}-2}}$ fluxio primæ speciei septimi ordinis posterioris Catalogi; collatisq; terminis exibunt $d=\frac{a}{2}$, $e=-2$, et $f=a$, adeoq; $z\left(=\frac{1}{z^{\eta}}\right)=x$, $\sqrt{ax-2xx}=v$, et $\frac{8s}{a}-\frac{4xv}{a}+v\left(=\frac{8de}{\eta ff}\text{ in } s-\tfrac{1}{2}xv-\frac{fv}{4e}\right)=t$. Quorum constructio est ut, ad Ellipsis centrum C acta recta dC, constituatur super AC parallelogrammum æquale sectori ACd, et duplum altitudinis ejus ponatur esse longitudo Curvæ AD.(736)

Exempl: 7.(737) Existente $A\beta=\psi$, & $\alpha\delta$ Hyperbola ad quam æquatio sit $\sqrt{-a+b\psi\psi}=\beta\delta$, actaq; δT tangente ejus; proponatur curva VdD cujus basis AB sit $\frac{1}{\psi\psi}$, & normaliter incedens BD longitudo quæ oritur applicando aream $\alpha\delta T\alpha$ ad unitatem linearem. Jam ut hujus VD longitudo determinetur quæro

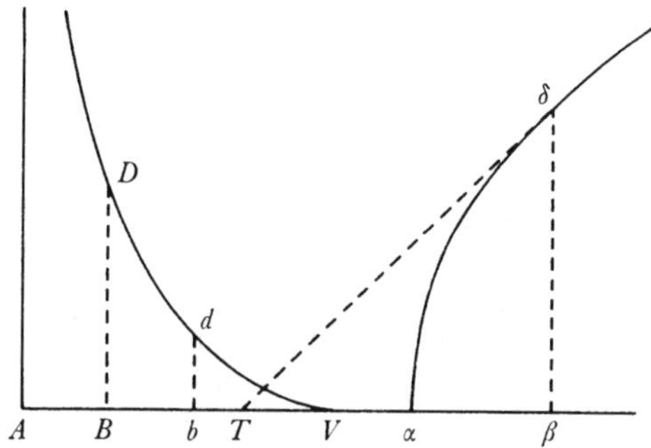

with the self-effacing phrase 'in simplicioribus vulgoq; celebratis figuris vix aliquid relatu dignum reperi quod evasit aliorum conatus nisi fortè longitudo Cissoidis ejusmodi censeatur', was communicated word for word to Leibniz in the *epistola posterior* of 24 October 1676 with the comment 'Demonstratio perbrevis est' (*Correspondence of Isaac Newton*, **2**: 117). Relatively ignorant as he was of contemporary research into the elementary differential geometry of curves, Newton was clearly hesitant of appearing too enthusiastic over a result which might be well known to contemporary foreign mathematicians (though to be sure, as we know in hindsight, the priority in rectifying the cissoid's general arc is unchallengeably his).

(734) An elliptical sinusoid, as we might name it. The manuscript at this point lacks a corresponding autograph figure but someone (apparently Jones) has pencilled in a repeat of Newton's diagram on his page 75g in accordance with the marginal reference 'Fig'. For

Example 6. Where Ad is an ellipse defined by the equation $\surd[az-2z^2]=y$, let the mechanical curve AD(734) be proposed whose property it is that, if Bd (or y) be produced till it meet this curve in D, BD is equal to the ellipse arc $\widehat{Ad}$. Now to determine this length the equation $\surd[az-2z^2]=y$ will yield Figure [on page 75g]

$$\tfrac{1}{2}(a-4z)/\surd[az-2z^2]=\dot{y},$$

and if 1 be added to its square there results $(a^2-4az+8z^2)/4(az-2z^2)$, the square of the fluxion of the arc $\widehat{Ad}$, while if to this 1 be again added there comes $a^2/4(az-2z^2)$, the root of which, $\frac{1}{2}a/\surd[az-2z^2]$, is the fluxion of the curve line $\widehat{AD}$. Here if z be taken outside the radical and in place of z^{-1} be written z^{η},(735) there will be obtained $\frac{1}{2}az^{-1}/\surd[az^{\eta}-2]$, a fluxion of the first species of the seventh order in the latter catalogue. On comparing terms there will prove to be $d=\frac{1}{2}a$, $e=-2$ and $f=a$, so that $z(=z^{-\eta})=x$, $\surd[ax-2x^2]=v$ and

$$8s/a-4xv/a+v(=[8de/\eta f^2]\times[s-\tfrac{1}{2}xv-\tfrac{1}{4}fv/e])=t.$$

The construction of these is: drawing the straight line dC to the ellipse centre C, on AC set up a rectangle equal to the sector ACd and put twice its altitude as the length of the curve AD.(736)

Example 7.(737) Where $A\beta=\psi$, let $\alpha\delta$ be a hyperbola having the equation $\surd[-a+b\psi^2]=\beta\delta$ and draw its tangent δT. Then let the curve VdD be proposed whose base AB is $1/\psi^2$ and perpendicular ordinate BD is the length which arises from dividing the area $\alpha\delta T\alpha$ by the linear unit. Now to determine the latter's

---

convenience we have inserted in the Latin text an appropriately modified version of that figure in which $\widehat{AdF}$ is an ellipse arc and the sinusoid AD is (correctly) shown to have an inflexion. It will be evident, on taking $AB=(z=)\,x$ and $BD=v$ as the perpendicular coordinates of the general point $D(x,v)$, that the curve will meet the vertical at F in some point E such that $EF=\widehat{AdF}$, that the arc ADE is symmetrical round its midpoint (at which it is consequently inflected), and that the lower part of the ellipse (not drawn) will correspondingly generate an arc $AD'E'$ in which each point D' is the mirror-image of the point D in the diameter $AF=\frac{1}{2}a$. Much like the circular sinusoid $v=\frac{1}{4}a\cos^{-1}(1-4x/a)$ which is generated in analogous manner from the circle $az-2z^2=2y^2$ (compare note (434)), the present curve will snake endlessly between the perpendiculars at A and F to AF, forming an infinite sequence of 'waves' congruent to $EDAD'E'$.

(735) That is, if -1 be written for η.

(736) In Leibnizian equivalent we would evaluate the sinusoid's arc

$$\widehat{AD}=t=\int_0^z \tfrac{1}{2}a/\surd[az-2z^2].dz \quad\text{as}\quad 8a^{-1}[s+\tfrac{1}{2}y(\tfrac{1}{4}a-z)],\quad\text{where}\quad (ABd)=s=\int_0^z y.dx.$$

(Note that Newton, in making consistent appeal to Ordo 7.1 of his 'catalogus posterior' on his page 81, has constructed an ellipse $v^2=ax-2x^2$ coincident with AdF by making $z=x$ and so $y=v$.) Evidently, since $BC=\frac{1}{4}a-z$ and therefore $\triangle BCd=\frac{1}{2}y(\frac{1}{4}a-z)$, we have

$$AC\times\widehat{AD}=2.[(ABd)+\triangle BCd]=2.(ACd).$$

(737) It would appear that this example was constructed *ad hoc* since Newton first began, in continuation, to write out the opening words of the following example: 'Proponatur Hyperbola ad quam æquatio est $\surd aa$ ...'.

fluxionem areæ $\alpha\delta T\alpha$ cum AB uniformiter fluit(738) & invenio esse $\frac{a}{4bz}\sqrt{b-az}$ posita $AB=z$ & fluxione ejus unitate. Nam est $AT=\frac{a}{b\psi}$(739) $=\frac{a}{b}\sqrt{z}$, ejusq; fluxio $\frac{a}{2b\sqrt{z}}$, cujus dimidium ductum in altitudinem $\beta\delta$ seu $\sqrt{-a+\frac{b}{z}}$ est fluxio areæ $\alpha\delta T$ descriptæ per tangentem δT. Quare fluxio illa est $\frac{a}{4bz}\sqrt{b-az}$, atq; hæc applicata ad unitatem fit fluxio incedentis BD. Hujus quadrato $\frac{aab-a^3z}{16bbzz}$ adde 1 quadratum fluxionis ipsius AB et prodit $\frac{aab-a^3z+16bbzz}{16bbzz}$, cujus radix $\frac{1}{4bz}\sqrt{aab-a^3z+16bbzz}$ est fluxio curvæ VD. Est autem hæc fluxio primæ speciei
||[113] sexti ordinis posterioris Catalogi, collatisq; terminis ||exeunt $\frac{1}{4b}=d$, $aab=e$, $-a^3=f$, $16bb=g$, adeoq; $z=x$, & $\sqrt{aab-a^3x+16bbxx}=v$ (æquatio ad unam Conicam sectionem, puta HG, cujus area $EFGH$ sit s, existente $EF=x$ & $FG=v$:) Item $\frac{1}{z}=\xi$ & $\sqrt{16bb-a^3\xi+a[a]b\xi\xi}=\Upsilon$ (æquatio ad aliam Conicam sectionem, puta ML, cujus area $IKLM$ sit σ, existente $IK=\xi$ & $KL=\Upsilon$:) Deniq; $\frac{2aabb\xi\Upsilon-a^3b\Upsilon-a^4v-4aabb\sigma-32abbs^{(740)}}{64b^4-2a^4}=t$.

Quare ut curvæ VD portionis cujuscunq; Dd longitudo noscatur, demitte db normalem ad AB fingeq; $Ab=z$ & exinde per jam inventa quære t, dein finge $AB=z$ et exinde etiam quære t & horum duorum t differentia erit longitudo Dd.(741)

(738) In other words, AB is taken as base variable.

(739) This is evident geometrically since $\beta\delta$ is the polar of T with respect to the hyperbola $\alpha\delta$ of centre A and consequently $AT\times A\beta = A\alpha^2$ (or a/b). Analytically, $AT = \psi-\phi.d\psi/d\phi$ where $\beta\delta = \phi = \sqrt{[b\psi^2-a]}$.

(740) A double error creeps in here. As we have seen in note (564) above, in Newton's value for the 'area' given in Ordo 6.1 of the 'Catalogus posterior' on his page 81 the terms '$-2dfgxv+4degv$' were accidentally omitted and the mistake has been carried through; correctly, the terms '$+16a^2b^2xv+32ab^3v$' should, after allowing for Newton's silent division by $\frac{1}{2}a^2/b$, be added to the present numerator. In the second place, Newton has silently divided through his original denominator by a^2/b; the present one, in line with the division of the numerator by only half this divisor, should be doubled to read '$128b^4-2a^4$'. Although both Horsley and Colson correct Newton's mistake in Ordo 6.1 of the latter catalogue of integrals,

length VD I seek the fluxion of the area $\alpha\delta T\alpha$ as AB flows uniformly[738] and find it to be $\frac{1}{4}a\surd[b-az]/bz$ on setting $AB=z$ and its fluxion unity. For $AT=a/b\psi$[739] $=(a/b)\surd z$ and so its fluxion is $\frac{1}{2}a/b\surd z$, half of which multiplied into the altitude $\beta\delta$ (that is, $\surd[-a+b/z]$) is the fluxion of the area described by the tangent δT. Hence that fluxion is $\frac{1}{4}a\surd[b-az]/bz$ and this divided by unity becomes the fluxion of the ordinate BD. To its square $\frac{1}{16}a^2(b-az)/b^2z^2$ add 1 (the square of the fluxion of AB) and there results $\frac{1}{16}(a^2b-a^3z+16b^2z^2)/b^2z^2$, whose root $\frac{1}{4}b^{-1}z^{-1}\surd[a^2b-a^3z+16b^2z^2]$ is the fluxion of the curve VD. This is a fluxion of the first species of the sixth order in the latter catalogue, and when terms are compared there proves to be $\frac{1}{4}b^{-1}=d$, $a^2b=e$, $-a^3=f$, $16b^2=g$, so that $z=x$ and $\surd[a^2b-a^3x+16b^2x^2]=v$ (the equation to a conic, say HG, whose area $EFGH$ is s, where $EF=x$ and $FG=v$); likewise, $z^{-1}=\xi$ and

$$\surd[16b^2-a^3\xi+a^2b\xi^2]=\Upsilon$$

(the equation to a second conic, say ML, whose area $IKLM$ is σ, where $IK=\xi$ and $KL=\Upsilon$). Finally, $\dfrac{2a^2b^2\xi\Upsilon-a^3b\Upsilon-a^4v-4a^2b^2\sigma-32ab^2s}{64b^4-2a^4}=t.$[740]

Therefore, to know the length of any portion Dd of the curve VD, let fall db normal to AB, suppose $Ab=z$ and seek t therefrom by what has now been found; then suppose $AB=z$ and again seek t from that: the difference of these two evaluations of t will be the length Dd.[741]

---

neither notices the present double confusion: compare their *Opera Omnia*, **1** (note (60)): 517 and *Method of Fluxions* (note (79)): 139 respectively.

(741) Some guile has gone into constructing this example. It will be clear that Newton wanted to give a rectification which involved recourse to one of the more complex integrals in his 'Catalogus posterior': for this, Ordo 6.1 is an obvious choice, yielding the rectification $t=\int^z Az^{-1}\surd[B+Cz+Dz^2].dz$ of a corresponding curve $y=\int^z Az^{-1}\surd[B+Cz+(D-A^2)z^2].dz$. Evidently, on making (with Newton) the simplification $D=A^2$, the curve

$$y=\int^z Az^{-1}\surd[B+Cz].dz$$

has to be constructed: that is, on substituting successively $z=1/u$ and $u=\phi^2$,

$$y=-\int^{1/\phi^2} 2A\phi^{-1}\surd[B+C\phi^{-2}].d\phi=\int^{1/\phi^2} 2A\surd[B\phi^2+C].d[1/\phi].$$

Clearly, if we construct the central conic $\psi^2=B\phi^2+C$, then $\phi-\psi.d\phi/d\psi=-C/B\phi$, so that the element of the area $T\alpha\delta$ will be $\frac{1}{2}\psi.d[\phi-\psi.d\phi/d\psi]=-\frac{1}{2}(C/B)\psi.d[1/\phi]$ and therefore $(T\alpha\delta)=(-\frac{1}{4}C/AB)\times y$. It is then natural to take $C=-a$, $B=b$ and $A=-\frac{1}{4}C/B=a/4b$, yielding $(T\alpha\delta)=y$.

EXEMPL 8.(742) Proponatur Hyperbola ad quam æquatio est $\sqrt{aa+bzz}=y$ et inde per Prob 1 elicietur $n=\frac{bz}{y}$ seu $\frac{bz}{\sqrt{aa+bzz}}$, cujus quadrato adde 1 & summæ radix erit $\sqrt{\frac{aa+bzz+bbzz}{aa+bzz}}=q$. Hanc fluxionem cùm non reperiatur in tabulis reduco in infinitam seriem, & primò per divisionem evadit $q=\sqrt{1+\frac{bb}{aa}zz-\frac{b^3}{a^4}z^4+\frac{b^4}{a^6}z^6-\frac{b^5}{a^8}z^8}$ &c[,](743) dein per extractionem radicis $q=1+\frac{bb}{2aa}zz-\frac{+4b^3+b^4}{8a^4}z^4+\frac{8b^4+4b^5+b^6}{16a^6}z^6$ &c. Et hinc per Prob 2 obtinetur t seu longitudo Hyperbolæ $=z+\frac{bb}{6aa}z^3-\frac{+4b^3+b^4}{40a^4}z^5+\frac{8b^4+4b^5+b^6}{112a^6}z^7$ &c.

Quod si Ellipsis $\sqrt{aa-bzz}=y$ proponatur debet signum ipsius b ubiq; mutari & habebitur $z+\frac{bb}{6aa}z^3+\frac{4b^3-b^4}{40a^4}z^5+\frac{8b^4-4b^5+b^6}{112a^6}z^7$ &c(744) pro longitudine ejus. Et posita insuper unitate pro b, emerget $z+\frac{z^3}{6aa}+\frac{3z^5}{40a^4}+\frac{5z^7}{112a^6}$ &c pro longitudine circuli: cujus seriei numerales coefficientes in infinitum inveniuntur multiplicando continuo per terminos hujus progressionis,

$$\frac{1\times1}{2\times3}\cdot\frac{3\times3}{4\times5}\cdot\frac{5\times5}{6\times7}\cdot\frac{7\times7}{8\times9}\cdot\frac{9\times9}{10\times11}\quad\&c.^{(745)}$$

||[114] ||EXEMPL. 8.(746) Proponatur deniq; Quadratrix VDE cujus vertex est V, existente A centro et AV semidiametro circuli interioris ad quem aptatur, atq; angulo VAE recto. Acta jam recta qualibet AKD secante circulum istum in K et Quadratricem in D, demissisq; ad AE normalibus KG, DB; dic AV a, AG z, VK x, et BD y, eritq; ut in superiore Exemplo, $x=z+\frac{z^3}{6aa}+\frac{3z^5}{40a^{[4]}}+\frac{5z^7}{112a^6}$ &c. Extrahe radicem

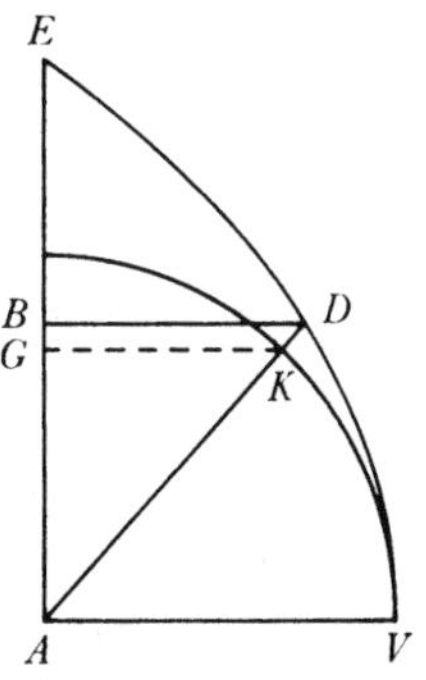

(742) This, the first of two concluding Examples '8' in which Newton abandons finite integration methods for expansions in infinite series, develops a remark in his earlier *De Analysi* (II: 216) that the quadrature of [q (or t) $=$] $\sqrt{[(1+ax^2)/(1-bx^2)]}$ 'dat longitudinem curvæ Ellipticæ', namely $y^2=(a+b)\,b^{-2}-(a+b)\,b^{-1}x^2$.

(743) Namely, on expanding $q^2=1+a^{-2}b^2z^2/(1+a^{-2}bz^2)$ as

$$1+a^{-2}b^2z^2(1-a^{-2}bz^2+a^{-4}b^2z^4-\ldots).$$

(744) That is, with $-b$ substituted for b in the previous result. Newton communicated an equivalent rectification of the ellipse $y^2=r^2-(r/c)\,x^2$ in his *epistola prior* of 13 June 1676 to Oldenburg for Leibniz, carrying the expansion through to $O(x^{13})$ (*Correspondence of Isaac Newton*, 2 (1960): 26); later, in the following paragraph, he added that 'Quæ autem de Ellipsi dicta sunt, omnia facilè accommodantur ad Hyperbolam: mutatis tantum signis

EXAMPLE 8.(742) Let there be proposed a hyperbola whose equation is $\surd[a^2+bz^2]=y$. From this by Problem 1 will be elicited $\dot{y}=bz/y$, that is, $bz/\surd[a^2+bz^2]$, to the square of which add 1 and the root of the sum will be $\sqrt{\dfrac{a^2+(b+b^2)\,z^2}{a^2+bz^2}}=\dot{t}$. Since this fluxion is not to be found in the tables I reduce it to an infinite series. First, by division there results

$$\dot{t}=\surd[1+(b^2/a^2)\,z^2-(b^3/a^4)\,z^4+(b^4/a^6)\,z^6-(b^5/a^8)\,z^8\ldots],^{(743)}$$

and then by extracting the root

$$\dot{t}=1+\frac{b^2}{2a^2}z^2-\frac{4b^3+b^4}{8a^4}z^4+\frac{8b^4+4b^5+b^6}{16a^6}z^6\ldots.$$

Hence, by Problem 2, is obtained t, that is, the hyperbola's length, equal to

$$z+\frac{b^2}{6a^2}z^3-\frac{4b^3+b^4}{40a^4}z^5+\frac{8b^4+4b^5+b^6}{112a^6}z^7\ldots.$$

But should the ellipse $\surd[a^2-bz^2]=y$ be proposed, the sign of b should everywhere be changed and there will be had

$$z+\frac{b^2}{6a^2}z^3+\frac{4b^3-b^4}{40a^4}z^5+\frac{8b^4-4b^5+b^6}{112a^6}z^7\ldots^{(744)}$$

for its length. When in addition b is put as unity, there will emerge

$$z+\tfrac{1}{6}z^3/a^2+\tfrac{3}{40}z^5/a^4+\tfrac{5}{112}z^7/a^6\ldots$$

for the length of a circle arc. The numerical coefficients of this series are found perpetually by continually multiplying by the terms of this progression:

$$\frac{1\times1}{2\times3},\quad\frac{3\times3}{4\times5},\quad\frac{5\times5}{6\times7},\quad\frac{7\times7}{8\times9},\quad\frac{9\times9}{10\times11},\quad\ldots.^{(745)}$$

EXAMPLE [9]. Lastly, let there be proposed the quadratrix VDE of vertex V, A being its centre, AV the radius of the inner circle to which it is adapted and the angle $V\widehat{A}E$ right. Drawing now any straight line AKD cutting that circle in K and the quadratrix in D and letting fall the normals KG, DB to AE, call $AV=a$, $AG=z$, $\widehat{VK}=x$ and $BD=y$. There will then, as in the previous example, be $x=z+\frac{1}{6}z^3/a^2+\frac{3}{40}z^5/a^4+\frac{5}{112}z^7/a^6+\ldots$. Extract the root z and there will emerge

ipsorum c...' (*ibid.*: 27). On adjusting Newton's 1676 expansion by the substitutions $r\to a$, $c\to a/b$ the next two terms in the present series prove to be

$$\frac{64b^5-48b^6+24b^7-5b^8}{1152a^8}+\frac{128b^6-128b^7+96b^8-40b^9+7b^{10}}{2816a^{10}}\quad\text{'\&c.'}$$

(745) This particular case, which yields the series expansion of $a\sin^{-1}(z/a)$, is developed at length from the circle's defining equation $y^2=a^2-z^2$ in the *De Analysi* of the year before. (See II: 232. The form of the coefficients is discussed shortly afterwards on II: 238.)

(746) Read '9' to preserve the numerical sequence.

z et emerget $z = x - \frac{x^3}{6aa} + \frac{x^5}{120a^4} - \frac{x^7}{5040a^6} +$ &c.[(747)] Cujus quadratum aufer de AK^q & residui radix $a - \frac{xx}{2a} + \frac{x^4}{24a^3} - \frac{x^6}{720a^5} +$ &c[(748)] erit GK. Jam cùm ex natura Quadratricis sit $AB = VK$ sive x, sitq; etiam $AG.GK::AB.BD(y)$,[(749)] divide $AB \times GK$ per AG et orietur $y = a - \frac{xx}{3a} - \frac{x^4}{45a^3} - \frac{2x^6}{945a^5}$ &c. Et inde per Prob: 1, $n = -\frac{2x}{3a} - \frac{4x^3}{45a^3} - \frac{4x^5}{315a^5}$ &c. Cujus quadrato adde 1 et summæ radix erit $1 + \frac{2xx}{9aa} + \frac{14x^4}{405a^4} + \frac{604x^6}{127575a^6}$ &c $= q$. Unde per Prob 2 obtinebitur t seu Quadratricis arcus $VD = x + \frac{2x^3}{27aa} + \frac{14x^5}{2025a^4} + \frac{604x^7}{893025a^6}$ [&c].[(750)]

§3. AN ADDENDUM ON THE THEORY OF GEOMETRICAL FLUXIONS[(1)]

From the original in the University Library, Cambridge[(2)]

[(3)][Hac methodo probandi curvas per æqualitatem vel datam rationem momentorum æquales esse vel datam rationem habere hic usus sum quòd cum methodis in his rebus usitatis affinitatem habeat: sed magis naturalis videtur quæ genesi superficierum ex fluendi motu] innititur: quæq; magis perspicua

(747) Inversion of the series expansion of $x = a\sin^{-1}(z/a)$ yields that of $z = a\sin(x/a)$. Compare the equivalent 'root extraction' in the *De Analysi* (II: 236).

(748) That is, $\sqrt{[AK^2 - AG^2]} = a\cos(x/a)$.

(749) Compare II, 2, 3: note (121). In terms of the perpendicular coordinates

$$AB(=\widehat{VK}) = x, \quad BD = y$$

the Cartesian equation of the present quadratrix is $y = x\cot(x/a)$, the right-hand side of which Newton proceeds to expand as an infinite series.

(750) This concluding example elaborates a remark in the *De Analysi* (II: 240) that 'Sic longitudo Quadratricis *VD*, licet calculo difficiliori, determinabilis est'. Newton communicated his present series evaluation of the length of the arc $\widehat{VD}$, together with the comparable expansions for the area *AVDB* (see II: 240) and the subtangent

$$y - x.dy/dx - a = (x^2/a)\,\text{cosec}^2(x/a) - a,$$

in his *epistola posterior* of 13 June 1676 to Oldenburg for Leibniz (*Correspondence of Isaac Newton*, **2** (1960): 28).

What further problems (if any) were scheduled to follow in the plan of the complete fluxional tract whose draft here abruptly terminates we can only hazard on the unsafe basis of the programme reproduced in Section 1 above (pages 28–30) and a late autobiographical remark (ULC. Add. 3968.9: 97^r), intended for insertion in an English draft of Newton's

$z = x - \frac{1}{6}x^3/a^2 + \frac{1}{120}x^5/a^4 - \frac{1}{5040}x^7/a^6 + \ldots$.[(747)] Take the square of this from AK^2 and the root, $a - \frac{1}{2}x^2/a + \frac{1}{24}x^4/a^3 - \frac{1}{720}x^6/a^5 + \ldots$,[(748)] of the remainder will be GK. Now since by the nature of the quadratrix $AB = \widehat{VK}$ (or x) and further $AG:GK = AB:BD$ (or y),[(749)] divide $AB \times GK$ by AG and there will arise

$$y = a - \tfrac{1}{3}x^2/a - \tfrac{1}{45}x^4/a^3 - \tfrac{2}{945}x^6/a^5 \ldots.$$

From this by Problem 1 comes $\dot{y} = -\frac{2}{3}x/a - \frac{4}{45}x^3/a^3 - \frac{4}{315}x^5/a^5 \ldots$, to whose square add 1 and the root of the sum will be

$$1 + \tfrac{2}{9}x^2/a^2 + \tfrac{14}{405}x^4/a^4 + \tfrac{604}{127575}x^6/a^6 \ldots = t.$$

Hence by Problem 2 will be obtained t, that is, the arc $\widehat{VD}$ of the quadratrix, equal to $x + \frac{2}{27}x^3/a^2 + \frac{14}{2025}x^5/a^4 + \frac{604}{893025}x^7/a^6 \ldots$.[(750)]

Translation

[(3)][I have here used this method of proving that curves are equal or have a given ratio by means of the equality or given ratio of their moments since it has an affinity to the ones usually employed in these cases. However, that] based on [the genesis of surfaces by their motion of flow appears a more natural approach,]

---

unpublished *Historia Methodi Infinitesimalis ex Epistolis antiquis eruta* (1717?), that 'There wanted that part wch related to the solution of Problemes not reducible to Quadratures'. See the introduction to the present Part 1.

(1) This brief systematic account of geometrical fluxions was, we presume, originally inserted in the preceding 1671 tract as a substantially augmented replacement for Newton's pages 92–4, but was evidently already detached from it by about 1710 when William Jones copied the latter text (under the title *Artis Analyticæ Specimina sive Geometria Analytica*) since he did not transcribe the present revise (compare §1: notes (2) and (3)). Following Jones' lead both Colson in his 1736 English version (§3: note (79)) and Horsley in 1779 in his first publication of the original Latin text (§2: note (60)) likewise omit this addendum—hardly by design in either case though, to be sure, insertion of this somewhat discursive geometrical paper would seriously unbalance the compact, carefully controlled flow of the essentially analytical tract which sired it. Apart from a brief, not very satisfactory notice by Alexander Witting in 1911 (in his 'Zur Frage der Erfindung des Algorithmus der Newtonschen Fluxionsrechnung', *Bibliotheca Mathematica*, $_3$**12** (1911–12): 56–60, especially 58) where Theorem 1 following is accurately located as the source of the celebrated 'fluxional' Lemma II of the second book of Newton's *Principia* (see note (11) below), the present manuscript has been ignored by modern scholars.

(2) ULC. Add. 3960.4: 33–46. The handwriting of this autograph manuscript is indistinguishable from that of the preceding tract, and we may safely assume that it was written when Newton came to revise his 1671 fluxional treatise in the winter of 1671/2.

(3) The text begins '—innititur:...' without indentation, but we have repeated the whole of the preceding paragraph from page 92 of the preceding tract (see §2: note (638)) to emphasize the continuity between the two.

et ornata evadet si fundamenta quædam pro more methodi syntheticæ præsternantur; qualia sunt hæc.

Axiomata

Ax. 1. Quæ fluxionibus æqualibus simul generantur sunt æqualia.

Ax. 2. Quæ fluxionibus in data ratione simul generantur, sunt in ratione fluxionum.

Nota, simul generari intelligo quæ tota eodem tempore generantur.(4)

Ax. 3. Fluxio totius æquatur fluxionibus partium simul sumptis.

Ubi nota quod profluxiones affirmativè ac defluxiones negativè ponendæ sint.(5)

|| *Ax. 4.* Fluxio major est quæ(6) majus producit.

Ax: 4. Momenta contemporanea sunt ut fluxiones.(7)

Theoremata.(8)

Th. 1. Positis quatuor perpetuò proportionalibus fluentibus quantitatibus: summa extremarum reciprocè ductarū in suas fluxiones æquatur summæ mediarum reciprocè ductarum in suas fluxiones.

Sint $A.B::C.D$ et erit $A\times \text{fl}:D+D\times \text{fl}:A=B\times \text{fl}:C+C\times \text{fl}:B$. Id quod eodem modo demonstrari potest ac solutio Prob. 1;(9) vel etiam sic. Esto M momentum ipsius A,(10) et cùm momenta fluentium sint ut fluxiones, erit $\frac{\text{fl}:B}{\text{fl}:A}M$ momentum ipsius B, $\frac{\text{fl}:C}{\text{fl}:A}M$ momentum ipsius C et $\frac{\text{fl}:D}{\text{fl}:A}M$ momentum ipsius D. Quare ubi A profluendo evadit $A+M$, B evadet $B+\frac{\text{fl}:B}{\text{fl}:A}M$, C evadet $C+\frac{\text{fl}:C}{\text{fl}:B}M$, ac D evadet $D+\frac{\text{fl}:D}{\text{fl}:A}M$;(11) hac lege scilicet ut hæ quantitates

(4) In other words, Newton presupposes that all the fluent magnitudes are simultaneously zero.

(5) A fundamental axiom: if $A = B\pm C$, then $\text{fl}(A) = \text{fl}(B)\pm\text{fl}(C)$.

(6) Newton has cancelled in continuation 'ex minori facit æquale aut ex æquali majus' (out of a lesser magnitude makes an equal one or out of an equal magnitude makes a greater one). Understanding that the revised axiom presupposes equal initial magnitudes, the two definitions are readily seen to be equivalent.

(7) A first draft in the margin at this point reads equivalently 'Fluxiones sunt ut momenta contemporanea fluxionibus istis generata'. (Newton has cancelled 'continuè generata' (continuously generated).) This fundamental observation opens the way to subsuming limit-increment arguments as fluxional ones, and conversely so.

(8) Universally in the sequel, both in footnotes and English translation, we have used the notation '$\text{fl}(X)$' to represent Newton's 'fluxio X' and its contracted form '$\text{fl}: X$'.

(9) Of the preceding fluxional tract, that is. Thus, on taking the fluxions of A, B, C, D to be a, b, c, d respectively and o to be the increment of 'time', it follows that $A\times D = B\times C$ and so

one which will come to be still more perspicuous and resplendent if certain foundations are, as is customary with the synthetic method, first laid. Such as these.

Axioms

Axiom 1. Magnitudes generated simultaneously by equal fluxions are equal.

Axiom 2. Magnitudes generated simultaneously by fluxions in given ratio are in the ratio of the fluxions.

Note: by simultaneous generation I understand that the wholes are generated in the same time.(4)

Axiom 3. The fluxion of a whole is equal to the fluxions of its parts taken together.(5)

Here note that increasing fluxions are to be set positive, decreasing ones negative.

|| *Axiom 4.* The greater fluxion is that which (6) produces the greater magnitude.

Axiom 4. Contemporaneous moments are as their fluxions.(7)

Theorems(8)

Theorem 1. Where four perpetually proportional fluent quantities are supposed, the sum of the products of each extreme and the other's fluxion is equal to the sum of the products of each of the middle ones with the other's fluxion.

Let there be $A:B = C:D$, and then

$$A \times \text{fl}\,(D) + D \times \text{fl}\,(A) = B \times \text{fl}\,(C) + C \times \text{fl}\,(B).$$

This can be demonstrated in the same manner as the solution of Problem 1,(9) or, alternatively, this way. Let M be the moment of A:(10) then, since the moments of fluents are as their fluxions, $[\text{fl}\,(B)/\text{fl}\,(A)]\,M$ will be B's moment, $[\text{fl}\,(C)/\text{fl}\,(A)]\,M$ that of C and $[\text{fl}\,(D)/\text{fl}\,(A)]\,M$ that of D. Hence when A, by flowing on, comes to be $A+M$, B will come to be $B+[\text{fl}\,(B)/\text{fl}\,(A)]\,M$, C will become $C+[\text{fl}\,(C)/\text{fl}\,(A)]\,M$ and D will be $D+[\text{fl}\,(D)/\text{fl}\,(A)]\,M$,(11) with the

---

$(A+oa)(D+od) = (B+ob)(C+oc)$; hence, on eliminating the equal products $A \times B$, $C \times D$ and dividing by o, we deduce in Fermatian style that $Ad+Da = Bc+Cb+O(o)$, from which Newton's result follows in the limit as o vanishes.

(10) In the terminology of the previous note $M = oa$. It follows directly that the moment of B is $ob = (b/a).oa$ and similarly for the moments of C and D.

(11) A cancelled continuation reads '& similiter ubi defluendo A evadit $A-M$, B evadet $B-\dfrac{\text{fl}:B}{\text{fl}:A}M$[,] C evadet $C-\dfrac{\text{fl}:C}{\text{fl}:A}M$ ac D evadet $D-\dfrac{\text{fl}:D}{\text{fl}:A}M$' (and similarly when A, by flowing back, comes to be $A-M$, B will come to be $B-[\text{fl}\,(B)/\text{fl}\,(A)]\,M$, C will become

$$C-[\text{fl}\,(C)/\text{fl}\,(A)]\,M$$

and D will be $D-[\text{fl}\,(D)/\text{fl}\,(A)]\,M$). This 'obvious' generalization is not, in fact, accurate unless the moment M is infinitely small: a condition which cannot be made at this stage of

etiamnum maneant proportionales. Duc ergo extrema et media in se & provenient

$$AD+\frac{\text{fl}:D}{\text{fl}:A}AM+DM+\frac{\text{fl}:D}{\text{fl}:A}MM$$
$$=BC+\frac{\text{fl}:C}{\text{fl}:A}BM+\frac{\text{fl}:B}{\text{fl}:A}CM+\frac{\text{fl}:B\times\text{fl}:C}{\text{fl}:A\times\text{fl}:A}MM.^{(12)}$$

Aufer æqualia AD & BC, & restantia æqualia duc in fl:A ac divide per M. Emerget enim

$$A\times\text{fl}:D+D\times\text{fl}:A+\text{fl}:D\times M=B\times\text{fl}:C+C\times\text{fl}:B+\frac{\text{fl}:B\times\text{fl}:C}{\text{fl}:A}M.$$

Ubi propter infinitam parvitatem momenti M rejectis terminis per id multiplicatis [prodit Theorema.][(13)]

Sint $AB.AD::AE.AC$,[(14)] et erit

$$AB\times\text{fl}:AC+AC\times\text{fl}:AB$$
$$=AD\times\text{fl}:AE+AE\times\text{fl}:AD.$$

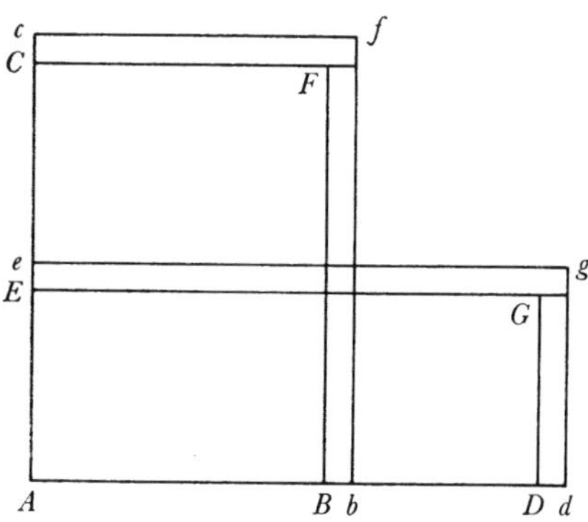

Nam augeantur hæ lineæ momentis suis Bb, Dd, Ee, Cc fluendo, & propter perpetuam earum proportionalitatem, adeoq rectangula ab extremis et medijs constituta perpetuo æqualia nempe $AF=AG$ & $Af=Ag$; augmenta rectangulorum istorum $BFCf$ & $DGEg$ æqualia erunt: hoc est

$$Ab\times Cc(Cf)+AC\times Bb(bF)=Ad\times Ee(Eg)+AE\times Dd(dG).$$

Cùm autem fluxiones sint ut momenta quantitatum ab istis continuò descripta. hoc est fl:$AC.Cc$::fl:$AB.Bb$. &c erit

$$Ab\times\text{fl}:AC.Ab\times Cc::A\times C\text{fl}:AB.AC\times Bb::$$
$$Ab\times\text{fl}:AC+AC\times\text{fl}:AB.Ab\times Cc+AC\times Bb.$$

---

argument since a subsequent division by the quantity M is required. (If in equal increments of 'time' o the variables A, B are augmented to $A+oa$, $B+ob$ and decrease to $A-ob$, $B-ob$, we deduce, on taking $B=f(A)$, that $f(A+oa)-f(A) = ob = f(A)-f(A-oa)$ and so

$$2f(A) = f(A+oa)+f(A-oa):$$

true only when $f(A) = B = kA+l$, where k, l are constants.) If we accept that Newton was conscious of this *non sequitur* in 1671, he had certainly forgotten it when he came to redraft the present theorem as Lemma II of the second book of his *Philosophiæ Naturalis Principia Mathematica* (London, 1687): 250–3, for in Case 2 of that lemma he tried, by an appeal to exactly

stipulation, namely, that these quantities still remain proportional. So multiply extremes and middles into each other and there will result

$$AD+\frac{\text{fl}\,(D)}{\text{fl}\,(A)}AM+DM+\frac{\text{fl}\,(D)}{\text{fl}\,(A)}M^2$$
$$=BC+\frac{\text{fl}\,(C)}{(\text{fl}\,A)}BM+\frac{\text{fl}\,(B)}{\text{fl}\,(A)}CM+\frac{\text{fl}\,(B)\times\text{fl}\,(C)}{[\text{fl}(A)]^2}M^2.^{(12)}$$

Take away the equals AD and BC, and the remainders, likewise equal, multiply by fl (A) and divide by M. There will emerge, in fact,

$$A\times\text{fl}\,(D)+D\times\text{fl}\,(A)+\text{fl}\,(D)\times M = B\times\text{fl}\,(C)+C\times\text{fl}\,(B)+\frac{\text{fl}\,(B)\times\text{fl}\,(C)}{\text{fl}\,(A)}\times M.$$

When, because of the infinite smallness of the moment M, terms multiplied by it are here rejected, [the theorem follows].(13)

Let $AB:AD = AE:AC$,(14) and there will be

$$AB\times\text{fl}\,(AC)+AC\times\text{fl}\,(AB) = AD\times\text{fl}\,(AE)+AE\times\text{fl}\,(AD).$$

For let these lines increase fluxionally by their respective moments Bb, Dd, Ee and Cc: then, because of their perpetual proportionality and so of the perpetual equality of the rectangles formed of the extremes and middles (specifically, $\Box AF = \Box AG$ and $\Box Af = \Box Ag$), the increments $BFC[c]f$ and $DGE[e]\,g$ of those rectangles will be equal; that is,

$$Ab\times Cc\ (\text{or}\ \Box Cf)+AC\times Bb(\Box bF) = \text{A}d\times Ee(\Box Eg)+AE\times Dd(\Box dG).$$

But since fluxions are as the moments of quantities continually described by them, that is, $\text{fl}\,(AC):Cc = \text{fl}\,(AB):Bb$ and so on, there will be

$$Ab\times\text{fl}\,(AC):Ab\times Cc = AC\times\text{fl}\,(AB):AC\times Bb$$
$$= (Ab\times\text{fl}\,(AC)+AC\times\text{fl}\,(AB)):(Ab\times Cc+AC\times Bb).$$

---

the present generalization, to do away with the inconvenient final step of making M infinitely small. Recasting his later argument in present terminology, we would 'deduce' that

$$AD-\frac{\text{fl}\,(D)}{\text{fl}\,(A)}AM-DM+\frac{\text{fl}\,(D)}{\text{fl}\,(A)}M^2 = BC-\frac{\text{fl}\,(C)}{\text{fl}\,(A)}BM-\frac{\text{fl}\,(B)}{\text{fl}\,(A)}CM+\frac{\text{fl}\,(B)\times\text{fl}\,(C)}{[\text{fl}\,(A)]^2}M^2,$$

from which Newton's desired result follows when it is subtracted from the previous (correct) exact equality and the remainder is divided by $2M/\text{fl}\,(A)$. Compare D. T. Whiteside, 'Patterns of Mathematical Thought in the later Seventeenth Century', *Archive for History of Exact Sciences*, **1** (1961): 179–388, especially 373.

(12) A following cancelled line displays the (incorrect) corresponding form, cited in the previous note, which results on substituting $-M$ for M in this equality.

(13) It will be clear that Newton here abandons this entirely valid analytical proof merely to recast it in geometrical form, and not at all because he is unsure of its cogency.

(14) Accordingly, BE is parallel to DC. As context requires, we have redrawn Newton's accompanying figure (whose original form may be seen in Plate II), making all its angles right.

Et eodem ratiocinio $Ad\times \text{fl}:AE+AE\times \text{fl}:AD$ erit ad $Ad\times Ee+AE\times Dd$ in eadem ratione. Quare $Ab\times \text{fl}\,AC+AC\times \text{fl}:AB$ & $Ad\times \text{fl}:AE+AE\times \text{fl}\,AD$, cùm sint in eadem ratione ad æqualia, etiam æqualia erunt.(15)

Sive $Ab+AC\times\dfrac{Bb}{Cc}=Ad\times\dfrac{Ee}{Cc}+AE\times\dfrac{Dd}{Cc}$. Adeoqꝫ cùm fluxiones sint ut momenta quantitatum ab istis continuò generata, hoc est $\dfrac{Bb}{Cc}=\dfrac{\text{fl}\,AB}{\text{fl}\,AC}$, $\dfrac{Ee}{Cc}=\dfrac{\text{fl}\,AE}{\text{fl}\,AC}$ & $\dfrac{Dd}{Cc}=\dfrac{\text{fl}\,AD}{\text{fl}\,AC}$, erit $Ab+AC\times\dfrac{\text{fl}\,AB}{\text{fl}\,AC}=Ad\times\dfrac{\text{fl}\,AE}{\text{fl}\,AC}+AE\times\dfrac{\text{fl}\,AD}{\text{fl}\,AC}$. Sive

$$Ab\times \text{fl}\,AC+AC\times \text{fl}\,AB=Ad\times \text{fl}\,AE+AE\times \text{fl}\,AD.$$

Decrescant jam rectangula Af et Ag donec in prima rectangula AF & AG redierint, & tunc Ab evadet AB atqꝫ Ad evadet AD. Quare in ultimo istius infinitè parvæ defluxionis momento,(16) hoc est in primo momento(16) fluxionis quadrangulorum AF et AG quando incipiunt augeri vel diminui, erit

$$AB\times \text{fl}:AC+AC\times \text{fl}:AB=AD\times \text{fl}\,AE+AE\times \text{fl}\,AD.$$

Q.E.D.(17)

Cor: 1. Positis tribus continuè proportionalibus, summa extremarum reciproce ductarum in suas fluxiones æquatur duplo mediæ ductæ in suam fluxionem. Sint $A.B::B.C$ et erit

* Theor 1. $$A\times \text{fl}:C+C\times \text{fl}:A(={}^*B\times \text{fl}:B+B\times \text{fl}\,B)=2B\times \text{fl}\,B.^{(18)}$$

Cor 2. Positis tribus continuè proportionalibus $A.B.C.$ si summa extremarum sit data quantitas, fluxio minoris extremæ(19) erit ad fluxionem mediæ ut duplum mediæ ad differentiam extremarum. $2B.\ C-A::\text{fl}\,A.\ \text{fl}\,B$. Nam cùm

* Ax 3. $A+C$ ex Hypothesi non fluat, *erit $\text{fl}:A+\text{fl}:C=0$, sive $\text{fl}\,C=-\text{fl}\,A$. adeoqꝫ

* Cor 1. $A\times \text{fl}:C=-A\times \text{fl}\,A$. Quare $\overline{C-A}\times \text{fl}\,A(=A\times \text{fl}:C+C\times \text{fl}\,A)^*=2B\times \text{fl}\,B$, hoc est $2B.\overline{C-A}::\text{fl}\,A.\text{fl}\,B$.

(15) The argument of this cancelled passage, replaced by the marginal revise which follows immediately upon it in our reproduction of the text, is completed by the two concluding sentences of this paragraph, 'Decrescant jam rectangula....Q.E.D.' Compare Plate I.

(16) Since (by Axiom 4) 'momenta contemporanea sunt ut fluxiones' we have here a first statement of the celebrated Newtonian *dictum* that the fluxions of quantities are in the 'prime' or 'ultimate' ratio of their evanescent moments: 'Fluxiones sunt quam proxime ut Fluentium augmenta æqualibus temporis particulis quam minimis genita, & ut accurate loquar, sunt in prima ratione augmentorum nascentium.... Eodem recidit si sumantur fluxiones in ultima ratione partium evanescentium'. (See the 'Introductio' to Newton's 1693 *Tractatus de Quadratura Curvarum*, first published as the second of two mathematical appendices to Newton's *Opticks: or, A Treatise of the Reflexions, Refractions, Inflexions and Colours of Light* (London, 1704): $_2$163–211, especially 166.)

(17) If, as in Newton's cancelled draft, we set $AB = A$, $AC = B$, $AD = C$ and $AE = D$ and denote their respective fluxions by a, b, c, d, then generally if

$$A:C = D:B \text{ (or } A\times B = C\times D)$$

it follows that $Ab+Ba = Cd+Dc$. This is, of course, the fundamental rule for determining the

And by the same reasoning $Ad \times \text{fl}\,(AE)+AE \times \text{fl}\,(AD)$ will be to

$$Ad \times Ee+AE \times Dd$$

in the same ratio. Hence

$$Ab \times \text{fl}\,(AC)+AC \times \text{fl}\,(AB) \quad \text{and} \quad Ad \times \text{fl}\,(AE)+AE \times \text{fl}\,(AD),$$

since they are in the same ratio to equals, will also be equal.(15)
In other words $Ab+AC \times (Bb/Cc) = Ad \times (Ee/Cc)+AE \times (Dd/Cc)$. Accordingly, since fluxions are as the moments of quantities continually generated by them, that is, $Bb/Cc = \text{fl}(AB)/\text{fl}\,(AC)$, $Ee/Cc = \text{fl}\,(AE)/\text{fl}\,(AC)$ and

$$Dd/Cc = \text{fl}\,(AD)/\text{fl}\,(AC),$$

there will be

$$Ab+AC \times [\text{fl}\,(AB)/\text{fl}\,(AC)] = Ad \times [\text{fl}\,(AE)/\text{fl}(AC)]+AE \times [\text{fl}\,(AD)/\text{fl}\,(AC)],$$

or $Ab \times \text{fl}\,(AC)+AC \times \text{fl}\,(AB) = Ad \times \text{fl}\,(AE)+AE \times \text{fl}\,(AD)$. Now let the rectangles Af and Ag diminish till they go back into the primary rectangles AF and AG: Ab will then come to be AB while Ad becomes AD. Hence at the last moment(16) of that infinitely decreasing fluxion—that is, at the first moment(16) of flux of the rectangles AF and AG when they start to increase or diminish—, there will be

$$AB \times \text{fl}\,(AC)+AC \times \text{fl}\,(AB) = AD \times \text{fl}\,(AE)+AE \times \text{fl}\,(AD).$$

As was to be proved.(17)

Cor. 1. Where three continued proportionals are supposed, the sum of the products of each of the extremes and the other's fluxion is equal to twice the middle one multiplied into its fluxion. Let $A:B = B:C$ and then

$$A \times \text{fl}\,(C)+C \times \text{fl}\,(A) = ({*}B \times \text{fl}\,(B)+B \times \text{fl}\,(\text{B}) \text{ or) } 2B \times \text{fl}\,(B).^{(18)}$$ * Th. 1.

Cor. 2. Where three continued proportionals A, B, C are supposed, if the sum of the extremes be a given quantity the fluxion of the lesser extreme(19) will be to that of the middle one as twice the middle to the difference of the extremes: $2B:(C-A) = \text{fl}\,(A):\text{fl}\,(B)$. For since by hypothesis $A+C$ is not fluent, therefore $^*\text{fl}\,(A)+\text{fl}\,(C) = 0$ or $\text{fl}\,(C) = -\text{fl}\,(A)$, so that $A \times \text{fl}\,(C) = -A \times \text{fl}\,(A)$. Hence * Ax. 3.
$(C-A) \times \text{fl}\,(A)\ (= A \times \text{fl}\,(C)+C \times \text{fl}\,(A)) = {*}2B \times \text{fl}\,(B)$, that is, * Cor. 1.

$$2B:(C-A) = \text{fl}\,(A):\text{fl}\,(B).$$

---

fluxion of a product: in particular, if $D = 1$ (and so $d = 0$), then the fluxion of $A \times B = C$ is $Ab+Ba = c$, which is virtually Newton's Corollary 7 below. The first five corollaries deal with the mildly interesting subgroup of cases in which (equivalently) $C = D$.

(18) That is, if $A \times C = B^2$, then $Ac+Ca = 2Bb$. Corollaries 2 and 3 following are the particular cases in which $A \pm C$ is constant, so that $a \pm c = 0$ and therefore $(C \mp A)\,a = 2Bb$; while Corollaries 4 and 5 deal with the cases where $A \pm B$ is constant, for which $a \pm b = 0$ and so $(2B \pm C)\,b = Ac$.

(19) The fluxion of the 'greater extreme' (C) will evidently be numerically the same as that of the lesser (A) but opposite in sign.

Cor 3. Sin differentia extremarum detur, fluxio alterutrius[20] extremæ erit ad fluxionem mediæ, ut duplum mediæ ad summam extremarum.

$$2B.A+C::\text{fl}A.\text{fl}B.$$

Demonstratur ut Cor 2.[21]

Cor 4. Quod si summa primæ et secundæ quantitatis detur, erit fluxio secundæ ad fluxionem tertiæ ut prima ad duplum secundæ auctum tertia.

$$A.2B+C::\text{fl}B.\text{fl}C.$$

* Ax 3. Nam cùm ex Hypothesi $A+B$ non fluat, *erit $\text{fl}A+\text{fl}B=0$ sive $\text{fl}B=-\text{fl}A$, adeoq; $C\times\text{fl}B=-C\times\text{fl}A$. Quare

$$A\times\text{fl}C(=2B\times\text{fl}B-C\times\text{fl}A)=\overline{2B+C}\times\text{fl}B;$$

hoc est, $A.2B+C::\text{fl}:\text{B}.\text{fl}:C.$

Cor: 5. Si deniq; differentia primæ et secundæ datur, erit fluxio alterutrius[20] ad fluxionem tertiæ ut prima ad duplum secundæ diminutum tertia.

$$A.2B-C::\text{fl}B.\text{fl}C.$$

Demonstratur ut Cor 4.[22]

Cor 6. Positis quotcunq; continuè proportionalibus, quarum una sit data quantitas & cæteræ fluentes: fluxiones fluentium erunt inter se ut fluentes illæ ductæ in numerum terminorum quibus distant a dato illo termino. Sint $A. B. C. D. E. F$ continuè proportionales et si datur C, erit

$$-2A.-B.D.2E.3F::\text{fl}A.\text{fl}:B.\text{fl}D.\text{fl}E.\text{fl}F.$$

Nam propter $C.\ D.\ E \div\!\!\div$, est $C\times\text{fl}E+E\times\text{fl}C=2D\times\text{fl}D$, per Cor 1. At ex Hypothesi $\text{fl}C=0$. Ergo $C\times\text{fl}:E=2D\times\text{fl}D$. hoc est $\text{fl}D.\text{fl}E(::C.2D)::D.2E.$

Iterum quia $D.E.F.\div\!\!\div$, erit $D\times\text{fl}:F+F\times\text{fl}:D=E\times2\text{fl}E.$ sive

$$D\times\text{fl}F=E\times2\text{fl}:E-F\times\text{fl}:D.$$

Sed e jam ostensis est $\text{fl}D.\text{fl}E(::D.2E)::E.2F.$ Ergo $F\times\text{fl}D=\frac{1}{2}E\times\text{fl}E.$ adeoq; $D\times\text{fl}F=\frac{3}{2}E\times\text{fl}E.$ et $\text{fl}:E.\text{fl}:F(::2D.3E)::2E.3F.$ Atq; ita in cæteris.[23]

Cor. 7. Si fluentes duæ quantitates se multiplicant fluxio Facti componitur ex fluxionibus factorum alterne ductis in factores. $\text{Fl}:AB=B\times\text{fl}A+A\times\text{fl}B.$ Nam $1.A::B.AB.$ Ergo per Th 1 [manifestum est].[24]

(20) The fluxional increase of two magnitudes differing by a constant is identical.

(21) Namely, 'cùm $A-C$ ex Hypothesi non fluat, erit $\text{fl}:A-\text{fl}:C=0$ sive $\text{fl}:C=\text{fl}:A$, adeoq; ... $\overline{C+A}\times\text{fl}:A=2B\times\text{fl}:B$' on replacing C by $-C$ in the previous argument.

(22) That is, on replacing B by $-B$ in the preceding corollary.

(23) If $W=C.X^p$ is the p-th term to the right of C, then its fluxion w will be $pC.X^{p-1}x$ (where x is the fluxion of X) and so in proportion to the fluxion, $qC.X^{q-1}$, of the q-th term to the right of C as 'fluentes illæ [$C.X^p$ and $C.X^q$ respectively] ductæ in numerum terminorum [namely, p and q] quibus distant a dato illo termino [C]'. Newton's proof of the basic result $w=pC.X^{p-1}x$ is essentially recursive: in modernized idiom, if for all integral n, $0\leqslant n<p$, $\text{fl}(C.X^n)=nC.X^{n-1}x$, then 'propter $C.X^{p-2}$, $C.X^{p-1}$ et $W=C.X^p\div\!\!\div$' it follows that

$$C.X^{p-2}\times w+C.X^p\times\text{fl}(C.X^{p-2}) = 2C.X^{p-1}\times\text{fl}(C.X^{p-1})$$

Cor. 3. But if the difference of the extremes be given, the fluxion of either[20] extreme will be to the fluxion of the mean as twice the mean to the sum of the extremes: $2B:(A+C) = \text{fl}(A):\text{fl}(B)$. This is proved as in Corollary 2.[21]

Cor. 4. If, however, the sum of the first and second quantities be given, the fluxion of the second will be to that of the third as the first quantity to twice the second increased by the third: $A:(2B+C) = \text{fl}(B):\text{fl}(C)$. For since by hypothesis $A+B$ is not fluent, therefore* $\text{fl}(A)+\text{fl}(B) = 0$ or $\text{fl}(B) = -\text{fl}(A)$, so that $C\times\text{fl}(B) = -C\times\text{fl}(A)$. Hence

* Ax. 3.

$$A\times\text{fl}(C)\ (= 2B\times\text{fl}(B)-C\times\text{fl}(A)) = (2B+C)\times\text{fl}(B);$$

that is, $A:(2B+C) = \text{fl}(B):\text{fl}(C)$.

Cor. 5. If, finally, the difference of the first and second quantities is given, the fluxion of either[20] will be to that of the third as the first to twice the second less the third: $A:(2B-C) = \text{fl}(B):\text{fl}(C)$. This is proved as in Corollary 4.[22]

Cor. 6. No matter how many continued proportionals are supposed, one of which is given and the rest are fluent, the fluxions of the fluents will be to one another as those fluents multiplied by the number of terms they are distant from that given term. Let A, B, C, D, E, F be in continued proportion, and if C is given, then

$$-2A:-B:D:2E:3F = \text{fl}(A):\text{fl}(B):\text{fl}(D):\text{fl}(E):\text{fl}(F).$$

For, because C, D, E are continued proportionals, by Corollary 1

$$C\times\text{fl}(E)+E\times\text{fl}(C) = 2D\times\text{fl}(D).$$

But by hypothesis $\text{fl}(C) = 0$. Therefore $C\times\text{fl}(E) = 2D\times\text{fl}(D)$; that is,

$$\text{fl}(D):\text{fl}(E)\ (= C:2D) = D:2E.$$

Again, because D, E, F are continued proportionals, therefore

$$D\times\text{fl}(F)+F\times\text{fl}(D) = E\times 2\text{fl}(E),\quad D\times\text{fl}(F) = E\times 2\text{fl}(E)-F\times\text{fl}(D).$$

But from what has already been shown $\text{fl}(D):\text{fl}(E)\ (= D:2E) = E:2F$. Consequently $F\times\text{fl}(D) = \frac{1}{2}E\times\text{fl}(E)$, so that $D\times\text{fl}(F) = \frac{3}{2}E\times\text{fl}(E)$ and

$$\text{fl}(E):\text{fl}(F)\ (= 2D:3E) = 2E:3F.$$

And so in the other cases.[23]

Cor. 7. If two fluent quantities are multiplied together, the fluxion of the product is composed of the fluxions of the factors each multiplied into the other factor: $\text{fl}(AB) = B\times\text{fl}(A)+A\times\text{fl}(B)$. For then $1:A = B:AB$ and the result is immediate by Theorem 1.[24]

---

and so $w = [2(p-1)-(p-2)]\,C^2.X^{2p-2}/C.X^{p-2}$. In seventeenth-century style Newton himself goes through this argument for $p = 2, 3$ and then leaves his reader to apply the pattern of proof thus established with an 'Atq; ita in cæteris'.

(24) The fundamental algorithm for finding the fluxion of the product of two magnitudes. Compare note (17).

Cor 8. Si fluens quantitas per fluentem quantitatem dividitur: fluxio Quoti prodit auferendo fluxionem divisoris multiplicatam per dividuum, a fluxione dividui multiplicata per divisorem & dividendo residuum per quadratum divisoris. $\text{Fl}:\frac{B}{A}=\frac{A\times \text{fl}\,B-B\times \text{fl}\,A}{AA}$. Nam $A.1::B.\frac{B}{A}$. Ergo per Th: 1, $A\times \text{fl}\frac{B}{A}+\frac{B}{A}\times \text{fl}\,A=1\times \text{fl}:B$, nam $B\times \text{fl}:1$ nihil est. Aufer utrobiꝗ $\frac{B}{A}\times \text{fl}\,A$ et residuum divide per A, et prodibit $\text{fl}:\frac{B}{A}=\frac{A\times \text{fl}\,B-B\times \text{fl}\,A}{AA}$.(25)

Cor 9. Fluxio radicis est ad fluxionem potestatis alicujus ut radix ad potestatem illam multiplicatam per numerum dimensionum. $\text{fl}:A.\text{fl}\,A^3::A.3A^3$. vel $\text{fl}:\sqrt{3}:A.\text{fl}:A::\sqrt{3}:A.3A$. & sic in alijs potestatibus. Patet per Cor: 6.(26)

Theor. 2. In triangulo quovis perpetim rectangulo(27) cujus latera quomodocunꝗ fluunt summa laterum ductorum in suas fluxiones æquatur hypotenusæ ductæ in fluxionem suam. Sit $AA+BB=CC$ et erit

$$A\times \text{fl}:A+B\times \text{fl}:B=C\times \text{fl}:C.$$

Nam per Cor 9 Th 1 est $\text{fl}:A.\text{fl}:AA(::A.2AA)[::]1.2A$, adeoꝗ

$$\text{fl}:AA=2A\times \text{fl}\,A.$$

Eadem ratione $\text{fl}\,BB=2B\times \text{fl}\,B$ & $\text{fl}\,CC=2C\times \text{fl}:C$. Quare cum $AA+BB=CC$ atꝗ adeo per Ax:1 & 3 $\text{fl}:AA+\text{fl}:BB=\text{fl}:CC$, erit

$$2A\times \text{fl}\,A+2B\times \text{fl}\,B=2C\times \text{fl}\,C.$$

Quod dimidiatum fit $A\times \text{fl}\,A+B\times \text{fl}\,B=C\times \text{fl}\,C$. Q.E.D.

Cor 1. Si crus(28) alterutrum sit data quantitas, erit fluxio alterius cruris ad fluxionem hypotenusæ ut hypotenusa ad crus illud alterum. Detur A, et erit $C.B::\text{fl}:B.\text{fl}:C$, nam $B\times \text{fl}\,B=C\times \text{fl}\,C$ propterea quòd $A\times \text{fl}\,A$ nihil sit.

Cor 2. Si hypotenusa datur, erit profluxio unius cruris ad defluxionem alterius ut illud alterum crus ad crus primum. Detur C, et erit $B.A::\text{fl}:A.\text{fl}:B$,(29) propterea quod $C\times \text{fl}:C$ nihil sit.

Theor 3. Si recta utcunꝗ moveatur per superficiem aliquam: fluxio superficiei isto motu generatæ erit ut fluxio alterius alicujus rectæ a dato puncto ad medium primæ rectæ demissæ, ducta in primam rectam. Describat recta AB superficiem

(25) The corresponding algorithm for finding the fluxion of the quotient of two magnitudes: essentially, Newton's proof determines $\text{fl}\,(B/A)$ as $[\text{fl}\,(B)-(B/A)\,\text{fl}\,(A)]/A$, that is, implicity as $\text{fl}\,(A^{-1}\times B)$ with $\text{fl}\,(A^{-1})$ evaluated as $-A^{-2}\,\text{fl}\,(A)$.

(26) Compare note (23). Corollary 6 establishes for integral p, q the result

$$\text{fl}\,(C.X^p):\text{fl}\,(C.X^q) = pC.X^p:qC.X^q.$$

Hence, on substituting $X^q = Y$, there follows $\text{fl}\,(C.Y^r):\text{fl}\,(C.Y) = rC.Y^r:C.Y$ for any rational index $r = p/q$.

Cor. 8. If a fluent quantity is divided by a fluent quantity, the fluxion of the quotient results by taking the fluxion of the divisor multiplied by the dividend from the fluxion of the dividend multiplied by the divisor and then dividing the residue by the square of the divisor: $\text{fl}\,(B/A) = (A\times\text{fl}\,(B) - B\times\text{fl}\,(A))/A^2$. For $A:1 = B:B/A$ and in consequence by Theorem 1 (since $B\times\text{fl}\,(1)$ is zero) $A\times\text{fl}\,(B/A)+(B/A)\times\text{fl}\,(A) = 1\times\text{fl}\,(B)$. Take away $(B/A)\times\text{fl}\,(A)$ from both sides and divide the residue by A: there will result

$$\text{fl}\,(B/A) = (A\times\text{fl}\,(B) - B\times\text{fl}\,(A))/A^2.$$ [25]

Cor. 9. The fluxion of a root is to that of some power of it as the root to that power multiplied by the number of the [power's] dimensions:

$$\text{fl}\,(A):\text{fl}\,(A^3) = A:3A^3, \quad\text{or}\quad \text{fl}(\sqrt[3]{A}):\text{fl}(A) = \sqrt[3]{A}:3A.$$

And so in the case of other powers. This is evident by Corollary 6.[26]

Theorem 2. In any perpetually right-angled triangle[27] whose sides are fluent in any way whatever, the sum of the sides multiplied into their fluxions equals the hypotenuse multiplied into its fluxion. Let $A^2+B^2 = C^2$ and there will be $A\times\text{fl}\,(A)+B\times\text{fl}\,(B) = C\times\text{fl}\,(C)$. For by Theorem 1, Corollary 9

$$\text{fl}\,(A):\text{fl}\,(A^2)\;(= A:A^2) = 1:2A,$$

so that $\text{fl}\,(A^2) = 2A\times\text{fl}\,(A)$. By the same reasoning $\text{fl}\,(B^2) = 2B\times\text{fl}\,(B)$ and $\text{fl}\,(C^2) = 2C\times\text{fl}\,(C)$. Hence since $A^2+B^2 = C^2$ and so, by Axioms 1 and 3, $\text{fl}\,(A^2)+\text{fl}\,(B^2) = \text{fl}\,(C^2)$, there will be $2A\times\text{fl}\,(A)+2B\times\text{fl}\,(B) = 2C\times\text{fl}\,(C)$. When this is halved it becomes $A\times\text{fl}\,(A)+B\times\text{fl}\,(B) = C\times\text{fl}\,(C)$. As was to be proved.

Cor. 1. If either leg[28] be a given quantity, the fluxion of the other one will be to that of the hypotenuse as the hypotenuse to that other leg. Let A be given and there will be $C:B = \text{fl}\,(B):\text{fl}\,(C)$, for $B\times\text{fl}\,(B) = C\times\text{fl}\,(C)$ seeing that $A\times\text{fl}\,(A)$ is zero.

Cor. 2. If the hypotenuse is given, the fluxion of one leg will be to the negative fluxion of the other as that other leg to the first one. Let C be given and there will be $B:A = \text{fl}\,(A):[-]\,\text{fl}\,(B)$, seeing that $C\times\text{fl}\,(C)$ is zero.

Theorem 3. If a straight line should move in any manner whatever over some surface, the fluxion of the surface generated by that motion will be as the fluxion of some other straight line let fall from a given point to the mid-point of the first line, multiplied into that first line. Let the straight line AB describe the

(27) An unlettered embryonic drawing of a right triangle which occurs in the manuscript at this point has been lightly cancelled and is accordingly not reproduced.

(28) Compare I: 25. In seventeenth-century terms, the 'legs' of a right triangle are those sides which together form the right angle.

(29) A preceding minus sign is to be understood.

ABC motu quocunq₃, & ad medium ejus punctum *D* erigatur perpendiculum *DE* ad datum quodvis punctum *E* terminatum: et fluxio superficiei *ABC* erit ut $AB \times \text{fl}:ED$. Sit enim *adb*, in momento temporis proximè sequenti, positio rectæ *AB* motu quocunq₃ prolabentis, et erit *ABba* momentum superficiei *ABC*, et *Dd* contemporaneum momentum rectæ *ED*. Quare cùm *ABba* æquetur $AB \times Dd$, & fluxiones sint ut momenta, erit fluxio generans *ABba* ut[(30)] *AB* in fluxionem generantem *Dd*. Q.E.D. Quod autem dixerim esse $ABba = AB \times Dd$: ad *A* et *B* erige ipsi *AB* perpendicula $A\alpha$, $B\beta$ occurrentia *ab* in α & β, et erit trapezium $AB\beta\alpha (= AB \times \overline{\tfrac{1}{2}B\beta + \tfrac{1}{2}A\alpha}) = AB \times Dd$. Et hoc trapezium non differt a momento *AB*[*ba*][(31)] nisi in spatijs $B\beta b$ & $Aa\alpha$, quæ ex Hypothesi[(32)] sunt infinitè minora dicto trapezio & ad eam collata instar nihili, vel puncti ad lineam.

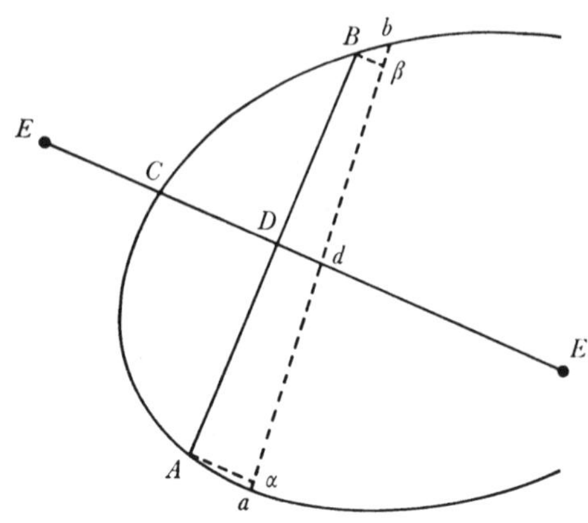

Schol.[(33)] Dixi fluxionem superficiei *ACB* esse ut $AB \times \text{fl}:ED$: sed quemadmodum Geometræ[(34)] statuere solent parallelogrammum æquale esse facto ex lateribus, quod (veriùs loquendo) est ut factum ex lateribus, sic ego ob usum commodiorem Theorematis hujus dicam in sequentibus esse $\text{fl}:ACB = AB \times \text{fl}:ED$.

Cor. Si in semicirculo *ABC* recta *AC* generet segmentum *ACE* gyrando circa terminum diametri *A*, tum ab altero diametri termino *B* demisso *BC* normali ad *AC*, erit fluxio segmenti *ACE* ut dimidium fluxionis perpendiculi *BC* ductum in *AC*: vel juxta Scholium $\text{fl}:ACE = \tfrac{1}{2}AC \times \text{fl}\,BC$.[(35)]

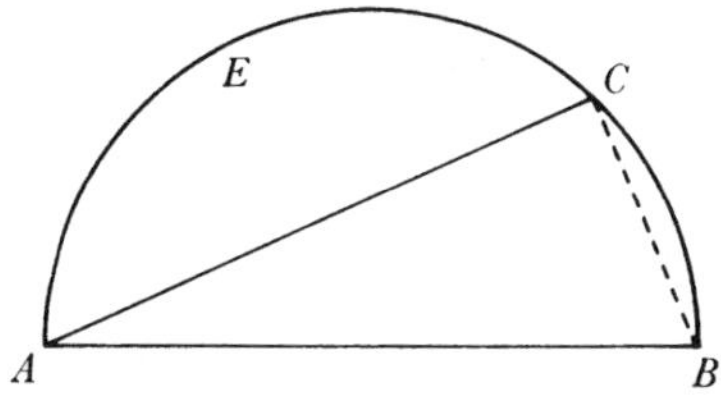

Theor 3.[(36)] Si Trianguli alicujus Basis et longitudine et positione detur, vertex autem sit ad rectam positione datam; demisso ab alterutro termino basis ad rectam illam positione datam perpendiculo, quod occurrat opposito cruri trianguli, erit fluxio ejus oppositi cruris ad fluxionem alterius cruris ut illud

(30) More precisely, read 'æqualis' (equal to).

(31) By a slip of his pen Newton here wrote '$AB\beta\alpha$'.

(32) Namely, that *ab* is indefinitely close to *AB*, so that $A\alpha$, $B\beta$, $a\alpha$ and $b\beta$ are all arbitrarily small.

(33) This appears in revised form below as the Scholium to Theorem 5.

(34) Newton apparently alludes to Cavalieri, John Wallis and others of the 'indivisibles' school of geometers who evaluated the area under a geometrical curve as the sum of (the infinite number of parallelograms of indefinitely small thickness formed by) 'all' parallel ordinates: evidently, when the areas of two such curves defined by coordinates set at equal

surface ABC by any motion whatever and at its mid-point D raise the perpendicular DE terminating at the arbitrary given point E: the fluxion of the surface ABC will then be as $AB \times \text{fl}\,(ED)$. For let adb be, in the next following instant of time, the position of the straight line AB gliding forward with any motion whatever: $(ABba)$ will then be the moment of the surface ABC, and Dd the contemporaneous moment of the straight line ED. Hence, since $ABba$ equals $AB \times Dd$ and fluxions are as their moments, the fluxion generating $ABba$ will be as[30] AB multiplied into the fluxion generating Dd. As was to be proved. But to justify my assertion that $ABba = AB \times Dd$, at A and B erect the perpendiculars $A\alpha$, $B\beta$ to AB meeting ab in α and β: then trapezium

$$AB\beta\alpha(= AB \times \tfrac{1}{2}[B\beta + A\alpha]) = AB \times Dd.$$

This trapezium does not differ from the moment $ABba$ except in the spaces $B\beta b$ and $Aa\alpha$ which are, by hypothesis,[32] infinitely less than the said trapezium and in comparison with it equal to nothing or like a point to a line.

Scholium.[33] I said that the fluxion of the surface ACB is as $AB \times \text{fl}\,(ED)$; however, just as geometers[34] are accustomed to set a parallelogram equal to the product of its sides when (more truthfully speaking) it is as the product of its sides, so I myself for more convenient application of this theorem will say in the sequel that $\text{fl}\,(ACB) = AB \times \text{fl}\,(ED)$.

Corollary. If in the semicircle ABC the straight line AC should generate the segment ACE in its rotation round the end-point A of the diameter, then, when from its other end B is let fall the normal BC to AC, the fluxion of the segment ACE will be as half the fluxion of the perpendicular BC multiplied into AC: or, following the scholium, $\text{fl}\,(ACE) = \tfrac{1}{2}AC \times \text{fl}\,(BC)$.[35]

Theorem 3.[36] If the base of some triangle be given in length and position, while its vertex lies on a straight line given in position, then when a perpendicular is dropped from one or other end of the base to that line given in position till its meet with the opposite leg of the triangle, the fluxion of this opposite leg

angles of ordination are compared, the constant sine of that angle may be set as unity and the parallelograms accordingly treated as rectangles. Of course, no geometer working in the classical tradition would dare at any time to assert generally 'parallelogrammum æquale esse facto ex lateribus'. See also note (42).

(35) In first draft Newton continued without break into his concluding paragraphs, 'Jactis hisce demonstrationum fundamentis...', which (much as on pages 92–4 of the preceding tract's Problem 9) apply the general theorems here established to determining the area 'under' branches of the Gutschoven quartic, the cissoid and the conchoid respectively. Subsequently he cancelled his first Theorem 3, perhaps intending it initially to be reinserted in revised form below, adding new Theorems 3–10/12 and 13 on the first six sides (37–42 in the modern pagination) of an inserted, quartered folio sheet.

(36) Originally numbered '4' and placed below the following Theorem, originally '3', but we have interchanged the position of the two theorems to accord with Newton's renumbering.

alterum crus ad partem hujus cruris inter verticem trianguli et perpendiculum illud situm. Sit AB basis trianguli, C vertex, DE locus verticis, et BD[37] perpendiculariter demissa ad DC occurrat AC in F, eritq $BC.FC::\text{fl}\,AC.\text{fl}\,BC$. Nam ab A ad DE demisso perpendiculo AD, erit (per Cor 1. Th 2) $BC.EC::\text{fl}\,EC.\text{fl}\,BC$, et $DC.AC$ (sive $EC.FC$) $::\text{fl}\,AC.\text{fl}\,DC=\text{fl}\,EC$.[38] Ergo ex æquo perturbatè $BC.FC::\text{fl}\,AC.\text{fl}\,BC$. Q.E.D.

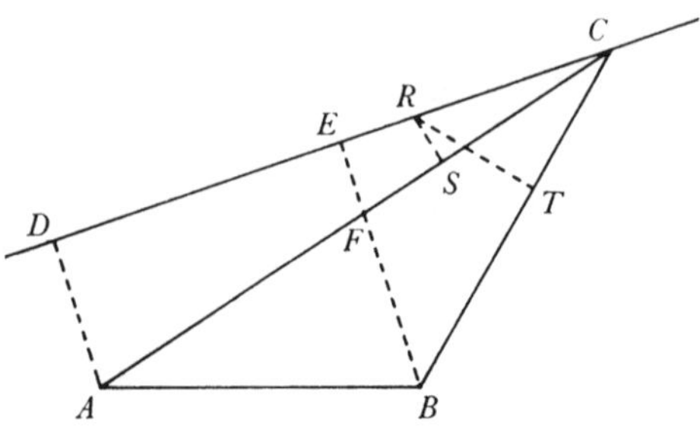

Cor 1.[39] Est fl: AC ad fl: BC ut cosinus anguli ACD ad cosin: BCD. Nam cosinus isti sunt ut BC ad FC.

Cor 2. Si punctum A infinitè distet a B, hoc est si AC sit ipsi AB parallela, age quamvis RQ occurrente[m] AC in Q, sitq RQ positione data, et demisso ad BC normali RT, erit

$$TC.SC::\text{fl}:BC.\text{fl}\,AC=\text{fl}:SC,$$

& propter datam rationem SC ad QC erit per ax 2 $SC.QC::\text{fl}:SC.\text{fl}:QC$. Ergo additis rationibus $TC.QC::\text{fl}:BC.\text{fl}:QC$. Q.E.D.[40]

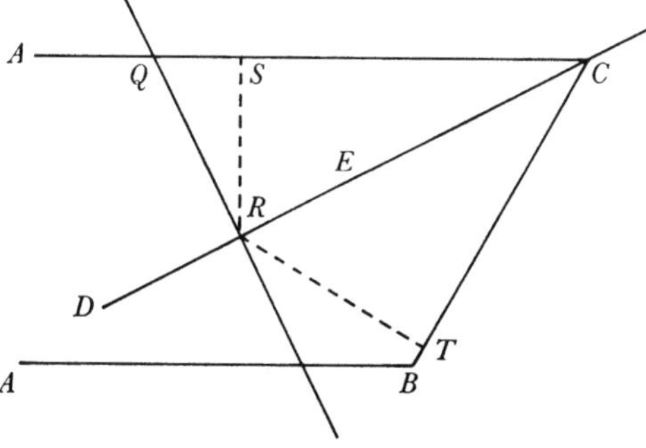

Theor. 4.[41] Si recta circa datum punctum gyrans, secet alias duas positione datas & ad commune punctum terminatas rectas: fluxiones earum quæ positione dantur, erunt ut illæ rectæ ductæ in conterminas partes lineæ gyrantis. Circa datum punctum A gyret recta AC, & inter gyrandum secet ea rectas positione datas DC ac DB in punctis C et B. Dico esse

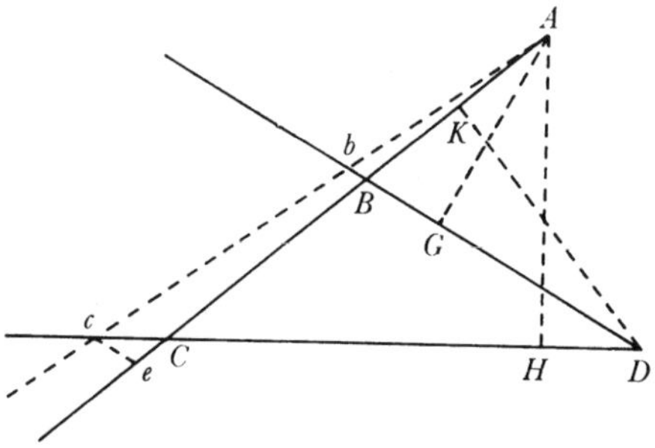

$$DB\times AB.DC\times AC::\text{fl}\,DB.\text{fl}\,DC.$$

Sit enim Abc positio rectæ gyrantis in proximo temporis momento, hoc est Cc momentum rectæ DC et Bb contemporaneum momentum rectæ DB; et ipsi DB parallela agatur ce occurrens AC in e: et propter sim: tri: CBD, Cec, erit $DC.DB::Cc.ce$. Dein propter sim: tri: Aec,

(37) Read 'BE'.

(38) $DC-EC = DE$, given in length and position as the orthogonal projection of AB on DC.

will be to that of the other leg as that other leg to the part of the former leg situated between the vertex of the triangle and that perpendicular. Let AB be the triangle's base, C its vertex, DE the locus of the vertex and $B[E]$ the line let fall perpendicularly to DC, meeting AC in F: then $BC:FC = \text{fl}(AC):\text{fl}(BC)$. For drop the perpendicular AD from A to DE and (by Th. 2, Coroll. 1) there will be $BC:EC = \text{fl}(EC):\text{fl}(BC)$ and

$$DC:AC \text{ (that is, } EC:FC) = \text{fl}(AC):\text{fl}(DC) \text{ (or fl}(EC)).^{(38)}$$

In consequence, by contraposed equals $BC:FC = \text{fl}(AC):\text{fl}(BC)$. As was to be proved.

Cor. 1.(39) Hence fl (AC) is to fl (BC) as the cosine of the angle $A\widehat{C}D$ to the cosine of $B\widehat{C}D$. For those cosines are as BC to FC.

Cor. 2. If the point A be infinitely distant from B (that is, if AC be parallel to AB), draw any RQ meeting AC in Q and let it be given in position: then, when the normal RT is let fall to BC, there will be $TC:SC = \text{fl}(BC):\text{fl}(AC)$, that is, fl (SC), and because of the given ratio SC to QC, by Axiom 2,

$$SC:QC = \text{fl}(SC):\text{fl}(QC).$$

Hence, on compounding the ratios $TC:QC = \text{fl}(BC):\text{fl}(QC)$. As was to be proved.(40)

Theorem 4.(41) If a straight line rotating round a given point should cut two other straight lines given in position and terminating at a common point, the fluxions of the lines given in position will be as those lines multiplied into the parts of the rotating line terminated at them. Let the straight line AC rotate round the given point A and during its rotation let it intersect the straight lines DC and DB given in position in the points C and B. I state that

$$DB \times AB:DC \times AC = \text{fl}(DB):\text{fl}(DC).$$

For let Abc be the position of the rotating line in the next instant of time, that is, let Cc be the moment of the line DC with Bb the contemporaneous moment of the line DB, and let ce be drawn parallel to DB meeting AC in e: then, because of the similar triangles CBD and Cec, $DC:DB = Cc:ce$, and again, because of

(39) A first cancelled continuation, evidently an early form of the following corollary, reads 'Si a quovis rectæ DC puncto R demittantur ad AC et BC perpendicula RS et RT: erit $SC.TC::\text{fl}AC.\text{fl}BC$. Nam $SC.TC::BC.FC$' (If from any point R of the line DC there be let fall to AC and BC the perpendiculars RS and RT, then $SC:TC = \text{fl}(AC):\text{fl}(BC)$. For $SC:TC = BC:FC$). It will be evident that $SC:RC = EC:FC$ and $RC:TC = BC:EC$, where $FC = EC\cos A\hat{C}D$ and $BC = EC\cos B\hat{C}D$.

(40) As in the preceding theorem, the unlocated point E in DC is evidently the foot of the perpendicular from B. It plays, of course, no part in the present argument.

(41) Originally '3': see note (36). In 1693, as we shall see in the seventh volume, Newton introduced a simplified version of this theorem into the 'Introductio' to his *Tractatus de Quadratura Curvarum*; compare his *Opticks* (note (16)): $_2$168.

ABb, erit $Ac.Ab::ec.Bb$, et additis rationibus

$$DC\times Ac.DB\times Ab::Cc.Bb::\text{(per ax 4) fl}:DC.\text{fl}:DB.$$

Coeant jam lineæ infinitè parùm distantes Ac & AC, et in momento concursus evadet $DC\times AC.DB\times AB::\text{fl}:DC.\text{fl}:DB$. Q.E.D.

Cor 1. Iisdem positis, et ab A demissis ad DB et DC normalibus AG et AH: erit primo $DC\times AC.DB\times BG::\text{fl}:DC.\text{fl}:AB$. Nam per Cor. 1 Th. 2, est $AB.BG::\text{fl}:DB.\text{fl}:AB$. sive $DB\times AB.DB\times BG::\text{fl}:DC.\text{fl}\,DB$. Ergo ex æquo $DC\times AC.DB\times BG::\text{fl}:DC.\text{fl}:DB$.

Cor 2. Erit secundo $DC\times CH.DB\times BG::\text{fl}\,AC.\text{fl}\,AB$. Nam per Cor 1 Th 2 est $CH.AC$ (vel $DC\times CH.DC\times AC$) $::\text{fl}\,AC.\text{fl}\,DC$. Et supra erat

$$DC\times AC.DB\times BG::\text{fl}\,DC.\text{fl}\,AB.$$

Ergo ex æquo $DC\times CH.DB\times BG::\text{fl}\,AC.\text{fl}\,AB$.

Hujusmodi alia Theoremata non inutilia proponi possent: ad fluxiones superficierum festinamus.

Theor. 5. Si recta quævis motu parallelo per aream aliquam, a duabus parallelis & positione datis rectis terminatam transferatur: erit fluxio areæ ut fluxio alterutrius rectæ parallelæ. Sint AB, DC rectæ parallelæ, & BC recta per spatium interjectum $ADCB$ in data inclinatione ABC translata et AD terminus a quo incipit transferri; et erit fl: $ADCB$ ut fl AB. Nam area BD est ut longitudo AB. Quare per Ax 2 fl: BD ut fl: AB.

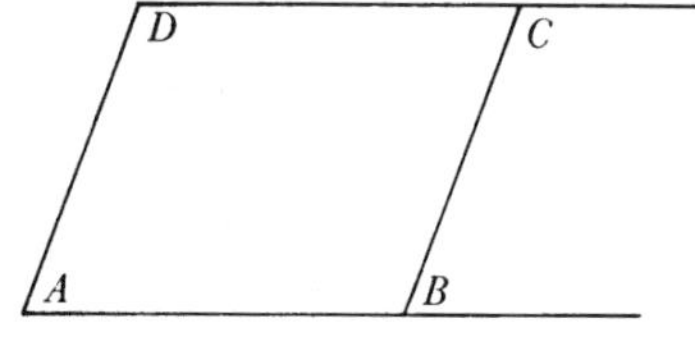

Schol. Si angulus ABC rectus sit, tum quemadmodum statui solet

$$BD=AB\times BC,$$

sic nos statuemus $\text{fl}\,BD=\text{fl}\,AB\times BC$ hoc est (per Cor 7 Th 1) $=BC\times\text{fl}\,AB$.(42) Sed hic sicut per $AB\times BC$ non intelligitur linea sed productum arithmeticum quod exprimit numerum unitatum in area BD ex Hypothesi quod unitas superficialis sit quadratum cujus latera sunt unitates lineares: sic in hoc Scholio per $BC\times\text{fl}\,AB$ non intelligitur fluxio linearis sed fluxio generans productum arithmeticum quod exprimit numerum̃ unitatum superficialium in BD ex Hypothesi quod momentum basis fluentis ductum in datam altitudinem parallelogrammi facit momentum parallelogrammi.

Theor 6. Si recta quævis motu parallelo transferatur per aream alijs duabus positione datis et non parallelis rectis terminatam, erit fluxio areæ ut fluxio alterutrius rectæ positione datæ ducta in rectam mobilem. Transferatur BC per spatium(43) CAB rectis AC AB

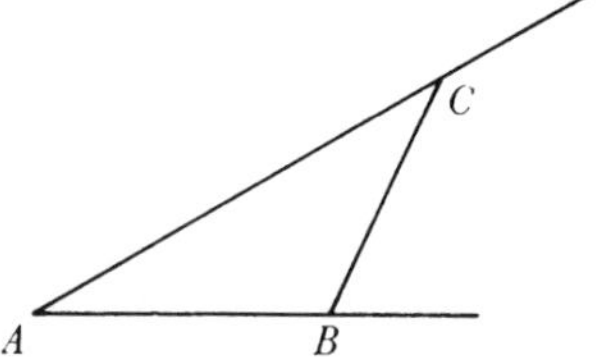

(42) Since $BC(=AD)$ is determined in length, its fluxion is zero. See also note (34).
(43) 'triangulum' (triangle), to be precise.

the similar triangles Aec and ABb, $Ac:Ab = ec:Bb$; so that, on compounding ratios, $DC \times Ac : DB \times Ab = Cc : Bb =$ (by Axiom 4) fl (DC) : fl (DB). Now let the lines Ac and AC, an infinitely little distance apart, coincide and at the moment of confluence there will come to be $DC \times AC : DB \times AB =$ fl (DC) : fl(DB). As was to be proved.

Cor. 1. With the same suppositions, on letting fall the normals AG and AH to DB and DC, there will be, first, $DC \times AC : DB \times BG =$ fl (DC) : fl (AB). For by Theorem 2, Coroll. 1 $AB:BG =$ fl (DB) : fl (AB), that is,

$$DB \times AB : DB \times BG = \text{fl}\,(DC) : \text{fl}\,(DB).$$

Therefore *ex æquo* $DC \times AC : DB \times BG =$ fl (DC) : fl (DB).

Cor. 2. Secondly, $DC \times CH : DB \times BG =$ fl (AC) : fl (AB). For by Theorem 2, Coroll. 1 $CH:AC$ (or $DC \times CH : DC \times AC$) $=$ fl (AC) : fl (DC), while above

$$DC \times AC : DB \times BG = \text{fl}\,(DC) : \text{fl}\,(AB).$$

Therefore *ex æquo* $DC \times CH : DB \times BG =$ fl (AC) : fl (AB).

Other theorems of this sort could be propounded which are not without use—but let us hurry on to surface fluxions.

Theorem 5. If any straight line be transported with a parallel motion over some area bounded by two parallel lines given in position, the fluxion of the area will be as the fluxion of either of the parallels. Let AB, DC be parallel lines, BC a straight line transported over the interposed space $ADCB$ at the given slope $\widehat{ABC}$ and AD the bounding line from which it begins to be conveyed: then fl $(ADCB)$ is as fl (AB). For the area $ADCB$ is as the length AB. Hence, by Axiom 2, fl $(ADCB)$ is as fl (AB).

Scholium. If the angle $\widehat{ABC}$ be right, then just as it is customary to set

$$(ADCB) = AB \times BC,$$

so we shall set fl $(ADCB) =$ fl $(AB \times BC)$, that is, (by Theorem 1, Coroll. 7) $BC \times$ fl (AB).[42] But as here by $AB \times BC$ is understood not a line but the arithmetical product which expresses the number of unities in the area $ADCB$ by the hypothesis that the surface unity is the square whose sides are linear unities, in the same way in this scholium by $BC \times$ fl (AB) is understood not a linear fluxion but a fluxion generating the arithmetical product which expresses the number of surface unities in BD by the hypothesis that the moment of its fluent base multiplied into the given altitude of a parallelogram makes the moment of the parallelogram.

Theorem 6. If any straight line be transported with a parallel motion over an area bounded by any two other non-parallel straight lines given in position, the fluxion of the area will be as the fluxion of either of the lines multiplied by the mobile line. Let BC be transported over the space[43] CAB bounded by the

positione datis terminatum: et erit fl: CAB ut $BC\times$ fl: AB. Etenim triangulum CAB est ut AB^q. ergo, per Ax 2, fl: tri: CAB est ut fl AB^q. Sed per Cor 6 & 9, Th: 1, fl AB^q est ut $2AB\times$ fl AB. hoc est cùm $2AB$ et BC sint in data ratione,(44) ut $BC\times$ fl AB.

Schol. Si ang ABC rectus sit, potest juxta Schol præcedens, poni

$$\text{fl}\,CAB = \tfrac{1}{2}BC\times \text{fl}:AB.$$

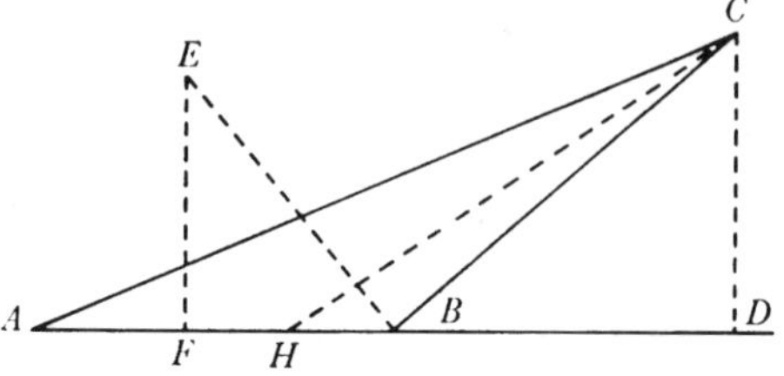

Theor 7. Si recta circa datum punctum gyrans, continuò terminetur ad aliam rectam positione datam: erit fluxio spatij a gyrante recta descripti, ut fluxio alterius rectæ. Gyrat recta CB circa punctum C sitq; CA terminus a quo incipit gyrare, et AB recta ad quam terminatur, et erit area ABC ut recta AB; adeoq; (per Ax 2) fl ABC ut fl AB.

Schol. Demisso ad AB normali CD, erit (juxta Schol Theor 5)

$$\tfrac{1}{2}CD\times \text{fl}:AB = \text{fl}\,ABC,$$

quia $\tfrac{1}{2}CD\times AB = ABC$.(45)

Theor 8. Iisdem positis, si ab alio insuper quovis puncto recta perpetim ducatur ad concursum priorum rectarum, et a primo puncto ad hanc rectam demittatur(46) linea perpendicularis terminata ad rectam positione datam: erit fluxio hujus novæ rectæ ducta in lineam perpendicularem, ut fluxio areæ a prima recta descriptæ.

A dato E agatur EB, et ad hanc perpendicularis CH occurrens AB in H, eritq; $CH\times$ fl: EB, ut fl: ABC. Nam demisso ad AB normali EF, erit per Cor: 1 Th: 2, fl AB(fl: FB) . fl: EB :: EB . FB. (hoc est propter sim. tri. EBF, HCD) :: HC . CD. Ergo $CD\times$ fl: $AB = HC\times$ fl: EB. Quare cùm CD datum sit, adeoq; $CD\times$ fl: AB ut fl: AB, sitq; etiam (per Th 7) fl AB ut fl ABC, erit $HC\times$ fl EB ut fl ABC.

Schol. Est et per Schol. sup.(47) $\tfrac{1}{2}HC\times \text{fl}\,EB(=\tfrac{1}{2}CD\times \text{fl}\,AB) = \text{fl}\,ABC$. Et insuper in eo temporis momento quo contingit angulum EBC rectum esse, est $\tfrac{1}{2}BC\times \text{fl}:EB = \text{fl}:ABC$ quia tunc HC et BC coincidunt.

Theor 9. Si recta circa datum punctum gyrans, secet alias duas positione datas rectas: superficierū inter datum punctum et rectas positione datas isto motu generatarū fluxiones erunt ut quadrata longitudinum generantium. Sit A datum punctum circa quod AC gyrat, sintq; BD et CD rectæ positione datæ, & AD principium a quo AC incipit gyrare, et erit fl: ADB . fl ADC :: AB^q . AC^q.

(44) Namely, $2\sin\hat{C}:\sin\hat{A}$.

(45) The length of CD is fixed and so its fluxion is zero.

(46) Newton first wrote 'agatur' (be driven).

(47) Read 'Scholium superius'.

straight lines AC, AB given in position: then fl (CAB) is as $BC \times$ fl (AB). For, indeed, the triangle CAB is as AB^2, and therefore, by Axiom 2, fl $(\triangle CAB)$ is as fl (AB^2). But by Theorem 1, Corollaries 6 and 9 fl (AB^2) is as $2AB \times$ fl (AB), that is, since $2AB$ and BC are in a given ratio,(44) as $BC \times$ fl (AB).

Scholium. If angle $A\widehat{B}C$ is right, according to the preceding scholium there may be put fl $(CAB) = \frac{1}{2}BC \times$ fl (AB).

Theorem 7. If a straight line rotating round a given point continuously terminate at another straight line given in position, the fluxion of the space described by the rotating line will be as that of the other line. Let the straight line CB rotate round the point C and let CA be the bound from which it begins to rotate with AB the straight line at which it terminates: the area ABC will then be as the line AB, so that (by Axiom 2) fl (ABC) is as fl (AB).

Scholium. When CD is let fall normal to AB, there will (following the scholium to Theorem 5) be $\frac{1}{2}CD \times$ fl $(AB) =$ fl (ABC) because $\frac{1}{2}CD \times AB = [\triangle]ABC$.(45)

Theorem 8. With the same suppositions, if further from any other point a straight line be drawn perpetually to the meet of the former lines and from the first point to this line there be let fall(46) one perpendicular terminating at the line given in position, the fluxion of this new line multiplied into the perpendicular one will be as the fluxion of the area described by the first straight line.

From the given point E draw EB and perpendicular to this CH meeting AB in H: then will $CH \times$ fl (EB) be as fl (ABC). For when EF is let fall normal to AB, by Theorem 2, Coroll. 1, there will be

fl (AB) (that is, fl (FB)) : fl $(EB) = EB:FB$

$=$ (because of the similar triangles EBF, HCD) $HC:CD$.

Therefore $CD \times$ fl $(AB) = HC \times$ fl (EB). Hence, since CD is given, so that $CD \times$ fl (AB) is as fl (AB), and also (by Theorem 7) fl (AB) is as fl (ABC), $HC \times$ fl (EB) will be as fl (ABC).

Scholium. By the previous scholium, in fact,

$$\tfrac{1}{2}HC \times \text{fl}\,(EB)\ (= \tfrac{1}{2}CD \times \text{fl}\,(AB)) = \text{fl}\,(ABC).$$

Moreover at the instant that the angle $E\widehat{B}C$ happens to be right,

$$\tfrac{1}{2}BC \times \text{fl}\,(EB) = \text{fl}\,(ABC)$$

since then HC and BC are coincident.

Theorem 9. If a straight line rotating round a given point should intersect two other straight lines given in position, the fluxions of the surfaces between the given point and the lines given in position which are generated by that motion will be as the squares of the generating lengths. Let A be the given point round which AC rotates, with BD and CD the straight lines given in position and AD the initial line from which AC starts to rotate: then

$$\text{fl}\,(ADB):\text{fl}\,(ADC) = AB^2:AC^2.$$

Sit enim *Abc* positio rectæ gyrantis in proximo temporis momento, et triangula infinitè parva *ABb*, *ACc* erunt momenta superficierum *ADB*, *ADC*, adeocꝫ ut ipsarum fluxiones. Sed per 15. 6. Elem. ista triangula sunt ut $AB \times Ab$ ad $AC \times Ac$:(48) Quæ ratio, si *Abc* retro volvatur donec redeat in *AC*, in ultimo ejus regressûs momento, hoc est in primo momento progressûs ubi *AC* incipit pergere ad *Ac* evadit AB^q ad AC^q. Q.E.D.

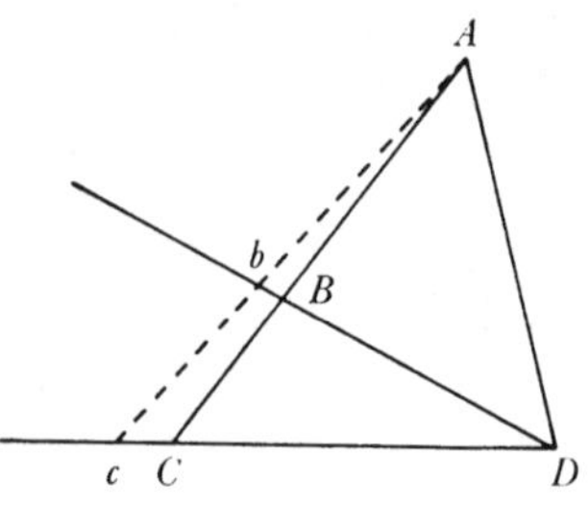

Theor: 10. Si recta positione data tangat curvā positione datam et utracꝫ secentur(49) ab alia utcuncꝫ motâ rectâ: fluxiones curvæ illius & tangentis ejus in eo temporis momento æquales erunt, quo mota illa linea secat utramcꝫ in puncto contactûs. Esto curva *RS*, Tangens ejus *AB*[,] punctum contactûs *C* et linea mobilis *DE*: dico fluxiones linearum *RC* et *AC* æquales evadere quando *DE* pertingit ad *C*. Nam in *RC* sumatur arcus infinitè parvus *Cc*, & cum hæc(49) juxta Hypothesin Archimedeam(50) pro recta haberi possit, produc eam utrincꝫ in directum, sitcꝫ ea producta *AB* propterea quod ipsa *AB* tantùm tangat curvam. *RS* itacꝫ et *AB* commune habent momentum *Cc*, & proinde eandem fluxionem dum *DE* transit per illud momentum. Q.E.D.

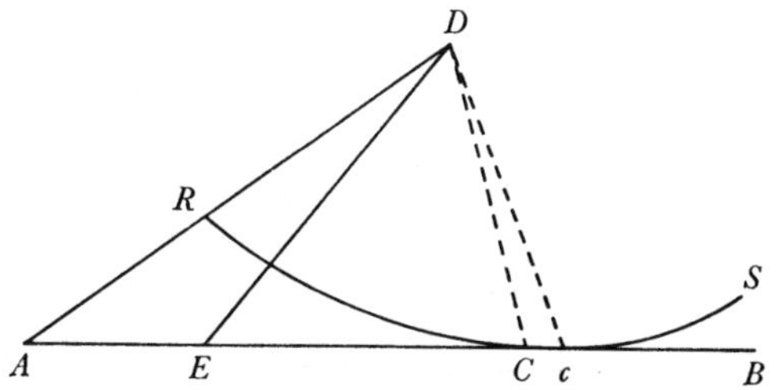

Cor. Hinc omnia quæ in Theor: 3 et 4 de fluxionibus rectarum positione datarum demonstrata sunt, conveniunt etiam fluxionibus curvarum quas rectæ illæ tangunt in intersectione cum linea mobili.(51)

Theor. 12.(52) Si recta illa *DE* circa datum punctum *D* convoluta, describat duas superficies quarum una *DRC* terminatur ad curvam *RC*, altera *DAC* ad tangentem curvæ *AC*: fluxiones illarum superficierum æquales erunt in eo

(48) For $\triangle ABb = \frac{1}{2}AB \times Ab \times \sin B\hat{A}b$ and $\triangle ACc = \frac{1}{2}AC \times Ac \times \sin C\hat{A}c$, with the angles $B\hat{A}b$, $C\hat{A}c$ coincident. Alternatively the result follows in more Euclidean terms by compounding the proportions $\triangle ABb:\triangle ACb = AB:AC$ and $\triangle ACb:\triangle ACc = Ab:Ac$.

(49) A slip of the pen for 'secetur' and 'hic' respectively.

(50) A curious phrase indeed, since this absolute limit-equation of the 'arcus infinitè parvus *Cc*' and the corresponding 'recta *Cc*' is alien to Archimedean as to all classical Greek geometrical thought: Archimedes himself would have argued here that the infinitesimal arc $\widehat{Cc}$ is bounded by the chord *Cc* and the two tangent segments drawn from *C* and *c* to their intersection and that *c* may be chosen sufficiently close to *C* to make the difference in magnitude of the latter bounds less than any arbitrarily small quantity. One might more reasonably qualify the 'hypothesis' as 'Robervallianam' since it is equivalent to postulating that the tangent at the point *C* is the instantaneous 'determination' (direction of motion) of the curve

For let Abc be the position of the rotating line in the next moment of time. The infinitely small triangles ABb, ACc will then be moments of the surfaces ADB, ADC and so as their fluxions. But by *Elements* VI, 15, those triangles are as $AB \times ab$ to $AC \times Ac$,(48) and this ratio, if Abc revolve backwards till it goes back again into AC, will, in the last instant of its return—that is, in the first moment of advance when AC starts to swing forward to Ac—, come to be AB^2 to AC^2. As was to be proved.

Theorem 10. If a straight line given in position be tangent to a curve given in position, and both be intersected by a second straight line moving in any way whatever, at that instant of time when the moving line intersects both in their point of contact the fluxions of the curve and its tangent line will be equal. Let the curve be RS, its tangent AB, the contact point C and the mobile line DE: I assert that the fluxions of the lines $\overset{\frown}{RC}$ and AC prove to be equal when DE arrives at C. For in $\overset{\frown}{RC}$ take the infinitely small arc $\overset{\frown}{Cc}$ and, since by Archimedes' postulate(50) this may be considered as straight, extend it either way in a straight line: this extended line must be AB for the simple reason that AB itself is tangent to the curve. Accordingly, $\overset{\frown}{RS}$ and AB have the common moment Cc and consequently, as DE traverses that moment, the same fluxion. As was to be proved.

Corollary. Hence everything which has been proved in Theorems 3 and 4 concerning the fluxions of straight lines holds true also for the fluxions of curves to which those straight lines are tangent at their intersection with a mobile line.(51)

Theorem 12.(52) If the straight line DE revolving round the given point D describes two surfaces, one of which (DRC) terminates at the curve RC, the other (DAC) at the curve's tangent AC, at that instant of time when the straight

at that point, but Newton in fact here merely reiterates a phrase of Barrow's that 'ob indefinitam sectionem curvula...pro recta haberi potest' (Isaac Barrow, *Lectiones Geometricæ* (London, 1670): Lectio XI, §1: 85). Two years before James Gregory, in making use of similar arguments, was conscious that they were not Archimedean: 'Demonstrandi methodo utor (ni fallor)...multo breviore quam Archimedea & non minus geometrica; utor quoque in propositionibus magis obviis methodo Cavalleriana, quæ etiam nullo negotio reducitur ad Archimedeam vel nostram' (*Geometriæ Pars Universalis* (Padua, 1668): Prooemium: [†2ᵛ/†3ʳ]).

(51) A favourite technique of Isaac Barrow's in his *Lectiones Geometricæ* (note (50)), used to reduce the problem of constructing a (linear) tangent to a given curve to the simpler one of constructing the tangent to a conic touching the given curve at the given point. In *Lectio* VIII, §XVIII (on pages 68–9), for instance, he in effect constructed the tangent at the point (X, Y) on the Gutschoven quartic $y^2 = x^4/(a^2-x^2)$ as the tangent at this point to the hyperbola $Yy = X^2x^2/(a^2-Xx)$ which there touches the quartic. In §§XVII/XIX of the same *Lectio* he gave comparable tangent constructions for the cissoid and the strophoid.

(52) Originally numbered '11'. In renumbering it Newton has added no new 'Theorema 11': perhaps this was to have been a revised form of the cancelled Theorem 3 above, although the present proposition is an evident generalization of Theorem 10 preceding and there seems no reason for separating them.

temporis momento quo recta circumacta transit per punctum contactus C. Nempe flDRC=flDAC quia tunc commune est utriusq; momentum CDc.

Cor: Hinc omnia quæ in Theor 7, 8 & 9 de fluxionibus superficierum rectis positione datis terminatarum demonstrata sunt, conveniunt etiam fluxionibus superficierum terminatarum curvis positione datis quas rectæ illæ tangunt.

Theor 13 (53) Si recta mobilis perpetuò tangat curvam, punctum contactûs in omni temporis momento erit centrum circa quod recta in illo momento volvitur.

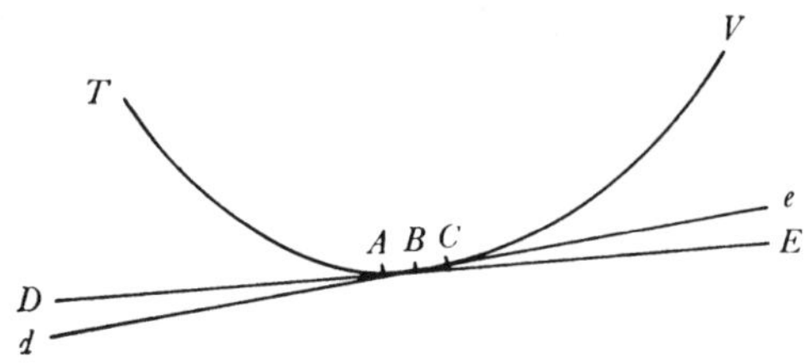

Concipe Curvam TV lineolis parvitate et multitudine infinitis constare quarum duæ sunto AB & BC. Hasce produc utrinq; in directum, nempe AB ad D et E et BC ad d et e, et manifestum est quod tangens mobilis in eo temporis momento quo volvitur de loco DE in locum de, convertitur circa punctum contactus B, propterea quod istud B sit communis intersectio locorum(54) DE ac de.

Cor 1. Hinc omnia quæ in Theor 3, 4, 7, 8, 9, 10, 11(55) de recta circa datum punctum ceu centrum volvente demonstrata sunt, conveniunt etiam rectæ perpetuò tangenti curvam lineam positione datam, si modò punctum contactûs circa quod recta illa in momento contactûs illius convolvitur, vicem centri dati gerere concipiatur.(56) Et proinde sigillatim

Cor 2. S(57)

(58)Jactis hisce demonstrationum fundamentis, methodus tenendi demonstrationes, uno et altero exemplo constabit.

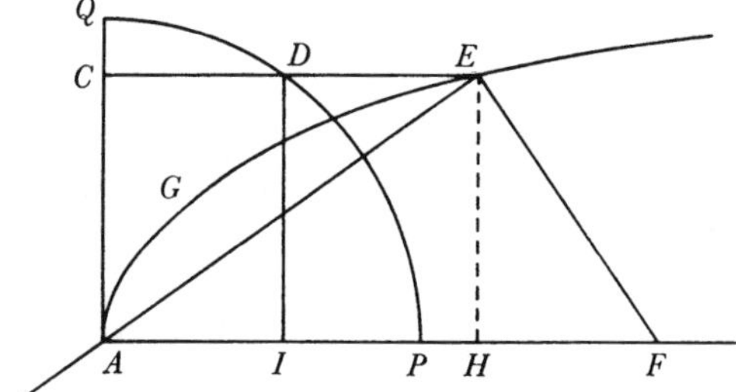

Proponatur itaq; constructio in Exemplo 2do(59) demonstranda. Per Cor. 2 Th. 1 est fl:ID ad fl:IP ut AI ad ID. Estq; AI ad ID ut ID ad CE ex natura Curvæ AGE. Et proinde $CE\times$fl:$ID=ID\times$fl:IP. Sed (per Schol. Th. 3) $CE\times$fl:ID=fluxioni

(53) Originally '12': compare the previous note.

(54) Newton perhaps intends 'linearum' (lines). Note that he presupposes that the line passes from DE into de uniquely by a rotatory movement; however, it may also 'slide' along the curve TAV with a motion compounded of both rotation and translation.

(55) Read '12': see note (52).

(56) An equivalent phrase 'pro centro habeatur' is cancelled.

line, as it goes round, passes through the contact point C the fluxions of those surfaces will be equal. Specifically, fl $(DRC) =$ fl (DAC) because then the common moment of both is (CDc).

Corollary. Hence everything which has been proved in Theorems 7, 8 and 9 concerning the fluxions of surfaces bounded by straight lines given in position holds true also for the fluxions of surfaces bounded by curves given in position to which those straight lines are tangent.

Theorem 13.(53) If a mobile straight line perpetually touches a curve, the point of contact at every moment of time will be instantaneously the centre round which that line revolves. Imagine that the curve $\widehat{TV}$ consists of line-segments infinite in their smallness and number, and let two of these be AB and BC. Extend these either way in a straight line—namely, AB to D and E, BC to d and e—and it is clear that the mobile tangent, in the instant of time in which it revolves from the position DE to the position de, turns round the point of contact B for the reason that the point B is the common meet of the positions(54) DE and de.

Cor. 1. Hence everything which has been proved in Theorems 3, 4, 7, 8, 9, 10 and 11(55) concerning a straight line revolving round a given point as its centre holds also for a straight line ever tangent to a curve given in position, provided that the point of contact round which that straight line turns at the moment of contact is conceived to take on the rôle of the given centre. And hence severally:

Cor. 2. ...(57)

(58)With these foundations for [constructing] proofs laid, the method of executing such proofs will be established by an example or two.

Accordingly, let the construction in Example 2(59) be proposed for proof. By Theorem 1, Corol. 2 fl (ID) is to fl (IP) as AI to ID, while from the nature of the curve AGE AI to ID is as ID to CE. Consequently,

$$CE \times \text{fl}\,(ID) = ID \times \text{fl}\,(IP).$$

But (by the scholium to Theorem 3) $CE \times$ fl (ID) = fluxion of area $ACEG$, and

(57) Newton halts his sentence, perhaps commencing 'S[i Trianguli alicujus]...', at its initial capital. Evidently he intended to add further corollaries corresponding 'sigillatim' to one or more of Theorems 3, 4, 7, 8, 9, 10 and 12.

(58) The manuscript of this revised version of pages 92–4 of the preceding tract, which (see note (37)) followed directly on the cancelled Theorem 3 above, has no illustrating figures, understanding those used previously in Examples 2–4. These are here repeated for the reader's convenience.

(59) The Gutschoven quartic whose area is evaluated analytically on page 87 of the preceding tract and then fluxionally on Newton's following page 92. The equivalent defining property of the quartic that the triangles ADI, FEH are congruent follows readily 'ex natura curvæ' and the triangle AEC is evidently similar to the latter and so to the former.

areæ $ACEG$, et $ID\times$ fl: IP = fluxioni areæ PDI. et proinde areæ illæ(60) per Ax: 1 æquantur. Q.E.D.

Proponatur denuò constructio qua Cissoidis area in Exemplo 3(61) determinatur. Ad hunc autem demonstrandam, lineæ punctim notatæ in schemate deleantur et agantur DQ, AE et Cissoidis Asymptoton QR.(62) Jam propter AQ, DQ, CQ ∺,(63) est (per Cor 2 Th 1) fl DQ . fl CQ :: DQ . $2CQ$. Et propter sim. tri. QDC, DEA, est

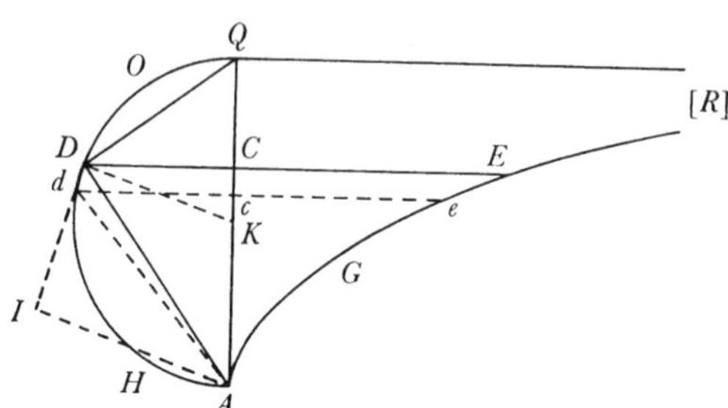

$$DQ\,.\,2CQ::ED\,.\,2AD.$$

Ergo fl DQ . fl CQ :: ED . $2AD$. &

$$ED\times \text{fl}\,CQ = 2AD\times \text{fl}:DQ$$

sive $=4\times\frac{1}{2}AD\times$ fl DQ. Sed per Cor. 1, Theor 3, est $\frac{1}{2}AD\times$ fl DQ = fluxioni generanti aream $ADOQ$. Et proinde per Ax: 2 area illa infinite longa $QREDO$ generatur quadrupla alterius $ADOQ$. Q.E.D.(64)

Deniq; ad demonstrandam constructionem areæ Conchoidalis in Exempl: 4;(65) Ubi demonstraveris ut supra(66) quod sit

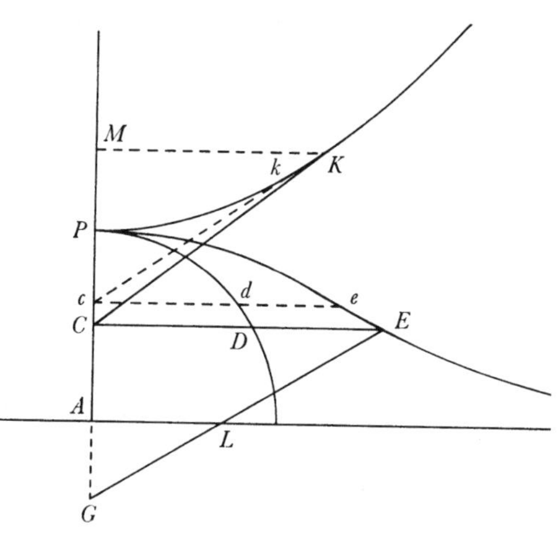

$$AG\,.\,AP::DE\,.\,MK,$$

sic procede. Est autem (per Cor. 2, Th. 3) $MK\times$ fl PC = fl: areæ PKC & $DE\times$ fl PC = fl areæ DPE. Ergo

$$\text{fl}\,PKC\,.\,\text{fl}\,DPE(::MK\,.\,DE)::AP\,.\,AG$$

Adeoq; per Ax. 2 Areæ PKC et DPE sunt in eadem ratione.(67)

---

(60) In comparable limit-increment terms

$$(ACEG) = \int_0^{ID} CE\,.\,d(ID) \quad \text{and} \quad (PDI) = \int_0^{IP} ID\,.\,d(IP)$$

with $d(ID):d(IP) = \text{fl}\,(ID):\text{fl}\,(IP)$.

(61) This fluxional evaluation of the cissoid's area, discussed analytically on page 87 of the preceding tract, is revised from the equivalent paragraph on Newton's page 91 following.

$ID \times$ fl (IP) = fluxion of area PDI. Consequently, those areas[60] are, by Axiom 1, equal. As was to be proved.

Again, let the construction by which the cissoid's area is determined in Example 3[61] be proposed. To prove this, let the broken lines in the figure be deleted and draw DQ, AE and the cissoid's asymptote QR.[62] Now, because AQ, DQ and CQ are in continued proportion, therefore (by Theorem 2, Corol. 1) fl (DQ) : fl $(CQ) = DQ : 2CQ$ and, because the triangles QDC, DEA are similar, $DQ : 2CQ = ED : 2AD$. Hence fl (DQ) : fl $(CQ) = ED : 2AD$ and

$$ED \times \text{fl}\,(CQ) = 2AD \times \text{fl}\,(DQ),$$

that is, $4 \times \frac{1}{2}AD \times$ fl (DQ). But by Theorem 3, Corol. 1 $\frac{1}{2}AD \times$ fl (DQ) is equal to the fluxion generating the area $ADOQ$. Consequently, by Axiom 2, the infinitely long area $QREDO$ is generated as four times the other one, $ADOQ$. As was to be proved.[64]

Finally, to prove the construction of the conchoidal area in Example 4,[65] when you have proved as above[66] that $AG : AP = DE : MK$, proceed thus. By Theorem 3, Corol. 3 $MK \times$ fl (PC) = fluxion of area PKC, while

$$DE \times \text{fl}\,(PC) = \text{fluxion of area } DPE.$$

Therefore fl (PKC) : fl (DPE) $(= MK : DE) = AP : AG$, so that, by Axiom 2, the areas PKC and DPE are in the same ratio.[67]

---

(62) For convenience we have added both the chord DQ and the asymptote QR to Newton's figure: these have, indeed, been added in pencil to the original inked diagram (on page 87 of ULC. Add. 3960.14) with a somewhat coarse flourish which is scarcely Newton's but may well be Jones'. (In the latter's transcript of the 1671 tract (see §1: note (21) both elements are incorporated in the copied figure.)

(63) Oughtred's notation for 'continuè proportionales'.

(64) Equivalently, $4\int_0^{DQ} \frac{1}{2}AD.d(DQ) = \int_0^{CQ} ED.d(CQ)$.

(65) Compare pages 87–8 and especially 91–2 of the preceding tract.

(66) See the bottom of page 91 in the preceding tract. Newton there shows that, since $AC \times AM = AP^2$, therefore $AG : GL = AC : (LE$ or$)$ $AP = AP : AM$ and consequently

$$AG : AP = (\sqrt{[GL^2 - AG^2]} \text{ or}) \ AL : (\sqrt{[AM^2 - AP^2]} \text{ or}) \ MK,$$

where by construction $AL = DE$ (compare §2: note (609)).

(67) Equivalently, since $(PKC) = \int_0^{PC} MK.d(PC)$ and $(DPE) = \int_0^{PC} DE.d(PC)$, we would argue that $(PKC) : (DPE) = MK : DE$, that is, $AP : AG$.

APPENDIX.
DAVID GREGORY'S 'TRACTATUS DE SERIEBUS'

[*c.* 1690].(1)

From the original in the University Library, Edinburgh(2)

||Pag: 1 || [56](3) ISAACI NEWTONI

TRACTATUS DE SERIEBUS INFINITIS ET CONVERGENTIBUS.

(4)Animadvertenti plerosque geometras posthabita fere veterum synthetica methodo analyticæ excolendæ plurimum incumbere et ejus ope tot tantasq; difficultates superasse ut pene omnia extra curvarũ quadraturas et similia quædam nondum penitus enodata, videantur exhausisse: placuit sequentia quibus campi analytici terminos expandere juxta ac curvarũ doctrinam promovere possem in gratiam discentiũ breviter conjungere.

Cum in numeris et speciebus operationes computandi similes sint neque differre videantur nisi in caracteribus quibus quantitates in istis definite in his indefinite designantur: demiror quod doctrinam de numeris decimalibus nuper inventam (si quadraturam Hyperbolæ per N. Mercatorem demas) nemini in mentem venerit speciebus itidem accommodare præsertim cum ad præclariora viam apperiat. Hujus autem de speciebus doctrinæ cum eodem modo ad

(1) The historical background to this manuscript tract, written out in the hand of David Gregory, is not known with any accuracy. Its first twenty-five pages are evidently based on or condense corresponding passages in Newton's 1671 fluxional tract (§§1/2 above) and from this the late H. W. Turnbull was encouraged to conclude, without any real proof, that the present paper accurately transcribes an abridgment (whose autograph original is no longer in existence) made by Newton himself. (See his edition of the *Correspondence of Isaac Newton*, **2** (1960): 43, note (12) and 158, note (66); **3** (1961): 9, note (1).) This conjecture does nothing to explain the incongruous content of the two final pages of Gregory's version (loose jottings, it would appear, from Newton's Kinckhuysen notes and his 1676 *epistolæ prior et posterior* to Leibniz) nor why these alone are written in English. The only pertinent, dated documentary evidence we have occurs in a letter of 30 January 1689/90 sent by John Craige to Colin Campbell (now in the Campbell collection in Edinburgh University Library), in which Craige wrote out the following 'Probl: 6' virtually word for word as 'a general Method for finding the Curvature of any given curve,...as I copied [it] out of Mr Newtons manuscript'. (An extract from the letter, which omits Craige's transcript of the problem itself, is reproduced in the *Correspondence of Isaac Newton*, **3** (1961): 8–9. In early 1685, when Craige met Newton during a visit to Cambridge, he was allowed to inspect certain 'manuscripta' of Newton's relating to the problem of quadrature and infinite series expansions, one of which must have been the 1671 tract: compare Craige's *Methodus Figurarum...Quadraturas determinandi* (London, 1685): 27, and *De Calculo Fluentium Libri Duo* (London, 1718): Præfatio: [b2^r].) The close similarity between the Craige and Gregory versions of this curvature problem, in contrast to the wide discrepancy between both and corresponding portions of Problem 5 of the 1671 tract, cannot be accidental, but it would now seem impossible to determine whether Craige quoted accurately from a (lost) Newtonian revise—which he perhaps took away with him to Scotland in

Algebram relata sit ac doctrina decimaliũ numerorũ ad vulgarem arithmeticam: operationes Additio, Substractio, Multiplicatio, Divisio et extractio Radicum exinde addisci possunt modo lector utriusq; et Arithmeticæ et Algebræ vulgaris ||[2] peritus fuerit, et noverit correspondentiam inter decimales ||numeros et terminos algebraicos in infinitum continuatos: scīz[5] quod singulis numerorum locis proportione decimali dextrorsum perpetuo decrescentibus correspondent singuli specierum termini secundum seriem dimensionum numeratorum vel denominatorum uniformi progressione in infinitum continuatam (prout factum in sequentibus) ordinati. Et quemadmodum commoditas decimalium in eo consistit ut fractiones omnes et radicales in eos reductæ quodammodo naturam integrorum induant: sic etiam infinitarum specierum commoditas est, quod per eas abstrusiorum terminorum genera (quales sunt fractiones a compositis quantitatibus denominatæ, compositarum radices, et radices affectarum æquationum) possunt ad simplicium genus reduci, ad infinitas nempe fractionum series numeratores ac denominatores simplices habentium, in quibus nullæ sunt aliorum difficultates propemodum insuperabiles. Imprimis itaq; reductiones aliarum quantitatum ad hujusmodi terminos, et methodos computandi minus obvias ostendam, dein hanc analysin ad solutiones problematum applicabo.

Reductiones per divisionem et extractionem radicum a sequentibus exemplis cum similibus operandi modis in Arithmeticâ decimali et speciosa collatis elucescet.

1685 and subsequently lent to Gregory—or whether he made a highly selective abridgment of Newton's original 1671 fluxional tract, which is here copied by Gregory without significant variance. However, the total lack of any hint that a hypothetical Newtonian revise ever existed must make the latter inference the more plausible one. Whatever be the truth, Gregory's script is here reproduced in full so that the reader may compare it with the allied portions of Newton's preceding fluxional treatise and then make up his own mind as to its provenance and importance. Since, with the exception of its last two pages, it does not differ significantly from its parent tract, we restrict our commentary for the most part to textual matters. Some trivial errors and omissions are adjusted within square brackets.

(2) David Gregory MSS, A56. The tract is paginated by Gregory himself.

(3) Gregory's cross-reference to his 'Index contentorum in M. S. 4to signato A' (now in Trinity College, Cambridge. R.2.85: 9). Unfortunately, the manuscript was there recorded as the last entry on a page whose bottom portion has crumbled away, leaving only the figures '56' readable: any useful comparison of the present title with the one there listed is consequently no longer possible. There can be little doubt, however, that this title is Gregory's, and it is, to be sure, strongly reminiscent of one given by him about October 1694 to a similar compendium, 'Isaaci Newtoni Methodus Fluxionum...' (St Andrew's. QA33 G8 D12), in which he incorporated other fragments of Newton's unpublished calculus researches.

(4) Though for want of the corresponding Newtonian autograph (see §1: note (2) above) we cannot be certain, these opening paragraphs appear to repeat word for word the introduction to Newton's 1671 tract.

(5) Read 'scilicet'. The contraction is a favourite one of Gregory's.

||[3] ||*Exempla reductionis per Divisionem.*(6)

Proposito $\frac{aa}{b+x}$ divido aa per $b+x$ in hunc modum.

$$b+x)aa+\ 0\ \left(\frac{aa}{b}-\frac{aax}{bb}+\frac{aaxx}{b^3}-\frac{aax^3}{b^4}+\frac{aax^4}{b^5}-\&c:\right.$$

$$\begin{array}{l} aa+\frac{aax}{b} \\ \hline 0-\frac{aax}{b}+0 \\ -\frac{aax}{b}-\frac{aaxx}{bb} \\ \hline 0+\frac{aaxx}{bb}+0 \quad [\&c] \end{array}$$

et prodit $\frac{aa}{b}-\frac{aax}{bb}+\frac{aaxx}{b^3}-$&c. quæ series in infinitum continuata tantum valet ac $\frac{aa}{b+x}$. Vel posito x primo [divisoris termino] hoc modo

$$x+b)aa(\text{prodibit}\,\frac{aa}{x}-\frac{aab}{x^2}+\frac{aabb}{x^3}-\&c.$$

Ad eundem modum fractio $\frac{1}{1+xx}$ reducitur ad $1-x^2+x^4-x^6+\&$. Vel ad $x^{-2}-x^{-4}+x^{-6}-x^{-8}+$&c. Et fractio $\frac{2x^{\frac{1}{2}}+x^{\frac{3}{2}}}{1+x^{\frac{1}{2}}-3x}$ ad

$$2x^{\frac{1}{2}}-2x+7x^{\frac{3}{2}}-13x^2+34x^{\frac{5}{2}}-\&c.$$

Ubi obiter notandum est quod usurpo x^{-1}, x^{-2}, x^{-3}, &c: pro $\frac{1}{x}, \frac{1}{x^2}, \frac{1}{x^3}$, &c: et $x^{-\frac{1}{2}}, x^{-\frac{2}{3}}, x^{-\frac{1}{4}}$ &c pro $\frac{1}{\sqrt{x}}, \frac{1}{\sqrt[3]{x^2}}, \frac{1}{\sqrt[4]{x}}$, &c. idque ob analogiam rei quæ deprehendi potest ex hujusmodi geometricis progressionib^s^ $x^3, x^{\frac{5}{2}}, x^2, x^{\frac{3}{2}}, x, x^{\frac{1}{2}}, x^0$ (sive 1), $x^{-\frac{1}{2}}, x^{-1}, x^{-\frac{3}{2}}, x^{-2}, x^{-\frac{5}{2}}$ [&c]. Ad hunc modum pro $\frac{aa}{x}-\frac{aab}{x^2}+\frac{aabb}{x^3}$ scribi potest
||[4] ||$aax^{-1}-aabx^{-2}+aabbx^{-3}$. Et sic vice $\sqrt{aa-xx}$ scribi potest $\overline{aa-xx}^{\frac{1}{2}}$: et $\overline{aa-xx}^2$ vice quadrati ex $aa-xx$ et $\left.\frac{abb-y^3}{by+y^2}\right|^{\frac{1}{3}}$ vice $\sqrt[3]{\frac{abb-y^3}{by+y^2}}$, et sic in alijs. Unde merito potestates distingui possunt in affirmativas et negativas, integras et fractas.

(6) Except for slight changes in notation and some rounding off in the central calculation (the work, presumably, of either Craige or Gregory), this repeats word for word the corresponding section on pages 2–3 of the 1671 tract (§§1/2 above).

Exempla reductionis per extractionem radicum.(7)

Proposito $aa+xx$ radicem ejus quadratam ut sequitur extraho.

$$aa+xx\left(a+\frac{xx}{2a}-\frac{x^4}{8a^3}+\frac{x^6}{16a^5}-\frac{5x^8}{128a^7}+\&c.\right.$$

$$\begin{array}{l} aa \\ \hline 0\ +xx \\ \qquad xx+\dfrac{x^4}{4aa} \\ \hline \qquad 0\ -\dfrac{x^4}{4aa} \\ \qquad\qquad -\dfrac{x^4}{4aa}-\dfrac{x^6}{8a^4}+\dfrac{x^8}{64a^6} \\ \hline \qquad\qquad\quad 0\ +\dfrac{x^6}{8a^4}-\dfrac{x^8}{64a^6}\quad [\&c] \end{array}$$

Et prodit $a+\frac{x^2}{2a}-\frac{x^4}{8a^3}+\frac{x^6}{16a^5}-\frac{5x^8}{128a^7}+\&c.$ Ubi notandū quod circa finem operis eos omnes terminos negligo quorum dimensiones transcendunt dimensiones ultimi termini ad quem cupio quotientem solummodo produci. Potest etiam ordo terminorum inverti ad hunc modū x^2+a^2 et radix erit

$$x+\frac{a^2}{2x}-\frac{a^4}{8x^3}+\frac{a^6}{16x^5}-\frac{5a^8}{128x^7}+\&c.$$

||[5] ||Sic ex $aa-xx$ radix quadrata est $a-\frac{x^2}{2a}-\frac{x^4}{8a^3}-\frac{x^6}{16a^5}-\&c.$ et ex $x-x^2$ radix quadrata est $x^{\frac{1}{2}}-\frac{1}{2}x^{\frac{3}{2}}-\frac{1}{8}x^{\frac{5}{2}}-\frac{1}{16}x^{\frac{7}{2}}$ &c. et ex $aa+bx-x^2$ est

$$a+\frac{bx}{2a}-\frac{x^2}{2a}-\frac{bbxx}{8a^3}-\&c.$$

et ex $\frac{1+axx}{1-bxx}$ est $\frac{1+\frac{1}{2}ax^2-\frac{1}{8}aax^4+\frac{1}{16}a^3x^6-\&c}{1-\frac{1}{2}bx^2-\frac{1}{8}bbx^4-\frac{1}{16}b^3x^6-\&c}$. [&] insuper divisione fit

$$\begin{array}{llll} 1+\frac{1}{2}bx^2 & +\frac{3}{8}b^2x^4 & +\frac{5}{16}[b^3]\,x^6 & +\&c. \\ \quad +\frac{1}{2}a & +\frac{1}{4}ab & -\frac{3}{16}ab^2 & \\ & -\frac{1}{8}aa & -\frac{1}{16}a^2b & \end{array}$$

Operationes vero per debitam præparationem non raro abbreviari possunt. Ut in allato exemplo ad extrahendam radicem ex $\frac{1+ax^2}{1-bx^2}$, si non eadem fuisset

(7) Except, again, for slight notational changes and some shortening of the main calculation, an insignificantly variant repeat of the corresponding section on pages 3–4 of Newton's tract in §2. Note that a marginal addition by Newton (see §2: note (6)) has here been incorporated into the main text: decisive proof that the present version is a later compilation.

numeratoris ac denominatoris forma, utramq; multiplicassem per $\sqrt{1-bx^2}$, et sic prodijsset $\dfrac{\sqrt{1\begin{smallmatrix}+a\\-b\end{smallmatrix}xx-abx^4}}{1-bxx}$ et reliquū opus perficeretur extrahendo radicem numeratoris tantū ac dividendo per denominatorem.

Ex hisce credo manifestū est quo pacto radices aliæ possunt extrahi et quælibet compositæ magnitudines (quibuscunque radicibus vel denominatoribus perplexæ ut hic videri[8] est $x^3+\dfrac{\sqrt{x-\sqrt{1-x^2}}}{\sqrt[3]{ax^2+x^3}}-\dfrac{\sqrt[5]{x^3+2x^5-x^{\frac{3}{2}}}}{\sqrt[3]{x+x^2}-\sqrt{2x-x^{\frac{2}{3}}}}$) in series infinitas simplicium terminorum reduci.

||[6] ||*De Reductione affectarum æquationum.*[9]

Propositis vero affectis æquationibus, modus quo radices earum ad hujusmodi series reduci possunt obnixius explicari debet, idq; cum earum doctrina quam hactens in numeris exposuerunt Mathematici, per ambages (superfluis etiam operationibus adhibitis) tradatur, ut in specimen operis in speciebus non debeat adhiberi. Imprimis itaque numerosam affectarum æquationum resolutionem compendiose tradam, dein speciosam similiter explicabo.

Proponatur æquatio $y^3-2y-5=0$, et fit 2 numers utcunque inventus qui minus quam decima sui parte differt a radice quæsita. tum pono $2+p=y$ et pro y substituo $2+p$ in æquationem, et inde nova prodit $p^3+6pp+10p-1=0$, cujus radix p exquirenda est ut quotienti addatur: nempe (neglectis p^3+6p^2 ob parvitatem) $10p-1=0$, sive $p=0{,}1$ ad veritatem proxime accedit; scribo itaq; 0,1 in quotiente. et suppono $0{,}1+q=p$ et hunc ejus fictitium valorem ut ante substituo et prodit $q^3+6{,}3q^2+11{,}23q+0{,}061=0$. et cum $11{,}[23]q+0{,}061=0$
||[7] veritatem appropinquet, sive ||ferè sit $q=-0{,}0054$ (dividendo nempe 0,061 per 11,23, donec tot eliciantur figuræ quot loca primis figuris hujus et principalis quotientis exclusivè intercedunt, quemadmodum hic duo sunt inter 2 et 0,005) scribo $-0{,}0054$ in inferiori parte quotientis, siquidem negativa sit. Et supponens $-0{,}0054+r=q$ hunc ut prius substituo. et sic operationem ad placitum produco pro more subjecti diagrammatis.

(8) Evidently Newton's 'videre' is meant.

(9) The following pages 6–16 of Gregory's text summarize pages 4–15 of Newton's 1671 tract.

(10) Compare page 5 of Newton's tract. Gregory has omitted the light cancellation strokes (see §2: note (13)) which modify the original scheme, and Newton's following explanation.

(11) That is, 'isthac'.

		$\left(\begin{matrix}+2{,}10000000\\-0{,}00544852\end{matrix}\right.$
		$+2{,}09455148$
$2+p=y.$	$+y^3$	$+8+12p+6p^2+p^3$
	$-2y$	$-4\ -2p$
	-5	-5
	Summa$=$	$-1+10p+6p^2+p^3$
$0{,}1+q=p$	$+p^3$	$+0{,}001\ +0{,}03q+0{,}3qq+q^3$
	$+6p^2$	$+0{,}06\ \ +1{,}2q\ +6qq$
	$+10p$	$+1\qquad +10q$
	-1	-1
	Summa$=$	$0{,}061+11{,}23q+6{,}3qq+q^3$
$-0{,}0054+r=q.$	$+q^3$	$-0{,}000000157464+0{,}00008748r-0{,}0162r^2+r^3$
	$+6{,}3q^2$	$+0{,}000183708\qquad -0{,}06804r\qquad +6{,}3r^2$
	$+11{,}23q$	$-0{,}060642\qquad +11{,}23r$
		$+0{,}061$
	Summa$=$	$-0{,}00005416\qquad +11{,}162r$
$-0{,}00004852+s=r.$		&c: (10)

||[8] ||Denique negativam partem quotientis ab affirmativâ subduco et oritur 2,09455148 quotiens absoluta. Præterea notandū est quod sub initio operis si dubitarem an $0{,}1=p$ ad veritatem satis accederet, vice $10p-1=0$, finxissem $6pp+10p-1=0$, et ejus radicis nihilo propioris primam figuram in quotiente scripsissem: et hoc modo secundam vel etiam tertiam quotientis figuram explorare convenit, ubi in æquatione secundariâ circa quam versaris quadratum coefficientis penultimi termini non sit decies major quam factus ex ultimo termino ducto in coefficientem termini antepenultimi. quinimo laborem plerumque minues, præsertim in æquationibus plurimarum dimensionum si figuras omnes quotienti addendas hoc modo (id est extrahendo minorem radicem ex tribus ultimis terminis æquationis ejus secundariæ) quæras: sic enim figuras duplo plures in quotiente qualibet vice lucraberis.

His in numeris sic ostensis consimiles operationes in speciebus explicandæ restant, de quibus convenit [sequentia] prænoscere, 1°: quod e speciebus coefficientibus aliqua præ reliquis (si sint plures) insignienda sit, ea nempe
||[9] quæ est aut fingi potest esse omnium ||longe minima vel maxima vel datæ quantitati vicinissima, cujus rei causa est, ut ob ejus dimensiones in numeratoribus vel denominatoribus terminorum [quotientis] perpetim auctas, illi termini continuo minores et inde quotiens radici propinquior evadat, sicut ante de specie x in exemplis reductionis per divisionem et extractionem radicum manifestum esse potest. Pro istac(11) vero specie in sequentibus ut plurimum usurpabo

etiam x vel z, quemadmodum et y, p, q, r, s &c pro specie radicali extrahendâ. 2°: Si quando fractiones complexæ intricatæ[12] vel surdæ quantitates in æquatione proposita [tolli debent] per methodos Analystis satis notas; quemadmodum si habeatur $y^3+\frac{b^2}{b-x}y^2-x^3=0$ multiplico per $b-x$, et ex facto

$$by^3-xy^3+b^2y^2-bx^3+x^4=0$$

valorem y elicio: vel possum fingere $y\times b-x=v$ et sic scribendo $\frac{v}{b-x}$ pro y oritur æquatio fractione libera. Et sic de cæteris.[13] 3°. æquatione sic præparatâ, opus ab inventione primi termini quotientis initium sumit, de quâ ut et consimili subsequentium terminorum inventione hæc esto regula generalis, cum species indefinita (x vel z) ad quem casum cæteri duo casus sunt reducibiles,[14]
||[10] parva esse fingitur. E terminis in quibus ||species radicalis (y, p, q vel r) non reperitur selige depressissimum respectu dimensionum indefinitæ speciei (x vel z) [&c] deinde alium terminum in quo sit illa species radicalis selige, talem nempe ut progressio dimensionum utriusq; præfatæ speciei a termino prius assumpto ad hunc terminum continuata, quam maxime potest descendat vel minime ascendat. Et si qui sint alij termini quorum dimensiones cum hâc progressione ad arbitrium continuata conveniant eos etiam selige: denique ex his selectis terminis tanquam nihilo æqualibus quære valorem dictæ speciei radicalis et quotienti appone.

Cæterum ut hæc regula magis elucescat placuit insuper ope sequentis diagrammatis exponere. descripto angulo recto BAC, latera ejus BA AC divide
||[11] in partes æquales, et inde normales erige distribuentes ||angulare spatium in æqualia quadrata vel parallelogramma quæ concipio denominata esse à

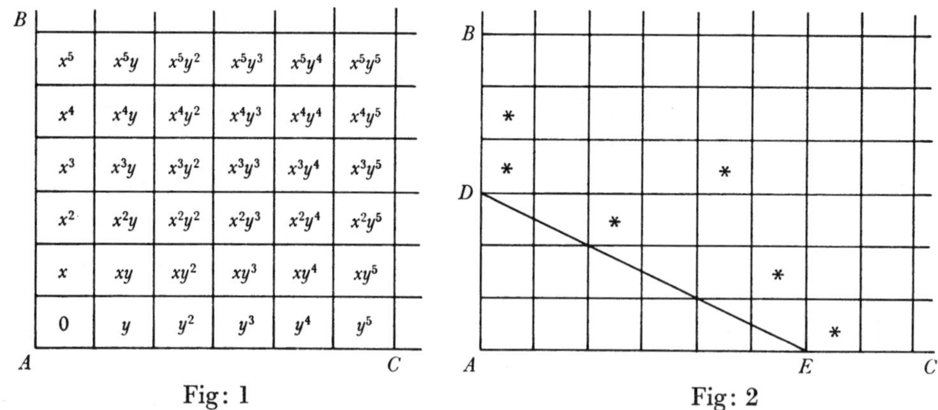

Fig: 1 Fig: 2

(12) See §2: note (21). The original tract has 'complexæ' written in over 'intricatæ', so cancelling it. In ignorance of Newton's convention, the present version plays safe and reproduces both! Newton himself would scarcely have made such a sloppy, illogical transcription and we must infer, surely, that he is not the author of the present abridgment.

dimensionibus specierum x et y prout vides in fig: 1. inscriptas, deinde cum æquatio aliqua proponitur parallelogramma singulis ejus terminis insignio nota aliquâ, et regula ad duo vel plura fortè ex singulis parallelogrammis applicata, quorū unum sit humillimum in columna sinistrâ juxta AB et alia ad regulam dextrorsum sita cæteraque omnia non contingentia regulam supra eam jaceant. Seligo terminos æquationis per parallelogramma contingentia regulam designatos et inde quæro quantitatem quotienti addendam. Sic ad extrahendam radicem [y] ex æquatione $y^6-5xy^5+\frac{x^3}{a}y^4-7a^2x^2y^2+6a^3x^3+b^2x^4=0$, parallelogramma hujus terminis respondentia signo nota aliquâ $*$ ut in fig: 2: dein applico regulam DE ad inferiorem e locis signatis in sinistrâ columna, eamq; ab inferioribus ad superiora dextrorsum gyrare facio donec aliū similiter vel
||[12] forte plura e reliquis signatis in sinistra columna cœperit attingere: ||Videoq; loca sic attacta esse x^3, x^2y^2, et y^6. E terminis itaque $y^6-7a^2x^2y^2+6a^3x^3$ tanquam nihilo æqualibus(15) quæro valorem y, et invenio quadruplicem, $+\surd ax$, $-\surd ax$, $+\surd 2ax$, et $-\surd 2ax$, quorum quemlibet pro initio quotientis accipere licet, prout e radicibus qu[a]mpiam extrahere decretum est.

Sic ex $y^5-byy+9bx^2-x^3=0$ seligo $-byy+9bx^2=0$, et inde obtineo $3x=y$ [pro] initiali termino quotientis. Et ex $y^3+axy+a^2y-x^3-2a^3=0$, seligo

$$y^3+a^2y-2a^3=0$$

et radicem ejus $+a$ ascribo quotienti. Et ex $x^2y^5-3c^4xy^2-c^5x^2+x^7=0$ seligo $x^2y^5+c^7=0$ quod exhibet $[\sqrt[5]{}-]\frac{c^7}{x^2}$ pro initio quotientis. et sic de cæteris.

Cæterum invento hoc termino si is contingat esse negativæ potestatis, æquationem per eandem indefinitæ speciei potestatem deprimo, eo ut non opus sit inter solvendum deprimere, et insuper ut regula de superfluis terminis elidendis mox tradenda apte poss[i]t adhiberi, si proposito

$$8z^6y^3+az^6y^2-27a^{[9]}=0$$

cujus quotiens exordiri debet à $-\frac{3a^3}{[2]z^2}$, deprimo per z^2 ut fiat

$$8z^4y^3+az^4y^2-27a^{[9]}z^{-2}=0.$$

antequam solutionem ineo.

(13) This non-committal phrase replaces some eleven lines of Newtonian text on pages 7–8 of the original tract.

(14) This phrase, a marginal addition in Newton's original text (see §2: note (26)), is here set by Gregory in a ludicrous place in the main sentence.

(15) Note the omission of a following parenthesis, 'et insuper si placet...', on page 9 of the original tract.

||[13] ||Subsequentes quotientum termini eadem methodo ex æquationibus secundarijs inter operandum prodeuntib^s eruuntur.(16)

His præmissis restat ut praxim resolutionis exhibeam.

Sit itaq; $y^3+a^2y+axy-2a^3=0$ æquatio resolvenda, et ex ejus terminis $y^3+a^2y-2a^3=0$ æquatione fictitia juxta tertium è præmissis elicio $y-a=0$, et scribo $+a$ in quotiente. deinde cum $+a$ non accuratè valeat y pono $a+p=y$, et pro y in terminis æquationis in margine descriptis substituo $a+p$ terminosq; resultantes (p^3+3ap^2+axp &c) rursum scribo in margine, ex quibus iterum juxta tertiū è præmissis excerpo terminos $4a^2p+a^2x=0$ pro æquatione fictitia quæ cum exhibeat $p=-\frac{1}{4}x$ scribo $-\frac{1}{4}x$ in quotiente: præterea cum $-\frac{1}{4}x$ non sit accurate p, pono $-\frac{1}{4}x+q=p$, et pro p in terminis marginalibus substituo $-\frac{1}{4}x+q$, terminosque resultantes ($q^3-\frac{3}{4}xq^2+3aq^2$ &c) iterum scribo in margine, ex quibus denuo juxta regulam præfatam seligo terminos $4a^2q-\frac{1}{16}ax^2=0$ pro æquatione fictitiâ quæ cum exhibeat $q=\dfrac{xx}{64a}$ scribo $+\dfrac{x^2}{64a}$ in quotiente, porro cum $\dfrac{x^2}{64a}$ non accurate valeat q pono $\dfrac{x^2}{64a}+r=q$ et procedo ut prius. ut indicat operationis diagramma.

[14] ||

		$\left(a-\dfrac{x}{4}+\dfrac{x^2}{64a}+\dfrac{131x^3}{512a^2}+\dfrac{509x^4}{[16]38[4]a^3}+\&c:\right.$
$a+p=y.$	$+y^3$	$a^3+3aap+3ap^2+p^3$
	$+axy$	$aax+axp$
	$+aay$	a^3+a^2p
	$-x^3$	$-x^3$
	$-2a^3$	$-2a^3$
$-\dfrac{x}{4}+q=p.$	$+p^3$	$-\dfrac{x^3}{64}+\dfrac{3x^2q}{16}-\dfrac{3xq^2}{4}+q^3$
	$+3ap^2$	$+\frac{3}{16}ax^2-\frac{1}{2}axq+3aq^2$
	$+axp$	$-\frac{1}{4}ax^2+axq$
	$+4a^2p$	$-a^2x+4a^2q$
	$+a^2x$	$+a^2x$
	$-x^3$	$-x^3$

(16) The remainder of Newton's paragraph (on page 9 of the 1671 tract) is omitted.

(17) The last line of Newton's scheme and a following three and a half pages of text (on pages 11–14 of the 1671 tract) are omitted.

(18) Newton's text is picked up again in the middle of its page 14.

(19) An 'improvement' not in Newton's original tract.

(20) Some further Newtonian observations on pages 15–16 of the original are omitted.

$\frac{x^2}{64a}+r=q.$		
	$+q^3$	$*$
	$-\frac{3}{4}xq^2$	$*$
	$+3aq^2$	$+\frac{3x^4}{4096a}+\frac{3}{32}x^2r+3ar^2$
	$+\frac{3}{16}x^2q$	$+\frac{3x^4}{1024a}+\frac{3}{16}x^2r$
	$-\frac{1}{2}axq$	$-\frac{x^3}{128}-\frac{1}{2}axr$
	$+4a^2q$	$\frac{ax^2}{16}+4a^2r$
	$-\frac{65}{64}x^3$	$-\frac{65}{64}x^3$
	$-\frac{1}{16}ax^2$	$-\frac{1}{16}ax^2$ [17]

||[15] ||[18] Hactenus indefinitam speciem supposui parvam esse, quod si datæ quantitati vicina supponatur, pro indefinit[è] parvâ differentia pono speciem aliquam, et hâc substitut[â] solvo ut ante. Quemadmodum in

$$\tfrac{1}{5}y^5-\tfrac{1}{4}y^4+\tfrac{1}{3}y^3-\tfrac{1}{2}y^2+y[+]a-x=0,$$

cognito vel ficto x esse ejusdem quantitatis ac a, pono z differentiam scribendo $a+z$ pro x et orietur æquatio[19] $\frac{1}{5}y^5-\frac{1}{4}y^4+\frac{1}{3}y^3-\frac{1}{2}y^2+y[-]z=0$ solvenda ut in præcedentibus. Sin autem species illa supponatur indefinitè magna; pro reciproc[o] ejus indefinite parvo pono speciem aliquam, quâ substitutâ solvo ut ante. Sic habita æquatione $y^3+y^2+y-x^3=0$, ubi x cognoscitur vel fingitur esse valde magnum, pro reciproce parvo $\frac{1}{x}$ pono z et substituto $\frac{1}{z}$ pro x oritur $y^3+y^2+y-\frac{1}{z^3}=0$, cujus radix invenitur $\frac{1}{z}-\frac{1}{3}-\frac{2}{9}z+\frac{7}{81}z^2+\frac{5}{81}z^3+\&c$, et x si placet restituto fiet $y=x-\frac{1}{3}-\frac{2}{9x}+\frac{7}{81x^2}+\frac{5}{81x^3}+\&c.$

||[16] ||Si quando ex aliqua harum trium suppositionum res non omnino aut non commode succedat ad aliam recurrendum [est]. Sic in

$$y^4-x^2y^2+xy^2+2y^2[-2y]+1=0.$$

cum primus terminus obtineri debet fingendo $y^4+2y^2-2y+1=0$, quæ tamen nullam admittit possibilem radicem, tento quid fiet aliter, quemadmodum si fingam x parum differre à $+2$ sive $2+z=x$, substituendo $2+z$ vice x prodibit æquatio $y^4-zzy^2-2zy^2-2y+1=0$ et quotiens ordi[e]tur à $+1$, vel si fingam x indefinite magnam esse, sive $\frac{1}{x}=z$, obtinebitur $y^4-\frac{y^2}{z^2}+\frac{y^2}{z}+2y^2-2y+1=0$. et $+z$ pro initio quotientis. Et hac ratione secundum varias hypotheses procedendo licebit varijs modis extrahere ac designare radices.[20]

||[17] ||VARIORUM PROBLEMATUM CURVAS SPECTANTIUM RESOLUTIO.(21)

PROBLEMA: I

Relatione quantitatum fluentium inter se datâ, fluxionum relationem determinare.(22)

Ad hoc et sequens problema omnes reliquorum difficultates reduci possunt, concipiendo spatiū a motu locali utcunqꝫ accelerato vel retardato describi: et hinc est quod in sequentibus considerem quantitates quasi generatæ essent per incrementum continuum, ad modum spatij quod mobile percurrendo describit. Quantitates sensim crescentes FLUENTES denominabo, ac designabo finalibus literis [v,] x y et z, et celeritates quibus singulæ a motu generante fluunt (quas FLUXIONES vel CELERITATES dicam) designabo l, m, n, et r respectivè: hisce præmissis rem e vestigio aggredior.

Æquationem quâ data relatio exprimitur dispone secundum dimensiones alicujus fluentis quantitatis puta x ac terminos ejus multiplica per quamlibet arithmeticam progressionem ac deinde per $\frac{m}{x}$, et hoc opus in qualibet fluenti quantitate seorsim institue, dein omniū factorū summam pone nihilo æqualem et habes æquationem desideratam.

||[18] ||*Exemplum 1.*

Si quantitatum x et y relatio sit $x^3-axx+axy-y^3=0$, terminos primo secundum x ac deinde secundū y dispositos multiplico ad hunc modum

$$\begin{array}{cccc} x^3 & -ax^2 & +axy-y^3=0. & \\ \frac{3m}{x} & \frac{2m}{x} & \frac{m}{x} & 0 \\ \multicolumn{4}{l}{3mx^2-2max+may} \end{array} \qquad \begin{array}{ccc} -y^3 & +axy & \begin{matrix}-ax^2\\+x^3\end{matrix} \\ \frac{3n}{y} & \frac{n}{y} & 0 \\ \multicolumn{3}{l}{-3ny^2+anx} \end{array}$$

et factorum summa [est] $3mx^2-2max+may-3ny^2+anx=0$, æquatio quæ dat relationem inter fluxiones m et n, nempe si assumas x ad arbitrium, æquatio $x^3-ax^2+axy-y^3=0$ dabit y, quibus determinatis erit

$$m:n::3y^2-ax:3x^2-2ax+ay.$$

---

(21) A Gregorian title which evidently summarizes the opening sentence on page 17 of the original.

(22) A summary of pages 17–20 of the 1671 tract. The opening paragraph reproduces essentially the second sentence and closing paragraph of the introductory to fluents and fluxions on pages 17–18, while the remaining text reproduces the initial paragraph together with Examples 1 and 5 (omitting the long 'Demonstratio' on pages 20–1) of Newton's Problem 1.

(23) Read 'z et x'. One more careless transcription. There is no 'H' in Gregory's figure.

Præparatio in Exempl: 2.

Pone BD ordinatam esse in angulo recto ad AB, et quod ADH sit curva quæ per relationem inter AB et BD æquatione qualibet exhibitam definitur. AB vero dicatur x, et curvæ area ADB ad unitatem applicata dicatur z, dein erige perpendiculum $AC=1$, et per C duc CE parallelam AB et occurrentem BD in E. Et concipiendo has duas superficies genitas esse per motū rectæ BED, manifestum erit quod earum fluxiones (hoc est fluxiones quantitatum $1\times z$, et $1\times x$ sive
||[19] x et z)(23) sunt inter se ut ||BD et BE lineæ generantes, est ergo $r.m::BD$ ad BE (sive 1) adeoque $r=m\times BD$. Et hinc est quod z in æquatione qualibet designante relationem inter x et aliam quamlibet fluentem quantitatem y involvi potest, et tamen fluxionum m et n relatio nihilominus inveniri.

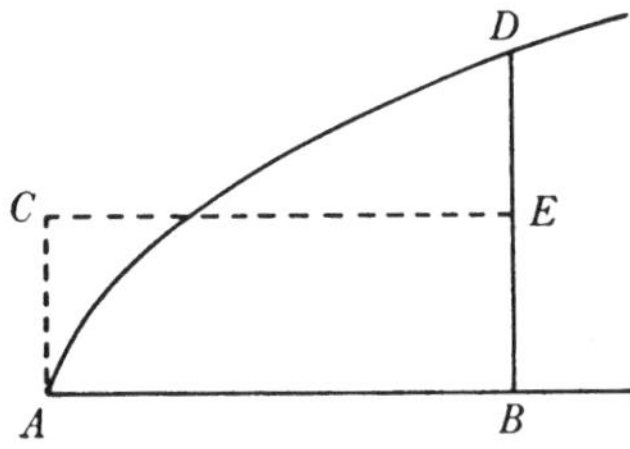

Exemplum: 2.

Quemadmodum si ponatur $z^2+axz-y^4=0$ pro designanda relatione inter x et y, ut et $\sqrt{ax-x^2}=BD$ pro curva determinandâ quæ pro[in](24) erit circulus, æquatio $z^2+axz-y^4=0$ sicut in præcedente dabit $2rz+arx+amz-4ny^3=0$ pro relatione celeritatum m, n et r, et præterea cum sit $r=m\times BD$ sive $m\times\sqrt{ax-x^2}$ pro eo substituo hunc valorem et orietur

$$\overline{2mz+amx}\times\sqrt{ax-xx}+amz-4ny^3=0$$

æquatio definiens relationem celeritatum m et n.

PROBL: 2.

Expositâ æquatione fluxiones quantitatum involvente, invenire relationem.(25)

Quo fluentes quantitates a se invicem clarius distinguantur, fluxionem quæ
||[20] in numeratore rationis ||disponitur, haud impropriè relatam quantitatem nominare possum et alteram ad quam refertur correlatam ut et fluentes quantitates ijsdem respectivè nominib^s insignire: et(26) possis imaginari correlatam quantitatem esse TEMPUS vel potius aliam quamvis æquabiliter

(24) Gregory's manuscript here reads 'prout'.

(25) A short excerpt from page 26 of the 1671 tract: namely, the 'solutio generalis' of Case 1 of Newton's Problem 2. We may guess that the discussion of fluxional equations on Newton's pages 27–40 was too advanced for Gregory at the time he penned the present version of the 1671 tract.

(26) Since the following Case 2 of Newton's Problem 2 is not here discussed, the phrase 'quo sequentia promptiùs intelligantur' is apparently omitted as irrelevant.

fluentem quantitatem qua tempus exponitur et mensuratur, et alteram sive relatam quantitatem esse Spatium quod Mobile ut=cunque acceleratam vel retardatam in illo tempore transigit.

Solutio.

Fluentem quantitatem quam unice æquatio complectitur suppone correlatam quantitatem, et æquatione perinde dispositâ (hoc est faciendo ut ex una parte habeatur fluxionis alterius ad hujus fluxionem relatio et valor ejus in simplicibus terminis ex altera) multiplica valorem rationis fluxionū per correlatam quantitatem, dein singulos ejus terminos divide per numerum dimensionum quib^s illa quantitas inibi afficitur et quod oritur valebit alteram fluentem quantitatem.

Sic expositâ $nn = mn + m^2x^2$, suppono x esse correlatam, et æquatione reductâ

||[21] $\frac{n}{m} = 1 + x^2 - x^4 + 2x^6$ &c. || [Jam] hunc valorem fluxionum n et m multiplico per x et oritur $x + x^3 - x^5 [+] 2x^7$ [&c], quod divido per numerum dimensionum quantitatis x et inde provenit $x + \frac{1}{3}x^3 - \frac{1}{5}x^5 + \frac{2}{7}x^7$ &c quod pono $= y$.

Probl: 3.

Propositæ alicujus curvæ aream determinare.(27)

Problematis resolutio in eo fundatur ut quantitatū fluentium relatio ex relatione fluxionum (per probl: 2) eliciatur. Et imprimis sit(28) recta BD cujus motu quæsita area $AFDB$ describitur super basi AB positione data erectè incedat, Concipe $\square ABEC$ a parte ejus $BE = 1$ interea describi, et posita BE fluxione parallelogrammi erit BD fluxio areæ quæsitæ. Sit ergo $AB = x$, erit etiam $ABEC(1 \times x) = x$ et $BE = m$, Dic insuper aream $AFDB = z$ et erit $BD = r = \frac{r}{m}$ quia $m = [1]$,(29) et proinde per æquationem definientem BD simul definitur fluxionum ratio $\frac{r}{m}$ et exinde per probl: 2: elicietur relatio fluentium x et z.

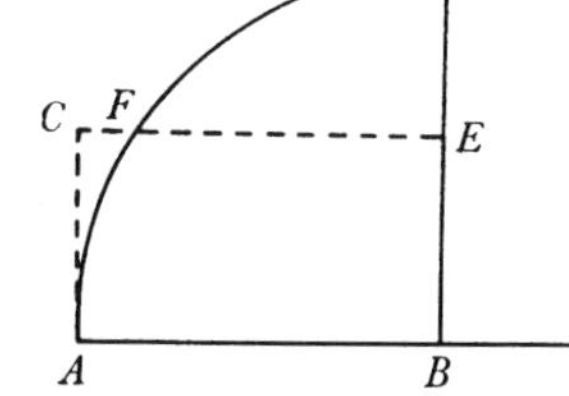

(27) Gregory reproduces the opening page (75a) of Newton's Problem 9, with some added examples from pages 75b and 75c.

(28) A careless version of Newton's 'si'.

(29) Gregory's text reads 'r'.

(30) Compare pages 75f–75i of Newton's 1671 tract above. Newton himself, however, does not compute any areas corresponding to $x = \frac{1}{4}$. Our 'proveniet' replaces Gregory's 'seu'.

(31) A summary version of the early portion of Newton's Problem 7 on page 71 of the preceding tract.

Exempla.

Detur $\frac{x^2}{a}=r$ vel $=\frac{r}{m}$ et per probl: 2: emerg[e]t $\frac{x^3}{3a}=z$.

‖[22] ‖Detur $\frac{x^3}{a^2}=r$, per probl: 2 emerget $\frac{x^4}{4aa}=z$.

Præterea sit $\sqrt{ax}=r$ æquatio ad parabolam et proveniet $\frac{2}{3}\sqrt{ax^3}=z$. Et sic de alijs.

Detur $r=\sqrt{aa+xx}$ æquatio nempe ad hyperbolam seu

$$r=a+\frac{x^2}{2a}-\frac{x^4}{8a^3}+\frac{x^6}{16a^5}\quad \&c:$$

inde per probl: 2: $z=ax+\frac{x^3}{6a}-\frac{x^5}{40a^3}+\frac{x^7}{112a^5}$ &c.

Ad eundem modum esto $r=\sqrt{aa-xx}$ æquatio ad circulum[:] orietur

$$z=ax-\frac{ax^3}{6a}-\frac{x^5}{40a^3}-\frac{x^7}{112a^5}\quad \&c:$$

Atque ita si detur $r=\sqrt{x-x^2}$ æquatio iterum ad circulū[: proveniet] $r=x^{\frac{1}{2}}-\frac{1}{2}x^{\frac{3}{2}}-\frac{1}{8}x^{\frac{5}{2}}$ &c: adeoque erit $z=\frac{2}{3}x^{\frac{3}{2}}-\frac{1}{5}x^{\frac{5}{2}}-\frac{1}{28}x^{\frac{7}{2}}-$ &c:

Facile est numeris has areas exprimere, ut si ponas $AB=x=\frac{1}{4}$ et fractiones ad decimales reducas.[30]

PROBL: 4.

Curvas pro arbitrio multas invenire quarum areæ per finitas æquationes designentur.[31]

Sit AB basis Curvæ et sit $AC=1$, et CE parallela AB, concipe has areas $ACEB$ et ADB a rectis BE et BD per AB delatis generari et earum fluxiones perpetim erunt BE et BD, quare □ $AE=x$ et curvæ aream[32] [$ADB=z$],
‖[23] ‖et fluxiones m et r erunt ut BE et BD, adeoq; posito $m=1=BE$ erit $BD=r$. Jam ad arbitrium assumatur æquatio pro definienda relatione z ad x, exinde per probl: 1 elicietur r, atque ita duæ erunt æquationes quarum posterior curvam definiet et prior aream ejus.

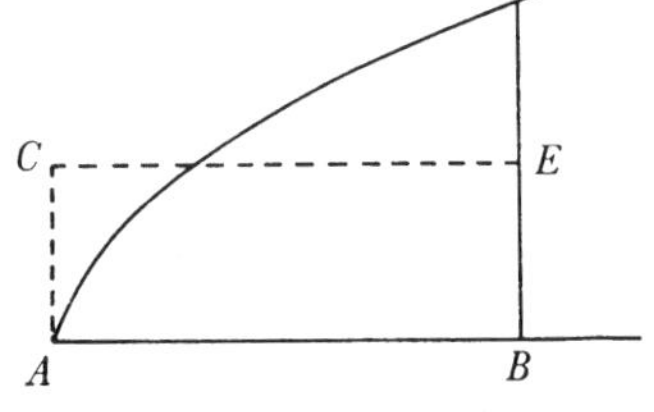

Ex: gr: pono $x^2=z$, inde per probl: 1: $2mx=r$.

2: assume $\frac{x^3}{a}=z$, inde per probl: 1: $\frac{3xx}{a}=r$.

3: Sumatur $\sqrt{ax^3}=z$, et provenit $\frac{3}{2}\sqrt{ax}=r$. Et sic de alijs.

(32) Read 'area'.

PROBL: 5.

Curvas pro arbitrio multas invenire quarum areæ ad aream datæ alicujus curvæ relationem habea[n]t per finitas æquationes designabilem.(33)

Sit FDH data curva atque GEI quæsita, earum applicatæ DB CE arearum quas transigunt fluxiones, erunt ut illæ applicatæ ductæ in velocitates incedendi hoc est in fluxiones basiū. Si[n]t ergo $AB=x$, $BD=v$, $AC=z$, $CE=y$, area $AFDB=s$ et area $AGEC=t$, [harum] fluxiones sint p et q, estque

$$m\times n:v\times y^{(34)}::p:q,$$

quare si $m=1$ et $v=p$ erit $ry=q$ seu $y=\frac{q}{r}$.

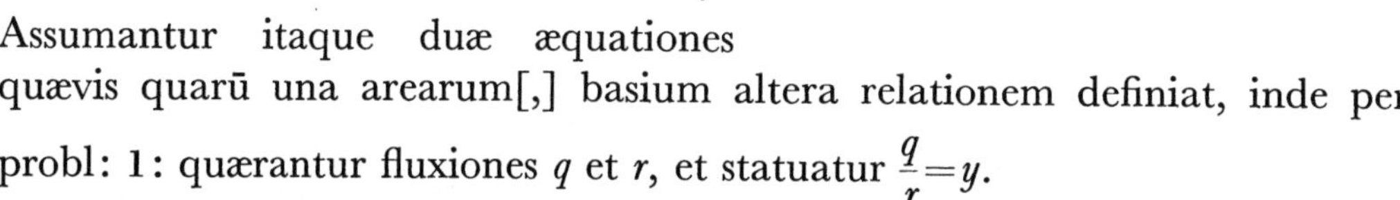

Assumantur itaque duæ æquationes quævis quarū una arearum[,] basium altera relationem definiat, inde per probl: 1: quærantur fluxiones q et r, et statuatur $\frac{q}{r}=y$.

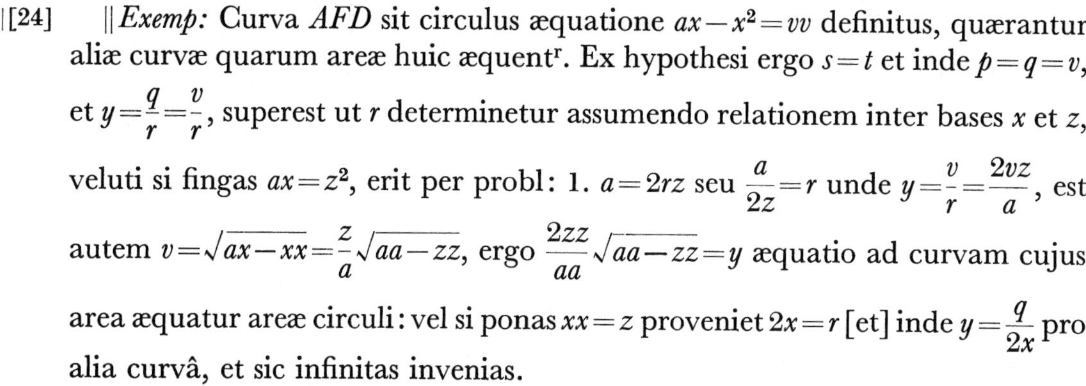

||[24] ||*Exemp:* Curva AFD sit circulus æquatione $ax-x^2=vv$ definitus, quærantur aliæ curvæ quarum areæ huic æquent^r. Ex hypothesi ergo $s=t$ et inde $p=q=v$, et $y=\frac{q}{r}=\frac{v}{r}$, superest ut r determinetur assumendo relationem inter bases x et z, veluti si fingas $ax=z^2$, erit per probl: 1. $a=2rz$ seu $\frac{a}{2z}=r$ unde $y=\frac{v}{r}=\frac{2vz}{a}$, est autem $v=\sqrt{ax-xx}=\frac{z}{a}\sqrt{aa-zz}$, ergo $\frac{2zz}{aa}\sqrt{aa-zz}=y$ æquatio ad curvam cujus area æquatur areæ circuli: vel si ponas $xx=z$ proveniet $2x=r$ [et] inde $y=\frac{q}{2x}$ pro alia curvâ, et sic infinitas invenias.

(33) A slightly variant repeat of the opening to Newton's Problem 8 (on page 72 of the 1671 tract).

(34) Read '$m\times v:r\times y$'.

PROBL: 6.

Curvæ alicujus ad datum punctum CURVATURAM *invenire.*(35)

In ejus resolutionem(36) hæc præmitto generalia. 1°. Circuli eadem est undique curvatura, et inæqualiū circulorum curvaturæ sunt reciproce proportionales diametris. 2°. Si circulus curvam ad partem concavam in dato puncto tangat, sitq; talis magnitudinis ut alius contingens circulus in angulo contactus inscribi nequeat, circulus ille ejusdem est curvitatis ac curva in isto puncto contactus. 3°. Itaque centrum Curvitatis ad aliquod Curvæ punctum est centrum tangentis circuli æqualiter incurvati, et sic radius Curvitatis est pars perpendiculi ad
||[25] ||centrum terminata. 4°. Et proportio curvitatis ad diversa ejus puncta e proportione Curvitatis circulorū æque curvorum innotescit.

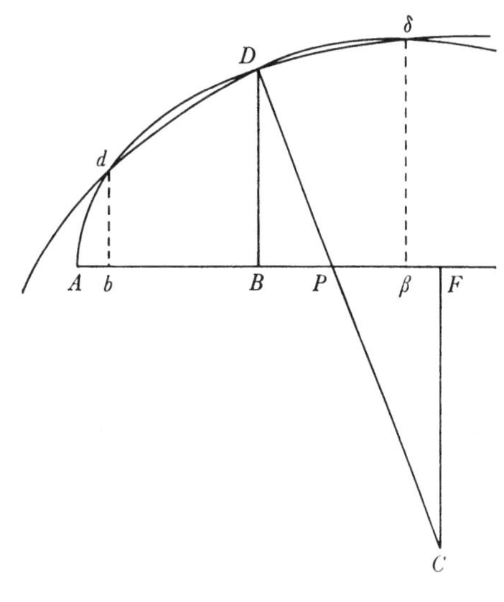

Problema itaq; huc redit ut inveniatur radius vel centrum curvitatis.(37) Centro itaque C et radio CD describatur circulus secans curvam in punctis δ, D, d, et demissis db, DB, $\delta\beta$ et CF perpendicularibus ad basim AB(38) dic, $AB=x$, $BD=y$, $AF=v$, $FC=t$, $DC=s$, et erit $BF=v-x$ ac $DB+FC=y+t$, quorum quadratorum aggregatum æquatur quadrato DC, hoc est $vv-2vx+xx+y^2+2ty+tt=ss$ (quam abbrevias fingendo $vv+tt-ss=qq$, unde $x^2-2vx+y^2+2ty+q^2=0$. postquam vero t, v, et q inveneris si s desideres fac $s=\sqrt{vv+tt-qq}$). Sit jam curva(39) $ax=y^2$, cujus ope ex præcedente extermina x,

(35) Excerpts from Problem 5 of Newton's tract. The first paragraph summarizes (from the tract's page 51) his general observations on curvature and a following sentence repeats his reduction of the problem to finding the radius or centre of 'curvity'. The remaining text reproduces (from pages 62–3) a somewhat altered version of Newton's method of constructing the circle of curvature at a point as that having 3-point contact at the point. We have noticed already (note (1) above) that John Craige in January 1690 communicated an all but identical repeat of the present text to Colin Campbell. To stress the congruence of Craige's and Gregory's texts, the few minor points of difference (punctuation and transpositions apart) are noted below.

(36) 'solutionem' (Craige); compare Newton's 'constructionem'.

(37) Craige here inserts the phrase 'In sequenti figura sit DC perpendicularis Curvæ $AD\delta$ cujus axis AF:'.

(38) Craige writes '...et CF ad basin AB normalibus'.

(39) Craige inserts 'aliqua data puta parabola cujus proprietas est' at this point.

unde prodit hæc æquatio

$$\begin{array}{llll} \frac{y^4}{a^2} - \frac{2v}{a}y^2 & + 2ty & + qq & = 0. \\ \quad + y^2 & & & \\ 4 \quad 2 & 1 & 0 & \\ 2^{(40)} \quad 1 & 0 & -1 & \\ \hline \frac{8y^4}{aa} - \frac{4v}{a}y^2 + 2y^2 & & & = 0, \end{array}$$

quæ tres habet radices æquales. Quare bis per arithmeticam progressionem multiplico[(41)] et provenit sive $v = \frac{2yy}{a} + \frac{a}{2}$. Unde colligitur $BF = 2x + \frac{a}{2}$.[(42)]

||[26] ||[(43)]To extract the square root of a binomial let b be the greater part and c y^e^ lesser of a binomial. I say $\sqrt{\tfrac{1}{2}\sqrt{bb-cc}+\tfrac{1}{2}b}+\sqrt{\tfrac{1}{2}b-\tfrac{1}{2}\sqrt{bb-cc}}$ [is the required root]. So $\sqrt[2]{33+\sqrt{800}} = 5+\sqrt{8}$. and $\sqrt[2]{116+12\sqrt{8}} = 6+\sqrt{80}$.[(44)]

To extract any root of a binomial let n denote y^e^ exponent of the root, b y^e^ greater and c the lesser part of the binomial, r a number not exceeding the root above $\frac{1}{2}$. I say the root of the binome[(45)] is $\dfrac{\frac{\sqrt[n]{bb-cc}}{r}+r}{2}+dd-\sqrt[n]{bb-cc}$,[(46)] in which I suppose $d = \dfrac{\frac{\sqrt[n]{bb-cc}}{r}+r}{2}$. I suppose only y^e^ whole numbers contained in $\frac{\sqrt[n]{bb-cc}}{r}+r$ to be divided by 2. I say this is y^e^ root when y^e^ rational parte is biggest but when it is least[(47)] the root is

$$r - \frac{\sqrt[n]{bb-cc}}{r} + dd + \sqrt[n]{bb-cc}^{(46)} \left[\text{in which } d = \frac{r - \frac{\sqrt[n]{bb-cc}}{r}}{2}\right].$$

So $\sqrt[3]{20+\sqrt{392}} = 2+\sqrt{2}$, for $n=3$, $20=b$, $\sqrt{392}=c$ and $r = 3\frac{1}{2} = \frac{7}{2}$. also

$$\sqrt[3]{25+\sqrt{968}} = 1+\sqrt{8}.^{(48)}$$

---

(40) Read '3' correctly. Interestingly, Gregory's text reveals that he transcribed the correct multiplier from Craige, only later cancelling it in favour of the incorrect '2' here reproduced!

(41) Craige here adds in parentheses (solely for Campbell?) 'Hudenius his method'.

(42) In his letter Craige adds the concluding phrase 'ideoq; habetur punctum F, a quo demissa normalis FC secabit curvæ perpendicularem DC in puncto C, quod est Centrum Curvitatis'. Compare Newton's concluding sentence on page 63 of his 1671 tract.

(43) The abrupt change from Latin to English marks the end of quotation from Newton's 1671 fluxional tract. These two final pages of Gregory's text are evidently loose jottings (by Craige or Gregory?) on other Newtonian mathematical papers. The extremely uncouth style and unconventional orthography ('yee', for example) is scarcely Newton's.

[49]Putting y[e] right sine x and radius r y[e] arch[50] shall be =

$$x+\frac{1\times1\times x^2}{2\times3\times r^2}\times A+\frac{3\times3\times x^2}{4\times5\times r^2}\times B+\frac{5\times5\times x^2}{6\times7\times r^2}\times C+ \ \&c.$$

Or if y[e] arch be x, y[e] right sine[50] is =

$$x-\frac{x^2}{2\times3[\times]r^2}\times A-\frac{x^2}{4\times5\times r^2}\times B-\frac{x^2}{6\times7\times r^2}\times C- \ \&c$$

and y[e] versed sine[50] $=\dfrac{x^2}{1\times2\times r}-\dfrac{x^2}{3\times4[\times r^2]}\times A-\dfrac{x^2}{5\times6\times r^{[2]}}\times B-$ &c.

||[27] ||OF ANGULAR SECTIONS.

Ane arch being given to find another in a given proportion, suppose y[e] diam: d, the chord of the given arch x, y[e] arch sought to y[e] given as n to 1, the Chord of y[e] arch[51] is $nx+\dfrac{1-nn}{2\times3\times dd}\times A[xx]+\dfrac{9-nn}{4\times5\times dd}\times B[xx]+\dfrac{25-nn}{6\times7\times dd}\times C[xx]+$&c: So suppose $n=3$, y[e] chord of y[e] arch triple y[e] arch of y[e] chord x is $3x-4x^3$ for $9-nn=0$ and consequently all y[e] rest of y[e] series breaks off. But if yee would trisect ane arch whose chord is c, $3x-4x^3=c$. and $x=$ to y[e] chord of y[e] sought arch. In like maner to find ane arch quintuple of a given one, its Chord is

(44) Taking $b = 33$, $c = 20\sqrt{2}$ (or $b^2-c^2 = 17^2$) and $b = 116$, $c = 48\sqrt{5}$ (with $b^2-c^2 = 44^2$) respectively. The rule is Kinckhuysen's (see II, **3**, 1, §1: note (37)), but the source for Gregory's present text is probably page 5 of Newton's *Observationes* (II: 374). The examples are to be found in both authors.

(45) Namely, $\sqrt[n]{[b+c]}$.

(46) A radical sign is omitted. Read '$+\sqrt{dd-\sqrt[n]{bb-cc}}$' and '$+\sqrt{dd+\sqrt[n]{bb-cc}}$' respectively.

(47) Note that n must in this case be odd.

(48) Examples of the two cases $b > c$ and $b < c$. In the former, $r = \frac{1}{2}([2\sqrt[3]{\{20+\sqrt{392}\}}]+1)$ so that $d = \frac{1}{2}[3\frac{1}{2}+2/3\frac{1}{2}] = 2$; in the latter, $r = \frac{1}{2}([2\sqrt[3]{\{25+\sqrt{968}\}}]+1)$ so that

$$d = \tfrac{1}{2}[4-7/4] = 1.$$

The rule is, of course, Descartes' (see II, **3**, 1, §1: note (39)) but, again, Gregory's probable source is Newton's manuscript *Observationes* on Kinckhuysen's restatement of it (see II: 382 ff.). The two numerical examples occur in Schooten's *Additamentum* to Descartes' *Geometria*.

(49) The remainder of Gregory's text consists of loose notes relating to the trigonometrical series expansions communicated by Newton in his *epistola prior* of 13 June 1676 to Leibniz. We know that Gregory had acquired by the early 1690's the Collins copy of that letter now in St Andrew's (QA33 G8 D3: 11–17), but the pertinent passages had already been published by John Wallis in Chapter XCV of his *Treatise of Algebra, both Historical and Practical* (London, 1685): 341–7.

(50) That is, $r\sin^{-1}(x/r)$, $r\sin(x/r)$ and $r(1-\cos(x/r))$ respectively. See the *Correspondence of Isaac Newton*, **2** (1960): 25 for the original Latin presentation of these expansions.

(51) Namely, $d\sin(n\sin^{-1}(x/d))$. Compare Newton's *Correspondence*, **2**: 25 for Newton's Latin exposition of this expansion. The numerical instances are evidently Gregory's additions.

$=5x+\frac{1-25}{2\times3\times dd}xxA+\frac{9-25}{4\times5\times dd}xxB$, and consequently to divide ane arch in 5 equall parts let c be the chord of ye given arch and x ye sought Chord shall be found by resolving this $c=5x+\frac{1-25}{2\times3\times dd}xxA+\frac{9-25}{4\times5\times dd}xxB$. So that its evident ye series breaks of when n is ane odd number.

Having A ye Chord of a given arch and B ye Chord of its half, (52)ye given arch is $=\frac{8B-A}{3}$ very near.(53)

---

(52) We omit an incomprehensible phrase 'z ye chord of the tenth of'.

(53) For, to quote Newton's own justification of this *Theorema Hugenianum* in his *epistola prior* (*Correspondence*, **2**: 29–30), 'finge arcum illum esse z, et circuli radium r; juxtaque superiora erit $A\ldots=z-\frac{z^3}{4\times6rr}+\frac{z^5}{4\times4\times120r^4}-$&c. et $B=\frac{1}{2}z-\frac{z^3}{2\times16\times6rr}+\frac{z^5}{2\times16\times16\times120r^4}-$&c. ... indeque emerget ... $\frac{8B-A}{3}=z$ errore tantum existente $\frac{z^5}{7680r^4}-$&c in excessu'.

(1) Nothing precise is known regarding the composition history of this group of papers. In content they are closely akin to Problem 8 of Newton's 1671 fluxional tract (**2**, §2 above) and may well have been intended, in revised form, to be added to that problem. Certainly, in what would appear to be Newton's first documented reference to them in his *epistola posterior* to Leibniz in late October 1676 this connexion is made: 'quando...Curva aliqua non potest geometricè quadrari sunt ad manus alia Theoremata pro comparatione ejus cum Conicis Sectionibus, vel saltem cum aliis figuris simplicissimis quibuscunq; potest comparari.... Pro trinomiis etiam et aliis quibusdam Regulas quasdem concinnavi' (Newton to Oldenburg, 24 October 1676 = *Correspondence*, **2** (1960): 117). Writing to Collins in amplification a fortnight later he was, in criticism of the Leibnizian 'method of Transmutations', somewhat diffidently insistent that 'all that can be done by it may be done better wthout it, by ye simple consideration of ye ordinatim applicatæ....The advantage of ye way I follow you may guess by the conclusions drawn from it wch I have set down in my answer to Mr Leibnitz: though I have not said all there. For there is no curve line exprest by any æquation of three terms, though the unknown quantities affect one another in it, or ye indices of their dignities be surd quantities (suppose $ax^\lambda+bx^\mu y^\sigma+cy^\tau=0$, where x signifies ye base, y ye ordinate, λ, μ, σ, τ ye indices of ye dignities of x & y, & a, b, c known quantities with their signes + or −)...but I can in less then half a quarter of an hower tell whether it may be squared or what are ye simplest figures it may be compared wth, be those figures Conic sections or others. And then by a direct & short way...I can compare them....The same method extends to æquations of four terms & others also but not so generally' (*ibid.* **2**: 179–80). We shall see in the sixth and seventh volumes that Newton (about 1685) drew on his present researches into trinomial and multinomial quadrature for an unpublished scholium intended for his *Principia*, and then subsequently (in late 1691) substantially revised them for the earliest version of his *De Quadratura Curvarum*. (Note that 'A manuscript by Newton on quadratures [?1676]' reproduced by H. W. Turnbull in his edition of *The Correspondence of Isaac Newton*, **2**: 171–4 is in fact a middle portion of the latter, and so to be dated as '1691'.)

(2) The date is conjectured on the basis of our assessment of Newton's handwriting style. That the language of these papers is English rather than Latin reinforces that intuitive opinion. On internal grounds (see note (1)) these researches are evidently a sequel to Problem 8 of the 1671 tract, while the autumn of 1676 is a firm post-date for their composition.

(3) Add. 3962.6: 77v. Much as in Problem 8 of the 1671 tract (compare 2, §2: note (414))

3

THE QUADRATURE OF CURVES DEFINED BY EQUATIONS OF A FINITE NUMBER OF TERMS[1]

[1671?][2]

From the original drafts in the University Library, Cambridge.

§1. FIRST CALCULATIONS.[3]

$$ax^m + bx^n y^o + cy^p = 0. \quad s = t. \quad p = q = y. \quad \frac{y}{r} = v.^{(4)}$$

$$xy = z. \quad y + lx = r. \quad \frac{y}{y+lx} = v. \quad yv + lxv = y. \quad yyv + lzv = yy.^{(5)}$$

$$x : y [::] 1 [:] z. \quad xz = y. \quad z + rx = l. \quad \frac{l-z}{x} = r. \quad \frac{xy}{l-z} = v.$$

$$\frac{xxz}{l-z} = v. \quad max^{m-1} \begin{matrix} +nbx^{n-1}y^o \\ +oblx^n y^{o-1} \end{matrix} + plcy^{p-1} = 0. \quad l = rx - z.^{(6)}$$

$$\frac{max^{m-1} + nbx^{n-1}y^o}{-obx^n y^{o-1} - pcy^{p-1}} = [l=] rx - z.$$

$$\frac{max^{m-2} + nbx^{n-2}y^o - obzx^{n-1}y^{o-1} - \dfrac{pczy^{p-1}}{x}}{[-]obx^n y^{o-1} - pcy^{p-1}} = r.$$

$$\frac{obx^n y^o + pcy^p}{max^{m-2} + nbx^{n-2}[y^o] - obzx^{n-1}y^{o-1} - pczx^{-1}y^{p-1}} = \left[-\frac{y}{r} = -\right] v.^{(7)}$$

---

s and t denote, in Leibnizian equivalent, the areas $\int y\,.\,dx$ and $\int v\,.\,dz$ respectively, while l, p, q, r are the fluxions of y, s, t, z with regard to the base variable x (whose fluxion is assumed to be unity). Note the double use of p, as the fluxion of s and as an arbitrary index.

(4) Newton's equation of s and t implies the equality of their fluxions $p = y$ and $q = vr$.

(5) Newton computes the effect of a second restricting condition $xy = z$. Presumably the fluxion l (or $\dot{y}$) is to be eliminated between this final equation and the trinomial's derivative $max^{m-1} + nbx^{n-1}y^o + l(obx^n y^{o-1} + pcy^{p-1}) = 0$, yielding the corresponding relationship between v, x and y.

(6) Read '$+z$', an error carried through the remaining computation. To compensate, all following terms containing z should be changed in sign.

(7) The relationship between v, x and y corresponding to the alternative restricting condition $xz = y$. As before, this is obtained by eliminating l between the trinomial's derivative and

§2. FIRST EXTENDED DRAFT.(1)

[1] Theorem 1. If $ax^\alpha+bv^\beta x^\epsilon+cv^\gamma x^\zeta+dv^\delta x^\eta$ &c$=0$ be an æquation defining any curve line ABC whose basis AB is x & ordinate BC is v & area ABC is s. Take τ for any number & make $x=z^\tau$ & $\frac{s}{\tau}$ shall be y^e^ area of y^e^ curve DEF whose Base DE is z & ordinate EF y & æquation

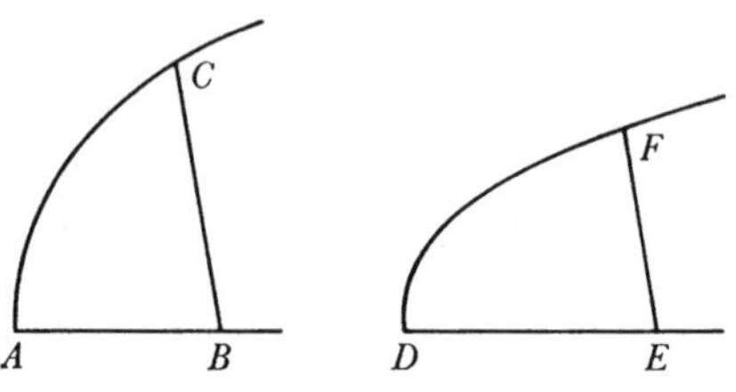

$$ay^\alpha z^{-\alpha\tau+\alpha}+by^\beta z^{-\beta\tau+\beta+\epsilon\tau}+cy^\gamma z^{-\gamma\tau+\gamma+\zeta\tau}+dy^\delta z^{-\delta\tau+\delta+\eta\tau}\ \&\text{c}=0.^{(2)}$$

Theorem 2.(3) If $\overline{ax^\epsilon+bx^\zeta+cx^\eta+dx^\theta+\&\text{c}}|^\lambda=v$. Then is $-s+vx$ the area of y^e^ curve $\lambda\epsilon az^\epsilon+\lambda\zeta bz^\zeta+\lambda\eta cz^\eta$ &c$\times\overline{az^\epsilon+bz^\zeta+cz^\eta\ \&\text{c}}|^{\lambda-1}=y$. Or if

$$\frac{-m+n}{\lambda}\times s+\frac{m}{\lambda}vx^{(4)}=\text{ area of the new curve}$$

the curve shall be

$$\frac{\epsilon m}{+n}az^\epsilon\frac{+\zeta m}{+n}bz^\zeta\frac{+\eta m}{+n}cz^\eta\frac{+\theta m}{+n}dz^\theta\ [\&\text{c}]\times\overline{az^\epsilon+bz^\zeta+cz^\eta+dz^\theta\ [\&\text{c}]}|^{\lambda-1}=y.$$

x and z being equal.(5)

Theorem 3. If $\overline{az^\epsilon+bz^\zeta+cz^\eta+dz^\theta\ \&c}|^\lambda=$areæ curvæ Tunc curva erit(6) $\lambda\epsilon az^{\epsilon-1}+\lambda\zeta bz^{\zeta-1}+\lambda\eta cz^{\eta-1}+\lambda\theta dz^{\theta-1}$ [&c] $\times\overline{az^\epsilon+bz^\zeta+cz^\eta+dz^\theta\ \&\text{c}}|^{\lambda-1}=y.^{(7)}$

$l=(z+rx$ or) $z+xy/v$. Since with neither of the restrictions $xy=z$, $xz=y$ is the relationship between v, x and y simple, Newton evidently saw no point in implementing the approach, passing on in the sequel to consider the 'simplest' curve with whose area the area $s=\int y.dx$ under the given trinomial 'may be compared'.

(1) Add. 3962.6: 74^r^–76^r^.

(2) Compare the opening paragraphs of Problem 8 of the 1671 tract. If, more generally, the curve AC is defined by the given Cartesian equation $f(x,v)=0$ and the area

$$ABC\ (\text{or } s=k\int v.dx)$$

is taken to be τ times the area DEF (or $k\int y.dz$), where $x=z^\tau$ and $A\hat{B}C=D\hat{E}F=\sin^{-1}k$, then $v=\tau y.dz/dx$ and hence $f(z^\tau, yz^{1-\tau})=0$ is the corresponding defining equation of the curve DF.

(3) The following text has been transposed in accordance with Newton's manuscript instruction.

(4) Despite Newton's thrice repeated observation at this point to the contrary,

$$\text{‘}\left(\text{Note }\frac{m}{\lambda}vx+\frac{n-m}{\lambda}s=\text{areæ}\right)\text{’},\quad\text{read}\quad\text{‘}\overline{-\frac{m}{\lambda}+n}\times s+\frac{m}{\lambda}vx\text{’}$$

here.

(5) Immediate conclusions from the equality of $s=\int v.dx$ to $vx-\int x(dv/dx).dx$.

(6) 'If ... = area of the curve, then shall y^e^ curve be.'

Theorem 4. If $\overline{az^\epsilon+bz^\zeta}|^\lambda \times \overline{cz^\eta+dz^\theta}|^\mu$ be the area, then the curve is

$$\overline{\lambda\epsilon az^{\epsilon-1}+\lambda\zeta bz^{\zeta-1}} \times \overline{\mu\eta cz^{\eta-1}+\mu\theta dz^{\theta-1}} \times \overline{az^\epsilon+bz^\zeta}|^{\lambda-1} \times \overline{cz^\eta+dz^\theta}|^{\mu-1}=y.^{(8)}$$

Theorem 5. If $gv^\alpha+cx^\epsilon=1$ or $v=\overline{\dfrac{1-cx^\epsilon}{g}}\Big|^{\frac{1}{\alpha}}$. Make $\dfrac{v^{\frac{1}{\pi}}}{x^{\frac{\epsilon}{\alpha\pi}}}=z$, or $\dfrac{v^\alpha}{x^\epsilon}=z^{\alpha\pi}$. And $xv-\dfrac{+\alpha+\epsilon}{\alpha}s$ shall be y^e area of y^e curve $\dfrac{\pi\epsilon z^{\pi-1}}{\overline{c+gz^{\alpha\pi}}|^{\frac{\alpha+\epsilon}{\alpha\epsilon}}}=y.^{(9)}$ assuming any number for π, & s for y^e area of y^e first curve.(10)

Theorem 6. If $x^\lambda \times \overline{a+bx^\eta+cx^{2\eta}}|^\pi=v$. & $x^\eta=z^\theta$. $\dfrac{\theta}{\eta}s$(11) is y^e area of

$$z^{\frac{\theta\lambda+\theta-\eta}{\eta}} \times \overline{a+bz^{[\theta]}+[c]z^{2[\theta]}}|^\pi=y.^{(12)}$$

---

[2] The Quadrature of all Curves whose Æquations consist of but three termes.

1. If y^e indefinite quantities be affected by one another the affection must be taken away thus. Let y^e æquation be $bv^\alpha+cv^\beta x^\epsilon+dx^\zeta=0$. And make $\dfrac{\alpha-\beta}{-\beta+\alpha+\epsilon}=\tau$ & $x^{\frac{1}{\tau}}=z$ & $\dfrac{s}{\tau}$ shall be y^e area of y^e curve $\overline{\dfrac{by^\alpha+cy^\beta}{-d}}\Big|^{\frac{\beta-\alpha-\epsilon}{\alpha\epsilon+\beta\zeta-\alpha\zeta}}=z.^{(13)}$ supposing its ordinatim applicata be y, or $[z]y-\dfrac{s}{\tau}$ shall be its area if y^e ordinate be z. You may

---

(7) This follows at once on taking the fluxion of the area $DEF = \int y\,.dz = (az^\epsilon+bz^\zeta+\ldots)^\lambda$. Newton henceforth assumes the angles at B and E to be right.

(8) This line is rightly cancelled since it invokes the untenable differential algorithm $d(\alpha\times\beta)/dz = (d\alpha/dz)\times(d\beta/dz)$, where $\alpha = (az^\epsilon+bz^\zeta)^\lambda$, $\beta = (cz^\eta+dz^\theta)^\mu$.

(9) The preceding numerator should read '$\pi z^{\pi-1}$' simply.

(10) On taking $v^\alpha = z^{\alpha\pi}x^\epsilon$, it follows that $\dfrac{dv}{dx} = x^{\frac{\epsilon}{\alpha}}\dfrac{d}{dx}(z^\pi)+\dfrac{\epsilon v}{\alpha x}$, so that

$$xv-s\ (\text{or } \textstyle\int v\,.dx) = \int x\,.dv$$

becomes equal to $\int x^{\frac{\epsilon}{\alpha}+1}.d(z^\pi)+\dfrac{\epsilon}{\alpha}s$; consequently, $xv-(1+\epsilon/\alpha)\,s = \int y\,.dz$ where

$$y = \pi z^{\pi-1}(x^{-\epsilon})^{-[1/\alpha+1/\epsilon]},$$

and it remains to substitute $c+gz^{\alpha\pi}$ for $x^{-\epsilon}$.

(11) Read '$\dfrac{\eta}{\theta}s$'.

(12) This follows straightforwardly on substituting $z^{\theta/\eta}$ for x in $s = \int v\,(dx/dz)\,.dz$ and equating $(\eta/\theta)\,s$ with $\int y\,.dz$.

(13) On f. 77^v Newton has entered a slightly revised version of the preceding: 'In Reg[ula] 1 [make] $\dfrac{\alpha-\beta+\epsilon}{\alpha-\beta}=\tau$. $x^\tau=z$ & τs y^e area of $\overline{\dfrac{by^\alpha+cy^\beta}{[-d]}}\Big|^{\frac{\beta-\alpha-\epsilon}{\alpha\epsilon+\beta\zeta-\alpha\zeta}}=z$ the base'.

also make $\frac{\alpha}{\alpha+\zeta}=\tau$. $x^{\frac{1}{\tau}}=z$ & $\frac{s}{\tau}$ y^e^ area of y^e^ curve $\frac{\overline{by^{\alpha-\beta}+dy^{-\beta}}\Big|^{\frac{\alpha+\zeta}{\alpha\epsilon+\beta\zeta-\alpha\zeta}}}{-c}=z$. Or make $\frac{\beta}{-\beta+\epsilon-\zeta}=\tau$.(14) $x^{\frac{1}{\tau}}=z$ & $\frac{s}{\tau}$ the area of y^e^ curve $\frac{\overline{cy^{\beta-\alpha}+dy^{-\alpha}}\Big|^{\frac{-\zeta-\beta+\epsilon}{\alpha\epsilon+\beta\zeta-\alpha\zeta}}}{-b}=z$.(15) The simplest is to be chosen viz y^t^ wch makes the least denominator of y^e^ index of y^e^ binomials dignity, unless it be a negative integer. for if that should be chosen it would hinder some of y^e^ following operations.(16)

Example. If y^e^ curve be $bv^3-cvvx-dx^4=0$. Then in y^e^ first case is $\tau=\frac{-1}{0}$ &c. In y^e^ second $\tau=-3$. $\frac{1}{x^3}=z$. & $-\frac{1}{3}s$ the area of $\frac{b}{c}y-\frac{d}{cyy}=z$ for $\frac{\zeta[+]\beta-\epsilon}{\alpha\epsilon+\beta\zeta-\alpha\zeta}$(17) the index of y^e^ dignity of y^e^ Binomiall is 1.(18)

Thus may all affected æquations(19) be reduced to pure binomialls. And y^e^ foundation of this reduction is y^e^ first Theorem.

2. In y^e^ next place if y^e^ index of y^e^ binomial be either negative or greater then an unit, it must be increased or diminished by an unit so often by the following Rule untill it come w^th^in y^e^ compas of an unit. Let y^e^ æquation be $\overline{az^{\epsilon}+bz^{\zeta}}\big|^{\lambda}=v$. And $\frac{vx-\epsilon s-s}{\zeta-\epsilon\times\lambda b}$(20) shall be y^e^ area of this new curve $x^{\zeta}\times\overline{ax^{\epsilon}+bx^{\zeta}}\big|^{\lambda-1}=y$. that is $\overline{ax^{\frac{\epsilon\lambda-\epsilon+\zeta}{\lambda-1}}+bx^{\frac{\zeta\lambda}{\lambda-1}}}\Big|^{\lambda-1}=y$. This is for diminishing y^e^ affirmative index wch must

(14) The preceding numerator should read '$-\beta$'.

(15) Three fundamental ways of reducing the given trinomial equation to an explicit form $z=F(y)$ by substituting z^{τ} for x and equating $s=\int v.dx$ to $\tau\int y.dz$ (so that $v=yz^{1-\tau}$ as above). In the first place, on setting $(\beta-\alpha)(1-\tau)+\tau\epsilon=0$ $\left(\text{or } \tau=\frac{\alpha-\beta}{\alpha-\beta+\epsilon}\right)$ and dividing through by $z^{\alpha(1-\tau)}$, the given trinomial may be replaced by the reduced corresponding equation $by^{\alpha}+cy^{\beta}+dz^{\tau\zeta-\alpha(1-\tau)}=0$, where $\tau\zeta-\alpha(1-\tau)=[(\alpha-\beta)\zeta-\alpha\epsilon]/[\alpha-\beta+\epsilon]$. Secondly, on putting $\alpha(1-\tau)-\tau\zeta=0$ $\left(\text{or } \tau=\frac{\alpha}{\alpha+\zeta}\right)$ and dividing through by $y^{\beta}z^{\zeta\epsilon}$, the given trinomial may be replaced by $by^{\alpha-\beta}+cz^{\beta(1-\tau)+\tau(\epsilon-\zeta)}+dy^{-\beta}=0$, in which

$$\beta(1-\tau)+\tau(\epsilon-\zeta)=[\beta\zeta+\alpha(\epsilon-\zeta)]/[\alpha+\zeta].$$

Lastly, on making $\beta(1-\tau)+\tau(\epsilon-\zeta)=0$ $\left(\text{or } \tau=\frac{-\beta}{-\beta+\epsilon-\zeta}\right)$ and dividing through by $y^{\alpha}z^{\tau\zeta}$, the given trinomial may be replaced by $bz^{\alpha(1-\tau)-\tau\zeta}+cy^{\beta-\alpha}+dy^{-\alpha}=0$, where

$$\alpha(1-\tau)-\tau\zeta=[\alpha(\epsilon-\zeta)+\beta\zeta]/[-\beta+\epsilon-\zeta].$$

(16) For a negative integer index implies the existence of non-zero algebraic terms in the denominator of the value of the ordinate, a threat to the possibility of an exact quadrature. A tentative continuation of the paragraph, cancelled abruptly, reads '& therefore one o[f?]'.

(17) The numerator should be '$\alpha+\zeta$', of course.

be repeated till it be an unit or less. But for increasing y^e^ negative make[(21)] $\overline{ax^{\frac{\epsilon\lambda+\epsilon-\zeta}{\lambda+1}}+bx^{\frac{\zeta\lambda}{\lambda+1}}}\Big|^{\lambda[+]1}=y$.[(22)] & y^e^ area of this curve shall be

$$\frac{yx}{\epsilon+1}+\frac{\lambda\epsilon-\lambda\zeta+\epsilon-\zeta}{\epsilon+1}s.^{(23)}$$

By this repeated y^e^ index may always [be] brought to be affirmative unles when λ is a negative integer for then it can only be brought to a negative unit.[(24)] If ϵ happen to be -1[(25)] change y^e^ order of y^e^ terms signified by $ax^{\epsilon}+bx^{\zeta}$ using y^e^ one for y^e^ other. And observe this also, to use[(26)] ax^{ϵ} for that term wch gives y^e^ simplest conclusion. You may first of all take away ζ by y^e^ first reduction, but it is not materiall.

3. Having reduced y^e^ æquation thus far, let it be $\overline{ax^{\eta}+bx^{\theta}}|^{\frac{\mu}{\nu}}=y$, where $\frac{\mu}{\nu}$ signifies y^e^ index of y^e^ dignity of y^e^ binomial reduced to y^e^ least fraction.[(27)] Let $\frac{p}{q}$ be y^e^ least fraction equal to $\frac{\eta\mu-\nu}{\theta\mu-\nu}$ & putting $p\nu-q\nu=t$ from $p+\frac{\nu}{[\mu]}$ & $q+\frac{\nu}{[\mu]}$ subduct y^e^ greatest multiplex of it that can be subducted from them both if they be both affirmative: or if one or both be negative add y^e^ least[(28)] multiplex of t

(18) There is something badly wrong with this example. For the given curve τ is, in the first case, $\frac{1}{2}$ and, in the second, $\frac{3}{7}$, while the 'index of y^e^ dignity of y^e^ Binomiall' is then -7. Newton's values hold for the revised trinomial $bv^3-cv^4x-dx^{-4}=0$.

(19) Consisting of but three terms, that is.

(20) The numerator should read '$vx-\lambda\epsilon s-s$'. The result follows at once by any equivalent to integration by parts, since $vx-s$ (or $\int v.dx$) $=\int x.dv$ with $x.dv/dx=\lambda[\epsilon v+(\zeta-\epsilon)by]$, so that $vx-(\lambda\epsilon+1)s=(\zeta-\epsilon)\lambda b\int y.dx$.

(21) Newton first wrote 'And for increasing y^e^ negative suppose also y^e^ given æquation b[e]'.

(22) That is, '$x^{-\zeta}\times\overline{ax^{\epsilon}+bx^{\zeta}}|^{\lambda+1}=y$'.

(23) Both denominators should read '$\lambda\epsilon+\epsilon-\zeta+1$', while a factor b is missing from the right-hand term. Some considerable guile, not at once apparent in Newton's bare statement of the result, has evidently been needed to achieve it, either by an equivalent to integration by parts or by inverting and suitably amending the previous result. It is not, in particular, wholly self-evident that the term yx should appear in the 'area', nor that its derivative $d(yx)/dx$ may be expressed in the form $(\lambda+1)(\epsilon y+[\zeta-\epsilon]bv)+(1-\zeta)y$.

(24) The integral $\int(Ax^{\alpha}+Bx^{\beta})^{-1}.dx$, $\alpha\neq\beta$, will in general require the logarithmic function for its exact evaluation, while Newton is here concerned only with rational algebraic quadrature.

(25) Read 'If $\lambda\epsilon+\epsilon-\zeta$ happen to be -1': Newton wishes to guard against the case where the denominators in his above expression become zero.

(26) A first cancelled continuation reads 'y^e^ index of y^e^ dignity of that term for ϵ'.

(27) Newton first continued: 'Then reduce $\eta\mu-\nu$ & $\theta\mu-\nu$ to their least common denominator wch reject & divide their numerators by y^e^ greatest common divisor, & supposing the difference of the products be t subduct y^e^ greatest multiplex of t that can be subducted from them both...'.

(28) A following word 'integer' is cancelled.

y[t] is requisite to destroy y[e] negative signe. let y[e] results be γ & δ & $\overline{az^\gamma+bz^\delta}|^{\frac{\mu}{\nu}}=v$ shall be y[e] simplest curve wth whose area the area of $\overline{ax^\eta+bx^\theta}|^{\frac{\mu}{\nu}}=y$ may be compared:(29) unless $2\gamma-t$ & $2\delta-t$ or $3\gamma-t$ & $3\delta-t$ or $4\gamma-t$ & $4\delta-t$ &c or $3\gamma-2t$ & $3\delta-2t$ &c be less then γ & δ, or y[t] γ & δ or $2\gamma-t$ & $2\delta-t$ or $3\gamma-t$ & $3\delta-t$ or $3\gamma-2t$ & $3\delta-2t$ &c be divisible by μ or by any factor of μ, so as to make a product less then γ & δ. For then y[e] least number that can be found by any of these ways must be used for γ & δ.(30) Excepting only when $\frac{\gamma\mu-\nu}{\gamma\nu-\delta\nu}$ or $\frac{\delta\mu-\nu}{\gamma\nu-\delta\nu}$ may be a simpler fraction then $\frac{\mu}{\nu}$. For y[n] those numbers are to be taken for γ & δ which make one of those two fractions the simplest that may be.(31)

Having thus found y[e] simplest curve(32) wth w[ch] y[e] propounded curve may be compared: it remains that we show how to compare them. And this is done thus. Make $\frac{\eta-\theta}{\gamma-\delta}=\tau$ & $x^\tau=z$(33) & $\frac{\mu\theta-\mu\delta\tau+\nu\tau-\nu}{\nu\tau}=\beta$.(34) & τs shall be y[e] area of y[e] curve $z^\beta\times\overline{az^\gamma+bz^\delta}|^{\frac{\mu}{\nu}}$ in w[ch] β will be either 0 or some multiplex of $\overline{\gamma-\delta}$. If 0 y[e] work is done; if some multiplex of $\gamma-\delta$, & also affirmative then let its area be P & y[e] area of y[e] curve $z^{\beta-\gamma+\delta}\times\overline{az^\gamma+bz^\delta}|^{\frac{\mu}{\nu}}=Y$ be Q & $\frac{\mu\gamma+\nu\beta-\nu}{\nu}aP+\frac{\mu\delta+\nu\beta-\nu}{\nu}bQ$

(29) Newton plans to find integers $k\mu$ and τ such that, when the substitution $x^\tau=z$ is made, there proves to be

$$(\tau s \text{ or}) \ \tau\int(ax^\eta+bx^\theta)^{\mu/\nu}.dx = \int z^{k\mu(\gamma-\delta)}(az^\gamma+bz^\delta)^{\mu/\nu}.dz.$$

At the outset, unfortunately, he mistakenly evaluates τs as $\int z^{1-1/\tau}(az^{\eta/\tau}+bz^{\theta/\tau})^{\mu/\nu}.dz$, but if we accept this we may restore his further analysis without trouble. On putting $\eta-\nu/\mu=p\tau$, $\theta-\nu/\mu=q\tau$ and $\nu(\gamma-\delta)=t$, it follows that $\gamma=p+\nu/\mu-kt$, $\delta=q+\nu/\mu-kt$ and accordingly $\gamma-\delta=p-q=(\eta-\theta)/\tau$. To find the 'simplest' curve $y=(az^{\gamma-kt}+bz^{\delta-kt})^{\mu/\nu}$ with whose area the area $s=\int v.dx$ may be 'compared', Newton adds the conditions that p and q be co-prime (to make τ as large as possible) and that the rational k be chosen to make γ and δ least positive in value.

(30) If the rational fraction $k=r/s$ is such that both $s\gamma-rt<\gamma$ and $s\delta-rt<\delta$ hold true, then the substitution $z=u^s$ will yield $s\int u^{s-1}(au^{s\gamma-rt}+bu^{s\delta-rt})^{\mu/\nu}.du$ as 'simplest' comparable integral. Similarly, where M is a factor of μ (that is, since $p\mu/q$ is integral while p, q are co-prime, M is a factor of s), the substitution $z=u^{s/M}$ will simplify the case where both $s\gamma-rt<M\gamma$ and $s\delta-rt<M\delta$ hold true.

(31) This somewhat mysterious sentence is explained below (see note (39)): Newton intends the reduced defining equation, say $(ax^\gamma+bx^\delta)^{\mu/\nu}=v$, to be simplified by means of his first rule. Unfortunately he has yet again made a small but consequential error, computing that the substitutions $x=z^\tau$, $v=yz^{\tau-1}$ will convert the area $\int v.dx$ into its equal

$$\tau\int y.dz=\tau(yz-\int z.dy).$$

Accepting this, we derive the erroneous defining equation $az^{\gamma\tau}+bz^{\delta\tau}=(yz^{\tau-1})^{\nu/\mu}$, which (on setting $\gamma\tau$ or $\delta\tau$ equal to $(\tau-1)\nu/\mu$, that is, $\tau=\nu/(\nu-\gamma\mu)$ or $\tau=\nu/(\nu-\delta\mu)$ respectively) assumes the 'simpler' alternative forms

$$z=([y^{\nu/\mu}-a]/b)^{(\gamma\mu-\nu)/\nu(\gamma-\delta)} \quad\text{and}\quad z=([y^{\nu/\mu}-b]/a)^{-(\delta\mu-\nu)/\nu(\gamma-\delta)}.$$

shall be $z^{\frac{\mu\beta-\mu\gamma+\mu+\nu-\nu\beta}{\mu}} \times \overline{az^\gamma+bz^\delta}|^{\frac{\mu}{\nu}}$(35). Whence Q being known P is also given. In like manner substituting $\beta-\gamma+\delta$ for β you may proceed to y^e^ area of $z^{\beta-2\gamma+2\delta}\times\overline{az^\gamma+bz^\delta}|^{\frac{\mu}{\nu}}$ & so on substracting at every operation $\gamma-\delta$ from β till β vanishing you come to y^e^ curve $\overline{az^\gamma+bz^\delta}|^{\frac{\mu}{\nu}}$ whose area therefore given will give y^e^ area of $z^\beta\times\overline{az^\gamma+bz^\delta}|^{\frac{\mu}{\nu}}$. But if β be negative change y^e^ terms signified by az^γ & bz^δ for one another so that $\gamma-\delta$ may be negative & proceede as before till β vanish. Thus have you a comparison wth y^e^ simplest curve.(36) Unless either $\frac{\gamma\mu-\nu}{\nu\gamma-\nu\delta}$ or $\frac{\delta\mu-\nu}{\nu\gamma-\nu\delta}$ be an integer or a simpler fraction then $\frac{\mu}{\nu}$, For then putting $\overline{az^\gamma+bz^\delta}|^{\frac{\mu}{\nu}}=y$(37) & its area D, if $\frac{\nu\mu-\nu}{\gamma\nu-\nu\delta}$ be simplest make $\frac{\nu}{\nu-\gamma\mu}=\tau.\ x^{\frac{1}{\tau}}=z$ & $\frac{D}{\tau}$ shall be y^e^ area of y^e^ curve $\overline{\frac{y^{\frac{\nu}{\mu}}-a}{b}}\Big|^{\frac{\gamma\mu-\nu}{\gamma\nu-\delta\nu}}=z$. or if $\frac{\delta\mu-\nu}{\gamma\nu-\delta\nu}$ be simplest make $\frac{\nu}{\nu-\delta\mu}=\tau$(38) & $x^{\frac{1}{\tau}}=z$. & $\frac{D}{\tau}$ shall be y^e^ area of $\overline{\frac{y^{\frac{\nu}{\mu}}-b}{a}}\Big|^{\frac{\delta\mu-\nu}{\gamma\nu-\delta\nu}}=z$: z now signifying

(32) That is, $[y=]\ z^{l(\gamma-\delta)}(az^\gamma+bz^\delta)^{\mu/\nu}$, where $l=k\mu$ is integral.

(33) A first cancelled continuation reads '& τs shall be y^e^ area of y^e^ curve whose ordinate is $\overline{az^{\frac{\mu\eta-\nu+\nu\tau}{\mu\tau}}+bz^{\frac{\mu\theta-\nu+\nu\tau}{\mu\tau}}}\Big|^{\frac{\mu}{\nu}}$. Then if y^e^ indexes of y^e^ dignities of z here be both affirmative'. (The numerators of these 'indexes of y^e^ dignities of z should, in fact, be $\mu\eta+\nu-\nu\tau$ and $\mu\theta+\nu-\nu\tau$ respectively. Compare note (29) and the following redraft.)

(34) The numerator should read '$\mu\theta-\mu\delta\tau-\nu\tau+\nu$'. See note (29) above.

(35) This is badly wrong. If for simplicity we set

$$az^\gamma+bz^\delta = \alpha \text{ (so that } z.d\alpha/dz = a\gamma z^\gamma+b\delta z^\delta)$$

and, with Newton, make $\int z^\beta\alpha^{\mu/\nu}.dz = P$ and $\int z^{\beta-\gamma+\delta}\alpha^{\mu/\nu}.dz = Q$, identification of

$$z^r\alpha^s = a(r+\gamma s)\int z^{\gamma+r-1}\alpha^{s-1}.dz+b(r+\delta s)\int z^{\delta+r-1}\alpha^{s-1}.dz$$

with $AP+BQ$ (where A, B are constants) yields $r=\beta-\gamma+1$, $s=\mu/\nu+1$,

$$A = a(r+\gamma s) = a(\beta+1+\gamma\mu/\nu),\quad B = b(r+\delta s) = b(\beta-\gamma+\delta+1+\delta\mu/\nu).$$

This error in detail does not, of course, affect the truth of the reduction scheme proposed by Newton.

(36) Newton has cancelled the clarification 'that is whose index of y^e^ dignity of y^e^ binomial is $\frac{\mu}{\nu}$'. A first draft of the following reads: 'If you would now try to reduce y^e^ curve to a binomial wth a simpler dignity put $\overline{az^\gamma+dz^\delta}|^{\frac{\mu}{\nu}}=y$ & make y y^e^ base & z y^e^ ordinate & repeat all y^e^ former reductions'.

(37) For consistency with the following read '$\overline{ax^\gamma+bx^\delta}|^{\frac{\mu}{\nu}}=v$'.

(38) Read '$\frac{\nu}{\delta\mu-\nu}=\tau$'. Similarly the numerator of the 'dignity' of the following binomial should be '$\nu-\delta\mu$'.

y^e^ ordinate.[39] And this curve is to be reduced by y^e^ precedent rules to y^e^ simplest form, that is so that y^e^ index of y^e^ dignity of y^e^ binomial be affirmative & less then an unit, & y^e^ indices of y be integer & y^e^ least integers that may be. Thus will y^e^ area at length come to a rectilinear figure if it be capable of a Geometric quadrature or to a conick section if it be comparable wth one of those, or to ye simplest figure it may be compared w^th^.[40] Every operation bringing ye figure to a degree more & more simple till it become capable of no further reduction.

---

§3. THE FINAL, AUGMENTED VERSION.[1]

[1] The Quadrature of all curves whose æquations consist of but three terms.

NB. x & z are y^e^ bases & A, B, C &c y^e^ propounded curves.[2]

Let y^e^ propounded curve be $bv^\alpha+cv^\beta x^\epsilon+dx^\zeta=0$. Make $\dfrac{\alpha-\beta+\epsilon}{\alpha-\beta}=\tau$, & $x^\tau=z$, & τA shall be y^e^ area of $by^\alpha+cy^\beta+dz^{\frac{\alpha\epsilon+\beta\zeta-\alpha\zeta}{\beta-\alpha-\epsilon}}=0$, supposing their ordinatim applicatæ be v & y & bases x & z. But if $\tau=0$ or inf[inity] or y^e^ index of z here be infinite, or a negative integer make $\dfrac{\alpha+\zeta}{\alpha}=\tau$, $x^\tau=z$, & τA shall be y^e^ area of $by^{\alpha-\beta}+dy^{-\beta}+cz^{\frac{\alpha\epsilon+\beta\zeta-\alpha\zeta}{\alpha+\zeta}}=0$. This second Rule is to be used also when ϵ is nothing & β is a negative integer. You may also make $\dfrac{-\epsilon+\beta+\zeta}{\beta}=\tau.\ x^\tau=z$ & τA

---

(39) See note (31) above. Observe that $D/\tau = \int y.dz = yz-\int z.dy$.

(40) The continuation 'wch may be acquiesced in if none of y^e^ former reductions will profit any further' is cancelled. Newton first concluded that 'Every operation brings the Figure to a simpler degree: & when none of these reductions will profit any further you may acquiesce [in y^t^ reduction]'.

(1) Add. 3962.6: 70^r^–72^r^.

(2) In the sequel Newton usually sets $A = \int v.dx$, $B = \int y.dz$.

(3) Compare §2: note (15). The substitution $x^\tau = z$ subject to the restriction that

$$\tau\int v.dx = \int y.dz \text{ (or } v = yz^{1-1/\tau})$$

converts the given defining equation $bv^\alpha+cv^\beta x^\epsilon+dx^\zeta = 0$ into the trinomial

$$by^\alpha z^{\alpha(1-1/\tau)}+cy^\beta z^{\beta(1-1/\tau)+\epsilon/\tau}+dz^{\zeta/\tau} = 0.$$

According as there is put $\alpha(\tau-1) = \beta(\tau-1)+\epsilon$, or $\zeta = \alpha(\tau-1)$, or $\beta(\tau-1)+\epsilon = \zeta$ (that is, as $\tau = 1+\epsilon/(\alpha-\beta)$ or $1+\zeta/\alpha$ or $1+(\zeta-\epsilon)/\beta$), this equation may be written as either

$$by^\alpha+cy^\beta+dz^\lambda = 0,\quad \lambda = [(\alpha-\beta)\zeta-\alpha\epsilon]/(\alpha-\beta+\epsilon),$$

or $by^{\alpha-\beta}+cz^\mu+dy^{-\beta} = 0$, $\mu = [\alpha\epsilon+(\beta-\alpha)\zeta]/(\alpha+\zeta)$, or lastly

$$bz^\pi+cy^{\beta-\alpha}+dy^{-\alpha} = 0,\quad \pi = [(\alpha-\beta)\zeta-\alpha\epsilon]/(\zeta-\epsilon+\beta).$$

Evidently, those alternatives for which τ is either zero or infinity are to be avoided, likewise those for which τ is negative.

y^e^ area of $cy^{\beta-\alpha}+dy^{-\alpha}+bz^{\frac{\alpha\epsilon+\beta\zeta-\alpha\zeta}{\epsilon-\beta-\zeta}}=0$.(3) And in one of these three cases y^e^ denominator of y^e^ index of y^e^ dignity of z divided by y^e^ numerator will always be an(4) affirmative integer if y^e^ curve be capable of a geometrick quadrature. Whence y^e^ valor of z being rational y^e^ quadrature is manifest without further transmutation. But if it admit of comparison with y^e^ area of y^e^ Hyperbola or Ellipsis, the numerator of y^e^ index will always in two of y^e^ three cases be y^e^ number 2.(5) In w^ch^ & all other more composed cases proceed thus

1. Let y^e^ base be x & y^e^ ordinate be $\overline{cx^{\epsilon}+dx^{\zeta}}\big|^{\theta}$, & making $1+\epsilon\theta=\tau$ & $x^{\tau}=z$, τB shall be y^e^ area of $\overline{c+dz^{\frac{\zeta-\epsilon}{\tau}}}\Big|^{\theta}$. Or making $1+\zeta\theta=\tau$ & $x^{\tau}=z$, τB shall be y^e^ area of $\overline{cz^{\frac{\epsilon-\zeta}{\tau}}+d}\Big|^{\theta}$.(7) Chose that case where τ is least, & may by y^e^ following operations be reduced to be least, unless it make y^e^ index of y^e^ dignity of z a negative integer: for that would hinder the following operation.

B signifies y^e^ area of $\overline{cx^{\epsilon}+dx^{\zeta}}\big|^{\theta}$.(6)

2. The ordinate being reduced to this form in wch one of y^e(8)^ terms c or d is cleared of its coefficient z: the next work is to diminish y^e^ index of its dignity by continuall adding or subducting an unit. Let the ordinate be $\overline{ax^{\eta}+b}\big|^{\theta}=v$, & $\dfrac{\eta A+A-vx}{\eta\theta b}$(9) shall be y^e^ area of $\overline{ax^{\eta}+b}\big|^{\theta-1}=y$. Or backwards $\overline{ax^{\eta}+b}\big|^{\theta+1}=y$ & it's area shall be $\dfrac{\eta A+\theta\eta A+yx}{\eta+1}$.(9) By repeating y^e^ first when y^e^ index θ is affirmative, or y^e^ last when it is negative y^t^ index may be always brought to be affirma-

(4) 'unit'. is cancelled.

(5) For the reduced trinomial to define a central conic it must be of the form

$$z^2 = Ry^2+Sy+T,$$

with R and either S or T non-zero. Possible cases are $(\lambda = \frac{1}{2})\,\alpha = 2$, $\beta = 0, 1, 2$ or $\beta = 2$, $\alpha = 1, 0$; $(\mu = \frac{1}{2})\,\alpha = 0$, $\beta = -2$ or $\alpha = 1$, $\beta = -1$; $(\pi = \frac{1}{2})\,\alpha = -2$, $\beta = 0$ or $\alpha = 1$, $\beta = 1$ or $\alpha = 0$, $\beta = 2$.

(6) Compare the preceding marginal note.

(7) An immediate application of the preceding reduction to the trinomial

$$cx^{\epsilon}+dx^{\xi}-v^{1/\theta} = 0,$$

where $B = \int v\,.dx$.

(8) 'nomina[l]' is cancelled.

(9) Read $\dfrac{\text{`}\theta\eta A+A-vx\text{'}}{\eta\theta b}$ and $\dfrac{\text{`}\eta bA+\theta\eta bA+yx\text{'}}{\theta\eta+\theta+\eta+1}$ respectively. These are exactly the same errors, notation apart, as in the first draft (see §2: notes (20) and (23)). If we define $az+b = \pi$ and $A(k,\theta) = \int z^k\pi^{\theta}.dz$, then, since $\dfrac{d}{dz}(z^{k+1}\pi^{\theta}) = (\theta+k+1)z^k\pi^{\theta}-\theta bz^k\pi^{\theta-1}$, it follows that $(\theta+k+1)A(k,\theta) = z^{k+1}\pi^{\theta}+\theta bA(k,\theta-1)$. Newton's (corrected) recursive forms follow readily on taking $k = \eta^{-1}-1$, since the substitution $z = x^{\eta}$ yields

$$A(\eta^{-1}-1,\theta) = \eta\int(ax^{\eta}+b)^{\theta}.dx.$$

tive & less then an unit, or if you please to be negative & nearer to nothing(10) then a negative unit. Both cases may be retained as simplest in their degree.

And by y^e same means an unit may be afterwards so often added to or subducted from $\frac{1}{\eta}$, that is y^e numerator of η to or from its denominator, till y^e denominator be less then y^e numerator whether negative or affirmative.(11) Which work being done you will have reduced y^e Figure to y^e simplest cases it can be reduced to while one of its terms is unaffected wth y^e unknown quantities. of which cases there are 4, one wth both indexes(12) affirmative one wth both negative & two wth one affirmative & y^e other negative. Chose.

Now though this is often(13) y^e simplest form where one of y^e terms is free from being coaffected,(14) yet sometimes y^e curve may be reduced to a simple degree by this meanes. Let y^e æquation be $ax^{\frac{\beta}{\gamma}}+b+cv^{\frac{\delta}{\epsilon}}=0$. & if either of y^e indexes be integer let it be $\frac{\delta}{\epsilon}$, ϵ being an unit & put θ for y^e greatest common divisor of β & δ, & making $x^{\frac{\theta}{\gamma}}=z$, $\frac{\theta}{\gamma}A$ shall be y^e area(15) [of] $az^{\frac{\beta+\gamma\delta-\theta\delta}{\theta}}+bz^{\frac{\gamma\delta-\theta\delta}{\theta}}+cy^{\delta}=0$.(16) But if neither of y^e indexes be integer let θ signify any common divisor of β & δ & η any number not greater then ϵ affirmative or negative, make $x^{\frac{\theta}{\eta\gamma}}=z$, & $\frac{\theta}{\eta\gamma}A$ shall be y^e area of $az^{\frac{\eta\gamma\delta-\theta\delta+\beta\eta\epsilon}{\theta\epsilon}}+bz^{\frac{\eta\gamma\delta-\theta\delta}{\theta\epsilon}}+cy^{\frac{\delta}{\epsilon}}=0$, where such numbers must be taken for η & θ as will make $\frac{\eta\gamma-\theta}{\epsilon}$ an integer, if it may be conveniently according to any of y^e 4 precedent cases.(17) But if not, then you may(18) put η & θ equall to an unit, & make $x^{\frac{1}{\gamma}}=z$ & $\frac{A}{\gamma}$ shal be y^e area of $z^{\gamma-1}\times\overline{\frac{az^{\beta}+b}{-c}}\Big|^{\frac{\epsilon}{\delta}}=y$.(19)

---

(10) This replaces the somewhat illogical 'less'.

(11) In the terminology of note (9) the basic reduction formula in this case is

$$(\theta+k+1)\,aA(k,\theta) = z^{k}\pi^{\theta+1}-kbA(k-1,\theta).$$

From this, on substituting $z = x^{\eta}$, there follows

$$\eta sa\int x^{r\eta-1}(ax^{\eta}+b)^{\theta}.dx = x^{(r-1)\eta}(ax^{\eta}+b)^{\theta+1}-\eta(r-1)\,b\int x^{(r-1)\eta-1}(ax^{\eta}+b)^{\theta}.dx,$$

where $r = k+1$ and $s = \theta+r$. Iteration of this recursive scheme yields a series expansion of the left-hand integral in decreasing powers of x which is structurally identical with a celebrated equivalent communicated by Newton to Leibniz in his *epistola posterior* of 24 October 1676 (see 2, §2: note (540)).

(12) Specifically, η and θ.

(13) The more optimistic adverb 'usually' is cancelled.

(14) That is, 'affected' by either of the coordinate variables (as, for instance, the term b in the equation immediately following).

[2] The Quadrature of many[20] Curves whose Æquations consist of more then three terms.

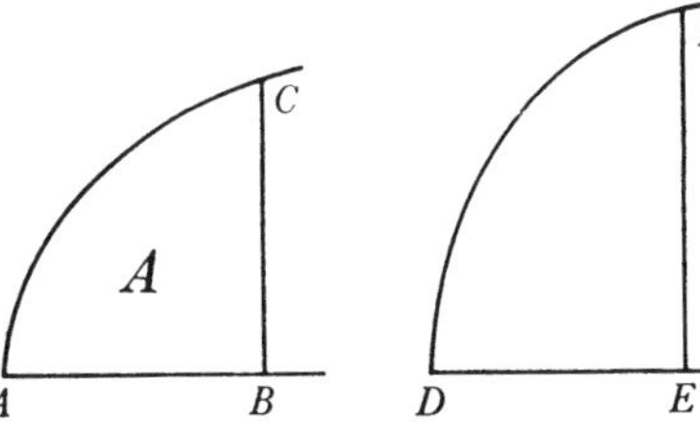

Let $cv^{\alpha}x^{\zeta}+dv^{\beta}x^{\eta}+ev^{\gamma}x^{\theta}+fv^{\delta}x^{\lambda}$ &c$=0$ be y^e^ æquation defining any Curve Line ABC, whose basis $AB=x$ & ordinate BC is v & area ABC is A. Take τ for any number whole or broken[,] affirmative or negative, & make $x^{\tau}=z$ & τA shall be y^e^ area of y^e^ curve DEF whose Base DE is z & ordinate $EF\,y$, & Equation[21] $cy^{\alpha}z^{\frac{\alpha\tau-\alpha+\zeta}{\tau}}+dy^{\beta}z^{\frac{\beta\tau-\beta+\eta}{\tau}}+ey^{[\gamma]}[z]^{\frac{\gamma\tau-\gamma+\theta}{\tau}}+fy^{\delta}z^{\frac{\delta\tau-\delta+\lambda}{\tau}}+\&c=0$. Substitute therefore either $\dfrac{\alpha-\zeta-\beta+\eta}{\alpha-\beta}$ or $\dfrac{\beta-\eta-\gamma+\delta}{\beta-\gamma}$ or $\dfrac{\alpha-\zeta-\gamma+\delta}{\alpha-\gamma}$ for τ,[22] & if in any of y^e^ three cases y^e^ indices of y^e^ dignities of z become y^e^ same in all terms of y^e^ æquation but one,[23] divide y^e^ æquation by y^e^ dignity of z in those terms & you will have z in a single dignity. Put v for z & x for y & let y^e^ æquation be $cx^{\mu}+dx^{\nu}+ex^{\xi}+\&c=v^{\pi}$.[24]

---

(15) Namely (in Leibnizian terms) $\int y.dz$.

(16) The substitution $x^{\theta}=z^{\gamma}$ subject to $\theta\int v.dx=\gamma\int y.dz$ determines $v=yz^{1-\gamma/\theta}$, and Newton's result follows by dividing through the derived trinomial $az^{\beta/\theta}+b+cy^{\delta}z^{\delta(1-\gamma/\theta)}=0$ by $z^{\delta(1-\gamma/\theta)}$.

(17) Here, similarly, substitution of $x^{\theta}=z^{\eta\gamma}$ subject to $\theta\int v.dx=\eta\gamma\int y.dz$ determines $v=yz^{1-\eta\gamma/\theta}$ and Newton's result follows on dividing through the derived trinomial by $z^{(\delta/\epsilon)(1-\eta\gamma/\theta)}$.

(18) Newton first concluded with the words 'take 1 for y^e^ greatest common divisor of β & δ, & make $z^{\frac{\eta\gamma-\theta}{\theta}}\times\overline{\dfrac{az^{\frac{\beta\eta}{\theta}}+b}{-c}}\Big|^{\frac{\epsilon}{\delta}}=y$'.

(19) The particularization $\eta=\theta=1$ yields at once the derived trinomial equation $z^{(\delta/\epsilon)(\gamma-1)}(az^{\beta}+b)+cy^{\delta/\epsilon}=0$.

(20) 'ALL' is cancelled!

(21) Much as before, the substitution $x=z^{1/\tau}$ subject to $\tau\int v.dx=\int y.dz$ determines $v=yz^{1-1/\tau}$ correspondingly.

(22) These values result from equating in pairs the numerators ($\alpha(\tau-1)+\zeta$, $\beta(\tau-1)+\eta$, $\gamma(\tau-1)+\theta$ respectively) of the indices of z in the first three terms of the preceding equation.

(23) A vast assumption on Newton's part, as he soon, no doubt, came to realize (compare note (20)). The present section of his paper could more accurately be entitled 'The Quadrature of Curves whose æquations consist of more then three terms when ye indices of ye dignities of one fluent quantity may be made y^e^ same in all terms of y^e^ æquation but one'.

(24) In first draft of the following cancelled passage Newton originally added a final sentence to the present paragraph: 'And in y^e^ next place you are to take away y^e^ dignity of z in any term ax^{μ} by putting $\dfrac{\mu+\pi}{\pi}=\tau$. $x^{\tau}=z$, & τA y^e^ area of $c+dz^{\frac{\nu-\mu}{\tau}}+ez^{\frac{\xi-[\mu]}{\tau}}+\&c=y^{\pi}$'. As before, the substitution $x=z^{1/\tau}$ subject to $\tau\int v.dz=\int y.dz$ determines $v=yz^{1-1/\tau}$. Newton then divides through by $z^{1-1/\tau}$, equating the index $(\mu-\pi[\tau-1])/\tau$ of the first term to zero.

The next work is to alter y^e^ dignity of v^{π} if $\frac{1}{\pi}$ be not affirmative & less then an unit. And this is done by y^e^ help of these two Theorems

1st If $\overline{ax^{\xi}+bx^{\eta}+cx^{\theta}+\&c}|^{\lambda}=v$. Then is $\frac{vx-A}{\lambda}$ the area of

$$\overline{\zeta ax^{\xi}+\eta bx^{\eta}+\theta cx^{\theta}+\&c}\times\overline{ax^{\xi}+bx^{\eta}+cx^{\theta}[+\&c]}|^{\lambda-1}[=y],$$

or more indefinitely $\frac{m}{\lambda}vx+\overline{n-\frac{m}{\lambda}}A$ is y^e^ area of

$$\overline{\begin{matrix}\zeta m\\+n\end{matrix}ax^{\xi}\begin{matrix}+\eta m\\+n\end{matrix}bx^{\eta}\begin{matrix}+\theta m\\+n\end{matrix}cx^{\theta}+\&c}\times\overline{ax^{\xi}+bx^{\eta}+cx^{\theta}+\&c}|^{\lambda-1}=y.$$

2. $kx^{\pi}\times\overline{ax^{\xi}+bx^{\eta}+cx^{\theta}[+\&c]}|^{\lambda}$ is y^e^ area of

$$\overline{\begin{matrix}\lambda\zeta\\+\pi\end{matrix}kax^{\xi+\pi-1}\begin{matrix}+\lambda\eta\\+\pi\end{matrix}kbx^{\eta+\pi-1}\begin{matrix}+\lambda\theta\\+\pi\end{matrix}kcx^{\theta+\pi-1}[+\&c]}$$
$$\times\overline{ax^{\xi}+bx^{\eta}+cx^{\theta}[+\&c]}|^{\lambda-1}[=y.]^{(25)}$$

But yet these Theorems profit not generally where y^e^ æquation is of more then 4 terms, & therefore I shall only describe their use in this case.(26)

Or

$$\overline{cx^{\mu}+dx^{\nu}+ex^{\xi}[+\&c]}|^{\frac{1}{\pi}}=v.$$

And y^e^ next work will be to reduce x within y^e^ radical, to y^e^ least dimensions by taking τ for a number whose denominator is y^e^ least common denominator of $\nu-\mu$ & $\xi-\mu$ &c & numerator y^e^ greatest common divisor of their numerators, & making $x^{\tau}=z$ & τA y^e^ area of $z^{\frac{1}{\tau}+\frac{\mu}{\pi\tau}-1}\times\overline{c+dz^{\frac{\nu-\mu}{\tau}}+ez^{\frac{\xi-\mu}{\tau}}[+\&c]}|^{\frac{1}{\pi}}=v$.(27) And then if $\frac{1}{\pi}$ be affirmative & greater then an unit, extract what is rational, till y^e^ index of y^e^ dignity of y^e^ radical remain less then an unit. & lastly bring y^e^ rational coefficient of y^e^ radical to its least terms by this Theorem

$$kx^{\pi}\times\overline{ax^{\xi}+bx^{\eta}+cx^{\theta}+\&c}|^{\lambda+1}$$

(25) The first theorem repeats, in corrected form (see §2.1: note (4)), Theorem 2 of the previous version. The latter one is a slightly generalized version of Theorem 3 which follows it. Both are derivable at once by differentiating and collecting terms. As usual, A denotes the area $\int v.dx$ and so $vx-A$ its complement $\int x.dv$.

(26) On cancelling the preceding passage Newton decides to extend a previous paragraph.

(27) Read 'y'. Newton no longer (compare note (24)) equates the index $(\mu/\pi-[\tau-1])/\pi\tau$ to zero, but chooses τ to make the remaining indices $(\nu-\mu)/\tau$ and $(\xi-\mu)/\tau$ 'simple'.

(28) The second of the preceding cancelled theorems, except that the index of the binomial is increased by unity.

(29) Newton first wrote 'y^e^ æquation expressing', no doubt intending to equate the value of the following ordinate to its base v.

is y^e^ area of

$$\overline{\begin{matrix}+\zeta\\+\lambda\zeta\\+\pi\end{matrix}kax^{\zeta+\pi-1}\begin{matrix}+\eta\\+\lambda\eta\\+\pi\end{matrix}kbx^{\eta+\pi-1}\begin{matrix}+\theta\\+\lambda\theta\\+\pi\end{matrix}kcx^{\theta+\pi-1}+\&c}\times\overline{ax^{\zeta}+bx^{\eta}+cx^{\theta}+\&c}|^{\lambda}\,[=y.]^{(28)}$$

Put therefore y^e^ ordinate of (29) y^e^ curve to be

$$\overline{gx^{\beta}+hx^{\gamma}+\&c}\times a+bx^{\zeta}+cx^{\theta}+\&c.^{(30)}$$

And comparing y^e^ terms, if β be y^e^ biggest index make $\beta=\zeta+\pi-1$, &

$$g=\zeta ka+\lambda\zeta ka+\pi ka$$

& subduct &c. And repeat this so long as it is convenient.(31)

---

[3] Besides these, there are some other rules. As for Binomials this. If $gv^{\alpha}+hx^{\eta}=1$, or $\overline{\dfrac{1-hx^{\eta}}{[+]g}}\Big|^{\frac{1}{\alpha}}=v$. Make $\dfrac{v^{\alpha}}{x^{\eta}}=z^{\alpha\pi}$ or $\dfrac{v^{\frac{1}{\pi}}}{x^{\frac{\eta}{\alpha\pi}}}=z$. And $xv-\dfrac{+\alpha+\eta}{\alpha}A$ shall be y^e^ area of $\dfrac{\pi\eta z^{\pi-1}}{\overline{h+gz^{\alpha\pi}}|^{\frac{\alpha+\eta}{\alpha\eta}}}$ (32) $=y$ assuming any number for π & A for y^e^ area of y^e^ first curve.(33)

---

(30) Read '$\ldots\overline{ax^{\zeta}+bx^{\eta}+cx^{\theta}+\&c}|^{\lambda}$'.

(31) This is, as we have seen (compare the integral tables reproduced in I: 346–63, 406–10 and on pages 244–54 above), a favourite Newtonian technique for integrating a given algebraic expression in finite terms, either exactly or in terms of other given integrals. In successive drafts of his *De Quadratura Curvarum* written during 1691–3 (to be reproduced in the sixth and seventh volumes) Newton considerably extended this approach, but such generalizations become rapidly unrealistic and devoid of interest.

(32) The numerator should read '$\pi z^{\pi-1}$' simply, an error copied from the previous version (see note (33), and compare §2: note (9)).

(33) Apart from trivial change of notation, this repeats Theorem 5 in §2.1 above. As usual, A denotes $\int v.dx$.

PART 2

MISCELLANEOUS RESEARCHES
(Early 1670's)

INTRODUCTION

In this second part we reproduce three autograph papers which have little in common except that Newton wrote them some time during the period 1670–3 covered by the present volume. Of their background and composition history contemporary evidence tells us almost nothing, but in default we may advance certain tentative conjectures as to their purpose which are founded on the structure and content of the documents themselves.

The first piece, essentially unoriginal and thoroughly schoolmasterish in tone, is a restyling and slight abridgment of the second book of Euclid's *Elements*, evidently taking as its source text Barrow's 1655 Cambridge edition which Newton had studied so intently almost a decade earlier.(1) The following paper, described none too accurately by Samuel Horsley in October 1777(2) as 'A fragment relating to the comparison of curved surfaces', is, if more sophisticated in its approach, scarcely less traditional in its theme: namely, evaluation of the surface areas and solid content of certain cylindrical, conical and spherical surfaces by an improved Archimedean technique. In the first Euclid's theorems are agreeably and succinctly restated and Newton's variant demonstrations are cleanly expounded, but the topic is too slight and restricted to afford creative outlet for an inventive talent—a mere compulsory skating of figures by a master of the art, in fact. In our guess the paper was intended as an undergraduate exercise, perhaps as a humdrum lecture from the Lucasian chair to students without aspiration to greater heights of mathematical proficiency. The latter too, we may surmise, was perhaps to form the basis of a lecture course which either came to nothing or has vanished utterly into oblivion. It is clear that Newton was none too happy in this classical world where the exhaustion method is the *sine qua non* of any proof equivalent to integration and the magnitude sought is to be confined within bounds which increase or decrease monotonically to that value as their common limit. Like Huygens before him,(3)

(1) See **1**: note (1) below, and compare I: 12, note (28).

(2) On a paper slip loose in ULC. Add. 3963.16.

(3) Namely, in his 1658 researches into the volume and surface area of various conoids of revolution. Having slaved away at the time-consuming task of recasting his original indivisibles arguments into Archimedean form Huygens gave up the frustrating uphill struggle with the words 'ad fidem faciendam apud peritos haud multum interest, an demonstratio absoluta [per methodum exhaustionum] tradatur an fundamentum ejus demonstrationis [per indivisibilia], quo conspecto non dubitent demonstrationem perfectam dari posse. Fateor tamen etiam in hac rite instituenda ut clara concinna omniumque aptissima sit, peritiam et ingenium elucere, uti in Archimedis omnibus operibus. Verum et prior et longe præcipua est inveniendi ratio ipsa, hujus cognitio potissimè delectat atque a doctis expetitur, quamobrem magis etiam hæc methodus sequenda videtur qua brevius clariusque comprehendi et ob oculos poni potest. Tum verò et nostro labori parcimus in scribendo, et aliorum in legendo, quibus vacare

halfway through he loses patience with the *longueurs* of the method, substituting a less rigorous approach by indivisibles.

The third manuscript, dealing with the simple harmonic motion induced in a vertical cycloidal arch by the action of terrestrial gravity (presumed constant), is a research paper of a different order of difficulty. From a strictly mathematical viewpoint it represents an elegant application of differential properties of the cycloid, notably concerning its tangent and its evolute, in ways pioneered—whether or not Newton knew it at the time of writing—by Huygens and Brouncker. Dynamically, it combines two fundamental concepts first expounded in general terms by Newton: the measure of the instantaneous velocity v in the direction of motion generated by the central force $f(x)$ as $\sqrt{\left[2\int_0^x f(x)\,.\,dx\right]}$, and the determining condition $f(x)\,.\,dx/ds = -ks$ for the isochronicity of the motion induced along the arc-length s. In his *Principia*[4] Newton was later to show that in a direct-distance force field ($f(x) \propto x$) the resulting tautochrone is a hypocycloid, but in the present manuscript he is content to repeat the Huygenian result that, where $f(x) = g$ (constant), then

$$v = ds/dt = (2gx)^{\frac{1}{2}} = -(g/2D)^{\frac{1}{2}} \sin (g/2D)^{\frac{1}{2}}\, t$$

and $d^2s/dt^2 = -(g/2D)\, s$ are the equations of the simple harmonic motion of period $2\pi(2D/g)^{\frac{1}{2}}$ in the cycloid of Cartesian equation

$$y = \sqrt{[x(D-x)]} + \tfrac{1}{2}D \sin^{-1}[1-2x/D],$$

which is therefore the simple tautochrone. Historically it is of some interest to establish the extent to which Newton is here indebted to previously published work. We may accept the evidence of Newton's handwriting, loosely confirmed by David Gregory twenty years later in a memorandum recording a guided tour by Newton through his early papers,[5] that the manuscript was composed about 1670. It follows that the present researches into cycloidal motion are effectively independent of the contemporary investigations of Huygens, Brouncker and Pardies, all of whom first published their (kinematical) proofs that the cycloid is the simple tautochrone in 1673.[6] It is our conjecture that Newton wrote his paper about the close of 1672, soon after reading James

tandem amplius non poterit, ut ingentem multitudinem Geometricorum inventorum quæ augetur in dies doctoque hoc sæculo in immensum porro exitura videtur, evolvant, siquidem prolixam illam ac perfectam veterum methodum scriptores usurpant' (*Œuvres complètes*, **14** (1920): 337). Newton would, no doubt, wholeheartedly have agreed.

(4) *Philosophiæ Naturalis Principia Mathematica* (London, $_1$1687): Liber 1, Propositio LI: 151–2.

(5) See 3: note (2) below.

(6) These investigations are outlined in the following appendix.

Gregory's stimulating essay on pendular and projectile motion[7] but before he acquired Huygens' *Horologium Oscillatorium* and Pardies' *Statique* in the middle of the following year.[8] The reader is, of course, free to form his own opinion of the extent of Newton's dependence, but is none the less invited to contrast the structure of Huygens', Brouncker's and Pardies' proofs (as summarized in the following appendix) with the considerably variant Newtonian equivalent reproduced in Section 3 below. Whatever judgement is made regarding the kinematical aspects of the present paper, however, the originality and absolute priority of its dynamical portion is beyond dispute. Of Newton's contemporaries Huygens alone came subsequently—in researches begun in late 1674 but not published till 1934[9]—to realize that a necessary and sufficient condition for motion in a curve to be isochronous is that the acceleration induced in the instantaneous direction of motion be proportional to the distance along the curve to the point of greatest velocity; and that in the case of simple gravity g this yields the tangential condition $g\,dx/ds \propto s$ which at once determines the tautochrone to be a cycloid.

APPENDIX: HUYGENS, BROUNCKER AND PARDIES ON CYCLOIDAL MOTION

[1659–1673][1]

The immense improvement in the accuracy in time-keeping of the mechanically driven pendulum clock during the seventeenth century was evidently the prime cause of the widespread interest at this time in the 'true' tautochrone, that is,

(7) 'Tentamina Quædam Geometrica De Motu Penduli et Projectorum', published in appendix to Patrick Mathers' *Great and New Art of Weighing Vanity* (Glasgow, 1672). Compare 3: note (15).

(8) Newton received both as gift copies: Huygens' *Horologium* from Oldenburg on 4 June 1673 (*Correspondence of Isaac Newton*, **1** (1959): 284) and Pardies' 'little but ingenious tract' from Collins in mid-September following (*ibid.*, **1**: 307).

(9) See 3: note (14) below.

(1) This outline is severely tailored to our present purpose of exhibiting possible antecedents for the manuscript on cycloidal motion reproduced in section 3 below: in particular we have not hesitated to reletter the original figures to conform with Newton's usage in that paper. For a wider treatment of the topic consult Vollgraff's editorial *avertissement* to Huygens' 1659 researches on the cycloid (*Œuvres complètes de Christiaan Huygens*, **16** (1929): 344–9: 'Tautochronisme de la cycloïde') and A. Ziggelaar's account of 'Les Premières Démonstrations du Tautochronisme de la Cycloïde' (*Centaurus*, **12** (1967): 21–37). Both are, however, less than fair to Brouncker and neither mentions Hooke, a provocative discussion of whose ideas on isochronism is given by R. S. Westfall in 'Hooke and the Law of Universal Gravitation' (*British Journal for the History of Science*, **3** (1967): 245–61, especially 252–5).

the curve in which a body falls 'naturally' in a vertical plane under the pull of terrestrial gravity (assumed constant over the whole depth of fall) so that the period of its double oscillation is independent of the amplitude of its swing. In 1638 Galileo had published his belief that small oscillations of the simple circular pendulum are effectively isochronous[(2)] but was contradicted in 1647 by Mersenne who, having tested the assertion experimentally, announced firmly that 'scies vibrationes eiusdem funependuli non esse æquales, nequidem proximè', a view echoed ten years later by Christiaan Huygens.[(3)] Nonetheless, for small swings the vibrations of a circular pendulum are approximately isochronous. Why should this be so? To answer this point Huygens set himself in late 1659 to compute the time of fall over a general circle arc.[(4)]

With some slight rephrasing his argument proceeds in the following way. If a body be supposed to fall under the pull (say g) of terrestrial gravity in the circle arc $P'\widehat{\pi'}C$ of radius $AC = A\pi' = b$ from rest at P' to the lowest point C at a depth $CV = c$ below it, then the instantaneous velocity v at π' is proportional to $\sqrt{VQ} = (c-x)^{\frac{1}{2}}$— in fact $v^2 = 2g(c-x)$ when suitable units are chosen—[(5)] and the slope dx/dy at π' is equal to $\pi'Q/AQ = y/(b-x)$, where $CQ = x$ and $Q\pi' = y = \sqrt{[x(2b-x)]}$ are Cartesian coordinates of the general point π'. On taking $\widehat{C\pi'} = s$ it follows that $ds/dx = b/\sqrt{[x(2b-x)]}$ and consequently that the time of fall

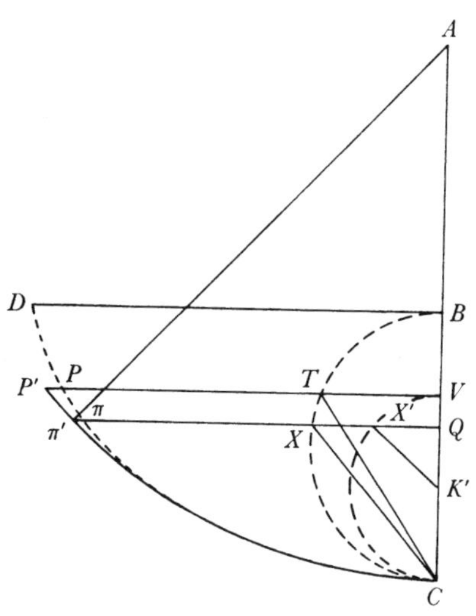

$t_{(\text{rest at } P')P' \to \pi'}$ is $\displaystyle\int_c^x \frac{b}{\sqrt{[2gx(2b-x)(c-x)]}} \cdot dx,$

an elliptical integral which is not exactly evaluable in terms of elementary functions other than by expanding the radical as an infinite series.[(6)] When,

(2) *Discorsi e Dimostrazioni Matematiche intorno a Due Nuove Scienze* (Leyden, 1638): 87–95 [= *Opere* (ed. A. Favaro), **8** (Florence, 1898): 130–9].

(3) See Marin Mersenne, 'Reflectiones Physico-Mathematicæ' [= *Novarum Observationum Physico-Mathematicarum... Tomus III* (Paris, 1647): 63–235]: Caput XIX, §2 'Funependuli vibrationes non esse isochronas': 153; and compare Huygens' *Horologium* (The Hague, 1658): 12.

(4) *Œuvres complètes*, **16** (1929): 392–413. The main researches are dated between 1 and 15 December 1659 (N.S.).

(5) This generalization of an equivalent Galileian 'assumption' for uniformly accelerated motion in an oblique straight line (see **3**: note (15) below) is here (implicitly) made for the first time. It is, of course, fundamental in any discussion of the tautochrone. Two years later Brouncker was to make the same generalization from Stevin's theorem on motion in inclined planes (see note (16)).

however, the arc $\widehat{CP'}$ of swing (and so CV) is small, so *a fortiori* is $CQ=x$ and therefore the time of fall $t_{(\text{rest at } P')P'\to\pi'} \approx \frac{1}{2}\sqrt{\frac{b}{g}}\int_c^x \frac{1}{\sqrt{[x(c-x)]}}\,.\,dx.$(7) As a result the time of a semi-oscillation (from rest at P' to C) is very nearly $\frac{1}{2}\pi(b/g)^{\frac{1}{2}}$, independent of the height $CV=c$ from which the body falls from rest to C. This theoretical justification of the practical accuracy of the simple circular pendulum at once suggested to Huygens a way of defining the true tautochrone $\widehat{P\pi C}$, evidently coincident with the circle arc $\widehat{P'\pi'C}$ in the immediate vicinity of C: for if, where now $CQ=x$, $Q\pi=y$ are the co-ordinates of the point π in the isochronous curve, the slope at π is exactly determined by $ds/dx=(\frac{1}{2}b/x)^{\frac{1}{2}}$ on putting $\widehat{C\pi}=s$, the time of fall to C from rest at P through the depth CV will be accurately $\frac{1}{2}\pi(b/g)^{\frac{1}{2}}$, independent of the choice of the starting-point P of motion. Accordingly, the tangent at π to the tautochrone will be parallel to CX, where X is the meet of the horizontal πQ with the semicircle drawn on the diameter $BC=\frac{1}{2}b$—the defining property, as Huygens well knew, of the cycloid $\widehat{D\pi C}$ generated by the point C of the semicircle $\widehat{BXC}$ as it rolls along beneath the horizontal DB.(8) In a way obvious after Wallis published Wren's rectification of the cycloid's general arc as twice that of the corresponding chord in the generating circle (here $\widehat{CP}=2CT$),(9) it remained only to show how the 'cheek' $\widehat{AD}$ of the cycloid's evolute (a congruent cycloid) could, by folding its upper portion AR along itself, be used to guide the end-point C of the thread AC instantaneously into coincidence with the general point π on the lower cycloid $\widehat{DPC}$.

(6) James Gregory in his 'Tentamina Quædam Geometrica De Motu Penduli et Projectorum' (appended to Patrick Mathers' *The Great and New Art of Weighing Vanity* (Glasgow, 1672)) was the first to give an equivalent series expansion of the integral. In his terms, on making $AV = b$, $VP' = c$, $VQ = a$ and $AC = r$ there results

$$t_{P'\to\pi'}\left(=\int_0^a \frac{r}{\sqrt{[2ga(c^2-2ab-a^2)]}}\,da\right) \propto \int_0^a \frac{r^{\frac{1}{2}}a^{-\frac{1}{2}}}{\sqrt{[1-2bc^{-2}a-c^{-2}a^2]}}\,da.$$

When the trinomial is expanded as an infinite series in a (taken to be small enough for convergence) it follows, upon integrating term by term and removing fractional powers of a by appropriately substituting $(a/r)^{\frac{1}{2}} = d/c$, that $t_{p'\to\pi'} \propto 2c^{-2}r^2d+\frac{2}{3}bc^{-2}ra^2/d\ \ldots$. Gregory's preliminary computations for this are scrawled in the margin of Collins' letter to him of 4 December 1669 (see H. W. Turnbull in the *James Gregory Tercentenary Memorial Volume* (London, 1939): 374–6).

(7) That is, $\frac{1}{2}(b/g)^{\frac{1}{2}}(\pi-\cos^{-1}[1-2x/c]) = \frac{1}{2}(b/g)^{\frac{1}{2}}\times V\hat{K'}X'$, where $\pi'Q$ meets the semicircle (of centre K') drawn on the diameter VC in X'. This corollary is made explicit in Huygens' immediately following synthetical restyling of the result (see note (10)).

(8) Compare **1**, 2, §2: note (307) above. The obvious source for Huygens' knowledge of this property of the cycloid's tangent was Schooten's *Commentarii in Librum II, O* of the Latin edition of Descartes' *Geometrie* (*Geometria* (Leyden, $_1$1649): 226 = Amsterdam ($_2$1659): 267).

(9) See **1**, 2, §2: note (308), and compare **3**: note (16) below.

Over the next few days following his initial discovery Huygens recast his argument in a manner which allowed him to make direct comparison of the time of fall over any specified portion of a cycloid's arc with that of uniform motion in a defined arc of a circle. In essence, if it is given that a body falls

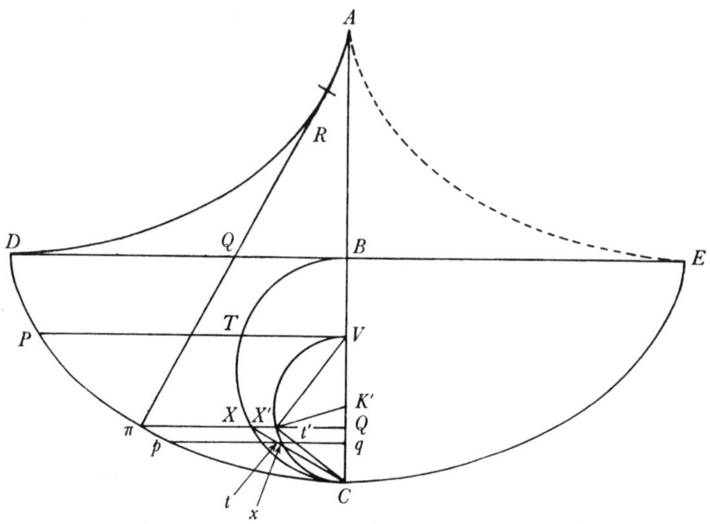

freely under gravity in the cycloid $\widehat{DC}$ from rest at P, then its instantaneous speed at point π after a fall through the depth VQ is proportional to $\sqrt{VQ}$: precisely, where g is the constant gravitational acceleration, it is $\sqrt{[2g \times VQ]}$, so that the time $t_{(\text{rest at } P)\pi\to p}$ of fall over the indefinitely small arc $\widehat{\pi p}$ is $\widehat{\pi p}/\sqrt{[2g \times VQ]}$. Let πQ and its parallel pq meet the semicircles on the diameters

(10) See Huygens' *Œuvres complètes*, **16**: 404–12, dated by him '15 Dec. 1659' (N.S.).

(11) *Horologium Oscillatorium sive De Motu Pendulorum ad Horologia aptato Demonstrationes Geometricæ* (Paris, 1673): 'Pars Secunda. De descensu Gravium & motu eorum in Cycloide', Propositions XVI–XXVI: 43–58 [= *Œuvres complètes*, **18** (1934): 158–87].

(12) In preface to his *Horologium Oscillatorium* (note (11)) he wrote (page 2): 'Hanc cum jam pridem amicis horum intelligentibus notam fecerimus (nam non multo post primam horologij editionem [in 1658; see note (3)] animadversa fuit) nunc eandem, demonstratione quàm potuimus accuratissima firmatam, omnibus legendam proponimus'.

(13) Thus, writing to Robert Moray on 10 February 1662 (N.S.) in connection with Brouncker's first, erroneous attempt to identify the cycloid as the tautochrone, Huygens remarked that 'Lors que j'estois l'hyver passé à Paris, Monsieur Auzout qui est fort bon geometre me donna aussi une demonstration du mesme Theoreme qui estoit pareillement fausse' (*Œuvres complètes*, **4** (1891): 51). In a letter to Oldenburg on 24 June 1673 (N.S.) Huygens made his attitude to second inventors clear: 'ce n'est pas grande chose...d'avoir fait la demonstration d'une proposition desia trouvé.... Le principal, et ce qu'il y a de plus difficile dans ces choses de Geometrie, c'est de les trouver comme sçavent tres bien ceux qui s'en meslent' (*ibid.*, **7** (1897): 314). Forty years later Newton was to use the same tone to Leibniz during the calculus priority squabble.

BC, VC in points X, X' and x as shown, and let CX, CX' intersect pq in points t, t'. In the limit therefore as the points π, p coincide (when $\widehat{\pi p}=Xt$) the time of fall over $\widehat{\pi p}$ from rest at P is $Xt/\surd[2g\times VQ]$, that is, on taking $AC=2BC$,

$$\frac{Xt}{X't'}\left(\text{or }\sqrt{\frac{BC}{VC}}\right)\times\frac{X't'}{Qq}\left(\text{or }\sqrt{\frac{VC}{QC}}\right)\times\frac{Qq}{\widehat{X'x}}\left(\text{or }\frac{X'Q}{X'K'}\right)\times\frac{\widehat{X'x}}{\surd[2g\times VQ]}=\sqrt{\frac{AC}{g}}\times\frac{\widehat{X'x}}{VC}$$

since $X'Q^2=VQ\times QC$ and $X'K'=\frac{1}{2}VC$. Consequently, on summing over all the infinitesimal arcs $\widehat{\pi p}$ it follows that $t_{(\text{rest at }P)P\to p}\propto\widehat{Vx}/VC$, and in particular that $t_{(\text{rest at }P)P\to C}\propto\widehat{VC}/VC=\frac{1}{2}\pi$, independently of the height VC of fall.(10) The proof eventually presented by Huygens to the world in 1673 is substantially the same, though the details of its argument are there conducted with the rigour befitting the exhaustion form in which it is set.(11)

Justifiably proud of this double proof of his fundamental result Huygens lost no time in conveying it in general terms to his friends(12) but kept its demonstration to himself. Quickly, news of his identification of the tautochrone as the common cycloid spread through the learned world and the implied challenge to repeat his proof was taken up by several contemporary mathematicians with varying success.(13) Huygens himself no doubt spoke of it during his visit to London in the spring of 1661 when he visited the Royal Society's apartments at Gresham College, meeting John Wallis and its President, Lord Brouncker. The latter, in any event, lost little time in seeking to confirm Huygens' result. A first, evidently erroneous demonstration communicated by him to Huygens in January 1662(14) we may pass over, but a far more interesting revised proof despatched a month later(15) falters only in its final argument. As a preliminary Brouncker supposes the vertical chain $S_0S_1\ldots S_x$ of x rectilinear links S_iS_{i+1}, each

(14) 'A Demonstration of the Equality of Vibrations in a Cicloid-Pendulum' (reproduced from the original in the Huygens collection at Leyden in the latter's *Œuvres complètes*, **4** (1891): 28–31). In so far as it is possible to attach precise meaning to such vague phrases as 'one degree of velocity is acquired from that continuance of motion or second impulse', Brouncker appears to confuse the tautochrone with the isochrone (curve of uniform vertical descent) and fails inevitably to identify the latter with the cycloid. In fact, as Leibniz was to show in his 'De Linea Isochrona, In qua grave sine acceleratione descendit' (*Acta Eruditorum* (April 1689): 195–8), the isochrone is a semicubical parabola. Replying to Moray on 10 February 1662 (N.S.) Huygens justly observed that '[Monsieur le Mylord Brouncker] conclud que la boule en descendant du point [P], et estant arrivé en h, elle y aura la mesme vistesse qu'elle avoit en [P] ce que je n'entens pas comment il peut concevoir, puis qu'en [P] elle n'avoit nulle vistesse, et qu'en descendant de là elle passe par tous les degrez de tardité, selon les principes que je suis' (*Œuvres complètes*, **4**: 51).

(15) *Œuvres complètes*, **4**: 88–91, enclosed with Moray's letter to Huygens of 14 March 1662 (N.S.) (*ibid.*, **4**: 85–7). Brouncker's more acceptable second proof was incorporated in the version deposited soon after in the archives of the Royal Society and first published by Thomas Birch in his *History of the Royal Society of London for improving of Natural Knowledge*, **1** (London, 1756): 70–4.

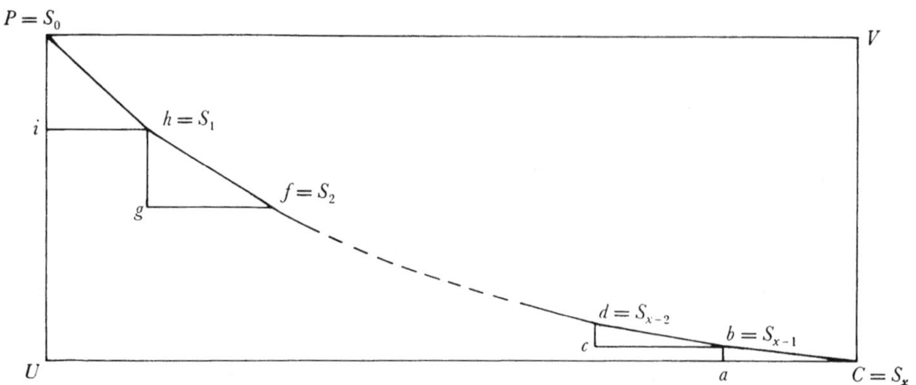

of unit length, to be jointed in such a way that the difference, s_i say, in vertical height between S_{i-1} and S_i is $(x-i+1)\,a$, where a is some constant: thus in the accompanying figure (where Brouncker's own lettering has been superimposed) $ab=s_x=a$, $cd=s_{x-1}=2a, \ldots, iP=s_1=xa$. It follows that, if the point $P(X, Y)$ is determined by the Cartesian co-ordinates $CV=X=\frac{1}{2}x(x+1)\,a$ and $VP=Y$ with the length (x units) of the chain $S_0S_1\ldots S_x=S$, then $\delta(CV)=\delta X=iP=xa$ where correspondingly $\delta S=hP=1$, so that $(\delta S/\delta X)^2=(1+1/x)\,D/X$ on taking $D=1/2a$. In the limit, therefore, as the divisions S_i in the chain become infinite in number while its length S is unchanged there results $dS/dX=(D/X)^{\frac{1}{2}}$, so that the chain coincides ultimately with the arc S of the cycloid

$$Y=\sqrt{[X(D-X)]}+\tfrac{1}{2}D\sin^{-1}[1-2X/D].$$

Now if, in the finite case, a body falls from rest at $P=S_0$ to $C=S_x$ (such that, it is understood, the speed of fall is unaffected by the instantaneous change in direction at each of the linking points S_i), then, on denoting the speed attained at S_i by v_i, there follows $v_i^2-v_{i-1}^2=2gs_i$ and so $v_p^2 \propto \sum_{1\leqslant i\leqslant p} s_i$, the vertical distance

(16) Brouncker's proof of this fundamental measure of the 'energy' of a fall path traversed from rest (under the gravitational pull g) is, most interestingly, wholly independent of Galileo's *Discorsi* (note (2)), taking its lead from the second corollary to Proposition XIX of Book 1 of Simon Stevin's work on statics (*De Beghinselen der Weeghconst* (Leyden, 1586) [= *The Principal Works of Simon Stevin* (ed. E. J. Dijksterhuis), **1** (Amsterdam, 1955): 178–9], studied by Brouncker in the French translation published in the fourth volume of Girard's *Œuvres Mathématiques de Simon Stevin de Bruges, où sont insérées les Mémoires mathématiques* (Leyden, 1634)). By Stevin's theorem it follows that, if a_p is the 'power of the weight' (component of gravity) which accelerates motion along $S_pS_{p+1} = 1$ uniformly, so that $t_p = 1/\frac{1}{2}(v_{p-1}+v_p)$ is the time of fall over that segment, then $s_{p+1}/s_p = a_{p+1}/a_p$ and therefore

$$\frac{s_{p+1}}{s_p} = \frac{(v_{p+1}-v_p)/t_{p+1}}{(v_p-v_{p-1})/t_p} = \frac{v_{p+1}^2-v_p^2}{v_p^2-v_{p-1}^2},$$

whence $v_p^2-v_{p-1}^2 \propto s_p$ (where the proportionality factor is in fact $2g$).

fallen from $P=S_0$ to S_p.(16) In consequence (on inserting the correct proportionality factor in Brouncker's argument)

$$v_p^2=(g/D)\sum_{1\leqslant i\leqslant p}(x-i+1)=(g/2D)\,(px-\tfrac{1}{2}(p-1)\,p)$$

and, finally, the total time of fall over the chain $S_0S_1\ldots S_x$ from rest at $P=S_0$ is $\sum_{1\leqslant p\leqslant x} 1/\tfrac{1}{2}(v_{p-1}+v_p)$, that is,

$$\sum_{0\leqslant p\leqslant x-1}\frac{2(D/g)^{\frac{1}{2}}}{\sqrt{[(p+1)\,x-\frac{1}{2}p(p+1)]}+\sqrt{[px-\frac{1}{2}(p-1)p]}}$$

$$=(2D/g)^{\frac{1}{2}}\sum_{1\leqslant s\leqslant x}\frac{\sqrt{[x(x+1)-(s-1)\,s]}-\sqrt{[(x+1)-s(s+1)]}}{s}$$

on taking $p=x-s$ and rationalizing the denominator. To prove that fall over the general cycloidal arc $\widehat{PC}$ is isochronous it remains to show that the limit of this aggregate as x becomes infinitely great attains a constant (finite) value. This Brouncker did not succeed in doing, though by an ingenious diagonal summation he arrived at the equivalent expression

$$(2D/g)^{\frac{1}{2}}\lim_{x\to\infty}\,(\sqrt{[x(x+1)]}-\sum_{1\leqslant s\leqslant x}\sqrt{[x(x+1)-s(s+1)]}/s(s+1)).$$

Huygens, it may be noted, upon receiving this unwieldy proof justifiably found it 'tres obscure' and asked for further enlightenment from Brouncker, apparently in vain.(17)

By the middle 1660's it had come to be widely realized that the isochronism of the cycloidal pendulum depended intrinsically on the proportionality of the length of its general arc $\widehat{PC}$ (or twice the corresponding chord

$$TC=\sqrt{[BC\times VC]}$$

(17) See Huygens' comments on the manuscript sent to him by Brouncker (*Œuvres complètes*, **4** (1891): 91), where he noted both 'Que c'est une nouuelle demonstration et qu'il a abandonnè la premiere' (see note (14)) and 'Que toute la demonstration depuis le commencement est aussi tres obscure'. Compare his subsequent letter of 24 June 1673 (N.S.) to Oldenburg where he affirmed that 'Milord Brouncker m'en envoya une demonstration, et puis une autre meilleure, mais qui ne laissoit pas d'avoir encore quelqu'obscuritè pour moy' (*ibid.*, **7** (1897): 313). At the end of his life his unfavourable response to Brouncker's proof had hardened into intolerance: 'cette autre [demonstration] qu'il m'avoit envoyée étoit fausse...aussi' (*ibid.*, **18** (1934): 667).

It will be obvious that in the limit as x becomes infinitely great, each portion of Brouncker's final expression tends to infinity. Yet the apparently unattainable was so nearly in his grasp, for on taking $\sigma = s/x$ the time of fall (from rest at P) over the cycloidal arch $\widehat{PC}$ is

$$(2D/g)^{\frac{1}{2}}\lim_{x\to\infty}\sum_{1\leqslant\sigma x\leqslant x}\sigma^{-1}(\sqrt{[1+1/x-\sigma(\sigma-1/x)]}-\sqrt{[1+1/x-\sigma(\sigma+1/x)]})$$

$$=(2D/g)^{\frac{1}{2}}\int_0^1\sigma^{-1}.d[-(1-\sigma^2)^{\frac{1}{2}}]=\tfrac{1}{2}\pi(2D/g)^{\frac{1}{2}}.$$

of the generating circle) to the square root of the depth VC of fall. Thus in November 1666 Robert Hooke gave an 'account of inclining pendulums' to the Royal Society in which, generalizing somewhat uneasily from the case of small swings 'very near of equal duration' in the circle arc $\widehat{P'C}$ (where the 'quantity of strength' or gravitational pull acting instantaneously in the direction of motion and hence the maximum speed (at C) varies as $P'V \approx \widehat{P'C}$), he proposed this criterion for the tautochrone by compounding the resulting (true) condition for the 'equality of duration of vibrations of differing arches' that the maximum speed $v_{(\text{from rest at } P) \text{ at } C} \propto \widehat{PC}$ with the Galileian condition that $v_{(\text{from rest at } P) \text{ at } C} = v_{(\text{from rest at } V) \text{ at } C} \propto \sqrt{VC}$.(18) But Hooke never applied his dubiously founded if exact criterion for isochronism of motion to specifying the shape of the tautochrone and it was left for Brouncker in 1673 to confirm its accuracy.

The immediate occasion was the appearance at Paris in spring 1673 of Ignace Gaston Pardies' work on statics, whose spare concluding 'pages vuides' were given over to his demonstration of the 'uniformité [du mouvement qui se feroit dans une Cycloïde], afin que quand M. Huygens aura publié sa démonstration, je puisse voir si j'ay esté assez heureux pour concourir avec un si grand homme'.(19) Huygens' own magisterial *Horologium Oscillatorium*(20) was published soon after, but before he saw it Brouncker had Oldenburg publish anonymously his 'Demonstration of the Synchronisme of the Vibrations made in a Cycloid, given by a Person of Quality'(21) in order to establish publicly his own priority over Pardies. There once more we find the cycloid introduced as

(18) See Birch's *History of the Royal Society* (note (15)), **2** (London, 1757): 126, and compare R. S. Westfall, 'Hooke and the Law of Universal Gravitation', *British Journal for the History of Science*, **3** (1967): 245–61, especially 253–5. Evidently, if a central force field $f(s)$ is such that $v_{\max\,(\text{at } s=0)}\left(\text{or } \sqrt{\left[2\int_0^s f(s)\,.\,ds\right]}\right) \propto s$, then $f(s) = d^2s/dt^2 \propto s$ and the periodic motion is simple harmonic.

(19) *La Statique ou La Science des Forces Mouvantes* (Paris, 1673, reissued 1674): 232. As we shall see, Pardies' proof is essentially a demonstration that motion in a cycloid under simple gravity is isochronous; unlike Huygens' more powerful approaches it yielded no explicit measure for the time of fall over a given arc. None the less Huygens was much impressed by Pardies' mechanical researches in general and upon the vibrating string in particular. Upon his death not long afterwards Huygens wrote to Oldenburg on 24 June 1673 (N.S.) that 'je ne puis m'empescher icy, de vous dire, que je regrette extresmement [s]a perte' (*Œuvres complètes*, **7** (1897): 314).

(20) See note (11).

(21) *Philosophical Transactions*: No. 94 (for 19 May 1673): 6032. The Latin title reads 'Nobilissimi cujusdam Angli Demonstratio Synchronismi Vibrationum peractarum in *Cycloide*; nunc juris publici facta ex occasione quam suppeditavit Rev. P. *Pardies*, de eodem Argumento Demonstrationem exhibens ad calcem libelli nuper ab ipso Gallicè editi de *Statica*'.

the limit of a chain $Cbd \dots hP$ made up of x equal links $Cb = bd = \dots = hP$, now of length a where $ab = b$, $cd = 3b$, ..., $iP = (2x-1)\,b$ and so $CV = x^2 b$; however, as the criterion for isochronism Brouncker now introduced out of the blue the condition that the vertical height CV be proportional to the square of the arc-length $\widehat{CP} = xa$. Unfortunately, as printed, Brouncker's sketchy argument was considerably garbled and the point of the existence criterion for the tautochrone was lost. As soon as Huygens read it he sought enlightenment from Oldenburg,(22) who in reply passed on the 'principle' underlying Brouncker's new proof of the isochronism of cycloidal motion:

'When y^e ꝑpendicular heights are in duplicate ꝑportions to ye Lengths of ye lines in wch the bullets fall (respectivè) ye times of ye descents are equal; as in a circle the proportion of [CV] to [CS] is 2plicat to y^e proport. of [CX] to [CY]; therefore the times of y^e bullets descent on [CX] and [CY] are equal. But so it is everywhere in the cycloid; therefore y^e time of descent from every part thereof is y^e same'.(23)

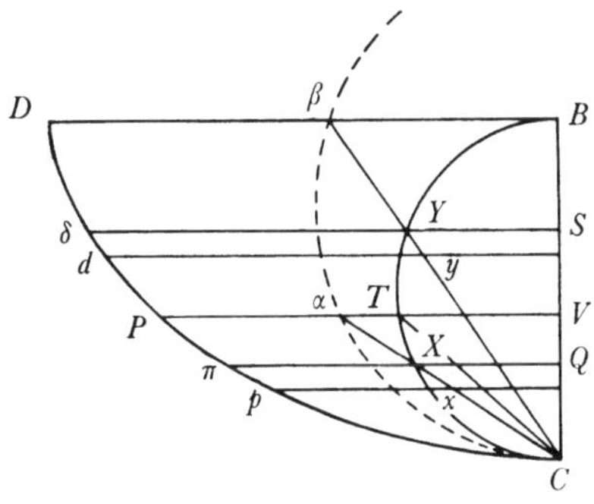

The reason for Brouncker's somewhat strange insistence on his priority over Pardies now becomes clear, for when flesh is added to the bones of this 'principle' the resulting proof is structurally identical with that propounded at the end of

(22) 'Pour ce qui est de la demonstration que vous avez mise dans vostre Journal touchant l'isochronisme de la cycloïde il n'y a pas moyen d'y rien comprendre de la maniere qu'elle y est, de sorte qu'il faut bien attendre qu'on l'explique d'avantage, et mesme les figures qui y sont, et dont pourtant il n'est fait aucune mention, semblent indiquer que cette explication n'est que différée. Je me souviens que lors que j'eus communiquè chez vous par lettres l'invention de cette proprietè de la cycloïde, Milord Brouncker m'en envoya une demonstration, et puis une autre meilleure, mais qui ne laissoit pas d'avoir encore quelqu' obscuritè pour moy, et ce sera cette derniere a ce que je puis juger qu'il aura voulu publier dans vostre Journal' (Huygens to Oldenburg 24 June 1673 (N.S.), published in his *Œuvres complètes*, **7** (1897): 313).

(23) Oldenburg's note on the verso of Brouncker's letter to him of 23 June 1673 (Royal Society, Letter Book B.1.11): the observation was duly communicated to Huygens four days later (*Œuvres complètes*, **7** (1897): 322). In his letter Brouncker wrote that 'As to what [Monsieur Hugens] is pleased to say concerning my Demonstration [see notes (13) and (22)], I doe acknowledge that to invent is much more than to demonstrate, & that likely in this case I had never thought of or don the latter if Mr Hugens had not done & made known that former, nor did I offer at it but for my own satisfaction untill he should be pleased to publish his. Nor doe I value myselfe at all upon it, & therefore would not yield to the printing it under my name, but did lette it passe on its way, not having leisure to explain it (though I think not it wants it so much) more for the method than the thing itselfe (having had the misfortune of a fire which has destroyed some papers I wish had been preserved) with intention to make use thereof hereafter if I have time, when perhaps something may be presented to public view worth the owning'.

the latter's *Statique*. Thus, if points δ, d, P, π, p are chosen in the cycloidal arc $\widehat{DC}$ such that $\widehat{DC}:\widehat{PC} = \widehat{\delta C}:\widehat{\pi C} = \widehat{dC}:\widehat{pC}$, then $t_{(\text{rest at } D)\delta\to d} = t_{(\text{rest at } P)\pi\to p}$. In proof, when the parallels DB, δS, dy, PV, πQ and px are drawn to the vertical BC as shown in the figure (where the semicircle on diameter BC meets πQ, δS, PV in X, Y, T respectively and CX, CY are extended to meet PV, px, DB, dy in the points α, x, β, y), it follows readily that

$$\widehat{PC}:\widehat{DC} = TC:BC = XC:YC = \alpha C:\beta C = xC:yC$$

and hence that the points α, β are on the same circle tangent at C to $B\widehat{T}C$. At once, by Galileo's theorems on the 'naturally' accelerated motion along inclined planes,(24) the speeds of fall attained at X, x and Y, y respectively from rest at α, β are in the proportion $\alpha C:\beta C = Xx:Yy$ and consequently the times over Xx and Yy are equal. Further, by Huygens' generalization of Galileo's 'assumption', the speed attained in an arbitrary curve at any level depends only on the vertical distance fallen from rest; accordingly, the speeds attained in the cycloidal arc $\widehat{DC}$ from rest at P, D are those attained at equal depths in the lines αC, βC from rest at α, β. In the limit, therefore as the pairs of points π, p; δ, d; X, x; Y, y coincide (when the vanishingly small arcs $\widehat{\pi p}$ and $\widehat{\delta d}$ become equal and parallel to Xx and Yy respectively), the time from rest at P over $\widehat{\pi p}$ is equal to that from rest at D over $\widehat{\delta d}$. Hence, on dividing the arcs $\widehat{Pp}$ and $\widehat{Dd}$ into vanishingly small, proportional corresponding sections, it follows that the time of fall over $\widehat{Pp}$ from rest at P is equal to that over $\widehat{Dd}$ from rest at D, independently of the depth VC of fall,(25) so that $\widehat{PC}$ and $\widehat{DC}$ are traversed in equal times. Brouncker's criterion for isochronism is an immediate corollary since

$$\widehat{\pi C}:\widehat{\delta C} = XC:YC = VC:SC.$$

How and when Newton first came to hear of Huygens' successful investigation of cycloidal motion is, it would appear, not recorded. Till the summer of 1673, when he received gift copies of both Huygens' and Pardies' published demonstrations of the isochronism of such motion,(26) his extant correspondence is silent on the topic. It is tempting to believe that he had already heard something of Huygens' researches and of Brouncker's partial success in repeating his proof that the tautochrone is the common cycloid, but this must remain

(24) In the tract 'De Motu Naturaliter Accelerato' inserted in the *Giornata Terza* of his 1638 *Discorsi* (note (2)).

(25) Compare Ziggelaar's 'Premières Démonstrations' (note (1)): 32–3. Pardies, in fact, constructs a division of the arcs $\widehat{PC}$ and $\widehat{DC}$ in geometrical proportion (when the time over each is the same) but this is unnecessary.

(26) See note (8) of the preceding introduction.

uncircumstantiated conjecture. As we have said, we believe that the internal structure of Huygens', Brouncker's and Pardies' demonstrations points unmistakably to the essential independence of Newton's equivalent arguments regarding free fall in a cycloidal arc.

1

THE SECOND BOOK OF EUCLID'S 'ELEMENTS' REWORKED[1]

[*c.* 1671][2]

From the originals in the University Library, Cambridge[3]

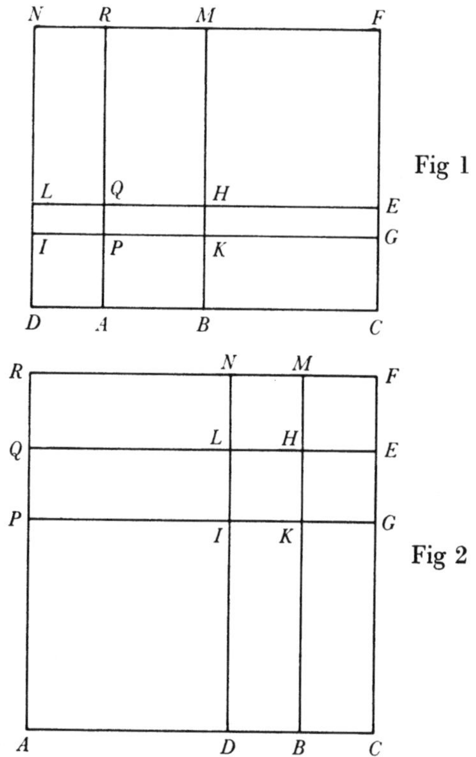

[1] [4]Sint AB, BC lineæ duæ; earum summa AC, differentia AD.[5] Super AB, AC, AD fac quadrata AH, AF, AI; quorum latera producta se mutuò secent in G, E, K, L, M, N, P, Q, R. Et priores decem[6] propositiones secundi Libri Elementorum sic enunciari ac demonstrari possunt.

Prop 1, 2, 3.[7] Rectangulum sub rectis duabus æquatur summæ rectangulorum sub una rectarum et partibus alterius, sive rectæ illæ utcunq; inæquales sint ut in Prop 1, sive æquales ut in Prop 2 sive una earum æquetur parti alterius ut in Prop 3.

$$AP\times AC(=AG=AK+BG)$$
$$=AP\times AB+AP\times BC.^{(8)}$$

Prop 4.[9] Quadratum ex summa duarum rectarum æquatur summæ

(1) Nothing certain is known regarding the background history of this piece. We may guess that this restyling and amelioration of the second book of Euclid's *Elements* was prepared by Newton for an intended commentary on Isaac Barrow's *Euclidis Elementorum Libri XV. breviter demonstrati* (Cambridge, 1655), perhaps for delivery as a set of more workaday lectures from the Lucasian chair in 1670 or 1671. As we have seen (I: 12, note (28)) Newton's own annotated copy of Barrow's edition (now Trinity College, Cambridge. NQ. 16.201) still exists. In the margin alongside 'Liber II.' Newton added in commentary on Propositions 2–11: '[II, 2] If $a=b+c$, y^{n} $aa=ab+ac$. [II, 3] & $ab=bb+cb$. [II, 4] & $aa=bb+2bc+cc$. [II, 5] &

$$\overline{a+b}\times\overline{a-b}=aa-bb.$$

Translation

[1] [4]Let AB, BC be two lines; AC their sum, AD their difference.[5] Upon AB, AC, AD construct the squares AH, AF, AI, and let their sides when produced intersect one another in G, E, K, L, M, N, P, Q, R. The first ten[6] propositions of the second book of the *Elements* may then be stated and proved in this manner.

Propositions 1, 2 and 3.[7] The rectangle contained by two straight lines is equal to the sum of the rectangles contained by one of the lines and the parts of the other, whether those lines be in any way unequal (as in Proposition 1) or equal (as in Proposition 2) or whether one of them be equal to one part of the other (as in Proposition 3).

$$AP\times AC(=\square AG=\square AK+\square BG)=AP\times AB+AP\times BC.^{(8)}$$

Proposition 4.[9] The square of the sum of two straight lines is equal to the sum

---

[II, 6] & $aa=bb+2bc+cc$. [II, 7] [&] $aa+cc=bb+2ac$. [II, 8] & $4ac+bb=aa+2ac+cc$. Or $aa+cc+2ac=bb+4ac$. [II, 9] $bb+cc=aa-2ab+2bb$. [II, 10] $aa+cc=bb+2bc+2cc$. [II, 11] $aa-ax=xx$.' The last entry is Newton's reminder that Euclid there constructs the problem of cutting a given line-length a into portions x and $a-x$ in 'golden' section. Compare T. L. Heath, *The Thirteen Books of Euclid's Elements Translated from the text of Heiberg with Introduction and Commentary* (Cambridge, $_2$1926): 403.

(2) This date, hazarded from Newton's handwriting, is confirmed by the fact that the first draft of this autograph manuscript is (see note (3)) entered on the verso of preliminary versions of Problems 1 and 2 of **3**, 2, §1 and of **3**, Appendix 1.6 below.

(3) The main revised text, given in [1], is Add. 3959.2: 22^r while additions in [2] are reproduced from the fuller draft in Add. 3970.11: 635^v/633^v.

(4) This first paragraph is not present in the draft version. The two accompanying diagrams illustrate the cases in which the point D lies outside or within the segment AB: in first draft a single figure had to serve both purposes.

(5) In either figure, that is, $DB = BC$.

(6) In first draft (see note (3)) Newton considered all fourteen propositions of Euclid's second book. His preliminary discussion of Propositions 11–14 is given in [2] following.

(7) 'Prop. I. Si fuerint duæ rectæ lineæ AB, AF, secetúrque ipsarum altera AB in quotcunque segmenta AD, DE, EB: rectangulum comprehensum sub illis duabus rectis lineis AB, AF, æquale est eis, quæ sub insecta AF, & quolibet segmentorum AD, DE, EB comprehenduntur rectangulis' (Barrow's *Euclid* (note (1)): 40). Newton's accompanying gloss on Barrow's figure is repeated from Add. 3970.11: 633^v.

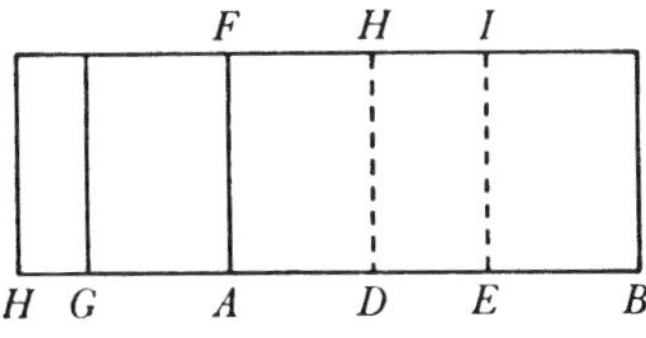

(8) Here and below, in our English version we honour the seventeenth-century practice of denoting squares and rectangles by a pair of opposite corners.

(9) 'Prop. IV. Si recta linea Z secta sit utcunque; Quadratum, quod à tota Z describitur, æquale est, & illis quæ à segmentis A, E describuntur quadratis, & ei, quod bis sub segmentis A, E comprehenditur, rectangulo' (Barrow's *Euclid*: 41). That is,

$$Z^2 \text{ (or } (A+E)^2) = A^2+E^2+2AE.$$

quadratorum ex lineis una cum duplo rectangulo ex ijsdem.

$$AC^q(=AF=AH+HF+HC+HR=AH+HF+2HC)=AB^q+BC^q+2ABC.$$

Prop 5 juxta fig 1 & Prop 6 juxta fig 2.(10) Rectangulum ex summa et differentia duarum rectarum æquatur differentiæ quadratorū ex lineis.

$$CAD(=AG=AK+BG=AK+PL=AH-IH)=AB^q-BC^q.$$

Prop 7.(11) Quadratum ex differentia duarum rectarum æquatur excessui summæ quadratorum ex lineis supra duplum rectangulū ex ijsdem.

$$AD^q(=AI=AH-DH+IH-QK=AH-2DH+IH)=AB^q-2ABD+BD^q.$$

Prop 8.(12) Differentia quadratorū ex summa et ex differentia duarum rectarum æquatur quadruplo rectangulo ex rectis.

$$AC^q-AD^q(=AF-AI=DE+LF+PN=DE+IM+PN=DE+PM$$
$$=2DE=4BE)=4ABC.$$

Prop 9 juxta Fig 1 et Prop 10 juxta Fig 2.(13) Summa quadratorum ex summa et ex differentia duarum rectarum æquatur duplo summæ quadratorum ex rectis.

$$AC^q+AD^q(=AF+AI=GM+AH+HR+GB+AI$$
$$=GM+AH+QK+DK+AI=GM+AH+AH=2EM+2AH)$$
$$=2AB^q+2BC^q.$$

---

(10) 'Prop. V. Si recta linea *AB* secetur in æqualia *AC*, *CB*, & non æqualia *AD*, *DB*, rectangulum sub inæqualibus segmentis *AD*, *DB* comprehensum, unà cum quadrato, quod fit ab intermedia sectionum *CD*, æquale est ei, quod à dimidia *CB* describitur, quadrato'

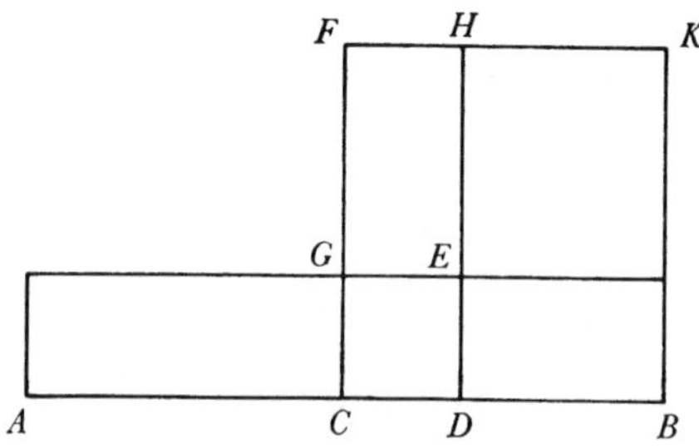

(Barrow's *Euclid*: 42). Prop. VI is the corresponding case where *D* no longer lies within the segment *CB*. Evidently, since in Newton's accompanying figure (reproduced from Add. 3970.11: 633v) $FG = DB$, the 'rectangle' $AD \times DB$ is equal to

$$(AC \text{ or}) \ CB\times FG+(CD \text{ or}) \ GC\times DB,$$

that is, to $CB\times FC-GC\times CD$.

of the squares of those lines together with twice the rectangle contained by them.

$$AC^2(=\square AF=\square AH+\square HF+\square HC+\square HR$$
$$=\square AH+\square HF+2\square HC)=AB^2+BC^2+2AB\times BC.$$

Proposition 5 (after figure 1) and Proposition 6 (after figure 2).[10] The rectangle of the sum and difference of two straight lines is equal to the difference of the squares of the lines.

$$CA\times AD(=\square AG=\square AK+\square BG=\square AK+\square PL=\square AH-\square IH)=AB^2-BC^2.$$

Proposition 7.[11] The square of the difference of two straight lines is equal to the excess of the sum of the squares of the lines over twice the rectangle contained by them.

$$AD^2(=\square AI=\square AH-\square DH+\square IH-\square QK$$
$$=\square AH-2\square DH+\square IH) = AB^2-2AB\times BD+BD^2.$$

Proposition 8.[12] The difference of the squares of the sum and the difference of two straight lines is equal to four times the rectangle contained by the lines.

$$AC^2-AD^2(=\square AF-\square AI=\square DE+\square LF+\square PN$$
$$=\square DE+\square IM+\square PN=\square DE+\square PM=2\square DE=4\square BE)=4AB\times BC.$$

Proposition 9 (after figure 1) and Proposition 10 (after figure 2).[13] The sum of the squares of the sum and the difference of two straight lines is equal to twice the sum of the square of the lines.

$$AC^2+AD^2(=\square AF+\square AI=\square GM+\square AH+\square HR+\square GB+\square AI$$
$$=\square GM+\square AH+\square QK+\square DK+\square AI=\square GM+\square AH+\square AH$$
$$= 2\square EM+2\square AH) = 2AB^2+2BC^2.$$

---

(11) 'Si recta linea Z secetur utcunque; Quod à tota Z, quódque ab uno segmentorum E, utraque simul quadrata, æqualia sunt illi, quod bis sub tota Z, & dicto segmento E comprehenditur, rectangulo, & illi, quod à reliquo segmento A fit, quadrato' (Barrow's *Euclid*: 43). In other words, $Z^2+E^2 = 2ZE+A^2$ (or $(Z-E)^2$).

(12) 'Si recta linea Z secetur utcunque; rectangulum quater comprehensum sub tota Z, & uno segmentorum E, cum eo quod à reliquo segmento A fit, quadrato, æquale est ei, quod à tota Z, & dicto segmento E, tanquam ab una linea $Z+E$ describitur, quadrato' (Barrow's *Euclid*: 44). That is, $4ZE+A^2$ (or $(Z-E)^2$) $= (Z+E)^2$.

(13) 'Prop. IX. Si recta linea AB secetur in æqualia AC, CB, & non æqualia AD, DB. quadrata, quæ ab inæqualibus totius segmentis AD, DB fiunt, simul duplicia sunt, & ejus, quod à dimidia AC, & ejus, quod ad intermedia sectionum CD fit, quadrati' (Barrow's *Euclid*: 44; compare Newton's diagram accompanying note (10) above).

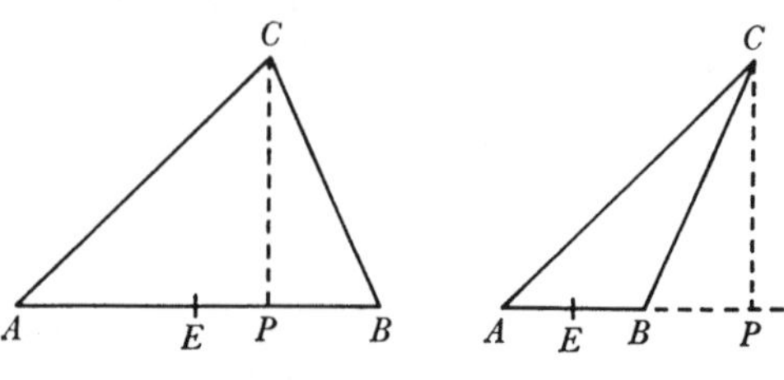

[2] [Prop 12 & 13.](14) Adhæc si bisecetur AB in E erit differentia quadratorum laterum $ACq - BCq$ æqualis duplo rectangulo sub basi AB, et perpendiculi(15) a medio basis distantia EP. Nam $ACq - BCq = APq - BPq = AP + BP$ in $AP - BP$. In priori triangulo $AP + BP$ est AB et $AP - BP$ seu $AB - 2BP$ est $2EB - 2BP$ seu $2EP$. In posterior $AP + BP$ seu $AB + 2BP$ est $2EB + 2BP$ id est $2EP$ & $AP - BP$ est AB. Ergo in utroq $AP + BP$ in $AP - BP$ est $2AB \times EP$. Quare $ACq - BCq = 2AB \times [E]P$. Q.E.D.(16)

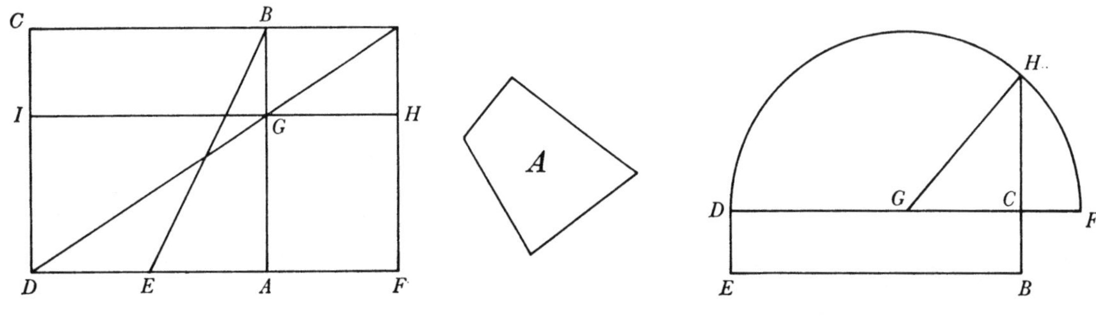

---

(14) 'Prop. XII [XIII]. In amblygoniis [oxygoniis] triangulis *ABC* quadratum, quod fit à latere *AC*[*AB*] angulum obtusum *ABC* [acutum *ACB*] subtendente, majus [minus] est quadratis, quæ fiunt à lateribus *AB*, *BC* obtusum [*AC*, *CB* acutum] angulum *ABC* [*ACB*] comprehendentibus, rectangulo bis comprehenso, & ab uno laterum *BC*, quæ sunt circa obtusum [acutum] angulum *ABC*[*ACB*], in quod...cadit perpendicularis *AD*, & ab assumpta exterius [interius] linea *BD*[*DC*] sub perpendiculari *AD* prope angulum obtusum *ABC* [ac tum *ACB*]' (Barrow's *Euclid*: 45 [46]). The lurking ambiguity in the opening phrase 'In oxygoniis [*sc.* 'acute-angled'] triangulis' of the enunciation of II, 13 (compare T. L. Heath's *Thirteen Books* (note (1)): 406–7) has evidently not occurred to Newton.

[2] [Propositions 12 and 13.][14] Further, if AB be bisected in E, the difference of the squares of the sides (AC^2-BC^2) will be equal to twice the rectangle contained by the base (AB) and the distance (EP) of the perpendicular[15] from the mid-point of the base. For

$$AC^2-BC^2=AP^2-BP^2=(AP+BP)(AP-BP).$$

In the first triangle $AP+BP$ is AB and $AP-BP$ or $AB-2BP$ is $2(EB-BP)$ or $2EP$. In the latter, $AP+BP$ or $AB+2BP$ is $2(EB+BP)$, that is $2EP$, while $AP-BP$ is AB. In either, therefore, $(AP+BP)(AP-BP)$ is $2AB\times EP$. Hence $AC^2-BC^2=2AB\times EP$. As was to be proved.[16]

---

(15) CP, that is.

(16) The two following diagrams illustrate the two remaining Propositions 11 and 14 of Euclid's second book: their proofs are not repeated here by Newton perhaps because they are constructions (and not theorems) or maybe because he felt he could not improve upon them. The left-hand figure (where, in extension of Euclid's, Newton has correctly extended DG through the meet of GB and FH) shows the division of the line AB at G in extreme and mean ratio (Barrow's *Euclid*: 45: 'Prop. XI. Datam rectam lineam AB secare in G, ut comprehensum sub tota AB, & altero segmentorum BG rectangulum, æquale fit ei, quod à reliquo segmento AG fit, quadrato'; compare VI, 30). Euclid constructs the square $ABCD$, bisecting DA in E and making $EF=EB$: at once $AB^2=EF^2-EA^2=(DF\times AF$ or) $AG(AB+AG)$, so that $AB\times GB=AG^2$. Evidently, on taking $AB=a$ and $AG=x$, there results $EB=\frac{1}{2}a\sqrt{5}$ and hence $AF=AG=\frac{1}{2}a(\sqrt{5}-1)$ is the positive root of the 'golden section' equation $a(a-x)=x^2$: compare note (1). The right-hand figure repeats that accompanying 'Prop. XIV. Dato rectilineo A, æquale quadratum invenire' (Barrow's *Euclid*: 47), where it is presumed that the given quadrilateral (A) has, by I, 45, been reduced to a triangle and thence to the rectangle $BCDE$ of equal area. Immediately, on taking $CF=CB$, the meet H of BC with the semicircle on diameter DF determines $CH=\sqrt{[DC\times CB]}$ to be the side of the square required.

2

RESEARCHES INTO THE ELEMENTARY GEOMETRY OF CURVED SURFACES(1)

[*c.* 1670?](2)

From an original draft in the University Library, Cambridge(3)

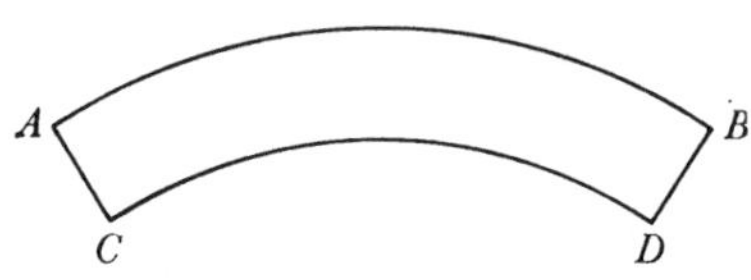

Ax 1. Superficies *ABDC* a curvis lineis æquidistantibus(4) *AB*, *CD* & eorum perpendiculis *AC*, *BD* comprehensa minor [est] parallelogrammo(5) $AB \times AC$, major parallelogrammo $CD \times AC$.

Ax 2. Solidum a curvis superficiebus æquidistantibus et ambitu quovis ⊥ lari comprehensū majus est parallelipipedo cujus basis æquatur(6) convexæ superficiei & altitudo intervallo superficierum, majus vero ℘pipedo cujus basis æqu. minori sup℘i et altitudo itidem interval. su℘erum.(7)

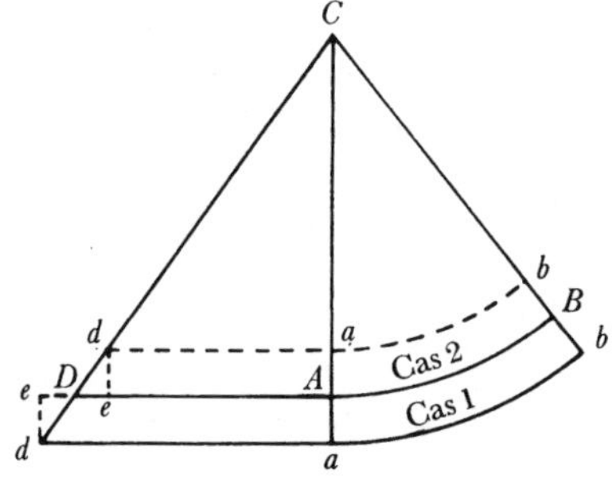

Prop. 1. Si Circuli cujusvis sectori *CAB* æquale triangulū rectangulum *CAD* super radio *CA* constituatur, ejus perpendiculum(8) *AD* æqualis est arcui sectoris *AB*.

Si negas pone $Ae = AB$, eiq; erige normalem *ed* occurrentem *CD* in *d*, ad *CA* demitte ⊥ *da* & centro *C* radio *Ca* describe arcum *ab* occurrentem *CB* in *b*. Est itaq; $eA \times Aa = AB \times Aa$.(9) Sed tri *CAD*. tri $Cad(::CA^q . Ca^q)::$sect *CAB*. sect *Cab*. Ergo cum sect $CAB =$ (Hyp)$\triangle CAD$, erit sect $Cab =$ ([33]. v[I])$\triangle Ca[d]$ & (ax 2.1)(10) limbus $BAab = DAad$. Sed in Cas 1 [ubi *AB* ponitur major *AD*] est

(1) This Archimedean discussion of the surface area and volume of various cylindrical and spherical segments is—particularly in its use of the classical exhaustion proof—scarcely typical of Newton's general approach to such problems but a glance at Plate III (which reproduces the manuscript's second page) will confirm that it is no mere effortless reworking of a passage in some previous commentary on Archimedes' *Sphere and Cylinder*. Once more we are tempted to think that the source of Newton's inspiration was Isaac Barrow, who in several undated 'Incerti temporis...Lectiones' appended to his Lucasian lectures on mathematics as later published (*Lectiones Mathematicæ XXIII...Habitæ Cantabrigiæ A.D. 1664, 1665, 1666* (London, 1683): 339–88) sought to expound the 'methodum...quâ Archimedes præclara

Translation

Axiom 1. The surface-area $ABDC$ comprehended by the equidistant(4) curves AB, CD and the perpendiculars AC, BD to them is less than the rectangle(5) $\widehat{AB}\times AC$, greater than the rectangle $\widehat{CD}\times AC$.

Axiom 2. The solid comprehended by equidistant curved surfaces and any perpendicular surround is greater than the parallelepiped whose base is equal to the(6) convex surface and whose height is the distance between the surfaces, but less than the parallelepiped whose base is equal to the lesser surface and whose height is likewise the distance between the surfaces.(7)

Proposition 1. If, equal to the sector CAB of any circle, there be set up on the radius CA the right triangle CAD, its perpendicular(8) AD is equal to the sector's arc $\widehat{AB}$.

If you deny it, take $Ae=\widehat{AB}$ and normal to it erect ed meeting CD in d, let fall da perpendicular to CA and with centre C, radius Ca describe the arc ab meeting CB in b. In consequence $eA\times Aa=\widehat{AB}\times Aa$. But triangle CAD: triangle Cad $(=CA^2:Ca^2)=$sector CAB: sector Cab. Accordingly, since sector $CAB=\triangle CAD$ (by hypothesis), therefore sector $Cab=\triangle Cad$ ([*Elements*], VI, 33) and so (by axioms 2, 1)(10) fringe $BAab=DAad$. But in Case 1, where $\widehat{AB}$ is put greater than

sua Theoremata, libris qui extant comprehensa, adinvenit'. We may perhaps go further and guess that Newton intended his present researches to form the nucleus of a similar lecture series of his own on Archimedean geometry. Other influences on Newton are hard to define. His introduction of the 'cylindrical wedge' (cuneus cylindricus) in Proposition 3 recalls momentarily to mind Grégoire de Saint-Vincent's investigation of parabolic 'hoofs' (*Opus Geometricum* (Antwerp, 1647): Book 9, Part 5: 1020–37), while the 'pumpkin' (figura peponiformis) he invokes in the proof of Proposition 6 has a Keplerian ring (see note (46)) but these allusions are surely no more than fortuitous.

(2) Hesitantly conjectured solely on the basis of Newton's present handwriting.

(3) Add. 3963.16: 184r–185v.

(4) The necessary restriction that they must also be convex is understood. It is made explicit in 'Ax 2' following.

(5) The continuation 'cujus basis est = AB & altitudo [AC]' (whose base is equal to $\widehat{AB}$ and height AC) is cancelled.

(6) Understand 'majori' (greater).

(7) It is interesting that Newton makes these two convexity lemmas axiomatic in his paper, but probably the parallel with the status accorded by Archimedes to the lambanomena of his *Sphere and Cylinder* is intentional. (Compare E. J. Dijksterhuis, *Archimedes*, Copenhagen (1956): 145–9.) Both may of course be justified in an obvious way by appeal to infinitesimals.

(8) 'altitudo' (height) is cancelled.

(9) We omit the continuation '⊐ (ax 1) $BAab$', which is true only of Case 1 and doubtless intended by Newton to be cancelled after he reorganized his further argument into two cases.

(10) Of Euclid's *Elements*, that is. See *Euclidis Elementorum Libri XV. breviter demonstrati, Operâ, Is. Barrow* (Cambridge, 1655): 6: '1. Quæ eidem æqualia, & inter se sunt æqualia'; '2. Et si æqualibus æqualia adjecta sunt, tota sunt æqualia'.

(ax 1)[11] $BAab \sqsubset AB \times Aa = eAad$. Ergo $DAad \sqsubset eAad$ pars toto. Q.F.N.[12] Similiter in Cas 2 ubi AB ponitur minor AD est $BAab$ (ax 1) $\sqsupset AB \times Aa$ et inde $DAad \sqsupset eAad$, totum parte. Q.F.N.[12] Non sunt ergo AD et AB inæquales. Q.E.D.

Cor. Hinc cognita longitudine arcus cognoscitur sector & vice versa.

Prop 2. Si super Sectore circuli CAB Sector Cylindricus & super æquali triangulo Prisma ejusdem altitudinis constituantur, Sector Cylindricus æquabitur Prismati & ejus convexa superficies AB lateri prismatis AD.

Cas. 1. Si negas sit N differentia Prismatis et sectoris Cylindrici & in sectore inscribe Polygonū[13] $CAfgB$ quod minus differat a sectore isto quam $\frac{1}{2}N$ & cujus etiam basis minus differet a base sectoris quam $\frac{1}{2}N$ applicatū ad altitudinem figuræ R. Utrumq; enim fieri potest per demonstrata ad Prop 2 & 10. XII.[14] Et differentiam basiū dic D. Cum itaq; Prismata CAD, $CAfgB$ sint ut bases (Cor 1. Pr 9 XII) hoc est ut triang CAD ad triang $CAD - D$[15] vel ductis omnib^s^ in alt[itudinem] fig[uræ] ut Prism CAD ad Prism $CAD - RD$. erit RD differentia[16] Prismatum CAD, $CAfgB$. Sed erat D minus quam $\frac{\frac{1}{2}N}{R}$ et proinde RD erit minus quam $\frac{1}{2}N$. Sed Sector Cylindr. etiam minus differt a Prismate $CAfgB$ quam $\frac{1}{2}N$ (Hyp.) Ergo minus differt ab altero Prismate CAD quàm N. Non est ergo N differentia eorum ut supponebatur & proinde non differunt.[17]

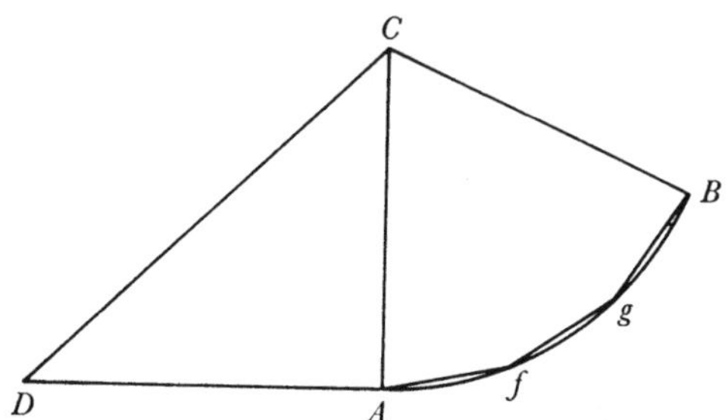

Cas 2. Contentis figurarum probatis æqualibus, superficies AB et AD ostendentur æquales simili argumentatione qua probavimus lineas AB et AD in priori Prop. æquales esse.

---

(11) Axiom 1 of Newton's present text.

(12) 'Q[uod] F[alsum] N[egatur]', presumably: a Newtonian variant on Barrow's usual rendering 'quæ [falsa] repugnant' (see his *Euclid* (note (10)): Liber XII, Prop. II: 293) of Euclid's ὅπερ ἐστὶν ἀδύνατον. Or it might be 'Q[uod] F[ieri] N[equit]'.

(13) More exactly, 'solidum polygonale' (a solid whose base is the 'polygon' *AfgBC*).

(14) To be precise, in these propositions (Barrow's *Euclid* (note (10): 292–3, 301–2) their author employs a repeated bisection of the circle arc $\widehat{AB}$ but Euclid's argument is general: since each arc *Af*, *fg*, ... may be taken arbitrarily small, so also the magnitude of the difference in both perimeter length and surface area of the sector *ABC* and inscribed 'polygon' *Afg* ... *BC* may be made as small as is desired.

(15) Newton first continued 'vel ut sector $CAB - \frac{N}{A}$' (or as sector $CAB - N/A$). Here for '$\frac{N}{A}$' read '$\frac{N}{R}$', a slip of Newton's pen which he himself has corrected several times in the sequel.

AD, by axiom 1[11] $BAab > \widehat{AB} \times Aa = eAad$, and therefore the part *DAad* is greater than the whole *eAad*: which is absurd. Likewise in Case 2, where $\widehat{AB}$ is put less than *AD*, by axiom 1 $BAab < \widehat{AB} \times Aa$, and thence the whole *DAad* is less than the part *eAad*: which is again absurd. Therefore *AD* and $\widehat{AB}$ are not unequal. As was to be proved.

Corollary. Hence when the length of the arc is known the sector is known, and conversely so.

Proposition 2. If a cylindrical sector be set up on the circle sector *CAB* and a prism of the same height on the equal triangle, the cylindrical sector will be equal to the prism and its convex surface *AB* to the prism's side *AD*.

Case 1. If you deny it, let *N* be the difference between the prism and the cylindrical sector and in the sector inscribe the polygon[13] *CAfgB*, which must differ from that sector by less than $\frac{1}{2}N$ while its base also must differ from the sector's base by less than $\frac{1}{2}N$ divided by the height *R* of the figure. (Both, to be sure, can be accomplished on the lines of the proofs of [*Elements*], XII, Propositions 2 and 10.)[14] Also call the difference of the bases *D*. Then since the prisms *CAD*, *CAfgB* are as their bases (XII, 9, Corollary 1), that is, as triangle *CAD* to triangle $CAD - D$[15] or, on multiplying everything into the figure's height, as prism *CAD* to prism $CAD - RD$, *RD* will be the difference[16] of the prisms *CAD*, *CAfgB*. But *D* was taken less than $\frac{1}{2}N/R$ and consequently *RD* will be less than $\frac{1}{2}N$. But (by hypothesis) the cylindrical sector also differs from the prism *CAfgB* by less than $\frac{1}{2}N$, and therefore differs from the other prism *CAD* by less than *N*. Therefore *N* is not their difference, as was supposed, and consequently they do not differ.[17]

Case 2. When the contents of the figures have been proved equal, the surfaces $\widehat{AB}$ and *AD* will be shown to be equal by an argument similar to that by which we demonstrated the equality of the lines $\widehat{AB}$ and *AD* in the previous proposition.

---

(16) This replaces 'excessus' (excess).

(17) Note how Newton's employment of the 'differentia' *N* allows him to reduce the prolixity of the classical exhaustion proof. The technique was pioneered in print by Blaise Pascal (who invoked a 'difference *Z*' in his 'Demonstration à la maniere des Anciens de l'Egalité des Lignes Spirale & Parabolique' [the final tract of his *Lettres de A. Dettonville contenant Quelques-vnes de ses Inuentions de Geometrie* (Paris, 1658)]: 12–16), but Newton presumably here borrows the method from James Gregory, who several times made analogous use of an arbitrarily small 'differentia α' in, for instance, his 'Analogia inter Lineam Meridianam Planispherii Nautici: & Tangentes Artificiales Geometricè Demonstrata, &c' (*Exercitationes Geometricæ* (London, 1668): 14–24). Compare D. T. Whiteside, 'Patterns of Mathematical Thought in the later Seventeenth Century' (*Archive for History of Exact Sciences*, **1** (1961):179–388, especially 346–8).

Cor.[18] Convexa superficies Cylindri est ad su*p*f[iciem] basis ut quadruplum altitudinis Cylindri ad diametrum basis.[19]

Schol. Potuit hæc Propositio facilius ostendi ducendo tam perimetrum basis quàm basem in altitudinem figuræ per motum localem: nam æquales lineæ in eandem altitudinem ductæ generant æquales superficies, & æquales superficies æqualia solida. Sed id nulli postulato vel principio Euclideo respondet.[20]

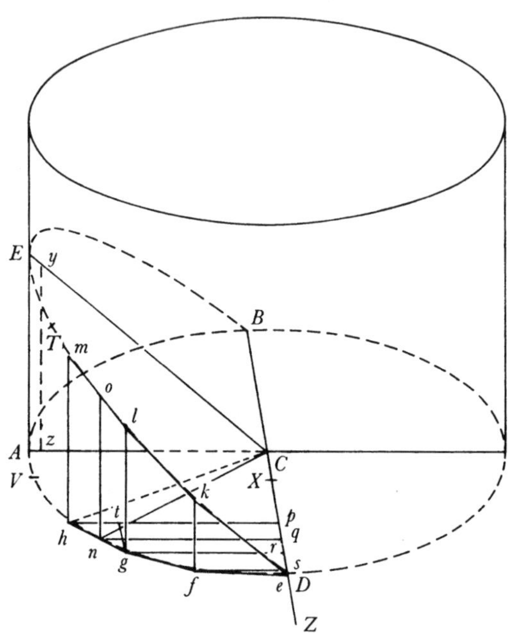

Lemma [1]. Si cylindrus rectus super Basi BAD[21] constitutus & polygonum circumscriptum[22] habens cujus basis est $efgh$, secetur per planum aliquod BED basem bisecans in BD,[23] & polygono occurrens in k, l, m; et a termino basis polygoni ad diametrū basis demittatur $\perp hp$: superficies polygoni $efk+kfgl+lghm$ æqualis erit rectangulo quod continetur a segmento diametri ep et altitudine abscissi solidi[24] cylindrici EA.

Nam Polygoni latus aliquod hg biseca in n[25] ubi circulum inscriptum tangit, et inde ad CD demitte $\perp nq$

(18) A cancelled first opening 'Schol' indicates that this corollary was added on the spur of the moment.

(19) On taking h and d to be respectively the height and base diameter of a cylinder, its curved surface is πdh and its base area $\frac{1}{4}\pi d^2$.

(20) Compare the opening 'Postulata' of Newton's contemporary 'Problems for construing æquations' (II: 452) and his preliminary observations on Problems 1 and 2 of his 1671 fluxional tract (1, 2, §2: Newton's page 17). The former tract is proof that he—as Viète before him in a parallel situation (*Supplementum Geometriæ*, Tours, 1593 [= *Opera Mathematica* (Leyden, 1646): 240–57])—had no compunction about transcending the arbitrary restrictions of classical Greek geometry when it suited his purpose.

(21) The continuation 'cujus centrum est C' (whose centre is C) is cancelled.

(22) Understand 'æquilaterum' (regular) as the phrase it replaces (see note (23)) makes clear. When Newton later inserted part of Proposition 3's proof as 'Lem. 2' (note (28)), he added the personal reminder 'Let this Lemma be done by inscription' to conform with its structure but never, in fact, made the necessary changeover. We have left the lemma in its present, inconsistent 'circumscribed' form: in the 'inscribed' equivalent the polygon $efgh$ will commence at D, having its vertices on the circle arc $\widehat{AD}$, and n will be the midpoint of the chord gh.

(23) A continuation 'et circa basem describatur polygonum æquilaterum $efgh$ a cujus angulis [ad diametrum demittantur perpendicula]' was cancelled when its gist was incorporated elsewhere.

(24) The equivalent 'segmenti resecti' is cancelled.

Plate III. The comparison of curved surfaces (**2**, 2).

Corollary.(18) The convex surface of a cylinder is to the surface-area of its base as four times the cylinder's height to its base diameter.(19)

Scholium. This proposition could have been shown more easily by spatial motion by drawing both the base perimeter and the base into the figure's height: for equal lines when drawn into the same height generate equal surfaces, and equal surface-areas generate equal solids. But that answers to no Euclidean postulate or principle.(20)

Lemma 1. If a right cylinder, set up on the base BAD(21) and having a circumscribed(22) polygon whose base is *efgh*, be cut by some plane *BED* bisecting the

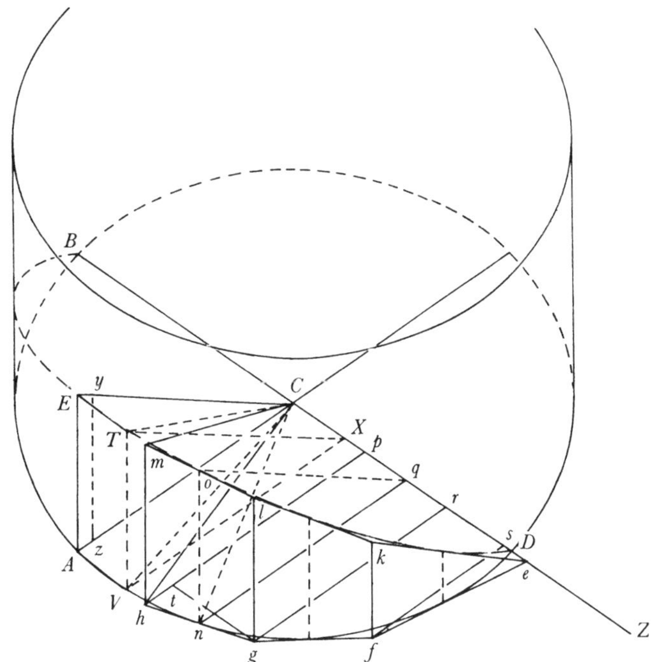

base in BD(23) and meeting the polygon in *k*, *l*, *m*, and if from the end-point of the polygon's base there be dropped *hp* perpendicular to the base diameter, then the polygon's surface $efk+kfgl+lghm$ will be equal to the rectangle contained by the segment *ep* of the diameter and the height *EA* of the cylindrical solid cut off.

For bisect some side *hg* of the polygon in *n*(25) (the point at which it touches the inscribed circle) and from there to *CD* let fall the perpendicular *nq*; also erect

(25) Newton had more than a little difficulty in identifying points correctly in his confused manuscript figure (see Plate III). In a first equivalent version 'Etenim a medio alicujus lateris polygoni, ut *n*' he confused the midpoint in turn with the end-points *h*, *g* of its chord. Below, in our reproduction of the text we have replaced all Newton's incorrect namings (*n*, *p*, *A*, *E*, *e*, *h*, *X*) of points by their correct identifications (*l*, *q*, *z*, *y*, *l*, *g*, *p* respectively) and in the English version have inserted a variant figure which, it is hoped, illustrates the elements of Newton's sketch more clearly.

& erige $\perp$ *no* plano[(26)] *BED* occurrens in *o*, et erit $no = \frac{1}{2}mh + \frac{1}{2}[l]g$. Sed figura *mhgl*, cum rectangula sit ad *h* & *g*, æqualis est parallelogrammo ad eandem basem *hg* & mediam altitudinem laterum $\frac{1}{2}mh + \frac{1}{2}lg$ constituto, [et] proinde æqualis $no \times hg$. Sed acto radio *nC*, et ad *hp* & *pe* demissis $\perp$ ^lis^ *gt*, *gr* propter sim. tri. *hgt*, $nC[q]$, erit $hg.gt(rp)::nC(AC).n[q]::EA.on$. Quare $rp \times EA = hg \times on$ hoc est = superficiei *mhgl*.[(27)] Et eodem modo, demisso perpendiculo *fs* demonstrabitur quod superficies *lgfk* sit $= sr \times EA$, et superficies $kfe = er \times EA$, Adeoq; summa omnium superficierum $mg + lf + kf[e]$ est $pr + rs + se \times EA$ hoc est $pe \times EA$. Q.E.D.

Lem. 2[(28)] [C]oncipe polygonum *efgh* inscribi[(29)] et a terminis planorū laterum erectorum *hm*, *gl*, *fk* plana duci ad centrum figuræ quibus figura resolvatur in pyramides $mhg[l]C$, $[l]gfkC$, $kfeC$ verticem habentes ad *C* & cum istarum omnium pyramidum æqualis[(30)] sit altitudo nempe *nC*, duc trientem altitudinis in summam basium $mg \times lf \times kfe$ hoc est (per Lemma [1]) in $EA \times e[p]$ et habebitur summa pyramidum hoc est solidū polygonum[(31)] *Cmhe*.

Prop 3. Si Cylindrus rectus super basi *BAD* constitutus secetur plano aliquo *BED* basem Bisecante in *BD*, & segmentum *BEDA* rursus secetur plano *CTV* ad basem perpendiculari & per centrum basis transeunte, a *V* ad *CD* demitte $\perp$ *VX*: segmentum secundum *CTVD* æquale erit parallelipipedo[(32)]

$$\frac{1}{3}EA \times AC \times DX.$$

[(33)]Si negas sit *N* differentia inter segmentum istu[d] & paralle[li]pipedum $\frac{1}{3}EA \times AC \times DX$, et in segmentū inscribe polyg[onum] *Dfghmlk*[(34)] laterū tam minutorum & multorum ut excessus quo superatur a segmento sit minor quam $\frac{1}{2}N$ ut et sinus versus *Az*[(35)] minor quam $\dfrac{N}{2AE \times XD}$. Utrumq; enim fieri potest per 1. x siquidem qualibet bisectione laterum Polygoni, et sinus versus, et spatia inter latera Polygoni & superficiem cylindri plusquam dimidio minuantur ut ex demonstratis ad Prop 10. & 16 l[ibri] 12 satis patet.[(36)] Ergo

(26) 'circulo' (circle) is here justly cancelled: the arc *BED* is elliptical!

(27) Subsequently, the sides of the circumscribed polygon are supposed to become infinite in number. Newton breaks completely away from the classical (Archimedean) tradition at this point, choosing to prosecute his further argument by an indivisibles approach. The rigorous approach by means of Archimedes' convexity lemmas would demand that the cylindrical surface element be bounded in its area by the trapezium *ghml* (the element of the 'polygonum circumscriptum') and a corresponding trapezium of an inscribed 'polygon', and that a *reductio ad absurdum* argument should prove that the difference between these areas is zero in the limit as the sides of the circumscribing and inscribed polygons increase in number to infinity.

(28) Originally this was the opening paragraph of Newton's proof of 'Prop 3' following, but it is placed here in accordance with his memorandum 'Let this be y^e^ 2^d^ Lem then add the Prop wth this demonstr. Si negas...' (compare note (33)).

(29) Newton assumes that the preceding lemma has been proved 'by inscription' (see note (25)).

the perpendicular no to the plane(26) BED, meeting it in o, and then will be $no = \frac{1}{2}(mh+lg)$. But the figure $mhgl$, since it is right-angled at h and g, is equal to the rectangle set up on the same base hg and with the mean height $\frac{1}{2}(mh+lg)$ of the sides, and consequently equal to $no \times hg$. But, on drawing the radius nC and letting fall the perpendiculars gt, gr to hp and pe, because of the similar triangles hgt, nCq it will be $hg:gt$ (or rp) $= nC$ (or AC)$:nq = EA:on$. Therefore $rp \times EA = hg \times on$, that is, equal to the surface $mhgl$.(27) And in the same manner, when the perpendicular fs is let fall, it will be proved that surface $lgfk = sr \times EA$ and surface $kfe = er \times EA$, with the result that the sum of all the surfaces $mg+lf+kfe$ is $(pr+rs+se) \times EA$, that is, $pe \times EA$. As was to be proved.

Lemma 2.(28) Conceive the polygon $efgh$ to be inscribed(29) and that from the boundaries of the erected plane sides hm, gl, fk planes are drawn to the figure's centre, by which the figure shall be resolved into pyramids $mhglC$, $lgfkC$, $kfeC$ having their vertex at C: then, since all those pyramids have an equal(30) height, namely nC, multiply one-third the height into the sum $mg+lf+kfe$ of the bases, that is (by Lemma 1) into $EA \times ep$, and there will be obtained the sum of the pyramids, that is, the polygonal solid(31) $Cmhe$.

Proposition 3. If a right cylinder set up on the base BAD be cut by some plane BED bisecting the base in BD and the segment $BEDA$ be again cut by the plane CTV perpendicular to the base and passing through the base's centre, let fall the perpendicular VX from V to CD: the second segment $CTVD$ will then be equal to the parallelepiped(32) $\frac{1}{3}EA \times AC \times DX$.

(33)If you deny it, let N be the difference between that segment and the parallelepiped $\frac{1}{3}EA \times AC \times DX$ and in the segment inscribe the polygon $Dfghmlk$(34) of sides sufficiently minute and numerous to make the segment's excess over it less than $\frac{1}{2}N$ and in addition the versed sine Az(35) less than $\frac{1}{2}N/(AE \times XD)$. (Both, to be sure, may be accomplished by [*Elements*] x, 1, seeing that by any bisection of the polygon's sides both that versed sine and the spaces between the polygon's sides and the cylinder's surface may be diminished by more than half: this is evident enough(36) from the proofs of xii, Propositions 10

(30) This replaces 'communis' (common).

(31) 'contentum' (content) is cancelled.

(32) Understood to be rectangular.

(33) See note (28). In the following paragraphs several occurrences of p have been silently corrected to X: compare note (25).

(34) The cancelled variant '*efghmlk*' relates to the original proof (here reproduced) of Lemma 1 by circumscribing the polygon. In the present inscribed form Newton correctly locates the end-point of the polygon at D (compare note (22)).

(35) Of $h\hat{C}n = n\hat{C}g$, that is, since (in the inscribed form) n is taken to be the midpoint of the chord gh, so that $zC = nC$ is equal to $AC\cos h\hat{C}n$.

(36) Contrast Newton's proof of Case 1 of Proposition 2 above.

$\frac{2}{3}Cyz \times XD$ ($=$ (per Lem 2) Polyg[ono])(37) minus differt a segm[ento isto] Cylindr[ico] quàm $\frac{1}{2}N$, At $\frac{2}{3}CAE \times XD$ differt a $\frac{2}{3}Cyz \times XD$ per $\frac{2}{3}AEyz \times XD$ quod minus est quam $\frac{2}{3}AE \times Az \times XD$ & multo magis minus quàm

$$AE \times Az \times XD$$

hoc est quàm $\frac{1}{2}N$ (supra).˙ Quare $\frac{2}{3}CAE \times XD$ differt a segm[ento] Cylind[rico] minus quam toto N. Falso igitur ponebatur N differentia et proinde differentia nulla est. Q.E.D.

Cor 1. Totus cuneus cylindricus $BEDA$ æquatur $\frac{2}{3}EAC \times BD$.(38)

Cor 2. Si cuneus [$BEDA$] secetur plano aliquo TVX parallelo EAC, fac $BX.CD::CD.CZ$ et erit $\frac{2}{3}TVX \times XZ =$ segmento cunei [$XTVD$].(39) Nam &c.

Prop 4. Stantibus jam positis superficies segmenti Cunei Cylindrici [TVD](40) erit $EA \times XD$.

Probatur ut Prop 1.(41)

Prop 5. Coni cujus basis ad axem erectus est(42) superficies convexa est ad basem ut latus ad radium basis.

Prob[atur] ut Prop 1.(43)

Cor. Si conus secetur planis duobus ad axem $\perp$[lis] superficies segmenti æqualis erit superficiei cylindri ejusdem altitudinis cujus basis diameter est perpendiculum(44) medio segmenti lateris ad axem erectum.

---

(37) Namely, *DfghVXTmlk*.

(38) That is, $\frac{1}{3}EA \times AC \times BD$.

(39) This is badly out. Since the triangles are similar,

$$\triangle EAC : \triangle TVX = AC^2 \text{ (or } CD^2) : VX^2 \text{ (or } BX \times XD)$$

and therefore $(XTVD)$ = 'second segment' $(CTVD)$ − pyramid $(CTVX)$

$$= \tfrac{2}{3}\triangle EAC \times XD - \tfrac{1}{3}\triangle TVX \times CX = \tfrac{1}{3}\triangle TVX(2CD^2/BX - CX).$$

We may consequently correct Newton's text to read

'fac $BX.CD::2CD.CZ$ et erit $\frac{1}{3}TVX \times XZ =$ segmento cunei $XTVD$'

(make $BX:CD = 2CD:CZ$ and then $\frac{1}{3}TVX \times XZ =$ wedge segment $XTVD$).

(40) The text reads '*mhz*'!

(41) For, by Lemma 1, in the limit as the sides of the polygon increase in number to infinity (and so the vertex e coincides with D) the cylindrical surface mhD is equal to $pD \times EA$ and the present proposition is the particular instance when mh coincides with TV.

(42) In a right cone, that is.

(43) By an analogous argument, if we erect the right pyramid $C'efgh$ ... of vertex C' on the regular polygon circumscribing the circle of diameter BD (in Newton's original Lemma 1), the element $C'hg$ of its raised surface is equal in area to $\frac{1}{2}C'n \times hg$. Accordingly, in the limit as each side hg becomes infinitely small (when the raised pyramidal surface coincides with the curved surface of the right cone of vertex C' standing on the base circle BnD), the cone surface

and 16.) Therefore $\frac{2}{3}Cyz \times XD$ (equal to the polygon[37] by Lemma 2) differs from that cylindrical segment by less than $\frac{1}{2}N$. But $\frac{2}{3}CAE \times XD$ differs from $\frac{2}{3}Cyz \times XD$ by $\frac{2}{3}AEyz \times XD$, which is less than $\frac{2}{3}AE \times Az \times XD$ and so much less still than $AE \times Az \times XD$, that is, less than $\frac{1}{2}N$ (by the above). Hence the difference between $\frac{2}{3}CAE \times XD$ and the cylindrical segment is altogether less than N. As a consequence the difference was falsely set to be N and there is therefore no difference at all. As was to be proved.

Corollary 1. The whole cylindrical wedge $BEDA$ is equal to $\frac{2}{3}EAC \times BD$.[38]

Corollary 2. If the wedge $BEDA$ be cut by some plane TVX parallel to EAC, make $BX:CD = CD:CZ$ and then $\frac{2}{3}TVX \times XZ =$ wedge segment $XTVD$.[39] For...

Proposition 4. With the previous suppositions still standing, the surface-area of the segment TVD[40] of the cylindrical wedge will be $EA \times XD$.

The proof is as in Proposition 1.[41]

Proposition 5. In a cone whose base is at right angles to its axis[42] the convex surface is to the base as its side to the base radius.

The proof is as in Proposition 1.[43]

Corollary. If a cone be cut by two planes perpendicular to its axis, the surface of the segment will be equal to the surface of a cylinder of the same height, the diameter of whose base is the perpendicular[44] raised on the mid-point of the segment's side as far as the axis.

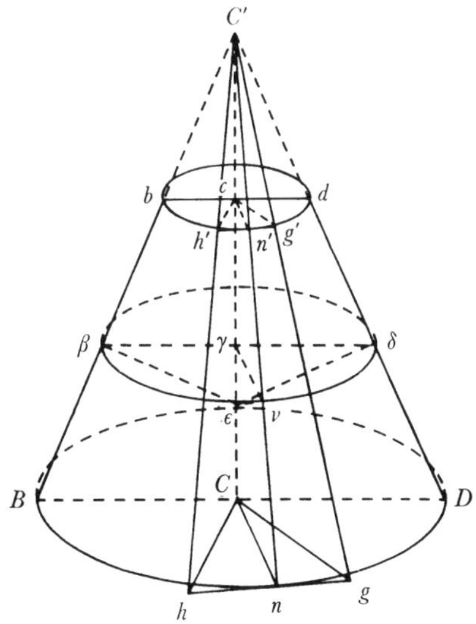

is equal to $\frac{1}{2}BC' \times \pi \times BD$ and so in proportion to the area ($\frac{1}{4}\pi \times BD^2$ or $\frac{1}{2}BC \times \pi \times BD$) of the base circle as BC' to BC.

(44) For 'diameter' read 'semidiameter' (radius): Newton first wrote 'semisum[m]a diametrorum sectionum' (the half-sum of the diameters of the sections) in place of 'perpendiculum'. If the circle $\beta\nu\delta$ is the mean segment between the end circles BnD, $bn'd$ of the conic

Prop 6. Sector sphæræ æqualis est(45) Cylindro basem haben[ti] circulum in sphæra maximū & altitudinem sinum versum sectoris.

Probatur per fig. Peponiformem.(46)

Cor. Tota sphæra(47) [&c]

Prop 7. Segmentum sphæræ(48) [&c]

Prop 8. Superficies sphærici segmenti æqualis est superficiei Cylindri eandem habentis alt[itudinem] & basem circulum in sphæra maximum.(49)

Cor. Hæe superficĉes æqualis est ⊙lo cujus radius est linea a vertice segmenti ad terminum basis ducta.(50)

Prop 9. Superficies Trianguli spherici.(51) [...]

---

frustum $BbdD$, and ϵ (in the axis $C'C$) is the meet of all perpendiculars to the cone's surface in the perimeter of that mean circle, then (by Proposition 5) the frustum's surface, that is, the difference of the cone surfaces $BnDC'$ and $bn'dC'$, is equal to

$$\pi(BC' \times BC - bC' \times bc) = \pi(\beta\gamma/\beta C')(BC'^2 - bC'^2) = 2\pi \times \beta\gamma \times Bb = 2\pi \times \beta\epsilon \times Cc$$

since $Bb:Cc = BC':CC' = \beta\epsilon:\beta\gamma$. The same result follows more directly since $\nu\epsilon(=\beta\epsilon)$ is normal to the element $hh'g'g = Bb \times \frac{1}{2}(hg + h'g')$ of the curved surface of the conic frustum.

(45) Newton first continued 'trienti alti[tudini]' (one-third the height). Read correctly 'duobus trientibus cylindri [habentis]' (two-thirds the cylinder).

(46) This is reminiscent of Keplers 'melo' (melon), one of the conoidal spindles (fusa conoidea) discussed in the 'Supplementum ad Archimedem' appended to his *Nova Stereometria Doliorum Vinariorum, In Primis Austriaci, figuræ omnium aptissimæ* (Linz (1605): D4v–H2v [= *Gesammelte Werke*, **9** (1960): 36–71, especially 60–1]). Here, of course, the 'pumpkin' is a

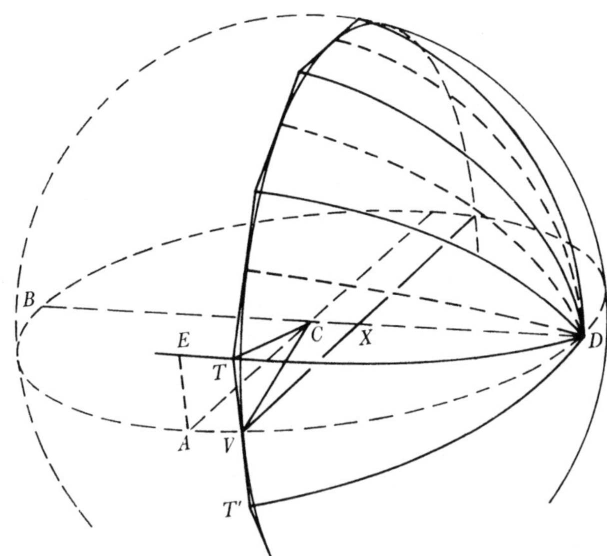

surface of revolution only when its component cylindrical wedges are of an infinitesimal thickness and the coincidence in name is very probably fortuitous. If we conceive the spherical sector formed by rotating $\widehat{DV}$ round the axis to be made up of an infinite number of infinitely

Proposition 6. A sector of a sphere is equal to[45] a cylinder having a great circle on the sphere for its base and the versed sine of the sector for its height.

The proof is by means of a pumpkin-shaped figure.[46]

Corollary. The whole sphere...[47]

Proposition 7. A segment of a sphere...[48]

Proposition 8. The surface of a spherical segment is equal to the surface of a cylinder having the same height and a great circle on the sphere for its base.[49]

Corollary. This surface is equal to a circle whose radius is the line drawn from the vertex of the segment to an end-point of its base.[50]

Proposition 9. The surface-area of a spherical triangle...[51]

---

thin cylindrical wedges $CTT'D$ (symmetrical about the plane BVD), then by Proposition 3, since EA is the element of the circle perimeter ($2\pi \times AC$) formed by rotating point A round the axis BD, the volume of the spherical sector is equal to

$$\int_{-AC}^{AC} \tfrac{1}{3}AC \times XD \times 2\pi . d(AC) = \tfrac{2}{3}\pi \times AC^2 \times XD.$$

In Newton's (corrected) form the latter is two-thirds the volume of a cylinder of base radius AC (that is, whose base is a great circle of the sphere obtained by rotating $\widehat{BAD}$ round BD) and of height XD, the versed sine $AC(1-\cos V\hat{C}D)$ of the sector.

(47) Doubtless 'Tota sphæra æqualis est duobus trientibus cylindri basem habentis circulum in sphæra maximum & altitudinem sphæræ diametrum' (The whole sphere is equal to two-thirds of a cylinder having a great circle on the sphere for its base and the sphere's diameter for its height). The result is Archimedes' (*Sphere and Cylinder*, I, 34).

(48) Since the volume of the cone formed by rotating the triangle CVX round BCD is $\frac{1}{3}\pi \times VX^2 \times CX$, the volume of the spherical segment obtained by revolving the circle segment VXD round BD is $\frac{1}{3}\pi(2AC^2 \times XD - VX^2 \times CX)$.

(49) By an analogous argument from Proposition 4 the spherical surface formed by rotating the arc $\widehat{DV}$ round BD is

$$\int_0^{AC} XD \times 2\pi . d(AC) = 2\pi \times AC \times XD,$$

where $2\pi \times AC$ is the perimeter of a great circle of the sphere. Again the result is Archimedes' (*Sphere and Cylinder*, I, 42).

(50) That is, VD. Newton first wrote equivalently 'cujus radius est chorda dimidij arcus ejus' (whose radius is the chord [$2AC \sin V\hat{C}D$] of its half arc). The corollary follows at once by observing that $2AC$ (or BD) $\times XD = VD^2$.

(51) In context Newton would appear to mean by a 'Triangulum sphericum' the triangle TVD cut off between the great circles $\widehat{DA}$, $\widehat{DE}$ and the small circle $\widehat{T'VT}$ (parallel to the great circle $\widehat{AE}$): if so, at once by Proposition 8 the area of the triangle TVD is $\widehat{AE}$ (or $AC \times \hat{D}$) $\times XD$. The 'spherical' triangle TVD formed by great circle arcs through T, V and D (taken two at a time) has, in contrast, the area $AC^2 \times \hat{D}$ by Harriot's theorem, but this result cannot be attained by Newton's present theorem. (On Harriot's proof that the area of a spherical triangle formed by great circles is proportional to the 'spherical excess' of its angles see J. A. Lohne, 'Thomas Harriot als Mathematiker', *Centaurus*, **11** (1965): 19–45, especially 27–31 and 44. Girard and Cavalieri found the result independently.)

3

HARMONIC MOTION IN A CYCLOIDAL ARC[1]

[Early 1670's][2]

From the original manuscript in the University Library, Cambridge[3]

[De Motu Gravis in Trochoide descendentis.][4]

In Fig: primâ[6].

Possunt hæ tres Figuræ ad unam reduci, ut factum vides in fig 2da.[5]

Sit DCE Trochoides ad $\frac{1}{2}$ circulum BCY pertinens quæ planum horizontale tangat in C insistens ei normaliter. Inqꝫ curva DC grave descendat a D ad C dilapsum per puncta δ, P, et π. Et agantur δYS, $P[TV]$, πXQ parallelæ ad DE &c.

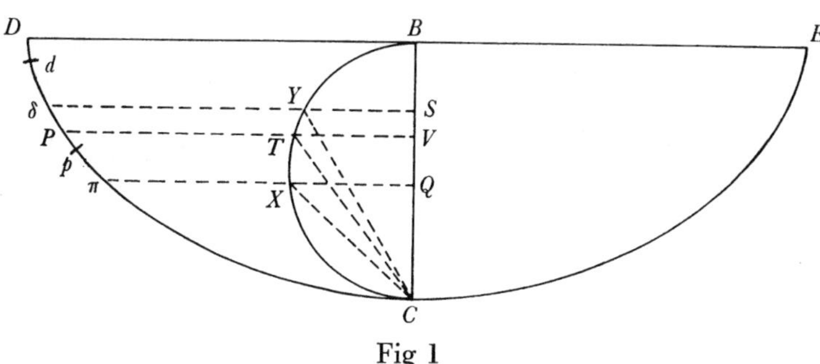

Fig 1

1. Dico quod gravitatis efficacia[7] sive descendentis acceleratio in singulis descensûs locis D, [δ], [P], &c est ut spatium describendum DC, δC, PC, &c. Scilicet obliquitas descensus minuit efficaciam gravitatis ita ut si gravia duo

(1) These Newtonian researches into the motion of a body under simple gravity in a vertical cycloidal pendulum have previously been published by A. R. and M. B. Hall in their *Unpublished Scientific Papers of Isaac Newton* (Cambridge, 1962): 170–80, and by J. W. Herivel in his *Background to Newton's Principia: A Study of Newton's Dynamical Researches in the Years 1664–84* (Oxford, 1966): 198–207.

(2) In one of his 'Adnotata Math: ex Neutono. 1694. Maio' made during a visit to Cambridge in the late spring of that year David Gregory recorded having seen this and an immediately preceding gravitational manuscript (ULC. Add. 3958.5: 87^r–88^r, reproduced in *The Correspondence of Isaac Newton*, **1** (1959): 297–9) in the phrase 'Vidi M.S. ante annum 1669...ubi omnia fundamenta suæ philosophiæ facta sunt gravitas scilicet [Lunæ] versus [Terram], planetarumque versus [Solem]....Vidi etiam in illo M.S. Equidiuturnitatem penduli intra Cycloides suspensi ante editum *Horologii Oscillatorii* Hugenii' (*Correspondence of Isaac Newton*, **3** (1961): 331). While not denying the independence of Newton's proof that the

Translation

[The Motion of a Heavy Body falling in a Cycloid](4)

In the first figure(6)

These three Figures may be reduced to one, as you see done in Figure 2.(5)

Let DCE be a cycloid pertaining to the semicircle BCY, which is to touch the horizontal plane at C, being ordinate at right angles to it. And in the curve DC let a heavy body descend from D to C, falling through the points δ, P and π. Moreover, let δYS, PTV and πXQ be drawn parallel to DE, and the rest.

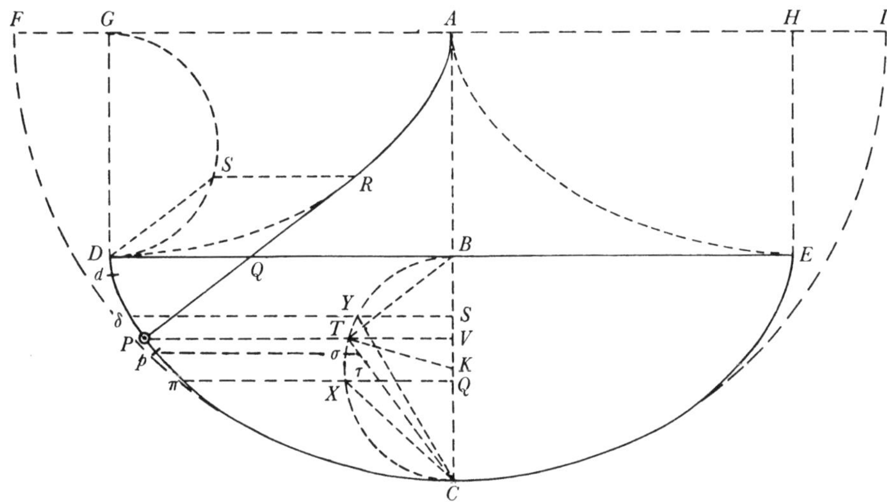

1. I assert that the effect of gravity,(7) that is, the acceleration of the falling body, at each point D, δ, P, ... of the fall path is as the space DC, δC, PC, ... to be described. Evidently the sloping of the fall path diminishes the effect of gravity in such a way that, if two heavy bodies are poised to descend to C, one

vertical cycloid is a tautochrone (see note (15) below), we prefer to trust the strong internal mathematical evidence that this piece is contemporary with his 1671 fluxional tract (reproduced in 1, 2) rather than follow Gregory's implied dating 'ante annum 1669'. Newton's memory was never good on exact dates and we should not trust to within a year or two a remark thrown out in conversation over twenty years afterwards.

(3) Add. 3958.5: 90^r–91^v, with accompanying figures (having two points Q) on 89^r.

(4) The manuscript text is untitled. In their *Unpublished Papers* (note (1)) A. R. and M. B. Hall suggest the equivalent heading 'Gravia in Trochoide Descendentia'.

(5) This marginal entry is here inserted from f. 89^r.

(6) In lettering this figure Newton originally entered the points δ, P, T, V as P, O, V, R respectively. The following text has been modified to accord with the revised form.

(7) The component of the 'force of gravity', as Newton named it on the first page of his Waste Book (see I: 456), which acts instantaneously in the direction of motion. Huygens in his contemporary researches into 'Newtonian' force (see note (14)) called the 'efficacia' of gravity its 'incitation' or 'incitatio'.

descensura sint ad C alterum B rectà per diametrū BC, alterum $[Y]$ obliquè per chordā YC: Minor erit acceleratio gravis Y propter obliquitatem descensus idq; in ratione YC ad BC ita ut ambo gravia simul perveniant ad C.(8) Est autem BC parallela curvæ in D, ac YC parallela ipsi in δ,(9) ideoq; acceleratio gravis in D est eadem cum acceleratione gravis descendentis in BC ut et acceleratio gravis in δ eadem cum acceleratione gravis descendentis per YC. Quare descendentis acceleratio in D est ad accelerationem ejus in δ ut BC ad YC, sive ut eorum dupla(10) DC ad δC. Q.E.O.

2.(11) Gravia in Trochoide descendentia, alterum a D, alterum a quolibet alio puncto P, simul pervenient ad C. Nam ut sunt longitudines DC PC, ita accelerationes sub initio motus in D et P: quare spatia primò descripta puta Dd & Pp(12) erunt in eadem ratione. Unde dividendo est $DC.PC::dC.pC$. Quare accelerationes in d et p permanent in eadem ratione, et etiamnum generabunt velocitates descendentium in eadem ratione. Adeoq; spatia(13) δC et πC erunt in illa ratione idq; continuò donec utrumq; simul in nihilum evanescat. Quare gravia simul attingent punctum C.(14)

Potuit etiam hoc inde ostendi quòd posito $DC.\delta C::PC.\pi C$ sit

$$D\delta.P\pi::\surd BS.\surd[V]Q$$

(8) A well-known theorem of Galileo's ('De Motu naturaliter Accelerato' [incorporated in the 'Giornata Terza' of his *Discorsi e Dimostrazioni Matematiche intorno à due nuoue scienze Attenenti alla Mecanica & i Movimenti Locali* (Leyden, 1638)]: Theorema VI) employed to similar effect by Huygens and James Gregory in their studies of pendular motion (see the appendix to the preceding introduction).

(9) Fermat's theorem; compare **1**, 2, §2: note (307).

(10) By Wren's 1658 rectification; see **1**, 2, §2: note (308).

(11) Newton first began this paragraph 'Grave a quolibet puncto P descendet ad C in eodem tempore ac si descendisset a D' (A heavy body will fall from any point P to C in the same time it would have taken to fall there from D).

(12) A following cancelled phrase, set in parenthesis, reads 'quantumvis parva cogitans' (supposing these to be arbitrarily small).

(13) '$D\delta$ $P\pi$ et eorū residua' ($D\delta$, $P\pi$ and their remnants) is cancelled.

(14) In modern analytical terms, on taking t to be the time of fall from rest at P to C and setting $CB = D$, $CQ = x$, $Q\pi = y = \surd[x(D-x)]+\frac{1}{2}D\sin^{-1}[1-2x/D]$, $\widehat{\pi C} = s = 2\surd[Dx]$ and $\widehat{PC} = S$, then Newton's 'acceleratio' (due to the component of constant gravity g acting instantaneously in the direction of the cycloidal fall path) is

$$d^2s/dt^2 = (-g\,dx/ds =) - (g/2D)\,s \propto s.$$

It follows that $\delta\delta' : \pi\pi' = \widehat{DC} : \widehat{PC}$, where a body falls from rest at D over $\delta\delta'$ in the same indefinitely small time that it falls from rest at P over $\pi\pi'$, and hence that after release from D, P it covers proportional lengths of the arcs $\widehat{DC}$, $\widehat{PC}$ in the same time interval, so reaching C after equal times of fall. In his final section below Newton will obtain an explicit expression for the time of fall from rest at P to C by an equivalent kinematical argument. Here he does not attempt to solve the differential equation $d^2s/dt^2 = -(g/2D)\,s$ of 'simple' harmonic motion in the form $s = S\cos(g/2D)^{\frac{1}{2}}t$ and so $v = ds/dt = -(g/2D)^{\frac{1}{2}}S\sin(g/2D)^{\frac{1}{2}}t$, yielding at once the

B straight down along the diameter BC, the other Y at a slope along the chord BC, the acceleration of the body Y will be decreased because of the slope of its path of fall and this in the ratio YC to BC: both heavy bodies as a result will reach C simultaneously.(8) But BC is parallel to the curve at D while YC is parallel to it at δ;(9) in consequence the acceleration of a heavy body at D is the same as the acceleration of one falling in BC, and likewise the acceleration of a heavy body at δ is the same as the acceleration of one falling along YC. Hence the acceleration of the falling body at D is to its acceleration at δ as BC to YC, that is, as their doubles(10) DC to δC. As was to be shown.

2.(11) Heavy bodies falling in the cycloid, one from D and a second from any other point P at random, will reach C simultaneously. For proportionally as the lengths DC, PC so are the accelerations at the start of motion at D and P; and consequently the spaces, say Dd and Pp,(12) described in the first instant will be in the same ratio. Hence *dividendo* it is $DC:PC = dC:pC$. Therefore the accelerations at d and p remain in the same proportion and will still continue to generate speeds of descent in the same proportion. As a result the spaces(13) δC and πC will be in that proportion, and continuously so till each vanishes simultaneously into nothing. Therefore the bodies will arrive simultaneously at the point C.(14)

The same conclusion could also have been demonstrated from the fact that, on assuming $DC:\delta C = PC:\pi C$, then

$$D\delta : P\pi = \surd(BS) : \surd(VQ)$$

---

corollary that its period $2\pi(2D/g)^{\frac{1}{2}}$ is independent of the choice of $\widehat{PC} = S$ and so of the initial point P. We will return to this theme in the sixth volume.

At a period when such exact scientists as Hooke and Wallis preferred to measure the effect of force by an increment of velocity (rather than the ratio of that increment to a corresponding increment of time) this dynamical approach by equating instantaneous acceleration in the direction of motion to the component of the force of 'gravitas' which provokes it was radically novel, indeed unprecedented. From it was to evolve the powerful general theory of central force orbits which Newton later presented in Proposition 41 of the first Book of his *Philosophiæ Naturalis Principia Mathematica* (London, 1687). In a somewhat more sophisticated dress the present application to pendular motion appeared in the following Proposition 53 of *Principia*'s Book 1: 'Proinde cum hæc [vis] sit ut via describenda..., accelerationes corporis vel retardationes in...partibus proportionalibus describendis, erunt semper ut partes illæ, & propterea facient ut partes illæ simul describantur. Corpora autem quæ partes totis semper proportionales simul describunt, simul describent totas' (p. 158). The parallel here with Huygens, who in his *Horologium Oscillatorium* in 1673 developed an essentially kinematic approach to simple harmonic motion (see note (20)), is not obvious but none the less real. In his study of 'Newtonian' forces about the beginning of 1674, unpublished till recently, he in fact propounded the present dynamical argument wholly independently of Newton. (See Huygens' *Œuvres complètes*, **18** (The Hague, 1934): 489.) This continual parallelism of much of Newton's and Huygens' mathematical thought is no accident: many of their contemporaries were educated in the same intellectual tradition and had access to the same published literature, but only James Gregory possibly had the same quality of mind and mathematical ability to explore the subtleties of the new science of continuous motion in time.

: : velocitas post descensum ad profunditatem BS. ad velocitatem post descensum ad profunditatem $[V]Q$.(15)

3. Itaq; si grave undulet in Trochoide undulationes quælibet erunt ejusdem temporis.

IN FIG: $\overline{\text{SCDÂ}}$.

Super Diametrum Trochoidis DE erige perpendiculum $BA = BC$ et a puncto A hinc inde describe duas semi-Trochoides AD AE, tangentes rectam DE in D

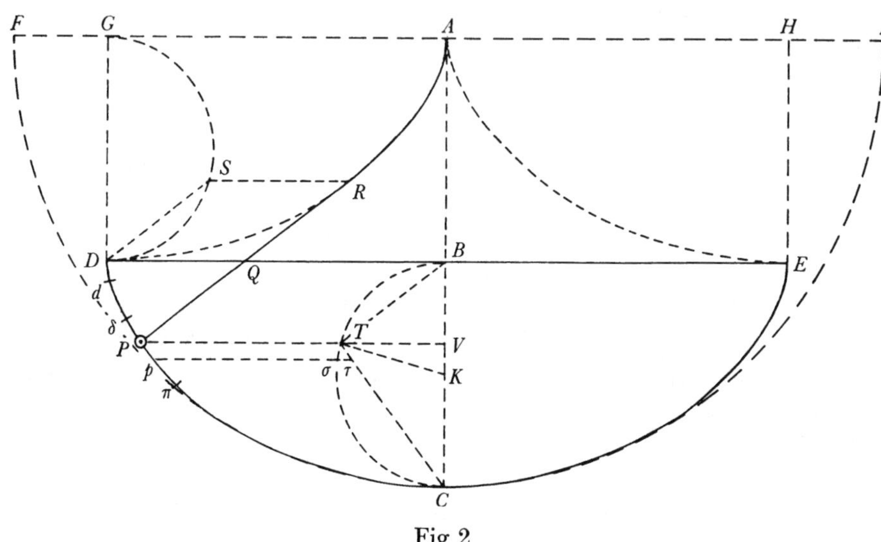

Fig 2

et E, adeoq; ejusdem magnitudinis cum Trochoide DCE. Jam puncto Q in BD ad arbitriū sumpto, fac arcus BT ac DS æquales longitudini DQ et comple parallelogramma $BQPT$ ac $DQRS$. Et constat(16)

---

(15) Since (in Figure 1) $DC:\delta C = PC:\pi C$ (and so $BC:YC = TC:XC$), at once

$$\frac{D\delta^2}{P\pi^2} = \frac{DC^2}{PC^2} = \frac{BC^2}{TC^2} = \frac{YC^2}{XC^2} = \left(\frac{BC^2-YC^2}{TC^2-XC^2}\text{ or}\right)\frac{BC-SC}{VC-QC}.$$

In modern terms (see previous note) the differential equation $d^2s/dt^2 = -k^2s$ (where $k^2 = g/2D$) yields $v/(S-s) = -k\sin kt/(1-\cos kt)$, constant for any given time t and so independent of $\widehat{PC} = S$. Newton's assumption that the 'velocitas post descensum ad profunditatem VQ' is the same after fall in the cycloidal arc $\widehat{P\pi}$ (from rest at P) as it is after fall vertically (from rest at V) to Q, and so, by Theorems I and II of Galileo's 'De Motu naturaliter Accelerato' (note (8)), is proportional to $\sqrt{VQ}$, raises an interesting historical point. By the Scholium to Galileo's Theorem II it readily follows, as Huygens argued in Proposition VIII of the 'Pars Secunda. De descensu Gravium & motu eorum in Cycloide' of his *Horologium Oscillatorium sive De Motu Pendulorum ad Horologia aptato Demonstrationes Geometricæ* (Paris, 1673): 33–4, that the terminal speed acquired by a body falling along a chain of line-segments in a vertical plane depends only on the depth fallen, provided that no speed is lost by the body at the 'corners' of the chain when the body instantaneously changes direction. If this is accepted Huygens'

or as the speed after fall to depth BS is to speed after fall to depth VQ.(15)

3. Accordingly, if a heavy body should oscillate in a cycloid, whatever the oscillations they will be isochronous.

In the second figure

On the diameter DE of the cycloid raise the perpendicular BA equal to BC, and from the point A on each side describe the two half cycloids AD and AE, tangent to the line DE at D and E and consequently of the same size as the cycloid DCE. Now, taking the point Q at random in BD, make the arcs $\widehat{BT}$ and $\widehat{DS}$ equal to the length DQ and complete the parallelograms $BQPT$ and $DQRS$. It is then settled(16)

---

conclusion is immediate: 'liquet...per curvam quamlibet lineam descendente mobili (nam curvas tanquam ex infinitis rectis compositæ essent hic considerare licet) semper eandem illi velocitatem acquiri si ab æquali altitudine descenderit: tantamque eam esse velocitatem, quantam casu perpendiculari ex eadem altitudine adipisceretur'. In his 'Tentamina Quædam Geometrica De Motu Penduli et Projectorum' which was published the previous year in appendix to Patrick Mathers' *Great and New Art of Weighing Vanity* (Glasgow, 1672) James Gregory had, however, anticipated a serious objection to this, arguing that 'mobilis, quod in diversis rectis movetur, impetum seu velocitatem mutari in rectarum occursu, ita ut velocitas in prima linea sit ad velocitatem in secunda, in ratione radij ad cosinum inclinationis mutuæ rectarum': the Huygenian argument is therefore valid only in the limit-case where the line-segments each become infinitely small and so the whole chain a 'linea curva [ubi] nullæ tales sunt inclinationes', in consequence of which 'in lineis curvis mobilia eâdem velocitate incedunt, qua in lineis rectis...sive horizonti perpendicular[ibus] sive...inclinat[is]'. When Newton came to thank Oldenburg on 23 June 1673 for a gift copy of Huygens' *Horologium Oscillatorium* he reiterated Gregory's objection: 'In ye Demonstration of ye 8th Proposition, *De descensu gravium*, there seems to be an illegitimate supposition, namely y^t^ ye flexures at B & C do not hinder ye motion of ye descending body. For in reality they will hinder it.... If this supposition be made becaus a body descending by a curve line meets with no such opposition, & this Proposition is laid down in order to ye contemplation of motion in curve lines: then it should have been shown that though rectilinear flexures do hinder, yet ye infinitely little flexures which are in curves, though infinite in number, do not at all hinder the motion' (*Correspondence of Isaac Newton*, **1** (1959): 290). Since Collins had already brought Mathers' book to Newton's attention in July 1672 (*ibid.*, **1**: 224)—and sent him a copy soon after?—we may infer that Newton in mid-1673 was familiar with Gregory's pendulum tract. The hypothesis that the present manuscript was provoked by Gregory's 'Tentamina' a little before June 1673 would both explain Newton's present easy acceptance of the 'energy' equation

$$v^2 = 2\int_0^{x=VQ} g(dx/ds).ds = 2g\times VQ$$

and allow his evident independence of Huygens in framing the variant kinematical approach to cycloidal motion developed below (see note (31)).

(16) See Example 4 of Problem 5 of Newton's 1671 fluxional tract (**1**, 2, §2 above). We should not make too much of Newton's originality in compelling a body to follow a cycloidal path by means of the congruent 'cheeks' of its evolute. When Robert Moray communicated Brouncker's 'Demonstration of the Equality of Vibrations in a Cicloid-Pendulum' with cycloidal 'Cheekes' used in an analogous way to make his 'Bullet' be 'carried in a Cicloid',

1. Quod QP normalit^r insistit Trochoidi DC in P, & quod QR tangit Trochoidem DRA in R.

2. Quod QP et QR in directum jacent propter parallelismum rectarum DS BT. Adeoq omnis recta perpendicularis ad Troch DC tanget Troch AD et contra.

3. Quod sit $PQ = TB = DS = \frac{1}{2}$curvæ DR, adeóq $2PQ$ sive $PR = DR$. Et recta $PR+$cu[r]va $RA =$toti curvæ $DRA =$Rectæ CA.

4. Quare si ARP sit filum datæ longitudinis, cui pondus P appenditur ita undulans intra trochoides AD et AE ut filum ab ipsis paululum prohibeatur ne(17) in rectum protendatur, quemadmodū videre est in parte AR, ubi se applicat ad Trochoidem: Tunc pondus P undulabit in Trochoide DCE, adeoq quamlibet utcunq longam vel brevem undulationem in eodem tempore perficiet.

5. Patet etiam quod undulationes in circulo FCI centro A descripto modò sint perbreves (puta 10^{gr}(18) hinc inde vel minus) sunt ejusdem temporis proximè ac in Trochoide DCE. Nam(19) undulatio in utroq casu fit circa centrū A nisi quod filum paululum incurvatur in uno casu; quæ curvatura quam parva sit ex eo percipies quod R tantum supra rectam DE esse debes imaginari quantum P cadit infra.(20)

In Figurâ tertiâ.

1. Stantibus jam ante positis cum gravis a D per P ad C descendentis velocitas in loco P est ut(21) radix altitudinis BV(22) hoc est linea BT: pro designanda illa velocitate exponatur eadem BT.(23)

2. Dein sit Pp particula spatij DC in ejusmodi particulas infinitè multas et æquales divisi, et agatur $p\sigma\tau$ parallela ad PTV secans semicirculum in σ et

---

Huygens replied forthwith on 10 February 1662 (N.S.) that 'La proprieté de la Cycloide, de ce que par son evolution, il se descrit une courbe pareille n'estoit pas difficile a demonstrer apres que Monsieur Wren a decouuert la dimension de cette ligne'. (See Huygens' *Œuvres complètes*, **4** (1891): 51; and compare **1**, 2, §2: note (308).)

(17) 'semper' (always) is cancelled.

(18) A more conservative following alternative 'vel 5^{gr}' (or 5°) is cancelled.

(19) Newton first continued 'punctum R tunc haud recedit a puncto A, et filum ['adhuc minus' is cancelled] vix omninò incurvatur' (then the point R hardly departs from the point A and the thread is [still less] scarcely at all curved).

(20) If (in Figure 1) VP, $Q\pi$ meet the circle quadrant $\widehat{CF}$ in the respective points P', π', then, on setting $CB = D$, $CV = X$, $CQ = x$, $QP' = y' = \sqrt{[x(4D-x)]}$ and $\widehat{C\pi'} = s'$, it follows that $dx/ds = y'/2D$ while

$$ds/dt = v_{(\text{rest at } P') \text{ at } \pi'} = v_{(\text{rest at } V) \text{ at } Q} = \sqrt{[2g(X-x)]},$$

so that the time $$t_{(\text{rest at } P')\, P' \to \pi'} = \int_X^x 2D/\sqrt{[2gx(4D-x)(X-x)]}\,.\,dx.$$

1. That QP stands normally to the cycloid DC at P, while QR is tangent to the cycloid DRA at R.

2. That QP and QR lie in a straight line due to the lines DS and BT being parallel. Consequently, every straight line perpendicular to the cycloid DC is tangent to the cycloid AD, and conversely so.

3. That $PQ = TB = DS =$ half the curve $\widehat{DR}$, so that $2PQ$, that is, $PR = \widehat{DR}$ and line $PR+$curve $\widehat{RA} =$ total curve $\widehat{DRA} =$ line CA.

4. Therefore, if $\widehat{ARP}$ be a thread of given length and the weight P be hung from it, swinging between the cycloids AD and AE in such a way that the thread is somewhat prevented from[17] straightening itself out by them (as can be seen in the portion AR where it folds itself along the cycloid), then the weight P will vibrate in the cycloid DCE and will consequently complete any vibration, however long or short, in the same time.

5. It is further evident that vibrations in the circle FCI described with A as centre, provided they be very short (say $10°$[18] to either side, or less), occur very nearly in the same time as in the cycloid DCE. For[19] the vibration in either case takes place around the centre A except that in one instance the thread is curved slightly inward—and how small this curvature is you will perceive from the circumstance that you should imagine R to be just as far above the line DE as P falls below it.[20]

In the third figure

1. With what has been previously supposed continuing to stand, since the speed at P of a heavy body falling from D by way of P to C is as the[21] root of the height BV,[22] that is, as the line BT, for denoting that speed let its exponent be this same line BT.[63]

2. Next, let Pp be a minute part of the space DC after it has been divided into infinitely many equal minute parts of the same sort, and draw $p\sigma\tau$ parallel to

---

Evidently, as Huygens showed in December 1659 (see the appendix to the preceding introduction), when X (and so x) is small in comparison with D

$$t_{P'\to\pi'} \approx \frac{1}{2}\sqrt{\frac{2D}{g}}\int_X^x 1/\sqrt{[x(X-x)]}\,.\,dx;$$

hence the total period ($4t_{P'\to C}$) of motion in the circle of radius $CA = 2D$ when the swing is small is approximately $2\pi(2D/g)^{\frac{1}{2}}$, independent of the height $CV = X$ of vibration. Compare note (14).

(21) 'quadra[ta]' (square) is cancelled.

(22) A first cancelled continuation reads 'pro designanda illa velocitate exponatur illius BV radix cujusmodi est BT si modò diameter [sit unitas]' (for denoting that speed let its exponent be the root of BV, such as BT provided the diameter be unity).

(23) In terms of the notation established in note (14) BT is in fact equal to

$$(D/2g)^{\frac{1}{2}} \times \text{speed } v_{(\text{rest at } D) \text{ at } P}.$$

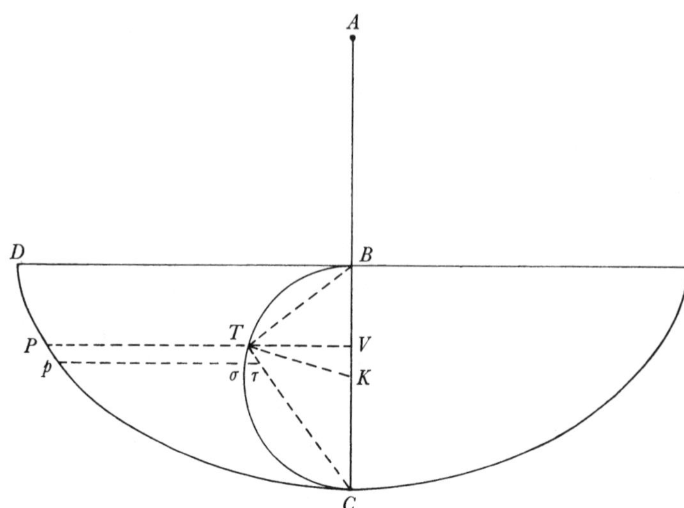

rectam TC in τ. Jam propter parvitatem, curvarum portiunculæ Pp ac $T\sigma$ pro rectis haberi possunt, adeoq; $P[p]$ ac $T\tau$ erunt æquales propter parallelismum et inde $T\tau$ erit datæ licet infinitè parvæ longitudinis; ductâq; semidiametro TK, triangula $T\sigma\tau$, TKB erunt similia, siquidem latera unius sunt perpendiculariter posita ad alterius latera correspondentia viz $T\sigma$ ad TK, $T\tau$ ad TB et $\sigma\tau$ ad BK. Quare est $BT.TK::T\tau.T\sigma$. sive $BT\times T\sigma = TK\times T\tau$. Adeoq; cum TK ac $T\tau$ pro datis habenda sunt, erunt BT ac $T\sigma$ reciprocè proportionalia. Cum itaq; tempus et velocitas quibus datum spatium ut Pp desc[r]ibitur sunt reciprocè proportionalia, et BT pro velocitate exponitur, exponi potest etiam $T\sigma$ pro tempore. Atq; ita si spatij Dp pars quælibet Pp describitur in parte tempo[ris] $T\sigma$, describetur totum spatium Dp in toto tempore $[B]\sigma$.(24)

3. Hinc posito quod semicircumferentia BTC designat tempus in quo spatium DC percurritur,(25) ut noscas in quo tempore pars DP describetur age $PT\parallel DB$ et arcus BT designet tempus.

4. Potest etiam tempus per angulum BCT, vel per inclinationem descensus Pp aut per longitudinem DQ designari.(26)

5. Cæterùm ut tempora perpendicularis descensûs cum temporibus descensûs in hâc curvâ conferantur, pone quòd grave a B per V ad C descendit, et cùm linearum BT quadrata sunt ut lineæ BV: exponatur BT pro designando tempore descensus ad V. Adeò ut si grave descendat ad C in tempore BC, descendet ad V in tempore BT.(27)

6. Jam cùm descensus æquè a D ac B sub initio sunt ad Horizontem per-

(24) Retaining the proportionality factor of the previous note we may restyle Newton's argument as follows: since the increments Pp, $T\sigma$, $T\tau$ are supposed to be vanishingly small

$$Pp = T\tau = T\sigma\times(BT/TK) = T\sigma\times(2/gD)^{\frac{1}{2}}\times v_{\text{at }P}$$

PTV intersecting the semicircle in σ and the straight line TC in τ. Now, because of their smallness, the minute portions $\widehat{Pp}$ and $\widehat{T\sigma}$ of curves can be taken for straight lines; accordingly Pp and $T\tau$ will be equal because they are parallel and hence $T\tau$ will be of given though indefinitely small length; and so, on drawing the radius TK, the triangles $T\sigma\tau$, TKB will be similar seeing that the sides of one are set perpendicularly to corresponding sides of the other, namely, $T\sigma$ to TK, $T\tau$ to TB and $\sigma\tau$ to BK. Therefore $BT : TK = T\tau : T\sigma$, or

$$BT \times T\sigma = TK \times T\tau.$$

Accordingly, since TK and $T\tau$ are to be taken as given quantities, BT and $T\sigma$ will be inversely proportional. And so, since the time and speed of description of a given space such as Pp are inversely proportional and BT is the exponent of the speed, $T\sigma$ can also be taken as the exponent of the time. And in this case if any part Pp of the space $\widehat{Dp}$ is described in the part $T\sigma$ of time, the total space $\widehat{Dp}$ will be described in total time $\widehat{B\sigma}$.(24)

3. Hence, supposing that the half circumference $\widehat{BTC}$ denotes the time in which the space $\widehat{DC}$ is covered,(25) to know in what time the part $\widehat{DP}$ shall be described draw $PT \| DB$ and the arc $\widehat{BT}$ will denote that time.

4. The time may also be denoted by the angle $B\hat{C}T$ or by the incline of the fall path Pp or alternatively by the length of DQ.(26)

5. Moreover, to compare the times of vertical descent with the times of fall in this curve, suppose that a heavy body falls from B through V to C and, since the squares of the lines BT are as the lines BV, let the exponent BT be used to denote the time of fall to V. As a result, should the heavy body fall to C in the time BC, it will fall to V in the time BT.(27)

6. Now since the paths of fall from D no less than from B are at the start

---

and therefore $\quad t_{(\text{rest at } D)\, P\to p} = Pp/v_{\text{at } P} = (2/gD)^{\frac{1}{2}} \times T\sigma,$

so that $\quad t_{(\text{rest at } D)\, D\to p} = (2D/g)^{\frac{1}{2}} \times \widehat{BT}/BC.$

(25) Newton first wrote in sequel more simply 'arcus BT designet tempus in quo spatium $\widehat{DP}$ per [curretur]' (the arc $\widehat{BT}$ will denote the time in which the space $\widehat{DC}$ shall be covered).

(26) For $B\hat{C}T = (\frac{1}{2}B\hat{K}T$ or) $\widehat{BT}/BC$ is the slope of the cycloid at P to the vertical, while the subnormal DQ to the cycloid $\widehat{DPC}$ at P is equal to $\widehat{BT}$.

(27) In explicit terms the time of vertical fall from B to V is

$$t_{(\text{rest at } B)\, B\to V} = \sqrt{[2BV/g]} = (2D/g)^{\frac{1}{2}} \times BT/BC,$$

and so (compare note (24)) in the terms of Newton's third paragraph will be denoted by BT. The equality of the two proportionality factors $(2D/g)^{\frac{1}{2}} \times 1/BC$, fundamental to Newton's further argument, is justified below (see note (29)) by the observation that in the 'first' moment of fall from the level DB the increments of arc $\widehat{BT}$ and chord BT (denoting respectively the time of fall in the cycloid from D to P and in the vertical from B to V) are equal.

pendicula[r]es, manifestum est quod utrumq; grave D ac B incipit æqualit[r(28)] descendere, etsi D confestim in obliquum fertur. Adeoq; lineæ per quas tempora descensuum designantur ita debent inter se constitui ut initialiter exhibeant æqualia tempora descensuum æque-altorum et postea rectè exhibebunt tempora descensuum æquè altorum ut fiunt sensim inæqualia. Et hinc patet arcum et ejus chordam BT (cum sint initialiter æqualia) non modo recte designare hæc tempora seorsim sed et inter se conferre.(29) Ita ut(30) posito quod grave defertur a D ad C in tempore BTC, non modò sequetur quod deferetur ad P in tempore BT, sed etiam quod decidet a B ad V in tempore BT vel a B ad C in tempore BC: et contra.

Tempora etiam descensus ab alijs curvæ punctis ut P exhinc noscuntur siquidem partes proportionales in æqualibus temporibus peraguntur.(31)

Sed præcipuum est quod ex dato tempore in quo pendulum datæ longitudinis vibrat, datur tempus in quo grave ad datam profunditatem descendet.(32) Nam si grave decidat a B ad C in tempore BC fac $BC^q . BTC^q :: BC . A\gamma$, ac decidet ab A ad γ in tempore BTC quod est unius semivibrationis. Posito autem $AC = 1,0000$, calculus dabit $A\gamma$ 1,2337, p cujus quadruplum 4,9348 descendet in tempore unius vibrationis, hoc est per 5(33) AC ferè, et per 19,7392 sive per $19\frac{3}{4}AC$ fere in vibratione replicata.(34)

Nota [1] quod motus gravis a D ad C descendentis persimilis est motui puncti in rota uniformiter mota(35) quod describit Trochoiden, respectu velocitatis.

2. Quod pendulum ex argento vivo confectum diutius perseverat in motu.(36)

---

(28) 'in eodem tempore' (in the same time) is cancelled.

(29) Compare note (27).

(30) To justify the following indicative tense we should perhaps read the equivalent conjunction 'Itaq;'.

(31) It is a weakness from a logical viewpoint that, in generalizing his present kinematical argument, he had to introduce the proof of isochronism developed previously on a dynamical basis. Compare note (14) above. Huygens in his *Horologium Oscillatorium* (see the appendix to the preceding introduction) achieved the same result by a single more sophisticated kinematical argument.

(32) A first continuation reads 'Nam si grave descendat a D ad C in tempore $BTC = 15708$, decidet ab B ad C in tempore $BC = 10000$. Et facto $BC^q . BTC^q :: BC . A\gamma = 24674$. decidet ['insuper' is cancelled] ad γ in tempore BTC, hoc est dum semissis undulationis peragitur, adeoq; in tempore unius undulationis decidet per $4A\gamma$ sive per spatium 98696' (For if a body descends from D to C in the time $B\widehat{TC} = 15{,}708$, it will fall down from B to C in the time $BC = 10{,}000$ and, on making $A\gamma = BC \times (B\widehat{TC}/BC)^2 = 24{,}674$, down to γ in the time $B\widehat{TC}$, namely the period of half a swing. Accordingly, in the time of one swing it will fall through the distance $4A\gamma$, that is, 98,696).

perpendicular to the horizon, it is obvious that each body D and B begins its descent on a par,(28) though D is immediately borne away at a slant. Accordingly, the lines by means of which the times of their descent are represented ought to be arranged in regard to each other in such a way as initially to indicate equal times of fall to equal depths: thereafter they will correctly indicate the times of fall to equal depths as soon as these become sensibly unequal. It is hence apparent that the arc $\widehat{BT}$ and its chord (since they are initially equal) correctly denote these times not only separately but also comparably so one with the other.(29) Consequently, supposing that a heavy body travels down from D to C in the time $B\widehat{TC}$, it will follow not only that it will travel down to P in the time $\widehat{BT}$ but also that it will fall from B down to V in the time $\overline{BT}$ or from B to C in the time BC; and conversely so.

Further, the times of descent from other points of the curve such as P are known from this, seeing that proportional parts are covered in equal times.(31)

But the important point is that, given the time in which a pendulum of given length vibrates, from it is given the time in which a heavy body will descend to a given depth.(32) For, should a body fall from B down to C in the time BC, make $BC^2 : B\widehat{TC}^2 = BC : A\gamma$ and it will fall from A down to γ in the time $B\widehat{TC}$ (of one half-vibration, that is). On taking $AC = 1{\cdot}0000$ computation will yield $A\gamma = 1{\cdot}2337$ and it will descend through four times this $4{\cdot}9348$ (that is, $5AC$ almost) in the time of one vibration and through $19{\cdot}7392$ (or $19\frac{3}{4}AC$ almost) in a vibration back and forth.(34)

Note that:

1. The motion of a heavy body descending from D to C is, with regard to its speed, closely similar to the motion of the point in the uniformly moving wheel(35) which describes the cycloid.
2. A pendulum constructed of quicksilver stays longer in motion.(36)

---

(33) The verbal equivalent 'quinquies' is cancelled.

(34) In a complete oscillation, that is. Evidently, since $B\widehat{TC}/BC = \frac{1}{2}\pi$, $A\gamma$ (not shown in Newton's figure but assumed to coincide in direction with AC) is equal to $\frac{1}{8}\pi^2 \times AC$. As an unstated corollary, the time of fall from rest at A to γ (the period of oscillation of the pendulum, namely) is equal to $\sqrt{[32A\gamma/g]} = 2\pi(AC/g)^{\frac{1}{2}}$. Compare note (14).

(35) As it rolls along beneath DBE at a uniform rate.

(36) Because, for a given shape of pendulum, the proportion of its surface area to its weight will be less and so it will be less affected by air resistance, presumably. Conversely, a heavier pendulum will be more affected by friction at A and the elasticity of the 'thread' $\widehat{ARP}$.

PART 3

RESEARCHES IN GEOMETRICAL OPTICS (*c.* 1670–1672)

INTRODUCTION

As we have seen, a little before Christmas in 1671 his colleague at Cambridge, Isaac Barrow, informed John Collins of Newton's intention 'to send up all the lectures he hath read since he was Professor to be printed here, which he sayth will be 20 Dioptrick Lectures, and some about infinite series'.(1) We will not here repeat the tangled story of the fate of that projected edition and the further attempts to publish its component treatises separately,(2) but in introducing the manuscript record of Newton's work in geometrical optics at this period it is pertinent not merely to outline the technical content of these 'dioptrick lectures' but also to examine the historical circumstances which gave birth to them and to detail the ties between these dioptrical researches and the tract on infinite series and fluxions(3) which Newton was also preparing at this time for the press.

In a somewhat loose compliance with the statutes of his Lucasian professorship(4) Newton in the autumn of 1674 delivered up to the University Librarian,

---

(1) See the introduction to Part 1 above. The quotation is from Collins' letter to James Gregory of 23 February 1671/2 (*James Gregory Tercentenary Memorial Volume* (London, 1939): 218).

(2) See pages 3–13 above and A. R. Hall's thoughtful discussion of 'Newton's First Book', *Archives internationales d'Histoire des Sciences*, **13** (1960): 39–61, especially 44–7.

(3) Reproduced as **1**, 2 above. See also **1**, Introduction: note (16).

(4) These are printed, from the signed and dated copy in the Lucasian papers (now ULC. Res. 1893: Packet E), in William Whiston's *Account of Mr. Whiston's Prosecution at, and Banishment from, the University of Cambridge...now Reprinted... With an Appendix: Containing Mr. Whiston's farther Account* ((London, 1718): 42–6: 'A Copy of Mr. Lucas's Statutes. Confirmed by the Royal Authority'). Under penalty of suspension of the professorial stipend the statutes decreed 'ut dictus Professor semel quotannis, proxime ante festum sancti Michaelis [that is, 29 September], non pauciorum quam decem ex illis, quas præcedente anno publice habuerit, Lectionum exemplaria nitide descripta Procancellario exhibeat, in publicis Academiæ archivis asservanda' (*ibid.*, 43). This proviso was never enforced rigorously. When, in a letter to the Vice-Chancellor (Baldroe) in mid-1669, Collins spoke of Barrow's 'Treatise of Opticks, wch he is prepared to deliver in...as his anniversary Lectures, according to the laudable Constitution or Iniunction laid upon your Mathematick Professor' (Shirburn Castle MS 101.H.2.1, reproduced in S. P. Rigaud, *Correspondence of Scientific Men of the Seventeenth Century*, **1** (Oxford, 1841): 137–8), the fact remained that Barrow had not yet deposited copies of any of his lectures given over the previous five and a half years. A letter from Collins to James Gregory a year later (24 December 1670) suggests that already publication of lectures by the Lucasian professor had been accepted by the executors, Robert Raworth and Thomas Buck, as an alternative to deposit for, Collins observed, 'Mr Barrow told me the Mathematick Lecturer there [in Cambridge] is obliged either to print or put 9 Lectures yearly in Manuscript into the publick Library, whence Coppies of them might be transcribed' (*Correspondence of Isaac Newton*, **1** (1959): 54). No doubt Newton between 1670 and 1674 satisfied the executors that he intended to publish his 'dioptrick lectures' and hence was excused having to deposit the statutory 'decem semel quotannis'.

Robert Peachey, the 'pars 1^{a}' and 'pars 2da' of a manuscript treatise of 'Optica', neatly divided in the margin in his own hand into four series of *lectiones*: one (of 8 lectures) commencing in 'Jan 1669',(5) a second (comprising 10 lectures) beginning in 'Octob: 1670', a third (also of 10 lectures) starting in 'Octob. 1671' and, finally, three lectures dated 'Octob 1672'.(6) Despite the apparently irrefutable authority of this marginal chronology we may no longer accept this simple one-to-one correspondence of passages in the 'Optica' with the lectures given by Newton in the public schools at Cambridge. As R. S. Westfall has recently pointed out, Newton kept by him till his death an earlier untitled version of these same 'lectures' divided into two series only: one (of 8 lectures) commencing likewise in 'Jan 1669', while the second (of 10 lectures) begins three months earlier in 'Julio 1670'.(7) Unfortunately, not only do the dates not exactly correspond but it is clear that in revising the earlier work into his deposited 'Optica' Newton considerably reorganized and augmented its content without attempting to make any allowance in the marginal division into 'lectiones'. To draw up even a rough concordance between the two is indeed an eye-opener: above all, the fourth section of Part 1 of the later 'Optica'(8) which Newton claimed to have expounded to an audience in the autumn of 1670 is entirely lacking in the earlier version, while lectures 3–8 of the first series (dated 'Jan 1669') of the latter deal with material postponed in the 'Optica' till October 1671. The conclusion is clear: except where the two proposed chronologies coincide (namely, in the opening lectures in January 1670) we can afford to follow neither without further corroboration. It is certainly tempting to distinguish the earlier (autograph) version of Newton's optical 'lectures' as a more accurate temporal and textual record of the *lectiones* actually delivered

(5) That is, 1669/70 in the Julian calendar. An equivalent marginal dating in ULC. Add. 4002 (see note (7)) led David Brewster to write on the latter's flyleaf 'An early form of the Lectiones Opticæ. P[age] 1 is dated at the top Jan 1669. As Newton was not created Lucasian Professor till Octr 29 1669 the Lectiones would appear to have been prepared for delivery in College or in the University otherwise than as professor'!

(6) In a note, dated 'Octr 21th 1674', on the flyleaf of the bound volume (ULC. Dd. 9.67) containing the 'Optica' Peachey recorded that 'These Lectures of M[r] Isaac Newton y[e] Mathematik Professor were delivered by him into ye hands of Dr Spencer ye Vice-chancellor of the University & by Mr Vice-chancellor delivered unto mee for to place in ye University Library according to ye order of ye Founder of that Lecture'. A concordance of this manuscript with the posthumous *Isaaci Newtoni...Lectiones Opticæ, Annis* MDCLXIX, MDCLXX *&* MDCLXXI. *In Scholis publicis habitæ: Et nunc primum ex MSS. in lucem editæ* ((London, 1729); see 1, §2: note (2)) was made by Edleston in his *Correspondence of Sir Isaac Newton and Professor Cotes* (London, 1850): xci/xcii.

(7) ULC. Add. 4002: 1–70 and 70–128 respectively (compare 1, §1: note (1) below). Westfall in his otherwise accurate criticism of the 'Optica' ('Newton's Reply to Hooke and the Theory of Colors', *Isis*, **54** (1963): 82–96, especially 83–4 and 95–6) erroneously asserts (p. 96) that 'The later manuscript...contains only one large section not found in the earlier one—"De variis Colorum Phænomenis" at the end'.

from the Lucasian chair during 1670, and to suggest that Newton took full advantage of the statutory stipulation that the deposited 'Optica' be a well polished transcript[9] of the lectures as given, but neither conjecture is confirmed by known evidence. If, however, we do not make too much of the *rapprochement*, the several links which can be traced between the 'Optica' and Newton's contemporary fluxional tract[10] indicate that the source manuscript subsequently published as *Isaaci Newtoni Lectiones Opticæ* was composed some time after the autumn of 1670, perhaps in the winter of the following year.[11]

These inconsistencies in the manuscript record do not seriously affect the accepted view of Newton's Lucasian lectures on optics. Whereas five years before Descartes had been the overriding influence on the development of his ideas, when Newton was appointed to his university professorship in succession to Isaac Barrow it was entirely natural that he should pursue a theme inaugurated by his predecessor in 'lectiones scholasticæ' which had just appeared in book form.[12] To be sure, the only other topics—pure mathematics and perhaps dynamics—which he then knew well enough to expound to a captive audience of Cambridge undergraduates would, as he must have realized, have been wholly above their heads. In his own optical lectures it had been Barrow's explicit intention not so much to promote new research into the subject as to epitomize existing knowledge of the catoptrics and dioptrics of white light (as developed by Alhazen, Kepler, Descartes, Scheiner, Maignan and others he named), presenting it in digestible form for students ignorant of the basic elements of the science.[13] While the logical organization of the work (developed

(8) Reproduced in its entirety in 1, §2 below.

(9) To be precise, an 'exemplar nitide descriptum': see note (4).

(10) See note (3).

(11) Compare note (6). It is not, of course, significant that the last three lectures in the 'Optica' are dated 'Octob 1672': the lectures must have been composed initially in advance of their delivery, while Newton's marginal dates and lecture divisions were evidently added at a single sitting some time before the manuscript was deposited.

(12) *Lectiones XVIII, Cantabrigiæ in Scholis publicis habitæ; In quibus Opticorum Phænomenωn Genuinæ Rationes investigantur, ac exponuntur....Ab Isaaco Barrow Socio Collegii S. Trinitatis, Matheseos Professore Lucasiano, necnon Societatis Regiæ Sodale* (London, 1669). The *imprimatur*, signed by the Vice-Chancellor (Baldroe), the Master of St John's and a fellow of Magdalene, is dated 22 March 1668/9.

(13) In his preliminary 'Epistola ad Lectorem' Barrow wrote that 'quale scriptum attrectas...non utique tibi [*sc.* the 'Benigne Lector' whom he addresses] soli elaboratum; non sponte productum; non diuturnâ meditatione subactos exhibens feriantis ingenii conceptus; at *Lectiones Scholasticas*; primùm officii necessitate expressas; tum subinde properantiùs effusas, ut absolveretur pensum, ac hora deflueret; demùm ad promiscui literarii populi instructionem comparatos, cujus intererat complura (qualis tibi videbuntur) leviora non prætermitti.... Enimverò, quò tibi satisfacerem, expediret scio multa detruncare, meliora substituere, pleraque transponere, omnia ad incudem limamque revocare; quæ tamen adniti,

axiomatically from six 'Hypotheses Opticæ primariæ et fundamentales [seu] leges...ab experientiâ confirmatæ')[14] was new and he had fresh things to say on many mathematical points, Barrow nowhere claimed to have tested his theoretical results experimentally and his few remarks on colour (playfully thrown away at the end of his twelfth lecture) were scarcely less shadowy than the Cartesian hypothesis of a secondary modification of white light they were meant to supplant.[15] Without discarding either the logical form or terminology[16] of Barrow's optical introduction Newton could not only expand certain mathematical passages in his predecessor's work but, more importantly, develop the experiments needed to support his own radical hypothesis that 'white' light is the compound effect of a congeries of elemental coloured rays, each with its own separate degree of refraction, and also discuss its application to the theory of the rainbow and the construction of improved 'catadioptricall' telescopes.

So it was that in mid-January 1670—somewhat nervously, we suspect, and no doubt for once in his life before a crowded auditorium—the new Lucasian professor began the highly polished opening to his first[17] lecture:

The recent invention of telescopes has so exercised most geometers that they appear to have left to others nothing in optics which is untrodden, nor any room for further discovery. Above all, since the discourses you heard not long since in this place[18] were packed with such a variety of optical matters and abundance of new things together with their finely accurate demonstrations, my efforts might perhaps appear to be for nothing and my labour useless should I undertake to treat this science yet again. But since I observe that geometers have a false idea of a certain property of light in regard to its refractions, while tacitly presupposing in their demonstrations a particular physical hypothesis[19] which is not well established, I judge my action will not be unacceptable

nec stomachi mei, nec otii fuit; sed...prout nata sunt emittere malui; quàm operosè lambendo aliam in formam...refingere. Quinimò postquam edendi propositum inii, seu fastidio correptus seu novandi subiturum studium fugitans, nè quidem horum magnam partem relegere sustinui'. As a study of the following text shows, there was a great deal of truth behind this front of modesty.

(14) *Lectiones XVIII* (note (12)): 7–19.

(15) See Barrow's *Lectiones XVIII* (note (12)): Lectio XII, §XVII: 85–6. Compare Percy H. Osmond, *Isaac Barrow: his Life and Times* (London, 1944): 128–30.

(16) How much Newton's early, intuitive ideas of the nature of a hypothetico-deductive system owed to Barrow's concept of an experimentally verifiable scientific hypothesis has never been adequately explored. (See, however, R. H. Kargon, *Atomism in England from Hariot to Newton* (London, 1966): 118–21.) In preference to the Cartesian index d/e which he had hitherto systematically used, in his optical lectures Newton went over to the Barrovian notation I/R for the constant ratio of the sine of incidence at an interface to that of refraction.

(17) Or so we conjecture. If Newton ever gave a separate inaugural lecture to mark his appointment to the Lucasian chair this has disappeared without trace.

(18) The Cambridge public schools, where Barrow had given his own eighteen optical lectures a little while before.

if I subject the principles of this science to a fairly severe examination, subjoining what I have thought over concerning them and also verified by multiple experiment to what my reverend predecessor last told you of in this room.

Those who are knowledgeable in dioptrics[(20)] imagine that perspective tubes might be brought to any desired degree of perfection, were it only practicable to impart any geometrical figure they wished to lenses as they are given their fine polish. For this purpose various instruments have been designed by means of which lenses might be ground into hyperbolic or even parabolic shapes;[(21)] but no one at the present time has succeeded in fashioning those figures exactly. This is, in fact, [as pointless as] to plough the seashore. So that they may no longer expend their efforts on a profitless undertaking, I dare to pledge them my word that, even though all should turn out extremely happily, it would by no means answer their expectations: for even though lenses be shaped according to the best figures for the purpose which can be contrived, they will still achieve no more than twice the performance of spherical ones polished to an equal perfection. I do not assert this as if to contend that writers on optics have been at fault in this matter, for in regard to what they aimed to demonstrate all they have said is accurate and very true. However, they have left something to be discovered by their successors, and that of highest importance: in refraction, namely, I find that a certain irregularity throws everything out, not only making conic figures little superior to spherical ones but determining spherical ones to be much less effective than they would be if this refraction were uniform.[(22)]

And so I plant my foot in dioptrics, not to survey it afresh in any systematic way, but merely, in the first place, to develop this property inherent in the nature of light, and then to show how considerably the perfection of dioptrics is impeded as a consequence of this property and in what way that obstacle may, in so far as the nature of the subject allows, be bypassed. Here I shall advance not a few things regarding both the theory and practice of telescopes and likewise microscopes,[(23)] showing that the height of

(19) The mechanical view (supported with some variation in detail by Descartes, Gassend, Boyle, Hooke, Barrow and others) that the spectrum colours are a secondary modification of the elemental light ray due to the varying transverse and rotary speeds of its component globules. Compare M. Roberts and E. R. Thomas, *Newton and the Origin of Colours* (London, 1934): 43–5; A. I. Sabra, *Theories of Light from Descartes to Newton* (London, 1967): 240–3.

(20) Newton refers particularly to Descartes and Hooke. Compare 1, §2: note (61) below.

(21) Compare I: 561–2, 568.

(22) See 1, §2: notes (60) and (61); compare the final scholium in Appendix 3 also.

(23) This topic is not explicitly discussed in either version (notes (6) and (7)) of Newton's optical lectures. The improved Newtonian reflector with refracting eye-piece, his 'catadioptricall Telescope', given by him to the Royal Society in autumn 1671 (and now still displayed there) is described in a manuscript reproduced in photocopy in Roberts and Thomas, *Newton and the Origin of Colours* (note (19)): Plates V and VI. From this (Wickins'?) Latin version Oldenburg drew the English 'Accompt of a New Catadioptrical Telescope invented by Mr. Newton...Professor of the Mathematiques in the University of Cambridge' which he published in the *Philosophical Transactions*, **8**: No. 81 (for 25 March 1672): 4004–7 [= I. B. Cohen, *Isaac Newton's Papers and Letters on Natural Philosophy* (Cambridge, 1958): 61–4]. An earlier autograph sketch of a purely 'catoptrical' reflector is reproduced as Plate II (facing page 76) in *The Correspondence of Isaac Newton*, **1** (1959), while an advanced catadioptrical

perfection in optics is (contrary to received opinion) to be sought in a mixture of dioptrics and catoptrics....[24]

With a final sentence 'And in the interim I shall present an extended explanation of the distinction between colours, together with their genesis by prisms and also from coloured bodies' Newton swung into his main text.

Here we can be interested only in those passages in his *Lectiones opticæ* which are, as Pemberton—or whoever it was that edited its 1728 English version—phrased it, 'in a manner purely Geometrical'.[25] Below, from the first, untitled version of Newton's lectures are reproduced three extracts dealing respectively with construction of the refraction point in a plane interface of a ray passing between two given points, definition of the locus—a Dioclean cissoid—of the Barrovian radiation centres of a point with respect to a given refraction point in a plane interface as the refractive index varies, and determination of the maximum dispersion between heterogeneous rays occurring (according to a model derived from prior prismatic experiments) when the density of the refractive medium varies from zero to infinity. From a mathematical viewpoint these are revealing illustrations, in an optical setting, of his technique for constructing the roots of algebraic equations by intersecting conics, of his deployment of limit-increments to effect a geometrical differentiation, and of his treatment of

telescope with a refracting object lens included is discussed in a contemporary 'Telescopij novi delineatio' (*ibid.*, **1**: 272–3, reproduced in 2, §1 below; compare also Appendix 1.6/7). Some practical disadvantages of a reflecting telescope in general were discussed by Newton in his letter to Oldenburg on 4 May 1672 (*ibid.*, **1**: 153–4). A brief reference to the possibility of a reflecting microscope occurs in an earlier letter to the same correspondent on 6 February 1671/2 (*ibid.*, **1**: 96). Newton's views on the practicality of an achromatic refracting telescope are discussed below, but see also 1, §2: note (61).

(24) Our translation from Newton's carefully phrased Latin original (ULC. Add. 4002: 1–2) reproduced in Appendix 1 below.

(25) *Optical Lectures Read in the Publick Schools of the University of Cambridge, Anno Domini, 1669. By the late Sir Isaac Newton* (London, 1728): Preface: vi. On Pemberton as possible editor see Appendix 2: note (1).

(26) Compare 1, §1: notes (6) and (20). On the other hand, taking Barrow's self-effacing preface for literal truth, Newton's eighteenth-century editors were prone to exaggerate the importance, and indeed number, of the two constructions which (see 1, §2: notes (8) and (25)) the former inserted in his *Lectiones* XVIII in 1669 as 'communicated by a friend [*sc.* Newton]'. (In his 'Epistola ad Lectorem' Barrow had written that 'postquam edendi propositum inii, seu fastidio correptus seu novandi subiturum studium fugitans, nè quidem horum magnam partem relegere sustinui; verùm, quod tenellæ matres factitant, â me depulsum partum amicorum haud recusantium nutriciæ curæ commisi, prout ipsis visum esset, educandum aut exponendum. Quorum unus (ipsos enim honestum duco nominatim agnoscere) D. *Isaacus Newtonus*, collega noster (peregregiæ vir indolis ac insignis peritiæ) exemplar revisit, aliqua corrigenda monens, sed & de suo nonnulla penu suggerens, quæ nostris alicubi cum laude innexa cernes'. John Collins was also named as 'Alter [amicus qui] ingente suo cum labore editionem procuravit'.) Compare I: 10, note (26) and Appendix 2: note (14) below.

the extreme values of a given function: in personal terms, the first two extracts afford an intimate insight into a side of Newton's character which eluded subsequent editors of his lectures—his deference at this period to Barrow, his predecessor in the Lucasian chair.(26) In sequel (and for the same reasons) we reproduce in full the last section of the 'Opticæ pars prima'. Their evident mathematical qualities apart, these nine propositions and their accompanying scholia (dealing with refraction at a spherical interface and hence at any general surface of revolution which it instantaneously osculates) lead at once into Newton's more specialized researches into the refraction of homogeneous light in lenses, his unrewarded efforts to construct an achromatic compound lens, and his definitive explanation of the sequence of colours in the general rainbow along with the magnitude of their angular radii. His investigation of the refractive power of thin lenses, ignoring both spherical and chromatic aberration,(27) and his conclusive mathematical formulation, in culmination to two thousand years of slow empirical advance and Descartes' primitive, numerical theory, of the structure of the n-ary bow(28) both confirm the acuteness of Newton's insight into contemporary dioptrical problems and afford one more provocative parallel with similar researches—unpublished in the seventeenth century—pursued by Christiaan Huygens.(29) We must be equally impressed how little it meant to Newton not to have studied in depth the published work

(27) Compare 2, §1 and Appendix 1. The construction of a general aplanatic lens 'by points' is broached in 2, §2 and Appendix 3.

(28) Compare 1, §2: Propositions 35, 36 and scholium, with a numerical discussion of the primary and secondary bows in Appendix 2. Carl B. Boyer has splendidly summarized the complex sequence of events in which the Newtonian codification was but a single highlight in his study of *The Rainbow: From Myth to Mathematics* (New York, 1959). Some complements, particularly with regard to still unpublished researches by Thomas Harriot, are given in J. A. Lohne's 'Regenbogen und Brechzahl' (*Sudhoffs Archiv für Geschichte der Medizin und der Naturwissenschaften*, **49** (1965): 401–15).

(29) See his *Œuvres complètes*, **13** (The Hague, 1916): passim, but especially the pages listed in the concluding table of *Matières traitées* under the heads of 'Lentilles', 'Aberration sphérique', 'Aberration chromatique' and 'Arc-en-ciel'. Compare also A. E. Bell, *Christian Huygens and the Development of Science in the Seventeenth Century* (London, 1947): chapter x, 'Huygens's Optical Studies': 165–75. Huygens' dioptrical researches, as collected in the 'Dioptrica' which he began in 1653 and continued to add to throughout his life, were first published (in incomplete form) in his *Opuscula Posthuma* in 1703. It is interesting to observe that the editor of Newton's *lectiones* asserted in 1729 that this publication provoked the latter to issue his own *Opticks* the following year: 'Newtonus...de [luce & coloribus] per multos annos prorsus sile[bat], vix tandem anno 1704 amicorum precibus victus, ut absolutissimum illud de Opticis Opus in publicum proferret; quo fortasse vel diutius privati essemus, nisi hæc Newtoni inventa insigni geometræ Hugenio adeo placuisse jam compertum esset, ut magnam partem libri ejus de Dioptricis in hujus principia extruxisset, qui Hugenii liber inter Opera ejus Posthuma anno præcedente in lucem prodiit: hinc sperandum esset imperitis istis nugatoribus silentium imponi' (*Lectiones Opticæ* (note (6)): Præfatio: vi).

of his predecessors, Descartes and Barrow excepted.(30) His calculation of the chromatic aberration of rays (radiating from a unique point) which are refracted at a spherical interface and his construction of a 'new' catadioptrical telescope in which the mirror is a coated lens placed so that its chromatic distortion is minimized(31) have, of course, no precedent.

On the topic, finally, of Newton's failure to construct an effectively achromatic refracting telescope we may be a little more expansive. Persuaded by an unfortunate, ambiguous phrase in his *Opticks* that 'the improvement of Telescopes of given length by Refractions is desperate',(32) his eighteenth-century successors—no less than the majority of modern scholars(33)—came to believe that Newton in the late 1660's turned to the construction of reflecting telescopes because he was convinced of the theoretical impossibility as well as the practical difficulty of constructing a colour-free refracting lens combination.(34) In November 1753, notably, the London optician John Dollond (apparently ignorant of the conflicting dispersion model proposed in his earlier *Lectiones*(35)) publicized his deduction from the modified variant theorem hesitantly presented by Newton in 1704 that in a thin achromatic glass-water lens 'the sum of the refractions will be $=0$', concluding that 'the aberration arising from the different refrangibility...cannot be corrected by any number of refractions whatsoever'.(36) Yet it is clear from his reply in June 1672 to Hooke's charge of 'laying aside the thoughts of improving Optiques by Refractions' that, although his 'successes on the Tryals I have made of that kind, which I shall now say have been less than I sometimes expected, and perhaps than he at present hopes for' had been unpromising, Newton was not yet discouraged; indeed

what I said...was in respect of Telescopes of the ordinary construction, signifying, that their improvement is not to be expected from the *well-figuring* of Glasses, as

(30) Conduitt's story that Newton as a freshman at Trinity in 1661 read 'Kepler's Opticks' unaided (I: 15, note (3)) is entirely unsupported. If there is any grain of truth in this anecdote the book referred to was probably the 1653 Cambridge edition of Kepler's *Dioptrice* (issued with Gassend's *Institutio Astronomica* and Galileo's *Nuncius Sidereus* for use, no doubt, as an undergraduate text). There is nothing to indicate any acquaintance on Newton's part with Kepler's more penetrating *Ad Vitellionem Paralipomena, quibus Astronomiæ Pars Optica traditur* of 1604.

(31) See 1, §2: Proposition 37 and 2, §1: 'Telescopij novi delineatio' respectively.

(32) *Opticks: Or, A Treatise of the Reflexions, Refractions, Inflexions and Colours of Light* (London, 1704): $_1$75. See Appendix 4: note (6) below.

(33) H. W. Turnbull is a rare exception. See his edition of *The Correspondence of Isaac Newton*, **1** (Cambridge, 1959): 104, note 16.

(34) Compare H. C. King's *History of the Telescope* (London, 1955): 67–9, 144–7. The construction of an adequate achromatic compound lens in practice depended less on acknowledging its theoretical possibility than on accurately grinding the lens surfaces and casting glass free from optical defect. In his assessment of 'The Quality of the Image produced by the Compound Microscope: 1700–1800' (*Historical Aspects of Microscopy* (Cambridge, 1967): 151–73, especially 170) by comparison of a dozen eighteenth-century microscope lenses S. Bradbury

Opticians have imagin'd; but I despaired not of their improvement by other constructions; which made me cautious to insert nothing that might intimate the contrary. For, although successive refractions that are all made the same way, do necessarily more and more augment the errors of the first refraction; yet it seem'd not impossible for *contrary* refractions so to correct each others inequalities, as to make their difference regular; and, if that could be conveniently effected, there would be no further difficulty. Now to this end I examin'd, what may be done not only by *Glasses alone*, but more especially by a Complication of divers successive *Mediums*, as by two or more Glasses or Crystals with Water or some other fluid between them; all which together may perform the office of *one Glass*, especially of the Object-glass, on whose construction the perfection of the instrument chiefly depends.(37)

It would be nearer the truth to say that Newton never succeeded in constructing an achromatic compound lens and in time came to doubt its practical feasibility. All his hard work was not, however, in vain for later, in his *Opticks*, he suggested how a compound lens might 'very much correct the errors of ye refractions...in so far as they arise from the sphericalness of the figure'.(38) Further argument on this point we must leave to others.

---

concludes—contrary to Newton's expectation—that even where complex eyepiece systems were used distortion was considerable and that 'it seems probable that it was spherical aberration rather than chromatic errors which proved to be the chief limiting factor'.

(35) See 1, §1: note (37) below.

(36) 'A Letter...concerning a Mistake in M. Euler's Theorem for correcting the Aberrations in the Object-Glasses of refracting Telescopes (*Philosophical Transactions* **48**. 1 (1753): § XLIII: 289–91; compare Euler's 'Sur la perfection des verres objectifs des lunettes', *Mémoires de l'Académie de Berlin*, **3** (1747): 274–96 [=*Opera Omnia* (3) **6** (Zurich, 1962): 1–21]). In his reply, while necessarily agreeing that if Newton's revised dispersion model (*Opticks* (note (32)): Book 1, Part II, Experiment VIII: $_{1}$94) were 'juste à la rigeur' an achromatic compound lens would be impossible, Euler pointed out that it is valid only 'à peu près', in which case no such theoretical objection is possible (*Philosophical Transactions* **48**: 294–5 [=*Opera* (3) **6**: 41–2]). Historians have since come unquestioningly to accept Newton's 'error' as axiomatic, usually in William Whewell's loosely equivalent form that 'the dispersion [spectrum-length] must be the same when the [mean] refraction is the same' (*History of the Inductive Sciences from the Earliest to the Present Times*, **2** (London, $_{1}$1837): 356; compare David Brewster's *Life of Sir Isaac Newton* (London, 1831): 50, 57–8, and his revised *Memoirs of the Life, Writings and Discoveries of Sir Isaac Newton*, **1** (Edinburgh, 1855): 85, 110–11).

(37) 'Mr. Isaac Newtons Answer to some Considerations upon his Doctrine of Light and Colors', *Philosophical Transactions*, **8**: No. 88 (for 18 November 1672): 5084–103, especially 5084–5 [=*Papers and Letters* (note (23)): 116–17]. The original letter is reproduced in *The Correspondence of Isaac Newton*, **1**: 172 (compare I: 575, note (60)), while a lengthier first draft (never sent) of Newton's 'Answer' to Oldenburg is given in 1, §2: note (61) below.

(38) See the concluding Appendix 4.

APPENDIX 1. NEWTON'S PROLOGUE TO HIS 'LECTIONES OPTICÆ'

[January 1670][1]

From an autograph manuscript in the University Library, Cambridge[2]

Inventio Telescopiorum nupera plerosq; Geometras ita exercuit, ut nihil in Optica non tritum, nullum inventioni præterea locum alijs reliquisse videantur. Et insuper cùm dissertationes quas hic non ita pridem audivistis, tantâ rerum Opticarum varietate, novorum copiâ, et accuratissimis eorundē demonstrationibus fuerint compositæ; frustranei fortè videantur conatus et labor inutilis, si ego scientiam hanc iterum tractandam suscepero. Verùm cùm Geometras in quadam lucis proprietate, quæ ad Refractiones spectat hucusq; hallucinatos videā, dum demonstrationibus suis hypothesin quandam Physicam haud benè stabilitam tacitè supponunt:[3] non ingratum me facturum judico, si principia Scientiæ hujus examini severiori subjiciam, et quæ ego de ijs simul excogitavi, et experientia multiplici habeo comperta, subnectam ijs, quæ Reverendus meus Antecessor his loci postrema dixit.

Imaginantur sibi[4] Dioptrices studiosi, quòd Perspicilla ad quemlibet perfectionis gradum perduci possent, si[4] modò vitris dum perpoliuntur, geometricam, quam vellent, figuram communicare concederetur. Et in eum finem instrumenta varia fuerunt excogitata, quibus vitra in figuras Hyperbolicas vel etiam Parabolicas contererentur; sed exacta figurarum istarum fabricatio nemini hucusq; successit. Scilicet aratur littus; et nè labores suos in negotio desperato diutiùs insumant, ijs audeo spondere, quòd licèt omnia fierent perquam[4] felicitèr, nihil minùs tamen quàm votis suis responderent: Etenim

(1) On this point we may accept Newton's marginal dating since both manuscript versions of the *lectiones* (see note (2)) read 'Jan 1669 Lect 1' in the margin alongside. It is completely reasonable that he would begin his lectures at the start of the first term following his appointment to the Lucasian professorship (on 29 October 1669).

(2) Add. 4002: 1–2, epitomized in a marginal subhead on the first page as 'Incepti ratio' (Reason for the undertaking). Significant variants in the revised 'Optica' (ULC. Dd. 9.67: $_1$1–2) are listed in following footnotes. A near contemporary vernacular rendering is given in (Pemberton's?) posthumous English edition of *Optical Lectures Read...Anno Domini, 1669. By the late Sir Isaac Newton* (London, 1728): 1–5; see Appendix 2. Our own more modern translation is incorporated in the preceding introduction.

(3) Later changed to read 'demonstrationes suas in hypothesi quadam Physica haud bene stabilita tacitè fundantes' (tacitly founding their proofs on a certain physical hypothesis not well established).

(4) Omitted in redraft.

(5) 'Sectionum' was later added for clarity.

(6) Newton afterwards changed the tense to the more logical 'figo'.

(7) 'eam' is understood and was indeed subsequently inserted.

vitra licèt efformentur secundum figuras in istum finem optimas, quæ possunt excogitari, tamen non duplo plus præstabunt quàm sphærica, æquali politurâ perfecta. Hæc autem non ideò loquor, quasi peccatum esse a scriptoribus Optices contenderem; illi enim omnia pro intentione demonstrationum suarum accuratè quidem et verissimè dixerunt; sed aliquid tamen idq maximi momenti reliquerunt posteris inveniendum. Scilicet in refractionibus irregularitatem quandam reperio, quæ omnia perturbat, et non solum efficit, ut figuræ Conicarū[5] Sphæricas non multùm superent, sed etiam ut sphæricæ multò minus præstent quàm præstarent, si dicta refractio esset uniformis.

Itaq in Dioptricâ pedem fixi[6] non ut[7] pertractarem de integro, sed tantùm, ut hanc de natura lucis proprietatis rimarer primò, deinde ut ostenderem quantum ex hac proprietate perfectio Dioptrices impeditur, & quo pacto incommodum istud, quatenus natura rei sinit, devitetur. Ubi nonnulla proferam quæ ad Telescopiorum juxta et Microscopiorum, tum Theoriam tum Praxin spectant; ostendens quod Optices summa perfectio (præter opinionem receptam) ex Dioptrica et Catoptrica mixtis petenda est. Ac interea discrimen colorum et eorum genesin a Prismatibus et corporibus etiam coloratis fusè explicabo.

APPENDIX 2. PEMBERTON'S(?)[1] DESCRIPTION OF NEWTON'S 'OPTICÆ PARS 1ª'.

From the preface to the 1728 English edition of the *Optical Lectures*[2]

It was as long ago as the Year 1666, when Sir *Isaac Newton* first found out his Theory of Light and Colours.[3] Upon Dr *Barrow*'s resigning to him the Professor-

(1) The editor concluded his preface with the observation that 'at the bottom of the Pages we have added here and there some very short Remarks, which, we presume, will not be altogether useless to such, as are not thoroughly informed of these Matters. By this our Labour we question not, but we shall merit the Thanks of all, who are curious in these Speculations. And we shall moreover proceed in the same manner to deserve well at their Hands; for we hope shortly to present the Publick with several Mathematical Pieces, that were long ago written by our great Author, though never yet printed; whereby will be given still farther Proofs, how early that Genius exerted itself, which was at length able to produce those divine Works, the *Principia*, and the *Opticks*'. The technical quality of these footnotes (compare 1, §2: notes (16) and (19)) and the announcement of a projected edition of Newton's early mathematical papers (compare **1**, Introduction: note (31) above) point to the editor being Henry Pemberton, though William Jones (or even James Wilson?) is an outside possibility.

(2) *Optical Lectures Read in the Publick Schools of the University of Cambridge, Anno Domini, 1669. By the late Sir Isaac Newton, then Lucasian Professor of the Mathematicks. Never before Printed. Translated into English out of the Original Latin* (London, 1728): Preface [dated 'June 29. 1727']: iii–xii.

(3) The source for this remark is evidently 'A Letter of Mr. *Isaac Newton*, Professor of the Mathematicks in the University of Cambridge; containing his New Theory about Light and

ship of the Mathematicks at Cambridge, he made A. 1669, this Discovery the Subject of his publick Lectures in that University. In 1671 he began to communicate it to the World, as also a Description of his Reflecting Telescope, in the *Philosophical Transactions.* About the same time he intended to publish his Optical Lectures, wherein these Matters were handled more fully; together with a Treatise of Series and Fluxions. But the Disputes, which were occasioned, by what he had already suffered to come abroad, deterred him from that Design. And hence he conceived so great an Horror for any thing, that looked like Controversy, that the constant Importunities of his Friends could not prevail upon him to print his Book of Optics until the Year 1704.(4) As to his Lectures, they were deposited, at the time they were read, amongst the Archives of the University. From whence many Copies have been taken, and handed about by the Curious in these matters.(5)

These Lectures are divided into two Sets or Parts. What is treated of in the last Part, relates to the Doctrine of Colours, and was left imperfect; but has been since published in the *Opticks* by Sir *Isaac* himself with great Improvements. The first Part is compleat and preparatory to the other: And as it contains but little in common, with what has been already printed, we have thought fit to make it now publick. The Reader will find in it Abundance of Particulars worthy their great Author, and such as will even at present appear entirely new.(5) It is divided into four Sections....But a short View of the Whole take as follows.

The first Section [*The Refrangibility of Rays is different*] gives us a very full and plain Account of the different Refrangibility in the Rays of Light, with the

---

Colors: sent by the Author to the Publisher [Oldenburg] from Cambridge, Febr. 6. 1671/72' (*Philosophical Transactions*, **8**: No. 80 (for 19 February 1671/2): 3075–87, especially 3075–6 [= *Correspondence of Isaac Newton*, **1** (1959): 92]): 'To perform my late promise to you, I shall without further ceremony acquaint you, that in the beginning of the Year 1666 (at which time I applyed my self to the grinding of Optick glasses of other figures than *Spherical*,) I procured me a Triangular glass-Prisme, to try therewith the celebrated *Phænomena* of *Colours*'. Compare J. A. Lohne, 'Isaac Newton: The Rise of a Scientist, 1661–1671' (*Notes and Records of the Royal Society of London*, **20** (1965): 125–39).

(4) The editor (perhaps William Jones?) observed similarly in preface to the full Latin edition of the following year (*Isaaci Newtoni, Eq. Aur. in Academiâ Cantabrigiensi Matheseos olim Professoris Lucasiani, Lectiones Opticæ, Annis* MDCLXIX, MDCLXX *&* MDCLXXI. *In Scholis publicis habitæ: Et nunc primum ex MSS. in lucem editæ* (London, 1729): Præfatio: v–xii, especially v–vi) that 'Continent [hæ prælectiones] inventa de luce & coloribus, quæ auctor detexit anno 1666; quorum specimen anno 1671 coram regiâ societate exhibitum fuit, eodemque anno in Transactionibus...Philosophicis in lucem editum; & sub idem tempus hunc librum edidisset ipse auctor, nisi ineptæ quorundam imperitorum cavillationes eum deterruissent. Imo adeo abhorrebat Newtonus ab hujusmodi altercationibus, ut de hoc argumento per multos annos prorsus sileret, vix tandem anno 1704 amicorum precibus victus, ut absolutissimum illud de Opticis Opus in publicum proferret'.

(5) The 'Præfatio' to the *Lectiones Opticæ* (note (4)) reads correspondingly (pages vi–vii): 'Liber de Opticis anno 1704 editus ab hoc tractatu haud parum differt. Multa quidem in

Experiments from whence it was deduced; and, amongst many other curious things relating thereto, an elegant Demonstration of the Case, where the Image of the Sun made through a Prism would be circular, provided the common opinion of Refractions was true. The Subject of the second Section [*Of the Measure of Refractions*] is the Measure of Refractions in transparent Substances, as well Fluid as Solid, and the comparing the Refractions of heterogeneal Rays; and these are performed not only in Mediums as contiguous to the Air, but when contiguous to one another; all which are illustrated by a Description of the Instruments for making the Experiments, and by Examples, together with suitable Demonstrations.

The other two Sections [*Of the Refractions of Planes/Curve Surfaces*] are in a manner purely Geometrical. In the first of them the Effects of the Refractions of Rays are considered, as they are incident upon one or two plane Surfaces. The first nineteen Propositions relate to the Refractions made by a single Plane. Of these the first eight treat of homogeneal Rays; containing some of the Principles of Dioptricks.(6) In the Scholium to the eighth Proposition(7) our Author has a curious Speculation concerning the apparent Place of the Image of an Object seen by Refraction. The rest of these Propositions are about the Divarications and Limits of heterogeneal Rays; as they are refracted at a Surface separating two Mediums, whose Densities are considered either as permanent, or the Density of any one of the Mediums is supposed to be varied;(8) amongst which occurs, at the Conclusion of Prop. XII.(9) what is very remarkable, that, in Rays

---

utroque inveniuntur eodem sensu, sed ratione diversâ tradita; in illo autem non pauca e præstantissimis inventis comparent, quæ hic non leguntur. Magna enim pars istius operis iis explicandis occupatur, quæ lumini contingunt, dum per tenues lamellas perlucidas transit; hujusmodi vero experimenta breviter tantum in fine harum prælectionum memorantur. Porro ex iis, quæ de hoc argumento in Transactionibus Philosophicis edita fuere, apparet auctorem nostrum in animo habuisse in hæc phænomena ulterius inquirere; sed inventa sua de his rebus vix perfecisse videtur intra duodecimum vel quindecimum annum, postquam hæc, quæ nunc edimus, lecta fuerant [*sc.* till about 1685; compare Appendix 3 below]. Quanquam autem his inventis liber noster caret; multa tamen præclara hic occurrunt, quæ in altero non habentur. Ibi enim auctor cavisse, quantum potuit, videtur, ne demonstrationes geometricas cum argumentis philosophicis immisceret; & ubi necesse fuit propositionem mathematicam proponere, ejus demonstratio vix unquam occurrit. Contra autem, hic omnia geometrica in hoc argumento necessaria fuse demonstrat; quæ forsitan in altero libro ideo omiserit, quoniam haud dubitavit, quin hæ prælectiones aliquando lucem essent visuræ; cum non modo publice Cantabrigiæ lectæ in archivis fuerant repositæ, sed etiam [see 1, §2: note (2)] alia exemplaria in amicorum manibus adservabantur'. In the fourth edition of Newton's *Opticks* (London, 1730) which appeared the following year cross-references to mathematical theorems in the 'Opticæ pars prima' were introduced in footnotes at appropriate points.

(6) Of these an early draft of Proposition 6 is reproduced in 1, §1.1.

(7) A first draft of this is reproduced as the opening of 1, §1.2 below.

(8) An early version of this is given in 1, §1.3.

(9) Compare 1, §1: notes (17), (20) and (28).

of every Sort refracted at the same Point of a plane Surface the Locus of the Centers of their Radiations is the vulgar Cissoid. From thence to the End of this Section is concerning the Affections of both homogeneal and heterogeneal Rays refracted by two Planes; which chiefly have Relation to the Experiments of the Prism, from whence our Author deduced his Theory of Light and Colours. And here is shewn, at Prop. XX. and XXI. that, if Rays diverge to a Prism, the homogeneal ones, after the double Refraction, will still continue to diverge, but some of the heterogeneal ones will converge; therefore at Prop. XXII. that of Rays so refracted from an Object to the Eye, some will fall gradually nearer to the Vertex of the Prism than others, as they are more and more refrangible; whence from Prop. X. are defined the Orders of Colours in the Image made by Refraction; at Prop. XXIII. XXIV. that the bigger the vertical Angle of the Prism, or the denser its Matter is, the Difference of the Refraction will be so much the greater, whence the Colours in the Image will become the more manifest; at Prop. XXV. XXVI. that the Rays so falling on the Prism, that the Refraction on each Side may be equal, in homogeneal Rays the Angle, which the incident and emerging Rays comprehend, will then become the greatest; but in heterogeneal Rays the difference of those Angles will become at that time the least. And at the last Proposition our Author sets down a Mechanical Solution of the following Problem; Rays being refracted from one given Point to another given Point by a Prism given in Position; to find the Angles comprehended by the heterogeneal Rays. He says, to perform this geometrically, would require such a Construction, as the Ancients called Linear,(10) or what could not be effected by the Help of the Conicks.

The last Section(11) regards Rays, as they are refracted by curve Surfaces. Its chief Contents are, at Prop. XXIX. XXX. XXXII. XXXIII. the finding both the principal Focus, and also that of every particular Ray, not only in Spheres, but in any curve Surface whatever;(12) at Prop. XXXI. the computing the Errors

(10) Evidently so, since finding the refraction point in a single plane interface requires (1, §1: note (5)) the solution of a quartic equation and hence an equation of sixteenth degree will determine the path of a light ray of given refractive index between two fixed points through a given prism.

(11) Reproduced as 1, §2 below from the manuscript 'Optica' deposited in the Cambridge archives in 1674 by Newton.

(12) On this point the editor of the 1729 *Lectiones Opticæ* (note (4)) is more specific, noting (pages viii–ix) that 'Hic nimirum demonstratur modus [Barrovii] focum superficierum sphæricarum inveniendi, & in omnibus aliis curvis locum habere ostenditur ope radii, qui dicitur, curvaturæ; item causticæ (quas vocant) a refractione ortæ hic determinantur. Has itaque causticas in superficiebus sphæricis & ipse Barrovius [see 1, §2: note (25)] determinavit. Hisce vero prælectionibus Newtonus in curvis omnibus radii curvaturæ earum ope has causticas exhibet. Hos quidem curvaturæ radios jam pridem consideraverat, & modum eos inveniendi in libro de fluxionibus, anno 1665[!] scripto, docuerat; idemque argumentum ulterius prosecutus est in alio libro anno 1671 scripto. Hoc apparet ex epistolâ ipsius ad

arising from the Figures of Optical Glasses; at Prop. XXXIV. the Invention of such Curves, as will accurately refract the Rays of Light to any given Focus; and at Prop. XXXV. XXXVI. the determining the Rain-bow. In all this Section our Author makes no mention of heterogeneal Rays, until he comes to the last Proposition, wherein he determines the Errors caused by the different Refrangibility of the Rays of Light. And from this Proposition compared with the thirty first he deduces this Conclusion, that the Imperfection of Optical Instruments is not owing, as has been all along thought, to the Unfitness in the Figure of the Glasses, but to the different Refrangibility in the Rays of Light.(13) This Consideration put our Author upon the noble Invention of the Reflecting Telescope; a very particular Account of which is given in his *Opticks*...

In the Preface of the learned Dr. *Barrow*'s *Lectiones Opticæ* printed in 1669, there is given a great Character of Sir *Isaac Newton*, at a time when he was altogether unknown to the World. And the Dr. is so candid as farther to declare, that he there altered several Things upon his Advice, and inserted some of his Inventions, as an additional Ornament to his own. What Dr. *Barrow* published then of our Authors without their Proofs, the Reader will find demonstrated in the following Discourse...(14)

Collinsium Dec. 10. 1672. inter cæteras in *Commercio epistolico* editâ [see **1**, Introduction, Appendix: Document E], & parte quâdam illius in editione ultimâ *Principiorum Philosophiæ* [*Philosophiæ Naturalis Principia Mathematica* (London, $_3$1726): Liber 2, Lemma II, Scholium: 246]; nec non ex tractatibus ipsis, qui adhuc inediti restant, quorum cum varia exempla dispergantur, nonnulli in lucem eos proferre polliciti sunt'. Compare note (1).

(13) See 1, §2: notes (60) and (61).

(14) Compare note (26) of the preceding introduction. In preface (page viii) to the 1729 publication of the *Lectiones opticæ* (note (4)) its editor was a little more appreciative of Barrow's originality: 'Quod spectat ad prima optices elementa, auctor noster hic ubique Barrovii prælectiones opticas sequitur, & quæ ille de omni luce scripserat, Newtonus persequitur ulterius, & applicat ad diversam radiorum refrangibilitatem; rem Barrovio ignotam, sed ab eo omnino probatam, quando ei ab auctore nostra explicata fuit, quod una ex epistolis D. Collinsii in *Commercio Epistolico* [(London, $_1$1712): 28] edita testatur, ubi Barrovius de his prælectionibus loquens, opus vocat, quo majus præsens ætas vix protulit. Hic etiam multæ propositiones demonstrantur, quæ auctor cum Barrovio communicavit, & in prælectionibus ejus, demonstrationibus prætermissis, editæ fuere'. As we have seen (**1**, Introduction, Appendix: Document D) the actual words of Collins' draft letter to Vernon on 26 December 1671 spoke of '20 Optick Lectures which Dr Barrow reckons one of the greatest performances of Ingenuity this age hath affoarded'.

1

EXTRACTS FROM NEWTON'S LUCASIAN LECTURES ON OPTICS[1]

[*c.* 1670–1671][2]

From the originals in the University Library, Cambridge

§1. REFRACTION AT A PLANE INTERFACE.[3]

[1][4] [Construction of the point of refraction.]

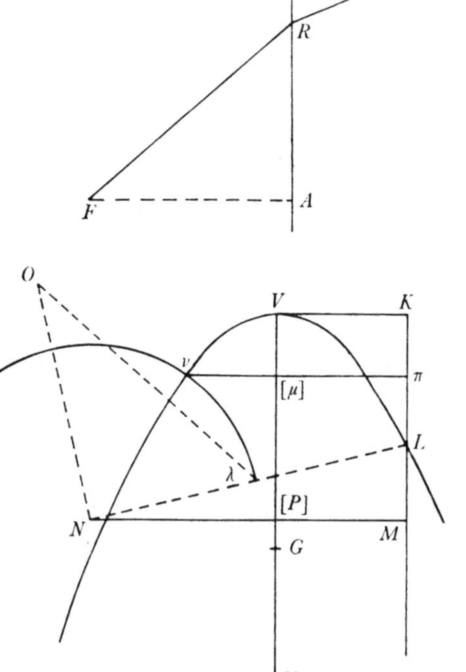

Prob. Dato puncto planum refringens irradiante, et alio etiam puncto, per quod refractus radius debet transire, positio radiorum, sive refringens punctum quæritur. Sit F punctum radios ejaculans, et R punctum ubi radius refringi debet ut postmodū transeat per datum X. Ab ijsdem F et X normales FA, et $X\alpha$ ad refringens planum demittantur. Et facto

$$\sqrt{II-RR}\,.\,I::AF\,.\,VG,$$

cum latere recto VG seorsim describatur Parabola Conica $VL\nu$, cujus vertex sit V, et axis VGH. Deinde ad axem erigatur a vertice normalis $VK=\frac{1}{2}A\alpha$ et agatur KM ad axem parallela, quæ secet Parabolam in L, et capiatur LM ejus longitudinis, ut sit $II-RR\,.\,RR::\frac{\alpha X^q}{2VG}\,.\,LM.$ eiq perpendiculariter insistens ad M ducatur MN, ut sit $VG\,.\,VG+LM::A\alpha\,.\,MN$, jungaturq LN et ad ipsam erigatur normalis $NO=A\alpha$, et in angulo ONL inscribatur $O\lambda=NL$, centroq N et intervallo $N\lambda$ describatur circulus, secans Parabolam in ν, unde $\nu\pi$ perpendiculariter ad LM demittatur. Deniq si huic $\nu\pi$ sumatur AR æqualis, erit R quæsitum punctum refractionis, quod radios FR, et RX positione determinat.[5] Q.E.F.

Translation

[1][4] [Construction of the point of refraction.]

Problem. Given a point irradiating a refracting plane [interface] and also another point through which a refracted ray shall pass, the position of the rays or the point of refraction is required.

Let F be the point throwing out the rays, and R the point at which a ray should be refracted so as afterwards to pass through the given point X. From these same points F and X let fall the normals FA and $X\alpha$ to the refracting plane. Then, making $\surd[I^2-R^2]:I = AF:VG$, with *latus rectum* VG describe the conical parabola LVv separately and let its vertex be V and axis VGH. Next, from the vertex to the axis raise the normal $VK = \frac{1}{2}A\alpha$ and draw KM parallel to the axis so as to intersect the parabola in L and take LM of length such that

$$(I^2-R^2):R^2 = \tfrac{1}{2}\alpha X^2/VG:LM.$$

Standing perpendicular to it at M draw MN so that

$$VG:(VG+LM) = A\alpha:MN,$$

join LN and to this raise the normal $NO = A\alpha$; then in the angle $O\widehat{N}L$ inscribe $O\lambda = NL$, with centre N and radius $N\lambda$ describe a circle intersecting the parabola in v, and from this let fall $v\pi$ perpendicular to LM. Finally, if AR be taken equal to this line $v\pi$, R will be the required point of refraction which determines the rays FR and RX in position.[5] As was to be done.

(1) These are taken from the two extant manuscript versions of the 'Lectiones opticæ' now known to exist: the earlier, autograph treatise (now ULC. Add. 4002) which Newton kept by him till his death, and the substantially augmented and rearranged revise, 'Opticæ pars 1ª/2ᵈᵃ' (ULC. Dd. 9.67, corrected and completed in his own hand), which he deposited in the University Library in 1674. The little evidence which can be gleaned concerning their composition has been presented in the preceding introduction.

(2) This firm dating, hazarded on the basis of documentary evidence discussed above, is substantiated by our assessment of Newton's handwriting style.

(3) Extracts from the section 'De radiorum semel refractorum Affectionibus' of the earlier (untitled) manuscript of the *Lectiones* dealing with the construction of the point of refraction at a plane interface, Newtonian 'centres of radiation' and dispersion in refractive media of varying density.

(4) Add. 4002: 99–100. The original Latin text has never been published, but J. A. Lohne included a commented German translation of the piece in his 'Fermat, Newton, Leibniz und das anaklastische Problem' (*Nordisk Matematisk Tidskrift*, **14** (1966): 5–25, especially 15–17).

(5) Since by the sine law of refraction there follows

$$I:R = \sin A\hat{F}R:\sin \alpha\hat{X}R = AR/FR:R\alpha/RX,$$

on making $AF = a$, $A\alpha = 2b$ and $\alpha X = c$ the problem reduces to constructing

$$AR = b+x \text{ (and so } R\alpha = b-x)$$

Cæterùm cùm hoc idem Problema a D[re] Barrow in Lect 5 rerum Opticarum eleganter solutum extet, potestis illum consulere, et ideò demonstrationem hujus constructionis brevitatis gratiâ prætermitto.[(6)]

---

such that $I:R = (b+x)/\sqrt{[a^2+(b+x)^2]}:(b-x)/\sqrt{[c^2+(b-x)^2]}$, that is,

$$(I^2-R^2)x^4+(a^2I^2-c^2R^2-2b^2[I^2-R^2])x^2-2b(a^2I^2+c^2R^2)x+b^2(a^2I^2-c^2R^2)+b^4(I^2-R^2) = 0.$$

The implicit restriction is made that the refractive index I/R shall be positive, and hence that R shall lie between points A and α: there will always be one solution—and may be three—in which a real point R taken outside the segment $A\alpha$ satisfies the condition

$$-IR = AR/FR:R\alpha/RX.$$

To solve this quartic Newton employs the geometrical approach elaborated in his 'Problems for construing equations' (II, **3**, 2, §2): namely, the real roots of the quartic are constructed as the ordinates x of the meets of the parabola $x^2 = \alpha y$ and the circle

$$x^2+y^2-2[b(a^2I^2+c^2R^2)/\alpha^2(I^2-R^2)]x+[(a^2I^2-c^2R^2)/\alpha^2(I^2-R^2)-2b^2/\alpha^2-1]\alpha y$$
$$+b^2(a^2I^2-c^2R^2)/\alpha^2(I^2-R^2)+b^4/\alpha^2 = 0,$$

in which for simplicity the parabola's latus rectum α is specified to be of length

$$VG = aI/\sqrt{[I^2-R^2]}.$$

In Newton's figure (to which we have added the points P and μ) $\mu\nu = x$ is constructed as the ordinate (corresponding to abscissa $V\mu = y$) of the meet $\nu(y, x)$ of the parabola νVL defined by the Cartesian equation $\mu\nu^2 = VG\times V\mu$ and the circle $\nu\lambda$ of centre N defined by

$$(\mu\nu-PN)^2+(V\mu-VP)^2 = N\lambda^2,$$

where, on making $KV = \frac{1}{2}A\alpha = b$ (so that $KL = KV^2/VG = b^2\sqrt{[I^2-R^2]}/aI$),

$$\dot{L}M = \frac{\alpha X^2}{2VG}\times\frac{R^2}{I^2-R^2} = \frac{c^2R^2}{2aI\sqrt{[I^2-R^2]}} \quad\text{and}\quad MN = \frac{A\alpha(VG+LM)}{VG},$$

$VP = KL+LM$, $PN = MN-KV$ and $N\lambda^2 = (O\lambda^2-ON^2 \text{ or})\ NL^2-A\alpha^2$, that is,

$$N\lambda = \sqrt{[LM^2+MN^2-A\alpha^2]}.$$

Evidently the construction yields $RA = b+x = (KV+\mu\nu \text{ or})\ \pi\nu$.

As J. A. Lohne has recently pointed out ('Dokumente zur Revalidierung von Thomas Harriot als Algebraiker', *Archive for History of Exact Sciences*, **3** (1966): 185–205, especially 186–7, 198) William Lower had obtained a similar quartic resolvent in terms of the variable line-segment AR about 1609, finding the refraction point at an interface between air and water (taking $I/R = 4/3$). Newton could have had no knowledge of this. Much later—probably in the very year that the present piece was composed—Sluse came across an equivalent manner of solution, and communicated its gist to Oldenburg on 25 July 1670 (N.S.) (see C. Le Paige, 'Correspondance de René-François de Sluse' [= *Bullettino di Bibliografia e di Storia delle Scienze Matematiche e Fisiche*, **17** (Rome, 1884): 427–554/603–726]: 643–4). Taking points $F(0, a)$ and $X(b, c)$ on the same side of the interface $A\alpha(y = 0)$, Sluse constructed the resulting refraction condition $I:R = (x-b)/\sqrt{[(x-b)^2+c^2]}:x/\sqrt{[x^2+a^2]}$ by the intersection of the two

However, since an elegant solution of this same problem by Dr Barrow is extant in Lecture 5 of his *Optical topics* you can consult that and I accordingly pass over my demonstration of this construction for brevity's sake.(6)

---

hyperbolas $xy = ab$ and $(I^2-R^2)(x-b)^2+c^2I^2 = R^2(y-a)^2$. (However, he made a gaffe in constructing the latter's upper vertex $(b, a-[I/R]c)$ and, since he drew only the lower branch of the former, missed one of the four possible meets.) Though Sluse's letter was not published for more than two centuries, it is worth remarking that little after receiving it Oldenburg allowed Collins to make a copy of it (now ULC. Add. 3971.1: 17r–18r) and that this (no doubt by way of Isaac Barrow) quickly found its way into Newton's possession. Nevertheless, the two methods—Sluse's and Newton's—are too dissimilar in detail to make the influence of the one upon the other likely.

(6) When Newton came to revise the present passage in his lectures, he in fact thought so highly of Barrow's 'elegant' solution that he substituted it for his own, making it Proposition 6 of Section 3 in the 'Pars prima' of his *Optica* (ULC. Dd. 9.67: $_1$41 [= *Optical Lectures Read in the Publick Schools of the University of Cambridge, Anno Domini, 1669. By the late Sir Isaac Newton*... (London, 1728): 101–2]). In his *Lectiones XVIII, Cantabrigiæ in Scholis publicis habitæ; In quibus Opticorum Phænomenωn genuinæ Rationes investigantur, ac exponuntur* (London, 1669): Lectio v §XII: 42 Barrow himself gave only a synthetic reformulation, but the preceding analysis is easily restored. From the basic defining condition $I:R = AR/FR:R\alpha/RX$, there follows readily $I^2:R^2 = AR^2/(FA^2+AR^2):R\alpha^2/(\alpha X^2+R\alpha^2)$, so that $I^2-R^2 = R^2\times\alpha X^2/R\alpha^2 - I^2\times FA^2/AR^2$ and consequently

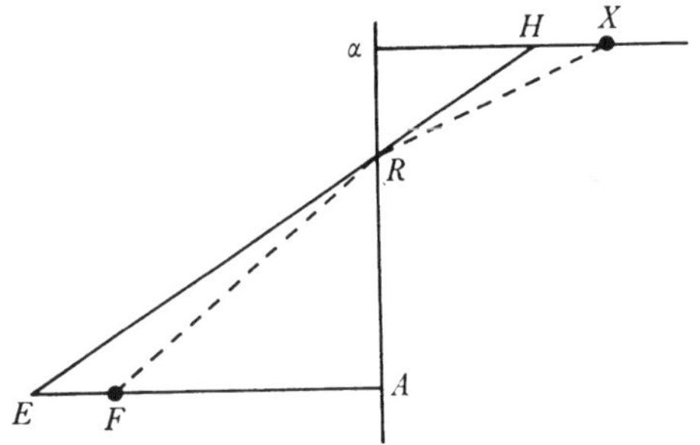

$$\frac{AR^2+\dfrac{I^2}{I^2-R^2}\times FA^2}{AR^2} = \frac{\dfrac{R^2}{I^2-R^2}\times \alpha X^2}{R\alpha^2}.$$

Hence, if E is taken in FA such that $EA/FA = I/\surd[I^2-R^2]$ and if ER meets αX in H, the line-segment $RH = \alpha X\times R/\surd[I^2-R^2]$ is determined in length and to find R we need only solve the Apollonian 'verging' problem of laying off the constant segment RH in the right angle $A\hat{\alpha}X$ so as to pass through the fixed point E. (There will be two or four real solutions according to the location of the point E with respect to the lines $A\alpha$, αX; compare I: 509.) In his own 'Lectiones opticæ' at this point Newton contented himself with the final remark: 'Cæterùm quo pacto data recta angulo recto interserenda sit, quæ per punctum datum transibit in Lect 5 Dris Barrow per Hyperbolæ et circuli intersectionem ostenditur'. In the 'Problema Lemmaticum' (Lectio v, §VII: 40, repeated in his *Lectiones Geometricæ: In quibus (præsertim) Generalia Curvarum Linearum Symptomata declarantur* (London, 1670): Lectio VI, §III: 46) Barrow does indeed give an extremely elegant construction of this Apollonian 'neusis' by the intersection of a circle and a rectangular hyperbola. In his letter to Oldenburg of 25 July 1670 (N.S.) (see previous note) Sluse commented that Barrow's 'Problema Lemmaticum...sub Vietæ postulato 1° in *Supplemento Geometriæ* [= (ed.) F. van Schooten, *Opera Mathematica* (Leyden, 1646): 240: 'Postulatum...Ad supplendum Geometriæ defectum'; compare II, **3**, 2, §2: note (17)] continetur' but Viète merely made the obvious suggestion that 'Nicomedes' first conchoid' be used to construct the 'verging'.

[2][7] [Centres of radiation.]

Prop 3.[8] Si radij homogenei FR, $F\rho$, manantes a puncto quodam lucido F, refringantur a qualibet plana superficie AR, eorum refracti RM, $\rho\mu$ ab invicem postea divergent.[9] --- --- --- --- ---

Schol: Si angulus $R\phi\rho$, quem refracti radij comprehendunt ponatur esse infinitè parvus, punctum istud ϕ erit limes determinans intersectiones radiorum utrinq; jacentium, quas cum radio $R\phi$ vel $\rho\phi$ efficiunt: ....[10] Porrò cum inter-

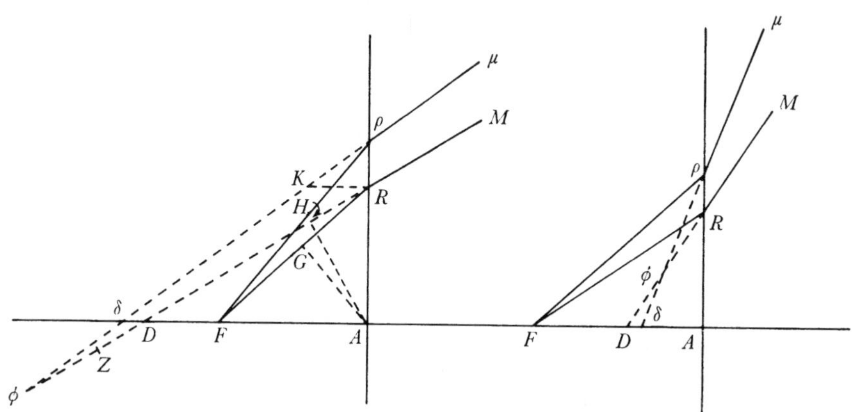

sectiones, quas radij utrinq; cum $\phi[R]$ efficiunt, sint eò densiores quo sunt viciniores puncto ϕ ac in illo puncto densissimæ: istud itaq; ϕ pro foco radij ϕR habendum est sive pro loco imaginis illuc per refractionem translatæ; habito scilicet ad eos solummodo radios respectu qui jacent in plano FAR, quod refringenti plano perpendiculariter insistit, transitq; per punctum radians F. Nam alij refracti quorum incidentes jacent in alijs planis per puncta F et R transientibus et obliquis ad refringens planum, radium $R\phi$ nec in puncto ϕ, nec ullibi omninò secabunt, si eos solummodò excipias quorum incidentes jacent in superficie conicâ cujus axis est AF, vertex F, et semi-angulus AFR; utpote qui omnes præfatum $R\phi$ in puncto D secabunt, quod in axe FA sit positum.[11] Et hujus itaq; $R\phi$ centra radiationis[12] præcipuè sunt duo, alterum ϕ a refractis jacentium in plano FAR effectum, et alterum $[D]$ a refractis jacentium in conicis superficiebus axe DFA angulisq; AFR, ADR descriptis.[13] Ad reliquos autem radios quod attinet, aliter circa FR quaquaversum positos, eorum refracti maximè appropinquant radio $R\phi$ alicubi inter D et ϕ. Adeò ut respectu oculi per cujus pupillæ centrum radius RM transit, locus imaginis per totum spatium ϕD diffundi debeat. Vel potiùs cùm spatium ϕD sit unici tantùm

(7) Add. 4002: 92–5, 101–2.

(8) That is, of the subsection 'de [radijs] similiter refrangibilibus' of the 'De radiorum semel refractorum Affectionibus'.

[2][7] [Centres of radiation.]

Proposition 3.[8] If the homogeneous rays FR, $F\rho$ as they spread from some point light-source F be refracted from any plane interface AR, their refracted rays RM, $\rho\mu$ will thereafter diverge from each other.[9]

--- --- --- --- ---

Scholium. If the angle $R\widehat{\phi}\rho$ contained by the refracted rays be supposed indefinitely small, that point ϕ will be a limit bounding the intersections of rays lying to either side made by them with the ray $R\phi$ or $\rho\phi$: ....[10] Furthermore, since the intersections made on either side with ϕR are the denser the closer they are to the point ϕ and densest of all at that point, that point ϕ is accordingly to be considered as the focus of the ray ϕR—in other words, as the place of the image there transported by refraction. To be sure, regard is had only for those rays lying in the plane FAR which stands perpendicular to the refracting plane and passes through the radiating point F. For other refracted rays, whose incident ones lie in other planes passing through the points F and R and at an angle to the refracting plane, will intersect the ray $R\phi$ neither at the point ϕ nor anywhere else, excepting only those whose incident rays lie in the cone-surface whose axis is AF, vertex F and half (vertex) angle $A\widehat{F}R$, since all these will intersect the aforesaid ray $R\phi$ in a point D set in the axis FA.[11] Accordingly, centres of radiation[12] of $R\phi$ are principally two in number: one, ϕ, brought about by the refractions of rays lying in the plane FAR, and a second, D, made by the refractions of rays lying in the cone-surfaces described with axis DFA and angles AFR, ADR.[13] As regards the remaining rays, otherwise located around FR in a haphazard way, their refractions mostly approach the ray $R\phi$ somewhere between D and ϕ, with the result that in respect of the eye through the centre of whose pupil the ray RM passes the image's location ought to be diffused throughout the whole space ϕD—or rather, since the space ϕD is the image of

(9) As Newton proceeds to show this is immediate since, by the sine law of refraction, $\sin A\widehat{F}\rho : \sin A\widehat{F}R = \sin A\widehat{\delta}\rho : \sin A\widehat{D}R$.

(10) In the half dozen lines following which are here omitted Newton discusses the positions of F and ϕ in regard to the interface $AR\rho$ which arise accordingly as the 'refractio fit e Medio rariori in densius [vel] E contra'.

(11) Since DFA is perpendicular to the refracting plane AR, any incident ray through F which meets the plane at a point on the circle through R of centre A will be refracted 'backwards' through D.

(12) 'foci' is cancelled.

(13) Newton means that the refractions of rays radiating from F in the cone formed by the side FR of the triangle AFR as it rotates round AFD will form a cone of light of vertex D formed by the rotation of DR round AD. He at once makes it clear that he is interested only in the 'plane' centres of radiation ϕ.

puncti F imago, debemus unicum aliquod in eo punctum quod lucis omnis ab eo versus oculum pergentis meditullium occupet, inter puncta D et ϕ in mediâ circiter distantiâ interjacens, pro sensibili imagine statuere. Puncti verò illius accurata determinatio, cùm omnium radiorum ab F versus oculi pupillam refractorum habenda sit æstimatio, problema solutu difficillimum præbebit nisi Hypothesi alicui, saltem verisimili, si non accuratè veræ innitatur assertio. Quemadmodum cùm radij æquè multi a termino D, alijsq; vicinis punctis, ac a termino ϕ alijsq; punctis similiter sibi vicinis versus oculum videantur profluere: locus imaginis ita debet in medio istorum terminorum statui, ut angulus quem radij duo a D et ϕ ad idem quodpiam pupillæ punctum convergentes includunt, a radio ab illo visionis loco(14) ad idem pupillæ punctum pergente quàm proximè semper bisecetur. Quâ Hypothesi admissâ nihil aliud agendum est, quàm ut fiat $M\phi+MD.MD::\phi D.DZ$, et erit Z locus visionis(14) puncti F quæsitus; posito nempe quòd M sit locus oculi. Nam cùm ponatur

$$M\phi+MD.MD::\phi D.DZ,$$

erit divisim $M\phi.MD::\phi Z.DZ$. Et proinde ductis tribus lineis a ϕ, D, et Z ad M vel potiùs ad punctum quodpiam huic M indefinitè vicinum, angulus quem externæ duæ continent, ab interjacente lineâ (per 3. 6. Elem)(15) quàm proximè semper bisecabitur. Cæterùm nè puncti ϕ positio gratis assumatur, placet illud insuper sequenti Methodo determinare, præsertim cùm in hac re videatur præcipuum.

Normalibus AG, AH a puncto A in radios dimissis, alterâ AG in incidentem radium FR et alterâ AH in refractum DR, factoq; $FG.DH::RF.R\phi$, punctum ϕ erit locus objecti F post refractionem visi habito unicè ad radios in plano FAR jacent[es] respectu. Scilicet cùm hoc punctum sit limes per interpositionem dirimens ac distinguens intersectiones radiorum utrinq; positorum, nè longâ propositionum serie ad hoc demonstrandum opus sit, illorum intersectiones finitis intervallis ab RD distantium vix respiciam, sed radij tantùm indefinitè propinquissimi intersectionem speculando determinabo propositum, siquidem ea ipsa (ut jam dictum est) sit punctum ϕ, quod quæritur. Et nè demonstratio hæc, quæ (dum nullis ferè fundamentis præmonstratis innititur) longiuscula futura est, vos(16) itaq; tædio afficiat, lubet ut in partes aliquot sive conclusiones distinguatur.(17)

(14) The 'locus imaginis' (image point), that is.

(15) Euclid's theorem that the internal bisector of a triangle divides the opposing side in the ratio of the other two adjoining ones (Barrow, *Euclidis Elementorum Libri XV. breviter demonstrati* (Cambridge, 1655): 116).

(16) Newton's captive Cambridge audience.

(17) The reader who finds the following argument somewhat tedious despite Newton's efforts may appreciate the following summary. Since

$$\frac{\rho F^2}{\rho\delta^2}=\frac{RF^2}{RD^2}=\frac{\rho F^2-RF^2}{\rho\delta^2-RD^2}=\frac{\rho A^2-RA^2}{(\rho A^2+A\delta^2)-(RA^2+AD^2)},$$

but a single point F, we ought, for the sensible image, to fix on some single point in it which, lying roughly midway between the points D and ϕ, may occupy a central place in all light proceeding from it in the direction of the eye. Since reckoning must be had of all rays refracted from F towards the eye's pupil, the accurate determination of that point will in truth provide a problem very difficult to solve unless the choice is asserted on the basis of some hypothesis which, if not accurately true, is at least plausible. For instance, since equally many rays appear to emanate in the direction of the eye from the end-point D and other nearby points as from the end-point ϕ and other points similarly near to it, the image's place ought to be set on mid-way between those end-points in such a way that the angle included by two rays converging from D and ϕ on the same arbitrary point of the pupil shall always be as closely as possible bisected by a ray going from that place of view(14) to the same point of the pupil. When this hypothesis is admitted, nothing else remains to be done but to make

$$(\mathrm{M}\phi + MD) : MD = \phi D : DZ$$

and then Z will be the required place of view(14) of the point F (supposing M to be the eye's place, that is). For since it is supposed that

$$(M\phi + MD) : MD = \phi D : DZ,$$

there will result, on dividing the ratios, $M\phi : MD = \phi Z . DZ$; and consequently, when three lines are drawn from ϕ, D and Z to M—or rather to any random point indefinitely near to M—, the angle contained by the two outside lines will always (by *Elements* VI, 3)(15) be exceedingly closely bisected by the intervening one. However, not to assume the position of the point ϕ out of hand, it is satisfying to determine it in addition by the following method, especially since in the present situation it seems particularly appropriate.

On letting fall the normals AG, AH from the point A onto the rays, one, AG, onto the incident ray FR and the second, AH, onto the refracted ray DR, and making $FG : DH = RF : R\phi$, the point ϕ will be the apparent place of the object F after refraction when regard is had uniquely to rays lying in the plane FAR. Specifically, since this point is a limit separating and distinguishing by its intervention the intersections of rays located on its either side, so that there be no need of a long sequence of propositions to prove this I shall take scarcely any notice of those of their intersections at a finite distance away from RD: rather I will accomplish my design by examining only the intersection of the indefinitely nearest ray, seeing that this meet (as has already been said) is the point ϕ required. And lest this demonstration, which (insofar as it is founded on almost no ground-work previously covered) will be a trifle long, should consequently weary you(16) by its tedium, it is agreeable to divide it up into several parts or conclusions.

Dico igitur imprimis quòd positis quorumlibet uniformium radiorum FR, $F\rho$, refractis RD, $\rho\delta$ secantibus perpendiculum FA in D ac δ: erit

$$R\rho\,.\,D\delta::\frac{FR^q}{AR+A\rho}\,.\,\frac{DR^q-FR^q}{AD+A\delta}.$$

In præcedentibus[18] enim ostensum est quòd FR & RD sunt ut sinus incidentiæ et refractionis, et sic $F\rho$ ad $\rho\delta$ habebit eandem rationem. Quare terminos quadrando erit $FR^q\,.\,RD^q::F\rho^q\,.\,\rho\delta^q$. et per conversam rationem

$$FR^q\,.\,RD^q-FR^q::F\rho^q\,.\,\rho\delta^q-F\rho^q.$$

rursus per conversam rationem, subintellectâ tamen permutatione, fit

$$FR^q\,.\,RD^q-FR^q::F\rho^q-FR^q\,.\,\rho\delta^q-F\rho^q-RD^q+FR^q.$$

Est autem $F\rho^q-FR^q=$[19] $A\rho^q-AR^q=R\rho^q+R\rho\times 2AR=R\rho\times\overline{AR+A\rho}$. Est etiam $\rho\delta^q-F\rho^q=A\delta^q-AF^q$ & $RD^q-FR^q=AD^q-AF^q$, adeoqꝫ

$$\rho\delta^q-F\rho^q-RD^q+FR^q=A\delta^q-AD^q=D\delta\times\overline{AD+A\delta}.$$

Quare est $FR^q\,.\,RD^q-FR^q::R\rho\times\overline{AR+A\rho}\,.\,D\delta\times\overline{AD+A\delta}$. Et applicando antecedentes ad $AR+A\rho$ et consequentes ad $AD+A\delta$ prodit

$$\frac{FR^q}{AR+A\rho}\,.\,\frac{RD^q-FR^q}{AD+A\delta}::R\rho\,.\,D\delta.$$

Q.E.O.

Porrò si radiorum FR, $F\rho$ distantia sit indefinitè parva; Dico quod erit

$$AD\times FR^q\,.\,AR\times\overline{RD^q-FR^q}::R\rho\,.\,D\delta.$$

Tunc enim segmenta $R\rho$, $D\delta$ pro infinitè parvis habenda sunt, sive lineæ AD, $A\delta$, ut et AR, $A\rho$ pro infinitè parùm differentibus; hoc est pro æqualibus. Evadit ergo $AR+A\rho=2AR$, et $AD+A\delta=2AD$. Et sic vice proportionis jam ante ostensæ oritur $\frac{FR^q}{2AR}\,.\,\frac{RD^q-FR^q}{2AD}::R\rho\,.\,D\delta$. Sive, multiplicando priorem rationem per $2AR\times AD$, est $AD\times FR^q\,.\,\overline{RD^q-FR^q}\times AR::R\rho\,.\,D\delta$.

---

therefore $$\frac{RF^2}{RD^2-RF^2}=\frac{\rho F^2-RF^2}{A\delta^2-AD^2}=\frac{\rho R(\rho A+RA)}{\delta D(\delta A+DA)},$$

so that $$\frac{\rho R}{\delta D}=\frac{RF^2(\delta A+DA)}{(RD^2-RF^2)(\rho A+RA)};$$

accordingly, since $\frac{\rho R}{KR}=\frac{\rho A}{\delta A}$, it follows that

$$\frac{R\phi}{D\phi}=\frac{KR}{\delta D}=\frac{RF^2(DA+\tfrac{1}{2}\delta D)(DA+\delta D)}{(RD^2-RF^2)(RA+\tfrac{1}{2}\rho R)(RA+\rho R)}.$$

In the limit, therefore, as ρ coincides with R (and so δ with D),

$$\frac{R\phi}{D\phi}=\frac{RF^2\times DA^2}{(RD^2-RF^2)\times RA^2}=\frac{RF^2\times DA^2}{(DA^2-FA^2)\times RA^2}$$

I assert, therefore, in the first place that, when the refractions RD, $\rho\delta$ of any uniform rays FR, $F\rho$ you please are supposed to intersect the perpendicular FA in D and δ, then $R\rho:D\delta = FR^2/(AR+A\rho):(DR^2-FR^2)/(AD+A\delta)$. For in the preceding(18) it was shown that FR and RD are as the sines of incidence and refraction, and so $F\rho$ will have the same proportion to $\rho\delta$. Hence, on squaring the terms, there will be $FR^2:RD^2 = F\rho^2:\rho\delta^2$ and, by conversion of the ratio, $FR^2:(RD^2-FR^2) = F\rho^2:(\rho\delta^2-F\rho^2)$, and yet again by conversion (with, however, an implicit permutation) there comes

$$FR^2:(RD^2-FR^2) = (F\rho^2-FR^2):(\rho\delta^2-F\rho^2-RD^2+FR^2).$$

But $F\rho^2-FR^2 =^{(19)} A\rho^2-AR^2 = R\rho^2+R\rho\times 2AR = R\rho(AR+A\rho)$. Also

$$\rho\delta^2-F\rho^2 = A\delta^2-AF^2 \quad\text{and}\quad RD^2-FR^2 = AD^2-AF^2,$$

so that $\rho\delta^2-F\rho^2-RD^2+FR^2 = A\delta^2-AD^2 = D\delta(AD+A\delta)$. Consequently

$$FR^2:(RD^2-FR^2) = R\rho(AR+A\rho):D\delta(AD+A\delta)$$

and on dividing through by $(AR+A\rho)/(AD+A\delta)$ there results

$$FR^2/(AR+A\rho):(RD^2-FR^2)/(AD+A\delta) = R\rho:D\delta.$$

As was to be shown.

Further, if the distance apart of the rays FR, $F\rho$ be indefinitely small, I assert that there will be $AD\times FR^2:AR(RD^2-FR^2) = R\rho:D\delta$. For then the segments $R\rho$, $D\delta$ are to be considered as infinitely small—in other words, the lines AD, $A\delta$ and also AR, $A\rho$ are to be taken as differing by an infinitely small amount, that is as equal. There results, therefore, $AR+A\rho = 2AR$ and $AD+A\delta = 2AD$. And thus in place of the proportion already shown before there arises $FR^2/2AR:(RD^2-FR^2)/2AD = R\rho:D\delta$; that is, on multiplying the first ratio by $2AR\times AD$, $AD\times FR^2:(RD^2-FR^2)\,AR = R\rho:D\delta$.

---

and so $$\frac{R\phi}{RD} = \frac{RF^2\times DA^2}{RF^2\times DA^2-(DA^2-FA^2)\times RA^2} = \frac{RF^2\times DA^2}{FA^2\times RD^2};$$

that is, on letting fall the perpendiculars AG, AH to FR, DR respectively,

$$\frac{R\phi}{RF} = \frac{DA^2/DR}{FA^2/FR} = \frac{DH}{FG}.$$

(18) Newton refers in parenthesis at this point to a preceding 'sect[io] 120' which we need not reproduce. It will be obvious that FR and DR are proportional to AR/DR and AR/FR, the sines of the refracted and incident angles ($A\hat{D}R$ and $A\hat{F}R$) respectively.

(19) Newton justifies this 'per sect[ionem] 119', but at once

$$F\rho^2-A\rho^2 = FA^2 = FR^2-AR^2.$$

Tertiò dico, quod est $AD^q \times FR^q . AR^q \times RD^q - AR^q \times FR^q :: R\phi . D\phi$. Nam erectâ RK ad AR normali, quæ secet radium $\rho\mu$ in K: est $A\rho . A\delta :: R\rho . RK$, sive

$$AR . AD :: R\rho . RK,$$

siquidem $A\rho$ et AR, nec non $A\delta$ et AD pro infinitè parùm differentibus habentur. Et priori ratione per $AD \times FR^q$ multiplicatâ divisâq; per AR orietur

$$AD \times FR^q . \frac{AD^q \times FR^q}{AR} :: R\rho . \rho K.$$

Quamobrem cùm supra inventum est

$$AD \times FR^q . AR \times RD^q - AR \times FR^q :: R\rho . D\delta,$$

si utriusq; permutatio subintelligatur, patebit esse

$$\frac{AD^q \times FR^q}{AR} . \rho K :: AR \times RD^q - AR \times FR^q . D\delta.$$

et multiplicando per AR permutandoq; fit

$$AD^q \times FR^q . AR^q \times RD^q - AR^q \times FR^q :: \rho K . D\delta.$$

Est autem $\rho K . D\delta :: R\phi . D\phi$. Quare et

$$AD^q \times FR^q . AR^q \times RD^q - AR^q \times FR^q :: R\phi . D\phi.$$

Dico deniq; Quòd est $FG . DH :: RF . R\phi$. Nam cùm sit

$$AD^q \times FR^q . AR^q \times RD^q - AR^q \times FR^q :: R\phi . D\phi,$$

erit divisim $AD^q \times FR^q . AD^q \times FR^q - AR^q \times RD^q + AR^q \times FR^q :: R\phi . RD$. At est $AD^q \times FR^q + AR^q \times FR^q = DR^q \times FR^q$, et $DR^q \times FR^q - AR^q \times RD^q = DR^q \times AF^q$. Quare $AD^q \times FR^q . DR^q \times AF^q :: R\phi . RD$. Ductisq; extremis et medijs in se invicem fit $AD^q \times FR^q \times RD = DR^q \times AF^q \times R\phi$. et applicando ad $FR \times DR^q$ oritur $\frac{AD^q \times FR}{DR} = \frac{AF^q \times R\phi}{FR}$. Quo in proportionalitatem resoluto prodit $\frac{AF^q}{FR} . \frac{AD^q}{DR} :: FR . R\phi$. Sed (per 8. 6. El) est $FR . AF :: AF . FG$; ut et

$$DR . AD :: AD . DH$$

et proinde $\frac{AF^q}{FR} = FG$ et $\frac{AD^q}{DR} = DH$, atq; adeò $FG . DH :: FR . R\phi$. Q.E.D.(20)

--- --- ---

(20) Newton began a following paragraph (here omitted) 'Sed videor actum agere, et his... paucis circa radios homogeneos...obiter notatis, ut eorum penitior cognitio habeatur, Lectiones, quas Vir Reverendus Dr Barrow de ijs fusè composuit, cōsulendas esse moneo.' His praise of Barrow was no mere formality for in revision in his 'Optica' (ULC. Dd. 9.67: $_1$41 [= *Optical Lectures* (note (6)): Book 1, Section 3, Proposition 8: 103–4]) he totally suppressed

In the third place I assert that $AD^2 \times FR^2 : AR^2(RD^2 - FR^2) = R\phi : D\phi$. For, when RK is raised normal to AR to intersect the ray $\rho\mu$ in K, there is

$$A\rho : A\delta = R\rho : RK,$$

that is $AR:AD = R\rho : RK$ seeing that $A\rho$ and AR and likewise $A\delta$ and AD are considered as differing by an infinitely small amount. When the first ratio is multiplied by $AD \times FR^2$ and divided by AR there will arise

$$AD \times FR^2 : AD^2 \times FR^2/AR = R\rho : \rho K.$$

As a consequence, since it was found above that

$$AD \times FR^2 : AR(RD^2 - FR^2) = R\rho : D\delta,$$

it will be evident (understanding a permutation in both) that

$$AD^2 \times FR^2/AR : \rho K = AR(RD^2 - FR^2) : D\delta$$

and, on multiplying by AR and permuting, there comes to be

$$AD^2 \times FR^2 : AR^2(RD^2 - FR^2) = \rho K : D\delta.$$

However, $\rho K : D\delta = R\phi : D\phi$. Consequently also

$$AD^2 \times FR^2 : AR^2(RD^2 - FR^2) = R\phi : D\phi.$$

I assert finally that $FG:DH = RF:R\phi$. For, since

$$AD^2 \times FR^2 : AR^2(RD^2 - FR^2) = R\phi : D\phi,$$

by division there will be

$$AD^2 \times FR^2 : (AD^2 \times FR^2 - AR^2 \times RD^2 + AR^2 \times FR^2) = R\phi : RD.$$

But $AD^2 \times FR^2 + AR^2 \times FR^2 = DR^2 \times FR^2$ and

$$DR^2 \times FR^2 - AR^2 \times RD^2 = DR^2 \times AF^2.$$

Hence $AD^2 \times FR^2 : DR^2 \times AF^2 = R\phi : RD$. On multiplying extremes and middles mutually into each other there comes $AD^2 \times FR^2 \times RD = DR^2 \times AF^2 \times R\phi$ and, on dividing through by $FR \times DR^2$, there arises

$$AD^2 \times FR/DR = AF^2 \times R\phi/FR.$$

When this is resolved as a proportion there results

$$AF^2/FR : AD^2/DR = FR : R\phi.$$

But (by *Elements* VI, 8) $FR:AF = AF:FG$ and again $DR:AD = AD:DH$; consequently $AF^2/FR = FG$ and $AD^2/DR = DH$, so that $FG:DH = FR:R\phi$. As was to be proved.(20)

--- --- --- --- ---

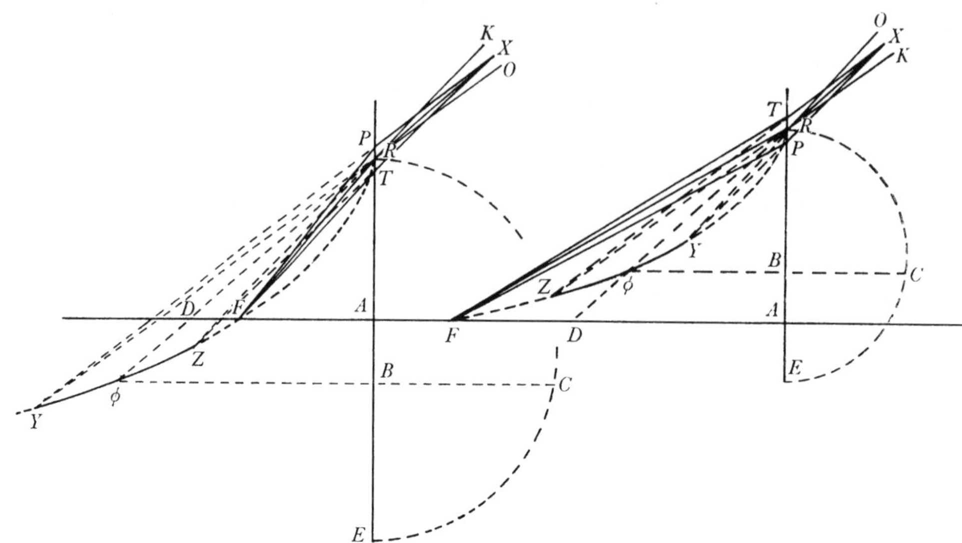

[21]Adhæc si describatur linea quædam curva $Y\phi Z$ in quâ foci[22] radiorum omnigenorum jacent secundū lineam *FR* incidentiū et ita refractorum in puncto *R* ut per totum angulum *KRO* divaricent: ista curva $Y\phi Z$ non malè assimilabitur objecto lucido cujus angulus visibilis ad oculū in *X* situm, sit *YXZ*, et distantia ab eodem oculo ad meditulliū ejus æstimata, ϕX. Sed notum est[23] quòd visibiliū apparentes magnitudines penè sunt reciprocè ut eorum distantiæ: Atqꝫ adeò longitudines ϕX penè sunt reciprocè ut anguli *YXZ*. ...[24]

Cæterùm curva prædicta in quâ radiorum omnis generis in puncto *R* refractorum radiationum centra locantur[25] est[26] Cissois vulgaris sive Dioclea,

his present limit-increment proof, eminently direct, however long-winded, in favour of the latter's 'Principale construendum Problema' (Barrow, *Lectiones XVIII* (note (6)): Lectio v, §XII: 42) with which his hearer—or intended reader?—was assumed to be familiar. Barrow's approach is by means of an ingenious application of his construction of the refraction point

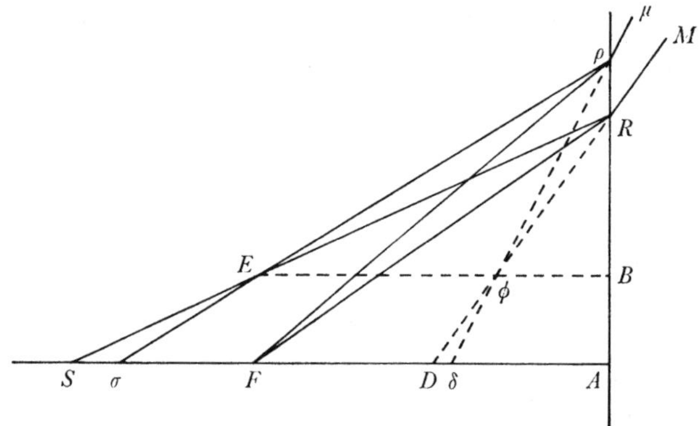

(corresponding to index I/R as before) at the interface *AR* when the given points *F*, ϕ are on the same side of it (compare note (6)): in this case two real refraction points, say *R* and ρ, will

[21]If moreover there be described a curve $Y\phi Z$ on which lie the foci[22] of rays of all kinds incident along the line FR and so refracted at the point R as to spread through the whole of the angle $K\widehat{R}O$, that curve $Y\phi Z$ will not awkwardly be likened to a lucid object whose visible angle to an eye situated at X would be $Y\widehat{X}Z$ and its estimated distance from the same eye to its central part ϕX. But it is known[23] that the apparent sizes of visible things are roughly inversely proportional to their distances; and in consequence the lengths ϕX are, roughly, inversely as the angles $Y\widehat{X}Z$....[24]

For the rest, the aforesaid curve on which are located[25] the radiation centres of rays of every kind refracted at R is [26] a common or Dioclean cissoid adapted

---

satisfy the problem. In specific terms, he constructs point E in the perpendicular ϕB to AR such that $EB:\phi B = I:\sqrt{[I^2-R^2]}$ and the points R, ρ in AR are located by drawing

$$RS = \rho\sigma = (R/\sqrt{[I^2-R^2]})\times FA$$

through E so that S, σ lie in ADF. Clearly, when the two points R and ρ coincide, then ϕ is the centre of radiation corresponding to the radiating point F: noting that in this case the length RS achieves a minimal value, Barrow deduces the defining condition

$$BR^3 = AB\times BE^2$$

(Lectio v, §§4/10). (Analytically, on taking $AB = \alpha$, $BE = \beta$, $BR = x$ and so

$$RS = (\alpha/x+1)\sqrt{[x^2+\beta^2]},$$

the condition that the derivative $d(RS)/dx$ be zero yields $x^3 = \alpha\beta^2$, so that $RS = (\alpha^{\frac{2}{3}}+\beta^{\frac{2}{3}})^{\frac{3}{2}}$.) In further argument, by the sine law of refraction $BR/\phi R:AR/FR = FR:DR = I:R$, so that

$$\frac{FR}{\phi R} = \frac{I}{R}\times\frac{AR}{BR} = \frac{I}{R}\left(1+\frac{BR^2}{EB^2}\right) = \frac{I}{R}\left(1+\frac{(I^2-R^2)\times BR^2}{I^2\times\phi B^2}\right)$$
$$= \frac{I^2\times\phi R^2-R^2\times BR^2}{I.R\times\phi B^2} = \frac{I^2\times DR^2\ (\text{or } R^2\times FR^2)-R^2\times AR^2}{I.R\times DA^2}$$
$$= \frac{R}{I}\times\frac{FR^2-AR^2}{DA^2} = \frac{FA^2/FR}{DA^2/DR},$$

which is Newton's present result.

(21) We proceed directly to Newton's construction of the locus of the centres of radiation (ϕ) of radiating point F as the refractive index varies. The omitted portion (Add. 4002: 96–100) is not relevant to the present extract. A little changed revision of the following paragraphs was included by Newton in his 'Optica' (ULC. Dd. 9.67: 46–7 [= *Optical Lectures* (note (6)): 119–21]).

(22) Compare note (12). Here too we should probably read 'radiationum centra' (centres of radiation).

(23) Notably by astronomers in assessing the distances of the sun and moon from the apparent diameters of their discs. Newton may well have in mind Gabriel Mouton's *Observationes Diametrorum Solis et Lunæ Apparentium* (Lyons, 1670) which had recently appeared. For small angles θ of observation, of course, $\tan\theta \approx \theta$.

(24) We omit a following exemplification which begins 'Instantiæ gratiâ, cùm punctū X in ipsâ refringenti superficie ad R existit apparens longitudo ipsius YZ erit angulus YRZ sive KRO...'. The points P and T are limits of refraction from F into X.

(25) 'cadant' (fall) is cancelled.

(26) A cancelled continuation reads 'si species forte desideretur' (if perchance its species be desired).

circulo accommodata cujus diameter RE est ad AR ut FR^q ad AF^q. Nam super diametro RE descripto circulo isto RCE, agatur quævis recta ϕBC normalis ad RE, circuloq; in C et curvâ in ϕ terminata. Et propter analoga latera similium triangulorum RAD, $RB\phi$, erit $AD^q . AR\times DR :: B\phi^q . BR\times\phi R$. Et applicando posteriorem rationem ad BR fiet $AD^q . AR\times DR :: \frac{B\phi^q}{BR} . \phi R$. Rursusq; ducendo consequentes rationum in $R\phi$ et applicando ad AR orietur

$$AD^q . DR\times R\phi :: \frac{B\phi^q}{BR} . \frac{R\phi^q}{AR}.$$

Est autem $\frac{AF^q}{FR} . \frac{AD^q}{DR} :: RF . R\phi$ ut priùs, et consequentibus in DR atq; antecedentibus in FR ductis, oritur $AF^q . AD^q :: FR^q . DR\times R\phi$ et vicissim

$$AF^q . FR^q :: AD^q . DR\times R\phi.$$

Quamobrem rationes eidem tertiæ[(27)] congruentes connectendo, habebitur $\frac{B\phi^q}{BR} . \frac{R\phi^q}{AR} :: AF^q . FR^q$, ducendoq; antecedentes rationum in BR et consequentes in AR prodibit $B\phi^q . R\phi^q :: AF^q\times BR . FR^q\times AR$ et insuper applicando posteriorem rationem ad AF^q fiet $B\phi^q . R\phi^q :: BR . \frac{FR^q\times AR}{AF^q}$. Sed cum posuerim $RE . AR :: FR^q . AF^q$, erit $\frac{FR^q\times AR}{AF^q} = RE$, et proinde $B\phi^q . R\phi^q :: BR . RE$, ac divisim $B\phi^q . R\phi^q - B\phi^q (BR^q) :: BR . BE$. Atqui ex naturâ circuli est BC media proportionalis inter BR et BE adeoq; est $BR . BE :: BR^q . BC^q$ et proinde

$$B\phi^q . BR^q :: BR^q . BC^q \quad \text{sive} \quad B\phi . BR . BC \div\!\!\div$$

Quod indicat curvam esse Cissoidem sicut ostendendū proposui.[(28)]

---

[3][(29)] [Extremes of dispersion in refractive media of varying density.]

Prop 6.[(30)] Heterogeneis radijs e Medio rariori in densius secundum eandem lineam in superficiem positione datam incidentibus, quo densius sit Mediū in

(27) Namely, '$AD^q . DR\times R\phi$'.

(28) Newton might well have established this result by a direct limit-increment argument (compare J. A. Lohne, 'The Increasing Corruption of Newton's Diagrams', *History of Science*, **6**, 1968: 69–89, especially 69–71). Thus, if in the preceding figure (on page 462) we suppose $A\hat{F}R = i$ and $A\hat{D}R = r$, where $\sin i/\sin r = I/R$, with their increments $R\hat{F}\rho = di$,

$$R\hat{\phi}\rho = dr \; (= [\tan r/\tan i] . di),$$

then $R\rho = FR.\sec i . di = \phi R.\sec r . dr$ and hence

$$\frac{\phi R}{FR} = \frac{\cos r}{\cos i}\times\frac{di}{dr} = \frac{\sin i/\cos^2 i}{\sin r/\cos^2 r}.$$

On taking $RE = a = FR.\sin i/\cos^2 i = FR^2\times AR/AF^2$, $\phi R = \rho$, $A\hat{R}\phi = \theta \; (= \frac{1}{2}\pi - r)$,

$$RB = x (= \rho\cos\theta) \quad \text{and} \quad B\phi = y (= \rho\sin\theta)$$

to a circle whose diameter RE is to AR as FR^2 to AF^2. For, having described that circle RCE on the diameter RE, draw any straight line ϕBC normal to RE and bounded by the circle at C and the curve at ϕ. Then, because they are corresponding sides of the similar triangles RAD, $RB\phi$, there will be

$$AD^2:AR\times DR = B\phi^2:BR\times\phi R.$$

On dividing the latter ratio by BR there will come $AD^2:AR\times DR = B\phi^2/BR:\phi R$, and again, on multiplying the denominators of the ratios by $R\phi/AR$, there will arise $AD^2:DR\times R\phi = B\phi^2/BR:R\phi^2/AR$. But, as previously,

$$AF^2/FR:AD^2/DR = RF:R\phi$$

and, when each ratio is multiplied by FR in its numerator and by DR in its denominator, there arises $AF^2:AD^2 = FR^2:DR\times R\phi$ and, on alternating,

$$AF^2:FR^2 = AD^2:DR\times R\phi.$$

Consequently, by joining ratios equal to the same third ratio,(27) there will be had $B\phi^2/BR:R\phi^2/AR = AF^2:FR^2$, and by multiplying the numerators of these ratios by BR and their denominators by AR there will result

$$B\phi^2:R\phi^2 = AF^2\times BR:FR^2\times AR,$$

while on dividing the latter ratio by AF^2 there will come in addition $B\phi^2:R\phi^2 = BR:FR^2\times AR/AF^2$. But since I set $RE:AR = FR^2:AF^2$, there will be $FR^2\times AR/AF^2 = RE$ and consequently $B\phi^2:R\phi^2 = BR:RE$, and then by division $B\phi^2:(R\phi^2-B\phi^2$ or$)\ BR^2 = BR:BE$. However, from the nature of a circle BC is the mean proportional between BR and BE, so that

$$BR:BE = BR^2:BC^2$$

and consequently $B\phi^2:BR^2 = BR^2:BC^2$: in other words $B\phi$, BR and BC are in continued proportion. This indicates that the curve is a cissoid, as I proposed to show.(28)

---

[3](29) [Extremes of dispersion in refractive media of varying density.]

Proposition 6.(30) When heterogeneous rays are incident on a surface given in position from a rarer medium into a denser one along the same line, the denser

---

the polar and Cartesian equations of the cissoidal locus ($\rho = a\sin^2\theta/\cos\theta$ and $y^2 = x^3/(a-x)$ respectively) are immediate. Evidently, when there is no refraction (that is, $I = R$ and so $i = r$) the point ϕ coincides with F, which is therefore in the cissoid $Y\phi Z$.

(29) Add. 4002: 115–20, extracted from a passage dealing with the dispersion ('differentia refractorum') of 'radii heterogenei sive dissimiliter refrangibiles' incident at the same angle at a plane interface between two media of varying density. These pages were reproduced with little change in Newton's 'Optica' as Proposition 17 of Book 1, Section 3 (ULC. Dd. 9.67: $_1$54–8 [= *Optical Lectures* (note (6)): 143–55]).

(30) Namely, of a final subsection 'De difformium radiorum affectionibus'.

quod radij incidunt eo major erit differentia refractionū ad certum usqꝫ terminum, et post eo minor perpetuò. Nam si Medium posterius densitate suâ valdè parùm superet anterius, ita ut refractiones indefinitè parvas efficiat, differentia refractionum erit etiam indefinitè parva, et proinde minor quàm foret si Medium posterius supponeretur densius ut refractiones evaderent majores. Quare aucta Medij posterioris densitate augebitur dicta refractionum differentia. Quod si densitas ejus in infinitum augeatur refractiones etiam quantum poterunt augebuntur, hoc est usqꝫ dum omnes refracti radij perpendiculariter emergant, angulis refractionum et eorum differentijs tunc prorsus evanescentibus. Quare differentia refractionum rursus diminuta est donec in nihilum evanuit.(31)

Schol. Etsi limitis ejus determinatio ubi(32) differentia refractionis evadit maxima, plus tædij et laboris administr[ar]e possit quam utilitatis, cùm tamen alicujus fortè momenti censeatur densitatem Medij cognoscere quod radijs in se refractis colores maximè conspicuos efficiat, non pigebit hunc insuper designare. Idqꝫ primò cùm incidentia fit obliquissimè.(33)

Cas: 1. Esto IX communis radiorum in superficiem AX quæcunqꝫ Media dirimentem obliquissimè incidentium via. Et eorum refracti ut ante(34) sunto $X\pi$ et $X\tau$.(35) Et agatur recta quævis $\pi\tau$ præfatæ superficiei parallela, quæ radijs istis occurrat in π ac τ; A quibus ad AX demissis perpendicularibus πC, τE, bisecetur CE in D et centro D distantia DX circulus describatur secans $C\pi$ in P et $E\tau$ in T, junganturqꝫ XP et XT. Dico quod cùm ea sit posterioris Medij densitas, ut radiorum secundum IX incidentium maximè refrangibiles(36) ad P et minimè refrangibiles(36) ad T refringat, tunc angulus PXT quam maximus evadet. Etenim utcunqꝫ Medium posterius ponatur densum, refracti radij ita

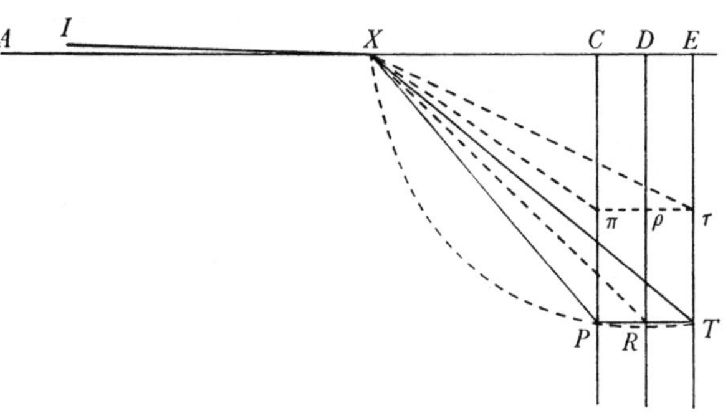

(31) It is assumed that the 'differentia refractionis' (dispersion) is a simply continuous function of the density of the refractive medium. As the latter density increases from zero to infinity, the dispersion increases correspondingly at first from zero but subsequently decreases back to zero again: an exemplary case of Newton's criterion in the opening paragraph of Problem 3 of the 1671 tract (**1**, 2, §2: Newton's page 41) for a function to attain a maximum value somewhere in an interval. The purpose of the following scholium is, with the aid of an appropriate geometrical model, to determine the maximum dispersion by means of both elementary and Fermatian techniques.

(32) Newton first continued equivalently 'dicta refractionis differentia fit maxima, metuo nè plus tædij...administret...' (the said refraction difference comes to be greatest will, I fear, result in more tedium...).

the medium into which the rays are incident the greater will be the difference between their refractions up to a certain limit and after that the less will it be perpetually. For if the latter medium exceeds the former very little in its density so that it effects indefinitely small refractions, the difference of the refractions will also be indefinitely small and consequently less than it would be if the latter medium were supposed denser so that the refractions would prove to be greater. Hence when the density of the latter medium is increased the stated difference of the refractions will increase. But if its density be increased infinitely the refractions too will increase as much as they can, that is, until all refracted rays emerge at right angles, when the refraction angles and their differences completely vanish. Hence the difference of the refractions has again diminished till it has dwindled to nothing.(31)

Scholium. Although the determination of the limit at which (32)the refraction difference turns out greatest may achieve more in the way of boredom and hard work than usefulness, it may perhaps be considered of some importance to find out the density of the medium which effects the most conspicuous colours when rays are refracted in it, and for that reason it will not be tiresome to specify this as well. First, then, when the incidence is at its most oblique.(33)

Case 1. Let IX be the common path of rays incident at their most oblique on the surface AX separating any media. As before(34) let their refracted rays be $X\pi$ and $X\tau$,(35) and draw any straight line $\pi\tau$ parallel to the above-mentioned surface, meeting those rays in π and τ. Letting fall the perpendiculars πC, τE from these points to AX, bisect CE in D and with centre D, radius DX draw a circle intersecting $C\pi$ in P and $E\tau$ in T, then join XP and XT. I assert that when the density of the latter medium is such as to refract the most refrangible(36) of the rays incident along IX to P and the least refrangible(36) of these to T, then the angle $P\widehat{X}T$ will prove to be at its largest. For, however dense the latter medium be supposed, the refracted rays will intersect the lines CP and ET in the

(33) When, that is, the incident ray IX just skims the surface of the interface AX.

(34) In the preceding proposition, lightly revised as Proposition 16 of Book 1, Section 3, of his 'Optica' (ULC. Dd. 9.67: $_1$52–4 [= *Optical Lectures* (note (6)): 137–42]), where Newton considers the complementary case in which the incident medium has varying density: 'non enim perinde est sive raritas vel densitas anterioris Medij, sive posterioris varietur, ut e præostensis pateat' (Add. 4002: 121). If XT, $X\tau$ in the present figure are taken to meet CP in M, μ respectively, Newton's modified condition in this case relating the extreme refracted rays is that $XM/XP = X\mu/X\pi$.

(35) $X\pi$ and $X\tau$ are the most and least refracted rays respectively. Newton first wrote 'sunto $X\pi$ purpuriformis et $X\tau$ rubiformis' but in a fit of positivism later cancelled the descriptive adjectives 'violet' and 'reddish'.

(36) Newton first wrote 'purpuriformes' and 'rubiformes' respectively. Elsewhere (compare previous note) for '[radij] extremam et viriditatem pingentes' (Add. 4002: 120) he has written 'extremi seu maxime difformes', while for '[radius] viriformis' is substituted 'mediocris refrangibilis' (mean refrangible ray).

lineas CP et $[E]T$ in punctis π ac τ secabunt ut recta $\pi\tau$ ipsi AX parallela sit.(37) Quare si ducatur linea $D\rho$, quæ lineas omnes $\pi\tau$ bisecet, centrum cujuscunꝗ circuli per puncta π ac τ transeuntis semper jacebit in eâdem $D\rho$. At ang: $\pi X\tau$ est ang: in segmento circuli per puncta π, τ, et X transeuntis; qui ideo erit maximus cùm ejusmodi circulus existit minimus, propterea quòd ratio subtensæ $\pi\tau$ ad circuli dimensiones tunc evadit maxima. Verùm iste circulus fit omnium minimus cùm centrum ejus cadit in D, siquidem pro semidiametro tunc habet XD minimam rectarum quæ ab X ad RD duci possunt. Est ergo ang $\pi X\tau$ tunc maximus cùm centrum circuli transeuntis per puncta π, τ, et X cadit in D. Adeoꝗ cùm circulus XPT et angulus PXT ejusmodi sint, liquet propositum.

Hinc obiter pateat hunc angulum PXT tunc etiam maximum evadere cùm talis est posterioris Medij densitas ut angulus refractionis mediocrium refrangibilium radiorum XR obliquissimè secundum IX incidentium sit semirectus;(38) et eo minorem perpetim fieri quò iste refractionis angulus a semirecto (excessu vel defectu) magis deviat....(39)

Cas: 2. Quòd si linea secundum quam incidunt radij non sit maximè obliqua, Problema emerget solidum;(40) sed lubet modum ostendere quo

(37) This is Newton's present dispersion hypothesis when the density of the refractive medium varies. In analytical terms, if the incident ray IX enters 'at its most oblique', being refracted at least, mean and most refracted angles r_B, r_G and r_R respectively, then the proportion $\cot r_B : \cot r_G : \cot r_R = XC:XD:XE$ is independent of the density of the refractive medium. Since Newton at this time was effectively restricted on an experimental basis to 'white' light incident in air and refracted into water, glass or 'christall', we may presume that the model was propounded on theoretical grounds. In particular, it is tempting to suppose that Newton here (in generalization of Descartes' analogy in his *Dioptrique* [= *Discours de la Methode*, Leyden, 1637: first *Appendice*]: 13–23: 'Discours Second de la Refraction') uses a modified Cartesian tennis racquet to hit a 'ball' of light down from the surface AX at X (which it grazes at a horizontal speed in proportion to its colour), so imparting a uniform velocity normal to it varying as the density. Some twenty years later—having meanwhile attempted experimental investigation of the problem—Newton abandoned this model in his *Fundamentum Opticæ* for the argument that 'Excessus sinuum refractionis supra communem sinum incidentiæ...sunt in eadem ratione' (ULC. Add. 3970.3: 411r [= *Opticks: Or, A Treatise of the Reflexions, Refractions, Inflexions and Colours of Light* (London, 1704): 94: 'The Excesses of the sines of refraction of several sorts of rays above their common sine of incidence when the refractions are made out of divers denser mediums immediately into one and the same rarer medium, are to one another in a given proportion']). In Newton's revised model, effectively, the proportion

$$(\operatorname{cosec} r_B - 1):(\operatorname{cosec} r_G - 1):(\operatorname{cosec} r_R - 1) = XD^2\ XC^2:1:XD^2\ XE^2$$

is independent of variations in the density of the refractive medium. Compare S. I. Vavilov, *Isaak N'juton: Lekcii po Optike* (Moscow–Leningrad, 1946): 281–3.

(38) Since, as Newton has just shown geometrically, the angle $P\hat{X}T$ of dispersion is greatest when $DX = DP = DT$, it follows that $D\hat{X}P > \frac{1}{4}\pi > D\hat{X}T$ and so for *some* 'mean' ray, say XR' it will be true that $D\hat{X}R' = \frac{1}{4}\pi$. However, since $XD = PD > RD$ and so $D\hat{X}R < \frac{1}{4}\pi$, R' cannot strictly coincide with R but will be located in PR such that $R'R \approx \frac{1}{2}CD^2/XC$. Since in practice the angle of dispersion is small, CD will be small in comparison to XC and so the distance $R'R$ is virtually negligible.

points π and τ such that the straight line $\pi\tau$ is parallel to AX.(37) Therefore, if the line $D\rho$ is drawn to bisect all lines $\pi\tau$, the centre of any circle whatever passing through the points π and τ will always lie on that same line $D\rho$. But $\pi\widehat{X}\tau$ is the angle in a segment of the circle passing through the points π, τ and X; and it will consequently be greatest when a circle of this kind turns out to be smallest, because the ratio of the subtense $\pi\tau$ to the circle's dimensions then turns out to be greatest. But that circle comes to be smallest of all when its centre falls at D, seeing that it then has for radius XD, the least of the straight lines which can be drawn from X to RD. Therefore the angle $\pi\widehat{X}\tau$ is then greatest when the centre of the circle passing through the points π, τ and X falls at D. Accordingly, since the circle XPT and the angle $P\widehat{X}T$ are so defined, the proposition is manifest.

It should hence be evident, by the way, that this angle $P\widehat{X}T$ also comes out to be greatest at the moment when the density of the latter medium is such that the refraction angle of the mean refrangible rays XR of rays incident at their most oblique along IX is half a right angle;(38) and that it becomes perpetually less the more that refraction angle deviates (by excess or defect) from half a right angle. ...(39)

Case 2. But if the line along which the rays are incident is not at its most oblique, the problem will emerge to be solid;(40) but I should like to show a way

(39) In this omitted portion Newton observed: 'Quemadmodum si refractiones ex aere in aquam, in vitrum et in crystallum peractæ conferantur, e calculo patebit quòd cùm angulus incidentiæ sit 90gr proximé, tunc angulus refractionis in aquam erit major semirecto, inq; vitrum erit minor. Quamobrem aqua minùs densa est et vitrum magis densū quàm ut efficiant angulum PXT maximum. Et proinde cùm crystallū sit adhuc densius efficiet istum PXT minorem quàm vitrum efficeret. Et sic vitrum etsi minùs refringat, in isthoc tamen casu heterogeneos radios in se refractos magis ab invicem dissipabit quàm cristallum, eoq; pacto colores in oppositam ejus superficiem projiciet magis distinctos'. In making experimental test of these theoretical deductions, however, he was honest enough to admit that 'hæc sunt expertu difficillima, quòd vitrum et crystallum densitate parùm differant, nec possint haberi satis crassa; et si possent, tum propter maximam crassitiem haud forent satis perspicua'. Note Newton's present belief that the refractive ratio increases monotonically with the density of a substance—evidently he was still unaware of the counter-examples to this principle (such 'sulphureous spirits' as oil of turpentine) which were later to cause him so much difficulty. Compare his suppressed 'Conclusion' to his *Opticks* (ULC. Add. 3970.3: 337r) where, having asserted that 'The refracting power of bodies *in vacuo* is proportional to their specific gravities', he was compelled to add: 'Note that sulphureous bodies *cæteris paribus* are most strongly refractive & therefore tis probable yt ye refracting power lies in ye sulphur & is proportional not to ye specific weight or density of ye whole body but to that of ye sulphur alone.'

(40) Constructible, that is, by intersecting 'loci solidi' or conics and so reducible analytically to a cubic or quartic equation. Here, most generally, if we suppose $\pi\hat{X}\tau$ to be the angle of dispersion of an incident ray which skims the interface AX, while $P\hat{X}T$ is that of the ray IX incident at the angle $\alpha = \sin^{-1}s$, then, on making $XE = a$, $XC = ka$ and $C\pi = E\tau = ax$, it follows that $\cos C\hat{X}P = s\cos C\hat{X}\pi = ks/\surd[x^2+k^2]$ and similarly $\cos C\hat{X}T = s/\surd[x^2+1]$, so that the problem reduces to maximizing $\cos^{-1}(ks/\surd[x^2+k^2]) - \cos^{-1}(s/\surd[x^2+1])$. Here, since

conditionibus ejus nonnihil mutatis, ad planum[(41)] reduci poterit. Sciendum est itaq; quod cùm inter extremos seu maximè difformes radios innumeri sint intermedij qui gradibus continuò successivis et infinitè parvis alij magis alijs refringuntur: [ideò] differentia refractionis extremorum radiorum conflata erit ex consimilibus intermediorum differentijs numero et parvitate infinitis. Jam cognitis proprietatibus istarum infinitē parvarum differentiarum possumus exinde de omnibus simul aggregatis, sive de differentijs finitè parvis quales intercedunt extremorum refractionibus, judicium proferre, præsertim cùm istæ differentiæ sint admodum exiguæ. Sic cognito quòd infinit[è] parvæ differentiæ augentur, diminuuntur vel simul maximæ evadunt aut minimæ: concludendum erit quòd omnium summa perinde augetur, diminuitur, vel maxima fit aut minima. Quod si non sint omnes simul maximæ vel minimæ, tamen summa pro maxima aut minima haberi potest cùm id accidit intermediæ parti.[(42)] Sic omnium colorum latitudo tunc maxima censeri potest, cùm id accidit viriditati.[(43)] Jam licèt Problema propositum cùm de differentijs finitè parvis agitur existat solidum, si tamen instituatur de differentijs infinitè parvis, ad planum reduci potest. Verùm huic solvendo nolo obnixè incumbere, sed breviter tantùm ostendam quo pacto calculus in hoc et ejusmodi alijs sit ineundus, ut ad æquationem perveniatur, ex quâ maximus angulorum infinitè parvorum possit elici....

Primò itaq; investiganda est regula vel Æquatio, quâ ex uno utcunq; refracto radio dato, refractus alter cum eo constituens angulum infinitè parvum cognosci poterit. Radijs e Medio densitate dato in Medium cujuslibet densitatis secundum obliquissimam lineam IX ut priùs incidentibus, sint XR et $X\rho$ refracti duo, quorum alter XR sit altero $X\rho$ paulo magis refrangibilis, differentiâ tamen infinitè parvâ. Et agatur lineola quævis $R\rho$ his in R et ρ occurrens, et refringenti[(44)] superficiei parallela. Ad quam superficiem normales etiam RD, $\rho\delta$

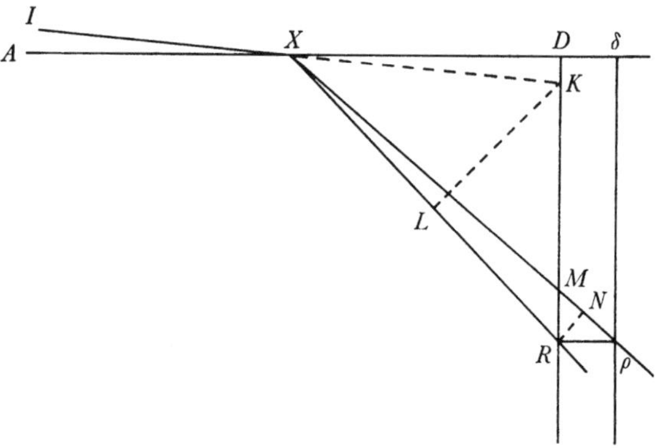

$C\pi = E\tau$ is a measure of the density of the refractive medium, x is variable and equation of the derivative to zero yields $x^6 - k^2(3-2s^2)x^2 - k^2(k^2+1)(1-s^2) = 0$, a cubic in x^2 which is readily constructible by means of intersecting conics (compare II, 3, 2, §2). The analytical equivalent $x^2 = k$ to the construction of Case 1 follows at once by taking $\alpha = \frac{1}{2}\pi\,(s = 1)$. When k is indefinitely close to 1 the factor x^2+1 may be eliminated from the cubic, leaving the quadratic condition $x^4 - x^2 - 2(1-s^2) = 0$ (compare note (46)).

(41) Reducible, that is, to intersecting 'loci plani' (circles and straight lines) and therefore in analytical terms to a quadratic equation.

how, when its conditions are somewhat altered, it may be reduced to a plane one.(41) It must accordingly be understood that, since between the extreme or most difform rays there are innumerable intermediate ones which are, some more than others, refracted in a continued succession of infinitely small gradations, the refraction difference of the extreme rays will in consequence be made up of the closely similar differences of the intermediate ones, infinite both in number and their smallness. Now, when the properties of those infinitely small differences are known, we can on that basis extend our judgement to cover them all when gathered together, that is, the finitely small differences such as fall between the refractions of the extreme rays, particularly since those differences are still slight. Thus, in the knowledge that the infinitely small differences increase, diminish or together come to be greatest or least, the conclusion shall be drawn that their total sum likewise increases, diminishes or becomes greatest or least. And even though they should not all be simultaneously at their maximum or minimum, nevertheless their sum can be considered a maximum or minimum when that happens in the case of an intermediate part.(42) Thus the breadth of the colours can be thought of as greatest at the moment when that happens in the case of green.(43) Now although, when finitely small differences are at issue, the problem proposed proves to be solid, if however it be undertaken for infinitely small differences it may be reduced to a plane one. But I am chary of pressing relentlessly on with its solution, and shall merely show briefly how the computation may be started in this instance and others of the kind in order to reach an equation from which may be derived the largest of the infinitely small angles....

In the first place, accordingly, a rule or equation must be discovered from which, given an arbitrary refracted ray, a second refracted ray forming an infinitely small angle with it may be ascertained. Where, as before, rays are incident from a medium given in density into a medium of any density along the most oblique line IX, let XR and $X\rho$ be two refracted rays, one of which XR is slightly more refrangible than the other $X\rho$, the difference being, however, infinitely small. Then draw any small line $R\rho$ (meeting these in R and ρ) parallel to the refracting surface. To this surface let fall also the normals RD, $\rho\delta$:

(42) Newton implicitly invokes the physically dubious postulate of uniform dispersion of the refracted rays: as he would have it, since these rays upon refraction are deviated through angles 'alike' varying with the density of the refractive medium, the maximum angle of dispersion occurs—at least, very nearly so—when the angle between two infinitely close 'mean' refracted rays attains its maximum. Of course, that this should be so is implicitly a definition of what constitutes the ray of 'white' incident light. With the somewhat crude experimental techniques at his disposal it is scarcely likely that Newton could have tested the physical accuracy of this postulate.

(43) We should presumably read 'radio mediocriter refrangibili' (mean refracted ray). Compare notes (35) and (36).

(44) The equivalent 'refractivæ' is cancelled.

demittantur, quas datam finitamq; distantiam ab X, ab invicem verò infinitè parvam habere finges, sed lineolam $R\rho$, cum radijs per R et ρ transeuntibus plus aut minùs ab XD vergere (quemadmodum in præcedentibus) concipies pro variâ posterioris Medij assumendâ densitate. Jam si recta DR secet radios $X\rho$ in M, et IX in K, cùm infinitè parvum triangulum $RM\rho$ sit simile triangulo DMX a quo triangulum KRX non nisi infinitè parvis differentijs RXM et DXK discrepat, quæ dissimilitudinem non inferunt, triangula etiam $RM\rho$ et RDX pro similibus haberi debent. Et proinde demissis perpendicularibus KL et RN, erit $XK.LR::R\rho.MN$. Adeoq; cùm sit

$$LR=\frac{XR^q-XK^q}{XR} \quad \text{(nam est } XR.KR(=\surd:\overline{XR^q-XK^q}:)::KR.LR.)$$

erit etiam $MN=\frac{XR^q-XK^q}{XR\times XK}$ in $R\rho$. Quæ differentia est inter XN sive XR et XM, et inde erit $XM=XR-:\frac{XR^q-XK^q}{XR\times XK}$ in $R\rho$. Inventa est itaq; relatio inter XK, XM, et XR cùm angulus IXA sit infinitè parvus: Quinetiam utcunq; IX obliqua ponatur, illæ XK, XM, et XR eandem relationem observabunt, siquidem reciprocè sint ut sinus incidentiæ et refractionis; et proinde inventa est etiam inter eas relatio pro quâvis obliquitate incidentis IX. Atq; ita cognitis vel utcunq; ad arbitrium assumptis XK et XR, inde XM simul cognoscitur. Quod primò determinandum proposui.

Quamobrem sit IX linea datum quemvis angulum AXI cum refringente superficie constituens; cæterisq; stantibus, erit $MN=\frac{XR^q-XK^q}{XR\times XK}$ in $R\rho$. Insuper est $RD(=\sqrt{XR^q-XD^q}).XD::MN.NR$. Atq; adeò est

$$NR=\frac{XR^q-XK^q \text{ in } R\rho\times XD}{XR\times XK\times\sqrt{XR^q-XD^q}}.$$

Quòd si NR dividatur per XR prodibit sinus anguli RXN respectu circuli cujus semidiameter sit unitas. Quare cùm angulus iste et sinus ejus sunt simul maximi,[45] ad maximum angulum determinandum quærenda erit maxima quantitas $\frac{NR}{XR}$, hoc est maximum $\frac{XR^q-XK^q \text{ in } R\rho\times XD}{XR^q\times XK\times\sqrt{XR^q-XD^q}}$. Sive (factâ per datum $\frac{R\rho\times XD}{XK}$ divisione) quærendum erit maximum $\frac{XR^q-XK^q}{XR^q\times\sqrt{XR^q-XD^q}}$. Id

(45) Newton has cancelled 'infinite p[arvi]'.

these you should conceive to have a given, finite distance from X but an infinitely small one from each other, while the line $R\rho$, along with the rays passing through R and ρ, you should imagine (as in the preceding) to verge more or less from XD according as you variously assume the density of the latter medium. Now, if the straight line DR cuts the rays $X\rho$ and IX in M and K, since the infinitely small triangle $RM\rho$ is similar to the triangle DMX, from which the triangle KRX varies only by the infinitely small differences RXM and DXM without effect on the similarity, the triangles $RM\rho$ and RDX ought also to be considered as similar. Consequently, on letting fall the perpendiculars KL and RN there will be $XK:LR = R\rho:MN$, so that, since

$$LR = (XR^2-XK^2)/XR$$

(for $XR:KR$ (that is, $\surd[XR^2-XK^2]$) $= KR:LR$), also $MN = \dfrac{XR^2-XK^2}{XR\times XK}R\rho$. This is the difference between XN (or XR) and XM, and hence

$$XM = XR-\frac{XR^2-XK^2}{XR\times XK}R\rho.$$

A relationship between XK, XM and XR when the angle $I\widehat{X}A$ is infinitely small has accordingly been found. But indeed, however obliquely IX be taken, those lines XK, XM and XR will keep the same relationship seeing that they are inversely as the sines of incidence and refraction, and consequently a relationship has also been found between them holding for any inclination of the incident ray IX. And so, when XK and XR are known or assumed arbitrarily at random, from them XM will at once be known. This I proposed to establish first.

Therefore let IX be a line forming any given angle $A\widehat{X}I$ with the refracting surface, other things remaining as before. Then $MN = \dfrac{XR^2-XK^2}{XR\times XK}R\rho$. Moreover RD (that is, $\surd[XR^2-XD^2]$)$:XD = MN:NR$, and so

$$NR = \frac{(XR^2-XK^2)\,R\rho\times XD}{XR\times XK\times\surd[XR^2-XD^2]}.$$

But if NR be divided by XR the quotient will be the sine of the angle $R\widehat{X}N$ with respect to a circle whose radius is unity. Hence, since that angle and its sine are at their greatest[(45)] simultaneously, to determine the maximum angle it will be necessary to seek the greatest quantity NR/XR, that is, the maximum of

$$\frac{(XR^2-XK^2)\,R\rho\times XD}{XR^2\times XK\times\surd[XR^2-XD^2]}:$$

in other words (on dividing out the given quantity $R\rho\times XD/XK$) it will be necessary to seek the maximum of $\dfrac{XR^2-XK^2}{XR^2\times\surd[XR^2-XD^2]}$. This can be done by

quod per Methodos de maximis et minimis satis notas fieri potest,[46] et prodibit

$$XR^{qq} = 3XK^q \times XR^q - 2XK^q \times XD^q.$$

Cujus æquationis constructio est ejusmodi.

A puncto quolibet incidentis radij IX demitte perpendiculum IA, et in eo sume $AF = AX$. Et XI producto ad B, ut sit $IB = \frac{1}{2}IX$, super BX describe semicirculum BEX cui inscribe $XE = XF$. Dein XB producto ad C ut sit $BC = BE$, super CX describe semicirculum CGX quem in G secet perpendiculum IG super diametro ejus ad I erectum. Deniq; centro X et intervallo GX describatur arcus GH secans AI productum in H, ducatur HX, et producatur versus R, eritq; XR ipsius IX refractus cùm tanta sit posterioris Medij densitas ut differentia refractionis RXM fiat omnium maxima.[47] Quo invento, densitas posterioris Medij talem refractionem efficientis facilè dabitur. Concip[i]es ergo radios XR et $X\rho$ esse mediocriter refrangibiles, diverso tamen gradu, et posterius Medium sic inventum, non modò inter istos sed et inter extremos seu maxime difformes radios maximam circiter[48] quam potest refractionis differentiam efficiet.

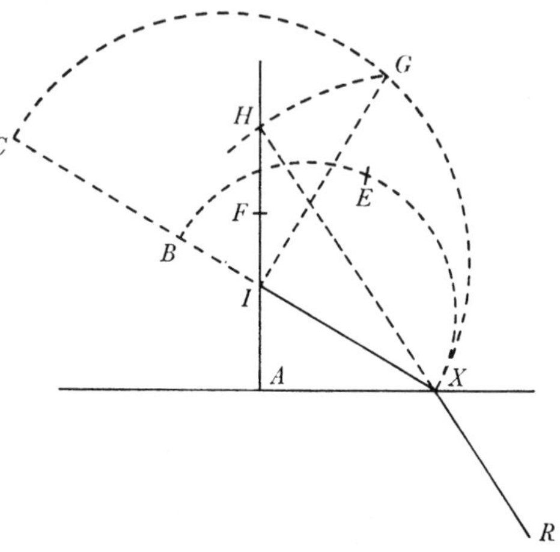

§2. REFRACTION AT A CURVED INTERFACE.[1]

De Refractionibus Curvarum Superficierum.[2]

Hæc de refractionibus planorum. De curvis et præsertim sphæricis superficiebus jam agendum est, quarum doctrinam respectu homogeneorum radiorum sequentibus propositionibus complecti conabimur.

(46) The 'ordinary' Huddenian technique soon to be expounded in Problem 3 of Newton's 1671 tract. (See **1**, 2, §2 above, and compare I: 214.) Here, since XR alone is variable, the condition for $(XR^2 - XK^2)/XR^2\sqrt{[XR^2 - XD^2]}$ to attain a local extreme value is that its derivative $XR^{-3}(XR^2 - XD^2)^{-\frac{3}{2}}(2XK^2[XR^2 - XD^2] - XR^2[XR^2 - XK^2])$ be zero. Evidently $XR = XD$ and $XR = \infty$ are minimum solutions between which the maximum defined by $2XK^2[XR^2 - XD^2] - XR^2[XR^2 - XK^2] = 0$ must exist. In the terminology of note (40), on taking $XD = a$, $XR/XK = \sqrt{[x^2+1]}$ and $XD/XK = s = \sin I\hat{X}A$ Newton's equation becomes

$$(x^2+1)^2 = 3(x^2+1) - 2s^2.$$

methods of maxima and minima which are well enough known,(46) and the result is $XR^4 = 3XK^2 \times XR^2 - 2XK^2 \times XD^2$. The construction of this equation is effected in some such manner as this.

From any point of the incident ray IX let fall the perpendicular IA and in it take $AF = AX$. Then, having produced XI to B so that $IB = \frac{1}{2}IX$, on BX describe the semicircle BEX and in it inscribe $XE = XF$. Next, with XB produced to C so that $BC = BE$, on CX describe the semicircle CGX and let the perpendicular IG raised on its diameter at I intersect it in G. Finally, with centre X and radius GX let there be described the arc GH intersecting AI produced in H, draw HX and produce it towards R: XR will then be the refracted ray of IX when the density of the latter medium is such that the refraction difference $R\hat{X}M$ becomes greatest of all.(47) When this is found, the density of the latter medium effecting such a refraction will easily be yielded. You should therefore imagine the rays XR and $X\rho$ to be averagely refrangible (though to a different degree), and the latter medium thus found will make the refraction difference not only between those but also between the extreme or most difform rays just about(48) the greatest it can be.

Translation

The Refractions of Curved Surfaces.(2)

This will do for the refractions of planes. We must now concern ourselves with curve surfaces, especially spherical ones, and we shall try to embrace their doctrine in respect of homogeneous rays in the following propositions.

---

(47) Newton's quartic has the solutions $XR = \pm\sqrt{[XK(\pm\sqrt{\{\frac{9}{4}XK^2 - 2XD^2\}} + \frac{3}{2}XK)]}$. To construct the greatest of these he puts $BC = \frac{3}{2}XI$ (that is, $BX/XA = \frac{3}{2}XK/XD$ since $XI:XA = XK:XD$) and $EX/XA = \sqrt{2}$, then $CB = EB = \sqrt{[BX^2 - EX^2]}$ (that is, $CB/XA = \sqrt{[\frac{9}{4}XK^2 - 2XD^2]}/XD$) and lastly $XG = \sqrt{[XI(CB+BX)]}$, so that XG (or XH) : $XA = XR:XD$ and therefore $H\hat{X}A = R\hat{X}D$.

(48) A wise late insertion. See note (42).

(1) This topic, nowhere discussed in the first version of Newton's optical lectures (see §1: note (1) above), had been broached in his first optical researches of 1664–66 (see I: 551–8, 572–6) but was first systematically treated in the 'Sectio 4ta' of the 'Pars 1a' of his revised 'Optica' (here reproduced) and his 'Theoremata Optica' of about the same date (2, §1 below).

(2) This portion (ULC. Dd. 9.67: $_1$64–77) of the 'Optica'—so Newton's marginal annotations would have it—was given out in lecture form during October 1670, though corresponding *Lectiones* 13–15 of its first version (ULC. Add. 4002: 95–113, partially reproduced in §1 above) relate to topics discussed in an earlier 'Sect. 3' of the present manuscript and are equally firmly dated as having been read in July of that year. These inconsistencies (and the still more important problem of determining which version of the 'lectiones opticæ' corresponds the more closely in form and content to the lectures actually given in the Cambridge *scholæ publicæ*) are discussed in the preceding introduction, but several points of similarity with

Prop. 28. Radij in curvam superficiem incidentis refractum ducere.

Nempe eadem est refractio radij a Curva ac est a plano contingente Curvam in puncto refractionis. Quære ergo refractum a contingente plano per Prop: 3.(3)

Prop. 29. Si radij seu paralleli seu ad punctum aliquod contermini se sphæræ objiciant refringendos, refractorum axi quamproximorum concursum sive focum(4) determinare.

Sit *A* punctum radios ejaculans versus sphæricam superficiem *BNP* centro *C* descriptam: E vertice et centro erige ad axem *AC* perpendiculares *BH* et *CI*:

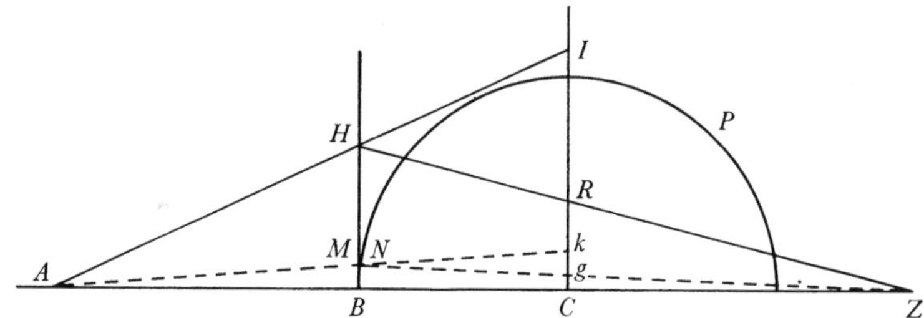

ipsisq; occurrentem in *H* et *I* age quamlibet *AI* per punctum *A*. Tum a puncto *C* versus *I* cape *CR* quæ sit ad *CI* ut sinus refractionis ad sinum incidentiæ, et age rectam *HR* occurrentem *AC* in *Z* et erit *Z* concursus refractorum quem determinare oportuit.

Sit enim *AN* radius axi vicinissimus incidens ad *N*, et occurrens *CI* in *k*. Age *NZ* occurrentem *CI* in *g*. Et, ut mos est, concipe infinitè parvum arcum *BN* æqualem esse *BM* segmento rectæ *BH* ad radium *Ak* terminato, et erit

$$CI.BH::Ck.BN,$$

ac $BH.CR::BN.Cg.$ et ex æquo $CI.CR::Ck.Cg.$ Hoc est *Ck* ad *Cg* ut est sinus incidentiæ ad sinum refractionis. Et proinde cùm anguli *CAk* et *CZg* ex Hypoth.

Newton's 1671 tract (compare **1**, Introduction: note (16)) suggest perhaps that the present revised text was not written till after the autumn of 1670. The bound volume containing this manuscript (mostly in Wickins' (?) amanuensis hand but augmented in places and corrected throughout by Newton himself) was, in accordance with statute, deposited in Cambridge University Library in October 1674 and is the unique source for the several secondary transcripts made (entirely legally) during Newton's lifetime—those, notably, by William Whiston (whose copy, taken about 1700, has now disappeared), Roger Cotes (now Trinity College, Cambridge. R.16.39, which 'Descripsi ex Autographo anno 1701/2. R.C.'), David Gregory (whose copy, made about May 1701, is now in private possession) and a 'Mr Meldrum' (British Museum. Sloane 3208.g.3, apparently a tertiary copy of about 1725 based on Gregory's transcript). In his Cambridge Lucasian lectures on 8 April and 2 June 1706 dealing with the rainbow Whiston introduced essentially unchanged proofs of Lemma 10 and Propositions 32, 33, 35, 36 below (demonstrated 'cum Cl. Newtono in Lectionibus suis Opticis MSS.') and these were subsequently published—the only extracts from the 'lectiones' to appear in Newton's lifetime—in his *Prælectiones Physico-Mathematicæ Cantabrigiæ in Scholis Publicis Habitæ. Quibus Philosophia Illustrissimi Newtoni Mathematica explicatius traditur, & facilius*

Proposition 28. To draw the refraction of a ray incident on a curved surface.

The refraction of a ray by a curve is, of course, the same as by a plane tangent to the curve at the point of contact. So seek the ray refracted by the tangent plane according to Proposition 3.(3)

Proposition 29. If rays which are either parallel or convergent to some point are cast upon a sphere to be refracted, to determine the meet or focus(4) of the refracted rays in the immediate vicinity of the axis.

Let A be the point throwing out rays towards the spherical surface BNP described with centre C. From its vertex and centre raise BH and CI perpendicular to the axis AC, and through the point A draw any line AI meeting these in H and I. Then from the point C in the direction of I take CR in proportion to CI as the sine of refraction to the sine of incidence and draw the straight line HR meeting AC in Z: Z will then be the meeting point of the refracted rays which it was required to determine.

For let the ray AN closest to the axis be incident at N and let it meet CI in k. Draw NZ meeting CI in g, and—as the convention is—conceive the infinitely small arc BN to be equal to the segment BM of the line BH terminated at M. Then $CI:BH = Ck:BN$ and $BH:CR = BN:Cg$, and so from equals

$$CI:CR = Ck:Cg;$$

that is, Ck is to Cg as the sine of incidence to the sine of refraction. Consequently, since by hypothesis the angles $C\widehat{A}k$ and $C\widehat{Z}g$ are infinitely small, so that Ck

demonstratur (Cambridge, 1710): 231–8. A year after Newton's death a first, English version of the 'Opticæ Pars 1ª' appeared as *Optical Lectures Read in the Publick Schools of the University of Cambridge, Anno Domini, 1669. By the late Sir Isaac Newton, then Lucasian Professor of the Mathematicks. Never before printed. Translated into English out of the Original Latin* (London, 1728), but for the original text—now published in full—the scholarly world had to wait another year till the appearance of *Isaaci Newtoni, Eq. Aur. in Academiâ Cantabrigiensi Matheseos olim Professoris Lucasiani, Lectiones Opticæ, Annis MDCLXIX, MDCLXX & MDCLXXI in Scholis publicis habitæ: Et nunc primum ex MSS. in lucem editæ* (London, 1729). The former was 'a faithful Translation of a very correct Copy [probably Gregory's], taken from the Latin original, as it was read in 1669' (Preface: xi), while the latter was edited in the first instance from an 'exemplum' which 'Newtonus olim Gregorio astronomiæ professori Saviliano dedit, a quo desumtum est illud [surely Meldrum's?], unde hæc editio impressa est, quod summâ fide & curâ descriptum fuisse cum eo Gregorii conferentes invenimus' and subsequently emended from 'exemplum illud Cantabrigiæ...asservatum...qu[od ipse Newtonus] in archivis Academiæ reposuerat'. The 1729 Latin text is the *editio princeps* on which all subsequent editions have been uniquely based: namely, by Castiglione (*Opuscula Mathematica, Philosophica et Philologica*, **2** (Lausanne and Geneva, 1744); repeated in the anonymously edited *Opera Omnia Optica* which appeared at Padua in 1749), by Horsley (*Opera quæ exstant Omnia*, **3** (London, 1782)) and—in Russian translation—by S. I. Vavilov (Moscow–Leningrad, 1946).

(3) The preceding Proposition 3 (in 'Sectio 3ª. De Planorum Refractionibus') constructs the refractions of incident rays by means of the sine law that the ratio of the 'sinus refractionis ad sinum incidentiæ' is constant for all angles of incidence (equal in modern terminology to the inverse of the refractive ratio).

(4) The 'focus principalis' (principal focus), that is.

sint infinitè parvi, adeoq Ck ad AN et Cg ad NZ perpendiculares vel saltem æquipollentes perpendiculis, erit NZ refractus ipsius AN. Q.E.D.

Coroll. 1.(5) Posito I ad R ut est sinus incidentiæ ad sinum refractionis erit $\frac{I}{R}AB.AC::BZ.CZ$. Est enim $\frac{I}{R}AB.AB(::I.R)::CI.CR$ et $AB.AC::BH.CI$, et ex æquo perturbatè $\frac{I}{R}AB.AC(::BH.CR)::BZ.CZ$.(6)

Coroll. 2. Si quando punctum A infinitè distet, seu parallelos radios ejaculetur, tum propter æquales BH et CI erit $I.R::BZ.CZ$.(7) Atq ita si refracti radij paralleli sint tum propter æquales BH et CR erit $I.R::AC.AB$.

Coroll. 3. Si e quatuor punctis A, B, C, et Z tria quævis dentur, potest quartum inveniri, ut e sequentibus exemplis patebit.

Exempl: 1. Dentur A, B, C, et quæratur Z. Scilicet est $\frac{I}{R}AB.AC::BZ.CZ$ adeoq divisim $\frac{I}{R}AB-AC.AC::BC.CZ$.

Exempl. 2. Si datis A, B, et Z, quæratur C. cùm sit $\frac{I}{R}AB.AC::BZ.CZ$ vicissim erit $\frac{I}{R}AB.BZ::AC.CZ$ et composite $\frac{I}{R}AB+BZ.BZ::AZ.CZ$.

Exempl. 3. Si datis A, C, et Z quæratur B, cum sit $\frac{I}{R}AB.AC::BZ.CZ$, sive $AB.\frac{R}{I}AC::BZ.CZ$. vicissim erit $\frac{R}{I}AC.CZ::AB.BZ$ et composite

$$\frac{R}{I}AC+CZ.CZ::AZ.BZ.$$

Possunt eadem determinari per ductum linearum. Veluti si datis A, B, et Z quæratur C, erige ad AZ normalem BH, cujusvis longitudinis et in ea cape $BI^{[\prime]}$ quæ sit ad BH ut I ad R. Junge AH et $I^{[\prime]}Z$ occurrentes in I, et IC normaliter demissa ad AZ incidet in punctum quæsitum C.(8)

Nota 1. Quod Z sit locus imaginis objecti A per refractionem exhibitæ cum spectatoris oculus in ipso axe ultra Z constituitur.

(5) Here, as universally in the first draft of his optical lectures, Newton employs Barrow's notation for the refractive ratio.

(6) And so (in agreement with his earlier derivation of this result by analytical means on I: 573–4) $CZ = AC \times BC/[(I/R)AB-AC]$. The present approach by geometrical limit-increments is far more elegant.

(7) As Barrow showed in his *Lectiones XVIII...In quibus Opticorum Phænomenωn genuinæ Rationes investigantur, ac exponuntur* (London, 1669: Lectio XII, §III: 81), in the equivalent form $I:R = NZ:CZ$ this is true generally for any ray AN incident parallel to the base BC, since $B\hat{C}N$ $(=\pi - N\hat{C}Z)$ and $C\hat{N}Z$ are the angles of incidence and refraction respectively, while $NZ:CZ = \sin N\hat{C}Z : \sin C\hat{N}Z$.

will be perpendicular to AN and Cg to NZ (or at least equivalently so), NZ will be the refracted ray of AN. As was to be proved.

Corollary 1.(5) On taking I to R as the sine of incidence to that of refraction, it will be $(I/R)\,AB:AC = BZ:CZ$. For $(I/R)\,AB:AB(=I:R) = CI:CR$ and $AB:AC = BH:CI$, and so from equals invertedly

$$(I/R)\,AB:AC(=BH:CR) = BZ:CZ.^{(6)}$$

Corollary 2. Should the point A ever be at an infinite distance, that is, if it cast out parallel rays, then because of the equality of BH and CI it will be $I:R = BZ:CZ$.(7) And similarly, if the refracted rays be parallel, then because of the equality of BH and CR it will be $I:R = AC:AB$.

Corollary 3. If any three of the four points A, B, C and Z be given, the fourth can be found—as will be evident from the following examples.

Example 1. Let A, B, C be given and Z sought. Plainly

$$(I/R)\,AB:AC = BZ:CZ$$

and so by dividing the ratios $(I/R)\,AB-AC:AC = BC:CZ$.

Example 2. If A, B and Z be given and C sought, since

$$(I/R)\,AB:AC = BZ:CZ,$$

by alternation $(I/R)\,AB:BZ = AC:CZ$ and by composition

$$(I/R)\,AB+BZ:BZ = AZ:CZ.$$

Example 3. If A, C and Z be given and B sought, since

$$(I/R)\,AB:AC = BZ:CZ \quad \text{or} \quad AB:(R/I)\,AC = BZ:CZ,$$

by alternation $(R/I)\,AC:CZ = AB:BZ$ and by composition

$$(R/I)\,AC+CZ:CZ = AZ:BZ.$$

The same may be ascertained by a linear construction. For instance, if A, B and Z be given and C sought, normal to AZ erect BH of any length you wish and in it take $BI^{[\prime]}$ in proportion to BH as I to R. Join AH and $I^{[\prime]}Z$ meeting in I, and IC let fall normal to AZ will cut it in the point C required.(8)

Note 1. Z is the place of the image of the object A produced by refraction when the viewer's eye is located on the axis itself beyond Z.

(8) Newton carelessly introduces a second point I without warning: to avoid confusion it is here denoted as $I^{[\prime]}$. A slightly changed version of this construction was inserted, as a 'Modus elegans ac expeditus...imaginem Geometricè designandi...ab amico communicatus', by Barrow at the end of Lectio XIV of his *Lectiones XVIII* (note (7)): 103–4.

2. Siquando refracti radij divergant, vel incidentes convergant, vel sint paralleli, similis erit problematis constructio mutatis tantum suo modo mutandis.

3. Si lux e puncto A emissa per pluras sphæricas superficies eundem Axem AC retinentes successivè transmittatur; Ad concursum post omnes refractiones determinandum, quære primò concursum radiorum post primam refractionem, deinde concursum eorundem post secundam refractionem, juxta ac si primariò emissi fuissent e puncto præcedentis concursûs. Et sic deinceps, donec ad ultimum concursum deventum sit. Atq; hoc pacto locus imaginis Objecti cujusvis per Telescopium sive Microscopium visi determinari potest.

4. Ope Coroll 3 Lentes ex sphæricis superficiebus confici possunt quæ Telescopijs modo quolibet designato constituendis inservient. Patet enim ex illo Corollario quòd non tantùm refractiones datarum Lentium investigari possunt, sed et Lentes delineari quæ datas refractiones peragent.

Lemma 9. Ad datam quamvis Curvam concursum Axis et vicinissimi perpendiculi determinare.

Sit BNn Curva, et ad quodvis ejus punctum n indeterminatè spectatum quære perpendiculum nc per notas methodos ducendi perpendicula Curvarum;(9) et simul invenies longitudinem Bc. Tum (demisso ad Bc normali nt) finge Bt vel nt infinitè parvam esse, seu nullam, et emerget longitudo BC cujus terminus est ad concursum axis cum vicinissimo perpendiculo.(10)

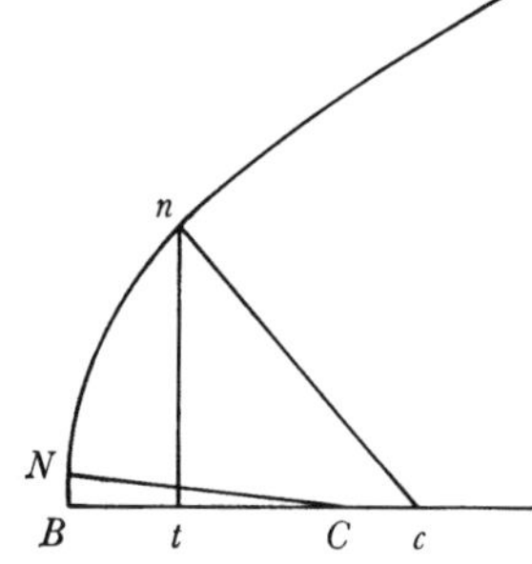

Exempl. 1. Sit BNn Parabola cujus latus rectum r; et Bt dic x, eritq; $Bc = x+\frac{1}{2}r$ ut notum est.(11) Pone jam $x = 0$ et restabit $\frac{1}{2}r$ pro longitudine BC ad verticem.

Exempl. 2. Sit BNn Ellipsis cujus latus rectum r et transversum q; eritq;, ut notum est, $Bc = x-\frac{r}{q}x+\frac{1}{2}r$. Jam pone $x = 0$ et restabit iterum(12) $\frac{1}{2}r$ pro longitudine BC ad verticem.

Nec secus in Curvis magis compositis procedendum est.

Prop. 30. Radijs in curvam quamvis superficiem quàm proxime perpendiculariter incidentibus, refractorum concursum seu focum determinare.

---

(9) Presumably Newton means Descartes' construction of the circle of centre c which touches the curve at n (see I, 2, 2, Historical Note) rather than Fermat's equivalent technique which constructs cn as the least or greatest line which can be drawn from c to the curve (compare I, 2, 4, §3: note (5)).

(10) It is assumed that the axis Bc is normal at B to the curve BNn. Since it is the limit-meet of perpendiculars to the curve in the immediate vicinity of B, the point C is evidently the curvature centre at B.

2. Whenever the refracted rays diverge or the incident ones converge or are parallel, construction of the problem will be similar merely by appropriate modification of what has to be modified.

3. If the light emitted from the point A be successively transmitted through several spherical surfaces having the same axis AC, to determine their meeting point after all the refractions seek first the meet of the rays after the first refraction, then—just as if they had originally been emitted from the previous meeting point—ascertain their meet after the second refraction, and so on until the last meeting point be reached. This way the location of the image of any object seen through a telescope or microscope can be determined.

4. With the aid of Corollary 3 lenses of service in the construction of telescopes of any design can be made up from spherical surfaces. For it is obvious from that corollary that not only may the refractions of given lenses be discovered but also lenses may be delineated which will execute given refractions.

Lemma 9. For any given curve to determine the meet of the axis and the nearest normal.

Let BNn be the curve and at any point n of it taken without particular regard seek the normal nc by the known methods of drawing normals to curves:(9) you will at once find the length of Bc. Then, letting fall nt perpendicular to Bc, suppose Bt or nt to be infinitely small, or in other words zero, and there will emerge the length of BC whose end-point is at the meet of the axis with the nearest normal.(10)

Example 1. Let BNn be a parabola of *latus rectum* r and call Bt x: then will Bc equal $x+\frac{1}{2}r$, as is known.(11) Now set $x = 0$ and there will remain $\frac{1}{2}r$ as the length of BC at the vertex.

Example 2. Let BNn be an ellipse of *latus rectum* r and main axis q: then, as is known, will $Bc = x-(r/q)\,x+\frac{1}{2}r$. Now set $x = 0$ and again(12) there will remain $\frac{1}{2}r$ as the length of BC at the vertex.

A like procedure should be followed in the case of more composite curves.

Proposition 30. Where the incident rays are, to a very close approximation, perpendicular to any curve surface, to determine the meeting point or focus of the refracted rays.

---

(11) The result is classical, namely Apollonius, *Conics*, v, 58 [= *Apollonii Pergæi Conicorum Libri* v. vi. vii.... *nunc primùm editi* (ed. G. A. Borelli) (Florence, 1661): 60]. The following Proposition 59 deals with the case of the ellipse (Newton's Example 2) correspondingly.

(12) In Cartesian terms $Bc = x+y\,dy/dx$ is the subnormal corresponding to the ordinate $tn = y$, and so in the case of the ellipse $y^2 = rx-(r/q)\,x^2$ Bc will have Newton's stated value. (The example is, in fact, Descartes'; see the latter's *Geometrie* [= *Discours de la Methode* (Leyden, 1637)]: Livre Second: 343/347.) To be sure, all conics $y^2 = rx\pm(r/q)\,x^2$ are equally curved at the origin, independently of the length of the *latus transversum* q.

Esto PBQ(13) curva quævis, A commune punctum, seu concursus incidentium radiorum, AB radius perpendicularis sive axis: et AN radius quàm proximè perpendicularis sive axi proximus. Sitq; NC ad curvam perpendicularis axiq;

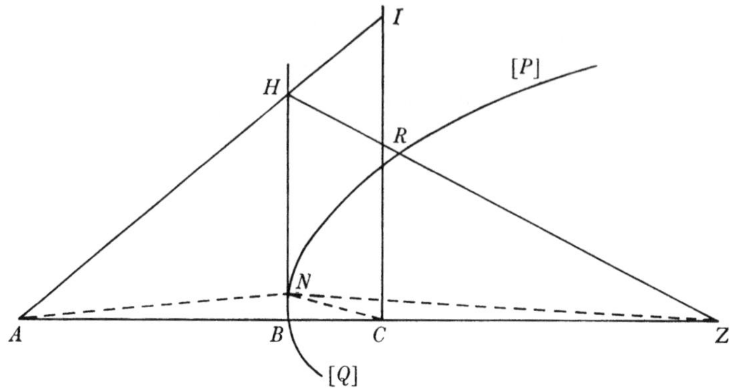

$A[B]$ occurrens ad C. Et puncto C per Lem. 9 invento, erige ad B et C perpendicula BH et CI, quibus in H et I occurrentem age quamvis AI: versus I cape CR quæ sit ad CI ut sinus refractionis ad sinum incidentiæ et recta HR occurret AB in quæsito refractorum concursu Z.

Probatur ad modum præcedentis Propositionis.(14) Et huic etiam consimilia Corollaria et Notæ competunt.

Prop. 31. Parallelis radijs in sphæram incidentibus refractorum ab Axe remotorum errorem a principali foco determinare.

Sit NBn sphæra, C centrum ejus, CB semi-diameter incidentibus radijs parallela, AN radius incidens et NK refractus ejus occurrens axi seu semidiametro CB in K. Et posito F principali foco, i.e. in quem radij prope axem

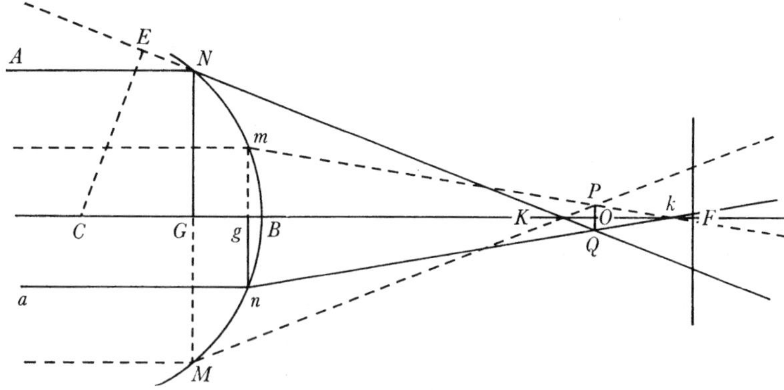

jacentes congregantur, quærendus erit error FK. Demitte ergo perpendiculares CE in NK et NG in CK, et dic $CB = a$, $GB = x$, et $CK = z$, atq; ex natura circuli erit $NG^q = 2ax - xx$, cui adde GK^q, hoc est $zz + 2x[z] - 2az + xx - 2ax + aa$,(15) et

Let PBQ(13) be any curve, A the common point or meet of the incident rays; AB the perpendicular ray (the axis, that is); and AN a ray exceedingly near to being perpendicular, that is, in the immediate vicinity of the axis. Further, let NC be normal to the curve, meeting the axis AB in C. Then, after finding the point C by Lemma 9, at B and C raise the perpendiculars BH and CI and draw any line AI to meet them in H and I. In the direction of I take CR in proportion to CI as the sine of refraction to the sine of incidence, and the line HR will meet AB in the required concourse Z of the refracted rays.

This is proved along the lines of the preceding proposition,(14) and to it also belong corollaries and notes which are closely similar.

Proposition 31. Where parallel rays are incident on a sphere, to determine how far refracted rays remote from the axis stray from the principal focus.

Let NBn be the sphere, C its centre, CB the radius parallel to the incident rays, AN an incident ray and NK its refraction meeting the axis—that is, the radius—CB in K. Then, on setting F as the principal focus (namely, the point at which rays lying close to the axis congregate), it will be required to assess the error FK. Let fall, therefore, the perpendiculars CE onto NK and NG onto CK, then call $CB = a$, $GB = x$ and $CK = z$. From the nature of the circle there will accordingly be $NG^2 = 2ax-x^2$ and, on adding GK^2 (that is

$$z^2+2xz-2az+x^2-2ax+a^2)^{(15)}$$

(13) The point P and the arc BQ, lacking in Newton's manuscript, have been added in the figure reproduced.

(14) For, to the order of BN^2, the arc $\widehat{BN}$ is indistinguishable from that of the osculating circle (of centre C).

(15) $GK = z+x-a$.

prodibit $NK^q = zz+2xz-2az+aa$. Jam cùm NG sit ad CE ut sinus incidentiæ ad sinum refractionis, sive ut I ad R et propter similitudinem triangulorum CEK et NGK, NK et CK sint in eadem ratione; erit

$$II.RR(::NK^q.CK^q)::zz+2xz-2az+aa.zz;$$

Adeoq; $IIzz = RRzz+2RRxz-2RRaz+RRaa$. et facta reductione

$$zz = \frac{2RRaz-2RRxz-RRaa}{RR-II}$$

extractaq; radice $z = \dfrac{RRa-RRx+R\sqrt{IIaa-2RRax+RRxx}}{RR-II}$. Ac radicali in infinitam seriem redacta[(16)] $z = \dfrac{Ra}{R-I} - \dfrac{RR}{IR-II}x - \dfrac{R^3}{2I^3a}xx - \dfrac{R^5}{2I^5a[a]}x^3$ &c. Jam cum per Coroll. 2 vel 3 ad Prop 29 sit $\dfrac{Ra}{R-I} = CF$ (id quod etiam innotescit ex valore z jam invento, fingendo esse $x=0$) ex hoc CF subduc inventum valorem z et restabit $\dfrac{RR}{IR-II}x + \dfrac{R^3}{2I^3a}xx + \dfrac{R^5}{2I^5a[a]}x^3$ &c pro valore erroris KF quem quærimus.

Coroll. 1. Si BG sive x ponatur valde exigua erit $\dfrac{RRx}{IR-II}$ quàm proximè æqualis KF. Tunc enim quantitates $\dfrac{R^3xx}{2I^3a} + \dfrac{R^5x^3}{2I^5aa}$ &c propter ascendentes potestates ejusdem x evadunt admodum exiguæ ut respectu termini $\dfrac{RRx}{IR-II}$ pro nullis haberi possint.

Coroll. 2. Quinetiam si statuas $NG=y$, erit $\dfrac{RRyy}{2IRa-2IIa} = KF$ circiter. Etenim est $NG^q = BG \times \overline{BC+BG}$ sive $= BG \times 2BC$ proximè, hoc est[(17)] $yy=2ax$ ferè vel $\dfrac{yy}{2a} = x$. et substituto $\dfrac{yy}{2a}$ pro x in valore ipsius KF emerget $\dfrac{RRyy}{2IRa-2IIa} = KF$.

Coroll. 3. Hinc errores KF sunt ut sagittæ GB vel ut quadrata semichordarum NG.[(18)]

Coroll 4. Si radius ANK detur positione et paralleli alicujus Axiq; proprioris et ad alteras axis partes incidentis radij an refractus nk ducatur secans Axem in k et hunc refractum NK in Q, et ad Axem demittatur normalis QO: linea KO

(16) $\sqrt{[I^2a^2-2R^2ax+R^2x^2]} = Ia - \dfrac{R^2}{I}x - \dfrac{R^2(R^2-I^2)}{2I^3a}x^2 - \dfrac{R^4(R^2-I^2)}{2I^5a^2}x^3 \ldots$ on extracting the square root algebraically. An editorial note at this point in the 1728 English printing of Newton's 'Optica' (*Optical Lectures* (note (2)): 182) is an interesting comment on contemporary knowledge of his early mathematical researches: 'The *Method of Series* was a very early Invention of our Author's, *viz.* in 1665. A particular Account of which may be seen in a Book called

to it, there will result $NK^2 = z^2+2xz-2az+a^2$. Now since NG is to CE as the sine of incidence to the sine of refraction (or as I to R) and again, because of the similarity of the triangles CEK and NGK, NK and CK are in the same ratio, it will be $I^2:R^2(=NK^2:CK^2) = (z^2+2xz-2az+a^2):z^2$, so that

$$I^2z^2 = R^2z^2+2R^2xz-2R^2az+R^2a^2.$$

On reduction $z^2 = \dfrac{2R^2(a-x)\,z-R^2a^2}{R^2-I^2}$ and with the root extracted

$$z = \frac{R^2(a-x)+R\sqrt{[I^2a^2-2R^2ax+R^2x^2]}}{R^2-I^2},$$

and so, when the radical has been brought to an infinite series,[16]

$$z = \frac{Ra}{R-I}-\frac{R^2}{I(R-I)}x-\frac{R^3}{2I^3a}x^2-\frac{R^5}{2I^5a^2}x^3\ldots.$$

Now since by Corollaries 2 and 3 to Proposition 29 $Ra/(R-I) = CF$ (which is also deducible from the value of z just now found by supposing $x = 0$), from this CF take away the value of z found and there will remain

$$\frac{R^2}{I(R-I)}x+\frac{R^3}{2I^3a}x^2+\frac{R^5}{2I^5a^2}x^3\ldots$$

for the value of the error KF we seek.

Corollary 1. If BG (that is, x) be taken exceedingly small, $[R^2/I(R-I)]\,x$ will to within a very close approximation be equal to KF. For then the quantities $[\frac{1}{2}R^3/I^3a]\,x^2+[\frac{1}{2}R^5/I^5a^2]\,x^3\ldots$ because of the rising powers of x come to be altogether so small that in regard to the term $[R^2/I(R-I)]\,x$ they can be considered as zero.

Corollary 2. Indeed, should you set $NG = y$, then $[\frac{1}{2}R^2/I(R-I)\,a]\,y^2 = KF$ just about. For $NG^2 = BG(BC+BG)$, that is $BG\times 2BC$ very nearly: in other terms,[17] $y^2 = 2ax$ or $\frac{1}{2}y^2/a = x$ almost. Then, when $\frac{1}{2}y^2/a$ is substituted for x in the value of KF, there will emerge $[\frac{1}{2}R^2/I(R-I)\,a]\,y^2 = KF$.

Corollary 3. Hence the errors KF are as the versed sines GB or as the squares of the half chords NG.[18]

Corollary 4. If the ray ANK be given in position and of some parallel ray an, incident nearer to the axis but on its further side, the refraction nk be drawn, intersecting the axis in k and the present refracted ray NK in Q, let fall the

Commercium Epistolicum Johannis Collins & aliorum de Analysi promota, first printed at *London* in 4[to]. *A*. 1712, and a second Edition with Additions in 8[vo]. *A*. 1722.' (Compare 1, Introduction: Appendix: note (1) above.)

(17) In effect, on replacing the circle arc $\widehat{BN}$ (of equation $y^2 = 2ax-x^2$) by the parabola (namely, $y^2 = 2ax$) of equal curvature at the point B of contact.

(18) By Corollaries 1 and 2 respectively.

evadet omnium maxima ubi radius *an* duplo minùs distat ab Axe circiter quàm radius alter AN. Demissa enim ad Axem normalis *ng* ponatur $=v$, $KO=s$, $GK=f$, et $KF=h$. et per Coroll. 3 hujus erit $yy.vv::KF.kF$, adeoq; $kF=\frac{hvv}{yy}$, quo a KF subducto restat $Kk=\frac{hyy-hvv}{yy}$. Præterea est $GK.GN::KO.QO$. Adeoq; $QO=\frac{ys}{f}$. Item $gn.gk(=GK$ proximè$)::QO.Ok$. Quare $Ok=\frac{ys}{v}$. Huic adde KO et iterum prodit $Kk=\frac{vs+ys}{v}$. Quamobrem est $\frac{vs+ys}{v}=\frac{hyy-hvv}{yy}$, factaq; divisione per $v+y$ et reducta æquatione prodit $s=\frac{hvy-hvv}{yy}$. Jam ut maximum s inveniatur multiplica terminos juxta Methodum Huddenij per dimensiones quantitatis indeterminatæ v, et emerget $0=\frac{hvy-2hvv}{yy}$, sive $y=2v$. Hoc est $NG=2ng$.(19)

Coroll 5. Et hinc KO ubi maximum est, æquatur quartæ parti ipsius KF circiter. Nam in valore ipsius s jam antè invento si scribas $2v$ pro y exoritur $\frac{1}{4}h=s$.

Coroll 6. Est etiam $OQ=\frac{Ry^3}{8Iaa}$.(20) Nam est

$$GK(=BF \text{ proximè}).GN::KO.OQ,$$

hoc est $\frac{Ra}{R-I}$(20) $.y::\frac{RRyy}{8IRa-8IIa}(=\frac{1}{4}KF).\frac{Ry^3}{8Iaa}$.(20)

Coroll 7. Si arcus BM sumatur æqualis BN, et $Bm=Bn$, ac radij ad puncta M et m refracti ducantur sibi occurrentes in P, constat esse spatium $PQ=\frac{Ry^3}{4Iaa}$(21) duplum nempe ipsius OQ. Et præterea constat refractos omnium radiorum in sphæricam superficiem inter N et M cadentium convergere in spatium hocce PQ, et itidem PQ esse minimum circulare spatium in quod possent congregari, adeoq; focum esse seu locum imaginis objecti parallelos radios in lentem ad usq; limites M et N apertam ejaculantis. Scilicet nulli radij possunt transilire hoc spatium, quia cùm OQ sit in data ratione ad KO, erit OQ simul maximum adeoq; punctum Q omnium versus F jacentiū remotissimum ab Axe in quo

(19) 'circiter' (roughly), that is. The 'error' *KF* roughly corresponds to the depth of focus of the spherical interface, while we might call *PQ* the diameter of the circle of 'confusion'. For Hudde's *methodus de maximis et minimis* see I, 2, 2, Historical Note. At this point in the 1728 English version (*Optical Lectures* (note (2)): 185) its editor (perhaps Pemberton?) added the footnote: 'Here our Author refers to *Hudden*'s Method *de Maximis et Minimis* printed *Anno* 1659 in *Carte's Geometry*; because his own *Method of Fluxions* [1, 2, §2] was not yet made publick, though he had written several small Tracts on this Subject in 1665, before he read these Lectures, and a larger one in 1671. These have been never yet printed, though many Copies of them in Manuscript are got abroad. The last is frequently mentioned in the *Commercium Epistolicum*'. (See 1, Introduction: Appendix.)

normal QO onto the axis: the line KO will prove to be greatest of all when the ray an is about half as distant from the axis as the other ray AN. For, letting fall the normal ng onto the axis, take it equal to v and $KO = s$, $GK = f$ and $KF = h$. Then by the present Corollary 3 there will be $y^2:v^2 = KF:kF$, so that

$$kF = hv^2/y^2;$$

and when this is taken from KF there remains $Kk = h(y^2-v^2)/y^2$. Further, $GK:GN = KO:QO$, so that $QO = ys/f$. Likewise

$$gn:gk \text{ (or } GK \text{ very nearly)} = QO:Ok,$$

and consequently $Ok = ys/v$. To this add KO and there again results

$$Kk = s(v+y)/v.$$

In consequence $s(v+y)/v = h(y^2-v^2)/y^2$, and after division by $v+y$ and reduction of the equation there results $s = h(vy-v^2)/y^2$. Now to find the maximum of s multiply the terms (according to Hudde's method) by the dimensions of the variable quantity v and there will emerge $0 = h(vy-2v^2)/y^2$ or $y = 2v$: that is, $NG = 2ng$.(19)

Corollary 5. And hence when KO is at its maximum it is about equal to one-quarter of KF. For if in the value of s already previously found you write $2v$ in place of y there arises $\frac{1}{4}h = s$.

Corollary 6. Also $OQ = \frac{1}{8}[R/Ia^2]\,y^3$.(20) For

$$GK \text{ (or } BF \text{ very nearly)} : GN = KO:OQ;$$

that is $Ra/(R-I)$(20)$:y = \frac{1}{8}[R^2/I(R-I)\,a]\,y^2$ (or $\frac{1}{4}KF$)$:\frac{1}{8}[R/Ia^2]\,y^3$.(20)

Corollary 7. If the arc BM be taken equal to BN and Bm equal to Bn, and the refracted rays at the points M and m are drawn meeting in P, it is settled that space $PQ = \frac{1}{4}[R/Ia^2]\,y^3$,(21) namely twice OQ. It is settled, moreover, that the refractions of all rays falling on the spherical surface between N and M converge in this space PQ, and that in the same way PQ is the least circular space in which they could converge and is accordingly the focus or position of the image of an object casting out parallel rays onto a lens uncovered to the limits M and N. Precisely, no rays can bypass this space because, since OQ is in given ratio to KO, OQ will simultaneously attain its maximum and accordingly the point Q is the remotest from the axis of all those lying towards F in

(20) Read '$\frac{RRy^3}{8IIaa}$', '$\frac{Ia}{R-I}$' and '$\frac{RRy^3}{8IIaa}$' respectively. Newton confuses $BF = Ia/(R-I)$ with $CF = (CB+BF$ or$)$ $Ra/(R-I)$, an error perpetuated in the 1728 published English version but corrected in the Latin *princeps* edition the following year.

(21) Compare note (20). This should be '$\frac{RRy^3}{4IIaa}$'.

radius quisquam concurrit cum externo radio NK. Neq; possunt in minus spatium congregari quia radij nk et mk secant externos radios in ipsissimis punctis P et Q quibus spatium PQ terminatur.

Coroll. 8. Si circuli NBM apertura augeatur vel minuatur error lateralis PQ erit ut y^3, sive ut cubus latitudinis aperturæ NM. Item si immutata apertura mutetur circuli magnitudo error PQ erit reciprocè ut aa sive ut CB^q, adeoq; ut BF^q siquidem CB et BF sint in datâ ratione. Sin vero et circuli magnitudo et apertura mutetur erit error ille PQ ut $\frac{y^3}{aa}$, sive ut $\frac{NM^{\text{cub}}}{BF^q}$ quemadmodum ex $\frac{Ry^3}{4Iaa}$[21] valore istius PQ constare potest.

Schol. Eodem ferè modo quo radiorum parallelè incidentium errores KF et PQ determinavimus consimiles divergentium vel convergentium errores, licèt calculo difficiliori, determinari possunt.[22]

Prop. 32. Si radij seu paralleli seu versus commune aliquod punctum inclinati[23] se sphæræ objiciant refringendos: refractorum extra axem sibi quam proximorum et in eodem plano cum incidentibus[24] jacentium concursum designare.

Sit AN incidens radius, NK refractus ejus et NV in plano trianguli ANK recta linea tangens sphæram ad N. Ad AN duc NR perpendicularem et occurrentem

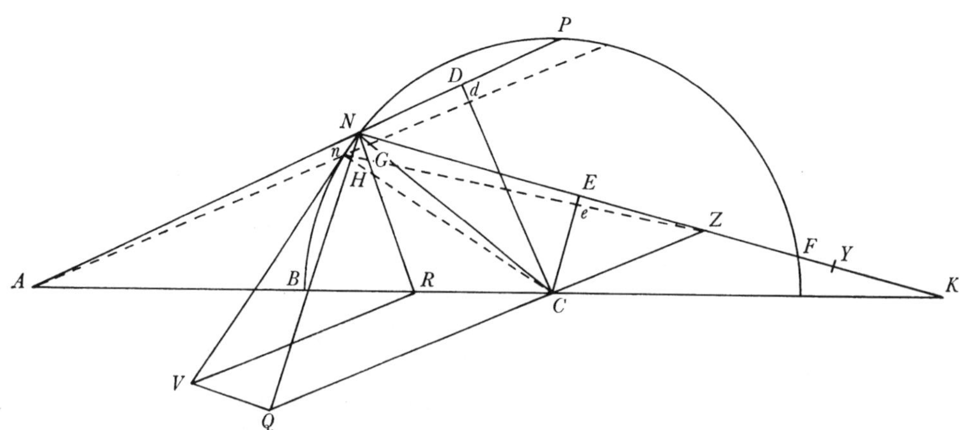

Axi AC in R, nec non RV parallelam et occurrentem tangenti NV in V. Item ad NK duc NQ perpendicularem et VQ parallelam convenientes in Q et age QC occurrentem NK in Z; eritq; Z concursus radiorum ipsi AN vicinissimorū.

Sit enim An alius ex incidentibus priori AN infinitè vicinus et occurrens NR in G. Age nZ occurrentem NQ in H et ad AN et NK e C centro sphæræ demitte normales CD et CE occurrentes An et nZ in d et e. Jam cum AN supponatur infinitè vicinus An arcus infinitè parvus Nn pro recta coincidente cum tangente

which any ray meets the outermost ray NK. Nor can they congregate in a lesser space because the rays nk and mk intersect the outermost rays in the very points P and Q which terminate the space PQ.

Corollary 8. If the aperture of the circle NBM increase or diminish, the lateral error PQ will be as y^3, in other words as the cube of the aperture's width NM. Likewise, if the aperture should be unchanged but the size of the circle alters, the error PQ will be inversely as a^2, that is CB^2, and so as BF^2 seeing that CB and BF are in given ratio. But if, indeed, both the circle's size and the aperture alter, the error PQ will be as y^3/a^2, that is as NM^3/BF^2: just as may be gathered from the value $\frac{1}{4}[R/Ia^2]\,y^3$[21] of PQ.

Scholium. In almost the same manner as we have determined the errors KF and PQ of parallel incident rays the closely similar errors of divergent or convergent ones may be ascertained, though the computation is more difficult.[22]

Proposition 32. If rays which are parallel or else converge at an angle towards some common point are thrown upon a sphere to be refracted, to specify the meet of the refracted rays lying outside the axis exceedingly close to one another and in the same plane as the incident ones.[24]

Let AN be an incident ray, NK its refraction and NV a straight line in the plane of the triangle ANK tangent to the sphere at N. To AN draw NR perpendicular, meeting the axis AC in R, and then parallel to it RV meeting the tangent NV in V. Likewise to NK draw the perpendicular NQ and the parallel VQ meeting one another in Q and extend QC till it intersects NK in Z: Z will then be the meeting point of [the refractions of] rays in the immediate vicinity of AN.

For let An be another of the incident rays indefinitely close to the first, AN, and meeting NR in G. Draw nZ meeting NQ in H, and to AN and NK from the centre C of the sphere let fall the normals CD and CE meeting An and nZ in d and e. Now since AN is supposed infinitely close to An, the infinitely small arc

(22) Compare I: 573–4. The excerpts which Whiston introduced into his 'prælectiones' in 1706 (see note (2)) begin with the following Proposition 32 (Whiston's first 'Problema' on his page 232).

(23) The equivalent phrase 'ad punctum aliquod contermini' is cancelled.

(24) To the modern reader this will be tautologous.

NV haberi potest, ac triangula NGn, NRV, ut et NHn, NQV pro similibus. Quare est $DC.Dd(::NR.NG::NV.Nn::NQ.NH)::EC.Ee$. Et conversè

$$DC.(DC-Dd)\,dC::EC.(EC-Ee)\,eC.$$

Et vicissim $DC.EC::dC.eC$. Est autem DC ad EC ut sinus incidentiæ ad sinum refractionis propterea quod NK sit refractus ipsius AN: adeoq; etiam dC ad eC est ut sinus incidentiæ ad sinum refracti[onis]. Et proinde cum anguli DAd et EZe sint infinite parvi [at]q; adeo Cd ad An et Ce ad nZ perpendiculares vel saltem perpendiculis æquipollentes erit nZ refractus ipsius An. Q.E.D.[25]

---

(25) An elegant construction of Barrow's problem (*Lectiones XVIII* (note (7)): Lectio XIII, §XXIV: 92) of defining the 'locum...imaginis in dato quovis refracto apparentis'. Barrow there derived the general diacaustic condition $NZ/EZ = AN\times DC\times NE/AD\times EC\times ND$ (here set by Newton as Corollary 2) by an equivalent limit-increment argument: namely, if nd and ne are continued to meet the circle again in p and f respectively, then in the limit as n passes into N (when all infinitely small circle arcs may be taken to be straight lines)

$$\frac{NZ}{EZ} = \frac{Nn}{\frac{1}{2}(Nn-Ff)} = \frac{Nn}{\frac{1}{2}(Nn+Pp)}\times\frac{\frac{1}{2}(Nn+Pp)}{Dd}\times\frac{Dd}{Ee}\times\frac{Ee}{\frac{1}{2}(Nn-Ff)} = \frac{AN}{AD}\times\frac{NC}{ND}\times\frac{DC}{EC}\times\frac{NE}{NC}$$

on substituting finite for equal infinitesimal ratios. Newton's preceding analysis, still preserved (ULC. Add. 4004: insert between 72^v, 73^r, reproduced in Appendix 1.1 below), is perhaps more revealing than his present synthetic justification of the construction. Barrow had already observed in the prefatory 'Epistola ad Lectorem' which introduces his *Lectiones XVIII* that 'D. *Isaacus Newtonus*...exemplar revisit,...de suo nonnulla penu suggerens, quæ nostris alicubi cum laude innexa cernes': true to his word he appended Newton's construction after his own researches into the problem with the words 'Subnectam & [Problematis constructionem] ab amico communicatam (aliâ methodo repertam ab ipso, concinneque demonstratam)' (§XXVI: 94).

In Proposition 33 following Newton adds the evident corollary to Barrow's diacaustic condition (formulated by him only for a circle) that, since only the curvature of the curve in the immediate neighbourhood of the point N concerns the position of the refraction centre Z of a ray of light AN incident upon any general interface BN, the Barrovian condition allows the construction of the diacaustic locus (Z), to which NZ is instantaneously tangent, on taking C to be instantaneously the centre of curvature at point N on the interface. It follows straightforwardly, on putting $D\hat{N}C = i$, $E\hat{N}C = r$, $NC = \rho$, $AN = x$ and $NZ = y$, that

$$\cot i\,(1-(\rho/x)\cos i) = \cot r\,(1-(\rho/y)\cos r).$$

The catacaustic condition $1/x+1/y = 2/\rho\cos r$ is an immediate corollary of taking $i = -r$. Barrow's (and indeed Newton's) priority in this matter was almost totally ignored by his contemporaries and has since been wholly forgotten. For the world at large the defining condition for the general diacaustic had to await Jakob Bernoulli's 'discovery' of it in 1693—generalizing earlier researches into the 'curva caustica' of a circle by Tschirnhaus (*Acta Eruditorum* (November 1682): 364–5; (February 1690): 68–73; (April 1690): 169–72), by Huygens (*Traité de la Lumière* (Leyden, 1690): Chapitre VI), by his brother Johann (*Acta Eruditorum* (January 1692): 30–5) and by himself (*ibid.* (March 1692): 110–16; (May 1692): 207–13)—to achieve public recognition. Jakob stated the construction of the general refraction centre in his short article concerning 'Curvæ Dia-Causticæ, earum relatio ad Evolutas, aliaque nova his affinia' (*Acta Eruditorum* (June 1693): 244–9, especially 244–6), but his proof

$\widehat{Nn}$ can be considered as a straight line coincident with its tangent NV and the triangles NGn, NRV and also NHn, NQV as similar. Accordingly

$$DC:Dd(=NR:NG=NV:Nn=NQ:NH)=EC:Ee,$$

and by conversion $DC:(DC-Dd \text{ or}) \, dC = EC:(EC-Ee \text{ or}) \, eC$, then by alternation $DC:EC = dC:eC$. But DC is to EC as the sine of incidence to that of refraction because NK is the refraction of AN: so also therefore is dC to eC as the sine of incidence to that of refraction. In consequence, since the angles $D\widehat{A}d$ and $E\widehat{Z}e$ are infinitely small and accordingly Cd and Ce are perpendicular to An and nZ (or equivalently so), nZ will be the refracted ray of An. As was to be demonstrated.(25)

of the result appeared only thirty years after his death in his *Opera* (**2** (Geneva, 1744): No. CIII, §XVIII 'Invenire relationem inter Evolutas & Diacausticas': 1077–80). From that article mathematicians took over not only Bernoulli's terminology ('*Dia-Causticas* voco Causticas per refractionem natas, reliquis ad distinctionem *Cata-Causticis* dictis, vel etiam *Causticis* simpliciter') but his personal claim to priority ('Solus Hugenius in Tractatu de Lumine schema nobis sistit integræ Dia-Causticæ, sed circularis tantum & per radios incidentes parallelos genitæ: Generalem vero Dia-Causticarum considerationem, earumque ad Evolutas relationem, primus, ni fallor, ego aggressus sum, nec irrito spero successu'). This failure to acknowledge Barrow's prior research into the topic is the more startling in that Huygens in his *Traité* (p. 122) had directed his reader to the *Lectiones XVIII* as a source of the passage which Bernoulli praises, while the latter had himself two years before singled out Barrow's *Lectiones Geometricæ: In quibus (præsertim) Generalia Curvarum Linearum Symptomata declarantur* (London, 1670) for a commendation which Newton, at the time of the calculus priority squabble, was to fling back at Leibniz time and again: 'qui calculum *Barrovianum* (quem decennio ante in Lectionibus suis Geometricis adumbravit *Auctor*, cujusque specimina sunt tota illa propositionum inibi contentarum farrago) intellexerit, alterum a Dn. *L*[*eibnitio*] inventum ignorare vix poterit, utpote qui in priori illo fundatus est, & nisi forte in differentialium notatione & operationis aliquo compendio ab eo non differt' (*Acta Eruditorum* (January 1691): 14). Soon after Jakob Bernoulli's article appeared L'Hospital wrote to his younger brother Johann, on 2 September 1693 (N.S.), that 'la relation que [M^r^ vôtre frere] donne entre les developpées et les caustiques par refraction est une decouverte tres curieuse et tres difficile: car je crois que vous vous ressouviendrez que vous ayant proposé autre fois la mesme chose vous en aviez abandonné le calcul parcequ'il étoit trop prolixe et trop ennuyeux, vous étant contenté d'indiquer la voye par laquelle on pouvoit y parvenir. C'est ce qui m'a fait juger que M^r^ vôtre frere avoit suivi quelque chemin plus court et apres avoir essayé inutilement diverses voyes, je crois enfin d'y estre tombé' (*Der Briefwechsel von Johann Bernoulli*, **1** (Basel, 1955): 184–5). The elegant limit-increment approach which he then outlined (*ibid.*: 185–6), published by him three years afterwards in his *Analyse des Infiniment Petits, pour l'Intelligence des Lignes Courbes* ((Paris, 1696): Section VII, §§132/133: 120–3), is the source for all subsequent research into diacaustics. Clifford Truesdell, typifying a recent swing in historical attitude, has asserted that 'In the case of every problem of major interest to which l'Hôpital has had a claim, a lesson or letter from [Johann] Bernoulli stands in the background' ('The New Bernoulli Edition', *Isis*, **49** (1958): 54–62, especially 60) but here is the exception in which L'Hospital caught Johann out when the latter accused him of 'un terrible calcul' and 'paralogisme' in his solution. (See his letters to Bernoulli of 7 October and 2 December 1693 (N.S.) [= *Briefwechsel*, **1**: 192–4, 197–9] and compare **1**, 2, §2: note (275) above.)

[(26)]Coroll. 1. Est $ND.NE$ (sive $NP.NF$)$::NR.NQ$. Nam actâ NC propter triang. NDC sim[ile] triang. NRV et triang NEC sim[ile] triang NQV, est $ND.NR(::NC.NV)::NE.NQ$ et inversè $ND.NE::NR.NQ$.

Hinc promptior emergit Problematis resolutio: nempe ad radios AN, NK erige normales NR, NQ quorum NR axi AC occurrat et NQ sit ad NR ut NF ad NP.[(27)] Dein Age QC quæ cum NK in quæsito Z conveniet.

Coroll 2. Est etiam $AN\times DC\times NE.AD\times EC\times ND::NZ.EZ$. Nam est $AD.AN::DC.NR$. et inde $NR=\frac{AN\times DC}{AD}$. Item $ND.NE::NR.NQ$ et inde $NQ=\frac{AN\times DC\times NE}{AD\times ND}$. Adeoq

$$AN\times DC\times NE.AD\times ND\times EC(::NQ.EC)::NZ.EZ.$$

Coroll 3. Si punctum radians A infinitè distet, sive parallelos radios ejaculetur: posito $I.R$::sin: incid:[.] sin: refract: erit $I\times NF.R\times NP::NZ.EZ$. In hoc enim casu AN et AD cum sint infinitè longæ pro æqualibus haberi debent. Atq adeò per Coroll 2 hujus erit $DC\times NE.EC\times ND::NZ.EZ$. Sed ex Hypothesi est $DC.EC::I.R$. et proinde $I\times NE.R\times ND(::NZ.EZ)::NP.NF$. Cæterùm de his vide plura in Lectionibus D[ris] Barrow.[(28)]

Notetur autem 1. Quod mutatis mutandis resolutio Problematis cuicunq casui facilè accommodatur sive radij incidentes divergant a puncto aliquo vel ad idem convergant vel incidant paralleli.[(29)]

(26) Newton's preliminary draft for these two Barrovian Corollaries (1 and 2) following is reproduced in Appendix 1.1 below.

(27) And so as their halves 'ut NE ad ND'.

(28) This particular case of the general proposition is, in fact, taken from Barrow's *Lectiones XVIII* (note (7)): Lectio XII, §VII: 82, where it is given direct proof: namely, if *nd* and

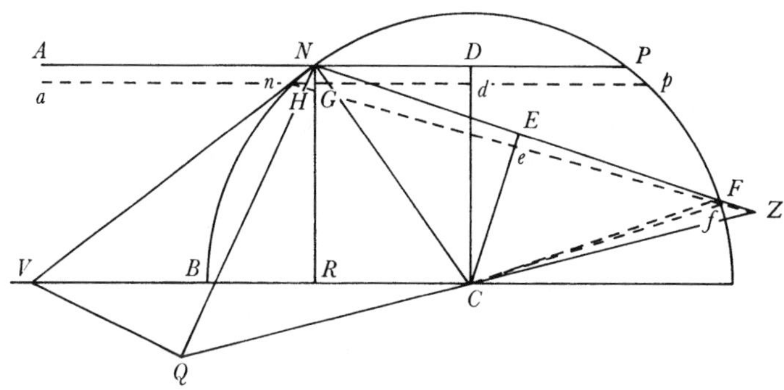

ne are continued to meet the circle in *p* and *f* respectively, in the limit as the parallel incident rays AN, *an* coincide

$$\frac{NZ}{EZ}=\frac{Nn}{\frac{1}{2}(Nn-Ff)}=\frac{Nn}{Dd}\times\frac{Dd}{Ee}\times\frac{Ee}{\frac{1}{2}(Nn-Ff)}=\frac{NC}{ND}\times\frac{CD}{CE}\times\frac{NE}{NC}.$$

[26]Corollary 1. It is $ND:NE$ (or $NP:NF$) $= NR:NQ$. For, on drawing NC, because the triangle NDC is similar to the triangle NRV and the triangle NEC to the triangle NQV, it is $ND:NR(= NC:NV) = NE:NQ$ and inversely

$$ND:NE = NR:NQ.$$

From this a readier solution of the problem emerges: namely, to the rays AN, NK erect the normals NR, NQ; of these let NR meet the axis AC [in R] and let NQ be to NR as NF to NP.[27] Then extend QC till it intersects NK in the required point Z.

Corollary 2. Further, $AN\times DC\times NE:AD\times EC\times ND = NZ:EZ$. For

$$AD:AN = DC:NR$$

and thence $NR = AN\times DC/AD$. Likewise $ND:NE = NR:NQ$ and thence $NQ = AN\times DC\times NE/AD\times ND$. Accordingly

$$AN\times DC\times NE:AD\times ND\times EC(= NQ:EC) = NZ:EZ.$$

Corollary 3. If the radiating point A be at an infinite distance, in other words if it throws out parallel rays, on setting I/R as the ratio of the sine of incidence to that of refraction there will then be $I\times NF:R\times NP = NZ:EZ$. For in this case since AN and AD are infinite in length, they should be considered as equal. Accordingly by the present Corollary 2 it will be

$$DC\times NE:EC\times ND = NZ:EZ.$$

But by hypothesis $DC:EC = I:R$ and consequently

$$I\times NE:R\times ND(= NZ:EZ) = NP:NF.$$

However, for more of these inferences see Dr Barrow's *Lectures*.[28]

But it should be noted that:

1. By making the necessary changes the solution of the problem is easily adapted to any case, whether the incident rays diverge from some point or converge to one or are incident parallel to one another.[29]

(Compare note (25).) The especial interest of this case is not at once apparent geometrically. Analytically, if we put $B\hat{C}N = D\hat{N}C = i$ and $E\hat{N}C = r$ (where $DC:EC = \sin i:\sin r = I:R$, constant) and denote their increments by di and dr respectively, it follows that $N\hat{C}n = di$ and (since $F\hat{C}K = E\hat{F}C - D\hat{N}E = 2r - i$) $F\hat{C}f = di - 2dr$, so that

$$NZ/EZ = \widehat{Nn}/\tfrac{1}{2}(\widehat{Nn} - \widehat{Ff}) = di/dr,$$

while $CD/ND = \tan i$ and $CE/NE = \tan r$. Hence the diacaustic condition

$$NZ/EZ = DC\times NE/EC\times ND$$

is equivalent to (Harriot's?) differential form of the sine law: namely, $di/dr = \tan i/\tan r$.

(29) In the extracts reproduced in his *Prælectiones* (note (2)) Whiston omits Newton's Notes 2 and 3 following.

2. Cum e radijs huic ANK proximis qui jacent in plano ANR conveniant in Z, qui vero in Conicâ superficie per revolutionem trianguli ANK circa latus AK generatâ jacent, conveniant in K; erit maxima radiorum ipsi ANK undiq; proximorum constipatio circa medium spatij KZ; puta ad Y.[30] Et proinde oculo in linea NK ultra K constituto, sensibilis imaginis objecti A per refractionem sphæricæ superficiei BN visi locus erit ad Y vel saltem intra limites K et Z. Nam locus ille non præcisè definitur.

3. Cum radij pluribus superficiebus successivè refringuntur et vicinorum post omnes refractiones concursum determines; primò quære concursum post primam refractionem, deinde concursum eorundem post secundam tanquam si primariò effluxissent e puncto præcedentis concursus: et sic deinceps ut ad Prop 29 dictum fuit.

Prop. 33. Radijs in quamcunq; curvam superficiem incidentibus refractorum sibi quàm proximorum, et in eodem plano cum incidentibus jacentium[31] concursum designare.[32]

Finge BNP jam non sphæram sed aliam quamcunq; curvam referre. Sitq; A commune punctum seu concursus incidentium radiorum, AN aliquis ex incidentibus, NK refractus ejus, et NC perpendicularis Curvæ ad punctum refringens. In hac NC quære intersectionem proximi alicujus perpendicularis (qualis nC) ad aliud proximum punctum refringens insistentis. Id quod alibi docebitur.[33] Sitq; ista intersectio C. Jam ductâ AC, demitte ad radios AN, NE normales CD, CE; ac erige NR, NQ; quorum NR occurrat AC in R. Sitq; NQ ad NR ut NE ad ND et acta QC conveniet cum refracto NK in desiderato proximorum refractorum concursu Z.

Probatur ad modum præcedentis Propositionis:[34] et huic etiam consimilia Corollaria et Notæ competunt.

Prop. 34. Figuram[35] determinare quæ radios Homogeneos sive parallelos sive ad commune aliquod punctum terminatos ita refringet ut refracti omnes ad aliud datum punctum accuratè conveniant.

(30) Compare the similar reasonings on possible definitions of the 'centre of radiation' of a point in §1.2 above.

(31) See note (24).

(32) See note (25) above. It will be obvious that in this general case likewise the refracted ray will be tangent to the diacaustic curve formed by the locus of Z with respect to fixed radiating point A as the incident point N varies along the general curve BN. This affords an immediate solution of Barrow's '*intricatissimum Problema* (cujúsque Solutio nullatenus aut laborem quem exigit, aut temporis jacturam compensabit) quo jubetur per datum punctum transeuntem refractum designare' (*Lectiones* XVIII (note (7))): Lectio XIII, §XXIII: 93): to find the point of incidence we need only produce the tangent from the given refraction point to the diacaustic till it intersects the given interface.

(33) By one or other of the curvature approaches expounded in Problem 5 of Newton's companion 1671 tract (1, 2, §2 above)!

2. Since those of the rays in the immediate proximity of the present one ANK which lie in the plane ANR meet at Z, while those lying in the cone-surface generated by the revolution of the triangle ANK round its side AK meet at K, the maximum compression of rays in proximity to ANK on either side of it will occur at about the middle of the space KZ: say, at Y.(30) In consequence, when the eye is placed in the line NK beyond K, the perceptible image of the object A viewed by way of a refraction in the spherical surface BN will be placed at Y, or at least within the confines of K and Z (for that place is not rigidly defined).

3. When rays are successively refracted at several surfaces and you are to determine the meeting point of neighbouring ones after all the refractions, first search out the meeting place after the first refraction; then seek their concourse after the second as though they had in the first instance streamed from the preceding meeting point; and so on in turn (as indicated in Proposition 29).

Proposition 33. Where rays are incident on any curved surface, to specify the meeting place of immediately neighbouring rays which lie in the same plane(31) as their incident ones.(32)

Suppose that BNP represents no longer a sphere but any other curve, and let A be the common point or concourse of the incident rays, AN one or other of them, NK its refracted ray and NC the normal to the curve at the refracting point. In this perpendicular NC seek its intersection with some immediately nearby normal (such as nC) erected on a second refracting point in the immediate vicinity—how to do so will be explained elsewhere(33)—and let that point be C. Now, after drawing AC, let fall the normals CD, CE to the rays AN, NE and erect NR, NQ perpendicular to them; of these let NR meet AC in R. Then let NQ be to NR as NE to ND, and when QC is extended it will intersect the refracted ray NK in the desired meeting point Z of the immediately neighbouring rays.

This is proved after the fashion of the preceding proposition(34) and to this one also pertain corollaries and notes which are closely similar.

Proposition 34. To determine the figure(35) which shall refract homogeneous rays, whether they be parallel or end at some common point, in such a way that all their refracted rays converge accurately upon some given point.

(34) See note (25) for details of the alternative analytical approaches proposed by Jakob Bernoulli and L'Hospital in 1693.

(35) The Cartesian oval $AN \pm (I/R)\, NZ =$ constant, defined by the differential relationship $\mp d(AN)/d(NZ) = I/R = \sin i/\sin r$ (constant), where $N\hat{T}S = i$ and $N\hat{T}Z = r$ are the respective angles of incidence and refraction. Newton had studied the problem of constructing the subnormal at a general point on this curve in some detail in the autum of 1664 (I, 3, Appendix 1, §§1/2).

Sit A concursus incidentium radiorum et Z refractorū, ac punctum aliquod B in recta AZ pro vertice curvæ ad arbitrium sumatur. Ab illo B capiantur in linea BZ versus medium densius BI cujusvis longitudinis et BR in ratione ad BI quam habet sinus incidentiæ ad sinum refractionis.(36) Centrisq; A et Z et intervallis AI et ZR describantur circuli se intersecantes in N et ipsius N locus erit curva quæ desideratam refractionem peraget.

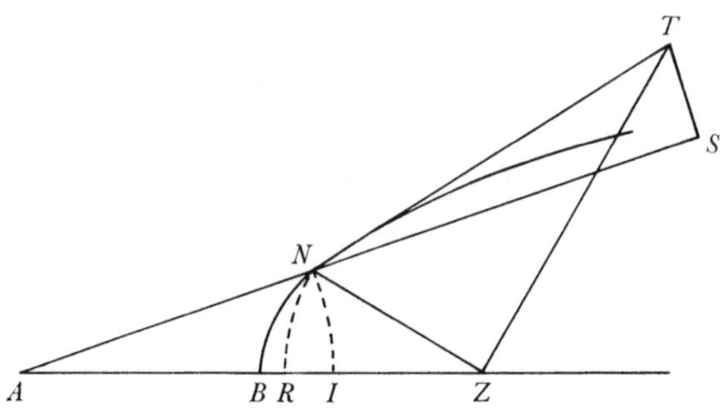

Quod ut pateat producatur AN ad S ut sit $NS.NZ::BI.BR$. et ad NS et NZ erigantur perpendiculares ST et ZT concurrentes in T et acta NT curvam tanget in N ut ex Methodo ducendi tangentes alibi(37) exposita constabit. Jam cum NS et NZ sint ut BI et BR, hoc est ut sinus incidentiæ et refractionis, et respectu sinus totiûs sive diametri NT sit NS sinus anguli NTS qui æquatur angulo incidentiæ radij AN, et NZ sinus anguli NTZ qui æquatur angulo refractionis radij NZ, patet esse NZ refractum ipsius AN. Q.E.D.

Nota 1. Potest etiam Curva huic usui inserviens describi quæ per datum quodvis punctum B extra axem AZ positum transibit. Scilicet agantur AB et

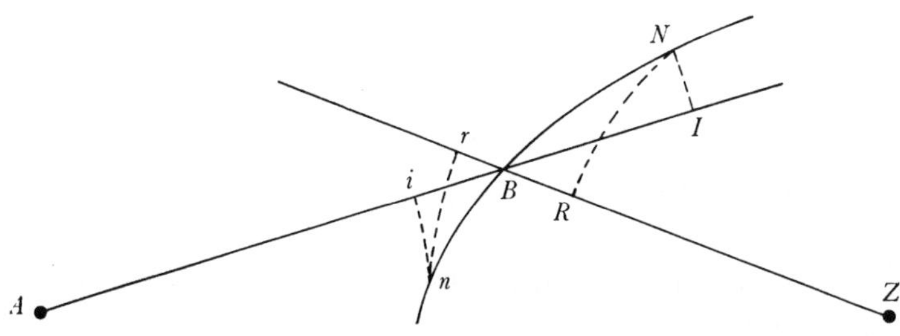

ZB et in ipsis capiantur BI et BR in ratione sinuum incidentiæ et refractionis. Et centris A et Z ac intervallis AI et ZR describantur circuli concurrentes in N eritq; N ad curvam quam oportet describere.(38)

2. Præfata Problematis resolutio mutatis mutandis se ad omnes casus extendit sive incidentes aut refracti radij convergant, divergant vel existant paralleli, sive refractio fiat e rariori Medio in densius, vel e densiori in rarius. Et quidem si radij ex neutra parte paralleli sint, i.e. si punctorum A et Z neutrum sit ad infinitam distantiam, Curva BN erit aliqua quatuor(39) Ellipsium quas Cartesius in hunc usum in Geometria descripsit. Sin alterutrum infinitè distet, ita ut radij punctum illud respicientes evadant paralleli, Curva erit

Let A be the meeting point of the incident rays, Z that of the refracted ones, and let some point B be assumed arbitrarily in the straight line AZ as the vertex of the curve. From that point B in the line BZ in the direction of the denser medium take BI of any length and BR to BI in the ratio of the sine of incidence to that of refraction.(36) Then with centres A and Z and radii AI and ZR describe circles intersecting one another in N, and the locus of N will be the curve which effects the desired refraction.

To make this evident, produce AN to S so that $NS:NZ = BI:BR$ and to NS and NZ raise the perpendiculars ST and ZT concurrent in T: then when NT is drawn it will touch the curve at N, as will be obvious from the method of drawing tangents expounded elsewhere.(37) Now since NS and NZ are as BI and BR, that is, as the sines of incidence and refraction, and since (with respect to the diameter NT as whole sine) NS is the sine of angle $N\widehat{T}S$ equal to the incident angle of the ray AN while NZ is the sine of the angle $N\widehat{T}Z$ equal to the refraction angle of the ray NZ, it is evident that NZ is the refraction of AN. As was to be proved.

Note 1. A curve serving this purpose can also be described passing through any given point B set outside the axis AZ. Specifically, draw AB and ZB and in them take BI and BR in the ratio of the sines of incidence and refraction. Then with centres A and Z and radii AI and ZR describe circles concurrent in N: N will be on the curve it is required to describe.(38)

2. On changing what is necessary to alter the solution of the aforesaid problem extends to all cases, whether the incident or refracted rays converge, diverge or be parallel, or whether the refraction should occur out of a rarer medium into a denser one or from a denser one into a rarer. To be precise, if the rays on neither side be parallel—that is, if neither of the points A and Z be at an infinite distance—the curve BN will be one or other of the four(39) ovals which Descartes has described for this purpose in his *Geometrie*. But if either one of those points be at infinity, so that the rays relating to that point prove to be

(36) Read 'quam habet sinus refractionis ad sinum incidentiæ' (of the sine of refraction to that of incidence).

(37) Namely, in Mode 3 of Problem 4 of the companion 1671 tract (**1**, 2, §2: note (228)).

(38) Newton in his accompanying figure in fact introduces two points I, i in AB and, correspondingly, R, r in ZB. His construction finds $(AN-AB):(ZN-ZB) = -BI/BR$, that is, $-I/R$ in the terminology of note (35), so that at once $AN+(I/R)\,NZ = AB+(I/R)\,BZ$, constant.

(39) Descartes distinguished four component 'ellipses' whose defining bipolar equation is (with regard to Newton's first figure) $AN+(I/R)\,NZ = AB+(I/R)\,BZ$, dividing the ovals for which the point B of intersection with AZ lies between the two latter points from those for which it does not, and separating the cases in which the refractive index I/R is positive from those in which it is negative in value. See Descartes' *Geometrie* [= *Discours de la Methode*, Leyden, 1637]: Livre Second: 252–7.

Conica sectio, uti notum est.[40] Et in hoc casu circulus RN vel IN propter infinitam centri distantiam evadet recta linea ipsi AZ ad R vel I perpendicularis.

Lemma 10.[41] E parallelis radijs ad circulum refractis radium illum determinare, cujus pars circulo inclusa datam habeat rationem ad partem refracti ejus eidem circulo inclusam.

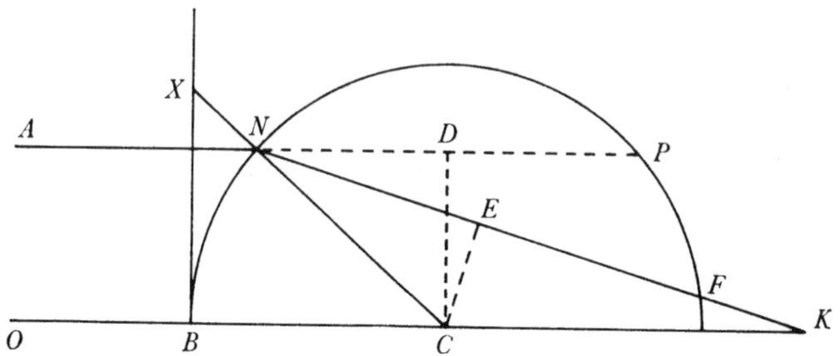

Sit AN radius incidens, NK refractus, NP et NF partes eorum circulo inclusæ, CD et CE perpendicula ad istas partes e centro circuli demissa, et BC semidiameter acta parallela AN. Sitq3 $CD.CE::I.R.$ et $NP.NF::p.q$. His positis ut innotescat punctum N quod radios AN & NK determinat, erige ad BC normalem BX cujus quadratum sit ad BC quadratum ut $\frac{qq-pp}{pp}$ ad $\frac{II-RR}{II}$ et acta CX secabit circulum in desiderato N. Est enim ex Hypoth: $p.q(::NP.NF)::ND.NE.$ et $I.R::CD.CE$; Quare $\frac{q}{p}ND=NE$, et $\frac{R}{I}CD=CE$. Porro cùm sit $ND^q+CD^q(=NC^q)=NE^q+CE^q$, aufer hinc inde ND^q+CE^q et restabit $CD^q-CE^q=NE^q-ND^q$. Hoc est substituendo valores CE et NE modò inventos, $CD^q-\frac{RR}{II}CD^q=\frac{qq}{pp}ND^q-ND^q$: et facta reductione $\frac{II-RR}{II}CD^q=\frac{qq-pp}{pp}ND^q$. Quo in proportionalitatem resoluto fit $\frac{qq-pp}{pp}.\frac{II-RR}{II}(::CD^q.ND^q)::BX^q.BC^q$. Q.E.D.

Prop. 35.[42] Sole sphæram pellucidam illustrante radiorum ejus post unam reflexionem emergentium maximam ad axem inclinationem determinare.[43]

(40) In this case the defining bipolar equation of the oval reduces to the focus-directrix property of a conic. See Corollary 2 of the 'Exemplum' of Mode 3 of Problem 4 in Newton's companion 1671 tract (**1**, 2, §2 above) and Descartes' *Dioptrique* [= *Discours de la Methode* (Leyden, 1637)]: Discours Huictiesme: 92–4, 103–5, where this corollary is given direct geometrical proof (compare I: 560, note (4)).

(41) Whiston's 'Problema alterum' (*Prælectiones* (note (2)): 234–5). Newton chooses a somewhat devious method of demonstrating that, given $\frac{\sin i}{\sin r}=\frac{I}{R}$ and $\frac{\cos i}{\cos r}=\frac{p}{q}$, then

$$\tan i=\sqrt{\frac{q^2/p^2-1}{1-R^2/I^2}}$$

parallel, the curve will be a conic, as is known.(40) In this case the circle RN or IN, because its centre is an infinite distance away, will prove to be a straight line perpendicular to AZ at R or I.

Lemma 10.(41) Of parallel rays refracted at a circle, to determine that one whose portion included by the circle shall have a given ratio to the portion of its refracted ray included by the same circle.

Let AN be the incident ray, NK its refraction, NP and NF the portions of these included by the circle, CD and CE perpendiculars let fall to those portions from the circle's centre, and BC a radius drawn parallel to AN. Again, let there be $CD:CE = I:R$ and $NP:NF = p:q$. With these suppositions, in order to ascertain the point N determining the rays AN and NK, to BC erect the normal BX whose square is to be to the square of BC as $(q^2-p^2)/p^2$ to $(I^2-R^2)/I^2$ and CX, when drawn, will intersect the circle in the desired point N. For by hypothesis $p:q(= NP:NF) = ND:NE$ and $I:R = CD:CE$; consequently

$$(q/p)\,ND = NE \quad \text{and} \quad (R/I)\,CD = CE.$$

Further, since $ND^2+CD^2(=NC^2) = NE^2+CE^2$, take away ND^2+CE^2 from either side and there will remain $CD^2-CE^2 = NE^2-ND^2$; that is, on substituting the values of CE and NE just now found,

$$CD^2-(R^2/I^2)\,CD^2 = (q^2/p^2)\,ND^2-ND^2;$$

and after reduction is made $[(I^2-R^2)/I^2]\,CD^2 = [(q^2-p^2)/p^2]\,ND^2$. When this is resolved as a proportion there comes

$$(q^2-p^2)/p^2:(I^2-R^2)/R^2(=CD^2:ND^2) = BX^2:BC^2.$$

As was to be proved.

Proposition 35.(42) Where the sun illuminates a transparent sphere, to determine the greatest angle of inclination to the axis of its rays which emerge after a single reflexion.(43)

(where in his figure $D\hat{N}C = i$, $E\hat{N}C = r$ are the incident and refracted angles respectively). Note the crucial step of equating CD^2-CE^2 (or $CN^2(\sin^2 i-\sin^2 r)$) with

$$NE^2-ND^2 \text{ (or } CN^2(\cos^2 r-\cos^2 i)).$$

(42) Whiston's 'Prob. (3.)' (*Prælectiones* (note (2)): 235–7). Somewhat confusingly, Newton's figure (which serves for both this and the following proposition) illustrates the theory of the secondary rainbow expounded in Proposition 36. We have introduced a modified diagram into our English translation.

(43) Descartes' classical defining condition for the angular radius of the primary rainbow: 'ayant pris la plume & calculé par le menu tous les rayons qui tombent sur les divers poins d'vne goutte d'eau, pour sçauoir sous quels angles aprés deux refractions & vne ou deux reflexions ils peuuent venir vers nos yeux, i'ay trouué qu'aprés vne reflexion & deux refractions, il y en a beaucoup plus qui peuuent estre veus sous l'angle de 41 à 42 degrés, que sous aucun moindre; & qu'il n'y en a aucun qui puisse estre vû sous vn plus grand' (*Meteores* [= *Discours de la Methode* (Leyden, 1637)]: Discours Huitiesme: 261).

Sit BNK sphæra proposita, $BC[Q]$ diameter sive axis incidentibus radijs parallelus, AN aliquis ex incidentibus, NF refractus ejus, FG reflexus et GR denuò refractus, et quærendus erit maximus angulorum quos RG cum axe BQ potest conficere.

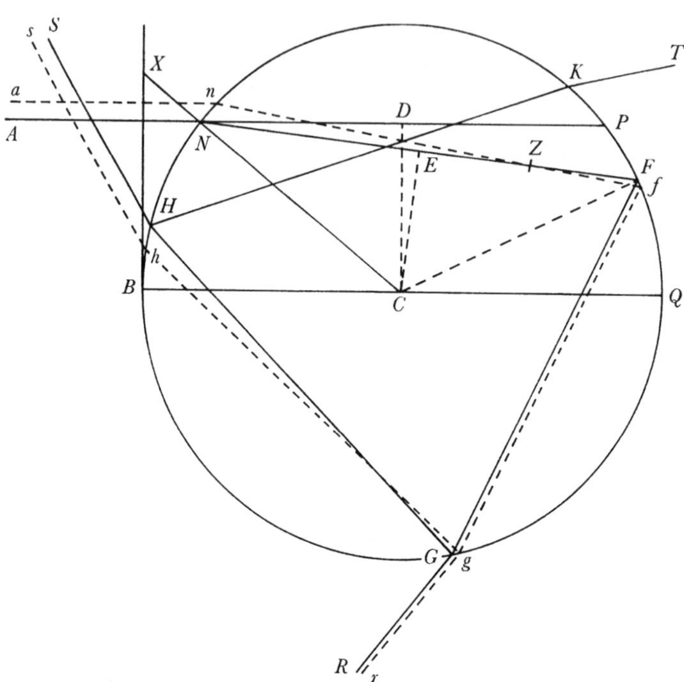

In quem finem advertendum est quod in eo solo casu ubi RG maximè inclinatur ad BQ, radij ipsi AN vicinissimi possunt emergere paralleli ad RG. Nam in alijs casibus ex emergentibus radijs sibi vicinissimis alij magis alijs continuò inclinantur ad BQ, adeoq; aliquantulum inclinantur ad se invicem.(44)

Advertendum est præterea quod radij emergent paralleli qui conveniunt ad punctum reflexionis.(45) Duc enim radium *an* ipsi AN parallelum et quàm

(44) Newton's proof is inadequate in that it does not consider whether other points in the arc $\widehat{PQ}$ might also reflect an incident refracted ray into a path which, after a second refraction, runs parallel to GR. This cannot, of course, happen.

(45) This is Barrow's reduction of the Cartesian test (note (43)) in his *Lectiones XVIII* (note (7)): Lectio XII, §XIII: 84. The proof there sketched essentially appeals to the convexity of the diacaustic formed by incident rays parallel to BC: if Z is the point where the diacaustic meets the circle (corresponding to the incident ray AN refracted along NZ), then it is an immediate corollary that the refractions of all incident rays parallel to AN meet its refraction NZ between N and Z or beyond Z accordingly as they are above or below AN, and therefore (in either case) intersect the circle between Z and Q. 'Exhinc apparet (id quod ab eximio *D. Slusio* monitum amicus mihi communicavit) potuisse *Cartesium* sine tabularum confectione suum *Iridis* angulum determinare' for, as Barrow went on to say, it follows at once that the point Z corresponds to maximum deviation ($\pi - 2Z\hat{C}Q$) between AN and RG. (The 'amicus' was John Collins, who communicated to Barrow in 1668 his copy of Sluse's letter to Oldenburg

Let BNK be the sphere propounded, BCQ its diameter or axis set parallel to the incident rays, AN some one of these incident rays, NF its refraction, FG its reflexion and GR its subsequent refraction: it will then be required to find the greatest angle which RG can make with the axis BQ.

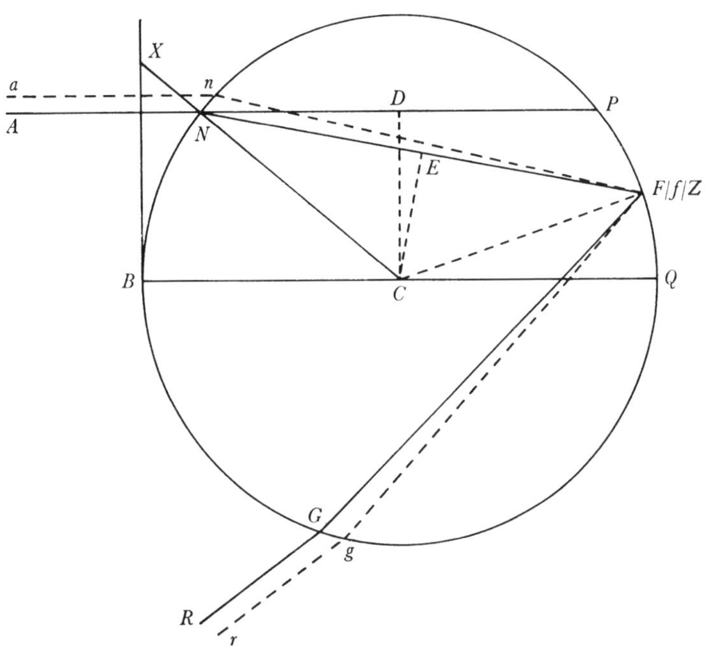

To this end it must be noticed that alone in the case where RG is most inclined to BQ can rays in the immediate vicinity of AN emerge parallel to RG. For in other cases of rays emerging in its immediate vicinity some of them are necessarily more inclined than others to BQ and so in turn are very slightly inclined to each other.(44)

It should further be noticed that those rays will emerge parallel which are coincident at the point of reflexion.(45) For draw the ray *an* parallel to AN and in

of 24 November 1667 (N.S.), reproduced in A. R. and M. B. Hall's *Correspondence of Henry Oldenburg*, **3** (Madison, Wisconsin, 1966): 594–6, see especially 596.) Newton's present 'proof' depends more directly on Descartes' criterion that when the deviation between AN and RG is a maximum, then parallels to AN in its immediate vicinity are refracted, reflected and again refracted parallel to RG. To make his somewhat loose appeal to symmetry watertight, note that, if *an* is refracted to f and then reflected to g so that gr is parallel to GR upon emergence, then (since angles of incidence and emergence are equal)

$$\widehat{Nn}\ (\text{or } \widehat{NF}-\widehat{nf}+\widehat{Ff}) = \widehat{gG}\ (\text{or } \widehat{FG}-\widehat{fg}-\widehat{Ff})$$

and therefore, because $\widehat{NF} = \widehat{FG}$, $\widehat{nf} = \widehat{fg}$, it follows that $\widehat{Ff} = 0$. Hence F, f (and so Z) coincide. In analytical terms (compare note (28)), where $D\hat{N}C = i$, $E\hat{N}C = r$, the deviation $\pi - 2F\hat{C}Q = \pi - 2(2r-i)$ is greatest when $NZ/EZ = di/dr = 2$.

proximum sitq; ejus refractus nf, reflexus fg ac iterum refractus gr. Et punctis F et f coincidentibus, cùm anguli NFn & GFg sint æquales et refractiones ad Nn et Gg similes, emergentes radij GR et gr æquè paralleli erunt ac incidentes AN et an.

Quærendus est itaq; radius AN cujus refractus cum refracto vicinissimi radij an concurrit ad F. Et quidem per Coroll. 3 Prop. 32 (demissis a centro sphæræ ad radios normalibus CD et CE, positoq; $I.R::CD.CE$) si radij isti ad quodvis punctum Z concurrant, erit $I\times NF.R\times NP(::NZ.EZ::NF.EF$, puncto nempe Z ad ipsum F juxta Hypoth: cadente$)::2.1$. Quare $I\times NF=2R\times NP$. et $I.2R::NP.NF$. Datur itaq; ratio NP ad NF et inde per Lemma 10 dabitur punctum N. Scilicet ad verticem circuli ducatur tangens BX cujus quadratum sit ad quadratum semidiametri BC ut $4RR-II$ ad $II-RR$ et agatur CX: hæc enim circulo occurret in N. Et ex invento N cætera nullo negotio determinantur.[46]

Coroll 1. Hinc fit $3RR.II-RR::CN^q.ND^q$. cùm enim sit

$$4RR-II.II-RR::BX^q.BC^q$$

componendo erit $3RR.II-RR(::CX^q.BC^q)::CN^q.ND^q$.

Coroll. 2. Est et $I.2R::ND.NE$. nam supra fuit $I.2R::NP.NF$.

Et ex his expeditior evadit problematis resolutio.

Schol. Una cum maxima inclinatione radij RG datur maximus arcuum FQ ad refractos NF terminatorum. Nam angulus FCQ quem FQ subtendit est æqualis angulo quem CF et AN comprehendunt, hoc est æqualis dimidio anguli, quem RG et AN vel BQ comprehendunt, et proinde arcuum FQ æquè ac angulorū ab RG et BQ comprehensorum maximus est qui radio AN in punctum jam inventum incidente definitur.[47]

(46) This last paragraph and the two following corollaries are a slight variation on Barrow's corresponding 'è dictis consectarium Theorema' (*Lectiones* XVIII (note (7)): Lectio XII, §XI: 83). In analytical equivalent, since at maximum deviation $\tan i/\tan r = 2$ (by notes (28) and (45)), on taking $\sin i/\sin r = I/R$, $ND/NE = \cos i/\cos r = I/2R$ and therefore

$$\frac{CN}{ND} = \sec i = \sqrt{\frac{2^2-1}{I^2/R^2-1}}, \quad \frac{CN}{CD} = \operatorname{cosec} i = \sqrt{\frac{2^2-1}{2^2-I^2/R^2}}.$$

Evidently, the primary bow is real only for $1 \leqslant I/R \leqslant 2$.

(47) This determining condition for the primary bow is, we now know, the historical discovery of Thomas Harriot about 1601 but was communicated by him only to a few close friends. (See J. A. Lohne, 'Regenbogen und Brechzahl', *Sudhoffs Archiv*, **49** (1965): 401–15, especially 408–9; and his 'Thomas Harriot als Mathematiker', *Centaurus*, **11** (1965): 19–45, especially 34–8. Harriot named $\widehat{FQ}$ the 'arcus egressionis'.) In ignorance, like all his contemporaries, of this prior accomplishment Christiaan Huygens in December 1652 achieved an equivalent result, but he too failed to publish the manuscript 'Tractatus de refractione et

its immediate neighbourhood, and let its refraction be *nf*, its reflexion *fg* and its second refraction *gr*. Then since, with the points F and f coincident, the angles $\widehat{NFn}$ and $\widehat{GFg}$ are equal and the refractions at Nn and Gg are congruent, the emergent rays GR and gr will be equally as parallel as the incident ones AN and *an*.

We are accordingly required to find the ray AN whose refraction meets the refraction of an immediately neighbouring ray *an* at F. Now to be sure by Corollary 3 of Proposition 32, if (on letting fall the normals CD and CE from the sphere's centre to the rays and on setting $I:R = CD:CE$) those rays be concurrent at any point Z, then $I\times NF:R\times NP(= NZ:EZ = NF:EF$ where, namely, the point Z according to hypothesis falls at F itself$) = 2:1$. Hence $I\times NF = 2R\times NP$ and $I:2R = NP:NF$. The ratio of NP to NF is accordingly given and from this fact (by Lemma 10) the point N will be given. Precisely, at the circle's vertex draw the tangent BX, letting its square be to the square of the radius BC as $4R^2-I^2$ to I^2-R^2, and draw CX: for this line will meet the circle in N. When N has been found the other details are determined without trouble.[46]

Corollary 1. Hence there comes $3R^2:[I^2-R^2] = CN^2:ND^2$. For since

$$[4R^2-I^2]:[I^2-R^2] = BX^2:BC^2,$$

by composition there will be $3R^2:[I^2-R^2]\ (= CX^2:BC^2) = CN^2:ND^2$.

Corollary 2. Also $I:2R = ND:NE$. For above there was $I:2R = NP:NF$. And from these there results a speedier solution of the problem.

Scholium. Along with the greatest incline of the ray RG there is given the greatest of the arcs FQ bounded by the refracted rays NF. For the angle $\widehat{FCQ}$ subtended by $\widehat{FQ}$ is equal to the angle contained by CF and AN, in other words to half the angle contained by RG and AN or BQ; and consequently the greatest of the arcs FQ no less than of the angles contained by RG and BG is fixed as that for which AN is incident at the point just now found.[47]

telescopiis' to which it was an appendix. In essence Huygens continued CF to meet the incident ray AN in (say) K, then maximized $\cos \widehat{FCQ} = \cos(2r-i)$ in terms of the variable

$$CK = x = a\sin r/\sin(i-r),$$

where a is the radius of the spherical 'raindrop'. On taking $\sin i/\sin r = I/R$ as before it follows that $\cos r = (I/R)^{-1}(a/x+\cos i) = (I/R)^{-1}[(I^2/R^2-1)x^2+a^2]/2ax$, and therefore

$$\begin{aligned}\cos(2r-i) &= (I/R)\cos r - 2a\cos^2 r/x + a/x\\ &= [(I^2/R^2-1)x^4+(I^2/R^2+2)a^2x^2-a^4]/2a(I^2/R^2)x^3.\end{aligned}$$

Equation of the derivative of this to zero yields, on neglecting $x = a$, $(I^2/R^2-1)x^2 = 3a^2$ and $\cos i = [(I^2/R^2-1)x^2-a^2]/2ax = \sqrt{[\tfrac{1}{3}(I^2/R^2-1)]}$, in agreement with Newton's present Barrovian result. (See Huygens' *Œuvres complètes*, **13** (The Hague, 1916): 146–50. On eliminating $I^2/R^2 = 1+3a^2/x^2$ in this maximum case Huygens at once derived the cubic condition

$$a\cos(2r-i) = (3a^2x^2+a^4)/(x^3+3a^2x),$$

Prop. 36.[48] Sole sphæram pellucidam illustrante, radiorum ejus post duas reflexiones emergentium minimam ad axem inclinationem determinare.

Sint AN et an radij duo incidentes sibi quàm proximi qui post duas reflexiones in F, f et G, g emergent secundum HS et hs. Et manifestum est[49] quòd in eo solo

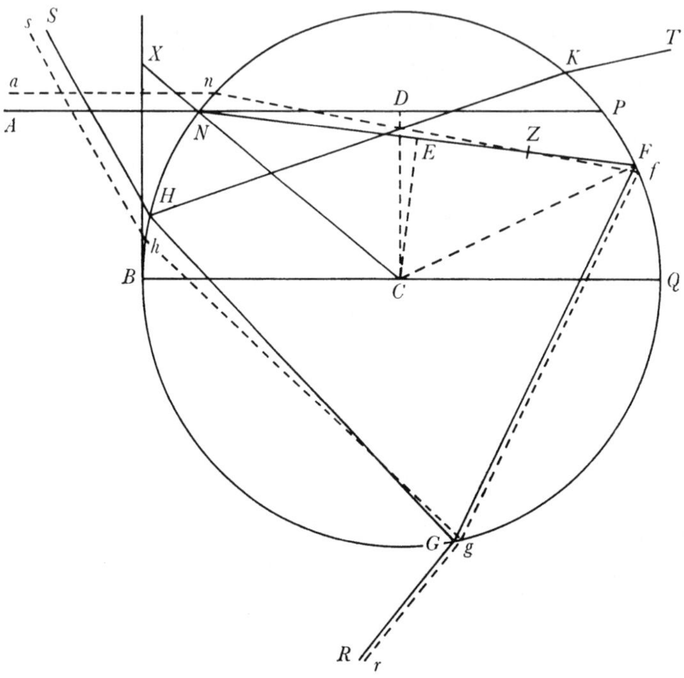

casu ubi acutus angulus quem BQ et SH comprehendunt minimus est, radij illi HS et hs possunt esse paralleli, uti supra[50] de radijs GR et gr dictum fuit. Et ubi hoc accidit radius etiam FG ad fg parallelus erit. Unde

$$2\,\text{arc}\,Ff(=\text{arc}\,Ff+Gg=\text{arc}\,FG-fg=\text{arc}\,NF-nf)=\text{arc}\,Nn-Ff,$$

adeoq3 $3\,\text{arc}\,Ff=\text{arc}\,Nn$. Et cùm NF dividatur in Z in ratione istorum arcuum ut patet,[51] erit $NZ=3ZF$, seu[52] $3EZ$. Cùm itaq3 per Coroll 3 Prop 32 sit $I\times NF.R\times NP::NZ.EZ$ sive $::3.1$, erit $I\times NF=3R\times NP$, sive $I.3R::NP.NF$. Datur itaq3 ratio NP ad NF et inde per Lem. 10 dabitur punctum N, ducendo nempe BX quæ circulum tangat in vertice B et cujus quadratum sit ad $BC^{\text{quadr.}}$ ut $9RR-II$ ad $II-RR$; et agendo CX quæ occurret periferiæ in N. Invento autem N cætera facilè determinantur.

Coroll. 1. Hinc est $8RR.II-RR::CN^q.ND^q$. Nam

$$9RR-II.II-RR::BX^q.BC^q$$

et componendo $8RR.II-RR(::CX^q.BC^q)::CN^q.ND^q$.

---

that is, $z^3-3t^2z-2st^2=0$ where $s=a\sec(2r-i)$, $t=a\tan(2r-i)$ and $z=x-s$.) Sluse, as we have seen (note (45)), independently attained the primary bow condition that $F\hat{C}Q$ must be a

Proposition 36.[48] Where the sun illuminates a transparent sphere, to determine the least angle of inclination to the axis of its rays emergent after two reflexions.

Let AN and an be two immediately neighbouring incident rays which after two reflexions at F, f and G, g emerge along HS and hs. It is obvious[49] that in that case alone where the acute angle comprised by BQ and SH is least can those rays HS and hs be parallel, as was observed above[50] with respect to the rays GR and gr. And when this happens the ray FG will also be parallel to fg. Hence $2\widehat{Ff}(=\widehat{Ff}+\widehat{Gg}=\widehat{FG}-\widehat{fg}=\widehat{NF}-\widehat{nf})=\widehat{Nn}-\widehat{Ff}$, so that $3\widehat{Ff}=\widehat{Nn}$. Since then NF is divided at Z in the ratio of those arcs—this is evident[51]—, it will be $NZ=3ZF$, that is[52] $3EZ$. Accordingly, since by Corollary 3 of Proposition 32 it is $I\times NF:R\times NP=NZ:EZ$, that is, as $3:1$, there will be

$$I\times NF=3R\times NP \quad \text{or} \quad I:3R=NP:NF.$$

As a result the ratio of NP to NF is given and from this, by Lemma 10, will be given the point N: namely by drawing BX, whose square shall be to BC^2 as $9R^2-I^2$ to I^2-R^2, to touch the circle at its vertex B and then extending CX to meet its circumference in N. When, however, N is found the other details are easily determined.

Corollary 1. Hence it is $8R^2:[I^2-R^2]=CN^2:ND^2$. For

$$[9R^2-I^2]:[I^2-R^2]=BX^2:BC^2$$

and by composition $8R^2:[I^2-R^2]\ (=CX^2:BC^2)=CN^2:ND^2$.

---

maximum, but the details of his rainbow researches remain unpublished. However, the broad outline which his compatriot Benedict Spinoza later published with a credit to 'Mijn Heer de Sluze' in his *Stelkonstige Reeckening van den Regenboog, Dienende tot naedere samenknoping der Natuurkunde met de Wiskonsten* ((The Hague, 1687): 15–19 [= *Opera quotquot reperta sunt* (ed. J. van Vloten and J. P. N. Land) **2** (The Hague, 1883): 507–20, especially 517–19]) shows Sluse's approach to have been modelled on Huygens', while his letter to Oldenburg of 20 August 1670 (N.S.) reveals that he was by then already *au fait* with the content of Lectio XII of Barrow's *Lectiones XVIII*. (The letter was first published by Le Paige in his edition of Sluse's 'Correspondance' [= *Bullettino di Bibliografia e di Storia delle Scienze Matematiche e Fisiche*, **17** (Rome, 1884): 427–554/603–726, especially 646–8] but by mid-1672 Newton already possessed a transcript of it in Collins' hand (now ULC. Add. 3971.1: 19^v–21^r), no doubt passed on to him by Barrow.)

(48) Excerpted by Whiston in 1706 as 'Prob. (4.)' of his *Prælectiones* (note (2)): 237–8.

(49) To the Cartesian reader at least. Compare Descartes' *Meteores* (note (43)): Discours Huitiesme: 261: 'ayant...calculé par le menu tous les rayons qui tombent sur les diuers poins d'vne goutte d'eau,...i'ay trouué aussy qu'aprés deux reflections & deux refractions, il y en a beaucoup plus qui vienent vers l'œil sous l'angle de 51 à 52 degrés, que sous aucun plus grand, & qu'il n'y en a point qui vienent sous vn moindre'.

(50) See note (44).

(51) For, since the infinitesimal arcs $\widehat{Nn}$ and $\widehat{Ff}$ are equally inclined to NF, the infinitely small triangles NZn, FZf are similar and so $\widehat{Nn}:\widehat{Ff}=NZ:(Zf\approx)\ ZF$.

(52) Since E bisects $NF=NZ$ (or $3ZF)+ZF$.

Coroll. 2. Est etiam $I.3R::ND.NE$ utpote cùm supra fuerit

$$I.3R::NP.NF.^{(53)}$$

Schol. Ad eundem modum maxima radij KT post tres reflexiones emergentis inclinatio ad axem, juxta ac maximus arcuum QG investigabitur. Scilicet in eo casu FG et fg convenient ad G(54) eritq3

$$\text{arc } Ff(=\text{arc } FG-fG=\text{arc } NF-nf)=Nn-Ff.$$

Et inde 2 arc Ff=arc Nn. et $NZ=2ZF$. Adeoq3 $4.1::NZ.EZ::$(per Coroll 3 ad Prop 32) $I\times NF.R\times NP$. Sive $I.4R::NP.NF$. Et proinde per Lem 10 $16RR-II.II-RR::BX^q.BC^q$. Unde consectatur esse

$$15RR.II-RR::CN^q.ND^q. \quad \text{et} \quad I.4R::ND.NE.$$

Atq3 ita si radij post quatuor reflexiones emergentis inclinatio minima desideretur determinabis faciendo ut sit $25RR-II.II-RR::BX^q.BC^q$. vel $24RR.II-RR::CN^q.ND^q$, et $I.5\text{R}::ND.NE$. Et sic præterea in infinitum.(55)

---

(53) These computations of the angular radius of the secondary rainbow have no parallel in Barrow's *Lectiones XVIII*, but they are in fact anticipated in the researches of Huygens in March 1667 (*Œuvres complètes*, **13** (The Hague, 1916): 163–8) and perhaps also of Sluse (compare Spinoza's *Regenboog* (note (47)): 17–18 [= *Opera*: 518–19]). Newton's results follow much as before from the secondary bow condition $NZ/EZ = di/dr = 3$ (the condition for the total deviation $2(i-r)+2(\pi-2r)$ to be minimal, though this is not immediately obvious for Newton's limit-increment argument). At once

$$\frac{CN}{ND} = \sec i = \sqrt{\frac{3^2-1}{I^2/R^2-1}} \quad \text{and} \quad \frac{CN}{CD} = \operatorname{cosec} i = \sqrt{\frac{3^2-1}{3^2-I^2/R^2}}.$$

In an attempted extension of his 1652 approach (note (47)) Huygens evaluated

$$\cos(3r-i) = -4a\cos^3 r/x+2(I/R)\cos^2 r+3a\cos r/x-I/R$$

as $[(I^2/R^2-1)^2x^6+(I^2/R^2-3)a^2x^4+(I^2/R^2+3)a^4x^2-a^6]/2a^2(I/R)^3x^4$, where as before

$$\sin i/\sin r = I/R \quad \text{and} \quad x = a\sin r/\sin(i-r).$$

Using standard Huddenian techniques he found the condition

$$(I^2/R^2-1)^2x^6-(I^2/R^2+3)a^4x^2+2a^6 = 0$$

for $\cos(3r-i)$ (and so the deviation) to attain an extreme value, but for a time did not succeed in factorizing it and never completed his examination of the secondary bow. (In fact, of its three factors $(I^2/R^2-1)x^2-2a^2$, $(I/R+1)x^2-a^2$ and $(I/R-1)x^2+a^2$ only the first yields a maximum for $\cos(3r-i)$, namely $x^2 = 2a^2/(I^2/R^2-1)$. Substitution of this in

$$\cos i = [(I^2/R^2-1)x^2-a^2]/2ax$$

yields $\cos i = \sqrt{[\frac{1}{8}(I^2/R^2-1)]}$ in agreement with Newton, while eliminating $I^2/R^2 = 1+2a^2/x^2$ from the sextic produces the quartic condition $a\cos(3r-i) = (6a^2x^2+a^4)/2(x^4+2a^2x^2)$.)

(54) By an argument analogous to that for the primary bow in Proposition 35. Compare the case $n = 3$ of the next note.

(55) We may give a general proof of these corresponding results for the tertiary and quaternary bows in the following way. (Compare Johann Bernoulli, 'Ordre du Calcul pour la détermination des Iris ou Arc-en-Ciels de toutes les Classes', published in his *Opera Omnia*,

Corollary 2. Also $I:3R = ND:NE$ for the reason that there was above

$$I:3R = NP:NF.^{(53)}$$

Scholium. In much the same way, just as the greatest of the arcs QG, will be discovered the greatest angle of inclination to the axis of a ray KT emergent after three reflexions. In this case, specifically, FG and fg meet at G(54) and it will be $\widehat{Ff}(=\widehat{FG}-\widehat{fG}=\widehat{NF}-\widehat{nf})=\widehat{Nn}-\widehat{Ff}$. From this $2\widehat{Ff}=\widehat{Nn}$ and $NZ=2ZF$, so that $4:1 = NZ:EZ = I\times NF:R\times NP$ (by Corollary 3 to Proposition 32) or $I:4R = NP:NF$. Consequently, by Lemma 10,

$$[16R^2-I^2]:[I^2-R^2] = BX^2:BC^2,$$

from which it follows that $15R^2:[I^2-R^2] = CN^2:ND^2$ and $I:4R = ND:NE$.

And thus, should the least incline of a ray emergent after four reflexions be desired, you will determine it by making $[25R^2-I^2]:[I^2-R^2] = BX^2:BC^2$ or $24R^2:[I^2-R^2] = CN^2:ND^2$ and $I:5R = ND:NE$. And so forth indefinitely.(55)

---

tam antea sparsim edita, quam hactenus inedita, **4** (Lausanne and Geneva, 1742): No. CLXXI. *Optica*, §III: 197–203.) Suppose that the infinitely close parallel incident rays AN, an are, after refraction at N, n, reflected p times at the points $F_i, f_i (i = 1, 2, \ldots, p)$ and subsequently emerge

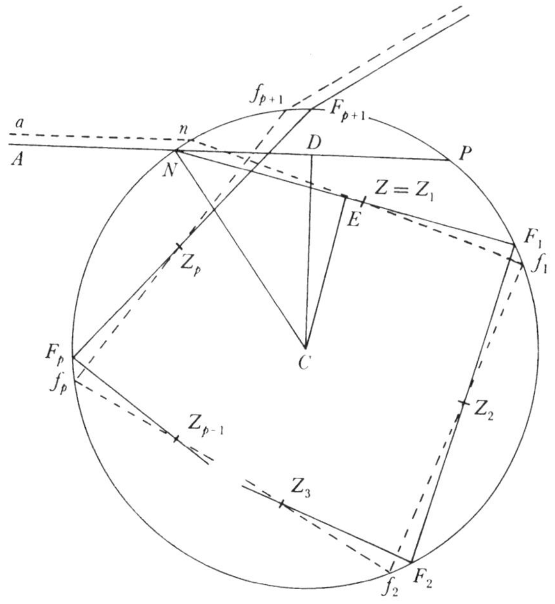

at F_{p+1}, f_{p+1} again parallel. If now we suppose the radius of the refracting sphere to be unity, since $\widehat{NF_1} = \widehat{F_1F_2} = \ldots = \widehat{F_pF_{p+1}}$ and $\widehat{nf_1} = \widehat{f_1f_2} = \ldots = \widehat{f_pf_{p+1}}$ it follows that

$$\widehat{Nf_p} = \widehat{Nn} + (\widehat{nf_p} \text{ or}) \, p\times\widehat{nf_1} = (\widehat{NF_p} \text{ or}) \, p\times\widehat{NF_1} + \widehat{F_pf_p}$$

and hence $\widehat{F_pf_p} = \widehat{Nn} - p(\widehat{NF_1}-\widehat{nf_1}) = \widehat{Nn} - p(\widehat{Nn}-\widehat{F_1f_1})$. In consequence, on taking $D\hat{N}C = i$, $E\hat{N}C = r$ the infinitesimal arcs have magnitudes $\widehat{Nn} = di$, $\widehat{F_1f_1} = -d(2r-i)$ (as in note (28))

Transactis refractionibus Homogeneorum radiorum jam restat ut Heterogeneos conferamus.(56) De horum ad plana refractionibus paulo fusiùs agebamus(57) eo ut Prismatum... affectiones innotescerent. Præcipuum verò quod circa curvas superficies jam determinandum occurrit, est quantitas erroris radiorum a quo oritur confusio sive indistincta visio objectorum quæ in Telescopijs per nimiam vitri objectum respicientis aperturam evenire solet. Et in hunc finem cum præmissa sit Prop 31 unde errores innotescunt qui in sphæricis superficiebus per ineptitudinem figuræ efficiuntur, sequentem jam subjungimus quâ errores ex inæquali refrangibilitate diversorum radiorum(58) orti determinari possunt.

Prop 37. Heterogeneis radijs in sphæram incidentibus erro[r]es ex inæqualibus radiorum similiter incidentium refractionibus progenitos determinare.

E puncto *A* in sphæram *NBM* centro *C* descriptam incidant secundum lineam aliquam *AN* radij duo maximè difformes quorum refracti sint *NF* et *Nf*

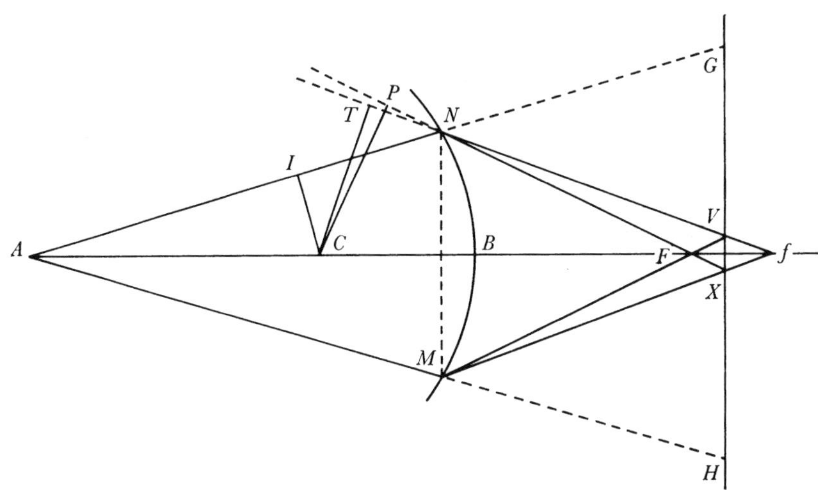

Axi occurrentes in *F* et *f*, et in illos demittantur perpendicula *CI*, *CP* et *CT*. Jam si accurata resolutio desideretur, refractiones radiorum *NF* et *Nf* seorsim computandæ sunt. Sed cùm arcus *NM* ponatur admodum exigua portio circuli, veritatem quam proximè assequemur assumendo angulos *CNI*, *CNP* et *CNT* ferè esse ut eorum sinus. Sit ergo *I* communis sinus incidentiæ, *P* sinus refractionis radiorum maximè refrangibilium ac *T* sinus ille minimè refrangi-

and so $\widehat{F_p f_p} = di - 2p.dr$. For the p-ary rainbow, by symmetry Z_p divides $F_{p+1}F_p$ in the same ratio as Z (or Z_1) divides NF_1; that is, $\widehat{Nn}:\widehat{F_1 f_1} = \widehat{F_{p+1} f_{p+1}}:\widehat{F_p f_p}$, so that

$$di:(di-2dr) = (di-2(p+1)\,dr):(di-2p.dr) \quad \text{or} \quad NZ/EZ = di/dr = p+1.$$

(In equivalent Cartesian terms this 'rainbow' condition is at once obtained by demanding that the total deviation $2(i-r)+p(\pi-2r)$ be maximum or minimum according as p is odd or even. This simplification was first introduced by Edmond Halley in his 'De Iride, sive de Arcu

Having dealt with the refractions of homogeneous rays, it now remains for us to talk briefly about[56] heterogeneous ones. We treated their refractions at planes somewhat copiously[57] in order to gain knowledge of the properties of prisms.... The important thing, however, which now suggests itself to be determined about curved surfaces is the quantity of stray of rays: this gives rise to the confusion or blurring of vision which usually happens in telescopes through an over-large aperture of the object glass. To that end, since from the premised Proposition 31 there may be known the errors produced in spherical surfaces by the unsuitability of their figure, we now adjoin the following theorem by which errors arising from the unequal refrangibility of different rays[58] can be determined.

Proposition 37. Where heterogeneous rays are incident on a sphere, to determine the errors engendered by unequal refraction of rays which are similarly incident.

From the point A onto the sphere NBM described with centre C let there be incident along some line AN two maximally difform rays, and let their refractions be NF and Nf meeting the axis in F and f. To these let fall the perpendiculars CI, CP and CT. Now were an accurate assessment desired, the refractions of the rays NF and Nf would have to be separately computed. But since the arc NM is taken to be an exceedingly minute portion of a circle, we shall pursue the truth to a high degree of approximation by assuming the angles $C\widehat{N}I$, $C\widehat{N}P$ and $C\widehat{N}T$ to be almost as their sines. So let I be the common sine of incidence, P the sine of the most refrangible rays and T that of the least refrangible ones.

Cœlesti, dissertatio Geometrica, qua methodo directâ Iridis utriusq; Diameter, data Ratione Refractionis obtinetur: Cum solutione Inversi Problematis, sive Inventione Rationis istius ex data Arcus Diametro', *Philosophical Transactions*, **22**, 1700–1 [1702]: No. 267 (for November/December 1700), §III: 714–24, especially 716.) The rest is immediate: on setting

$$\sin i/\sin r = I/R,$$

we deduce $\tan i/\tan r = p+1$ (note (28)) and so $\cos i/\cos r = I/(p+1)R$; hence, since

$$(I/R)^2 = \sin^2 i+(p+1)^2\cos^2 i,$$

$$\frac{CN}{ND} = \sec i = \sqrt{\frac{(p+1)^2-1}{I^2/R^2-1}} \quad \text{and so} \quad \frac{CN}{CD} = \operatorname{cosec} i = \sqrt{\frac{(p+1)^2-1}{(p+1)^2-I^2/R^2}}.$$

(Compare Halley's 'De Iride': 219; and Jakob Hermann, 'Méthode Géométrique & Générale de déterminer le Diamétre de l'Arc-en-Ciel, quelque Hypothèse de la Refraction qu'on suppose dans l'eau, ou dans toute autre liqueur transparente. Et le Diamétre de l'Arc-en-Ciel étant donné par Observation, en trouver la Raison de la Refraction', *Nouvelles de la Republique des Lettres* (June 1704): §IV: 658–71, especially 663–8.)

(56) It is just possible that the reading is 'conseramus' (grapple with).

(57) In the preceding 'Sectio 3ª. De Planorum Refractionibus' (ULC. Dd. 9.67: 39–64 [= *Optical Lectures* (note (2)): 97–172]), not here reproduced.

(58) Newton has cancelled the continuation 'similiter incidentium' (similarly incident).

bilium, eritq; ang: CNI. ang: $CNP::I.P$, et ang: CNP. ang: $CNT::P.T$. ac divisim ang: INP. ang: $CNP::P-I.P$, et ang: CNP. ang: $PNT::P.P-T$. et ex æquo ang: INP. ang $PNT::P-I.P-T$.

Sume jam arcum BM æqualem arcui BN, et radiorum secundū AM incidentium duc refractos MF, $M[f]$ prioribus occurrentes in V et X, age VX, et produc donec occurrat incidentibus radijs ad G et H. et patet VX esse latitudinem minimi spatij in quod omnes radij congregari possunt. Estq;

$$GX.VX(::\text{ang}\,GNX.\text{ang}\,VNX\ \text{proximè}::\text{ang}\,INP.\text{ang}\,PNT)::P-I.P-T.$$

Et $GH+VX(2GX).VX::2[P]-[2]I.P-T$. ac divisim

$$GH.VX::P+T-2I.P-T.$$

Unde datis P, T, et I dabitur ratio GH ad VX.(59)...

Schol. Ope hujus et Prop 31 errores homogeneorum radiorum quæ in sphæricis superficiebus per figuræ ineptitudinem eveniunt cum heterogeneorum erroribus conferi possunt et constabit hosce longe majores esse in parvis sphærarum portionibus: Atq; adeò heterogeneitatem Lucis et non ineptitudinem figuræ sphæricæ in causa esse quod Telescopia in majorem perfectionis gradum nondum promota habeamus.(60)... In Microscopijs quidem errores homogeneorum radiorum ex sphæricâ superficie vitri objectivi propter aperturam bene magnam enormes oriuntur et admodum sensibiles Adeo ut illa vitra si secundū conicam aliquam sectionem debitè formarentur, paulò perfectiora

(59) In a following numerical example (here omitted) Newton takes over a preceding result that the limits of dispersion of white light refracted from glass into air are given by $I:P:T = 44\frac{1}{2}:69\frac{1}{2}:68\frac{1}{2}$ and so computes $GH:VX = (P+T-2I):(P-T) = 49:1$ 'circiter'.

(60) We omit Newton's following numerical example which justifies this remark. Briefly, he asks the reader to conceive that NBM represents the object glass of a telescope whose front surface NM is plane, so that 'parallel' rays (from a distant object) entering from the left are refracted only at the rear, spherical surface NBM. He next sets the latter's radius $CB = 120''$ and the aperture $NM = 2''$. By Corollary 7 of Proposition 31 he then finds the 'error lateralis homogeneorum radiorum ortus ab ineptitudine figuræ Sphæricæ PQ' to be

$$\tfrac{1}{4}(R/I)\,(\tfrac{1}{2}NM)^3/CB^2$$

—this should be $\frac{1}{4}(R/I)^2(\frac{1}{2}NM)^3/CB^2$ (compare note (21)) but Newton's main point is unaffected—or about $\frac{17}{633,600}''$ on taking I/R (the mean refractive index of light passing from glass into air) to be 11/17 'circiter'. In contrast, when a comparable estimate is made of the effects of colour dispersion in the same lens (taking $I:P:T = 44\frac{1}{2}:69\frac{1}{2}:68\frac{1}{2}$), since the point A is taken to be at infinity it follows that $GH = NM = 2''$ and therefore (by note (59)) the 'error lateralis heterogeneorum ab invicem in eodem loco concursûs' $VX = \frac{2}{49}''$, that is, some 1500 times greater than PQ, 'Tanta sane disproportio, ut PQ respectu VX pro nullo haberi possit. Error quidem VX ... tantus est ut miror quod objecta per ejusmodi Telescopia tam distinctè videri possint. Sed alterius generis error PQ ... longé minor est quàm qui potest esse sensibilis, et proinde negligendus et indistincta visio erroribus ex heterogeneitate Lucis exortis solummodo tribuenda'. When shortly afterwards Newton wrote to Oldenburg (on 6 February

Then will it be $C\widehat{N}I : C\widehat{N}P = I : P$ and $C\widehat{N}P : C\widehat{N}T = P : T$, so that by dividing the ratios $I\widehat{N}P : C\widehat{N}P = [P-I] : P$ and $C\widehat{N}P : P\widehat{N}T = P : [P-T]$, and from equals $I\widehat{N}P : P\widehat{N}T = [P-I] : [P-T]$.

Now take the arc BM equal to the arc BN and of rays incident along AM draw the refractions MF, Mf meeting the previous ones in V and X, join VX and extend it to meet the incident rays at G and H. It is then evident that VX is the breadth of the least space in which all the rays can congregate. Here $GV : VX (\approx G\widehat{N}X : V\widehat{N}X = I\widehat{N}P : P\widehat{N}T) = [P-I] : [P-T]$ and

$$GH + VX \text{ (or } 2GX) : VX = [2P-2I] : [P-T],$$

and so by dividing the proportion $GH : VX = [P+T-2I] : [P-T]$. Hence when P, T and I are given the ratio of GH to VX will be given.(59)...

Scholium. With the aid of Proposition 31 and the present one the errors of homogeneous rays occurring in spherical surfaces through the inappropriateness of their figure can be compared with the errors of heterogeneous ones, and it will then be admitted that the latter are far greater over small portions of spheres. As a consequence the heterogeneity of light and not the unsuitability of a spherical figure is the reason why we have not yet advanced telescopes to a greater degree of perfection.(60)...In microscopes, to be sure, the errors of homogeneous rays arising from the spherical surface of the object glass due to its extremely large aperture are enormous and quite perceptible—so much so that, if those glasses were to be fashioned appropriately in the shape of some conic

1671/2) summarizing the progress of his optical researches over the preceding half dozen years, he referred explicitly to the present passage of his optical lectures, rather misleadingly antedating it by five years:

'When I understood [that Light consists of Rays differently refrangible]...I saw that the perfection of Telescopes was hitherto limited, not so much for want of glasses truly figured according to the prescriptions of Optick Authors [see note (61)]...as because that Light itself is a *Heterogeneous mixture of differently refrangible Rays*. So that were a glass so exactly figured as to collect any one sort of rays into one point, it could not collect those also into the same point, which having the same Incidence upon the same Medium are apt to suffer a different refraction. Nay, I wondered that seeing the difference of refrangibility was so great as I found it, Telescopes should arrive to that perfection they are now at. For measuring the refractions in one of my Prismes I found that supposing the common *sine* of Incidence upon one of its planes was 44 parts, the *sine* of refraction of the utmost Rays on the red end of the Colours made out of the glass into the Air would be 68 parts, and the *sine* of the utmost rays on the other end 69 parts: So that the difference is about a 24th or 25th part of the whole refraction. And consequently the object-glass of any Telescope cannot collect all the rays which come from one point of an object so as to make them convene at its *focus* in less room then in a circular space whose diameter is the 50th part of the Diameter of its Aperture; which is an irregularity some hundreds of times greater then a circularly figured *Lens* of so small a section as the Object glasses of long Telescopes are, would cause by the unfitness of its figure were Light uniform. This made me take *Reflections* into consideration...finding them regular' (*Correspondence of Isaac Newton*, **1** (1959): 95).

evaderent. Sed methodus tamen me non latet corrigendi errores illos absq; conicis sectionibus et efficiendi ut vitra e sphæricis superficiebus formari possint quæ radios homogeneos satis accuratè refringent, ne dicam quæ longe accuratiùs refringent obliquos radiorum penicillos quàm vitra alijs quibuscunq; figuris terminata.(61) Adeo ut Sphæricas superficies usibus dioptricis præ cæteris omnibus accommodatas esse censeam.

(61) It was (compare I: 551–4) Descartes who, in the 'Discours Huictiesme' of his *Dioptrique* [= *Discours de la Methode* (Leyden, 1637): $_2$89–121: 'Des Figures que doiuent auoir les cors transparens pour detourner les rayons par refraction en toutes les façons qui seruent a la veuë'] had pointed out that conics accurately refract rays of white light parallel to their major axis so as to make them converge on their further focus. This led Robert Hooke, for example, to assert in his *Micrographia: or some Physiological Descriptions of Minute Bodies made by Magnifying Glasses. With Observations and Inquiries thereupon* (London, 1665) that 'if Glasses could be made of those kind of Figures, or some other, such as the most incomparable Des Cartes has invented, and demonstrated in his Philosophical and Mathematical Works, we might hope for a much greater perfection of Opticks then can be rationally expected from spherical ones' (Preface: [e2r]). In February 1672 Oldenburg had passed Newton's letter of the 6th (note (60)) on to Hooke for comment. The latter, in his reply on 15 February (*Correspondence of Isaac Newton*, **1**: 110–14, especially 111), had been quick to pick up Newton's pessimism regarding the future of the refracting telescope, 'since it is not improbable, but that he that hath made soe very good an improvement of telescopes by his own tryalls upon Reflection, would, if he had prosecuted it, have done more by Refraction. And that Reflection is not the only way of improving telescopes, I may possibly hereafter shew some proof'. In a draft of his letter to Oldenburg of 11 June 1672 (*Correspondence*, **1**: 171–82) Newton answered Hooke's charge at length (ULC. Add. 3970.3: 445r/447v, partially reproduced in *Correspondence*, **1**: 191–2), the first part only of which was sent (*ibid.*, **1**: 172):

'I know not whence [Mr Hook] should conclude that I have wholly layd aside the thoughts of improving Telescopes & Microscopes by refractions. What I said in my former letter was in respect of the ordinary construction of those instruments, intimating that their improvement was not to be expected from the well figuring of glasses. But of other constructions I have ever since my first acquaintance wth optiques had considerations, & principally about regulating the refractions of the object glasses. For although successive refractions wch are all made the same way do necessarily more & more augment the errors of the first refraction: yet it seemed not impossible for contrary refractions so to correct each others inequalities as to make their

or other, they would prove to be slightly more perfect. However, a method does not escape me for correcting those errors without using conics and of succeeding in being able to fashion glasses of spherical surface which shall refract homogeneous rays accurately enough—not to say which shall refract oblique pencils of light far more accurately than glasses bounded by any other types of figure.(61) As a result it is my view that spherical surfaces above all others are suited to dioptrical purposes.

---

difference regular. And if so, there would then be no further difficulty Now to this end I have not onely examined what may be effected by one similar Object-glasse, but more especially by a complication of divers successive mediums, as by two or more glasses or chrystalls wth water or some other fluid between them, all wch together may perform the office of ane object-glasse. [Newton has cancelled a continuation 'But what the results in Theory or by tryalls have been I may possibly find a more seasonable occasion to declare. At present I shall onely tell you that they are inferior to reflexions.'] But though this way faile[d] my expectations as to regulating the refractions of heterogeneall rays yet as to homogeneall rays I found something in it wch Opticians have hitherto sought for in vain: Namely that those rays whether parallel diverging or converging may by such successive Mediums of sphericall figures be much more exactly refracted to or from a point then they can by one similar object glasse of the like sphericall figures....And this way I should prefer before the use of the conic sections because here the rays will be as accurately refracted to points that are on all sides distant from the principall focus, as to the principall focus it selfe, wch cannot be effected by the conic sections.

'If this be made use of for the Object-glasse of a Microscope the advantage will be very sensible although the Objects be illuminated wth heterogeneall light such as is the immediate light of the Sun or a Candle. But it will be much more sensible by illuminating them wth homogeneall light of any convenient colour such as may in some measure be copiously cast on it in an obscure Room wth Prisms & Lenses. For so the Microscope will beare a larger aperture & deeper charge, & perhaps wthout glareing admit of a stronger light to be cast on the Object.

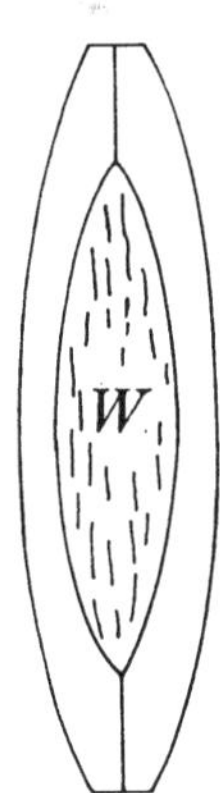

'As for Telescopes it would be necessary to make their Object glasses more flat as you see them represented in this figure where *W* designes the double convex water between the two convexo-concave glasses. [Compare 1: 575–6; and Appendix 4 below.] But since it is not in o^{r} power to illuminate remote Objects wth homogeneall light, I cannot foresee any considerable advantage that will accrew to these instruments by this Method.'

2

MISCELLANEOUS RESEARCHES INTO REFRACTION AT A CURVED INTERFACE[1]

[c. 1671]

From autograph drafts in the University Library, Cambridge

§1. THE THEORY AND CONSTRUCTION OF SPHERICAL LENSES.

[1] THEOREMATA OPTICA.[3]

[4]Si radius divergens a puncto dato A vel convergens ad punctum idem A incidit in Sphæram CVD ad punctum D, sit Sphæræ centrum C,[5] & secet AC

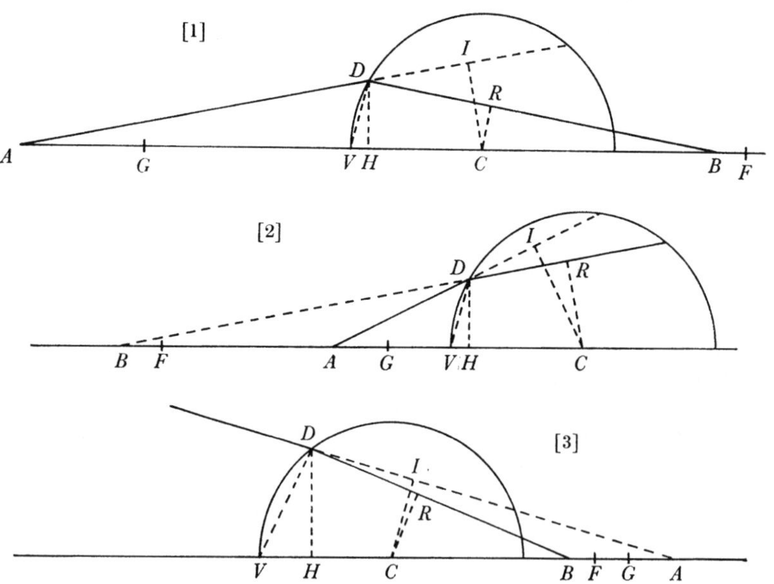

(1) These papers, dealing with the theory of the (thin) spherical lens and its application to the construction of a telescope in which chromatic distortion is minimized, together with the fabrication of a general lens which accurately refracts monochromatic light radiating from a point source to a point focus, are evidently related to the last section (reproduced in 1, §2 above) of the first part of the manuscript 'optical lectures for 1671–2' (ULC. Dd. 9.67, self-styled as 'Optica') deposited by Newton in 1674 in the archives of Cambridge University. Indeed, it is tempting to look upon them as fragments of an intended sequel in which, following

Translation

[1] Optical Theorems(3)

(4) If a ray diverging from the given point A or convergent to it falls upon the sphere CVD at the point D, let the sphere's centre be C(5) and let AC produced

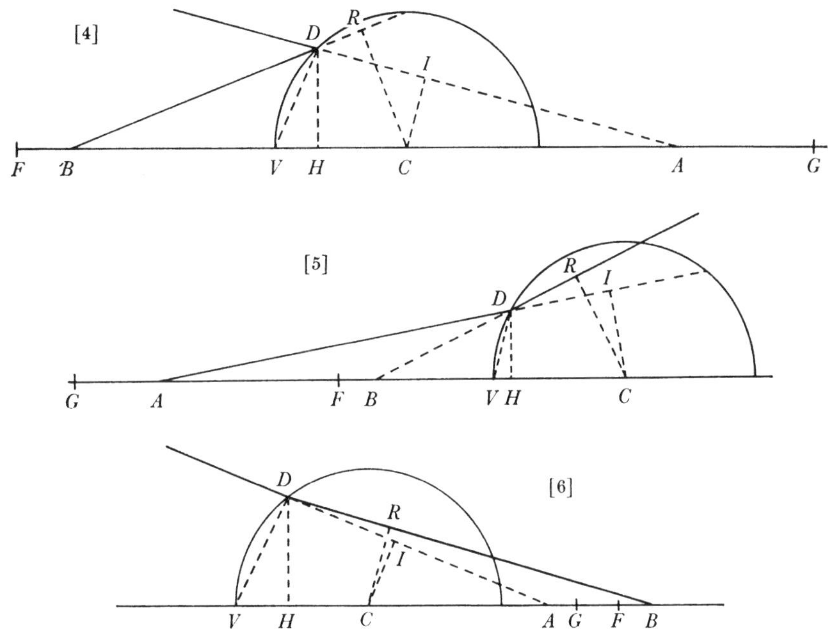

on Isaac Barrow's *Lectiones XVIII* (1, §2: note (7))—particularly its fourteenth lecture (pages 96–104)—and his own adumbration of the topic in Propositions 30 and 31 of the 'Optica', the theory of the refracting 'glass' lens was discussed at length. We might even conjecture that this topic was one of the subjects touched upon in his Lucasian lectures for 1672–3, of which (despite much recent speculation) nothing definite is known.

(2) Apart from the internal evidence discussed in the previous note, Newton's handwriting style fully confirms this date.

(3) Add. 4004: 71^r–72^v/72 bis^r.

(4) The generalization (to the case where the radiating point A is at a finite distance) promised by Newton in the scholium to Proposition 31 of his 'Opticæ Pars Prima' (see 1, §2: note (28)). On taking $AV = u$, $FV = v$ and $CV = r$ the basic refractive condition here established, $AC \times VF = (I/R)\,AV \times CF$, becomes $(u-r)(-v) = (I/R)\,u(r-v)$; that is,

$$1/r - 1/u = (I/R)\,(1/r - 1/v).$$

The six accompanying figures illustrate the various possibilities which arise: namely, Figures 1–3 represent (for $I > R$) the cases $1/r + 1/u = (I/R)\,(1/r - 1/v)$, $1/r + 1/u = (I/R)\,(1/r + 1/v)$ and $1/r - 1/u = (I/R)\,(1/r - 1/v)$; and Figures 4–6 correspondingly denote (for $I < R$) the cases $1/r - 1/u = (I/R)\,(1/r + 1/v)$, $1/r + 1/u = (I/R)\,(1/r + 1/v)$ and $1/r - 1/u = (I/R)\,(1/r - 1/v)$. The cases $1/r - 1/u = (I/R)\,(1/r + 1/v)$ with $I > R$, and $1/r + 1/u = (I/R)\,(1/r - 1/v)$ with $I < R$ are impossible.

(5) 'in recta AB situm' (sited in the line AB) is cancelled.

producta Sphæram in V et radium refractum DR in B: a punctis D et C ad AB, AD, BD demitte normales DH, CI, CR; sitq; sinus incidentiæ ad sinum refractionis seu CI ad CR ut I ad R; et facto $R, AC.I, AV::CF.VF$. erit F focus,(6) seu locus imaginis puncti A radios quaqua versum emittentis.

2do. A puncto V versus A cape VG ad VA ut est R ad I et error radij refracti DR a loco imaginis in axe AV, seu distantia punctorum B et F erit

$$\frac{AC,\ FC,\ FG,\ VD^q}{2AV^q,\ CV^q}, \quad \text{sive} \quad \frac{AC,\ FC,\ FG,\ VH}{AV^q,\ CV}$$

quamproxime.(7)

3. Ubi punctum [A] infinite distat ita ut radius incidens parallelus sit axi, pro AC scripto AV, et pro FG scripto $\frac{R,\ VA}{I}$ (nam hæc jam sunt æquipollentia:)(8) error BF fiet $\frac{\frac{R}{I}FC,\ VD^q}{2CV^q}$, vel $\frac{\frac{R}{I}FC,\ VH}{CV}$, vel $\frac{RR}{II-IR}VH$.(9)

4. Si radius non refringitur sed reflectitur a superficie Sphærica VD, eadem regula obtinet si modo ponatur $I.R::1.-1$,(10) et perinde capiatur VG ad contrarias partes VA fiatq; ipsi VA æqualis. Erit enim adhuc error

$$FB = \frac{AC,\ FC,\ FG}{AV^q,\ CV}VH,$$

vel $AV^q \times CV.ACF \times FG::VH.FB$.

Hujus autem Theorematis inventio talis est.(11) [Patet quod]

$$AD^q = AV^q + 2AC,\ VH. \quad DF^q = VF^q - 2CF,\ VH.^{(12)}$$

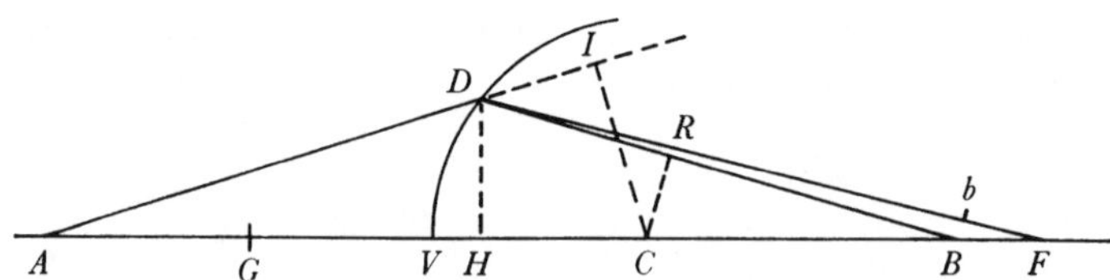

Et extractis radicibus, $AD = AV + \frac{AC}{AV}VH - \frac{AC^q}{2AV^{cub}}VH^q$ &c.

$$DF = VF - \frac{CF}{VF}VH - \frac{CF^q}{2VF^{cub}}VH^q \quad \text{\&c.}$$

(6) 'focus principalis' (principal focus), that is. Explicitly $CF = \frac{R \times AC \times CV}{I \times AV - R \times AC}$ (compare 1, §2: Proposition 29, Corollary 1).

intersect the sphere in V and the refracted ray DR in B. From the points D and C let fall the normals DH, CI, CR to AB, AD, BD and take the sine of incidence to the sine of refraction, that is, CI to CR, as I to R. Then, on making

$$R \times AC : I \times AV = CF : VF,$$

F will be the focus(6) or place of the image of the point A emitting rays in any direction.

2. Take VG (in the direction from V to A) in proportion to VA as R to I: then the error of the refracted ray DR from the place of the image in the axis AV, that is, the distance between the points B and F, will be

$$\frac{AC \times FC \times FG \times VD^2}{2AV^2 \times CV^2} \quad \text{or} \quad \frac{AC \times FC \times FG \times VH}{AV^2 \times CV}$$

very approximately.(7)

3. When the point A is infinitely distant, resulting in the incident ray being parallel to the axis, on writing AV in place of AC and FG in place of $R \times VA/I$ (for these are now equivalent)(8) the error BF will become

$$\tfrac{1}{2}(R/I)\, FC \times VD^2/CV^2, \quad \text{or} \quad (R/I)\, FC \times VH/CV, \quad \text{or} \quad [R^2/I(I-R)]\, VH.^{(9)}$$

4. If the ray is not refracted but reflected by the spherical surface VD, the same rule will hold provided only there be put $I/R = -1$(10) and, correspondingly, VG be taken the opposite way to VA and equal to it. For the error FB will still be equal to $(AC \times FC \times FG/AV^2 \times CV) \times VH$ and so

$$AV^2 \times CV : AC \times CF \times FG = VH : FB.$$

The method, however, of finding the theorem is this.(11) Evidently

$$AD^2 = AV^2 + 2AC \times VH \quad \text{and} \quad DF^2 = VF^2 - 2CF \times VH.^{(12)}$$

On extracting their roots $AD = AV + (AC/AV)\, VH - \tfrac{1}{2}(AC^2/AV^3)\, VH^2 \ldots$, while

$$DF = VF - (CF/VF)\, VH - \tfrac{1}{2}(CF^2/VF^3)\, VH^2 \ldots.$$

(7) Newton 'proves' this below. A scarcely relevant continuation 'Sed regula prior ni fallor melior est' (But the former rule, unless I am mistaken, is better) is cancelled.

(8) In the limit as A passes to infinity the ratio AC/AV becomes unity.

(9) Since $VD^2 = VH \times 2CV$ and $FC/CV = R/(I-R)$. The result was established analytically in 1, §2: Proposition 31, Corollary 1.

(10) Compare I: 576, note (61), last paragraph.

(11) The preliminary worksheets for the following computation are reproduced in Appendix 1. 4/5 below. Note that Newton has borrowed his terminology of 'fl[uxio]' and 'defl[uxio]' from **1**, 2, §3 above.

(12) $AD^2 = AV^2 + DV^2$ (or $2VC \times VH$) $+ 2AV \times VH$ and $DF^2 = VF^2 + DV^2 - 2VF \times VH$.

[Quare] fluxio ipsius $AD = \frac{AC}{AV}\text{fl}: VH - \frac{AC^q, VH}{AV^{cub}}\text{fl}\, VH$ &c. Defluxio ipsius $DF = -\frac{CF}{VF}\text{fl}\, VH - \frac{CF^q, VH}{VF^{cub}}\text{fl}\, VH$ &c. [adeoq;]

$$\frac{R}{I}\text{fl}: AD + \text{defl}: DF = -\frac{R, AC^q}{I, AV^{cub}} - \frac{CF^q}{VF^{cub}} \text{ in } VH, \text{ fl}\, VH \text{ \&c.}$$

Quod si nihil esset radij omnes accuratè refringerentur ad focum F. Tunc enim AD et DF fluerent in data ratione, juxta ea quæ Cartesius in Optica(13) probavit: Sed quia nihil non est, obliquitatis error superficiei VD erit ut illud $\frac{R}{I}\text{fl}\, AD + \text{defl}: DF$. Et ut error ille sive defluxio a legitima obliquitate(14) ita error angularis radij refracti. Jam vero est $\frac{R, AC}{I, AV} = \frac{CF}{VF}$ Ergo error angularis radij refracti est ut $\frac{AC}{AV^q}[+]\frac{CF}{VF^q}$ in $\frac{CF, VH}{VF}$, fl VH. vel etiam ut $\frac{I, CF}{R, AV, VF} + \frac{CF}{VF^q}$ in $\frac{CF, VH}{VF}$, fl VH seu ut

$$VF + \frac{R}{I}AV \text{ in } \frac{CF^q, VH}{VF^c, AV}\text{fl}\, VH \quad \text{seu} \quad \frac{FG, CF^q, VH}{AV, VF^{cub}}$$

posito fl $VH = 1$ & $VF + \frac{R}{I}AV = FG$. Datur autem ratio $\frac{CF}{VF}$ ad $\frac{AC}{AV}$. Ergo substituto posteriore fiet error ille ut $\frac{FG, CF, AC, VH}{AV^q, VF^q}$. Duc in VF^q(15) et error in axe FB erit ut $\frac{FG, CF, AC, VH}{AV^q}$ quando circuli radius determinatur. Divide per radium circuli(15) et fiet $\frac{FG, CF, AC, VH}{CV, AV^q}$ ut error BF in omni casu. Dato igitur

(13) Read 'Geometria' (*Geometrie*). See **1**, 2, §2: note (228). In this case, since

$$AD + (I/R)\, DF = AV + (I/R)\, VF,$$

constant, it follows at once (in Newton's terminology) that fl: $AD = -(I/R)$ fl: DF.

(14) Literally 'legitimate obliquity': Newton first wrote 'a legitima figura' (from the legitimate figure). This step in his argument is a makeshift one, for he neither defines what he means by 'error angularis' (presumably $F\hat{D}B$) nor states why it should be proportional to the 'error obliquitatis' $[(R/I) \times \text{fl}: AD + \text{fl}: DF]/\text{fl}: VH$.

(15) This successive multiplication (by VF^2) and division (by CV) come like a bolt out of the blue! It is not obvious, following Newton's argument, why the factor VF^2/CV should transform the 'error angularis' into the 'error in axe FB'. The reader may appreciate the following modernized recasting of it.

Since

$$AD = AV + \frac{AC}{AV}.VH - \frac{1}{2}\frac{AC^2}{AV^3}.VH^2 + O(VH^3)$$

Accordingly, $\text{fl}\,(AD) = (AC/AV - AC^2 \times VH/AV^3 \ldots)\,\text{fl}\,(VH)$, while

$$-\text{fl}\,(DF) = -(CF/VF + CF^2 \times VH/VF^3 \ldots)\,\text{fl}\,(VH),$$

so that

$$(R/I)\,\text{fl}\,(AD) - \text{fl}\,(DF) = -(R \times AC^2/I \times AV^3 + CF^2/VF^3 \ldots)\,VH \times \text{fl}\,(VH).$$

If this were zero, all rays would be refracted accurately to the focus F, for then the fluents AD and DF would be in given ratio and so satisfy Descartes' test in his *Dioptrique*.(13) But since it is not zero, the skew error of the surface VD will be as that quantity $(R/I)\,\text{fl}\,(AD) - \text{fl}\,(DF)$; and that error, in other words the fluxional departure from the correct skewness,(14) will be as the angular error of the refracted ray. Now to be sure $R \times AC/I \times AV = CF/VF$ and the angular error of the refracted ray is therefore as $\left(\frac{AC}{AV^2} + \frac{CF}{VF^2}\right) \times \frac{CF \times VH}{VF}\,\text{fl}\,(VH)$, or in addition as $\left(\frac{I \times CF}{R \times AV \times VF} + \frac{CF}{VF^2}\right) \times \frac{CF \times VH}{VF}\,\text{fl}\,(VH)$, or as

$$\left(VF + \frac{R}{I}AV\right) \times \frac{CF^2 \times VH}{VF^3 \times AV}\,\text{fl}\,(VH),$$

that is, $FG \times CF^2 \times VH/AV \times VF^3$ on setting $\text{fl}\,(VH) = 1$ and

$$VF + (R/I)\,AV = FG.$$

But the ratio of CF/VF to AC/AV is given and so, when the latter replaces it, the error will become as $FG \times CF \times AC \times VH/AV^2 \times VF^2$. Multiply by VF^2(15) and the error in the axis FB will be as $FG \times CF \times AC \times VH/AV^2$ when the circle's radius is determined. Divide by the circle's radius(15) and in every case

$$FG \times CF \times AC \times VH/CV \times AV^2$$

will come to be as the error BF. When, therefore, that error is given in one case,

---

and $$DB = VB - \frac{CB}{VB}.VH - \frac{1}{2}\frac{CB^2}{VB^3}.VH^2 - O(VH^3),$$

therefore $$\frac{R}{I}.\frac{d(AD)}{d(VH)} + \frac{d(DB)}{d(VH)} = \left(\frac{R}{I}.\frac{AC}{AV} - \frac{CB}{VB}\right) - \left(\frac{R}{I}.\frac{AC^2}{AV^3} + \frac{CB^2}{VB^3}\right).VH + O(VH^2) = 0$$

for any ray AD near to AV which is refracted along DB. Hence in the limit as D passes into V (and so B into F) $$\frac{CF}{VF} = \frac{R}{I}.\frac{AC}{AV} \quad \text{and so} \quad CF = \frac{R \times AC \times CV}{I \times AV - R \times AC};$$

accordingly, since

$$CF/VF - CB/BV = CV/BV - CV/VF = CV \times BF/VF \times VB,$$

in this limit $$\frac{CV \times BF}{VH \times VF^2} = \frac{R}{I}.\frac{AC^2}{AV^3} + \frac{CF^2}{VF^3}$$

errore illo in uno casu datur in omni. At in eo casu ubi est radius incidens axi parallelus datur error ille idem cum quantitate $\frac{FG, CF, AC}{CV, AV^q}$ VH ergo semper idem est cum hac quantitate.

(16)PROBL. [I.]

HABITA LENTE PLANO-CONVEXA, INVENIRE TUM CONVEXITATEM, TUM REFRACTIONEM VITRI.

Sit Lens *RS*, ejus superficies plana *RTS*, convexa *RVS*, axis *KF*. Lentis superficie plana solem respiciente, observentur imaginum solarium a radijs tum trajectis tum reflexis in charta obversa distinctissimè pictarum loci duo(17) *F*

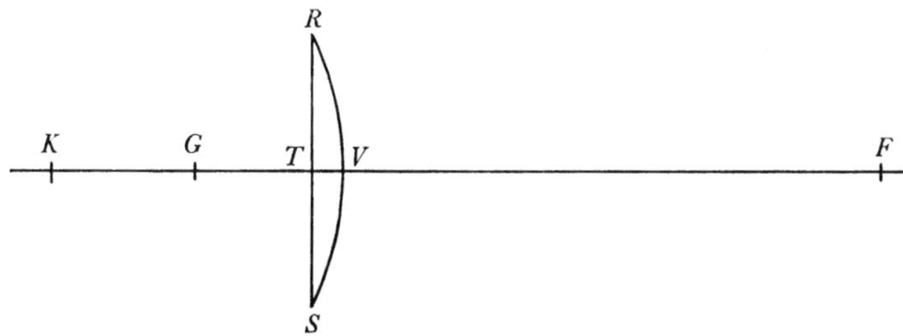

et *G*; *F* locus imaginis trajectæ, *G* locus reflexæ: et mensurentur quam accuratissime distantiæ *VF*, *TG*, ut et crassities vitri *TV*. Dein fac(18) ut $VF+2TV$ ad $VF-2GT$, ita sinus incidentiæ ex ære in vitrum ad sinum refractionis, ita *KT* ad *GT* et erit $2KV$ radius circuli *RVS*.(19)

---

and therefore, to $O(VH)$,

$$\frac{BF}{VH} = \left(\frac{R}{I}\cdot\frac{AC^2}{AV^3}+\frac{CF^2}{VF^3}\right)\cdot\frac{VF^2}{CV} = \left(\frac{I}{R\times AV}+\frac{1}{VF}\right)\cdot\frac{CF^2}{CV}$$

$$= \left(VF+\frac{R\times AV}{I}\right)\cdot\frac{I}{R\times AV}\cdot\frac{CF^2}{CV\times VF} = \frac{GF\times CF\times AC}{CV\times AV^2},$$

since $R\times AV/I = GV$ and $I\times CF/R\times VF = AC/AV$. Hence

$$BF = (AC\times VH/AV^2\times CV)\,CF\times FG.$$

On taking $AV = a$, $VH = x$, $CV = r$ and $AC = b = a-r$, it follows that $CF = Rbr/(Ia-Rb)$ and therefore, since

$$FG = VF+(R/I)\,AV = \frac{Iar}{Ia-Rb}+\frac{Ra}{I},\quad \text{at last}\quad BF = \frac{Rb^2}{Ia}\left(\frac{R}{Ia-Rb}+\frac{I^2r}{(Ia-Rb)^2}\right)x \quad\text{to}\quad O(x^2).$$

Since Newton had computed this 'error in axe' five years before (see I: 573–4, where the Cartesian refractive index e/d replaces the present Barrovian equivalent I/R), we may well incline to think that here he is merely concerned to contrive its synthetic proof without worrying too much about steps in his argument.

(16) The two following applications to the computation of the refractive power of given lenses are drafted on ULC. Add. 3970.11: 635r. Significant variants are noted in following footnotes.

it is given in all. But in the case where the incident ray is parallel to the axis that error is given to be identical with the quantity

$$(FG \times CF \times AC/CV \times AV^2)\ VH$$

and is therefore always identical with this quantity.

(16)Problem I

Where a plano-convex lens is had, to find both its convexity and the refraction of glass.

Let the lens be RS, its plane surface RTS, its convex one RVS, its axis KF. With the plane surface facing the sun, observe the two positions F and G of the solar images most sharply depicted by both transmitted and reflected rays on an opposing sheet of paper,(17) F being the position of the transmitted image, G that of the reflected one; then measure as precisely as possible the distances VF, TG and also the thickness TV of the glass. Next make(18) $VF+2TV$ to $VF-2GT$ as the sine of incidence from air into glass to that of refraction and as KT to GT, and $2KV$ will be the radius of the circle RVS.(19)

(17) In first draft Newton wrote 'observentur loci imaginis solaris ubi radij trajecti tum reflexi [accuratissime] congregantur et imaginem solis in charta obversa distinctissime terminatam pingunt' (observe the positions of the solar image where the transmitted and also the reflected rays most closely crowd together, painting an image of the sun with sharpest definition on a held up sheet of paper). An uncompleted revised version of this is cancelled on the present manuscript.

(18) In repetition of his draft, Newton first continued 'fac $VF-2GT.GT::TF+2TV.KT$, et erit $2KV$ semidiameter circuli RVS. Et sinus incidentiæ ex aere in vitrum erit ad sinum refractionis ut KT ad GT vel ut KF ad VF' (make $VF-2GT$ to GT as $TF+2TV$ to KT, and $2KV$ will be the radius of the circle RVS. And the sine of incidence from air into glass will be to the sine of refraction as KT to GT or as KF to VF).

(19) Here an incident ray parallel to the base CTV meets the plane interface RT of the lens (say in a) and passes on without deviation till it meets the second, spherical interface RV (of radius CV) in b; it is there partially refracted to F, partially reflected (through K) till it

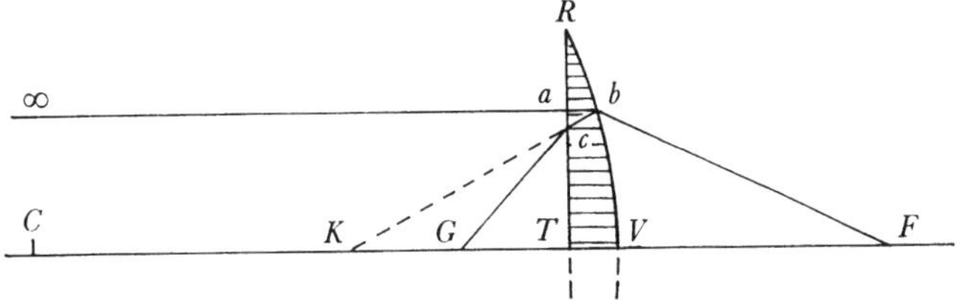

meets the plane interface again in c, where it is refracted to G. Hence, on taking the refractive index from air into glass as I/R, $CV/VF = (I-R)/R$, while correspondingly $CV = 2KV$ and $GT/KT = R/I$. Accordingly, $VF = [2R/(I-R)]KV = [2I/(I-R)]GT+[2R/(I-R)]TV$, so that $I(VF-2GT) = R(VF+2TV)$.

PROBL: [2.]

HABITA LENTE QUAVIS CONVEXO-CONVEXA, VEL ETIAM CONVEXO-CONCAVA CUJUS CONCAVITAS SIT CONVEXITAT[E] MULTO MINOR, INVENIRE TUM REFRACTIONEM VITRI, TUM CONVEXITA[TEM] LENTIS.

Sit Lens *RS*, superficies magis convexa *RVS*, minus convexa vel concava *RTS*, axis *kF*, vertices *V* ac *T*. Lentis hujus superficie minus convexa vel concava *RTS* solem directe respiciente, observentur quam accuratissimè solaris imaginis in charta obversa distinctissimè pictæ tam trajectæ locus *F* quam reflexæ locus *G*, et mensurentur distantiæ *VF*, *TG*, et crassities vitri *TV*. Dein altera Lentis superficie *RVS* solem respiciente observetur qu[am accuratissimè] locus imaginis reflexæ *H* et mensuretur distantia *VH*, quæ est imaginis illius a vitro. Biseca *TV* in *X*.(20) Et fac $\frac{1}{HX}-\frac{2}{FX}=A$. Et $\frac{1}{GX}+A=B$. Et $\frac{1}{B}=gX$. Et Len[s] plano-convexa ex consimili vitro confecta cujus vertices sint *T*, *V*, et convexitas(21) versus *F* sita æque[t] summæ convexitatum *RTS*, *RVS* in fig. 1 vel differentiæ convexitatis et concavitatis in fig. 2, projicere[t] solis imaginem refractam ad locum priorem *F*, reflexam vero ad locum [*g*] quam proximè.

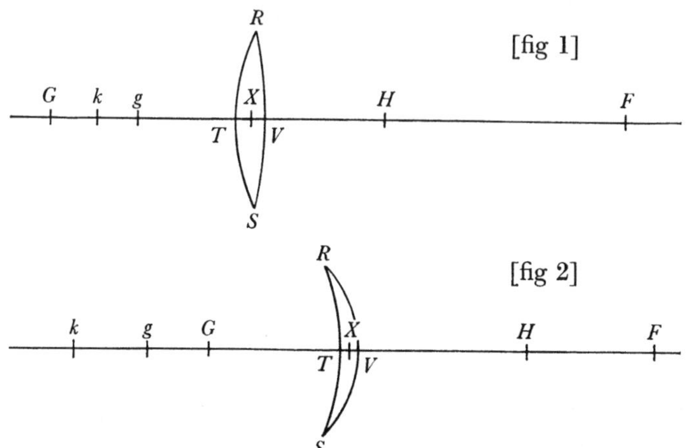

Unde † si fiat (juxta Problema prius) $VF-2gT.gT::VF+2TV.kT$ erit sinus incidentiæ ex ære in vitrum ad sinum refractionis ut *kT* ad *gT*, vel ut *kF* ad *VF*. Sit ista ratio *I* ad *R*, et erit $\frac{4I-2R}{R\times A}$(22) radius circuli *RT*[*S*]. Sit iste *C*. Fac $\frac{1}{2[k]V}$(23) $-\frac{1}{C}=D$ et erit $\frac{1}{D}$ radius circuli alterius(24) [*RVS*, posito quod superficies *RTS* sit convexa. Sin autem (fig. 2) illa superficies sit concava, *C* erit negativus] et poni debet $\frac{1}{2[k]V}$(23) $+\frac{1}{C}=D$.(25)

† *juxta problema præcedens erit* $VF-2gT$ *ad* $VF+2TV$ *ut sinus refractionis ad sinum incidentiæ.*

(20) In his draft Newton here continued 'vel melius seca in *X* ita ut sit $TX.XV::4.9$. vel forte $TX^q.VX^q::[G]T.VH$' (or, better, cut it in *X* so that $TX:XV = 4:9$. or maybe

Problem 2

Where any doubly convex lens is had, or also a convexo-concave one whose concaveness shall be much less than its convexity, to find both the refraction of the glass and the convexity of the lens

Let the lens be RS, its more convex surface RVS, its less convex or concave one RTS, its axis kF, its vertices V and T. With the less convex or concave surface RTS of this lens directly facing the sun, observe both the place F of the transmitted solar image and the place G of the reflected one when these are most sharply depicted on an opposing sheet of paper, and then measure the distances VF, TG together with the thickness TV of the glass. Next, with the other surface RVS of the lens facing the sun, observe as precisely as possible the place H of the reflected image and measure the distance VH of that image from the glass. Bisect TV in X[20] and make $1/HX-2/FX=A$, $1/GX+A=B$ and $1/B=gX$. Then a plano-convex lens fabricated from exactly similar glass, with its vertices at T, V and its convexity,[21] located towards F, equal to the sum of the convexities RTS, RVS in figure 1 or to the difference between the convexity and concaveness in figure 2, would project a refracted image of the sun to the previous position F, but a reflected one to the position g very approximately. Hence if (according to the previous problem)† there be made $[VF-2gT]:gT=[VF+2TV]:kT$, the sine of incidence from air into glass to the sine of refraction will be as kT to gT or as kF to VF. Let that ratio be I to R and then $(4I-2R)/R\times A$[22] will be the radius of the circle RTS. Let that be C, make $1/2kV$[23]$-1/C=D$ and $1/D$ will be the radius of the other circle[24] [RVS, supposing the surface RTS to be convex. But if (figure 2) that surface be concave, C will be negative] and there ought to be put $1/2kV$[23]$+1/C=D$.[25]

† *According to the preceding problem it will be $VF-2gT$ to $VF+2TV$ as the sine of refraction to the sine of incidence.*

$TX^2:VX^2 = GT:VH$). The reason for this proposed division of TV is not obvious and Newton's cancellation of it evidently mirrors his dissatisfaction with its usefulness. See note (25).

(21) The effective curvature of the lens, measured by the sum of the reciprocals of the radii of the spherical interfaces RVS, RTS. See note (25).

(22) Read '$\frac{2}{A}$' simply. Newton carries his error into the following example (see note (28)).

(23) The manuscript reads '$\frac{1}{2gV}$'.

(24) The bottom of the manuscript page (f. 72ʳ) has wholly disintegrated and the following parenthesis is our restoration.

(25) We may paraphrase Newton's argument as follows (compare note (19)). Given any lens RTV (convexo-convex in Newton's figure 1, concavo-convex in his figure 2), we may suppose that a ray which enters horizontally from the left is refracted at a in the direction of F' to b, where it is partially refracted a second time to F, partially reflected in the direction of

Exempli gratia.(26) In Telescopij cujusdam(27) vitro objectivo observabam $VF = 13^{\text{ped.}}\ 11^{\text{digit.}}$ $VH = 6^{\text{ped.}}\ 9\frac{13}{16}^{\text{digit}}$. $TG = 2^{\text{ped}}\ 4\frac{13}{16}^{\text{dig}}$. et $TV = \frac{2}{9}^{\text{dig}}$. seu $VF = 167^{\text{dig.}}$ $VH = 81{,}8125^{\text{dig}}$. $TG = 28{,}8125^{\text{dig}}$. $TV = 0{,}2222^{\text{dig.}}$ &c. Adeoq; $XF = 167^{\text{d}}{,}1111$. $XH = 81^{\text{d}}{,}9236$. $XG = 28^{\text{d}}{,}9236$. Unde prodit $A = 0{,}0002384^{\text{dig}}$. $B = 0{,}0348122^{\text{dig}}$. $gX = 28{,}7256^{\text{dig}}$. $VF + 2TV = 167{,}4444$ &c.

$$VF - 2gT = 109{,}771.$$

Ergo 167,444. 109,771::$I.R$. vel in minoribus numeris 29.19::$I.R$ aut magis accuratè 90.59 [vel] 61.40::$I.R$. [Nec non] $\dfrac{4I - 2R}{RA} = 17161^{\text{dig}} = 1430^{\text{ped}}$.(28) Unde circuli alterius RVS semidiameter erat quasi $7^{\text{ped}}\ 4^{\text{dig}}$. Atq; hæc ita se habebant in vitro objectivo Telescopij D^ris^ Babington.(29)

In altero Telescopio quod erat in archivis Academiæ,(30) mensuravi distantiam imaginis trajectæ a vitro objectivo $VF = 14^{\text{ped}}\ 3^{\text{dig}} + \frac{9}{10}^{\text{dig}}$(31)

G' to c, where it is refracted to G. Correspondingly, a ray entering horizontally from the right is refracted at α, reflected at β and again refracted at γ to H. For convenience, take the re-

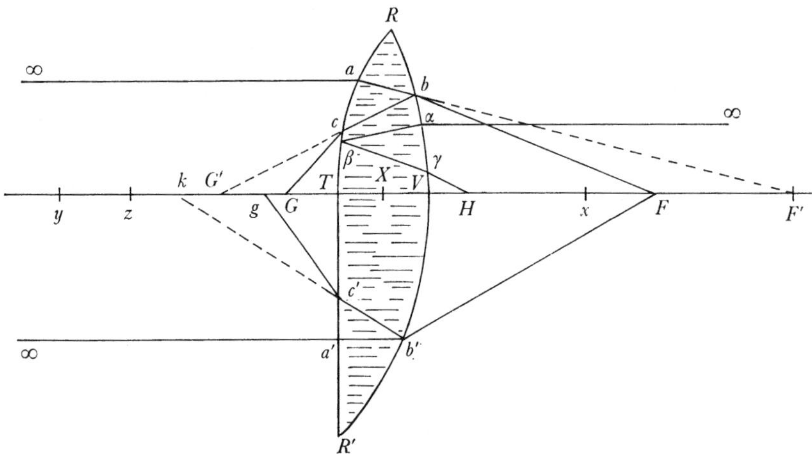

fractive index of the outer medium to the material of the lens to be μ—Newton, following Barrow, denotes this by I/R—and, where X is some convenient mean point within the axis TV of the lens, suppose $XF' = u$, $XF = v$, $G'X = v'$ and $GX = w$ and that the radii of the interfaces RT, RV are $Tx = r$, $yV = s$, while $TX = \epsilon$, $XV = \epsilon'$. First,

$$\frac{1}{r} = \mu\left(\frac{1}{r} - \frac{1}{u+\epsilon}\right) \quad \text{and} \quad \frac{1}{s} + \frac{1}{u-\epsilon'} = \mu^{-1}\left(\frac{1}{s} + \frac{1}{v-\epsilon'}\right)$$

defines the passage of the ray incident at a passing through b to F after two refractions. For small ϵ, ϵ'

$$-\frac{\mu-1}{r} + \frac{\mu}{u} \approx \frac{\mu}{u^2}\epsilon \quad \text{and} \quad \frac{\mu-1}{s} + \frac{\mu}{u} - \frac{1}{v} \approx \left(\frac{1}{v^2} - \frac{\mu}{u^2}\right)\epsilon'$$

are very nearly equal, and $1/v = (\mu-1)/\rho + O(\epsilon, \epsilon')$, where $1/\rho = 1/r + 1/s$. (This approximation becomes still more close when $\epsilon/\epsilon' \approx u^2/\mu v^2 - 1$. Compare note (20).) When the thickness of the lens is wholly negligible, its refractive power is equal to that of the plano-convex lens $R'TV$, where the radius of the spherical interface $R'V$ is $zX = \rho$; that is, a horizontal ray

For instance,[26] in the object glass of a certain telescope[27] I observed $VF = 13$ ft. 11 in., $VH = 6$ ft. $9\frac{13}{16}$ in., $TG = 2$ ft. $4\frac{13}{16}$ in. and $TV = \frac{2}{9}$ in.; that is (in inches) $VF = 167$, $VH = 81{\cdot}8125$, $TG = 28{\cdot}8125$ and $TV = 0{\cdot}2222$, so that $XF = 167{\cdot}1111$, $XH = 81{\cdot}9236$, $XG = 28{\cdot}9236$. From these there results $A = 0{\cdot}0002384$, $B = 0{\cdot}0348122$, $gX = 28{\cdot}7256$, $VF + 2TV = 167{\cdot}4444$,

$$VF - 2gT = 109{\cdot}771.$$

Therefore $I:R = 167{\cdot}444:109{\cdot}771$, or in smaller numbers $I:R = 29:19$ or more accurately $I:R = 90:59$ or $61:40$. Further

$$(4I - 2R)/R \times A = 17161 \text{ in.} = 1430 \text{ ft.}^{(28)}$$

From this the radius of the other circle *RVS* was virtually 7 ft. 4 in. This was indeed the case in the object glass of Dr Babington's telescope.[29]

In the instance of another telescope which was in the university archives[30] I measured the distance *VF* of the image transmitted by the object glass to be 14 ft. $3\frac{9}{10}$ in.[31]

---

entering from the left will pass into the new lens, without being deviated at a' and then be refracted at b' to F. Newton then supposes that at b' it is partially reflected in the direction of k and then refracted at c' to g. Now the reflexion of the first ray at b and subsequent refraction at c is determined by $2/s = 1/v' - 1/u$ and $1/r + 1/v' = \mu^{-1}(1/r + 1/w)$, so that, since

$$(\mu - 1)/r = \mu/u, \quad 1/GX \text{ (or } 1/w) = 2(\mu - 1)/r + 2\mu/s.$$

Similarly, $1/HX = 2(\mu - 1)/s + 2\mu/r$. Hence, since $1/FX = (\mu - 1)/\rho$, $A = 1/HX - 2/FX = 2/r$ and $B = 1/GX + A = 2\mu/\rho = 1/gX$. Further, on taking (with Newton) $r = C$ and $s = 1/D$, it follows that $1/C + D = 1/\rho = (1/2kX \approx)\, 1/2kV$. The value of the refractive index

$$\mu \text{ (or } I/R) = kT/gT = (VF + 2TV)/(VF - 2gT)$$

follows, as Newton says, immediately by Problem 1. (His erroneous value for A, namely $A = (4\mu - 2)/s$, appears to have arisen from evaluating $-2/FX = -2(\mu - 1)(1/r + 1/s)$ as $-2(\mu - 1)/r + 2(\mu - 1)/s$!)

(26) Preliminary calculations for this example (on ULC. Add. 3970.11: 633r/635r) show that the measurements were indeed taken from an actual telescope, but are here slightly rounded off. For instance, the length of *VF* was originally taken to be '13f $11\frac{1}{6}$ inch', while that of *TG* was averaged from '$28\frac{7}{8}$. $28\frac{3}{4}$ vel potius $28\frac{13}{16}$ inches' (*sic*). The 'crassities vitri' *TV* remains unchanged.

(27) Babington's; see note (29).

(28) Read '$\frac{2}{A} = 8388^{\text{dig}} = 699^{\text{ped}}$' ($2/A = 8388'' = 699'$). Compare notes (22) and (26).

(29) An example of where a close empirical agreement with theory does not vindicate the latter. Here $C = r$ is so large that Newton's error in computing its value (note (28)) will not appreciably affect the radius $1/D = s = 1/(1/2kV - 1/C)$ of the interface *RVS*. Humphrey Babington, Newton's senior at Trinity and briefly Vice-master of the college in 1690, was an old acquaintance (see I: 8, note (21)). We need read no expression of an interest in optics (otherwise unrecorded) into his possession of the telescope.

(30) Presumably Newton means that it was deposited in the University Library. The Registry would hardly have a use for it.

(31) Newton breaks off, leaving his experimental data incomplete.

[2] TELESCOPIJ NOVI DELINEATIO.(32)

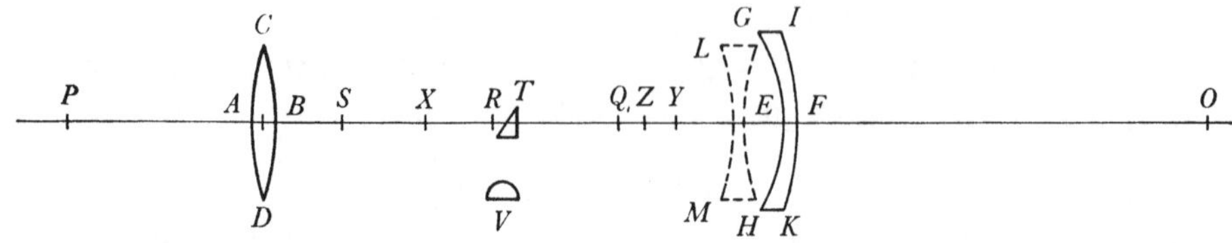

Vitrum objectivum CD parallelos radios refringat versus O. Imago O per refractionem concavæ superficiei GEH transferatur ad P, et inde per reflexionem superficiei specularis(33) ad Q; et inde per refractionem secundam superficiei GEH ad R ubi a speculo(34) obliquo T detorquetur per vitrum oculare perexiguum V ad oculum.

Sit imaginis translatio angularis ab O ad P et a P ad S tanta quanta corrigendis(35) vitri objectivi refractionibus erroneis ab inæquali refrangibilitate ortis sufficit et erit angularis translatio imaginis a Q ad R tanta quanta est a P ad S, et punctum S invenietur faciendo ut sit $BE.EO::EO.ES$.(36)

Sit X centrum circuli specularis IFK et Y centrum circuli refringentis concavi GEH. Et quoniam imaginis angulares translationes PX, XQ æquales sunt, ut et PS, QR; erunt etiam translationes SX, RX æquales: adeoq; si fiat $ES.SX::ER.RX$, vel $ES+SX.SX::EX.RX$, ex dato puncto X habebitur ultimæ imaginis locus R, e cujus regione consistet oculus.

Sit insuper Y centrum superficiei concavæ GEH, et quoniam est

$$EP.EQ::PX.QX, \quad \text{et} \quad I\times OE.R\times OY::EP.YP.$$

et $I\times ER.R\times YR::QE.QY$: inde derivabitur hæc conclusio. Fac

$$\frac{2I\times EO}{R\times EX}=a. \quad \frac{EO}{ER}=b. \quad b\times RX=c. \quad \frac{OX-c}{a-b+1}=XY.\text{(37)}$$

et habebitur circuli GEH centrum Y.(38) Ubi nota quod usurpo $\frac{I}{R}$ pro ratione

(32) ULC. Add. 4004: 73r, first published by H. W. Turnbull in *The Correspondence of Isaac Newton*, **1** (1959): 272–3. A variant draft of this piece, which evidently applies the preceding general theory of the lenses, is reproduced as Appendix 1.7 below, while Appendix 1.8 exhibits a worked example. Other related computations on ULC. Add. 3970.9: 631r/636v are discussed in note (39).

(33) Namely, *IFK*.

(34) Perhaps because of the tendency of an ordinary plane mirror to tarnish, Newton has inserted a 45° prism in his figure at *T*. Since the outer medium is implicitly assumed to be air, all near-horizontal rays will not noticeably be deviated on entering the (glass) prism and will suffer total internal reflexion at the opposite interface, emerging vertically downwards in the direction of *V* without significant chromatic distortion.

(35) Rather mysteriously Newton first wrote 'delendis' (...destroy).

(36) See note (38) below.

[2] OUTLINE OF A NEW TELESCOPE(32)

Let the object glass CD refract parallel rays towards O. Now let the image O be transferred by refraction in the concave surface GEH to P, from there by reflexion in the mirror surface(33) let it pass to Q, and from there by a second refraction in the surface GEH to R, at which point it is deflected downwards by the slanting mirror(34) T through the exceedingly minute eyepiece V to the eye.

Let the angular translation of the image from O to P and from P to S be as much as is sufficient to correct(35) errant refractions arising in the object glass from unequal refrangibility: the angular translation of the image from Q to R will then be as much as it is from P to S, and the point S will be found by making $BE:EO = EO:ES$.(36)

Let X be the centre of the mirror-circle IFK and Y the centre of the concave refracting circle GEH. Then because the angular image translations PX, XQ are equal, and so too are those PS, QR, the translations SX, RX will also be equal. Accordingly if there be made $ES:SX = ER:RX$ or

$$[ES+SX]:SX = EX:RX,$$

from the given point X will be had the position R of the final image and the eye shall take its stand in line with it.

Furthermore, let Y be the centre of the concave surface GEH. Then because $EP:EQ = PX:QX$ and $I\times OE:R\times OY = EP:YP$ and again

$$I\times ER:R\times YR = QE:QY,$$

there will thence be derived this conclusion: make $2I\times EO/R\times EX = a$, $EO/ER = b$, $b\times RX = c$, $(OX-c)/(a-b+1) = XY$(37) and there will be obtained the centre Y of the circle GEH.(38) Here note that I employ I/R for the sine

(37) This sentence was added at a late stage when Newton noticed a mistake in his computation on ULC. Add. 3970.11: 634v; see Appendix 1.6: note (26) below. A first, cancelled sentence reads (in line with the uncorrected draft on f. 633r) 'Fac $ER.EX::OR.P$, et $\frac{2I}{R}EO.OX::P.XY$' (Make $ER:EX = OR:P$ and $[2I/R]EO:OX = P:XY$).

(38) Newton's ingenious idea is to construct a reflecting telescope, in which the effective area of the mirror $GHKI$ is increased by a thin convex 'object glass' $ADBC$: to prevent tarnishing the mirror surface IFK is the coated convex face of a thin concavo-convex lens, the

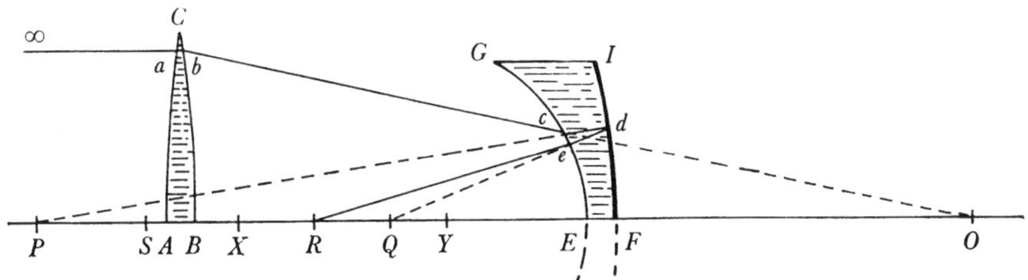

sinus incidentiæ ex aere in vitrum ad sinum refractionis: et suppono insuper vitri crassitiem EF ad instar nihil esse.(39)

---

§2. THE PROBLEM OF TWOFOLD REFRACTION RESOLVED.(1)

PROBLEMA.

(2)Data quavis refringente superficie CD quæ radios a puncto A divergentes quomodocunq; refringat: invenire aliam superficiem EF quæ refractos omnes DF convergere faciet ad aliud punctum B.

---

centres of whose spherical surfaces IFK, GEH are respectively X, Y. (The function of the doubly concave lens LM and the corresponding point Z is not clear, nor are they referred to in the accompanying text.) Newton then seeks to determine the relative position and curvature of the mirror-lens combination $GEFI$ such that chromatic aberrations are minimized when a ray of 'white' light, incident parallel to the axis $ABEF$, is refracted by the object lens ABC (say at a and b) through O, meeting the concave surface of the mirror in c, where it is refracted (in line with P) to d on the mirror surface, then reflected (through Q) to e, there finally to be refracted to R. On taking the refractive index from air into glass to be I/R and assuming the thicknesses AB, EF to be negligible, from Newton's preceding general theorem (note (6)) it follows that the refraction at c, reflexion at d and second refraction are respectively determined by

$$\frac{1}{EY}+\frac{1}{OE}=\frac{I}{R}\left(\frac{1}{EY}-\frac{1}{EP}\right),\quad \frac{1}{EP}+\frac{1}{EQ}=\frac{2}{EX}\quad\text{and}\quad \frac{1}{EY}-\frac{1}{EQ}=\frac{R}{I}\left(\frac{1}{EY}-\frac{1}{ER}\right).$$

Hence, when EP and EQ are eliminated,

$$\frac{1}{OE}-\frac{1}{ER}=\frac{2(I/R-1)}{EY}-\frac{2(I/R)}{EX}.$$

This path is equivalent to a simple reflexion in the mirror IF from S to R, where

$$1/ES+1/ER = 2/EX,$$

that is, $1/EP+1/EQ$. (Since the distance of the incident ray $\infty\, a$ from the axis is assumed to be small, very nearly $PS = QR$, so that in Newton's phrase the angles subtended at the mirror by these 'translations' will effectively be equal.)

If now the radii of the surfaces of the object lens ABC are r and s, then, on taking $OE = x$ and $1/\rho = 1/r+1/s$, $1/(x+EB) = (I/R-1)/\rho$, so that

$$\frac{1}{x}+\frac{1}{ES}\left[\text{or } 2(I/R-1)\left(\frac{1}{EY}-\frac{1}{EX}\right)\right]=\frac{2\rho}{x+EB}\left(\frac{1}{EY}-\frac{1}{EX}\right).$$

The condition that, for slight changes in the magnitude of the refractive index I/R and so of $OE = x$, the position of the final image R (and hence of the point S) remain invariant is that this equation in x should have a double root: which, on equating its derivative to zero, yields OE^2 (or x^2) $= EB\times ES$, as Newton requires (note (36)).

Finally, since $1/OE-1/ER = 2(I/R-1)/EY-2(I/R)/EX$, on taking $2(I/R)\,OE/EX = a$ and $OE/ER = b$ we may conclude that $1-b = (a\times EX/OE-2)\,OE/EY-a$ or

$$EY = (a\times EX-2OE)/(a-b+1);$$

of incidence from air into glass to the sine of refraction: in addition I suppose the thickness EF of the glass to be tantamount to nothing.(39)

Translation

PROBLEM

(2)Given any refracting surface CD which is to refract rays diverging from the point A in any manner whatever, to find a second surface EF which will make all refracted rays DF converge on some point B.

---

that is, on making $b \times RX$ (or $b \times EX - OE$) $= c$,

$$XY = \frac{(-b+1)EX+2OE}{a-b+1} = \frac{OX-c}{a-b+1}.$$

(39) In a tabulated scheme on ULC. Add. 3970.11: 631ʳ (for which there are additional calculations on f. 636ᵛ following) Newton has taken a worked example in which $I/R = 61/40$, $BO = 12$, $BE = 8$ and $EX = 3-i$, $i = 0, \frac{1}{2}, 1, 1\frac{1}{2}, 2$, computing (with some little difficulty and error in arithmetical detail) the five corresponding values of $SX[= EX-ES]$,

$$RX[= EX(2-EX)/(4-EX)], \quad ER[= 2EX/(4-EX)],$$

$a, b, c, XY, EY[= EX-XY], EP$ and EQ. The refractive index is a close approximation of that for air into glass, but whether Newton ever attempted to fabricate this improved design of reflecting telescope is not known. A related numerical example (reproduced in Appendix 1.7 below) takes the near equal 14/9 for its refractive index.

(1) This isolated 'Problema' in Newton's Waste Book (ULC. Add. 4004: 90ʳ) evidently, both by its handwriting and its content, belongs with the 'Theoremata Optica' reproduced in §1. Essentially—and with a brevity belying its subtlety—the piece gives a 'plane' construction by points of Descartes' problem (*Geometrie* [= *Discours de la Methode* (Leyden, 1637): $_2$297–413]: 363–8: 'Commēt on peut faire vn verre autant conuexe ou concaue, en l'vne de ses superficies, qu'on voudra, qui rassemble a vn point donné, tous les rayons qui vienent d'vn autre point donné') of finding a lens, one of whose faces is a given surface of revolution, such that all rays of (homogeneous) light radiating from a given point are accurately refracted by it to a second point. Five years before, as we saw in the first volume (I: 558–9, note (25)), Newton had attempted to find the Cartesian defining equation of the latter face corresponding to the particular cases where the given interface is one or other species of conic, but had failed to make any headway. The present general solution by points was repeated without structural change in his *soi-disant* 'lectures' for 1684–5 at Cambridge (see Appendix 3 below), and published in that form two years afterwards in his *Philosophiæ Naturalis Principia Mathematica* (London, $_1$1687: 233–4). Shortly after, Christiaan Huygens incorporated a similar solution, without acknowledging Newton's priority, in Chapitre VI of his *Traité de la Lumiere, Où sont expliquées les causes de ce qui luy arrive dans la Reflection, & dans la Refraction* (Leyden, 1690: 113–15). Accepting—unlike Newton—the Fermatian postulate of 'least time' that the speeds of light in incident and refracted media are proportional to the constant ratio d/e of the sines of incidence and refraction at their interface, Huygens was able to deduce thereby from the basic relation $AD+(d/e)DF+FB = AC+(d/e)CE+EB$ governing the transmission of light from point A to point B that all paths $ADFB$ are traversed in the same time (p. 113). A variant

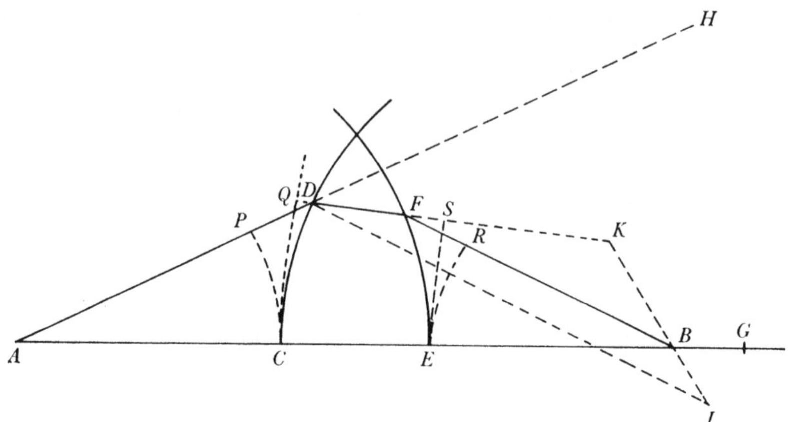

Junge AB. Eam secent refringentes superficies in C et E. Et posito d ad e(3) ut sinus incidentiæ ex medio ACD in medium EDC(4) ad sinum refractionis, produc AB ad G ut sit $BG.CE::d-e.e$, & AD ad H ut sit $AH=AG$, et DF ad K ut sit $DK.DH::e.d$. Junge KB et centro D radio DH describe circulum occurrentem KB productæ in L. Ipsi DL parallelam age BF et erit F punctum superficiei EF quæ radios omnes ADF convergere faciet ad punctum B. Nam defluxio DF est ad fluxionem $AD+BF$ ut e ad d: adeoq͡ȝ

$$CE-DF.AD+BF-AC-BE::e.d.^{(5)}$$

Haud secus Problema resolvitur ubi tres sunt vel plures refringentes superficies.(6)

---

solution on Newtonian lines which introduces the common diacaustic of the points A and B (see note (5)) was given by Guillaume de L'Hospital in his *Analyse des Infiniment Petits, pour l'Intelligence des Lignes Courbes* ((Paris, 1696): Section VII, §144:) and parroted eight years later by Charles Hayes in his *Treatise of Fluxions: Or, An Introduction to Mathematical Philosophy. Containing A full Explication of that Method by which the most Celebrated Geometers of the present Age have made such vast Advances in Mechanical Philosophy* ((London, 1704): Sect. x, Prop. V: 256).

(2) There is some indication in the manuscript that the circle arcs PC, RE and the diacaustic involutes QC, SE together with the points P, Q, R, S were added by Newton at a later date, presumably about mid-1685 when he revised his 'Problema' for publication (see note (1)).

(3) A curious reversion to the Cartesian notation for the refractive index. Compare a similar usage in Appendix 1.2–5/8 below.

(4) The more concrete 'ex aere in [vitrum?]' (from air into [glass?]) is cancelled.

(5) Newton's argument may be clearer if we convert it to an equivalent limit-increment one. If we suppose that the indefinitely close incident ray Ad is also refracted (at d and f) into B and draw $D\delta$, $D\delta'$ and $F\phi'$, $F\phi$ normal to Ad, df, fB respectively, then $\delta d/\delta' d = \phi f/\phi' f = d/e$ and therefore, since $d(AD) = Ad-AD = \delta d$, $d(DF) = df-DF = -(\delta' d+\phi' f)$ and

$$d(FB) = fB-FB = \phi f,$$

at once $d(AD+FB)/-d(DF) = d/e$, so that $AD+(d/e)DF+FB = AC+(d/e)CE+EB$, constant. Newton's construction of the second refraction point F corresponding to any point D in

(90)

Aug: de Doct. Christ. l. 2. c. 36. Eloquentia.

Præceptis eloquentiæ persuaderi etiam falsa, non tamen ideo est facultas ipsa culpabilis sed ea male utentium perversitas.

c. 37. Dialectica.

Plerumque accidit ut facilius homines res eas assequantur propter quas assequendas præcepta dial. discuntur quam talium præceptorum nodosissimas et spinosissimas disciplinas. Tanquam si quispiam præcepta dare volens ambulandi moneat non esse levandum pedem posteriorem nisi cum posuerit priorem deinde ... Ita plerumque citius ingeniosus videt non esse ratam conclusionem quam præcepta ejus capit: tardus autem non eam videt, sed multo minus quod de illa præcipitur.

Problema

Data quavis refringente superficie CD quæ radios a puncto A divergentes quomodocunque refringat: invenire aliam superficiem EF quæ refractos omnes DF convergere faciet ad aliud punctum B.

Junge AB. Eam secent refringentes superficies in C et E. Et posito d ad e ut sinus incidentiæ ex medio ACD in medium EDC ad sinum refractionis, produc AB ad G ut sit BG . CE :: d−e . e, & AD ad H ut sit AH = AG, et DF ad K ut sit DK . DH :: e . d. Junge KB et centro D radio DH describe circulum occurrentem KB productæ in L. Ipsi DL parallelam age BF et erit F punctum superficiei EF quæ radios omnes ADF convergere faciet ad punctum B. Nam fluxio DF est ad fluxionem AD + BF ut e ad d: adeoque CE − DF . AD + BF − AC − BE :: e . d.

Haud secus Problema resolvitur ubi tres sunt vel plures refringentes superficies.

Plate IV. The problem of twofold refraction to a point resolved (**3**, 2, §2).

Join AB and let it cut the refracting surfaces in C and E. Now, supposing d to e[3] to be as the sine of incidence from the medium ACD into the medium EDC[4] to the sine of refraction, produce AB to G so that $BG:CE = [d-e]:e$, AD to H so that $AH = AG$, and DF to K so that $DK:DH = e:d$. Join KB and with centre D, radius DH describe a circle meeting KB produced in L. Parallel to DL draw BF and F will be a point on the surface EF which shall make all rays ADF converge on the point B. For the negative fluxion of DF is to the fluxion of $AD+BF$ as e to d, so that

$$[CE-DF]:[AD+BF-AC-BE] = e:d.^{(5)}$$

The problem is resolved in like fashion where there are three or more refracting surfaces.[6]

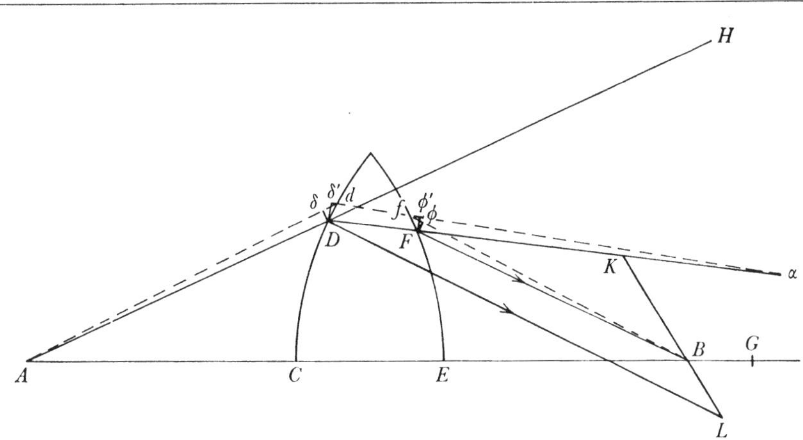

the given interface CD is easily justified. On taking $BG = (d/e-1)\,CE$, we have to construct $AD+(d/e)\,DF+FB = AG$, that is, $(d/e)\,DF+FB = DH$ on making $AH = AG$. Take point K in DF such that $FK = (e/d)\,FB$ and immediately $DK = (e/d)\,DH$, which fixes K in the refraction of the given incident ray AD through D. If therefore L is taken in KB such that $DL = DH$, then $DK:(DH$ or$)\ DL = e:d = FK:FB$, so that the refraction point F is fixed in DK as the latter's meet with the parallel to DL through B. For Newton's use of 'fluxio' and 'defluxio' compare **1**, 2, §3. It will be evident that the diacaustic (not shown) of the point A with regard to the interface CD must, if the conditions of the problem are to be satisfied, coincide with that of the point B with regard to the interface EF, and that the curves CQ, ES (here added in afterthought) are the involutes to this common diacaustic which pass through the points C, E respectively.

(6) If the ray passing between points A and B is refracted, following the indexes e_{i-1}/e_i, at the n points D_i, $i = 1, 2, \ldots, n$, then the path is to be constructed from the condition that $AD_0 + \sum_{1\leqslant i\leqslant n} [(e_{i-1}/e_i)\,D_{i-1}D_i] + D_nB = \text{constant}$.

APPENDIX 1. MISCELLANEOUS OPTICAL CALCULATIONS

[*c.* 1670][(1)]

From original drafts in the University Library, Cambridge

[1][(2)]

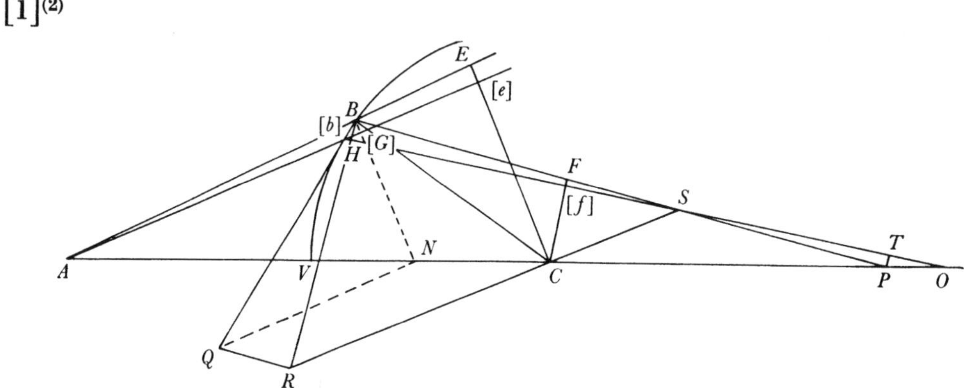

[Est] $BE.BC.BF::BN.BQ.BR$. [hoc est] $BE.BF::BN.BR$ [adeoq3] $\frac{FBN}{BE}=[BR=]\frac{BS, FC}{FS}$.[(3)]

---

[2][(4)] $VF=a$. $FC=c$. $VC=r$. $CS=z$. $AB=y$. $CB=x$.

$$rr=xx+yy.$$

motus de $\left.\begin{matrix}x\\y\\z\end{matrix}\right\}$ est $\left\{\begin{matrix}n.\\p=1.\\q.\end{matrix}\right.$ [Posito $Cd.Ce::d.e$ erit]

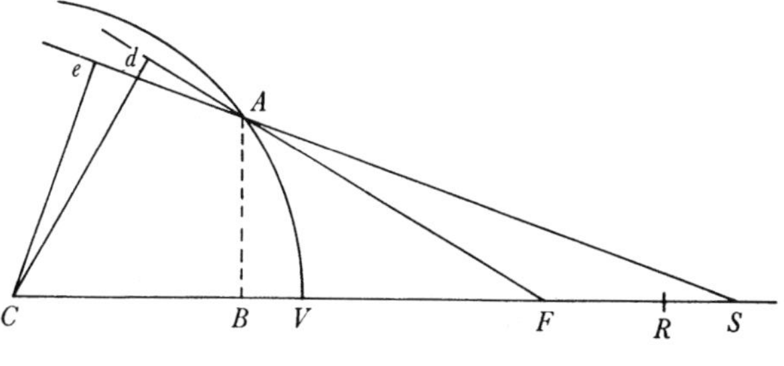

$$AF=\sqrt{rr+cc-2cx}:AB=y::CF=c:Cd=\frac{cy}{\sqrt{rr+cc-2cx}}.$$

---

(1) Preliminary sheets dealing with the refraction of (homogeneous) light at a spherical interface, together with related drafts of passages in 1, §2 and 2, §1 above. Rather than have lengthy footnotes, we have for the most part chosen to elaborate the logical development of their somewhat abruptly expressed argument by means of textual insertions within square parentheses. The date suggested is based on their uniformity of handwriting style and their evident close relationship with the preceding papers.

(2) From the verso of a slip inserted between folios 72 and 73 of Newton's Waste Book (Add. 4004).

& $AS = \sqrt{rr+zz-2zx}:[AB=]y::[CS=]z:\dfrac{zy}{\sqrt{rr+zz-2zx}} = eC$. [necnon]

$$d \times eC = e \times dC.$$

[ergo] $dz\sqrt{rr+cc-2cx} = ec\sqrt{rr+zz-2zx}$. [Hoc est quadrando et redigendo terminos in ordinem] $\begin{array}{l} eecczz-2eeccxz+eeccrr=0 \\ -ddcc \\ -ddrr \\ +2ddcx \end{array}$ [adeoqꝫ capiendo fluxiones]

$$2eeccqz-2ddccqz-2ddrrqz+4ddcxqz-2eeccxq+2ddczzn-2eecczn=0.$$

Sit jam $r=x$. erit $y=0$. [itaqꝫ] $AF=a:[AB=]y::[FC=]c:\dfrac{cy}{a}=Cd$. & $\dfrac{ecy}{da} = \left[\dfrac{e}{d} \times Cd\right] = Ce$. &

$$Ce-AB = \frac{ecy-day}{da}:CV=r=CB::[Ce=]\frac{ecy}{da}:\frac{ecr}{ec-da}=z=CR.$$

Quare per prior[m] Æquat[m(5)]

$$ee\dot{c}\dot{c}q-ddccq-ddrrq+2ddcrq-ee\dot{c}\dot{c}q+ecdaq-eeccn+ddczn=0$$

[sive] $ddaaq-decaq = \dfrac{ddeccrn-e^3c^3n+deeccan}{ec-da} \overset{(6)}{=} \dfrac{ddeccrn-e^3c^3n+deec^3n-deeccrn}{ec-da}$.

---

(3) Clearly a working draft of Corollaries 1 and 2 of Proposition 32 in the last section of the 'Opticæ Pars Prima' (1, §2). Where the ray AB radiating from A is refracted at the spherical surface VB (of centre C) to the refraction centre S, Newton as before draws BN perpendicular to AB to meet AC in N, then NQ perpendicular to BN to meet the tangent at B in Q, and lastly R is located in the parallel to BS through Q such that RB is perpendicular to BS, while CE, CF are the normals from C drawn to AB and BS. His stated result determines the Barrovian diacaustic condition $BS/FS = AB \times EC \times BF/AE \times FC \times BE$ since $BN:EC = AB:AE$. On inserting the points b, e, f, G (as shown) in Newton's figure, it is evident that the quadrilaterals $BCFE$ and $BQRN$ are each similar to the differential quadrilateral $BbHG$, so that the truth of the first proportion $BE:BC:BF = BN:BQ:BR$ is immediate. Further, since

$$\frac{BH}{Ff}\left(\text{or } \frac{BS}{FS}\right) = \frac{BH}{Bb}\left(\text{or } \frac{BR}{BQ}\right) \times \frac{Bb}{BG}\left(\text{or } \frac{BQ}{BN}\right) \times \frac{BG}{Ee}\left(\text{or } \frac{BN}{EC}\right) \times \frac{Ee}{Ff}\left(\text{or } \frac{EC}{FC}\right),$$

it follows that $BS/FS = BR/FC$ and hence that the line RC meets BF in the refraction point S, as Newton constructs it. Compare 1, §2: note (25).

(4) Add. 3958.2:32[r]. Newton seeks the 'error' RS between the meet of the refraction at A in the spherical interface VA (of centre C) of a ray incident from F and the corresponding 'focus principalis' R. Note Newton's several lapses from Latin into English in this worksheet, and also his use of a cancelling superscript dot to eliminate the terms $\pm e^2c^2q$ (not to be confused with the (unusual) literal fluxions n, p, q of x, y, z).

(5) After dividing through by $2z$ with its term '$-2eeccxq$' misread as '$-2eecczq$'! This mistake nullifies the remaining calculation. See note (7).

(6) Since $a = c-r$.

[hoc est] $daq \times da - ec = \dfrac{dehccrn + eehc^3n}{ec - da}$ [posito $h = d - e$]. But [ex natura circuli $xx + yy = rr$ fit] $nx + py = 0$. Or $\dfrac{-py}{r} = n$. & $rdaq \times \overline{da - ec}|^2 = ehccpy \times dr + ec$. [Ergo quia quantitatis] $RS = v$ [fluens est] $drav \times \overline{da - ec}|^2 = \dfrac{ehccyy \times dr + ec}{2}$ [erit] $v = RS = \dfrac{ehccyy \times dr + ec}{2dra \times \overline{da - ec}|^2}$. Or [posito $\sigma = dr + ec$ & $\pi = da - ec$] $RS = v = \dfrac{ehcc\sigma yy}{2da\pi\pi}$.(7)

---

[3](8) [$CA = a$. $CV = r$. $CG = x$. $GR = y$. $CD = z$. Ex natura circuli erit $xx + yy = rr$. Tum posito $CE . CF :: d . e$ fit

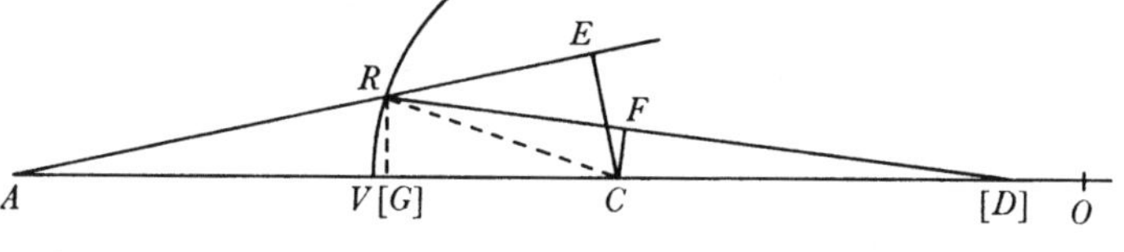

$AR = \sqrt{aa + rr - 2ax}$. $RD = \sqrt{zz + rr + 2zx} :: e, CA . d, CD$. Hoc est quadrando et redigendo terminos in ordinem fit] $zz = \dfrac{2xz + rr}{\dfrac{dd - ee}{ee} + \dfrac{ddrr - 2ddax}{eeaa}}$. [seu]

$$zz = \frac{2xz + rr}{\dfrac{dd}{ee} - \dfrac{2dxd}{eea} + \dfrac{rrdd}{eeaa} - 1}. \text{ [adeoq}\mathfrak{z}] \; z = \frac{x \mp \sqrt{xx + \dfrac{rrdd}{ee} - \dfrac{2rrddx}{eea} + \dfrac{r^4dd}{eeaa} - rr}}{\dfrac{dd}{ee} - \dfrac{2ddx}{eea} + \dfrac{ddrr}{eeaa} - 1}. \text{ [Fac]}$$

---

(7) Having computed $(c^2e^2 - c^2d^2 - d^2r^2 + 2cd^2x)z^2 - 2c^2e^2xz + c^2e^2r^2 = 0$ as the condition for the incident ray FA to be refracted back through S, Newton seeks to find the 'error' RS from the focus R by an ingenious argument, but unfortunately chooses a wrong base variable $CB = x$ and then clouds his mistake by a small slip in his calculation (see note (5)). On differentiating the basic equation there results, correctly,

$$\frac{q}{n}\left(\text{or } \frac{dz}{dx}\right) = -\frac{cd^2z^2 - c^2e^2z}{(c^2e^2 - c^2d^2 - d^2r^2 + 2cd^2x)z - c^2e^2x},$$

where $\lim_{x \to r}(z) = CR = Z = cer/(ce - ad)$. Newton then wishes to argue, on taking the new variable $BV = X = r - x$, that we may evaluate the coefficient α in the expansion of z as $Z + \alpha X + O(X^2)$ by evaluating the limit as $x \to r$ $(X \to 0)$ of

$$-\frac{q}{n}\left(\text{or } \frac{dz}{dX}\right) = \frac{c^2e^2 - cd^2Z}{c^2e^2 - a^2d^2 - c^2e^2r/Z} + O(X),$$

where $VF = a = c - r$. His error is then, in effect, to neglect the final term

$$-c^2e^2r/Z = -ce(ce - ad)$$

$x = r - v$.(9) [fit] $z = \dfrac{r - v \pm \sqrt{\dfrac{-rr}{rr} - 2vr + vv + \dfrac{rrdd}{ee} - \dfrac{2r^3dd}{eea} + \dfrac{r^4dd}{eeaa} + \dfrac{2rrddv}{eea}}}{\dfrac{dd}{ee} - \dfrac{2ddr}{eea} + \dfrac{ddrr}{eeaa} + \dfrac{2ddv}{eea} - 1}$. Fac

[$v=0$, erit CO] $= \dfrac{eeaar \pm dear, AV}{dd, AV^q - ee, AC^q}$ [posito $AV = a - r$]. [Unde] Error

$$[DO] = \frac{-eeaav \pm ea \times \dfrac{ddrav - eeaav}{d, AV}}{dd, AV^q - ee, AC^q + 2ddav}.$$

Fac [$AV = w$, erit CD] $= \dfrac{eear \pm dear, AV \pm \dfrac{deaarv}{AV} \mp \dfrac{e^3a^3v}{d, AV}.^{(10)}}{dd, AV^q - ee, AC^q + 2ddav}$ [adeoq; posito $v=0$ fit CO] $= \left(\dfrac{eeaar \pm dearw}{ddww - eeaa} = \right) \dfrac{ear}{dw - ea}$. [Unde] Error

$$[DO] = \frac{-2ddavz + \dfrac{ddeaarv - eeea^3v}{d, w}^{(11)}}{d^2w^2 - e^2a^2 \times dw + ea}$$

$$= \frac{ddear - 2d^3wz - e^3aa}{\overline{dw - ea}|^2} \times \frac{av}{dw}.^{(12)}$$

[Vel facilius posito] $\frac{e}{d}a = f$. [&] $AV = t = a - r$. [fit]

$$f\sqrt{zz + rr - 2zx} = fz + fr - \frac{fzx}{z+r} = z\sqrt{aa - 2ar + rr + 2ax} = z \times a - r + \frac{ax}{a-r}$$

---

in the denominator. Much as in the direct computation of

$$z = \frac{ce(ce(r-X) + \sqrt{[a^2d^2r^2 + 2cr(d^2r - ce^2)X + c^2e^2X^2]})}{c^2e^2 - a^2d^2 - 2cd^2X}$$

(compare I: 573–4 with e and d interchanged and c replaced by b), since

$$c^2e^2 - a^2d^2 - c^2e^2r/Z = ad(ce - ad),$$

on integrating there results

$$RS = z - Z = -\frac{c^2e}{ad}\left(-\frac{e}{ce-ad} + \frac{d^2r}{(ce-ad)^2}\right)X + O(X^2).$$

(8) Extracted from the upper verso of the Waste Book insert detailed in note (2) and from Add. 3970.11: 634r. Newton attempts a second time to estimate the refractive error (here DO) of a spherical interface, now by straightforward computation (compare note (7)). A trivial error in computation (note (10)) vitiates his efforts.

(9) That is, Newton sets $VG = x$.

(10) A term '$-eeav$' here omitted from the numerator spoils Newton's remaining computations.

(11) The second factor '$dw + ea$' in the denominator is *de trop*.

(12) To $O(v^2)$, that is. The first denominator should read '$d^2w^2 - e^2a^2$'.

[adeoq;] $zz = \frac{2ffrz - 2ffxz + ffrr}{aa - 2ar + rr + 2ax - ff}$. [Unde radicem extrahendo]

$$z = \frac{ffr - ffx + far - frr + \frac{far - f^3}{f + a - r}x^{(13)}}{aa - 2ar + rr - ff + 2ax} = \frac{fr}{a - r - f} + \frac{\frac{far - f^3}{f + a - r}x - \frac{2far}{a - r - f}x - ffx}{a - r + f \times a - r - f}$$

[hoc est, posito] $N = \frac{fr}{t - f}$. $z = N + \frac{\frac{far - f^3}{a - r + f}x - aNx - ffx}{a - r + f \times a - r - f}$ [sive]

$$z = N + \frac{\frac{far - f^3}{t + f} - aN - ff}{tt - ff}x = N + \frac{far - fft - 2f^3 - faN - taN}{\overline{t + f} \times \overline{t + f} \times \overline{t - f}}$$

[ubi est] $N = CD$.

---

[4][14] [Sit focus F, erit] d, AV, $FE = e$, FV, AE.[15] $AB^q = AV^q + 2AE \times VG$. $BF^q = VF^q - 2VG \times EF$. [Fac] $d.e :: AV.VN :: AB.BM$. [prodit] $\frac{e, AB}{d} = BM$.[16]

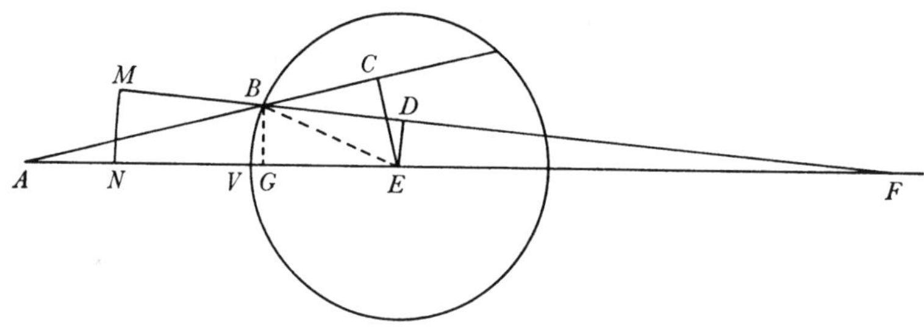

[Jam quia FN proxime æquat $FM = FB + BM$, erit $FN - \frac{e, AB}{d} = BF$ seu] $FN^q - 2\frac{e}{d}AB, FN + \frac{ee}{dd}AB^q = BF^q$. [Radices extrahendo fit]

$$AB = AV + \frac{AE, VG}{AV} - \frac{AE^q, VG^q}{2AV^{\text{cub}}} \text{ [\&c]}. \quad BF = VF - \frac{EF, VG}{VF} - \frac{EF^q, VG^q}{2VF^c} \text{ [\&c]}.$$

[adeoq; fluxiones capiendo]

(13) The denominator of the last term in the numerator should read '$a - r$' simply. The superfluous f is carried through the remaining calculation (in which a factor '2' disappears).

(14) Add. 3970.11: 638^r, a first attempt to evaluate the spherical error by the geometrical approach elaborated in the 'Theoremata Optica' (see 2, §1: note (11)). The change from Cartesian to Barrovian denotation (d/e to I/R) of the refractive index is probably not significant but may just possibly support our present ordering in sequence of Newton's worksheets. The curve MN in the figure is the involute through N of the diacaustic (not shown) of the radiating point A with respect to the spherical interface VB of centre E.

$\text{fl}\,AB = \frac{AE, \text{fl}\,VG}{AV} - \frac{AE^q, VG, \text{fl}\,VG}{AV^{\text{cub}}}$ [&c]. $\text{fl}\,BF = -\frac{EF}{VF}\text{fl}\,VG - \frac{EF^q, VG}{VF^c}\text{fl}\,VG$ [&c].

[Cum tamen sit $e\frac{AE}{AV} = d\frac{EF}{VF}$, erit defluxio quantitatis $e, AB + d, BF$ seu] $\frac{e, AE^q}{AV^c} + \frac{d, EF^q}{FV^c}$ in VG in fl VG = fl erroris. [hoc est] ut $\frac{d, FE, AE}{AV^q, FV} + \frac{d, EF^q}{FV^c}$ vel ut $\frac{AE}{AV^q} + \frac{EF}{FV^q} \times \frac{FE}{FV}$. vel ut $\frac{d, FE}{e, FV, AV} + \frac{FE}{FV^q} \times \frac{FE}{FV}$ vel ut $\frac{d, FV + e, AV \times FE^q}{AV \times FC^c}$ vel ut $\frac{\overline{d, FV + e, AV}}{FV} \times \frac{AE^q}{AV^c}$. vel ut $\frac{d, FV + e, AV \text{ in } AE, FE}{AV^q, FV^q}$ (17) ita error angularis. Et ut $\frac{d, FV + e, AV \text{ in } AE, FE}{AV^q}$ ita error in axe ab F. [Hæc igitur prodit regula.]

Cape $d.e :: AV.VN$ et erit $\frac{FN, AE, FE}{AV^q}$ ut error in axe ab F, si datur BG et GE. Item $\frac{FN, AE, FE, BG^q}{AV^q, VE^q}$ ut error in axe absoluta. Vel ut

$$\frac{FN \times \text{ang}\,BEG^{\text{quad}} \times \text{ang}\,BEC \times \text{ang}\,BED}{\text{ang}\,A^{\text{quad}}}\,^{(18)}.$$

---

[5](19)

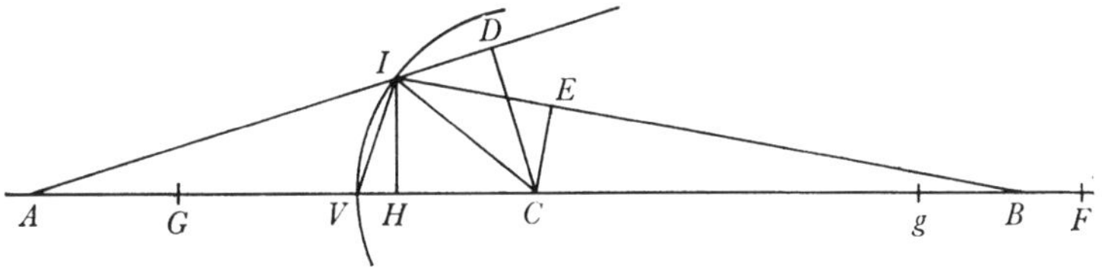

Si radius exiens a puncto A incidit in sphæram $V[I]$ ad I et refringitur ad B: posito sphæræ centro C, Axe AC, vertice V, ac demissis ad AC, AI, IB normalibus IH, CD, CE, sectaꝗ AV in G ut sit $CD.CE :: AV.GV$, fac $d \times AC.e, AV :: BC.BV$

(15) Compare 2, §1: notes (4) and (6).

(16) See note (1), and compare 2, §2: note (5).

(17) Since $AE/AV = d \times FE/e \times FV$ (note (15)).

(18) This alternative expression for the 'error in axe' is mysterious. Since 'ang A' ($B\hat{A}V$) is assumed to be small, so also is $B\hat{E}G$ and accordingly $B\hat{E}C \approx B\hat{E}D \approx \frac{1}{2}\pi$ while

$$B\hat{E}G/B\hat{A}V \approx BG/AV.$$

However, the lack of dimensional equivalence between numerator and denominator suggests that Newton abandoned it without completing (or correcting) it.

(19) Add. 3970.11: 638r/638v, a first revise of the preceding which, after substitution of Barrow's I/R for the Cartesian index d/e and some further revision, became the opening paragraphs of the 'Theoremata Optica' (2, §1). The manuscript text has no accompanying figure: that here reproduced is our restoration made from the figures of the final version.

et erit B focus. Et error BF a foco erit ut $\frac{GB, BC, CA, VI^q}{AV^q, CV^q}$ seu ut $\frac{GBC, CA, VH}{AV^q, CV}$ proxime. Et si radius AI parallelus est axi AV,(20) ita ut AV, AC et VG infinitæ sint et $CA = VA = \frac{d}{e}BG$[,] erit $\frac{\frac{d}{e}BC, VH}{CV}$ ut error ille. Si radius AI reflectitur Fac $AC.AV::FC.FV$ et cape F inter V et C. Dein produc AV ad G ut sit $VG = AV$ vel $AG = 2AV$ et error BF erit ut $\frac{BG, AC, BC, VH}{AV^q, CV}$.

Et vicissim si radius exiens a G refringitur in I et perger[e] debet ad focum [g] erit error ejus ab A ut $\frac{Fg, BC, CA, VH}{FV^q \times CV}$.

[Nam] $AD^q = AV^q + 2VH, AC$. $DF^q = VF^q - 2VH, CF$.

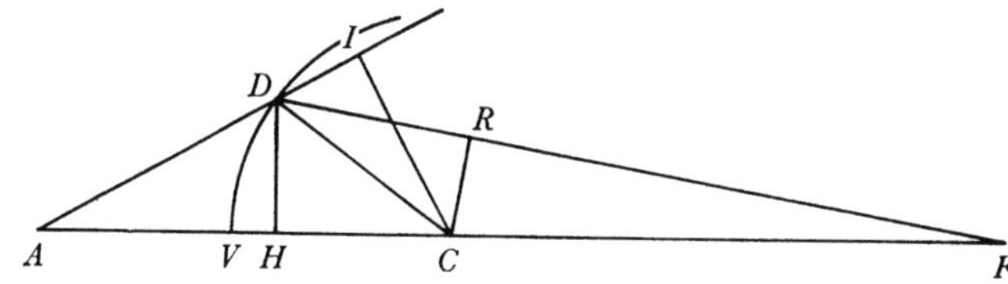

[Earum fluxiones]

$$2AD, \text{fl}\,AD = 2AC, \text{fl}\,VH. \quad 2DF, \text{fl}\,DF =^{(21)} 2CF \times \text{fl}\,VH.$$

[Erit etiam] $I.R::\text{fl}\,AD = \frac{AC}{AD}\text{fl}\,VH. \ \frac{R, AC}{I, AD}\text{fl}\,VH\left[=\frac{R}{I}\text{fl}\,AD. \text{ Necnon}\right]$

$\frac{CF}{DF}\text{fl}\,VH =^{(21)} \text{fl}\,DF. \left[\text{Quare } \frac{R}{I}\text{fl}\,AD + \text{fl}\,DF =\right] \frac{R, AC, DF - I, AD, CF}{I, AD, DF}\text{fl}\,VH.$

---

(20) In the limit, that is, when the point A has passed to infinity.

(21) A necessary minus sign is omitted.

(22) For criticism of the inadequacies of this argument see 2, §1: note (15).

(23) Add. 3970.11: 633r/634v, a preliminary worksheet for the 'Telescopij novi delineatio' reproduced above (see 2, §1: notes (32) and (39)). The structure of the 'telescope' is unchanged except for some renaming of points (note the two points F, R, K in Newton's present figure): an incident ray parallel to the axis $ABCD$ is refracted by the 'object glass' $AFBG$ through the (rightmost) point F till it meets the concave surface HCI of the lens–mirror combination $HILK$, where it is refracted (in line with P) to the mirror-face KDL, then reflected back (through Q) to the interface HCI and so once again refracted to R. The point K is defined by $KC = DF^2/BD$ (note (24)), while the (leftmost) point R is some suitable mean point between A and B (compare 2, §1: note (25)).

(24) The condition for minimum variation of the position of R when the refractive index of the incident (assumed homogeneous) light varies slightly; compare 2, §1: note (38).

$$AD = AV + \frac{AC}{AV} \times VH - \frac{AC^q}{2AV^3} VH^q \text{ [\&c]}. \quad DF = VF - \frac{CF}{VF} VH - \frac{CF^q}{2VF^3} VH^q \text{ [\&c]}.$$

$$\left.\begin{aligned} \frac{R}{I}\,\mathrm{fl}\,AD &= \frac{R}{I} \times \frac{AC}{AV} \times VH - \frac{R, AC^q}{2I, AV^3} VH^q \text{ [\&c]} \\ +\mathrm{fl}\,DF &= -\frac{CF}{VF} \times VH - \frac{CF^q}{2VF^3} VH^q \text{ [\&c]} \end{aligned}\right| = \frac{R}{I}\,\mathrm{fl}\,AD + \mathrm{fl}\,DF \text{ [hoc est]}$$

$$= \frac{-R, AC^q}{2I, AV^{cub}} VH^q - \frac{CF^q}{2VF^c} VH^q \text{ [quoniam] } \frac{R, AC}{I, AV} = \frac{CF}{VF}.$$

Error [igitur] ut $\frac{AC}{2AV^q} - \frac{CF}{2VF^q}$ in $\frac{CF}{VF} VH^q$ [&c].(22)

---

[6](23)

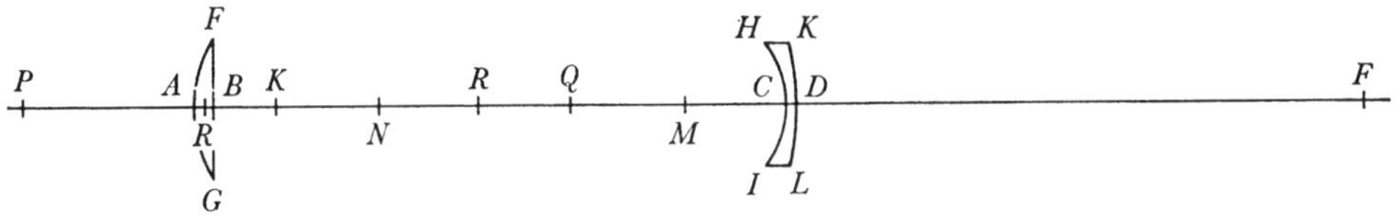

F, P, Q, R imagines succesivæ, N centrum superficiei KDL, M centrū superficiei HCI.

$DP.DQ::PN.QN.$ $DK.DR::KN.NR.$ $\frac{BF-DF}{DF} = \frac{DF}{CK}$ [id est]

$$BD.DF::DF.CK.^{(24)}$$

Ang PCK = ang QCR. Tota refractio Vitri HCI valere debet ang $CFI + CK[I]$ estq ang $CFI + CPI + QIR$.

Primo itaq fac $BD.DF::DF.CK$ et habebitur tota refractio vitri HCI. Dein fac $DK.KN::DR.NR$ et habebitur ultimus focus R e regione oculi.

Postea fac $DP.DQ::PN.QN.$ et $R, FM.I, FD::MP.CP.$ Et

$$I, QM.R, QD::MR.DR. \quad \text{vel} \quad QM.QD::R, MR.I, DR.$$

$I, FD-R, FM.I, FD::CM.CP.$ et $I, DR-R, MR.I, DR::MD.DQ.$ [Ergo]

$$\frac{I, FD, DM}{I, FD-R, FM} (=DP). \quad \frac{I, DR, DM}{I, DR-R, FR} (=DQ)::$$

$$\frac{R, FM, DN - I, FD, MN}{I, FD - R, FM} (PN). \frac{I, DR, MN - R, FR, DN}{I, DR - R, FR} [=QN].^{(25)}$$

---

(25) The factor 'FR' in the three terms '$-R$, FR' here should be ($FR-FM$ or) 'MR', as Newton subsequently corrected his manuscript to read (see next note). For clarity's sake we have separated the first, erroneous state of the following from its corrected version since Newton himself corrected his manuscript neither adroitly nor consistently.

[Hoc est] $FD.DR::R, FM, DN-I, FD, MN.\ I, DR, MN-R, FR, DN.$ [sive] $R, DR, DN, FM=2I, DR, FD, MN-R, FD, FN, DN.$ [adeoq;]

$$R, DN, FN, FR+R, DN, DR, MN=2I, DR, FD, MN \text{ [seu]}$$

$$MN=\frac{DN, FN, FR}{\frac{2I}{R}DR, DF-DR, DN}.\ \text{Cape ergo } DR.DN::FR.P.\ \text{et}$$

$$\frac{2I}{R}DF.FN::P.MN.^{(26)}$$

$$\frac{I, FD, DM}{I, FD-R, FM}(=DP).\frac{I, DR, DM}{I, DR-R, MR}(=DQ)::$$
$$\frac{R, FM, DN-I, FD, MN}{I, FD-R, FM}(PN).\frac{I, DR, MN-R, MR, DN}{I, DR-R, MR}[=QN]$$

[Hoc est] $FD.DR::R, FM, DN-I, FD, MN.\ I, DR, MN-R, MR, DN.$ [sive] $R, DR, DN, FM=2I, DR, FD, MN-R, FD, MR, DN.$ [adeoq;]

$$2I, DR, FD, MN=R, DN, DR, FM+R, DN, FD, MR$$
$$=[R, DN \text{ in}]\ DR, FN-DR, MN+FD, MN-FD, NR.$$

$$\text{[seu] } MN=\frac{DR, FN-FD, NR \text{ in } DN}{\frac{2I}{R}DR, DF+DN, DF-DN, DR}=\frac{DR, FN-DF, NR}{\frac{2I, DR, DF}{R, DN}+DF-DR}.^{(27)}$$

Datis lente objectiva *FG*, speculari superficie *KDL* et earũ distantia *AC* Fac $BD.DF::DF.CK$ et habebitur tota refractio quæ fieri debet in superficie *HCI* ad corrigendos errores vitri objectivi, nimirũ quanta sat esset ad transferendã Imaginẽ ab *F* ad *K*. cujus refractionis mensura statui potest $\frac{1}{CF}+\frac{1}{CK}$. Transferatur autem imago prima ab *F* ad *P* per refractionem superficiei *HCI*, dein a *P* ad *Q* per reflexionem superf[iciei] *KDL*, Postea a *Q* ad *R* per secundã refractionem superficiei *HCI* et [angularis] translatio ab *F* ad *P* et a *Q* ad *R* tanta esse debet quanta foret translatio ab *F* ad *K*. Vel translatio a *Q* ad *R* tanta quanta ab *P* ad *K* si modo transferretur ad *K*. Porro, fac $DK.KN::DR.NR.$ et habebitur locus ultimæ imaginis *R*, e regione oculi. Tertiò, fac

$$DR.DN::FR.P \quad \text{et} \quad \frac{2I}{R}DF.FN::P.MN$$

(26) The error underlying this first, mistaken result was not detected by Newton till after he had transferred its equivalent to the 'Telescopij novi delineatio' (see 2, §1: note (37)).

(27) Compare 2, §1: note (38). In line with that result, on taking $2(I/R)\,DF/ND = a$, $DF/DR = b$ and ($DF\times NR/DR$ or) $b\times NR = c$ there results $NM = (FN-c)/(a-b+1)$.

(28) This construction is taken from the erroneous first result found above! See notes (26) and (27).

et habebitur M centrum superficiei HCI.(28) Quarto si CM sumatur major vel minor sit ea Q et erit $\frac{1}{CM}-\frac{1}{Q}$ in $\frac{2I-R}{R}$(29) $=\frac{1}{x}+\frac{1}{y}$ in $\frac{I-R}{R}$. vel

$$I-R\,.\,2I-R::\frac{1}{CM}-\frac{1}{Q}\,.\,\frac{1}{x}+\frac{1}{y}.^{(30)}$$

---

[7](31)

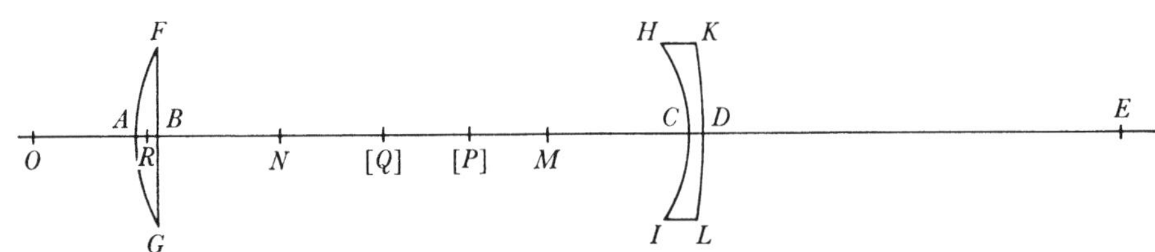

[Pone] $ME\,.\,CE::d\,.\,y.$ $MO\,.\,CO::e\,.\,y.$(32) [erit] $\frac{d,CE}{ME}=\frac{e,CO}{MO}$. [adeoq;] $d,CE,\overline{CO-CM}=e,CO,\overline{MC+CE}$. [Item] $\frac{d,CQ}{MQ}=\frac{e,CP}{MP}$. [adeoq;]

$$d,CQ,\overline{CP-CM}=e,CP,\overline{CQ-CM}.$$

[Posita autem] $DO\,.\,DR::DQ\,.\,DP$. [erit] $DO\,.\,DN::DN\,.\,DP$. [seu] $\frac{DN^q}{DR}=DQ$.(33) [Quibus inventis fit] $\frac{d,QCP-e,QCP}{d,CQ-e,CP}=CM=\frac{d,ECO-e,ECO}{d,CE+e,CO}$. [vel posito] $d-e=f$. [prodit] $\frac{f,QCP}{d,CQ-e,CP}=\frac{f,ECO}{d,CE+e,CO}=CM$. [hoc est]

$$-df,CE,CP,CQ-ef,CO,CP,CQ+df,CE,CO,CQ-ef,CO,CP,CE=0.$$

---

(29) An unexplained sentence at the head of the page reads '$R.2I-R$[::] rotatio radij circuli ad rotationem radij lucis' ($R:2I-R$ is as the rotation of the circle's ray (or radius?) to the rotation of the light ray).

(30) Since the meaning of x and y is not explained, it is difficult to understand this last sentence. If they denote the radii of the faces FAG, FBG (the latter of which is drawn plane in Newton's figure), in this case $(I/R-1)(1/x+1/y) = 1/BE$ and hence

$$1/BE:(1/CM-1/Q) = 2(I/R)-1,$$

apparently (see note (29)) the ratio of the 'rotatio radij lucis' to the 'rotatio radij circuli'. We can offer no suggestion as to what Newton meant by this. Was he perhaps misled by the same sort of error as he committed in his (later?) 'Theoremata Optica' (see 2, §1: note (22))?

(31) The recto side of an insert between folios 72 and 73 of Newton's Waste Book (Add. 4004), whose verso is reproduced in sections 1 and 3 of this appendix (see notes (2) and (8)). Newton seeks to throw his theory of the 'new telescope' outline in 2, §1 and the preceding section of this appendix into a more manageable form for computation. Note Newton's reversion to the Cartesian notation d/e for the refractive index with (compare I: 573) $d-e=f$.

(32) This is to leave y free.

(33) The condition for minimum variation of the point Q as the refractive index varies slightly. See note (24) and 2, §1: note (38).

[adeoq$\!\!\!\beta$] $d, f, CE, CQ, PO = e, f, CO, CP, QE = e, f, DN^q, QE$. [Ergo]

$$PO = \frac{e, DN^q, QE}{d, CE, CQ} = R. \quad \text{[Quare]} \sqrt{\frac{RR+4DN^q}{4}} : \pm\tfrac{1}{2}R = \frac{CO.}{CP.}^{(34)}$$

[Exempli gratia. Posito $d=14.e=9$. cape $DN=4\frac{1}{2}$. $RE=14$.(35) fiet]

$$7)20\tfrac{1}{4}(DQ\left[=2\frac{6\frac{1}{4}}{7}.\ QE=9\frac{6\frac{1}{4}}{7}.\ \text{etiam}\right]$$

$$\frac{9,20\frac{1}{4},9\frac{6\frac{1}{4}}{7}}{14,7,2\frac{6\frac{1}{4}}{7}} = \frac{89}{14} = 6\tfrac{1}{3} = PO = R. \text{ [necnon] } \sqrt{30\tfrac{1}{4}} \pm 3\tfrac{1}{6} = \frac{CO}{CP} = 5\tfrac{1}{2} \pm 3\tfrac{1}{6} = \frac{8\frac{2}{3}.}{2\frac{1}{3}.} \text{ [id est]}$$

$CO=8\frac{2}{3}$. $CP=2\frac{1}{3}$. [Unde] $CM = \dfrac{5,60\frac{2}{3}}{98+78} = \dfrac{303\frac{1}{3}}{176}\Big|^{(36)} \dfrac{455}{264} \times 12^{\text{inch}} = \dfrac{455}{22} = 20,6.$ [hoc est] $CM = 20\frac{2}{3}$ inches.(37)

---

(34) Since $CP+PO = R$ and $CP \times PO = DN^2$, it follows that CP, PO are the roots of the quadratic $x^2 - Rx + DN^2 = 0$.

(35) As the sequel makes clear, these are measured in feet as units.

(36) That is, on multiplying top and bottom by 3/2 and then converting feet to inches.

(37) Compare 2, §1: note (39).

(1) In a footnote at the end of the scholium to Proposition 36 of its last section, the editor of the posthumous English version of Newton's 'Opticæ Pars Prima' (ULC. Dd. 9.67: $_1$1–77, partially reproduced in 1, §2 above) correctly observed that 'The Use of the two last *Propositions* [35 and 36] is to determine the Rain-bow, as see our Author's *Opticks* [London, 1704: $_1$126–34:] Book I. Part II. Prop. IX' (*Optical Lectures Read in the Publick Schools of the University of Cambridge, Anno Domini, 1669. By the late Sir Isaac Newton* (London, 1728): 205). In substantiation of this remark, in the present appendix we reproduce two autograph passages from Newton's optical papers which deal with the formation of the primary and secondary bows. The first, an unpublished excerpt from his 'chemical' notebook, is irrefutable proof that Descartes' *Meteores*, in its Latin translation at least, was the source for Newton's ideas on the rainbow; the second, a too little known first Latin account of about 1671 drawn from the 'Opticæ Pars 2da', complements the mathematical theory established in the 'Pars Prima'.

(2) Add. 3975: [Newton's page] 14, to be dated, by the handwriting and immaturity of style, about 1665–66. These are evidently Newton's first thoughts on the rainbow: in particular he has not yet attained a satisfactory explanation of the colour sequence in the two bows which alone are commonly seen. Following Descartes' lead he adopts the 'large glasse Globe... filled wth water' as a convenient large-scale dioptrical model of the individual action of the raindrops which together produce the rainbow, for 'it is now agreed upon, that this Bow is made by refraction of the Sun's Light in Drops of falling Rain' (*Opticks: Or, A Treatise of the Reflexions, Refractions, Inflexions and Colour of Light* (London, 1704): $_1$126).

(3) Namely, because they are the least refracted (visible) coloured rays.

(4) *Renati Des Cartes Specimina Philosophiæ: seu Dissertatio de Methodo Recte regendæ Rationis, & Veritatis in Scientiis investigandæ: Dioptrice, et Meteora. Ex Gallico translata, & ab Auctore perlecta, variisque in locis emendata* (Amsterdam, [$_1$1644 →] $_2$1650): 277–96: Caput VIII. 'De Iride'. Sectio IX is headed 'Quomodo in Iride [colores] producantur; & quomodo ibi lumen ab umbra terminetur. Cur primariæ Iridis semidiameter 42 gradibus major esse nequeat, nec secundariæ

APPENDIX 2. NEWTON'S EXPLANATION OF THE RAINBOW

[*c.* 1666/*c.* 1671][(1)]

From the originals in the University Library, Cambridge

[1][(2)] 51 If y^e Sun *S* shine upon a large glasse Globe *abd* filled wth water And if you hold your eye very neare to y^e globe, y^e rays *bp* will appeare coloured

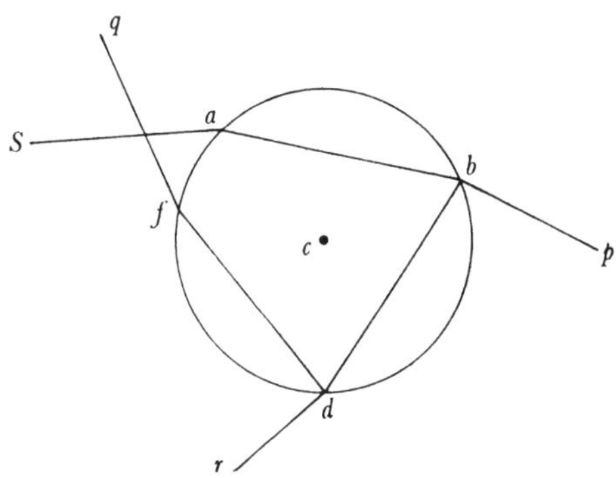

redd[(3)] & y^e farther you hold yor eye from y^e glasse y^e lesse they appeares coloured, untill y^e colour vanish. But y^e rays *rd* & *fq* appeare coloured at w^t distance so ever yor eye bee placed from ye Globe. The like you may observe by letting ye colours fall on a piece of paper.

52 Though one termination of light trajected through ye Prisme will not make both blews & reds; yet in this globe it doth (see Cartesij Meteora Cap 8 sec 9[(4)]) For ye rays *rd* & *fq* make all sorts of blews & reds; indeed by ye rays *bp* ye red is very distinct but ye blew is scarce discernable.

53 The colours of ye Rainbow must bee explicated by ye rays *rd* & *fq* (vide

---

semidiameter 51 gradibus minor. Curque illius superficies exterior magis determinata sit quàm interior...'. There is no evidence that Newton, whose knowledge of French was never good, read the *Meteores* in its original language at any time. (Compare I: 549.) A quarter of a century afterwards he wrote that 'This Explication of the Rain-bow is yet further confirmed by the known Experiment (made by *Antonius de Dominis* and *Des-Cartes*) of hanging up any where in the Sun-shine a Glass-Globe filled with Water, and viewing it in such a posture that the rays which come from the Globe to the Eye may contain with the Sun's rays an angle of either 42 or 50 degrees. For if the Angle be about 42 or 43 degrees the Spectator...shall see a full red Colour in that side of the Globe opposed to the Sun..., and if that Angle become less...there will appear other Colours, yellow, green and blue successively in the same side of the Globe. But if the Angle be made about 50 degrees...there will appear a red Colour in that side of the Globe towards the Sun, and if the Angle be made greater...the red will turn successively to the other Colours yellow, green and blue' (*Opticks* (note (2)): $_1$132–3).

Cartesij Meteora Cap 8 sec 1, 2, 3, 9, 10, 11, 12, 15)(5) For y^e bow may bee mad by drops of water forcibly cast up into y^e aire.(6)

---

[2](7) Superest jam mirum illud cœlestis arcûs spectaculum ad cujus explicationem Cartesius viam stravit, Huic enim debetur quod in guttis aquæ pluvialis decidentibus efformari cognoscimus.(8) Quemadmodum ex eo constat quòd nunquam videtur nisi cœlo pluente, quod Sole pluviam decidentem illustrante in vicis(9) nonnumquam apparuit quasi non in cœlo collocatus sed in aere vicino super oppositarum domuum parietibus effixus vel potius interjectus, quod aqua per artificium aliquod in altum sparsim ejaculata iridem ostendit et quod gramen rore matutino quasi guttulis minutissimis conspersum colores etiam iridis exhibet. Huic etiam debetur ingeniosissima de refractionibus guttæ earumqꝫ limitibus inventio.(10) Sed causam physicam minus fœliciter aggressus est. Hanc itaqꝫ ut(11) intelligatis concipite radium *AN* in globum *NFG* ad *N*

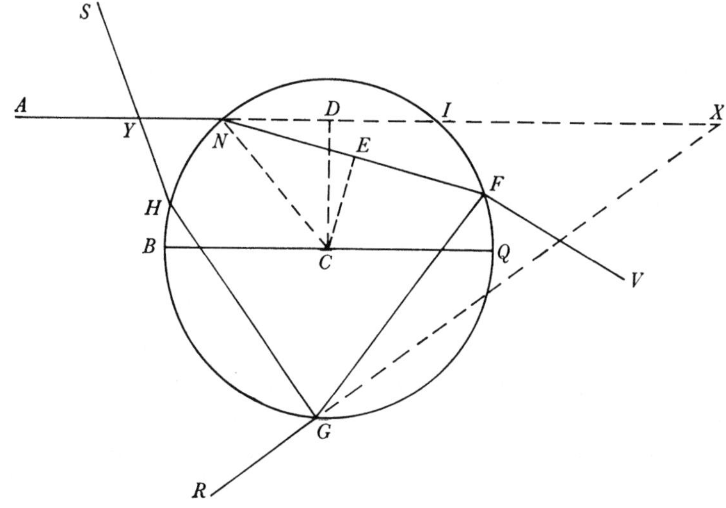

(5) These *sectiones* are entitled as follows: 'I. Non in vaporibus, nec in nubibus, sed tantùm in aquæ guttis Iridem fieri'; 'II. Quomodo [Iridis] causa ope globi vitrei aquæ pleni detegi possit'; 'III. Iridem interiorem & primariam oriri ex radiis qui ad oculum perveniunt post duas refractiones & unam reflexionem: exteriorem autem sive secundariã ex radiis post duas refractiones & duas reflexiones ad oculum pervenientibus: quò fiat ut illâ fit debilior'; 'IX. Quomodo [colores] in Iride producantur...' (see note (4)); 'X. Quomodo ista Mathematicè demonstrentur'; 'XI. Aquæ calidæ refractionem minorem esse quàm frigidæ atque idcirco primariam Iridem paulo majorem, & secundariam minorem exhibere....'; 'XII. Cur pars exterior primariæ Iridis, & contra interior secundariæ sit rubra'; 'XV. Quomodo aliæ prodigiosæ Irides varias figuras habentes, possint arte exhiberi'.

(6) 'This Bow never appears but where it Rains in the Sun-shine, and may be made artificially by spouting up Water which may break aloft, and scatter into Drops, and fall down like Rain' (*Opticks* (note (2)): $_1$126).

incidere et inde versus *F* refringi ubi rursus vel refringitur versus *V* vel forte reflectitur ad *G*, et si posterius eveniat tunc iterum in *G* vel refringitur ad *R* vel reflectitur ad *H*, et sic deinceps: ita ut ex radijs globum ingredientibus aliqui, ut *NFV*, statim egrediantur nullam reflexionem passi, alij, ut *FGR*, post unam reflexionem, et alij, ut *GHS*, post duas, alijq; post tres vel etiam plures.(12) Jam verò cùm guttæ pluviales respectu distantiæ ab oculo spectatoris sint admodum exiguæ ut physice pro punctis haberi possunt, non opus est ut earum magnitudines omnino consideremus, sed angulos tantum quos incidentes cum emergentibus radijs comprehendunt: Nam ubi anguli illi maximi sunt vel minimi, emergentes radij sunt solito confertiores et quia diversis radiorum generibus diversi competunt anguli maximi vel minimi, singula ad diversas plagas confertissimè tendentia in ijsdem prævalebunt ad colores proprios exhibendos.(13) Anguli itaq; maximi vel minimi quos singulorum generum

(7) Extracted from Newton's 'Opticæ Pars 2da. De Colorum Origine' (Dd. 9.67: $_1$1–101): 'Sect: 2da. De varijs Colorum Phænomenis. 4. De Phænomenis Lucis per Media sphæricè terminata transmissæ. Deq; Iride': 96–100. Newton here presents his improved version of the way the colours of the rainbow are seen by the human eye. In the early 1690's he inserted an augmented English paraphrase of this section in his 'first Book of Opticks' (ULC. Add. 3970.3: 119r–124r [= *Opticks* (note (2)): $_1$126–34]).

(8) Twenty years afterwards Newton had a somewhat clearer (but still cloudy) picture of pre-Cartesian researches into the theory of the rainbow: '...that this Bow is made by refraction of the Sun's Light in Drops of falling Rain...was understood by some of the Ancients, and of late more fully discovered and explained by the Famous *Antonius de Dominis* Archbishop of *Spilato*, in his Book *De Radiis Visus & Lucis*, published by his Friend *Bartolus* at *Venice*, in the Year 1611, and written above twenty Years before. For he teaches there how the interior Bow is made in round Drops of Rain by two refractions of the Sun's Light, and one reflexion between them, and the exterior by two refractions and two sorts of reflexions between them in each Drop of Water, and proves his Explications by Experiments made with a Phial full of Water, and with Globes of Glass filled with Water, and placed in the Sun to make the Colours of the two Bows appear in them. The same Explication *Des-Cartes* hath pursued in his *Meteors*, and mended that of the exterior Bow. But whilst they understood not the true origin of Colours, it's necessary to pursue it here a little further' (*Opticks* (note (2)): $_1$126–7).

(9) 'in built-up areas'.

(10) See note (5).

(11) Understand 'vos', Newton's captive Cambridge audience.

(12) 'For understanding therefore how the Bow is made, let a Drop of Rain...be represented by the Sphere *BNFG*, described with the Center *C*, and Semi-diameter *CN*. And let *AN* be one of the Sun's rays incident upon it at *N*, and thence refracted to *F*, where let it either go out of the Sphere by refraction towards *V*, or be reflected to *G*; and at *G* let it either go out by refraction to *R*, or be reflected to *H*; and at *H* let it go out by refraction towards *S*, cutting the incident ray in *Y*; produce *AN* and *RG* till they meet in *X*, and upon *AX* and *NF* let fall the perpendiculars *CD* and *CE*.... Parallel to the incident ray *AN* draw the Diameter *BQ*' (*Opticks* (note (2)): $_1$127).

(13) 'Now it is to be observed, that...when by increasing the distance *CD*, these Angles [*AXR* and *AYS*] come to their limits, they vary their quantity but very little for some time together, and therefore a far greater number of the rays which fall upon all the points *N* in the Quadrant *BL*, shall emerge in the limits of these Angles, then in any other directions. And

emergentes radij cum incidentibus possunt constituere determinandi sunt ut horum Phænomenων rationes rectè percipiamus.

Scilicet in coroll: 1 et 2 prop. 35[14] ostensum est emergentem radium GR ad incidentem AN minime inclinari cum sit $3RR.II-RR::CN^q.ND^q$. et $I.2R::ND.NE$. posito nempe I ad R ut sinus incidentiæ ad sinum refractionis. Et exhinc inventis ND et NE dabitur positione RG.[15]

Sit e.g. pro radijs maximè refrangibilibus sinus incidentiæ ad sinum refractionis, sive I ad R, ut 185 ad 138, prout in aqua pluviali proximè comperi, et erit $57132.15181(::3RR.II-RR)::CN^q.ND^q$. Adeoq; $ND=\sqrt{\frac{15181}{57132}CN^q}$ seu $=\frac{5155}{10000}CN$ unde per tabulam sinuum datur arcus NI $62^{gr}.\ 4^{min}$. Præterea cùm sit $I.2R::ND.NE$, hoc est $185.276::\frac{5155}{10000}CN.NE$; erit $NE=\frac{7691}{10000}CN$. et inde etiam per tabulam sinuum datur arcus NF $100^{gr}.\ 32^{min}$. Subduc jam duplum arcus NF ex aggregato arcus NI et arcus 180^{gr} sive semicirculi, et restabit $41^{gr}.\ 0^{min}$ pro inclinatione radij RG ad radium AN, sive pro angulo AXR; productis nempe AN et RG donec in X conveniant. Et hic angulus est sub quo intimus sive cæruleus limbus Iridis hujus apparere debet.

Ad eundem modum pro radijs minimè refrangibilibus posito sinu incidentiæ ad sinum refractionis ut 183 ad 138, uti dimensus sum; invenietur

$$ND=\frac{5028}{10000}CN, \quad \text{et} \quad NE=\frac{7583}{10000}CN;$$

indeq; per Tabulam sinuum arcus NI erit $60^{gr}.\ 22^{min}$, & arcus NF $98^{gr}.\ 38^{min}$, adeoq; angulus AXR $43^{gr}.\ 6^{min}$, sub quo extimus sive rubeus hujus Iridis limbus apparebit. Itaq; maxima ejus semidiameter est $43^{gr}.\ 6^{min}$, et minima $41^{gr}.\ 0^{min}$, et Orbitæ latitudo sive crassities $2^{gr}.\ 6^{min}$ circiter, vel potiùs $2^{gr}.\ 37^{min}$, addita

further it is to be observed, that the rays which differ in refrangibility will have different limits of their Angles of emergence, and by consequence according to their different degrees of refrangibility emerge most copiously in different Angles, and being separated from one another appear each in their proper Colours' (*Opticks*: $_1$128–9).

(14) Of the 'Opticæ Pars Prima', that is. See 1, §2: note (46).

(15) '...let the sine of incidence out of Air into Water be to the sine of refraction as I to R. Now if you suppose the point of incidence N to move from the point B continually...the Arch QF will first increase and then decrease, and so will the Angle AXR which the rays AN and GR contain; and the Arch QF and Angle AXR will be biggest when ND is to CN as $\sqrt{II-RR}$ to $\sqrt{3RR}$, in which case NE will be to ND as $2R$ to I' (*Opticks*: $_1$127–8).

(16) In the corresponding passage in his *Opticks* (note (2): $_1$129) Newton takes the refractive index of red light to be 108/81, and that of blue light 109/81. It follows that the outer (red) perimeter of the primary bow has an angular radius of 42° 2′, while the inner (blue) perimeter has a radius of 40° 17′, so that (allowing 30′ for the sun's apparent diameter) the bow width is computed to be 2° 15′, slightly less than the present value.

(17) Of the 'Opticæ Pars Prima', that is. See 1, §2: note (53).

diametro Solis.(16) Sed cùm colores in extremitatibus ad utrumqꝫ limbum debiliores sint quàm quæ propter nubium conterminarum splendorem videri possint, sensibilis ejus crassities duos gradus vix excedet.

Haud secus determinantur exterioris Iridis dimensiones. Nam ostensum est in Coroll: 1 et 2 Prop 36,(17) emergentem radium HS ad incidentem AN maximè inclinari cum sit $8RR.II-RR::NC^q.ND^q$. et $I.3R::ND.NE$.(18) Quamobrem pro radiorum maximè refrangibilium sinubus I et R substitutis numeris 185 et 138, ut supra; obtinebuntur $ND=\frac{3167}{10000}CN$, et $NE=\frac{7064}{10000}CN$. Et inde per Tabulam sinuum arcus NI 36^{gr}. 48^{min}, et arcus NF 89^{grad}. 53^{min}. Atqꝫ adeò angulus $AYS=52^{grad}$ 51^{min} qui erit minima(19) semidiameter Iridis hujus. Et similiter pro radiorum minimè refrangibilium sinubus I et R substituendo numeros supra positos 183 & 138, emergent $ND=\frac{3079}{10000}CN$, et $NE=\frac{6965}{10000}CN$. Unde per Tabulā sinuum eliciuntur arcus NI 35^{gr}. 52^{min}, et arcus NF 88^{gr}. 18^{min}. Adeoqꝫ angulus AYS erit 49^{gr}. 2^{min}, Iridis nempe minima semidiameter. Quamobrem si a maximo semidiametro 52^{gr}. 51^{min} auferatur minima 49^{gr}. 2^{min}, et residuo addatur semidiameter Solis 31^{min} emerget hujus Iridis crassities 4^{gr}. 20^{min}.(20) Sed propter majorem hujus quàm interioris Iridis obscuritatem colores vix ultra crassitiem trium et semissis videri posse conjicio.(21)

(18) '...if you suppose the point of incidence N to move from the point B continually... the Angle AYS which the rays AN and HS contain will first decrease, and then increase and grow least when ND is to CN as $\sqrt{II-RR}$ to $\sqrt{8RR}$, in which case NE will be to ND as $3R$ to I' (*Opticks*: ${}_1127$–8).

(19) Read 'maxima'.

(20) In the corresponding passage in his *Opticks* (note (2): ${}_1129$) Newton supposed I/R to vary in the range 108/81 to 109/81, computing the related angular radii of the inner (red) and outer (blue) perimeters of the secondary bow to be 50° 57′ and 54° 7′. On allowing half a degree for the sun's apparent diameter there resulted 3° 40′ for the bow width, scarcely more than the $3\frac{1}{2}$° girth he here conjectures to be visible.

(21) Twenty years later Newton added a paragraph summarizing the scholium to Proposition 36 of his 'Opticæ Pars Prima' (see 1, §2: note (55)), observing that 'the Angle which the next emergent ray (that is, the emergent ray after three reflexions) contains with the incident ray AN will come to its limit when ND is to CN as $\sqrt{II-RR}$ to $\sqrt{15RR}$, in which case NE will be to ND as $4R$ to I, and the Angle which the ray next after that emergent, that is, the ray emergent after four reflexions, contains with the incident will come to its limit, when ND is to CN as $\sqrt{II-RR}$ to $\sqrt{24RR}$, in which case NE will be to ND as $5R$ to I; and so on infinitely, the numbers 3, 8, 15, 24, &c being gathered by continual addition of the terms of the arithmetical progression 3, 5, 7, 9, &c. The truth of all this Mathematicians will easily determine' (*Opticks* (note (2)): ${}_1128$). He was subsequently realistic enough to add that 'The Light which passes through a Drop of rain after two refractions, and three or more reflexions, is scarce strong enough to cause a sensible Bow' (*ibid.*: ${}_1134$). Whether anyone before 1700 knew even where to look for the tertiary rainbow (round the sun at an angular radius of some 40°) is highly dubious. See C. B. Boyer's 'The Tertiary Rainbow: An Historical Account' (*Isis*, **49** (1958): 141–54) and his penetrating historical study of *The Rainbow: From Myth to Mathematics* (New York, 1959): 233–68, especially 251–3.

Jam deniq; ut harum Iridum rationes conspectui distinctè exhibeam, sunto E, F, & G guttæ per aëra utcunq; sparsæ; ZE, ZF, & ZG radij solares parallelè incidentes in guttas; EM, EN, & EO radij diversè refrangibiles e gutta E post unam reflexionem emergentes; atq; FN, FO, FP, & GO, GP, GQ consimiles

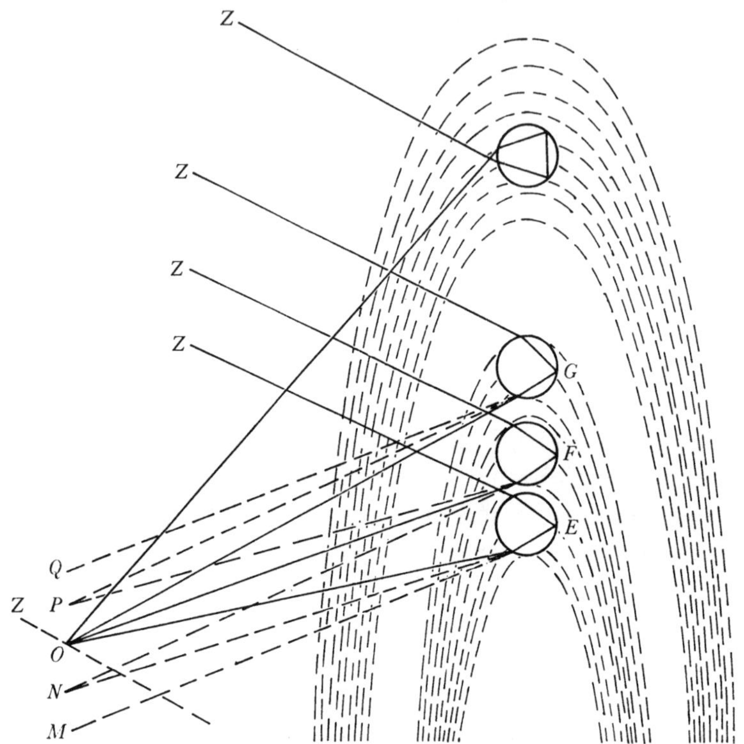

radij emergentes e guttis F ac G: nempe EO, FP, GQ maximè refrangibiles, EM, FN, GO minimè refrangibiles &c. Jam si spectantis oculus ad O consistat, ex Hypothesi manifestum est quod e radijs quos gutta E post unam reflexionem emittit soli maximè refrangibiles seu rubiformes(22) (qualis EO) impingent in oculum, reliquis ut EN et EM propter minorem refractionem præterlabentibus. Et proinde rubor(23) ad E conspicietur. E radijs autem quos gutta G post unam reflexionem emittit maximè refrangibiles qualis GQ præteribunt oculum propterea quod radio EO paralleli sunt, & alterius generis radij puta minimè refrangibiles seu cæruliformes(22) (qualis FO) in eum impingent, unde cæruleus color(23) apparebit in G. Et simili discursu gutta F in medio inter E ac G posita

(22) Interchange these to read 'cæruliformes' and 'rubiformes' respectively. Compare the following note.

(23) Read *vice versa* 'cæruleus color' and 'rubor'. Compare note (22).

(24) In Alexander's dark band, that is.

(25) Namely, (bluish) white.

radios mediocriter refrangibiles, ut FO, in oculum immittent reliquis, ut FN, FP, utrinq praeterlabentibus: indeq viriditas cernetur ad F. Eademq est ratio guttarum omnium ad easdem cum his guttis apparentes distantias ab axe OZ qui per Solem & oculum transit, positarum: et proinde ad distantias illas colores undiq apparebunt, hoc est arcus variegatus cujus interior limbus cæruleo, exterior rubro, & mediæ partes medijs coloribus tingantur, existente angulo OGQ, sive GOE, hoc est latitudine arcûs duorum circiter graduum juxta ea quæ jam ante ostendi. Estq similis discursus de arcu exteriori nisi quod ordo colorum propter contrariam inflexionem radiorum contrarius evadat. Guttæ autem quæ extra hos arcus ex una parte(24) sitæ sunt radios omninò nullos post unam vel duas reflexiones duasq refractiones in oculum immittent, ex altera autem parte omnigenos permistos eosq ferè insensibiles, et proinde nulla hujusmodi phænomena exhibere possunt, sed cœlum in illis locis colore solito(25) apparebit.

APPENDIX 3. THE PROBLEM OF TWOFOLD REFRACTION IN THE 'PRINCIPIA'.(1)

[Summer? 1685](2)

From an amanuensis'(3) draft in the University Library, Cambridge(4)

Prop. XCVII Prob. XLVII

Posito quod sinus incidentiæ in superficiem aliquam sit ad sinum emergentiæ in data ratione, quodq incurvatio viæ corporum ad superficiem illam fiat in

(1) When Newton came to draft the first book of his *Principia* in 1685 he added a final Sectio XIV 'De motu corporum minimorum quæ viribus centripetis ad singulas magni alicujus corporis partes tendentibus', the first three propositions of which have to do with a stream of corpuscles subject, in their passage through horizontally layered media, to a constant 'vis attractionis' acting vertically upwards (see his *Philosophiæ Naturalis Principia Mathematica* (London, $_1$1687): 227–31). In particular, where the total thickness of this interim layer is arbitrarily small, he was able to deduce that the corpuscle stream followed a parabolic path and so obeyed at its entrance and exit a sine law of 'refraction' (Proposition XCIV), while if this 'interface' were thick enough the parabolic stream of particles would be bent back and so emerge on the same side at an angle of 'reflection' equal to that of incidence (Proposition XCVI). In the present preliminary version of the following scholium Newton was then able to introduce the light ray as a model of such a corpuscle stream in the following way:

'Harum attractionum haud multum dissimiles sunt lucis reflexiones et refractiones factæ secundum datam [co]secantium rationem, ut invenit Snellius, et per consequens secundum datam sinuum rationem ut postmodum exposuit Cartesius. Namq radii in aere existentes... in transitu suo prope corporum vel opacorum vel transparentium angulos (quales sunt cultrorum aut fractorum vitrorum acies) incurvantur circum corpora quasi attracti in eadem... ut ipse observavi.... Cùm autem talis incurvatio radiorum fit in aere extra cultrum, debebunt etiam radij qui incidunt in cultrum, priùs incurvari in aere quàm cultrum attigunt. Fit igitur

spatio minimo quod ut punctum considerari possit;[(5)] requiritur superficies quæ corpuscula omnia de loco dato successivè manantia convergere faciat ad alium locum datum.

Sit *A* locus de quo corpuscula divergunt, *B* locus in quem convergere debent, *CDS* curva linea quæ circa axem *AB* revoluta describat superficiem quæsitam, *D*, *E* ipsius puncta duo quævis, et *EF*, *EG* perpendicula in corporis vias *AD*, *DB*. Accedat punctum *D* ad punctum *E*, et lineæ *DF* qua *AD* augetur ad lineam *DG* qua *DB* diminuitur ratio ultima erit eadem quæ sinus incidentiæ ad sinum emergentiæ. Datur ergo ratio incrementi lineæ *AD* ad decrementum lineæ *DB* et propterea si in *AB* sumatur ubivis punctum *C* per quod curva *CDE* transire debet, et capiatur *CM* ipsius *AC* incrementum [ad] ipsius *BC* decrementum *CN* in data illa ratione, centrisq; *A*, *B* et intervallis *AM*, *BN* describantur circuli duo se mutuò secantes in *D*: punctum illud *D* tanget curvam quæsitam *CDE*, eandemq; ubivis tangendo determinat. Q.E.I.[(6)]

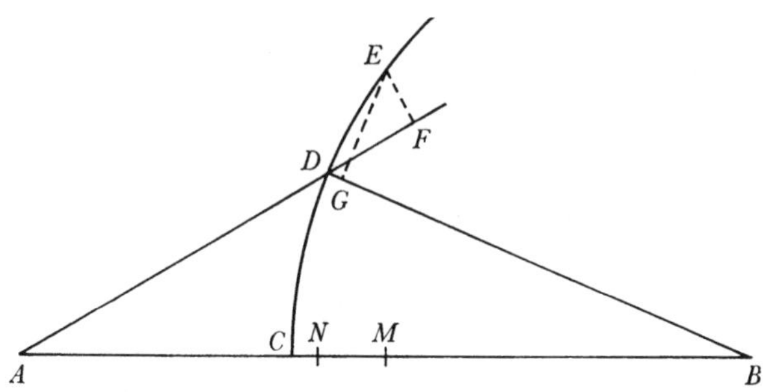

refractio non in puncto incidentiæ sed paulatim per continuam incurvationē radiorum factam partim in aere antequam attingunt vitrum, partim (ni fallor) in vitro postquam illud ingressi sunt:... Refractio igitur quia incipit in aere non fit per resistentiam vitri, et quia perseverat in vitro non fit per resistentiam aeris, sed attractioni jam expositæ similior est. Si radij in medio quovis resistentiam sentirent hi (sive motus sint sive mota corpuscula) perpetuo retarderentur et redderentur debiliores, omninò contra experientiam. Utrum verò reflexio et refractio fiant per attractiones disputet qui volet. Malim Propositione una et altera de motu, inventionem figurarum usibus opticis inservientium docere.' [In the published version (*Principia*: 232) the latter portion was considerably toned down to read 'Igitur ob analogiam quæ est inter propagationem lucis & progressum corporum, visum est Propositiones sequentes in usos opticos subjungere; interea de natura radiorum (utrum sint corpora necne) nihil omnino disputans, sed trajectorias corporum trajectoriis radiorum persimiles solummodo determinans'.]

The concluding propositions (XCVII and XCVIII) here announced have only the most tenuous connexion with the *Principia*'s basic theme—corpuscular force of one sort or another. We find it hard to resist the conclusion that Newton merely took this opportunity of publishing his elegant solution by points of the problem of twofold refraction (2, §2 above) just to have it available in printed form. A similar criticism might be levelled against many of the geometrical propositions in the preceding Sectio v (*Principia*: 70–103).

(2) Handwriting apart, this date is narrowly defined by the fact that the present piece is clearly a later portion of the second of the two incomplete drafts (now ULC. Dd. 9.46) of Book 1 of the *Principia* which Newton subsequently deposited in Cambridge University Library as his 'Lucasian lectures for 1684 and 1685'. We will return to this point in the sixth volume.

Corol. 1. Faciendo ut punctum A vel B in infinitum abeat vel migret ad alteras partes puncti C, habebuntur figuræ illæ omnes quas Cartesius in Optica et Geometria ad refractiones exposuit. Quarum inventionem cum Cartesius maximi fecerit et studiose celaverit(7) visum fuit his paucis exponere.

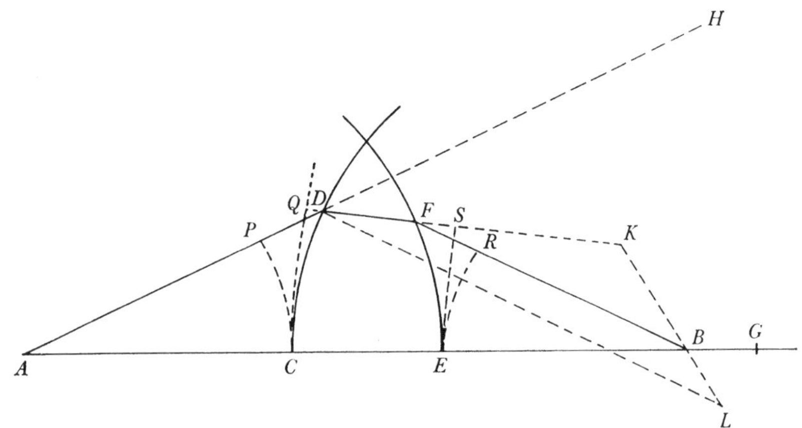

Corol. 2. Si corpus in superficiem quamvis CD, secundum lineam rectam AD lege quavis ductam incidens, emergat secundum lineam quamvis rectam $D[K]$, et a puncto C duci intelligantur lineæ curvæ CP, CQ(8) ipsis AD, $D[K]$ semper perpendiculares, erunt incrementa linearum PD, QD atq; adeo lineæ ipsæ PD, QD incrementis istis genitæ, ut sinus incidentiæ et emergentiæ; et contra.

(3) Humphrey Newton. Newton himself, however, has in places heavily corrected the manuscript in his own hand.

(4) Add. 3970.3: 428 bis^r/3970.9: 615^r–617^r, published in revised form in his *Philosophiæ Naturalis Principia Mathematica* (London, $_1$1687): 231–5. The text lacks diagrams: those here reproduced are copied from the published version.

(5) See note (1): here the (curvilinear) interface is taken to have infinitely small thickness so that the parabolic refraction of the stream of light corpuscles is reduced to a point.

(6) A rearranged but essentially unchanged version of Proposition 34 of the preceding 'Opticæ Pars Prima' (see 1, §2: note (35)).

(7) Compare 1, §2: notes (39) and (40). Newton here cancelled 'studiose' for 'studiosissime' at some subsequent time, but reverted to 'studiose' in the published version (*Principia*: 233). Even without the superlative he is somewhat less than fair to Descartes on this point, for it is probable that the latter—with Mydorge's assistance?—came upon the particular cases of the refracting ellipse and hyperbola in much the same, geometrical way as he discussed them in his *Dioptrique* before achieving the general case of the refractive (Cartesian) oval in his later *Geometrie*. (On Mydorge's possible anticipation of the hyperbolic case of the conic refracting parallel incident rays to the further focus see J. A. Lohne, 'Zur Geschichte des Brechungsgesetzes', *Sudhoffs Archiv für Geschichte der Medizin und der Naturwissenschaften*, **47** (1963): 152–72, especially 164–5.)

(8) $\widehat{CP}$ will be a circle arc of centre A (since A is fixed) while $\widehat{CQ}$ will be the involute through C of the diacaustic (not shown) of A with respect to the refracting interface CD. Compare 2, §2: note (2).

Prop. XCVIII Prob. XLVIII

Iisdem positis et circa axem AB descripta superficie qualibet attractiva CD regulari vel irregulari, per quam corpora de loco dato A exeuntia transire debent; requiritur alia superficies attractiva EF quæ corpora illa ad locum datum B convergere faciet.

Juncta recta AB secet superficiem primam in C et secundā in $[E]$, puncto $[E]$ utcunq; assumpto. Et posito M ad N(9) ut sinus incidentiæ in superficiem primam ad sinum emergentiæ utq; sinus emergentiæ ex superficie secunda ad sinum incidentiæ, produc AB ad G ut sit BG ad CE ut $M-N$ ad N, et AD ad H ut sit AH æqualis AG, et DF ad K ut sit DK ad DH ut N ad M. Junge KD et centro D intervallo DG describe circulum occurrentem KB productæ in L; Ipsiq; DL parallelā age BF; et punctum F tanget lineam EF, quæ circa axem AB revoluta describet superficiem quæsitam.(10)

Nam concipe lineas CP, CQ ipsis AD, DF respective et ER, ES ipsis FB, $[FD]$ ubiq; perpendiculares esse, et erit (per Corol. 2 Prop. XCVII) PD ad QD ut M ad N adeoq; ut DL ad DK vel FB ad FK, et divisim ut $DL-FB$ seu $HP-PD-FB$ ad FD seu $FQ-QD$, et composite ut $HP-FB$ ad FQ id est (ob æquales HP et CG, QS et CE) ut $CE+BG-FR$ ad $CE-FS$. Verum (ob proportionales BG ad CE et $M-N$ ad N) est etiam $CE+BG$ ad CE ut M ad N; adeoq; divisim FR ad FS ut M ad N, et propterea per Corol. 2 Prop. XCVII superficies EF cogit corpus in se secundum lineam DF incidens pergere in linea FR ad locum B. Q.E.F.

Schol.

Eadem methodo pergere liceret ad superficies tres vel plures.(11) Ad usus autem opticos maximè accommodatæ sunt figuræ sphæricæ. Si Perspicillorum vitra objectiva ex vitris duobus sphæricè figuratis et aquam inter se claudentibus conflentur, fieri potest ut a refractionibus aquæ corrigantur errores refractionum

---

(9) Note the new notation M/N for the refractive index. A few years later when Newton came to write his 'Fundamentum Opticæ' (ULC. Add. 3970.3: 409[→ 302]/410/415/394–8/425–6/407–8/405–6/403–4/401–2/399–400/419[+422]–21/411–14/423–4/417–18 [*sic*!]) he reverted to Barrow's index I/R, maintaining this subsequently in his *Opticks* (London, 1704: drafts and printer's copy exist in ULC. Add. 3970.3).

(10) Compare 2, §2: note (5). Newton implicitly evokes the property in his following proof that, since $\widehat{CQ}$ and $\widehat{ES}$ are involutes to the same curve (the diacaustic—not shown—of A with respect to $\widehat{CD}$, coincident with that of B with regard to $\widehat{EF}$), therefore $CE = QS$. Compare Lemma 3 to Problem 9 of the October 1666 fluxional tract (I: 433). It follows that

$$CE-DF = QD+FS = (N/M)(PD+FR) = (N/M)(AD-AC+FB-EB)$$

and therefore $AD+(M/N)DF+FB = AC+(M/N)CE+EB$, constant. Newton's construction readily results much as in 2, §2.

(11) See 2, §2: note (6).

quæ fiunt in vitrorum superficiebus extremis.(12) Talia autem Vitra Objectiva vitris Ellipticis et Hyperbolicis præferendæ sunt non solum quod facilius et accuratius formari possint, sed etiam quod penicillos radiorum extra axem vitri sitos accuratius refringant. Verum tamen diversa diversorum radiorum refrangibilitas impedimento est quo minus Optica per figuras vel sphæricas vel alias quascunq; perfici possit.(13)

APPENDIX 4. CORRECTING FOR SPHERICAL DISTORTION BY MEANS OF A COMPOUND LENS.(1)

[*c*. 1692?](2)

From an autograph draft in the University Library, Cambridge(3)

Now were it not for this different refrangibility of rays Telescopes might be brought to a greater perfection then we have yet described, by composing the

---

(12) See I: 575–6 and Appendix 4 following.

(13) A following incomplete, cancelled sentence reads 'Errores inde oriundi sunt (in Telescopijs) longè majores quàm qui ex figuris minus ['sphæricis' is cancelled] aptis oriri solent: et vitio vertendum est si quis ignorata errorum causa principali [laborem omnem suam in cæteris corrigendis collocaverit?]'. An 'impediment' can, of course, be overcome and Newton never suggested otherwise. Indeed, in a further sentence added subsequently in his own hand he was careful to add that 'Nisi corrigi possint errores illinc oriundi, labor omnis in cæteris corrigendis ['frustra' is cancelled] imperitè collocabitur' (published in *Principia*: 235).

A final following paragraph (also cancelled) reads 'Hactenus exposui motus corporum in spatijs liberis'. This echoes the scholium to Problem 5 of the revised version of his 'De Motu' (ULC. Add. 3965.7: 50r): 'Hactenus motum corporum in medijs non resistentibus exposui, id adeo ut motus corporum cœlestium in ætheri determinarem....'. The suggestion is strong that Newton at this time intended to discuss projectile motion in a resisting medium (Section I of Book 2 in the published *Principia*) as an appendix to his 'De Motu Corporum Liber Primus' in the style of 'De Motu', no doubt continuing with the 'De Corporum Liber Secundus' (ULC. Add. 3990; compare I: xx, note (12)).

(1) This extract from a preliminary draft of Newton's 'first Book of Opticks' evidently amplifies the first, uncommunicated version of his letter to Oldenburg on 11 June 1672, where he remarked that 'though this way [of 'regulating' the refractive dispersion of a simple telescopic object glass 'by a complication of divers successive mediums, as by two or more glasses or chrystalls wth water or some other fluid between them, all wch together may perform the office of one object-glasse'] faile[d] my expectations as to regulating the refractions of heterogeneall rays yet as to homogeneall rays I found something in it wch Opticians have hitherto sought for in vain: Namely that those rays whether parallel diverging or converging may by such successive Mediums of sphericall figures be made more exactly refracted to or from a point then they can by one similar object glasse of the like sphericall figures....And by making trialls wth severall liquors differing in their refractive densities [placed in the 'sphericall cavity' between 'two like & equall convexo-concave glasses something lesse then hemi-sphericall & cemented to either side of a flat Ring of Copper...in such manner, that their superficies may constitute two concentrick spheres'], it will be found that some of those wch

Object-glass of two glasses *ABED* & *BEFC* alike convex on y[e] outsides *AGD* & *CHF* & alike concave on y[e] insides *BME*, *BNE* with water in the cavity *BMEN*. Let the sine of incidence out of glass into air be as *I* to *R* & out of water into air as *K* to *R* & by consequence out of glass into water as *I* to *K* & let the diameter of the sphere to w[ch] all the convex side *AGD* or *CHF* is grownd be *D* & the diameter of the sphere to w[ch] y[e] concave side *BME* or *BNE* is ground be to *D* as the cube root of $KK-KI$ to the cube root of $RR-RI$:(4) & the refractions on y[e] concave side of

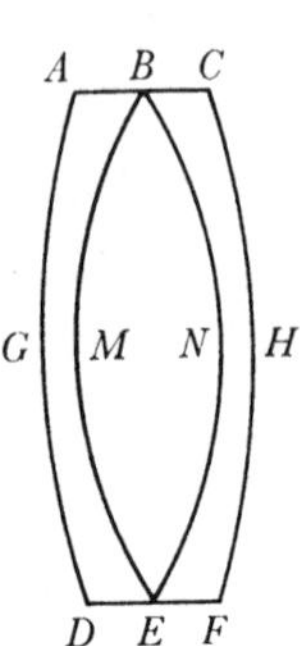

nearly equall rain water in density will make parallel homogeneall rays converge towards a point wth a very great though not Geometrically exactnesse....As to Telescopes it would be necessary to make their Object glasses more flat...' (ULC. Add. 3970.3: 447[v], reproduced in *Correspondence of Isaac Newton*, **1** (1959): 191–2). Compare **1**, §2: note (61) above.

(2) This conjectural dating, suggested by several points of circumstantial evidence in the surrounding context, is confirmed by Newton's handwriting.

(3) Add. 3970.3: 471[r], subsequently lightly revised (on 3970.3: 74[r]) in the marked printer's copy. The extract appeared publicly a decade later—along with a crucial printer's error (see note (4))—in Newton's *Opticks: Or, A Treatise of the Reflexions, Refractions, Inflexions and Colours of Light* (London, 1704): $_1$74–5.

(4) This appeared in Newton's *Opticks* (note (3)) in 1704 as '$RK-RI$', a printer's mistake which none of the numerous editions of the book has ever corrected—indeed, to our knowledge this false reading has never previously been detected, although 'Newton's' printed proportion at once (since $K \neq I$) reduces lamely to $K:R$, a manifest impossibility. The reader will draw his own conclusions regarding the depth and accuracy of past scholarly criticism of Newton's geometrical optics.

(5) Newton has left us no preliminary worksheet as a guide to the analysis which brought forth this result. If we take C, C' (in the axis *GMNH*) to be the centres of the respective spherical interfaces $\widehat{AGD}$, $\widehat{BME}$, we may suppose he intended the compound lens to correct for spherical distortion in such a way that all rays incident from F are refracted (at a in $\widehat{AG}$ from air into

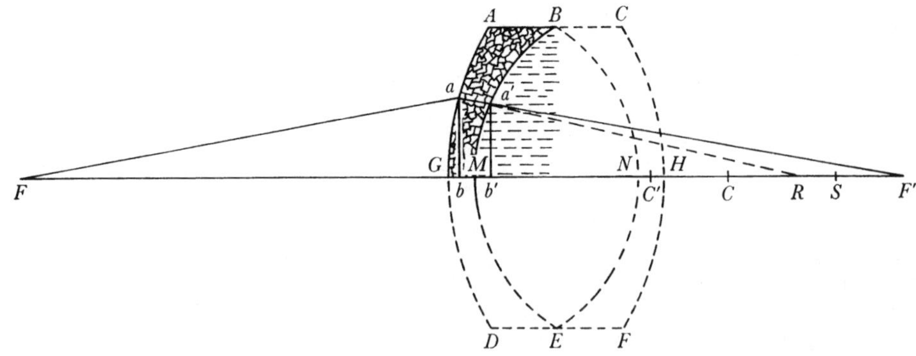

glass and at a' in $\widehat{BM}$ from glass into water) to the point F' 'wth a very great though not Geometrically exactnesse'. The condition (necessary and sufficient) for this to occur instantaneously for any incident ray Fa is that, where Fa is refracted through R, a ray incident from F' shall also be refracted at a' through R: in particular, when a and a' are indefinitely close to the axis, that F (with respect to the interface $\widehat{AG}$) and F' (with respect to $\widehat{BM}$) shall share the

y^e^ glasses will very much correct the errors of y^e^ refra[c]tions on the convex sides, so far as they arise from the sphericalness of the figure.(5) And by this meanes might Telescopes be brought to sufficient perfection were it not for y^e^ different refrangibility of several rays.(6)...

---

common principal focus S. In analytical terms (compare Appendix 1.3: note (8) above), on taking $FG = a$, $MF' = \alpha$, $GC = r$ (that is, $\frac{1}{2}D$) and $MC' = \rho$, then

$$GR = \frac{aRr}{a\mu+rI} - \frac{(a-r)^2 I}{2aRr}\left(\frac{I}{a\mu+rI} + \frac{R^2 r}{(a\mu+rI)^2}\right) y^2 + O(y^4)$$

and
$$MR = \frac{\alpha K\rho}{\alpha\lambda-\rho I} - \frac{(\alpha+\rho)^2 I}{2\alpha K\rho}\left(\frac{I}{\alpha\lambda-\rho I} - \frac{K^2\rho}{(\alpha\lambda-\rho I)^2}\right) y'^2 + O(y'^4),$$

where $I:K:R$ are in proportion as the refractive indices of glass, water and air respectively and $\lambda = K-I$, $\mu = R-I$, while $ab = y$ and $a'b' = y'$ are the vertical distances of points a and a' from the axis $GMNH$. When R passes into S, y and y' become indefinitely small, so that (on neglecting the thickness of the glass meniscus $ABED$)

$$(GS \text{ or}) \ aRr/(a\mu+rI) \approx (MS \text{ or}) \ \alpha K\rho/(\alpha\lambda-\rho I):$$

this determines the distance of F' from the lens in relation to that of F from it. The corresponding condition that R is the common refraction point of F and F' determines that

$$2RS/Iy^2 \approx 2RS/Iy'^2,$$

and so very nearly

$$\frac{(a-r)^2}{aRr}\left(\frac{I}{a\mu+rI} + \frac{R^2 r}{(a\mu+rI)^2}\right) \approx \frac{(\alpha+\rho)^2}{\alpha K\rho}\left(\frac{I}{\alpha\lambda-\rho I} - \frac{K^2\rho}{(\alpha\lambda-\rho I)^2}\right).$$

When a and α are both large in comparison with r, ρ and the refractive indices K/I and R/I, it follows that $I/R\mu r$ and $I/K\lambda\rho$ are approximately equal, so that $r/\rho \approx K(K-I)/R(R-I)$. In the telescopic case, in fact, if F is the object, then a will be 'infinity' so that, more accurately, $\dfrac{I}{\mu Rr} \approx \dfrac{(\alpha+\rho)^2}{\alpha K\rho}\left(\dfrac{I}{\alpha\lambda-\rho I} - \dfrac{K^2\rho}{(\alpha\lambda-\rho I)^2}\right)$, where $Rr/\mu \approx \alpha K\rho/(\alpha\lambda-\rho I)$: elimination of α between these equalities yields a cubic equation of some complexity connecting the radii r, ρ of the spherical interfaces $\widehat{AG}$, $\widehat{BM}$. Newton's suggested simplification of this as $2r/2\rho = \sqrt[3]{[K\lambda/R\mu]}$ (or $R\mu r^3 - K\lambda\rho^3 = 0$) we find theoretically baffling. Is it perhaps an empirical modification of the loose equality $r/\rho \approx K\lambda/R\mu$? We have not tested the physical accuracy of Newton's proportion by experiment.

(6) Compare 1, §2: note (61) and Appendix 3: note (13) above. Here, much oversimplifying his own historical position of twenty-five years earlier, Newton continued: 'by reason of this different refrangibility, I do not yet see any other means of improving Telescopes by Refractions alone than that of increasing their lengths, for which end the late contrivance of *Hugenius* seems well accommodated. For very long Tubes are cumbersome...and shake by bending so as to cause a continual trembling in the Objects, whereby it becomes difficult to see them distinctly.... Seeing therefore the improvement of Telescopes of given lengths by Refractions is desperate; I contrived heretofore a Perspective by reflexion, using instead of an Object Glass a concave Metal' (*Opticks* (note (3)): $_1$75. On Huygens' 'aerial' telescope see the review of his *Astroscopia Compendiaria, Tubi optici molimine liberata* (The Hague, 1684) in *Philosophical Transactions*, **14** (1684): No. 161 (for 20 July 1684): 668–70: 'Christiani Hugenii *Astroscopia compendiaria*.... Or the description of an Aerial Telescope'. Compare also his *Œuvres complètes*, **8** (1899): 488, note 2; and H. C. King's *History of the Telescope* (London, 1955): 54–6, 63–4.

APPENDIX

NEWTON'S MATHEMATICAL CORRESPONDENCE 1670–1673

NEWTON'S MATHEMATICAL CORRESPONDENCE, 1670–1673

In seeking to illuminate the background to the autograph mathematical papers here reproduced, we have quoted extensively from the letters which then passed between Newton, Collins and others of their contemporaries. Together with his somewhat spare autobiographical remarks such correspondence sheds a uniquely informative light on the social and psychological factors governing the creation and diffusion of many of Newton's more important mathematical writings of the period—notably his *De Analysi*, his *Observationes* on Kinckhuysen's *Algebra* and his 1671 tract on infinite series and fluxions.(1) As we might expect, these letters also treat of certain mathematical topics (arising out of queries put to him by his correspondents) which are not elsewhere discussed by Newton, though it is important to notice that these problems are all resolved by applying general techniques developed by him in his earlier researches and that the methods of solution are not, in themselves, essentially novel. To obtain a fully balanced assessment of Newton's mathematical achievement during the period 1664–73, as we can now know it, account must necessarily be taken of these epistolary offshoots of his mainstream researches. Accordingly, it seems appropriate at the present point in our edition of his mathematical papers to offer, in minor complement, an outline of these additional topics. As fundamental documentary source we shall draw upon the authoritative text of Newton's early letters recently established by H. W. Turnbull in his Royal Society edition, employing for ease of quotation the convention of referring to it (in square brackets) by volume and page number only. Thus, in the case of '[**1**: 229–32]' understand a reference to Newton's letter to John Collins of 20 August 1672, reproduced in *The Correspondence of Isaac Newton*, **1** (Cambridge, 1959): 229–32.

Newton's extant mathematical correspondence begins effectively in January 1670 with a letter (his first?) to Collins [**1**: 16–20] about six months after Barrow had communicated to him the tract *De Analysi per æquationes numero terminorum infinitas*.(2) Formalities apart, he there considered, 'without any Indication of his method' [**1**: 53], two 'Problemes' proposed a few weeks earlier by Collins, one seeking the sum of p terms of a 'musicall' series $a/(b+c_i)$, the second concerning 'y[e] resolution of Equations by tables'. On the latter topic Newton was curtly discouraging [**1**: 19–20]:

---

(1) See the introductions to II, **2**, 3 and **3**, 1, and to **1**, 2 above.

(2) Compare II: 165–70.

There may bee such Tables made for Cubick equations; & consequently shall serve for those of foure dimensions too: But scarcely for any others. Indeed could all Equations bee reduced to three termes only, tables might be made for all: but that's beyond my skill to doe it, & beleife that it can bee done. For those of three dimensions there needs but one column of figures bee added to ye ordinary tables of Logarithms, & ye construction of it is pretty easy & obvious enough. If you please I will some time send you a specimen of its composition & use, but I cannot perswade my selfe to undertake ye drudgery of making it.

In fact, it took Collins another two and a half years before he succeeded in prying out of Newton his way of 'approximating ye roots of affected Æquations by Gunters line'. At length, in reply on 13 July 1672 to a (lost) letter of Collins which renewed his appeal for information, Newton agreeably answered that 'If I ever applyed Gunters Sector to the resolving of affected æquations it hath now slipt out of my memory. Possibly it might be Gunters line wch being set upon 3 or 4 severall rulers is of ready use for finding ye 2 or 3 first figures of any affected æquation but there is no difficulty in y^e invention'. Collins was quick to accept [**1**: 226] Newton's 'profer' that 'if it be y^e same wch you meane, you may command it', and hence it was that on the following 20 August [**1**: 230] the latter finally described his nomogram in the 'instance' of the cubic equation $x^3-7x^2+16x=12$:

to resolve w^{ch}, place three of Gunters Rulers *BF*, *CG*, & *DH* parallel & equally distant from one another, & to any line *AE* wch crosseth them all apply y^e number 1 of the third ruler *DH*, y^e number 7 of y^e second *CG*, & y^e number 16 of y^e first *BF*, accordingly as y^e coefficients of y^e æquation are. Then in ye cross line taking ye point *A* as far from *B* as *B* is from *C* apply any ruler to that point *A*, about wch whilst you turne it slowly observe the numbers where it cuts the other rulers untill you see y^e summ of y^e numbers on the first & third ruler equall to y^e sum of y^e resolvend 12 & the number on y^e other ruler, wch when it happens the number on y^e 3^d ruler shall be y^e cube of y^e desired root. Thus in this case you see $8+32=12+28$, & therefore 8 ... is y^e cube of y^e desired root.

Evidently, if in any position the rotating ruler cuts off the length $\log(x^3)$ on the third logarithmic ruler *DH*, it will cut off corresponding lengths $\log(16x)$ and $\log(7x^2)$ on the others, and Newton's procedure is effectively to find a position of the rotating ruler which constructs $x^3+16x=7x^2+12$, as closely as possible, by inspection. Here, since $2^3+16.2$ is constructed equal to 7.2^2+12, $x=2$ is a root of the given cubic equation.(3) As Newton adds, 'The application of this to any æquations of higher dimensions is obvious': we merely transpose the equation so that all terms on its either side have positive coefficients and then attempt to balance the corresponding 'numbers' cut off by the rotating line on a set of

(3) A disadvantage of Newton's machine will be at once apparent: namely, its inability to distinguish between multiple roots. Here $x=2$ is a double root of the equation.

equally spaced 'Gunter's' rulers, either rectilinear or concentric round the point A. By scaling the ruler corresponding to the term in x^i in the ratio $1/i$, it is possible 'also so to proportion y^e^ rulers BF, CG, DH, &c y^t^ y^e^ line AK may be carried over them with parallel motion', independently of the distances BC, CD, ...(4)

The problem of summing a general harmonic series was more immediately attractive to Newton, and in achieving its practical solution he displayed considerable ingenuity. He saw at once in January 1670 [**1**: 16–17] that, since $\sum_{0\leqslant i\leqslant p-1} a/(b+c_i)$ may be expressed equivalently in the form

$$a\frac{d}{db}\Big[\prod_{0\leqslant i\leqslant p-1}(b+c_i)\Big]\Big/\prod_{0\leqslant i\leqslant p-1}(b+c_i),$$

the essential difficulty was to elaborate an adequate technique for determining the continued product $\prod_{0\leqslant i\leqslant p-1}(b+c_i)$. As we would expect, Newton concentrated his attention on the particular case for which $c_i=(i+h)c$, and particularly the symmetrical form in which $p=2n+1$, $h=-n$. In the latter instance the problem is to determine the 'aggregate'

$$\sum_{-n\leqslant i\leqslant +n} a/(b+ic)=a\frac{d}{db}\Big[\prod_{-n\leqslant i\leqslant +n}(b+ic)\Big]\Big/\prod_{-n\leqslant i\leqslant +n}(b+ic),$$

and to 'much faciliate y^e^ multiplication of denominators together' Newton tabulated the coefficients of the continued product $\prod_{-n\leqslant i\leqslant +n}(b+ic)$ for $n=1, 2, 3, 4, 5$ [**1**: 17].

For large p the complexities involved in summing p terms of the harmonic series $a/(b+ic)$ by this direct method rapidly become prohibitive, and in January 1670 Newton went on to consider solutions 'by approximation', understanding (as he made clear a year later) that 'the difference of the denominators [c] beares a lesse proportion to the denominator [b]' for convergence [**1**: 69]. In the general non-symmetrical case he states [**1**: 17–18] the 'progression' ('in w^ch^...the farther you proceede, the nearer you approach to truth')

$$\sum_{0\leqslant i\leqslant p-1} a/(b+ic)=(a/b)\,(p-q\gamma+r\gamma^2-s\gamma^3\ldots+(-1)^l\alpha_l\gamma^l\ldots),$$

where $\gamma=c/b$ and, on taking $\alpha_l=\sum_{0\leqslant i\leqslant p-1}(i^l)$, $q=\alpha_1$, $r=\alpha_2$, $s=\alpha_3$, ... More conveniently, however, when p is odd 'its better to put b for the Denominator of y^e^ middle terme of y^e^ propounded series...making n the number of termes from y^e^ said middle terme either way' [**1**: 18]. In this case he derives for the

(4) See James Wilson, *Mathematical Tracts of the late Benjamin Robins, Esq*;..., **2** (London, 1761): Appendix: 348–50. Compare also I, **3**, 3, §1: note (6).

'desired Aggregate' $\sum_{-n\leqslant i\leqslant n} a/(b+ic)$ a 'progression wanting each other terme & also converging much more towards the truth[5] then the former', $(a/b)(p+r'\gamma^2+t'\gamma^4\ldots+\beta_{2l}\gamma^{2l}\ldots)$, where (as before) $\gamma=c/b$ and also, on setting $\beta_{2l}=2\sum_{1\leqslant i\leqslant n}(i^{2l})$, $r'=\beta_2$, $t'=\beta_4$, ... We may conjecture[6] that the recursive developments of the coefficients α_l, β_{2l} stated by Newton for $l=1, 2, 3, 4$ were derived by an application (knowingly so, or not) of Pascal's 'unica ac generalis methodus ad summam potestatum cujuslibet progressionis inveniendam':[7] namely, the various α_l, $l=1, 2, 3, \ldots$, may be successively determined by appropriate elimination between the various

$$p^k=\sum_{0\leqslant i\leqslant p-1}[(i+1)^k-i^k]=p+\binom{k}{1}q+\binom{k}{2}r\ldots+k\alpha_k,\quad k=2, 3, \ldots, l;$$

while in a similar way the various β_{2l}, $l=1, 2, 3, \ldots$, are found in turn by appropriate elimination between

$$p((4m+1)^k-1)=-2n+\sum_{1\leqslant i\leqslant n}[(2i+1)^{2k+1}-(2i-1)^{2k+1}]$$
$$=4\binom{2k+1}{2}r'+16\binom{2k+1}{4}t'\ldots+2^{2k}(2k+1)\beta_{2k},$$

$k=1, 2, 3, \ldots, l$, where $m=n(n+1)=\frac{1}{4}(p^2-1)$. In the symmetrical case, on replacing $r'=\beta_2$, $t'=\beta_4$, ... by their values we obtain the infinite series expansion $\sum_{-n\leqslant i\leqslant n} a/(b+ic)=p(a/b)(a_0+\frac{1}{3}a_1 m\gamma^2+\frac{1}{5}a_2 m^2\gamma^4+\ldots)$, in which $\gamma=c/b$,

$$p=2n+1,\quad m=n(n+1)\quad\text{and}\quad a_0=1,\quad a_1=1,\quad a_2=1-\tfrac{1}{3}m^{-1},$$

$a_3=1-m^{-1}+\frac{1}{3}m^{-2}$, ... As a corollary [**1**: 18] Newton essayed a rule: 'a series of fractions being propounded: [if] you would have their aggregate ... not erring from truth above $\frac{1}{e}$ part of an unit ... make a rude guesse how many times $\frac{b}{nc}$ multiplyed into it selfe will bee about y^e^ bignesse of $\frac{5ae}{2b}$ more or lesse. And omit all those termes of the progression where b is of more then soe many dimensions'. In other words, if $\frac{5}{2}(a/b)(n\gamma)^{2k}\approx e^{-1}$ roughly, then only the first

(5) The terminology is James Gregory's, here used by Newton for the first time in extant manuscript: compare **1**, 2, §2: note (72) above.

(6) Compare [**1**: 21, note (8)]. See also J. E. Hofmann, *Studien zur Vorgeschichte des Prioritätstreites zwischen Leibniz und Newton um die Entdeckung der höheren Analysis*. I. *Materialien zur ersten mathematischen Schaffensperiode Newtons (1665–1675)* [= *Abhandlungen der Preussischen Akademie der Wissenschaften (1943)*. *Math.-naturw. Klasse* 2 (Berlin, 1943)]: 22–3.

(7) Sketched, for the instance $\sum_{1\leqslant i\leqslant 4}(3i+2)^3$, in Blaise Pascal's *Potestatum Numericarum Summa* (published posthumously in his *Traité Du Triangle Arithmetique, Avec Quelques Autres Petits Traitez sur la mesme Matiere* (Paris, 1665): $_4$34–41, especially 36–7).

$k+1$ terms of the series $(2n+1)(a/b)\sum_{1\leqslant i\leqslant\infty}[(2i-1)^{-1}a_i m^i\gamma^{2i}]$ are needed to evaluate $\sum_{-n\leqslant i\leqslant n} a/(b+ic)$ to within an error of e^{-1}. This implies the approximation $(2n+1)\sum_{(k+1)\leqslant i\leqslant\infty}[(2i-1)^{-1}a_i m^i\gamma^{2i}]\approx\frac{5}{2}(n\gamma)^{2k}$, closely true for Newton's first example (in which $\gamma=3/59$, $n=3$ and so $m=12$) but unrealistic in general.[(8)]

Newton concluded these observations on the summation of a harmonic series in his January 1670 letter with the remark that 'This Probleme much resembles y^e^ squaring of the Hyperbola: That being only to find y^e^ aggregate of fractions infinite in number & littlenesse, wth one common numerator to denominators whose differences are equall & infinitely little.[(9)] And as I referred all the series to y^e^ middle terme, the like may bee done conveniently in y^e^ Hyperbole' [**1**: 19]. The truth of this comparison of the sum of the harmonic series $\sum_i a/(b+ic)$ with the area under the hyperbola $xy=a/c$ is evident geometrically. Equivalently, as p tends to infinity in the preceding expansions, it follows that $\alpha_i/p^i=1/i$ and $a_i=1$, so that (on setting $\gamma=c/b$ as before) the corresponding limit forms of $\sum_{0\leqslant i\leqslant p-1} a/(b+ic/p)$ and $\sum_{-n\leqslant i\leqslant n} a/(b+ic/n)$ are

$$(a/c)\,(\gamma-\tfrac{1}{2}\gamma^2+\tfrac{1}{3}\gamma^3\ldots)=(a/c)\log(1+\gamma)$$

and $2(a/c)\,(\gamma+\frac{1}{3}\gamma^3+\frac{1}{5}\gamma^5\ldots)=(a/c)\log([1+\gamma]/[1-\gamma])$. This analogy fermented in Newton's mind during the next few weeks, and on 18 February he again wrote to Collins: 'I now see a way...how y^e^ aggregate of y^e^ termes of Musicall progressions may bee found...by Logarithms, but y^e^ calculations for finding out those rules would bee still more troublesom, and I shall rather stay till you have leisure to doe me the favour of communicating w^t^ you have already composed on y^t^ subject' [**1**: 27]. With that chilly response Collins had to be content for more than seventeen months till, after a long break in their correspondence, Newton again took up the topic on 20 July 1671 with the words: 'There having some things past between us concerning musicall progressions, & as I remember you desiring mee to communicate somthing wch I had hinted

(8) Compare [**1**: 22, note (10)]. Since n is taken to be large, it may be that Newton assumed $m\approx n^2$ and $a_i\approx 1$ for $i\geqslant k+1$, and then contrived some argument (perhaps by a comparison with the infinite geometrical progression $\sum_{(k+1)\leqslant i\leqslant\infty}[(n\gamma)^{2(i-k)}]=n^2\gamma^2/[1-n^2\gamma^2]$) which convinced him of the justice of the approximation $\sum_{(k+1)\leqslant i\leqslant\infty}[(2i-1)^{-1}(n\gamma)^{2(i-k)}]\approx 5/2(2n+1)$.

(9) This is reminiscent of the definition of the natural logarithm developed by Pietro Mengoli in Books 4 and 5 of his *Geometria Speciosa* (Bologna, 1659): namely, $\log(n/m)$ is the function which satisfies the inequality $\sum_{rm\leqslant i\leqslant(rn-1)}(1/i)>\log(n/m)>\sum_{(rm+1)\leqslant i\leqslant rn}(1/i)$ for all positive integers r. (See D. T. Whiteside, 'Patterns of Mathematical Thought in the later Seventeenth Century', *Archive for History of Exact Sciences* **1** (1961): 179–388, especially 224–5.) However, it is unlikely that Newton ever read Mengoli's book.

to you about it, w[ch] I then had not (nor have yet) adjusted to practise: I shall in its stead offer you somthing else w[ch] I think more to y[e] purpose' [**1**: 68]. The rule he then communicated is not at once obvious: to compute the sum, say $S = \sum_{0 \leqslant i \leqslant p-1} a/(b+ic)$, of 'any musicall progression...propounded whose last terme is $\frac{a}{d}$' and 'where the difference of the denominators beares a less proportion to the denominator of the first terme', Collins is directed to

choose any convenient number e (whither whole broken or surd) which intercedes these limits $\frac{2mn}{b+d}\left[\text{that is, } \frac{2mn}{m+n}\right]$ & $\sqrt{mn}$; supposing $b-\frac{1}{2}c$ to bee m, & $d+\frac{1}{2}c$ to bee n.(10) And this proportion will give you the aggregate of the termes very neare the truth. As y[e] Logarithm of $\frac{e+\frac{1}{2}c}{e-\frac{1}{2}c}$ to y[e] Logarithm of $\frac{n}{m}$, so is $\frac{a}{e}$ to y[e] desired summe.... The ground of this rule I believe you will easily apprehend by contemplating y[e] Hyperbola, what relation its area beares to musicall progressions.

It is evident that Newton's choice of the form of his proportion preceded his choice of the harmonic and geometric means of m and n as bounds to e. In the first instance it follows readily from the inequality

$$\frac{a}{c}\log\left(\frac{b+(p-1)c}{b-c}\right) > S > \frac{a}{c}\log\left(\frac{b+pc}{b}\right)$$

(visually apparent from considering bounds to the area under the rectangular hyperbola $xy = a/c$, $b-c < x < b+pc$) that $S \approx (a/c)\log(n/m)$, where $m = b-\frac{1}{2}c$ and $n = b+(p-\frac{1}{2})c = d+\frac{1}{2}c$. Hence if ϵ is chosen such that

$$\log(n/m) = (n-m)/\epsilon,$$

that is, since $n-m = pc$ and $\frac{c}{\epsilon} \approx \log\frac{\epsilon+\frac{1}{2}c}{\epsilon-\frac{1}{2}c}$, such that $\log\frac{n}{m}\Big/\log\frac{\epsilon+\frac{1}{2}c}{\epsilon-\frac{1}{2}c} \approx p$, Newton's proportion is an immediate consequence of $S \approx ap/\epsilon$. Subsequently, we may conjecture, he sought to bound the quantity e which made the near-equality $S \approx \frac{a}{e}\log\frac{n}{m}\Big/\log\frac{e+\frac{1}{2}c}{e-\frac{1}{2}c}$ even more nearly true. It would not be hard for him to prove that the harmonic mean, $\epsilon_H = 2mn/(m+n)$, and geometric mean

$$\epsilon_G = \surd[mn],$$

(10) In a first draft on the back of Collins' letter of 5 July (ULC. Add. 3977.9) to which he was replying, Newton wrote 'Suppose e = a meane proportion twixt b & d or any integrall or broken number interceding these limits $\frac{2\sqrt{ad}}{a+d}$ & $\sqrt{ad}$ [*sic*]', and subsequently 'Suppose e is a meane proportion twixt $b-\frac{1}{2}c$ & $d-\frac{1}{2}c$ or that it is any convenient integrall or broken number by guesse w[ch] differs not considerably from it. suppose it intercede the limits $\sqrt{bd}$ & $\sqrt{\overline{b-c}\times\overline{d-c}}$ very neare it'. In the latter it would appear that d is taken to be $a+pc$.

are both close lower bounds to ϵ,(11) from which it would follow that e is very nearly equal to both ϵ_H and ϵ_G. Newton's final verification, we may be sure, of the accuracy of these suggested approximate values of e would have been by means of the series expansions of $\sum_i a/(b+ic)$ which he had developed the year before. For simplicity suppose $p=2n+1$ and take $\frac{1}{2}(b+d)=b+nc=w$, and there results

$$S=\sum_{-n\leqslant i\leqslant n} a/(w+ic)=(ap/w)\,(1+\tfrac{1}{12}(p^2-1)\,c^2/w^2+\tfrac{1}{240}(p^2-1)\,(3p^2-7)\,c^4/w^4\ldots),$$

$$\log\,(n/m)=\log\,([w+\tfrac{1}{2}pc]/[w-\tfrac{1}{2}pc])=(cp/w)\,(1+\tfrac{1}{12}p^2c^2/w^2+\tfrac{1}{80}p^4c^4/w^4\ldots)$$

and $\log\,([e+\frac{1}{2}c]/[e-\frac{1}{2}c])=(c/e)\,(1+\frac{1}{12}c^2/e^2\ldots)$. On substituting these values in Newton's proportion and reducing, there follows

$$1+\tfrac{1}{12}c^2/e^2\ldots=\frac{1+\frac{1}{12}p^2c^2/w^2+\frac{1}{80}p^4c^4/w^4\ldots}{1+\frac{1}{12}(p^2-1)\,c^2/w^2+(\frac{1}{80}p^4-\frac{1}{24}p^2\ldots)\,c^4/w^4\ldots},$$

that is, $e^2=w^2(1+\frac{1}{12}p^2c^2/w^2\ldots)/(1+\frac{1}{2}p^2c^2/w^2\ldots)$ and hence

$$e=w(1-\tfrac{5}{24}p^2c^2/w^2\ldots).$$

In contrast, $\epsilon_H=w(1-\frac{1}{4}p^2c^2/w^2)$ and

$$\epsilon_G=w\surd[1-\tfrac{1}{4}p^2c^2/w^2]=w(1-\tfrac{1}{8}p^2c^2/w^2\ldots),$$

so that e is very nearly equal to the first of two arithmetic or geometric means between ϵ_H and ϵ_G.(12) Newton's replacement of natural logarithms by ones to base 10 in his final rule is not significant since only the ratio of $\log\,(n/m)$ to $\log\,([e+\frac{1}{2}c]/[e-\frac{1}{2}c])$ is to be computed.

In his letter to Collins of 8 February 1669/70, in addition to observing that 'Mr Barrow shewed mee some of your papers in wch I was much pleased at Monsieur Cassinis invention for finding y^e^ Apogæa & excentricitys of y^e^ Planets',(13) Newton outlined a rule 'sent you...to consider of w^t^ use it may bee'

(11) This result is implicit in Propositions 1, 2 and 32 of James Gregory's *Vera Circuli et Hyperbolæ Quadratura* (Padua, 1667), a book Newton knew well (see **1**, 2, §2: note (72) above).

(12) Compare Hofmann's *Studien* (note (6)): 42–4.

(13) This refers to a paper in Collins' hand, 'A new Geometricall and direct Method for finding the Apogæa Excentricities and Anomalies of the Motion of the Planets by Monsr Cassini' (ULC. Add. 3971.4: 77^r^ – 78^r^), which is his English translation ('Of my owne version out of French')—presumably that read to the Royal Society on 27 January—of Cassini's 'Nouvelle Maniere Geometrique & directe de trouver les Apogées, les Excentricitez, & les Anomalies du Mouvement des Planetes' (*Journal des Sçavans* (1669), issue for 'Lundy 2 Septemb.': Art. III). (Compare T. Birch, *History of the Royal Society*, **2** (London, 1756): 417.) The essence of the paper is a description of Cassini's method of determining the elements of planetary orbits 'in the Elliptick Hypothesis [of Seth Ward], supposing that one focus of the Ellipsis is the Center of the apparent Motion, and the other the Center of the meane Motion'. As such it was, soon afterwards, heavily critized by Nicolaus Mercator in the *Philosophical*

for solving an annuity problem previously communicated to him by Collins: namely, 'To know at what rate (N per cent) an Annuity of B is purchased for 31 yeares at y^e price A'. In solution Newton proposed a logarithmic algorithm, 'not exact but yet soe exact as never to faile above 2^d or 3^d at the most when the rate is not above 16^li per cent':

$$\frac{6\log_{10}(31B/A)}{100-50\log_{10}(31B/A)}=\log_{10}x, \quad \text{`putting } \frac{100+N}{100}=x\text{'}.$$

However, 'if the rate bee above 16 or 18^lib per cent, or wch is all one if A [bee lesse then] $6B$, then this rule $\frac{A+B}{A}=[x]$ will not err above 2^s. You may try the truth of these rules by the equation $x^{32}=\frac{A+B}{A}x^{31}-\frac{B}{A}$;... & working in logarithms'. The last equation causes no difficulty since it merely restates the standard annuity formula $Ax^{31}=B(1+x+x^2+\ldots+x^{30})=B(x^{31}-1)/(x-1)$. Here, when x is large, evidently $x^{31}-1\approx x^{31}$ and so $A\approx B/(x-1)$, with an error in the rate (when $A=6B$ or $x\approx 1{\cdot}16$) of £0·16, roughly. Newton's rule for smaller values of x is less easily explicable. The suggestion [**1**: 25, note (2)] that he evolved it by plotting the graph of $Y=\log_{10}(31B/A)$ against $X=\log_{10}x$ and identifying the resulting curve as (very nearly) the hyperbola

$$50XY-100X+6Y=0$$

is scarcely a typical Newtonian approach to problems of this sort and indeed, to our knowledge, has no documented parallel in his early papers—Newton's usual method is to manipulate the given conditions of the problem analytically, making use (if at all) of the roughest of sketches. While we cannot be hard and fast in restoring his present argument, it is here tempting to suppose that Newton was familiar (perhaps by inverting the expansion of $\log(1+y)=l$) with the series $(y=)\,x-1=l+\frac{1}{2}l^2+\frac{1}{6}l^2\ldots$, $l=\log x$.(14) Substitution of this in $B/A=x^k(x-1)/(x^k-1)$ yields at once

$$kB/A=1+\tfrac{1}{2}(k+1)\,l+\tfrac{1}{12}(k+1)\,(k+2)\,l^2+\tfrac{1}{12}(k+1)^2\,l^3\ldots,$$

so that $\log(kB/A)=\frac{1}{2}(k+1)\,l-\frac{1}{24}(k^2-1)\,l^2+O(l^4)$. On replacing common by natural logarithms and transposing, Newton's rule becomes

$$\log(31B/A)=100l/(6+50\mu l), \quad \mu=1/\log 10;$$

Transactions (**5** (1670): No. 57: 1168–75) in an account which may well have given Newton his first awareness of Kepler's areal law of planetary motion. See D. T. Whiteside, 'Newton's Early Thoughts on Planetary Motion: a fresh Look', *British Journal for the History of Science*, **2** (1964): 117–37, especially 131, note 48.

(14) Compare Hofmann's *Studien* (note (6)): 24–5. See also a note by Augustus de Morgan in the *Assurance Magazine and Journal of the Institute of Actuaries* **8**, 1859: 61–9.

correctly, $\log(31B/A) \approx 16l/(1+2\frac{1}{2}l)$ and the correspondence is close for $l \approx 2/25(16\mu-5) \approx 0{\cdot}041$ or $x \approx 1{\cdot}042$. The agreement remains narrow for Newton's example ($A = 1200$, $B = 100$) in which $x \approx 1{\cdot}074$.

Whatever his derivation of the logarithmical rule, Newton was pleased enough with it. Indeed, his enthusiasm temporarily overcame his native caution sufficiently to allow him to write to Collins a fortnight afterwards, on 18 February 1669/70 that 'if [my solution of the annuity Probleme] will bee of any use you have my leave to insert it into the *Philosophical Transactions*', but adding at once as his caution reasserted itself 'soe it bee w^th^out my name to it. For I see not what there is desirable in publick esteeme.... Of that Problem I could give exacter solutions, but that I have noe leisure at present for computations' [**1**: 27].

In the same letter Newton also enclosed a reply to a letter from Michael Dary (both now lost) in which, he told Collins, he sent his 'thoughts of w^t^ [Mr Dary] desired'. From a letter of Collins to Gregory a month later [**1**: 28] it is clear that the main reflexion forwarded by Newton in that enclosure was his series expansion for the area, $\int_{-B}^{B} \surd[R^2-x^2]\,.\,dx$, of the central zone of width $2B$ of a circle of radius R 'sent...to compare with...Darys approaches'.(15) Writing to Gregory again, on 24 December 1670, Collins had the broad outline of a novel extension to this basic result to communicate, for 'by a double quadrature...he doth the second Segm^ts^ of a Sphere and Sphæroid (and other round Solids) the particular Precepts for which I desired Dr Barrow to procure me' [**1**: 55]. It was evidently Newton's intention that the 'second Segments' of the ellipsoid

$$x^2/r^2 + y^2/s^2 + z^2 = R^2$$

(a spheroid for $r = s$) should be computed as the double integral

$$2\int_0^a \int_0^b \surd[R^2 - x^2/r^2 - y^2/s^2]\,.\,dy\,dx$$

(where the integrand is first to be expanded as a doubly infinite series of ascending powers of x^2 and y^2) but subsequent events suggest that he did not, at this time, perform the calculation in any detail. A year and a half later, to be sure, having learned in general terms of an equivalent result of Gregory's for the volume of the spheroid $x^2/c^2 + (y^2+z^2)/r^2 = 1$, which had been sent down from Scotland a full year previously,(16) Newton wrote back on 13 July 1672 that

(15) Newton, of course, had come upon this expansion in 1664–5 (see I, **1**, 3, §3: note (58)).

(16) In Gregory's letter to Collins of 17 May 1671: see the *James Gregory Tercentenary Memorial Volume* (London, 1939): 187–91, especially 188–9. Gregory, who had come upon the binomial expansion independently a few months earlier, was pessimistic (as Newton was to

'Mr Gregorys Problem of finding ye solidity of the second segments of a Sphere & yours of finding the surfaces of inclined round solids may be solved divers ways by infinite series, as I find by considering them in generall, but I foresee the calculations are intricate & unpleasant wch has made me neglect them, not thinking them worth transmitting to you' [**1**: 215]. This spurred Collins in reply not only to thank Newton 'for your kind Profer about sending up a Calculation for the 2^d^ Segm^ts^ of a Sphære or Sphæroid, which indeed may be of great use to Guagers (though tedious) were it but to compare with such false approaches as they are driven to use, to find how much they are erroneous' but at last to pass on a 'Coppy' of 'What Mr Gregory hath sent me about it',(17) adding that 'I rather wish his severall Series had been distinct than compounded and that as a Parabola is throughout Proportionall to a Sphære, he had found the genius of that Curve that is throughout Proportionall to a Segment of a Sphære' [**1**: 226].

Both of Collins' points were dealt with by Newton on 20 August 1672 in an unusually informative letter [**1**: 229–30]. 'Since your last', he began,

I have tryed the calculation for finding by an infinite series the content of ye second segments of an ellipsoid. The first series y^t^ I met with was this....(18) Which upon comparison proved ye same wth Mr Gregories & therefore I have exprest it in ye same letters. I tryed two or thre[e] others but could fine none more simple. Wherefore since I understand your designe is to get a rule for guageing vessells, this Problem having so bad success for y^t^ end I shall in its stead present you wth this following expedient.

Newton then pictured a 'Parabolick spindle' of length H, let us say $x^2+y^2=(R-\rho z^2)^2$, $-\frac{1}{2}H\leqslant z\leqslant\frac{1}{2}H$, lying on its side with its main axis (that of z) horizontal and filled with 'liquor' up to some known level, say $R+h$. For small H this truncated paraboloidal 'expedient' evidently closely approximates in gauge the true 'ellipsoid' $x^2+y^2=R^2-2R\rho z^2$ (a spheroid, in fact) cut off between the planes $z=\pm\frac{1}{2}H$. To approximate the spindle's volume Newton then states a rule which 'is not exact but approaches the content of y^e^ Parabolick spindle exactly enough for practice when the top of y^e^ liquor buts upon y^e^ end of

prove to be) concerning its present application. 'This method of infinite series', he told Collins, 'hath no good success in the second segments of round solids, at least so far as I can improve them, yet such as it is ye shall have it', while in afterthought he added that 'This series is nothing but a congeries of other serieses, all of them being infinite, yet is the best I can have to this purpose; if Mr Newton know any other, I hope ye will inform me. If ye would have it agree to the sphere, ye shall only put c in the place of r, which will render it more simple. I can give such a series as this for the second segment of any round solid; and if ye like this, I shall give you with the next a series for the second segments of an hyperbolic spindle, which I imagine is of greater consequence than anything else for guaging'.

(17) Specifically, a single sheet headed 'Out of Mr Gregories Letter of the 17th of May 1671' (now ULC. Add. 3968.41: 1^r^). Apart from the Gregorian series itself Collins transcribed for Newton the sentences quoted in note (16). Why Collins delayed more than a year before

y^e vessell.(19) If y^e vessel be just half full tis exact; if more then half full, tis something to[o] little; if lesse y^n half full, too much': namely, 'the whole quantity of the liquor in y^e vessel' is very nearly $(\frac{2}{3}\alpha_h+\frac{1}{3}\beta_h-\frac{1}{15}\gamma)\,H$, where

$$\alpha_h=\int_{-R}^{h}\pi\surd[R^2-x^2]\,.\,dx=R^2(\sin^{-1}[h/R]+\tfrac{1}{2}\pi)+h\surd[R^2-h^2],$$

$\beta_h=\int_{-(R-\frac{1}{4}\rho H^2)}^{h}\surd[(R-\frac{1}{4}\rho H^2)^2-x^2]\,.\,dx$ and $\gamma=\pi(\frac{1}{4}\rho H^2)^2$ are the areas, respectively, of the central cross-section of the liquid in the vessel, of its surface 'butting' upon the vessel's end and of the circle whose radius is the difference of their radii. In exact terms the liquid content, say L_h, of the vessel when filled to the height $R+h$ is, on taking $R_z=R-\rho z^2$, $\int_{-\frac{1}{2}H}^{\frac{1}{2}H}\int_{R}^{h}y\,.\,dx\,dz$; that is,

$$L_h=2\int_0^{\frac{1}{2}H}[R_z^2(\sin^{-1}[h/R_z]+\tfrac{1}{2}\pi)+h\surd\{R_z^2-h^2\}]\,.\,dz$$

As Newton observes, when h is zero $(\alpha_0=\frac{1}{2}\pi R^2,\ \beta_0=\frac{1}{2}\pi(R-\frac{1}{4}\rho H^2)^2)$

$$L_0=\pi\int_0^{\frac{1}{2}H}R_z^2\,.\,dz=\tfrac{1}{2}\pi H(R^2-\tfrac{1}{6}R\rho H^2+\tfrac{1}{80}\rho^2H^4)$$

is exactly equal to $H(\frac{2}{3}\alpha_0+\frac{1}{3}\beta_0-\frac{1}{15}\gamma)$, while in general

$$L_h-L_0\approx H[2h(R-\tfrac{1}{12}\rho H^2)-\tfrac{1}{3}h^3(1/R+\tfrac{1}{12}\rho H^2/R^2+\tfrac{1}{80}\rho^2H^4/R^3)],$$

$\alpha_h-\alpha_0\approx 2hR-\frac{1}{3}h^3/R$ and $\beta_h-\beta_0\approx 2h(R-\frac{1}{4}\rho H^2)-\frac{1}{3}h^3/(R-\frac{1}{4}\rho H^2)$, where h and H are both small, so that $L_h-H(\frac{2}{3}\alpha_h+\frac{1}{3}\beta_h-\frac{1}{15}\gamma)\approx\frac{1}{360}\rho^2h^3H^5/R^3$. Since H and R are assumed to be positive the last quantity, for small h and H, takes its sign from h in agreement with Newton's final remark.

In the latter half of his letter Newton gave Collins a tantalizing glimpse of his organic method of describing conics, presuming it to be 'not...an unpleasing speculation to your Mathematicians [viz. Gregory and Dary] to find out ye Demonstration' and adding, with regard to 'a set of Problems for construing æquations...set down rudely', that 'How ye afforesaid descriptions of ye Conick sections are to be applyed to these constructions I need not tell you' [**1**: 231]. Since this portion is reproduced *verbatim* in the second volume(20) we will not here examine it.

---

communicating Gregory's result is not known. Apart from a brief exchange of one letter each in July 1671 Collins and Newton apparently did not correspond between September 1670 and late April 1672.

(18) Taking (as Gregory suggested in May 1671) $c=r$, Newton here gives the first ten terms in the series expansion of $2\int_0^a\int_0^b\surd[r^2-x^2-y^2]\,.\,dy\,dx$, the volume of the 'second segment' $(-a\leqslant x\leqslant a,\ 0\leqslant y\leqslant b)$ of the sphere $x^2+y^2+z^2=r^2$.

(19) That is, when $|h|<R-\frac{1}{4}\rho H^2$.

(20) See II: 156–9.

One final point in Newton's mathematical correspondence up till the end of 1673 is worthy of note. Writing to Collins (after a break of several months) on 17 September of that year he related [**1**: 307] that

to decide the controversy between him & Mr Gunton... Mr Dary... desires my opinion about the relation of ye lines one to another which are drawn from ye center of an Ellipsis to the angular points of a polygon inscribed into ye same Ellipsis, wch consists of 24 equall sides. As it appeares to me, their relation cannot be accurately known without an Equation wch is 4 times decompounded of affected cubic equations & twice of quadratick ones, & by consequence would ascend to 324 dimensions: To compute wch would be a Herculean labour, & when done, it would be unmanageable. I doubt therefore y^t^... recours must be had to some mechanicall examination... but I leave it to you....

Evidently Dary's problem dealt with the particular instance in which four vertices of the inscribed equilateral 24-gon coincide with the end-points of the ellipse's axes.(21) In this case, given the ellipse defined by the Cartesian equation $y^2 = b^2(a^2 - x^2)$, say, in the quadrantal arc bounded by $P_0(a, 0)$ and $P_6(0, ab)$ five points $P_i(x_i, y_i)$, $i = 1, 2, 3, 4, 5$, are to be inserted such that

$$P_0P_1 = P_1P_2 = P_2P_3 = P_3P_4 = P_4P_5 = P_5P_6,$$

for then at once the 'semidiameter' through point P_i will have the length $\surd[x_i^2 + y_i^2] = \surd[(1-b^2)\,x_i^2 + a^2b^2]$. It is easy to show that, if the points P_{i-1} and P_i, $i = 1, 2, 3, 4$, are known, then the abscissa x_{i+1} of the next point P_{i+1} will be determined as the unique real root x of the cubic

$$[(1-b^2)\,(x + x_{i-1}) - 2x_i]\,[(x - x_{i-1})\,((1-b^2)\,(x + x_{i-1}) - 2x_i) + 2y_{i-1}y_i] + 4b^2y_i^2(x + x_{i-1}) = 0.$$

In geometrical terms, equivalently, the circle of centre P_i and radius $P_{i-1}P_i$ will meet the ellipse in three points (other than P_{i-1}), of which only P_{i+1} will be real. Similarly, the condition $P_0P_1 = P_5P_6$ yields the irreducible quartic equation $[(1-b^2)\,(x_1^2 - x_5^2 + a^2) - 2ax_1]^2 - 4a^2b^4(a^2 - x_5^2) = 0$ connecting x_1 and x_5. Newton's argument, we may conjecture, would then assert that, since each x_{i+1}, $i = 1, 2, 3, 4$, is given in terms of x_{i-1} and x_i by an equation of third degree, therefore x_5 will be determined in terms of x_1 by an equation of $3^4 = 81$ dimensions and so, when x_5 is eliminated by the quartic equation connecting it with x_1, the latter will be one of the (six) positive real roots of an equation of

(21) Dary placed this restriction upon a simpler version of the problem noted a few weeks earlier by Collins: namely, to inscribe an equilateral octagon in a given ellipse and so determine the length of its oblique 'semidiameter' (see S. P. Rigaud, *Correspondence of Scientific Men of the Seventeenth Century*, **2** (Oxford, 1841): 116). Collins at that time observed 'Mr Dary hath occasion to use this proposition in guaging' (in measuring the content of coopered barrels, oval in cross-section?).

$4 \times 3^4 = 324$ dimensions. In fact, the equation connecting x_1 and x_2 is quadratic in the former, while that connecting x_{i-1}, x_i and x_{i+1} ($i = 2, 3, 4$) contains the surd term $2y_{i-1}y_i = 2b^2\surd[(a^2 - x_{i-1}^2)(a^2 - x_i^2)]$, and this will boost the degree of the final eliminant.(22) Exact determination of x_1 is, in general, impossible and even its 'mechanicall' evaluation is forbidding, so that Newton's pessimism is well founded.

For the rest Newton's correspondence is concerned with optical matters(23) and with correcting certain errors of understanding on Collins' part regarding the roots of algebraic equations.(24) On these we need not dwell.

(22) Compare [**1**: 308, note (2)].

(23) The mathematically interesting passages in these letters are discussed in connexion with Newton's dioptrical writings at appropriate points in Part 3 above.

(24) Compare II: 282–4. Collins believed that 'one generall rule' could be applied to 'both kinds of Cubick Æquations, to wit as well those that are solved by meane Proportionalls as those that require Trisection' in July 1670 [**1**: 33]. Newton crushingly replied on 27 September [**1**: 43] that 'I can not...bee convinced y^t two meane proportionalls may bee found by trisecting an arch, or contrarily....I cannot therefore yet bee convinced that any one problem can be solved both those ways, wch if it could, it would bee noe hard matter to take away both ye two middle terms of any cubick æquation. Which whoever performes I shall esteem as a great Apollo & admire as much as if hee had squared ye circle, because I judg both impossible. And my reason is this that æquations to what termes soever they are reduced their reall roots never becom imaginary nor their imaginary roots reall (though indeed their true roots may become false & false ones true)'.

INDEX OF NAMES